Fundamentals of Enzymology

Fundamentals of Enzymology

SECOND EDITION

BY

NICHOLAS C. PRICE

Reader in Biochemistry, University of Stirling

AND

LEWIS STEVENS

Senior Lecturer in Biochemistry, University of Stirling

Oxford New York

OXFORD UNIVERSITY PRESS

Oxford University Press, Walton Street, Oxford OX2 6DP
Oxford New York
Athens Auckland Bangkok Bombay
Calcutta Cape Town Dar es Salaam Delhi
Florence Hong Kong Istanbul Karachi
Kuala Lumpur Madras Madrid Melbourne
Mexico City Nairobi Paris Singapore
Taipei Tokyo Toronto
and associated companies in
Berlin Ibadan

Oxford is a trade mark of Oxford University Press

Published in the United States
by Oxford University Press Inc., New York

First published 1982
Second Edition published 1989
Reprinted 1995, 1996

A catalogue record for this book is available from the British Library

Library of Congress Cataloging in Publication Data
Price, Nicholas C.
Funamentals of enzymology/by Nicholas C. Price and Lewis Stevens.—2nd ed.
p. cm.
Includes bibliographies and index.
1. Enzymes. I. Stevens, Lewis. II. Title.
[DNLM: 1. Enzymes. QU 135 P946f]
QP601.P83 1989 574.19'25—dc19 89–3155
ISBN 0 19 855296 3 (Pbk)

Printed in Malta by
Interprint Limited

For
Margaret, Jonathan, Rebekah, and Naomi, and
Evelyn, Rowena, Catherine, and Andrew.

Preface to the second edition

In preparing the second edition, we have attempted to highlight the major developments in enzymology in the seven years since the first edition appeared. Undoubtedly the most significant advances have been in the application of recombinant-DNA technology to the production of enzymes (Chapters 2 and 11), to the indirect determination of amino-acid sequences (Chapter 3) and to the analysis of enzyme structure and function using site-directed mutagenesis (Chapters 3 and 5). In many ways these advances can be said to have ushered in a 'golden age' of enzymology.

Other major advances have been made in the techniques of enzyme purification and analysis (Chapters 2 and 3), in the identification of catalytic domains within multienzyme proteins (Chapter 7), in understanding the organization of enzymes within membranes (Chapter 8), and in the application of purified enzymes in clinical analyses (Chapter 10).

We have been greatly encouraged by the response of colleagues to the first edition and hope that the new edition will continue to serve as a valuable guide for students and teachers of enzymology.

Stirling
August 1988

N.C.P.
L.S.

Preface to the first edition

Over the last few years a number of books that have dealt comprehensively with the kinetics, structures, and mechanisms of enzymes have appeared. In general, the emphasis of these books has been on understanding the behaviour of isolated enzymes and much less attention has been focused on enzymes in their natural environment in an organism. This book aims to restore the balance to some extent by discussing the properties of enzymes in systems of increasing complexity from isolated enzymes to enzymes in the cell. It should be of value to undergraduate students of biochemistry and related biological sciences. We have assumed that the reader will have an elementary knowledge of biochemistry such as might be gained from an introductory course at university.

In order to keep the length of the book within manageable limits, certain topics have been discussed only in outline. However, we have aimed to explain the principles involved in these topics in sufficient detail so that a student can pursue them if he or she wishes in the many books, research articles, and reviews referred to at the end of each chapter. We feel that it is important for undergraduates to make use of the original literature because biochemistry is such a rapidly expanding research-based subject.

In this book we have adopted the SI system of units and followed, as far as possible, the recommendations of the International Union of Biochemistry regarding the nomenclature of enzymes and substrates. Some of the recommended names may not be familiar to many biochemists, e.g. the enzyme RNA nucleotidyl transferase is more widely known as RNA polymerase. In such cases the more commonly used name is given as well as the recommended name in the Appendix at the end of the book, which also gives the reactions catalysed and Enzyme Commission numbers of all the enzymes mentioned in the book.

Stirling
January 1981

N.C.P.
L.S.

Acknowledgements

It is particularly appropriate in a book on biological catalysts to thank the many people who helped to bring this work to fruition. We should like to thank our colleagues in the department for many useful discussions and especially to pay tribute to Dr Evelyn Stevens who improved the style of much of the book. Professors Michael Rossmann and David Blow and Drs Heiner Schirmer and Georg Schulz generously supplied original photographs of enzyme structures. Dr Hal Dixon gave us some valuable advice on nomenclature, and Dr Linda Fothergill helped with some points relating to amino-acid sequencing. Dr Neil Paterson gave some very helpful comments and advice on current medical applications of enzymology. We should also like to thank Mrs May Abrahamson for her excellent typing. In one sense, all these people acted as more than catalysts; their intervention improved the book, rather than leaving it unchanged. The errors are, of course, our responsibility.

Acknowledgements for second edition

We would like to thank Khlayre Mullin, June Watson, and Marianne Burnett for their help in preparing the second edition, and Professor David Blow for supplying a photograph of the structure of tyrosyl-tRNA synthetase.

Contents

A note on units

In this book we have tried, wherever possible, to use SI units. These are based on the metre-kilogram-second system of measurement. Multiples of the basic units by powers of 10 are as follows.

Prefix		Abbreviation	Prefix		Abbreviation
$10 (10^1)$	deca	da	10^{-1}	deci	d
10^2	hecto	h	10^{-2}	centi	c
10^3	kilo	k	10^{-3}	milli	m
10^6	mega	M	10^{-6}	micro	μ
10^9	giga	G	10^{-9}	nano	n
10^{12}	tera	T	10^{-12}	pico	p

We do not use compound prefixes, thus 10^{-9} metre (m) = 1 nm, not 1 mμm. SI units for the various physical quantities mentioned in this book are listed below.

Quantity	SI unit
Time	*second* (s).
Length	*metre* (m). (Supplementary units retained for convenience are dm, cm)
Mass	*kilogram* (kg). Note that multiples are based on 1 gram (g), i.e. mg, μg rather than μkg, nkg (see above compound prefix rule)
Volume (given in units of length cubed)	*cubic metre* (m^3). For convenienoe the litre ($1 l = 1\ dm^3$) and millilitre ($1\ ml = 1\ cm^3$) are retained
Amount of substance	*mole* (mol). This quantity contains 1 Avagadro number of basic units (e.g. electrons, atoms, or molecules)
Concentration	*mol dm*$^{-3}$ used instead of molar (M) *mol kg*$^{-1}$ used instead of molal (m)
Temperature	*kelvin* (*K*) note $0\,°C = 273.15\ K$
Force	*newton* (N) ($1\ m\ kg\ s^{-2}$)
Pressure	*pascal* (Pa) ($1\ Nm^{-2}$). In this text, however, we have retained the atmosphere unit (atm)

Quantity	SI unit
	since the definition of standard states is then less cumbersome, being under a pressure of 1 atm rather than 101.325 kPa. (1 atm = 760 mm Hg = 101.325 kPa)
Energy	*joule* (J) (1 $m^2 kg s^{-2}$). One calorie = 4.18 J. The gas constant, $R = 8.31\ J\ K^{-1}\ mol^{-1}$ (1 electron volt = 96.5 $kJ\ mol^{-1}$)
Electric charge	*coulomb* (C) (1 ampere second). Faraday constant = 96 500 $C\ mol^{-1}$
Frequency	*hertz* (Hz) (1 s^{-1})
Viscosity	the units are $kg\ m^{-1}\ s^{-1}$ (Note that 1 centipoise = $10^{-3}\ kg\ m^{-1}\ s^{-1}$).

The main implications of these units for biochemists are as follows.

(i) Temperatures are quoted in kelvin (or absolute). Thus 25 °C is 298.15 K (in practice, this is given as 298 K).
(ii) Enthalpy, internal energy, and free-energy changes are given in J mol^{-1}.
(iii) Entropies are quoted in $J\ K^{-1}\ mol^{-1}$.
(iv) Molar concentrations are given in terms of mol dm^{-3} rather than M.
(v) Enzyme activities are quoted in terms of katals (the amount of enzyme catalysing the transformation of 1 mol substrate per second). Specific activities are quoted in terms of katal kg^{-1}.
(vi) The curie (Ci) is redundant as a unit of radioactivity; 1 Ci = 3.7 $\times 10^{10}$ disintegrations s^{-1}. (The unit disintegrations s^{-1} is given the name *becquerel* (Bq)).

For fuller discussions of SI units the following may be consulted:

Quantities, units, and symbols (2nd edn). The Royal Society, London (1975).
Physicochemical quantities and units (2nd edn). M. L. McGlashan. Royal Institute of Chemistry, London (1971).

List of abbreviations

The following standard abbreviations, which are in common use in biochemistry, are used in the text. Any others are defined where used.

A	absorbance
ACP	acyl-carrier protein
ADP	adenosine 5′-pyrophosphate (adenosine 5′-diphosphate)
AMP	adenosine 5′-phosphate (adenosine 5′-monophosphate)
ATP	adenosine 5′-triphosphate
cAMP	3′:5′ cyclic AMP
cDNA	complementary DNA
CoA	coenzyme A
CTP	cytidine 5′-triphosphate
dATP	2′-deoxyadenosine 5′-triphosphate
dGDP	2′-deoxyguanosine 5′-pyrophosphate (2′-deoxyguanosine 5′-diphosphate)
dGTP	2′-deoxyguanosine 5′-triphosphate
DNA	deoxyribonucleic acid
dTMP	thymidine 5′-phosphate (thymidine 5′-monophosphate or 2′-deoxyribosylthymine 5′-monophosphate)
dTTP	thymidine 5′-triphosphate (2′-deoxyribosylthymine 5′-triphosphate).
dUMP	2′-deoxyuridine 5′-phosphate (2′-deoxyuridine 5′-monophosphate)
EDTA	ethylenediaminetetra-acetate
FAD	flavin-adenine dinucleotide (oxidized form)
$FADH_2$	flavin-adenine dinucleotide (reduced form)
g	gravitational field, unit of (9.81 m s^{-2})
GDP	guanosine 5′-pyrophosphate (guanosine 5′-diphosphate)
GTP	guanosine 5′-triphosphate
h.p.l.c.	high-performance liquid chromatography
i.r.	infrared
ITP	inosine 5′-triphosphate
mRNA	messenger RNA
M_r	relative molecular mass
NAD^+	nicotinamide-adenine dinucleotide (oxidized form)
NADH	nicotinamide-adenine dinucleotide (reduced form)
$NADP^+$	nicotinamide-adenine dinucleotide phosphate (oxidized form)
NADPH	nicotinamide-adenine dinucleotide phosphate (reduced form)
NMN	nicotinamide mononucleotide
n.m.r.	nuclear magnetic resonance
RNA	ribonucleic acid
r.p.m.	revolutions per minute

rRNA	ribosomal ribonucleic acid
tris	2-amino-2-hydroxymethyl-propane-1, 3-diol (tris(hydroxymethyl) methylamine)
tRNA	transfer RNA
UDP	uridine 5′-pyrophosphate (uridine 5′-diphosphate)
UMP	uridine 5′-phosphate (uridine 5′-monophosphate)
UTP	uridine 5′-triphosphate
u.v.	ultraviolet

Nucleotide sequences

A sequence of nucleotides can be written as, e.g., A C G C U C where each letter signifies a nucleotide.* The convention is to write the order so that the phosphodiester link runs 3′ to 5′ from left to right. A nucleoside 5′-phosphate is written as pX, and a 3″-phosphate as Xp. The sequence above could also be written as ApCpGpCpUpC. To indicate that there is a phosphate, pyrophosphate or triphosphate group at the 5′-end, we would write

pA C G C U C	5′-phosphate
ppA C G C U C	5′-pyrophosphate
pppA C G C U C	5′-triphosphate

* The nucleotide is taken to be a ribonucleotide unit, unless the context makes it clear that it is a deoxyribonucleotide (e.g. in DNA). In a DNA sequence, the letter T refers to a thymidine phosphate (2′-deoxyribosylthymine phosphate) unit.

1 Introduction

1.1 Aims of the book

In this book we have tried to give a broad account of enzymology and have aimed to put current knowledge into perspective. Studies of enzymes have as their ultimate goal an understanding of the crucial role that these catalysts play in the metabolic processes of living organisms. Because of the complexity of such processes, it is at least necessary to gain an insight into the properties of enzymes in simpler systems, i.e. as isolated entities studied in the test tube or spectrophotometer cuvette. The chapters in the book follow a progression from the properties of isolated enzymes to the behaviour of enzymes in increasingly complex systems, leading up to the cell. We have included some discussion of the importance of enzymes in medicine and industry to emphasize that enzymology is not a purely academic subject but has increasingly wide applications.

1.2 Historical aspects

Enzymes are catalysts (i.e. they speed up the rates of reactions without themselves undergoing any permanent change). Each reaction taking place in the cell is catalysed by its own particular enzyme so that in a given cell there are a large number of enzymes. It is difficult to make a precise estimate of the number of different enzymes in each cell but it seems that a bacterial cell such as *Escherichia coli* makes about 3000 different proteins and a higher eukaryote cell about 50 000 of which the majority are enzymes. In the absence of enzymes many of these reactions would not occur even over a time period of years and life as we know it could not exist.

The word *enzyme* is derived from the Greek meaning 'in yeast' and was first used by Kühne in 1878.* At the time it was used to distinguish between what were referred to as 'organized ferments' (meaning whole microorganisms) and 'unorganized ferments' (meaning extracts or secretions from whole organisms). The term *enzyme* was thus intended to emphasize that catalytic activity was 'in yeast', i.e. a manifestation of an extract or secretion rather than of the whole organism. Although the term specified yeast, it was to be used for all 'unorganized ferments'. Of course, the catalytic activity of enzymes in microorganisms has been utilized by man for many thousands of years in processes such as fermentation and cheese-making but this was very much purely a practical use. It was only when it was shown that enzyme activity could be

* It was pointed out by Fruton[1] that the term was used as far back as the twelfth century by the Armenian philosopher, Theorianus—although obviously not in a biochemical context.

expressed without the need for an intact cellular structure that the study of enzymes could proceed along the paths already established in the study of chemistry. In this respect Büchner's demonstration (1897) that filtrates of yeast extracts could catalyse fermentation was highly significant.

Emil Fischer in 1894 had performed some classical studies on carbohydrate metabolizing enzymes in which he demonstrated the specificity shown by an *enzyme* for its *substrate* (the molecule acted on by the enzyme). On the basis of his experiments, Fischer proposed the 'lock and key' hypothesis to describe this interaction and for many years this proved to be a fruitful way to picture the binding of enzyme to substrate (Fig. 1.1)

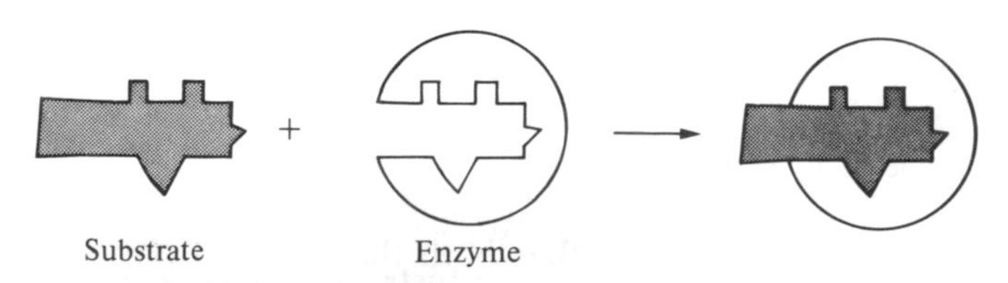

FIG. 1.1. Fischer's 'lock and key' hypothesis to explain enzyme specificity.

At this time (1890s) the chemical nature of enzymes was not clear. In fact this point was only established many years later after a number of enzymes had been crystallized and been shown to consist entirely of protein (i.e. made up of amino acids linked by amide bonds, see Chapter 3, Section 3.3). The first enzyme to be crystallized (in 1926 by Sumner) was urease, which catalyses the hydrolysis of urea to yield carbon dioxide and ammonia. Actually, it is now known that if Sumner had had more sensitive methods of analysis he would have found that the preparation contained a small amount of nickel (approximately 0.1 per cent by weight) which is essential for catalysis.[2] Perhaps in retrospect it was fortunate that he did not have such sensitive methods, since otherwise the nature of enzymes might not have been settled for many more years.

The development of the ultracentrifuge by Svedberg (also in the 1920s) allowed very high centrifugal fields capable of sedimenting macromolecules* to be generated. These studies showed that proteins in solution generally consist of homogeneous molecules of definite M_r (in the case of enzymes the M_r values range between about 10^4 and 10^7) rather than of colloidal suspensions. The description of enzyme structures in precise chemical terms was then a realistic possibility: this was first achieved in 1960 when the amino-acid sequence of ribonuclease[3] (an enzyme catalysing the hydrolysis of ribonucleic acid) was deduced. In 1965 the three-dimensional structure of lysozyme (an enzyme cleaving certain bacterial cell walls) was deduced by the technique of X-ray crystallography (see Chapter 3, Section 3.5.1), and for the first time a mechanism of action could be postulated in precise structural terms. We now know the

* The term macromolecule is generally reserved for molecules of relative molecular mass $(M_r) > 10\,000$

amino-acid sequence and three-dimensional structure of many enzymes and generalizations and comparisons between them can be undertaken.

During the late 1950s and the 1960s a number of observations were made that suggested that enzymes show considerable flexibility. In 1958 Koshland[4] proposed the 'induced fit' theory to account for the catalytic power and specificity shown by enzymes (see Chapter 5, Section 5.6). It also became clear that the catalytic activity of certain enzymes could respond to changes in physiological conditions. Monod and his colleagues[5,6] proposed their 'allosteric model' to explain in a quantitative way how the activity of certain enzymes can be regulated by the binding of small molecules (*effectors*) and this provided a basis for understanding many features of the control of enzymes in the cell. An important feature of models for allosteric enzymes in general is that they postulate that the binding of effectors to the enzymes induces structural changes in the enzymes (Chapter 6, Section 6.2.2).

The first chemical synthesis of an enzyme (ribonuclease) from amino-acid precursors was reported in 1969.[7] Although this represented a considerable achievement, it should be noted that both the chemical purity and catalytic activity of the preparation were rather low. The chemical synthesis can, perhaps, be said to represent the final proof that enzymes are no different qualitatively from other non-biological catalysts.

The application of recombinant DNA techniques to the study of enzymes has produced some remarkable new insights (see Chapter 3, Section 3.4.1 and Chapter 5, Sections 5.4.5 and 5.5). It has proved possible to alter catalytic activity and specificity in a rational manner by introducing mutations at defined positions using site-directed mutagenesis. This has helped in understanding the mechanism of enzyme action and has also opened the prospect of designing enzymes with specific required properties.[33] For example, the specificity of lactate dehydrogenase has been changed to that of malate dehydrogenase by introducing three particular mutations at the active site (see Chapter 5, Section 5.5.4.5).

Recently it has become evident that catalytic activity can be shown to a limited extent by biological molecules other than the 'classical' enzymes. In the process of enzyme catalysis the substrates are converted via a high-energy transition state to the eventual products (see Chapter 5, Fig. 5.1). A number of reports have shown that antibodies raised to stable analogues of the transition states of hydrolytic reactions (see Chapter 5, Section 5.3.4) can act as moderately effective catalysts for those reactions.[30–32] Fragments of RNA can also act as catalysts for reactions involving hydrolysis of RNA.[8,9] The observation of these 'ribozymes' has important implications for theories concerning the evolution of catalytic function.[10]

For further details on the history of enzymology see references 11 and 12.

1.3 Remarkable properties of enzymes as catalysts

Enzymes display a number of remarkable properties when compared with other types of catalyst. The three most important are their high catalytic power, their specificity, and the extent to which their catalytic activity can be regulated by a variety of naturally occurring compounds. These three properties will be illustrated below.

1.3.1 Catalytic power

Enzymes may increase the rate of a reaction by as much as 10^{14}-fold. There are not many examples where a direct comparison can be made between the rates of an enzyme-catalysed reaction and the reaction occurring under similar conditions of temperature, pH, etc. but in the absence of enzyme. This is because in the absence of an enzyme the rates may be too low to be measured easily. Where the comparison has been made, very high rate enhancements have been found,[13] e.g. hexokinase $> 10^{10}$, phosphorylase $> 3 \times 10^{11}$, alcohol dehydrogenase $> 2 \times 10^{8}$, and creatine kinase $> 10^{4}$. In other instances, where enzymatic and non-enzymatic catalysts are compared, the former catalyse much higher rates and in some cases do so at significantly lower temperatures (Table 1.1). The optimum conditions for enzyme catalysis are almost invariably moderate temperatures, and pHs that are not extreme. The contrast between an enzyme-catalysed reaction and that catalysed by a non-enzymatic catalyst is well illustrated by the process of nitrogen fixation (i.e. reduction of N_2 to ammonia). Nitrogenase* catalyses this reaction at temperatures around 300 K and at neutral pH. The enzyme is a complex system comprising two dissociating protein components, one of which contains iron and the other iron and molybdenum.[15] Several molecules of ATP are hydrolysed during the reduction, although the exact stoichiometry is still uncertain. By contrast, in the industrial synthesis of ammonia from nitrogen and hydrogen the conditions used are as follows:

TABLE 1.1
Examples of the catalytic power of enzymes

Substrate	Catalyst	Temperature (K)	Rate constant k $(\text{mol dm}^{-3})^{-1}\,\text{s}^{-1}$
Amide (hydrolysis)			
benzamide	H^+	325	2.4×10^{-6}
benzamide	OH^-	326	8.5×10^{-6}
benzoyl-L-tyrosinamide	α-chymotrypsin	298	14.9
Urea	H^+	335	7.4×10^{-7}
(hydrolysis)	urease	294	5.0×10^{6}
$2H_2O_2 \rightarrow 2H_2O + O_2$	Fe^{2+}	295	56
	catalase	295	3.5×10^{7}

Data taken from reference 14.

* Only certain prokaryotes can carry out nitrogen fixation and they may be either symbionts or non-symbionts. In terrestrial ecosystems, symbiotic fixation appears to exceed non-symbiotic and the most outstanding example of symbiotic fixation is that between bacteria of the genus *Rhizobium* and the roots of leguminous plants.

temperatures between 700 and 900 K, pressures between 100 and 900 atmospheres, and the presence of an iron catalyst, often promoted by traces of oxides of other metals.[16] The basis of catalytic power is discussed in Chapter 5, Section 5.3.

1.3.2 Specificity

Most enzymes are highly specific both in the nature of the substrate(s) they utilize and also in the reaction they catalyze. The range of specificity varies between enzymes. There are some enzymes that have relatively low specificities (*bond specificity*), e.g. certain peptidases, phosphatases, and esterases, which will utilize a wide range of substrates provided they contain the required chemical bond, i.e. peptide, phosphate ester, and carboxylate ester, respectively, in these three examples. Low specificity is more commonly encountered with degradative enzymes but is only very rarely observed with biosynthetic enzymes. The role of the former may be that of digestion, where wider specificity would be more economical. An intermediate set of enzymes show *group specificity*, e.g. hexokinase. This enzyme will catalyse the phosphorylation of a variety of sugars provided they are aldohexoses. However, many enzymes show *absolute* or *near-absolute specificity*, in which they will only catalyse at an appreciable rate the reaction with a single substrate (or a single pair of substrates in a bimolecular reaction), e.g. urease will only catalyse the reaction with urea, or with very similar analogues at a very much lower rate. The quantitative definition of specificity in kinetic terms is described in Chapter 4, Section 4.3.1.3.

The terms *group specificity* and *absolute specificity* can be readily appreciated in relation to low-M_r substrates, but when considering macromolecular substrates the position may be somewhat different in the sense that the active site of the enzyme can only interact with a part of the macromolecule. This is simply due to the relative dimensions of the active site and the macromolecule. A group of enzymes that has been studied extensively in the past decade is the restriction endonucleases. These enzymes generally recognize a sequence of four to six base pairs in DNA and then cleave the phosphodiester links in both strands but not necessarily in opposite positions. There are now known at least 400 of these enzymes differing in specificity.[17] In a sense, each enzyme shows absolute specificity for the region of the substrate that is in contact with the active site, although it can act on any DNA molecule or fragment that contains the appropriate sequence.

Another distinct feature of many enzyme-catalysed reactions is their stereospecificity; this is well illustrated in the case of NAD^+- and $NADP^+$-requiring dehydrogenases. It has been demonstrated by use of suitably labelled substrates that dehydrogenases catalyse the transfer of hydrogen from the substrate on to a particular side of the nicotinamide ring; these are designated A-side and B-side dehydrogenases (Fig. 1.2). In addition, almost all dehydrogenases act on either NAD^+ or $NADP^+$. The basis of these specificities is clear in the case of those dehydrogenases whose three-dimensional structures are known, e.g. liver alcohol dehydrogenase, lactate dehydrogenase, and glyceraldehyde-3-phosphate dehydrogenase[19] and is discussed further in Chapter 5, Section 5.5.4.3.

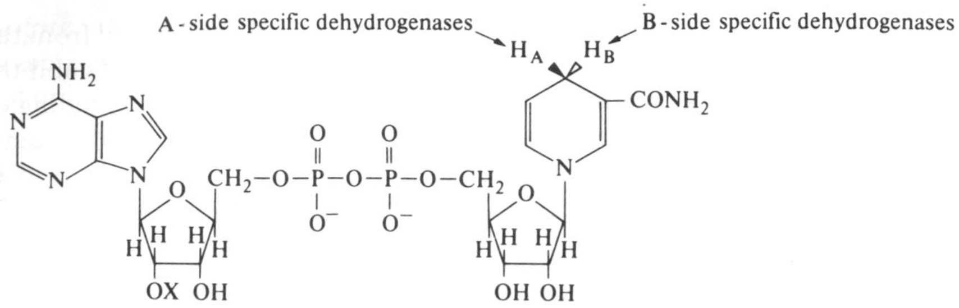

Reduced nicotinamide adenine dinucleotide (NADH), X = H, and reduced nicotinamide adenine dinucleotide phosphate (NADPH), X = phosphate.

A-side dehydrogenases	**B-side dehydrogenases**
Alcohol dehydrogenase (NAD^+)	Glycerol 3-phosphate dehydrogenase (NAD^+)
Lactate dehydrogenase (NAD^+)	3-Hydroxybutyrate dehydrogenase (NAD^+)
Malate dehydrogenase (NAD^+)	Glucose dehydrogenase (NAD^+)
Cytochrome b_5 reductase (NAD^+)	Glyceraldehyde-3-phosphate dehydrogenase (NAD^+)
Shikimate dehydrogenase ($NADP^+$)	Homoserine dehydrogenase ($NADP^+$)
3-Oxyacyl [acyl-carrier-protein] reductase ($NADP^+$)	Glucose-6-phosphate dehydrogenase ($NADP^+$)

FIG. 1.2. Stereospecificity of NAD^+- and $NADP^+$-requiring enzymes. For a more complete list see reference 18

There are a number of other examples of prochirality in enzyme-catalysed reactions.[20] These reactions occur in such a manner that although the reactant does not have a chiral* centre the reaction occurs in a stereospecific manner. A good example is the reaction catalysed by fumarate hydratase in which fumarate is converted to malate. It can be shown by carrying out the reaction in the presence of tritiated water (^3H_2O) that ^3H addition occurs in a stereospecific manner. There are several other examples of this phenomenon.

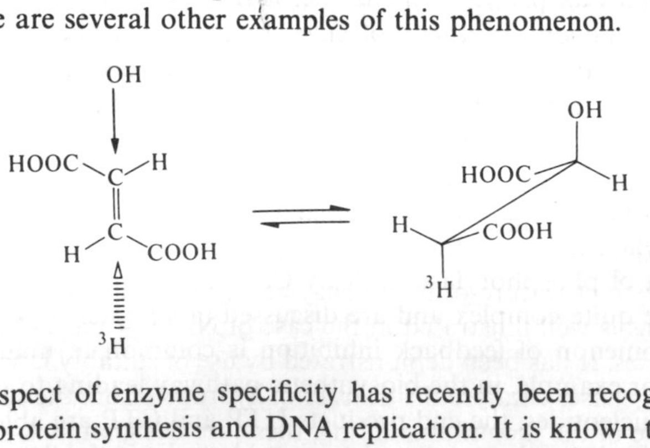

A further aspect of enzyme specificity has recently been recognized and is important in protein synthesis and DNA replication. It is known that the error rate in DNA replication *in vivo* is as low as one mistake in $10^8 - 10^{10}$ nucleotides polymerized and that the overall error rate in transcribing DNA and then translating the mRNA into protein (see Fig. 9.1) is about 1 in 10^4 amino-acid

* A compound is said to be chiral if it cannot be superimposed on its mirror image.

residues incorporated. This specificity is higher than would be expected from the relative energies of interaction between structurally similar amino acids and the aminoacyl-tRNA synthetases and is only possible because of a proof reading or editing mechanism.[21] The essence of this mechanism in the case of aminoacyl-tRNA synthetases will be described briefly here (a fuller account is given in Chapter 5, Section 5.5.5). The charging of tRNA catalysed by aminoacyl-tRNA synthetases occurs in two steps

amino acid + ATP + Enzyme → Enzyme-aminoacyl-AMP + pyrophosphate

Enzyme-aminoacyl-AMP + tRNA → aminoacyl-tRNA + AMP + Enzyme

The enzyme has to recognize a specific amino acid and also a specific tRNA. The latter is recognized very precisely since it is a large molecule and thus makes many contacts with the enzyme. The amino acids, which are very much smaller, are not selected so precisely. It is possible to monitor both the first and second steps in the reaction and this shows that significant errors do occur in the formation of the aminoacyl-adenylate intermediates but not in the formation of aminoacyl-tRNAs. However, any incorrect aminoacyl-adenylates formed rapidly become hydrolysed. There is evidence that the enzyme has a second site, distinct from that on which the synthesis occurs, which enables it to hydrolyse off mismatched amino acids, thus acting as an editing mechanism. A similar mechanism may also occur to ensure accurate DNA replication. The enzyme DNA nucleotidyltransferase (DNA polymerase) III also possesses exonuclease activity which acts in editing DNA replication[21] (see also Chapter 7, Section 7.11.3).

1.3.3 Regulation

The third important property of enzymes is that their catalytic activity may be regulated by small ions or other molecules. An example is the enzyme phosphorylase, which catalyses the first step in the breakdown of glycogen in skeletal muscle. This is an important enzyme that enables the carbohydrate reserves to be degraded in order to generate ATP required for muscle contraction. The onset of muscle contraction is triggered by release of Ca^{2+} from the sarcoplasmic reticulum (see Chapter 8, Section 8.4.4) and this also brings about the activation of phosphorylase to ensure the continued production of ATP. The mechanisms of regulation of phosphorylase both by Ca^{2+} and also by hormones such as adrenalin are quite complex and are discussed in Chapter 6, Section 6.4.2.

The phenomenon of feedback inhibition is common in many biosynthetic pathways. For example, in the biosynthetic pathway leading to the synthesis of pyrimidine nucleotides, the end products UTP and CTP are able to inhibit the first enzyme in the pathway; thus they are able to limit the flow of metabolites into that pathway and so regulate their own biosynthesis. This regulation is effected through changes in the catalytic activity of early enzymes in the pathway, carbamoyl-phosphate synthase and aspartate carbamoyltransferase; this topic is discussed in Chapter 6, Sections 6.2.2.1 and 6.3.1.

1.4 Cofactors

The demonstration of enzyme crystallinity and analysis of these crystals has shown that enzymes are proteins with chain lengths generally of 100–2500 amino acids. Enzymes such as chymotrypsin (a protein-hydrolysing enzyme) or triose-phosphate isomerase (an enzyme catalysing the interconversion of two phosphorylated three-carbon compounds) are active without needing any other factor present. However, many enzymes require a non-protein component for activity; this is termed a *cofactor*. One group of cofactors is metal ions. Thus carboxypeptidase (an enzyme catalysing the hydrolysis of proteins from the C-terminal end) requires zinc for activity. Removal of zinc (e.g. by EDTA) leads to inactivation of the enzyme: activity can be restored by addition of zinc, or to a lesser extent, by other metals. In this case, it appears that zinc forms an integral part of the enzyme structure. The kinases (enzymes that catalyse the transfer of the γ-phosphoryl group of ATP to some acceptor molecule) have a requirement for magnesium ions, but the evidence suggests that, in these cases, the metal ion is actually bound to substrate rather than to the enzyme (i.e. the true substrate is Mg-ATP^{2-} rather than ATP). Some examples of metal ion requirements among enzymes are given in Table 1.2.

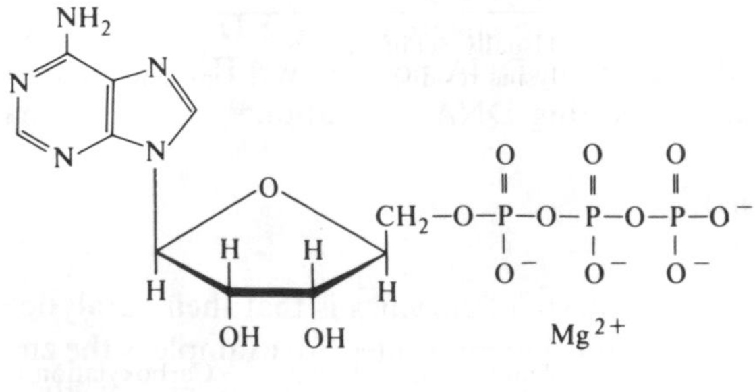

Mg-ATP^{2-}

The second major class of cofactors are organic cofactors (many of which are derivatives of B vitamins). Thus most carboxylases (enzymes involved in the incorporation of CO_2) requires biotin, which is covalently linked to its enzyme. Similarly, in the case of aminotransferases (enzymes involved in interconversions of oxo-acids and amino acids), pyridoxal phosphate is bound to the enzyme as a Schiff base. Some examples of organic cofactors are listed in Table 1.3.*

* NAD(P)$^+$ is often listed as a cofactor, and is generally known as a *coenzyme*. In fact it functions as a normal substrate for dehydrogenases and in our view is best considered as a substrate. It does, however, differ from most cell metabolites in that it becomes alternately reduced and oxidized. In this respect it does resemble redox cofactors. The same arguments apply to coenzyme A which can also be regarded as a substrate. Pantotheine, however, which is covalently attached to certain fatty-acid synthase systems (see Chapter 7, Section 7.11.1) is a cofactor. The problems of nomenclature of cofactors, coenzymes, and prosthetic groups is discussed further in reference 22.

TABLE 1.2
Examples of the metal ion requirements of some enzymes

Metal	Enzyme
Na	Intestinal sucrose α-D-glucohydrolase
K	Pyruvate kinase (also requires Mg)
Mg	Kinases (e.g. hexokinase, pyruvate kinase), adenosinetriphosphatases (e.g. myosin adenosinetriphosphatase)
Fe	Catalase, peroxidase, nitrogenase
Zn	Alcohol dehydrogenase, carboxypeptidase
Mo	Xanthine oxidase, nitrogenase
Cu	Cytochrome *c* oxidase, amine oxidase

TABLE 1.3
Examples of organic cofactors

Cofactor	Linkage to apoenzyme	Type of reactions catalysed by holoenzyme
Pyridoxal phosphate 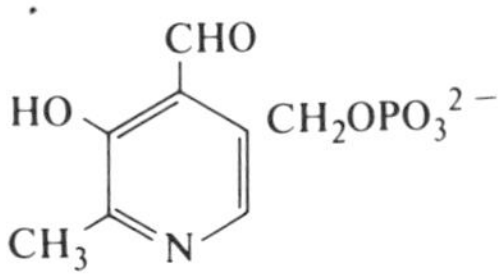 CHO HO $CH_2OPO_3^{2-}$ CH_3 N	Usually Schiff base to lysine residue	Transamination, decarboxylation, racemization
Biotin 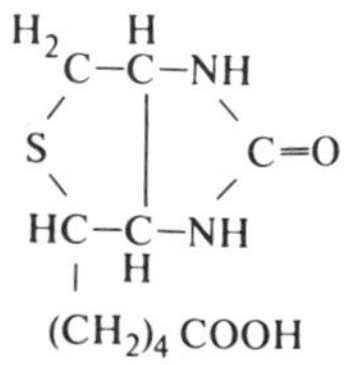 H_2 H C–C–NH S C=O HC–C–NH H $(CH_2)_4$ COOH	Amide bond to lysine residue	Carboxylation reactions, e.g. acetyl-CoA carboxylase, pyruvate carboxylase
Lipoic acid 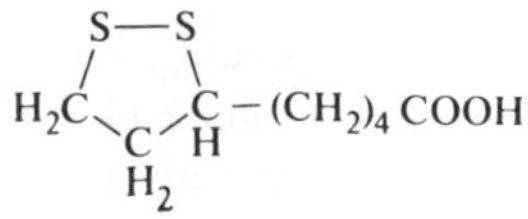 S—S H_2C C – $(CH_2)_4$ COOH C H H_2	Amide bond to lysine residue	Acyl transfer, e.g. pyruvate dehydrogenase and 2-oxoglutarate dehydrogenase systems

TABLE 1.3 (*Contd.*)

Cofactor	Linkage to apoenzyme	Type of reactions catalysed by holoenzyme
Thiamin diphosphate	Non covalent binding dissociation constant $\approx 10^{-6}$ mol dm^{-3}	Decarboxylation of 2-oxo-acids, e.g. pyruvate dehydrogenase and 2-oxoglutarate dehydrogenase systems
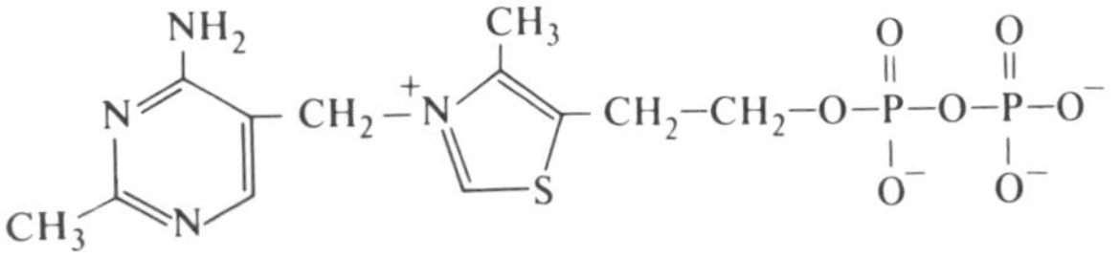 NH_2; N; CH_3; N; $CH_2-\overset{+}{N}$; CH_3; S; $CH_2-CH_2-O-P(=O)(O^-)-O-P(=O)(O^-)-O^-$		
Flavin nucleotides: flavin adenine dinucleotide (FAD) and flavin mononucleotide (FMN)	Non-covalent in, e.g. amino acid oxidase; covalent link in succinate dehydrogenase	Redox reactions, e.g. xanthine oxidase (FAD), succinate dehydrogenase (FAD), glucose oxidase (FAD), NADPH-cytochrome reductase (FMN)
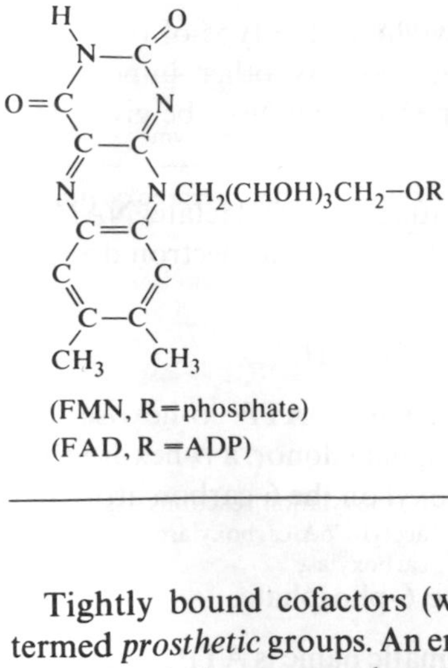 H; O; N—C; O=C; N; C—C; N; N $CH_2(CHOH)_3CH_2$—OR; C=C; C; C; C—C; CH_3; CH_3 (FMN, R=phosphate) (FAD, R=ADP)		

Tightly bound cofactors (which cannot be removed by dialysis) are often termed *prosthetic* groups. An enzyme containing a cofactor or prosthetic group is termed a *holoenzyme*: when the cofactor is removed, the resulting protein is termed an *apoenzyme*. Any small molecule or other species which can reversibly bind to an enzyme is termed a *ligand* (this can include substrates, inhibitors, metal ions, etc.).

1.5 Nomenclature and classification of enzymes

1.5.1 General classification

As in the development of organic chemistry, many enzymes were given 'trivial' names before any attempt was made at a system of nomenclature. In general the 'trivial' names consisted of the suffix '-ase' added to the substrate acted on (as in the case of urease) or implied something about the reaction catalysed (as in the case of lactate dehydrogenase, which catalyses the dehydrogenation of lactate to yield pyruvate). However, some trivial names are distinctly unhelpful in this respect, e.g. old yellow enzyme, catalase, papain, trypsin, rhodanese, etc.* The present-day accepted nomenclature of enzymes is that recommended by the Enzyme Commission (set up in 1955 by the International Union of Biochemistry in consultation with the International Union of Pure and Applied Chemistry). Enzymes are named according to certain well-defined rules (for details see references 23 and 24). The six major types of enzyme-catalysed reactions are:

1. oxidation–reduction reactions, catalysed by *oxidoreductases*;
2. group transfer reactions, catalysed by *transferases*;
3. hydrolytic reactions, catalysed by *hydrolases*;
4. elimination reactions in which a double bond is formed, catalysed by *lyases*;
5. isomerization reactions, catalysed by *isomerases*;
6. reactions in which two molecules are joined at the expense of an energy source (usually ATP), catalysed by *ligases* (also known as *synthetases*).

The full systematic name of an enzyme not only shows the type of reaction catalysed but describes the substrate(s) acted on and any other important information. One of the enzymes from each major group will now be given to illustrate the classification scheme.

Group 1, e.g. *lactate dehydrogenase.* The full systematic name is L-lactate: NAD^+ oxidoreductase, which denotes the fact that L-lactate acts as the electron donor and NAD^+ as the electron acceptor:

$$\text{L-Lactate} + NAD^+ \rightleftharpoons \text{pyruvate} + NADH.$$

Group 2, e.g. *hexokinase.* The full systematic name is ATP: D-hexose 6-phosphotransferase, indicating that ATP is the phosphate donor, a D-hexose the phosphate acceptor and that transfer is to the hydroxyl on the 6-carbon atom of the hexose:

$$ATP + \text{D-Hexose} \rightleftharpoons ADP + \text{D-hexose 6-phosphate}.$$

Group 3, e.g. *adenosinetriphosphatase.* The full systematic name is ATP phosphohydrolase, indicating that ATP, the substrate, is hydrolysed so as to split the bond which allows release of orthophosphate:

$$ATP + H_2O \rightleftharpoons ADP + \text{orthophosphate}.$$

Group 4, e.g. *fructose-bisphosphate aldolase.* The full systematic name is D-fructose 1,6-bisphosphate D-glyceraldehyde 3-phosphate-lyase, which indicates that the substrate D-fructose 1,6-bisphosphate is cleaved in such a way as to yield

* Old yellow enzyme is now termed NADPH: acceptor oxidoreductase, catalase is H_2O_2:H_2O_2 oxidoreductase, papain and trypsin do not have systematic names because of the wide variety of substrates acted upon, and rhodanese is termed thiosulphate sulphurtransferase.

D-glyceraldehyde 3-phosphate as one of the products:

$$\text{D-Fructose 1,6-bisphosphate} \rightleftharpoons \underset{\text{3-phosphate}}{\text{D-glyceraldehyde}} + \underset{\text{phosphate.}}{\text{dihydroxyacetone}}$$

Group 5, e.g. *triosephosphate isomerase.* The full systematic name is D-glyceraldehyde 3-phosphate ketol-isomerase, indicating that the aldose D-glyceraldehyde 3-phosphate is isomerized to the ketose dihydroxyacetone phosphate:

$$\text{D-Glyceraldehyde 3-phosphate} \rightleftharpoons \text{dihydroxyacetone phosphate.}$$

Group 6, e.g. *isoleucyl-tRNA synthetase.* The full systematic name is L-isoleucine: $tRNA^{Ile}$ ligase (AMP-forming) and it indicates that L-isoleucine becomes bonded to a specific tRNA acceptor ($tRNA^{Ile}$) and that in the process ATP is split to AMP and pyrophosphate*:

$$\text{ATP} + \text{L-isoleucine} + \text{tRNA}^{\text{Ile}} \rightleftharpoons \text{AMP} + \text{pyrophosphate} + \text{L-isoleucyl-tRNA}^{\text{Ile}}.$$

Many of the systematic names of enzymes are rather cumbersome and for everyday use shorter names can be used (e.g. lactate dehydrogenase, hexokinase, etc.). These shorter names are known as *recommended names.*

Enzymes are further classified by being assigned an Enzyme Commission (EC) number consisting of four parts (a, b, c, d).

The first number (a) indicates the type of reaction catalysed and can take values from 1 to 6 according to the classification of reaction types given above.

The second number (b) indicates the subclass, which usually specifies the type of substrates or the bond cleaved more precisely. In the case of oxidoreductases for example, this shows the type of chemical grouping acting as an electron donor, whereas with hydrolases it would show the type of bond that is being broken.

The third number (c) indicates the sub-subclass, allowing an even more precise definition of the reaction catalysed, in terms of the type of electron acceptor (oxidoreductases) or the type of group removed (lyases), etc.

The fourth number (d) indicates the serial number of the enzyme in its sub-subclass.

Following these rules the EC numbers of lactate dehydrogenase, hexokinase, adenosinetriphosphatase, fructose-bisphosphate aldolase, triosephosphate isomerase, and isoleucyl-tRNA synthetase are 1.1.1.27, 2.7.1.1, 3.6.1.3, 4.1.2.13, 5.3.1.1, and 6.1.1.5, respectively. All the enzymes mentioned in this book are given their recommended names, together with their EC numbers, and the reactions catalysed are listed in the Appendix at the end of the book. (There are a few enzymes that have not yet been assigned EC numbers because the reactions they catalyse have not been precisely determined.)

* The Enzyme Nomenclature Committee of the IUB has recommended that Group 6 enzymes be termed ligases and that the use of the term synthetase be discouraged (because of the possibility of confusion between synthetase and synthase).

It should be noted that the system of nomenclature and classification of enzymes is based only *on the reaction catalysed* and takes no account of the origin of the enzyme (i.e. from the species or tissue it derives). Enzymes catalysing the same reaction but isolated from different species may well have substantially different amino-acid sequences and may well act via different catalytic mechanisms, but these points will not be distinguished in the classification system. For example, the adenosinetriphosphatases from the inner mitochondrial membrane and from the sacroplasmic reticulum are not distinguished in this classification scheme, although the former is concerned with the transport of protons across the inner mitochondrial membrane and the latter with the transport of Ca^{2+} across the sacroplasmic reticulum (for details see Chapter 8, Section 8.4.4). Both catalyse hydrolysis of ATP. Another example is the two classes of fructose-bisphosphate aldolases discussed in Chapter 5, Section 5.5.3. Therefore, for a more exact specification of an enzyme the source should also be mentioned, and if necessary the isoenzyme type. The meaning of the term isoenzyme is discussed below.

1.5.2 Isoenzymes

Within a single species there may exist several different forms of enzyme catalysing the same reaction. These could differ from one another in terms of amino-acid sequence, some covalent modification (e.g. phosphorylation of serine hydroxyl groups), or possibly in terms of three-dimensional structure (conformational changes), etc. The term *isoenzyme* should be restricted to those forms of an enzyme that arise from *genetically determined differences in amino-acid sequence* and should not be applied to those derived by modification of the same amino-acid sequence.* Thus, for example, heart muscle malate dehydrogenase (which catalyses the oxidation of malate to oxaloacetate) occurs both in the cytoplasm and in the mitochondria; these two forms of the enzyme are termed *isoenzymes* of malate dehydrogenase. Similarly, in serum or in muscle extracts, lactate dehydrogenase occurs in a variety of forms (see Chapters 5 and 10, Sections 5.5.4.1 and 10.2); these are also referred to as isoenzymes. It is recommended that the naming of isoenzymes should be based on the extent of migration in an applied electric field using the technique of *electrophoresis* (see Chapter 2, Section 2.6.2.2) rather than on the basis of tissue distribution (e.g. brain type, muscle type, etc.) since this distribution can vary between different species (or in a single species or tissue with the stage of development). Electrophoresis is recommended as the basis of classification because it is a widely used technique with high resolving power. Isoenzymes are numbered starting with the species having highest mobility towards the anode (see for example Fig. 10.1).

1.5.3 Multienzyme systems

Multienzymes are proteins that exhibit more than one catalytic activity and are described in some detail in Chapter 7. For the purposes of nomenclature, the

* See reference 25 for recommendations on nomenclature of isoenzymes.

Enzyme Commission recommendation is that where more than a single catalytic activity is to be ascribed it should be referred to as a *system*. Thus, for example, the unit that contains all the catalytic entities for the synthesis of fatty acids is referred to as the fatty-acid synthase system. However, since each enzyme-catalysed reaction has an EC number and recommended name, it follows that multifunctional enzymes will have more than one EC number and position in the classification scheme. Two examples that are discussed later in the book will serve to illustrate this. The debranching enzyme (see Chapter 6, Section 6.4.2) that catalyses the removal of 1,6-branches in glycogen is a single polypeptide chain having two catalytic activities, amylo-1,6-glucosidase and 4-α-D-glucanotransferase and thus appears twice in the classification scheme as EC 3.2.1.33 and EC 2.4.1.25. Similarly, homoserine dehydrogenase and aspartate kinase activities are also associated with a single polypeptide chain (see Chapter 7, Section 7.11) and are numbered EC 1.1.1.3 and EC 2.7.2.4.

A further problem arises with the nomenclature of multienzyme systems, namely, that some consist of several polypeptide chains each having a distinct catalytic activity and associated with one another by non-covalent bonds, whilst others, like the two examples just described, exist as single polypeptide chains having multiple catalytic sites. The Recommendations (1984) of the Nomenclature Committee of the IUB did not include terminology for the systems just described, although there has been some debate on the subject.[26-29] The system adopted in this book is described in Chapter 7, Section 7.2.

1.6 The contents of this book

In the next five chapters we shall be discussing the behaviour of isolated enzymes, dealing in turn with isolation methods (Chapter 2), structural characterization (Chapter 3), kinetics (Chapter 4), catalytic action (Chapter 5), and control of activity (Chapter 6). These are areas in which there is now a considerable body of knowledge and understanding. Methods for isolation and characterization of enzymes are now well-established procedures. It is probable that many more enzymes will be isolate and characterized in the future using the techniques presently available, together with further refinements. We may therefore anticipate that in the next decade, the three-dimensional structures and the mechanism of action of many more enzymes will become known. Ultimately, of course, we want to know how enzymes behave in living cells. This involves in part a synthesis of the information obtained from the study of isolated enzymes, but it also requires detailed knowledge of the molecular morphology of the cell, which in turn requires methods for making measurements on intact cells. There are a large number of unanswered questions in this area, such as, what is the state of proteins in the cytosol? Are they just part of an intracellular soup or do they have an organization that so far has not been discerned? What are the concentrations of free ligands such as metal ions in cells? It is relatively straightforward to determine the total concentration of a ligand in cells but very difficult to ascertain the extent to which it is bound to different macromolecules. Many of these questions require new methods of investigation. The recent use of ^{31}P nuclear magnetic resonance (n.m.r.) to measure compartmentation of phosphorus-containing metabolites in intact tissues is such a development (see Chapter 8, Section 8.3.1.3). Some of these aspects are considered in Chapters 7 and 8 on multienzymes and enzymes in the cell, respectively.

As already mentioned, there are a large number of different enzymes in each cell. The amounts of certain enzymes are relatively constant throughout the life of a cell, whereas others vary with physiological conditions. How these amounts are controlled is the topic of Chapter 9 on enzyme turnover.

Of the 2477 enzymes listed in the 1984 classification scheme over 300 are commercially available in a purified or partially purified state. This has meant that enzymes can be used in some cases as reagents either on an analytical or on a preparative scale. The current developments in genetic engineering suggest that many more will be available on a larger scale in the future. These developments have assumed importance both in medicine and in industry and a discussion of them is therefore included in the final two chapters of the book.

References

1. Fruton, J. S., *Trends Biochem. Sci.* **3**, N281 (1978).
2. Dixon, N. E., Gazzola, C., Blakeley, R. L., and Zerner, B., *J. amer. chem. Soc.* **97**, 4131 (1975).
3. Hirs, C. H. W., Moore, S., and Stein, W. H., *J. biol. Chem.* **235**, 633 (1960).
4. Koshland, D. E. Jr. *Proc. natn. Acad. Sci. USA* **44**, 98 (1958).
5. Monod, J., Changeux, J.-P., and Jacob, F., *J. molec. Biol.* **6**, 306 (1963).
6. Monod, J., Wyman, J., and Changeux, J.-P., *J. molec. Biol.* **12**, 88 (1965).
7. Gutte, B., and Merrifield, R. B., *J. amer. Chem. Soc.* **91**, 509 (1969).
8. Uhlenbeck, D. C., *Nature, Lond.* **328**, 596 (1987).
9. Zaug, A. J., Been, M. D., and Cech, T. R., *Nature, Lond.* **324**, 429 (1986).
10. North, G., *Nature, Lond.* **328**, 18 (1987).
11. Dixon, M., in *Chemistry of life* (Needham, J., ed.) p. 15. Cambridge University Press (1971).
12. *FEBS Lett.* Vol. 62 Suppl., Enzymes: 100 years (1976).
13. Koshland, D. E., *J. cell. comp. Physiol.* **47**, Suppl. **11**, 217 (1956).
14. Laidler, K. J., and Bunting, P. S., *The chemical kinetics of enzyme action* (2nd edn), p. 256. Clarendon Press, Oxford (1973).
15. Mortenson, L. E., and Thornley, R. N. F., *A. Rev. Biochem.* **48**, 387 (1979).
16. Shreve, R. N., and Brink, J. A., *Chemical process industries* (4th edn), pp. 276–80. McGraw-Hill, New York (1977).
17. Old, R. W., and Primrose, S. B., *Principles of gene manipulation: an introduction to genetic engineering* (3rd edn). Blackwell, Oxford (1985).
18. You, K., Arnold, L. J., Allison, W. S., and Kaplan, N. O., *Trends Biochem. Sci.* **3**, 265 (1978).
19. Branden, C.-I., and Ekland, H., in *Dehydrogenases requiring nicotinamide coenzymes* (J. Jeffrey, ed.), p. 63. Birkhauser, Basel (1980).
20. Goodwin, T. W., *Essays Biochem.* **9**, 103 (1973).

21. Fersht, A. R., *Trends Biochem. Sci.* **5**, 262 (1980).
22. Bryce, C. F. A., *Trends Biochem. Sci.* **4**, N62 (1979).
23. Dixon, M., Webb, E. C., Thorne, C. J. R., and Tipton, K. F., *Enzymes* (3rd edn). Longman, London (1979).
24. *Enzyme nomenclature*, Recommendations of the Nomenclature Committee of the International Union of Biochemistry. Academic Press, Orlando (1984).
25. Nomenclature of multiple forms of enzymes, *Biochem. J.* **171**, 37 (1978).
26. Karlson, P., and Dixon, H. B. F., *Trends Biochem. Sci.* **4**, N275 (1979).
27. Welch, G. R., and Gaertner, F. H., *Trends Biochem. Sci.* **5**, VII (1980).
28. Von Dohren H., *Trends Biochem. Sci.* **5**, VIII (1980).
29. Karlson, P., and Dixon, H. B. F., *Trends Biochem. Sci.* **5**, VIII (1980).
30. Lerner, R. A., and Tramontano, A., *Trends Biochem. Sci.* **12**, 427 (1987).
31. Jacobs, J., Schultz, P. G., Sugasawara, R., and Powell, M., *J. amer. chem. Soc.* **109**, 2174 (1987).
32. Hansen, D. G., *Nature, Lond.* **325**, 304 (1987).
33. Wells, J. A., and Estell, D. A., *Trends Biochem. Sci.* **13**, 291 (1988).

2
The purification of enzymes

2.1 Introduction

In this chapter we shall discuss the isolation of enzymes, concentrating initially on the principles underlying the more important separation methods employed in this work. We shall then illustrate how these methods are used in practice by considering eight specific examples of purification of enzymes; the examples have been chosen to give a broad coverage of the methods and problems involved in such procedures. However, it is necessary to deal first with some more general questions regarding the need for, and strategy underlying, purification procedures.

2.2 Why isolate enzymes?

It is not difficult to appreciate that if we hope to gain a detailed understanding of the behaviour of an enzyme in a complex system (be it a subcellular organelle such as the mitochondrion, a cell, or a whole organism) we must first try to understand its properties in as simple a system as possible. In many cases this simple system would consist of a solution of enzyme in a medium containing only small ions, buffer molecules, cofactors, etc. However, in some cases, e.g. enzymes that are bound to cell membranes (see Chapter 8, Section 8.4), the isolated enzyme may be inactive in the absence of phospholipid or detergent, and our simple system would need to contain these additional components.

From studies of the isolated enzyme we can learn about its specificity for substrates, kinetic parameters for the reaction (see Chapter 4), and possible means of regulation (see Chapter 6). All these features would be useful in understanding the role of the enzyme in the more complex systems (see Chapters 7 and 8). In addition, enzymes pose some extremely intriguing questions about the structure of large molecules (see Chapter 3) and mechanisms of catalysis (see Chapter 5). Detailed studies of these aspects are only possible if we have been able to purify enzymes so as to remove contaminating enzymes and other large molecules. The ready availability of isolated enzymes has been of considerable value in a number of medical and industrial applications (see Chapters 10 and 11, respectively).

2.3 Objectives and strategy in enzyme purification

2.3.1 Objectives

The aim of a purification procedure should be to isolate a given enzyme with the *maximum possible yield*, based on the percentage recovered activity compared with the total activity in the original extract. In addition the preparation should possess the *maximum catalytic activity*, i.e. there should be no degraded or other inactivated enzyme present, and it should be of the *maximum possible purity*, i.e. it should contain no other enzymes or large molecules. In the early days of enzyme purification, crystallinity was taken to be proof of purity but there are now known to be a number of crystalline enzymes[1] that are impure, and indeed many present-day purification procedures do not involve a crystallization step (e.g. RNA nucleotidyltransferase; Section 2.8.4). Of course, crystallization is necessary if the three-dimensional structure of an enzyme is to be elucidated by X-ray crystallography (see Chapter 3, Section 3.5.1).

The catalytic activity of a preparation is determined by a suitable assay procedure (Chapter 4, Section 4.2) in which the rate of disappearance of substrate or the rate of appearance of product is determined under defined conditions of substrate concentration, temperature, pH, etc. The units of activity are usually expressed either as μmol substrate consumed, or product formed, per minute ('units' or 'international 'units') or as mol substrate consumed, or product formed, per second ('katal' in the SI system). There is no way of predicting the catalytic activity of a purified enzyme under a given set of conditions and so purification is carried out until the *specific activity* of the preparation (i.e. units per mg or katal per kg*) increases to reach a constant value that is not increased by further purification steps. The criteria used in assessing the purity of a preparation involve the application of analytical methods to detect contaminants: these are dealt with in more detail in Section 2.7.

2.3.2 Strategy

The steps involved in the purification of an enzyme can be discussed in terms of the type of 'flow sheet' shown in Fig. 2.1. Within this general scheme, the procedure to be adopted for a given enzyme will involve choices of: (i) source of enzyme; (ii) methods of homogenization; and (iii) methods of separation. These choices are discussed in the following sections (2.4 to 2.6). The progress of a purification is recorded in a purification table (see Section 2.8.1).

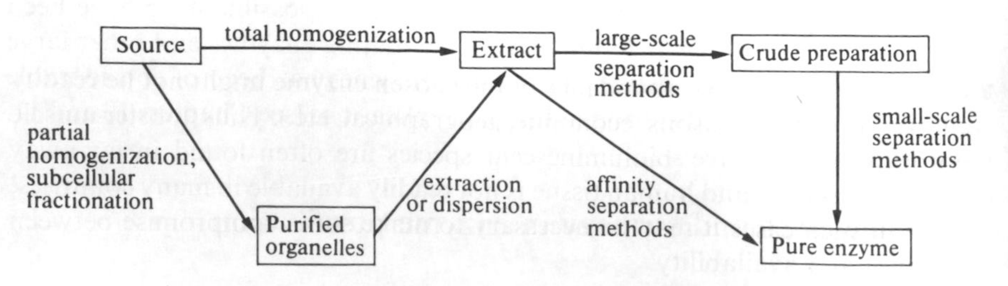

FIG. 2.1. Steps involved in the purification of enzymes.

* It should be noted that 60 units per mg corresponds to 1 katal per kg.

2.4 Choice of source

A number of factors influence the choice of starting material in an enzyme purification. Among the most important factors are the following.

2.4.1 Abundance of enzyme

In the absence of other factors it is obviously a good idea to choose a source in which the required enzyme occurs in large amounts. Thus, lactating mammary gland is an excellent source of enzymes such as acetyl-CoA carboxylase involved in fatty-acid biosynthesis,[2] and kidney is a good source of hydrolytic enzymes such as alkaline phosphatase, an enzyme that catalyses the hydrolysis of phosphate esters at alkaline pH.

In the case of microorganisms it is often possible to increase the level of a given enzyme by genetic or environmental manipulations.[3,4] Thus synthesis of β-D-galactosidase can be induced in *E. coli* by growing the bacteria in a medium containing lactose as the sole source of carbon. Increasingly, recombinant-DNA techniques are being employed to construct over-producing strains; for this approach to be successful it is necessary to isolate the gene coding for the enzyme of interest and to find a suitable system in which to express that gene. Details of these techniques can be found in reference 5. An example of the usefulness of the recombinant-DNA approach involved studies[6] on the enzyme 3-enolpyruvoylshikimate-5-phosphate synthase (EPSP synthase) in *E. coli*, which is involved in the biosynthesis of aromatic amino acids. The gene for this enzyme was isolated and inserted into the multicopy plasmid[5] pAT153. The modified plasmid was used to transform *E. coli* to give a new strain that produced approximately 100-fold more EPSP synthase than the wild-type strain; 4.8 mg pure enzyme could be isolated from 18g cells as a result of a 50-fold purification.[6] A second example of the recombinant-DNA approach is provided by glutathione reductase (Section 2.8.7). It should be pointed out that considerable problems can be encountered in attempting to express eukaryotic genes in prokaryotes[5] and that in some cases proteins that have been over-expressed in *E. coli* can form insoluble aggregates; recovery of active protein from these aggregates has not always been possible.[7]

2.4.2 Availability

A source with reasonable abundance of the chosen enzyme might not be readily available for various reasons (economic, geographical, etc.). Thus, lobster muscle is notoriously expensive, bioluminescent species are often found in far away 'exotic' locations,[8,9] and human tissue is not readily available in many countries! Clearly, in such cases it may be necessary to make some compromise between abundance and availability.

2.4.3 Comparative studies

It may be the case that a certain enzyme has been studied in one species or within one tissue in a species; in such cases it can be highly profitable to examine the corresponding enzyme in another species or another tissue. In Chapter 3, Section 3.4.9.5 we shall see how comparisons of a given enzyme from different species can be used to explore aspects of enzyme evolution. A knowledge of the properties of the different isoenzymes (see Chapter 1, Section 1.5.2) of lactate dehydrogenase is of use in discussing the role of these enzymes in heart and muscle.[10]

2.4.4 Subcellular location

If the enzyme catalysing a given reaction occurs at only one location within the cell* (e.g. succinate dehydrogenase and some other tricarboxylic acid-cycle enzymes occur only in the mitochondria), then the whole tissue can be homogenized or extracted in the purification procedure (Fig. 2.1), and subcellular fractionation is not essential. In some cases, however, a given enzyme may exist in a number of locations in the cell and some subcellular fractionation may well be necessary before the purification can proceed. Thus, in the case of mitochondrial adenosinetriphosphatase (Section 2.8.3) it is important to obtain a reasonably pure preparation of mitochondria from the chosen source (beef heart, a tissue rich in mitochondria) in order to avoid contamination from other adenosinetriphosphatases, for instance that involved in the transport of Na^+ and K^+ across the plasma membrane of the cell. Subcellular fractionation to prepare the constituent organelles of a cell is normally achieved by a succession of centrifugation steps with the larger organelles (nuclei, mitochondria) being sedimented first (see Chapter 8, Section 8.2.1).

Having taken these and possibly other factors into account, the choice of source of enzyme can be made. We now proceed to look at the steps in the purification procedure outlined in Fig. 2.1. Further details of the various methods can be found in reference 11.

2.5 Methods of homogenization

A number of methods for breaking open cells and extracting the contents are available. The choice of method usually depends on the type of tissue or organism which is used as the source of enzyme.

2.5.1 Mammalian tissue

The lack of a rigid cell wall makes homogenization of most mammalian tissues relatively easy. The tissue is often cut up into small pieces or minced prior to homogenization by either a rotating Teflon pestle in a glass mortar (the Potter–Elvehjem homogenizer) or a high-speed blender. Extraction is performed with isotonic† solutions in those cases where it is important to avoid the rupture

* A discussion of the subcellular locations of enzymes will be found in Chapter 8, Section 8.2.

† An isotonic solution is one which exerts the same osmotic pressure as the cell contents, containing approximately 0.3 mol dm^{-3} dissolved species (e.g. 0.3 mol dm^{-3} sucrose or 0.15 mol dm^{-3} KCl).

of subcellular organelles. This is particularly important in the case of cells from tissue such as liver which possess a large number of lysosomes; these are subcellular organelles containing a number of hydrolytic enzymes such as proteases (see Chapter 8, Section 8.2.4). In other cases a medium of low ionic strength such as 0.01 mol dm^{-3} KCl is often used for extraction. It may also be necessary in particular cases to add protease inhibitors or other substances (e.g. the reducing agent dithiothreitol) to prevent damage to the enzyme of interest during the extraction process.

The use of the pressure homogenization technique, in which a large pressure difference is created between the interior of a cell and its surroundings, has been discussed by Avis.[12]

2.5.2 Plant, fungal, and bacterial material

The rigid cell wall surrounding the cells in these types of material usually necessitates the use of harsher methods of extraction, e.g. grinding with abrasives such as alumina or sand, freezing and thawing, long periods of blending, or the addition of glass beads during blending. Alternatively, a French press, in which the cells are forced at high pressure through a narrow orifice, can be used. In this case, the mechanical shear forces lead to cell rupture.

However, it is often possible to disrupt the cell walls of plants, fungi, and bacteria by the use of appropriate hydrolytic enzymes. Protoplasts (cells completely or partially lacking a cell wall) may be prepared from Gram-positive bacteria, e.g. *Bacillus subtilis* by incubation with lysozyme.[13] By a modification of this method that includes the addition of EDTA, protoplasts have also been prepared from the Gram-negative bacterium *E. coli.*[14] Protoplasts have been prepared from fungi, e.g. yeast, *Neurospora* and *Aspergillus*, using enzyme extracts that generally contain a mixture of chitinase and 3-glucanases.[15] An alternative method may be to use a fungal mutant that lacks the normal cell wall, e.g. the slime mutant of *Neurospora*,[16] and which can therefore be readily disrupted by solutions of lower osmotic pressure.

Particular problems can arise with plant tissues,[17] because during homogenization the release of the contents of the vacuoles (usually acidic and containing proteases) can cause damage to enzymes. Addition of buffer can protect against such damage. In addition, plant tissues often contain certain phenolic compounds that are readily oxidized to form dark pigments that can be harmful to enzymes. The pigments can be removed by adsorption on polymers such as poly(vinylpyrrolidone) (Section 2.8.2).

2.6 Methods of separation

A successful separation of one substance from another depends on there being some property by which the substances may be distinguished. The principal properties of enzymes that can be exploited in separation methods are size, charge, solubility, and the possession of specific binding sites. Methods that

TABLE 2.1
Principal separation methods used in purification of enzymes

Property	Method	Scale
Size or mass	Centrifugation	Large or small
	Gel filtration	Generally small
	Dialysis: ultrafiltration	Generally small
Polarity		
(a) Charge	Ion-exchange chromatography	Large or small
	Chromatofocusing	Generally small
	Electrophoresis	Generally small
	Isoelectric focusing	Generally small
(b) hydrophobic character	Hydrophobic chromatography	Generally small
Solubility	Change in pH	Generally large
	Change in ionic strength	Large or small
	Decrease in dielectric constant	Generally large
Specific binding sites or structural features	Affinity chromatography	Generally small
	Affinity elution	Large or small
	Dye–ligand chromatography	Large or small
	Immunoadsorption	Generally small
	Covalent chromatography	Generally small

The term 'large-scale' is used to indicate that amounts of protein greater than about 100 mg can be handled at that particular step in the purification procedure.

exploit differences in these properties are listed in Table 2.1, and are described in the following sections (2.6.1 to 2.6.4).

2.6.1 Methods that depend on size or mass

2.6.1.1 Centrifugation[18]

Large molecules such as enzymes can be sedimented by the high centrifugal fields (up to 300 000 times gravity) generated by an ultracentrifuge. Although the rate at which any particular enzyme will sediment depends on a variety of factors including the size and shape of the molecule and the viscosity of the solution, it is found that in general the higher the M_r value the greater the rate of sedimentation. This method is not widely used in purification procedures to separate one enzyme from another because only small volumes (a few cm^3) can be dealt with in ultracentrifuges that operate at high centrifugal fields. However, centrifugation is very widely used to remove precipitated or insoluble material in the course of an isolation, e.g. to remove cell debris after homogenization or to collect enzyme that has been precipitated by the addition of ammonium sulphate (see Section 2.6.3.2). When centrifugation is being used in this way, only relatively low centrifugal fields (5000 to 50 000 *g*) are required and consequently volumes up to several dm^3 can be handled.

The use of differential centrifugation in the preparation of subcellular organelles is described in Chapter 8, Section 8.2.1.

2.6.1.2 Gel filtration[19,20]

In gel filtration, the separation between molecules of different sizes is made on the basis of their ability to enter the pores within the beads of a beaded gel. The most widely used types of gel are Sephadex (cross-linked dextrans), Bio-Gel P (cross-linked polymers of acrylamide), Sephacryl (cross-linked polymers of dextran and acrylamide) and Sepharose (agarose, which can be cross-linked for greater rigidity). Small molecules that can enter the pores of the beads are retarded as they pass down a column containing the gel; large molecules that are unable to enter the pores pass through the column unimpeded. By varying the size of the pores (which is controlled by the degree of cross-linking in the preparation of the beaded gel), it is possible to change the range of M_r values which can be fractionated. Sephadex G-100, for example, can fractionate globular proteins in the M_r range 4000–150 000.* If the shape of the enzyme is markedly non-spherical, it will be eluted from the column in an unexpected position, i.e. not at that expected on the basis of its M_r (see Chapter 3, Section 3.2.2). In order to minimize any non-specific interactions between the enzyme and the gel, it is important that extremes of ionic strength should be avoided.[20]

Gel filtration can be carried out on a large scale, but since large columns are rather time-consuming to run and the gels required to fill them are expensive, the method finds most application in the later (small-scale) stages of enzyme purifications.

2.6.1.3 Dialysis[21,22] *and ultrafiltration*[23,24]

A dialysis membrane such as cellophane acts as a sieve with holes large enough to permit the passage of globular proteins with M_r values up to about 20 000 but not of larger molecules. It is possible to change the pore sizes by various mechanical and chemical treatments.[25] Although dialysis is not generally useful for the separation of enzymes from each other, it is widely used during purification procedures to remove salts, organic solvents, or inhibitors of low M_r from solutions of enzymes.

In ultrafiltration, small molecules and ions pass through the dialysis membrane under the influence of an applied pressure (generally N_2 gas at about 4 atm pressure). This leads to concentration of an enzyme solution, which can be useful in reducing the volume of a sample during the purification procedure.

2.6.2 Methods that depend on polarity

2.6.2.1 Ion-exchange chromatography[26,27]

Ion-exchange chromatography depends on the electrostatic attraction between species of opposite charge. Ion exchangers usually consist of modified derivatives of some support material such as cellulose, Sephadex, etc. as shown in Fig. 2.2.

* The corresponding ranges for Sephacryl S-300 and Sepharose 6B are 10 000 to 800 000 and 10 000 to 4 000 000, respectively.

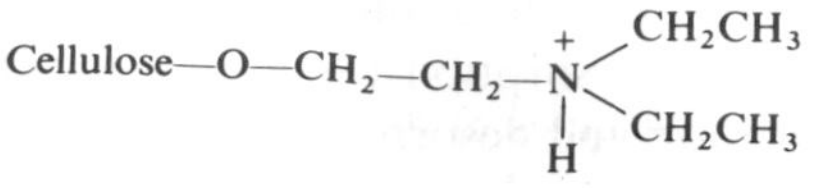

$$\text{Cellulose—O—CH}_2\text{—CH}_2\text{—}\overset{+}{\underset{|}{N}}\text{H}(\text{CH}_2\text{CH}_3)_2$$

DEAE-cellulose

(**d**iethyl**a**mino**e**thyl-cellulose)

possesses a p$K_a \approx 10$,
will bind negatively charged
species and is therefore an
anion exchanger

$$\text{Cellulose—O—CH}_2\text{—CO}_2^-$$

CM-cellulose

(**c**arboxy**m**ethyl-cellulose)

possesses a pK_a of ≈ 4,
will bind positively charged
species, and is therefore a
cation exchanger

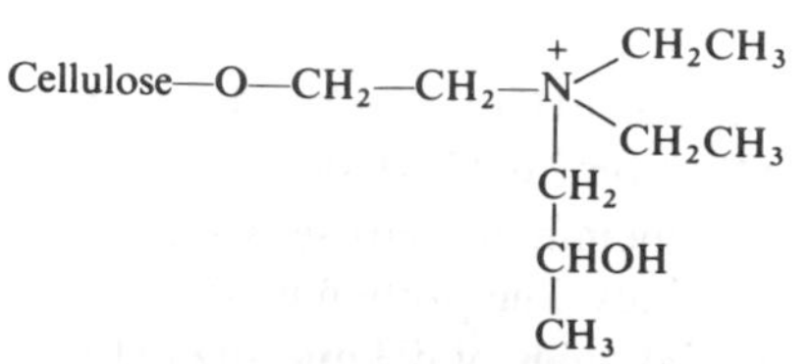

$$\text{Cellulose—O—CH}_2\text{—CH}_2\text{—}\overset{+}{N}(\text{CH}_2\text{CH}_3)_2\text{—CH}_2\text{—CHOH—CH}_3$$

QAE-cellulose
(**q**uaternary **a**mino **e**thyl-cellulose)
(more correctly, diethyl [2-hydroxypropyl]-
aminoethyl-cellulose),
a strongly basic ion-exchanger that
acts as an *anion exchanger*

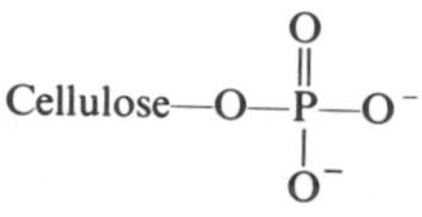

$$\text{Cellulose—O—P(=O)(O}^-\text{)—O}^-$$

Phosphocellulose,
possesses a second p$K_a \approx 6.5$,
will act as a *cation*
exchanger, but in
some respects acts also as
an affinity adsorbent[27]
(see Section 2.6.4.1)

FIG. 2.2. Ion exchangers commonly used in purifications of enymes.

During a purification procedure, the enzyme is usually applied to an ion exchanger in a solution of low ionic strength and at a pH at which the appropriate interaction will occur (i.e. the enzyme and the ion exchanger have opposite charges). Desorption of the bound species can be brought about either by changing the pH, and thus altering the charges, or, more commonly, by increasing the ionic strength of the solution so that the increased concentration of cations or anions will compete with the enzyme for the binding sites on the ion exchanger. Use of a gradient of increasing ionic strength permits the separation of proteins in a mixture on the basis of their ability to bind to the ion exchanger. When modified forms of Sephadex (such as DEAE-Sephadex) are used, separation is made on the basis of charge and (to a limited extent) size (see Section 2.6.1.2). It should be pointed out that DEAE-Sephadex shrinks as the ionic strength of the solution increases; this can cause problems during ion-exchange chromatography. The agarose-based ion exchangers such as DEAE-Sepharose do not suffer from this problem.

Ion-exchange chromatography can be performed on a large or a small scale. On a large scale it is often convenient to work in a batchwise manner, i.e. to adsorb the enzyme by adding the ion exchanger to a solution and then to pour the material into a column for controlled desorption using an ionic-strength gradient. On a small scale, both adsorption and desorption are performed in a column.

The degree of purification effected by an ion-exchange step is generally up to about 10-fold (as judged by the specific activity), but examples are known where much better results have been achieved[28] (see also Section 2.8.6). Ion-exchange chromatography finds very wide application in present-day purification procedures. In Section 2.6.4.2 we shall describe the technique of affinity elution from ion exchangers.

In general the desorption of proteins from ion-exchangers by changes in pH has not been of great use in protein purification, partly because the buffering properties of the bound proteins distort the shape of the applied pH gradient. However, the relatively new technique of 'chromatofocusing' overcomes most of these problems.[29] In chromatofocusing, the pH gradient is formed by using mixtures of ampholyte-type buffers (see Section 2.6.2.3) that are of high buffering capacity. The ion-exchanger is poly(ethyleneimine) agarose, which has a wide range of titrable groups of differing pK_a values. The protein mixture is adsorbed on the ion-exchanger at the upper end of the range of pH and the pH gradient is generated on the column by addition of the acid form of the ampholyte buffer. As the pH falls, proteins are eluted from the column roughly in the order of their isoelectric points (see Section 2.6.2.3) and the technique is claimed to be of high resolving power. The costs of the materials involved make working on a large scale expensive, and thus chromatofocusing is probably most useful at the later stages of a purification procedure.

2.6.2.2 *Electrophoresis*[30]

Electrophoretic separation is based on the differential movement of charged molecules under the influence of an applied potential difference. The rate of movement of a species is governed by the charge it carries and also by its size and shape. In order to minimize convection currents, the buffer (electrolyte) solution is soaked into a support (paper, cellulose powder, starch, or polyacrylamide gels). The position of a protein on the support can be determined by use of a stain such as Coomassie Blue which binds to proteins. Although the technique is normally performed on a small, analytical scale (a few milligrams or less), it is possible to use the method on a large preparative scale.[31] After preparative electrophoresis, the separated enzymes can be eluted out of the support or run off the bottom of a column in sequence.

2.6.2.3. *Isoelectric focusing*[32, 33, 34]

Isoelectric focusing is based not on the rate at which charged species move in an electric field but rather on their equilibrium position in a pH gradient. The pH gradient is established by applying a potential difference to a gel soaked with a solution containing a mixture of ampholytes that are polyamino acids with different charge properties and hence different isoelectric* points (pI values) (Fig. 2.3). When the potential difference is applied, a negatively charged molecule will migrate towards the anode until it encounters the acid; the acid will tend to neutralize the charge so the molecule will stop. The reverse argument holds for positively charged species. By this means the ampholytes will arrange themselves in the gel so that there is a gradual increase of pH from anode to cathode.

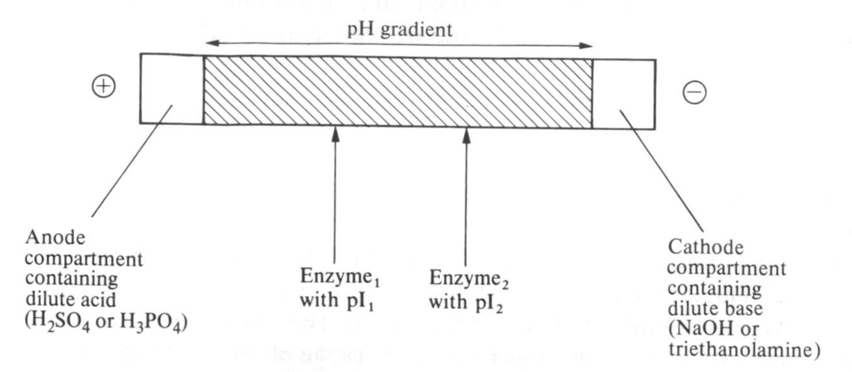

FIG. 2.3. Separation of enzymes by isoelectric focusing.

An enzyme applied to the gel will then migrate to the position in the gel where the pH equals its isoelectric point. For example if the enzyme were at a position where the pH is above its pI, then its net negative charge would ensure that it moved towards the anode until it reached its pI. It is possible to separate species that differ only slightly (as little as 0.01 pH units) in isoelectric points, and thus this method can be a very powerful tool in purification procedures. The method is generally used on an analytical scale, although it is possible to scale up for preparative work (up to gram quantities).[32]

2.6.2.4 *Hydrophobic interaction chromatography*

In an aqueous environment, hydrocarbons and other non-polar molecules will tend to associate with each other rather than with water molecules (see Chapter 3, Section 3.5.5.4 for a further discussion of this 'hydrophobic' effect). The surfaces of proteins contain a high proportion of charged and polar amino-acid side-chains (see Chapter 3, Section 3.5.4.5) but usually contain at least some non-polar side-chains. It is the presence of these non-polar side-chains that accounts for the ability of many proteins to bind to non-polar molecules. Such interactions are exploited in the technique of hydrophobic-interaction chromatography.[35, 36] In this technique, proteins are adsorbed on matrices such

* The isoelectric point of a compound is the pH at which it carries no net charge and hence does not move under an applied potential difference.

as octyl- or phenyl-Sepharose, usually at high ionic strength at which the hydrophobic interactions are stronger. Desorption of proteins can be brought about by decreasing the ionic strength and, if necessary, by addition of organic solvents or non-ionic detergents to weaken the hydrophobic interactions further. In the purification of 3-enolpyruvoylshikimate-5-phosphate synthase from *E. coli* (Section 2.4.1), hydrophobic-interaction chromatography on phenyl-Sepharose was employed to give a 5-fold purification with 80 per cent recovery of activity. The enzyme was adsorbed at 0.8 mol dm^{-3} $(NH_4)_2SO_4$ in a 0.1 mol dm^{-3} tris-HCl buffer at pH 7.5 and eluted using a decreasing gradient of $(NH_4)_2SO_4$ (0.8–0.0 mol dm^{-3}) in this buffer.[6]

2.6.3 Methods based on changes in solubility

The solubility of a compound in a given solvent depends on the balance of the forces between solute and solute and those between solute and solvent. If the former predominate the compound will be insoluble, whereas if the latter are predominant the compound will be soluble. In the course of enzyme purifications it is possible to alter the balance between these forces and hence precipitate the enzyme of interest or remove contaminating enzymes. Three of the most important ways of changing the solubility of enzymes are (i) to change the pH; (ii) to change the ionic strength; (iii) to decrease the dielectric constant. These methods can be applied on a large scale and are often used in the initial stages of a purification procedure.

2.6.3.1 Change in pH[37]

An enzyme is least soluble at its isoelectric point, since at this pH there will be no repulsive electrostatic forces between the enzyme molecules. Adjustment of the pH to the appropriate value can therefore be used to precipitate an enzyme; this method is employed in the case of adenosinetriphosphatase (Section 2.8.3). A change in pH can also be used to precipitate unwanted enzymes and proteins (see the case of adenylate kinase; Section 2.8.1). It is important to check that the enzyme of interest is not inactivated by exposure to these changes in pH.

2.6.3.2 Change in ionic strength[38]

Large charged molecules are generally only slightly soluble in pure (deionized) water; addition of ions promotes solubility by helping to disperse the charge carried by the large molecule. This phenomenon is known as salting in. However, if the ionic strength is increased beyond a certain point, the charged molecule will be precipitated (salting out). Although the theoretical basis for this behaviour is not too well understood, it is probable that a major factor is that at very high concentrations of salt, the concentration of water is significantly decreased, leading to a decrease in solute–solvent interactions and hence in solubility. The salt of choice in enzyme purification procedures is ammonium sulphate, which has the advantages of cheapness, high solubility in water (a saturated solution at

298 K (25°C) has a molarity of approximately 4 mol dm^{-3}) and lack of harmful effects on most enzymes.* Each enzyme will usually begin to precipitate at a certain concentration of ammonium sulphate and this forms the basis of an initial fractionation procedure. For instance, if it is known from a preliminary experiment that an enzyme, X, is precipitated at around 50 per cent saturation† ammonium sulphate, then the procedure might involve addition of ammonium sulphate to 45 per cent saturation, centrifugation to remove the unwanted precipitated protein followed by addition of more ammonium sulphate to the supernatant to bring the concentration up to 55 per cent. The precipitate containing X is removed by centrifugation.

In general, the degree of purification that can be achieved by ammonium sulphate fractionation is less than 10-fold, but the method is often used in the initial stages of a purification to reduce the volume of solution to be dealt with since the precipitated enzyme can be redissolved in a small volume of buffer. In at least two cases, however, it is possible to obtain pure enzyme by ammonium sulphate fractionation (glyceraldehyde-3-phosphate dehydrogenase[39] and fructose-bisphosphate aldolase[40] from rabbit muscle).

Ammonium sulphate is also commonly used in crystallization of enzymes.[41] The concentration of salt is raised until slight turbidity is observed; when the solution is left to stand, crystals of the enzyme should form. Some of the larger crystals (with dimensions of the order of 1 mm) may be suitable for X-ray crystallography (see Chapter 3, Section 3.5.1).

2.6.3.3 *Decrease in dielectric constant*[42, 43]

Addition of a water-miscible organic solvent (such as ethanol or acetone) will decrease the dielectric constant of a solution and hence increase electrostatic forces. Although there can be rather complex effects on both solute–solute and solute–solvent forces, in general the net effect is to cause precipitation of large, charged molecules such as enzymes. This method can be used on a large scale as an initial fractionation procedure. It must be remembered, however, that addition of organic solvents can sometimes lead to the inactivation of enzymes (see Chapter 3, Section 3.7.1) and it is normally important to work at low temperatures to minimize such inactivation.

An alternative method of precipitating proteins is to add the neutral water-soluble polymer poly(ethylene glycol).[43, 44] The polymer is thought to act by removing water from the hydration spheres of proteins.[44] Poly(ethylene glycol) preparations of average M_r 4000 or 6000 are commonly used and can give highly successful purifications (see Section 2.8.2). Difficulties can arise in removing residual traces of poly(ethylene glycol) from protein preparations but this is often unnecessary.[43]

* It should be noted that ammonium sulphate is a weak acid. When it is added to a weakly buffered solution, ammonia or another base should be added to maintain the pH at 7 or above.

† Concentrations of ammonium sulphate are usually expressed as percentages of the concentration required to saturate the solution.

2.6.4 Methods based on the possession of specific binding sites or structural features

Enzymes normally display highly specific interactions with their substrates and other ligands (see Chapter 1, Section 1.3.2). Advantage can be taken of this specificity in the so-called affinity separation methods of enzyme purification discussed below.

2.6.4.1 Affinity chromatography[45-49]

In affinity chromatography, a molecule such as a substrate or competitive inhibitor (see Chapter 4, Section 4.3.2.2) that interacts specifically with the enzyme of interest is linked covalently to an inert matrix (such as agarose). When a mixture is passed down a column containing the affinity matrix, only this enzyme is retained and other enzymes and proteins are washed away. The bound enzyme can be desorbed by a pulse of substrate, which will compete for the binding sites on the enzyme, or by changing the pH or ionic strength of the solution in such a way as to weaken the binding of enzyme to the column (Fig. 2.4).

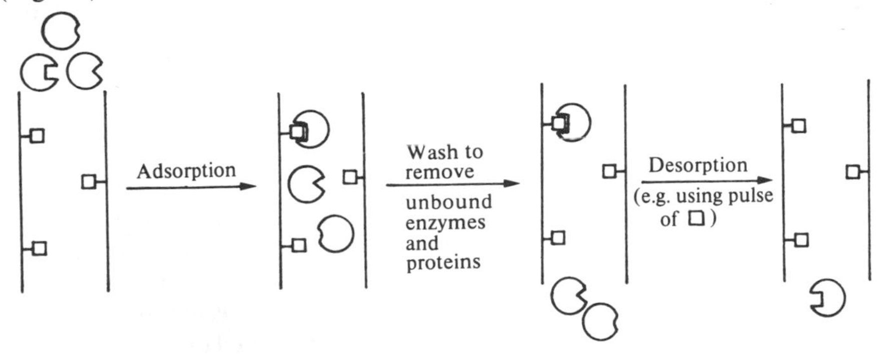

FIG. 2.4. Schematic illustration of enzyme purification using affinity chromatography.

This simple method is, in principle, capable of purifying an enzyme in a single step from a crude extract and a number of such purifications have been reported (see the example of Staphyloccal nuclease[50, 51] outlined in Fig. 2.5).

There are, however, a number of problems associated with the use of affinity chromatography in enzyme purifications; these include

1. Attaching a suitable substrate analogue or inhibitor to the matrix can be a difficult task and the reactions involved in coupling (Fig. 2.5) have not been completely characterized.
2. Linking of the ligand to the matrix may interfere with the binding to the enzyme (see Section 2.8.5) and lead to a loss of specificity in the interaction. In this connection, the need to interpose spacer arms between ligand and matrix to prevent interference by the matrix in the interaction has been recognized for some time.[52] However, the most widely used spacer (1,6-diaminohexane)

The enzyme acts on substrates of the general type pTp - X and is competitively inhibited by pdTp . The compound

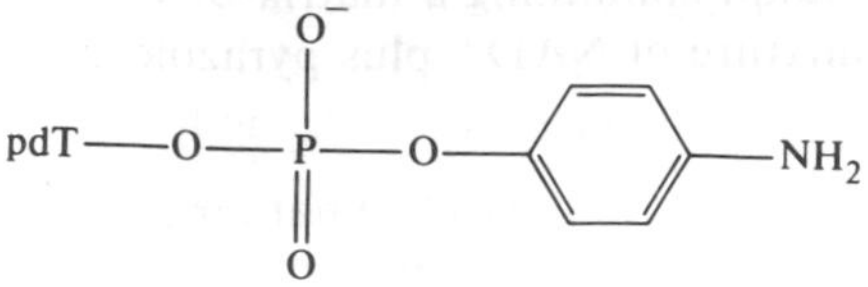

which acts as a powerful competitive inhibitor of the enzyme was attached to a cyanogen bromide-activated agarose gel via the free amino group:

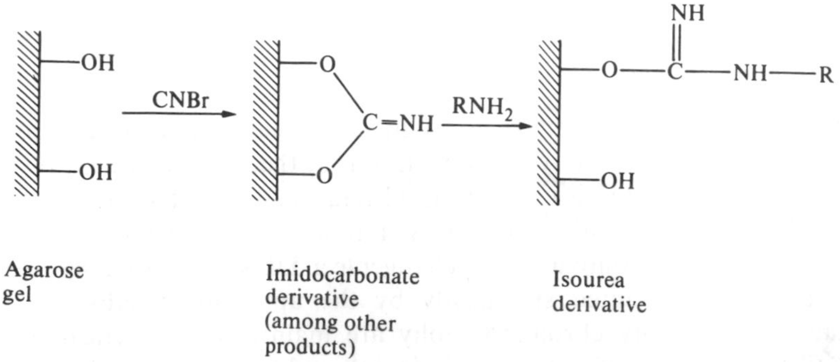

The nuclease was retained by the affinity column and could be subsequently eluted at low pH. Purification could be achieved in a single step from a crude cell extract.[51]

FIG. 2.5. Use of affinity chromatography in the purification of Staphyloccal nuclease.

can itself participate in non-specific hydrophobic interactions (see Chapter 3, Section 3.5.5.4) with enzymes,[53] so the use of alternative hydrophilic spacers has been suggested.[54]

3. For affinity chromatography to work succcessfully the strength of the interaction between matrix-bound ligand and the enzyme must be in the correct range.[55, 56] If it is too weak, with a dissociation constant greater than about 2 mmol dm^{-3} (see Section 2.8.7), the enzyme will not be retarded by the column. However, if the interaction is too strong (dissociation constant smaller than about 10^{-8} mol dm^{-3}), removal of the bound enzyme may only be possible under harsh conditions that lead to inactivation of the enzyme. Information obtained from inhibition studies of an enzyme in solution can be of considerable value in the successful design of affinity chromatography systems (see the example of hexokinase in Section 2.8.8).
4. Special problems are posed by enzymes that catalyse reactions involving more than one substrate. Thus it would be expected that all NAD^+-dependent dehydrogenases would 'recognize' matrix bound NAD^+. The differences between the various dehydrogenases occur in the binding of the second substrate, and in order to exploit these differences we normally have to

pay particular attention to the method of elution from the column. In the case of horse liver alcohol dehydrogenase, for example, specific elution of the enzyme from a column containing a matrix-bound AMP analogue could be obtained with a mixture of NAD^+ plus pyrazole, a competitive inhibitor of the enzyme.[57]

In spite of these problems, affinity chromatography has made a very significant contribution to the purification of enzymes and also of other biological molecules such as antibodies that possess specific binding sites. The technique is most likely to be successful when full use is made of information from solution studies (see Section 2.8.8).

2.6.4.2 *Affinity elution*[58]

Affinity elution is complementary to affinity chromatography in that in the former technique the specificity of interaction is at the stage of *desorption* from the chromatographic support material, whereas in the latter the specificity occurs at the stage of *adsorption*. In affinity elution, the enzyme of interest (along with others) is adsorbed onto an ion exchanger (e.g. DEAE-cellulose, see Section 2.6.2.1), and then is eluted specifically by the appropriate substrate. The advantages over affinity chromatography are mainly of a practical nature; namely (i) the often complex task of designing and attaching the appropriate ligand to the matrix is not required, and (ii) columns of high capacity are more readily available, since ion-exchangers are much cheaper than affinity matrices. Affinity elution was used in the early 1960s in the purification of fructose-bisphosphatase from rabbit liver,[59] but the technique has been developed more extensively by others, including Scopes,[60, 61] who has devised a method for the separation of all the enzymes involved in the glycolytic pathway (see Chapter 6, Section 6.4.1) in a single multiple-operation scheme. Thus, for instance, a mixture of glycolytic enzymes already partially fractionated by ammonium sulphate (those precipitating between 45 and 65 per cent saturation) is loaded on to a CM-cellulose column at pH 6.5; elution with phosphoenolpyruvate (0.5 mmol dm^{-3}) yields pyruvate kinase, elution with D-fructose 1,6-bisphosphate (0.2 mmol dm^{-3}) yields fructose-bisphosphate aldolase.[60, 61] An analogous procedure has been reported for purification of the tricarboxylic acid-cycle enzymes from beef heart.[62]

2.6.4.3 *Dye-ligand chromatography*[63]

In the mid 1970s it became clear that the dye Cibacron Blue F3G-A (Fig. 2.6) (which had been used to label high-M_r dextran preparations to provide a calibration for gel-filtration columns) could bind to a number of enzymes such as dehydrogenases and kinases.[64] These enzymes possess a common structural feature known as the nucleotide-binding domain (see Chapter 3, Section 3.5.3.3). Although the molecular basis for the apparent specificity has not been completely clarified,[65,66,67,68] the interaction has proved a useful tool in protein

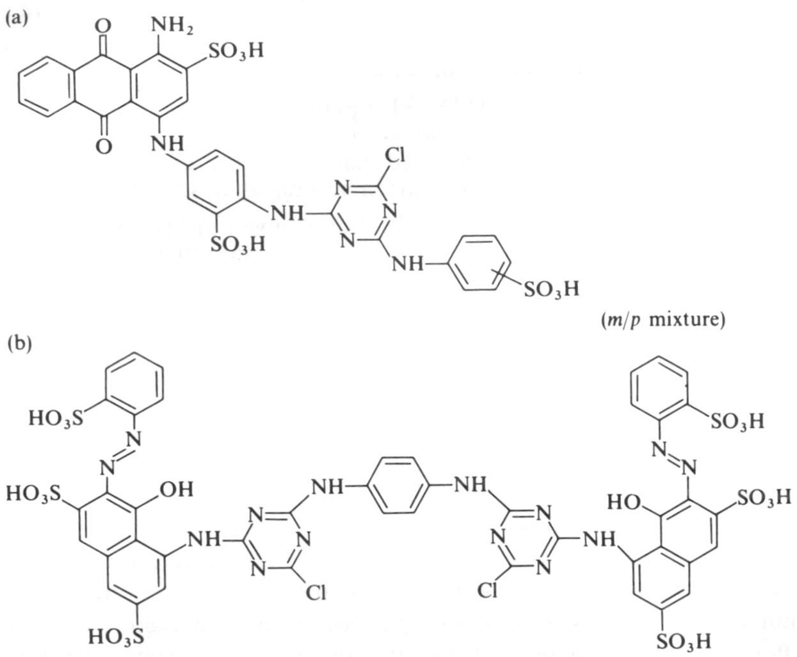

FIG. 2.6. Structures of two dyes used in dye-ligand chromatography. (a) Cibacron Blue F3G-A; (b) Procion Red HE-3B. These dyes can be linked to matrices such as agarose by reaction of the triazinyl chloride groups.

purification. The dye can conveniently be coupled to an agarose matrix via the triazinyl chloride group and the enzymes can be eluted by addition of appropriate substrate or ligand or merely by increasing the ionic strength.[64]

Since the early work on Cibacron Blue, a large number of triazine dyes have been examined as potential tools for protein purification. The Procion dye series (produced by Imperial Chemical Industries) have been particularly well studied* and many of them show high degrees of selectivity for particular classes of proteins. Thus Procion Red HE-3B (Fig. 2.6) linked to Sepharose binds to $NADP^+$-dependent dehydrogenases (such as glucose-6-phosphate dehydrogenase) in preference to NAD^+-dependent dehydrogenases (such as malate dehydrogenase) and can be used for purification of the former type from crude extracts of yeast.[69] By contrast NAD^+-dependent dehydrogenases are bound preferentially by Cibacron Blue F3G-A. Hey and Dean[70] have advocated the use of a 'tandem' strategy for protein purification in which a crude extract is first passed through a dye column that does not retain the protein of interest, and then through a second dye column that does retain the protein. The first column

* It should be pointed out that structural information about many of these dyes is not widely available, often for commercial reasons.

thus acts to remove a number of contaminating proteins that might interfere with the second step.

As well as varying the structure of the dye, the conditions of adsorption and elution can be varied.[63] In particular it is important to note that under certain conditions the immobilized dyes can act as effective cation exchangers. To minimize these effects, it is recommended that the operations should be carried out at an ionic strength $\geqslant 0.1$ and at pH $\geqslant 7$.[63] An example of the use of dye-ligand chromatography in protein purification is given in Section 2.8.7.

2.6.4.4 *Immunoadsorption chromatography*[71]

The high specificity shown in antibody–antigen interactions can often provide a suitable basis for purification procedures. If a small amount (ca. 0.1 mg) of a pure enzyme from one species is available, it can be used to raise antibodies in another species (usually a rabbit, goat, or mouse). A range of antibodies will be produced that vary in affinity for the enzyme because they recognize different structural features (epitopes) of the enzyme. After purification, the antibodies can be coupled to a matrix such as CNBr-activated Sepharose (see Fig. 2.5) and subsequently used to separate the enzyme antigen from a complex mixture. Desorption can be brought about by a change in pH, an increase in ionic strength or other treatments which weaken the antibody-antigen binding. (The desorption step can be the most difficult part of the entire procedure because the enzyme activity can be destroyed during the harsh conditions required to weaken the strong interactions.) The application of monoclonal antibody techniques[72, 73] has helped to overcome a number of the problems previously encountered; thus it is often possible to choose antibodies that have an appropriate affinity for the enzyme antigen and to produce an indefinite supply of such antibodies. The monoclonal-antibody technique has been used to purify aromatic L-amino acid decarboxylase from bovine brain.[74] In this case the previously purified enzyme from bovine adrenal medulla was used as the antigen for monoclonal-antibody production. The antibody was immobilized on an activated cross-linked agarose matrix and used to provide a 25-fold purification of enzyme from extracts of bovine brain which had been already subjected to ion-exchange chromatography. Elution of enzyme from the immobilized antibody was brought about by washing with a solution containing 50 mmol dm^{-3} acetic acid and 10 per cent (v/v) ethylene glycol.

2.6.4.5 *Covalent chromatography*

In covalent chromatography, a covalent bond is formed between molecules in the mobile phase and the stationary-phase matrix, in contrast to the non-covalent interactions described earlier in this section. The principal applications have been in the separation and purification of cysteine-containing peptides and proteins, via the disulphide exchange reactions shown in Fig. 2.7(a).

A commonly used matrix is activated thiol-Sepharose 4B in which a 2-thiopyridyl group is attached via a glutathione spacer arm in order to minimize

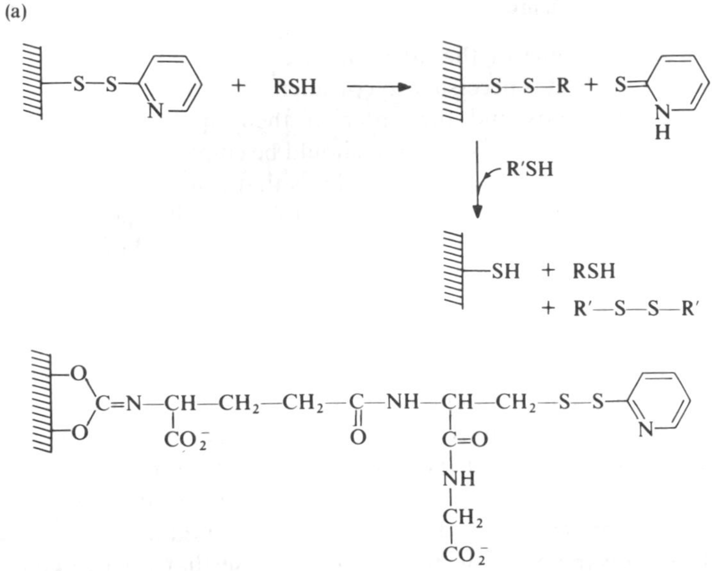

(a)

$$\text{Matrix}-S-S-\text{(2-pyridyl)} + RSH \longrightarrow \text{Matrix}-S-S-R + S{=}\text{(pyridine-2-thione, NH)}$$

$$\downarrow R'SH$$

$$\text{Matrix}-SH + RSH + R'-S-S-R'$$

(b)

$$\text{Matrix}{<}^{O}_{O}{>}C{=}N-\underset{CO_2^-}{CH}-CH_2-CH_2-\underset{O}{\overset{\|}{C}}-NH-\underset{\underset{\underset{\underset{CO_2^-}{|}}{CH_2}}{\underset{|}{NH}}}{\underset{C=O}{CH}}-CH_2-S-S-\text{(2-pyridyl)}$$

FIG. 2.7. Covalent chromatography. (a) Reaction scheme. In the first step a protein or peptide containing a cysteine group (RSH) reacts with the immobilized 2-thiopyridyl group to form a mixed disulphide. In the second step, a thiol reducing agent such as 2-mercaptoethanol or dithiothreitol (R′SH) is added to liberate RSH. (b) Partial structure of activated thiol-Sepharose.

steric interference to the disulphide exchange reactions[75] (Fig. 2.7(b)). The technique has been used to purify the cysteine-containing protease, papain, from dried papaya latex[75] and can also be adapted to separate a mixture of cysteine-containing proteins. By eluting with successively more powerful reducing agents (e.g. L-cysteine, reduced gluthathione, 2-mercaptoethanol, and finally dithiothreitol) separation of protein-disulphide isomerase from glutathione-insulin transhydrogenase could be achieved.[76]

Activated thiol-Sepharose can also be used to immobilize cysteine-containing proteins for affinity chromatography. This approach is described in Section 2.8.5 in connection with the purification of cytochrome *c* oxidase. A further application of covalent chromatography is mentioned in Chapter 3, Section 3.4.8, for the purification of a cysteine-containing peptide.

2.6.5 Other methods

In Sections 2.6.1 to 2.6.4 we have discussed the principal methods employed in enzyme purification procedures. There are, however, a number of other methods, including heat denaturation,[77, 78] fractional adsorption on calcium phosphate gels and hydroxyapatite,[79, 80, 81] and concentration by freeze-drying (lyophilization).[82] Further details about these methods can be found in the references given.

2.6.6 Choice of methods

Having described some of the advantages and disadvantages of the various separation methods, it is necessary to consider some of the factors that influence the choice of methods and the order of their application in a particular purification procedure. At the outset it should be emphasized that there is rarely only one method or combination of methods that can be used to purify a given enzyme. The actual sequence of methods employed will depend on a variety of factors such as (i) the scale of the preparation and the yield of enzyme required; (ii) the time available for the preparation; and (iii) the equipment and expertise available in the laboratory.

In general, methods based on changes in solubility (Section 2.6.3) are more suitable in the earlier (large-scale) stages of a purification, whereas methods involving column chromatography (e.g. ion-exchange chromatography) or electrophoresis are more appropriate in the latter (small-scale) stages.

In some cases, especially if there is the possibility of proteolysis during the procedure, it may be desirable to aim for as quick a purification as possible at the expense of a lower yield. If so, the use of methods based on solubility would be favoured because these are more rapid than those that involve column chromatography. Recently matrices have been developed to allow column chromatography procedures to be carried out more rapidly under high-performance liquid chromatography (h.p.l.c.) conditions. Gel filtration can be performed using Superose, a rigid cross-linked agarose-based matrix (Section 2.8.7). Other beaded hydrophilic matrices are available for ion-exchange chromatography and chromatofocusing (Section 2.6.2.1). Although these matrices can give superior resolution compared with conventional materials, it should be noted that they are expensive and require specialized apparatus. They are thus generally used in the final stages of purification procedures.

Despite the great advances made in recent years by the advent of affinity techniques (Section 2.6.4), there are a number of enzymes that are still conveniently prepared by 'old fashioned' techniques such as precipitation by ethanol or ammonium sulphate (Section 2.6.3). In these cases, which are generally enzymes present in large amounts, it is probably not worth the trouble to devise new procedures such as affinity chromatography unless there is evidence that the enzyme prepared by the traditional methods is heterogeneous or partially inactivated.

Whatever method or combination of methods is finally decided upon, the progress of the purification should be recorded in a purification table (see the example given in Section 2.8.1).

2.7 How to judge the success of a purification procedure

As explained in Section 2.3.1, the aim of a purification procedure is to obtain enzyme of the *maximum possible purity and maximum catalytic activity*. We shall now consider some methods to test these properties.

2.7.1 Tests for purity

Many of the methods of separation described in Section 2.6 can be used on an analytical scale to check that an enzyme preparation is homogeneous, although we should remember that an analytical test can only really be used to demonstrate the presence of impurities rather than prove their absence. Some of the more commonly employed analytical methods are listed in Table 2.2

TABLE 2.2
Some commonly employed analytical methods to check the purity of enzyme preparations

Method	Comments
Ultracentrifugation (Chapter 3, Section 3.2.1)	Not very satisfactory for detecting small (<5%) amount of impurities. Problems arise from associating–dissociating systems (Chapter 3, Section 3.2.4)
Electrophoresis (Section 2.6.2.2)	Should be run at a number of values of pH (as two enzymes might move together at a single pH)
Electrophoresis in the presence of sodium dodecylsulphate (Chapter 3, Section 3.2.3)	A good method for detecting impurities that differ in terms of subunit M_r. Very useful to detect proteolysis in preparations. Problems arise from enzymes composed of non-identical subunits since these give multiple bands (Chapter 3, Section 3.6.1.5)
Isoelectric focusing (Section 2.6.2.3)	A very sensitive method for detecting impurities. Sometimes artefacts can arise suggesting apparent heterogeneity[83, 84, 85]
N-terminal analysis (Chapter 3, Section 3.4.4.1)	Should indicate the presence of a single polypeptide chain. Some enzymes have a blocked N-terminus (Chapter 3, Section 3.4.7.1); others consist of more than one chain held together by disulphide bonds (e.g. chymotrypsin)

In using these methods we should be aware that the impurities may only constitute a small amount (1 per cent or so) of the total protein and may be missed altogether. It is often easier to test for the presence of contaminating enzymes by measurements of catalytic activity (e.g. testing for the presence of lactate dehydrogenase in a preparation of pyruvate kinase) than to use analytical methods to try to indicate their absence. However, if the preparation of the enzyme of interest appears to be homogeneous in a number of these analytical tests we can be reasonably confident that it is pure.

2.7.2 Tests for catalytic activity

In testing for the catalytic activity of a preparation it is important to check that the assay conditions (see Chapter 4, Section 4.2) are optimal, i.e. that any necessary activators or cofactors are present and that inhibitors are absent. It is also worthwhile to investigate the conditions under which the enzyme is stable during storage; in some cases a reducing agent such as dithiothreitol or 2-mercaptoethanol may be necessary to maintain cysteine side-chains in a reduced state; in others storage at low temperatures, e.g. in 50 per cent (v/v) glycerol solution at 255 K (−18°C), may serve to minimize processes leading to inactivation. The possibility of degradation (e.g. caused by traces of proteases) during long-term storage should also be borne in mind (see the example in Section 2.8.6).

2.7.3 Active-site titrations

Even if an enzyme preparation is shown to be homogeneous by the various analytical methods listed in Table 2.2, there is still a possibility that some of the enzyme may be in an inactive form. In a number of cases it has proved possible to estimate the amount of active enzyme by a technique known as active-site titration.[86] In this method we observe the rapid release of a product of reaction between the enzyme and a substrate (or pseudosubstrate), when the intermediate breaks down slowly, if at all, to regenerate enzyme (Fig. 2.8). Inactive enzyme does not give rise to any product, so the concentration of product formed gives the concentration of active enzyme.

(a) **General scheme**

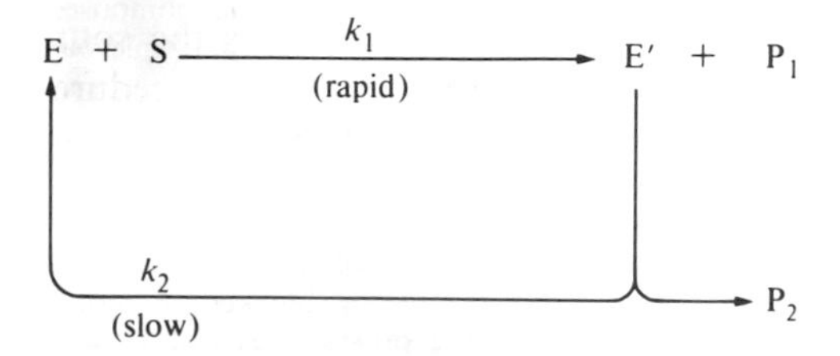

If k_2=0, the concentration of P_1 produced gives the concentration of active enzyme directly. If k_2 is less than k_1, but not equal to zero, it is still possible to calculate the concentration of active enzyme (see reference 91)

(b) **The concentration of active enzyme in a preparation of trypsin**

The reagent, *p*-nitrophenyl-*p′*-guanidinobenzoate (PNGB) reacts with

NO_2–C₆H₄–O–C(=O)–C₆H₄–NH–C($=\overset{+}{N}H_2$)(NH_2)

PNGB

the reactive serine side chain of trypsin to form an acyl enzyme derivative (see Chapter 5, Section 5.5.1.3) and liberate *p*-nitrophenol which can be detected by its yellow colour ($\lambda_{max} \approx 410$ nm). Measurement of the concentration of the *p*-nitrophenolate ion gives the concentration of the active enzyme directly. In this case, the acyl enzyme derivative does not break down to regenerate enzyme.

$E{-}CH_2{-}OH + PNGB \xrightarrow{-H^+} NO_2{-}C_6H_4{-}O^- + E{-}CH_2{-}O{-}C(=O){-}C_6H_4{-}NH{-}C(=\overset{+}{N}H_2)NH_2$

(stable)

FIG. 2.8. Active-site titration of an enzyme.

An example of the use of the method was the demonstration that commercial preparations of the protease trypsin contained only 59 and 72 per cent of active enzyme[87] (Fig. 2.8); the inactive contaminants could be removed by affinity chromatography. Active-site titration methods have been developed for a number of other enzymes, including chymotrypsin,[86] papain,[88] isoleucyl-tRNA synthetase,[89] and ornithine decarboxylase.[90]

2.8 Examples of purification procedures

In this section, eight examples of enzyme purification procedures will be outlined to show how the various separation methods described in Section 2.6 are applied in actual cases. The examples illustrate the wide variety of sources, methods, and problems encountered.

2.8.1 Adenylate kinase from pig muscle[92]

Adenylate kinase catalyses the reaction between adenine nucleotides

$$\mathrm{ATP} + \mathrm{AMP} \underset{}{\overset{\mathrm{Mg^{2+}}}{\rightleftharpoons}} 2\,\mathrm{ADP}$$

and is very widely distributed in nature. Pig muscle is chosen as the source because this enzyme gives crystals suitable for X-ray diffraction. The procedure is summarized in the flow sheet shown in Fig. 2.9; the progress of the purification is recorded in a purification table (Table 2.3).

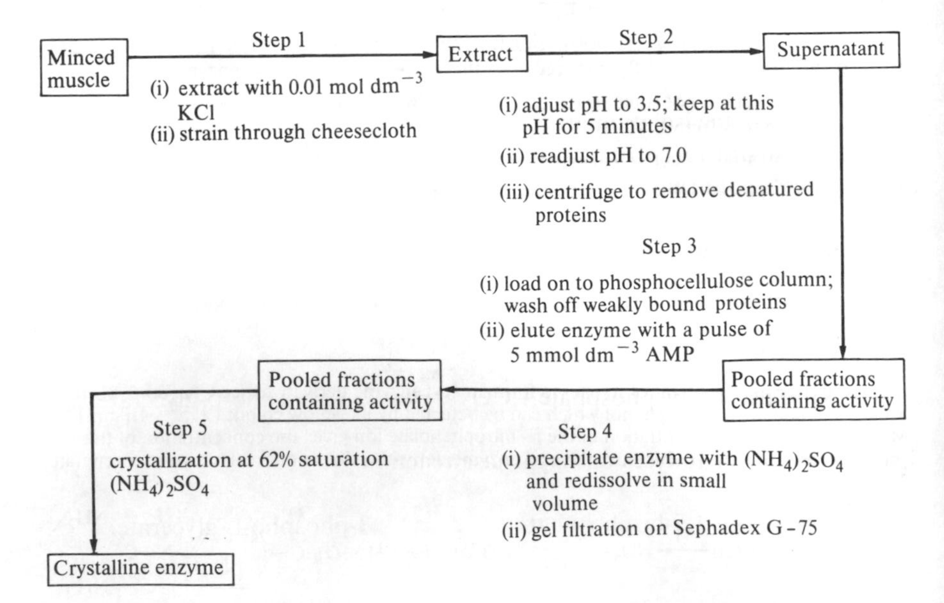

FIG. 2.9. Purification of adenylate kinase from pig muscle. All steps are performed at 273–278 K (0–5 °C).

TABLE 2.3
Purification table for adenylate kinase (see Fig. 2.9) starting with 6 kg muscle

Step	Total volume (cm^3)	Total protein (mg)	Total activity (katal)	Specific activity (katal kg^{-1})	Yield (%)	Purification factor per step
1 (extraction)	16 600	435 000	0.0413	0.095	(100)	(1.0)
2 (pH)	15 700	112 000	0.0365	0.325	88.3	3.42
3 (phosphocellulose)	1380	1716	0.0223	13.02	54.0	40.0
4 (gel filtration)	211	462	0.0200	43.17	48.4	3.32
5 (crystallization)	—	344	0.0160	46.5	38.7	1.08

The yield is calculated from the amount of activity at each step relative to the amount in the initial extract.

The purification factor of a step is calculated on the basis of the increase in specific activity after that step (e.g. 0.325 to 13.02 in Step 3 is a 40-fold increase).

Particular points to note in this purification scheme are:

1. The cells are easily disrupted and the enzyme is extracted in solutions of low ionic strength.
2. Advantage is taken in step 2 of the exceptional stability of adenylate kinase at low pH; this enables many contaminating proteins to be removed. The stability of the enzyme is probably associated with the large amount of regular secondary structure; see Chapter 3, Section 3.5.6.3.
3. Chromatography on phosphocellulose (step 3) is made very effective by the use of affinity elution (see Section 2.6.4.2) using AMP.
4. Gel filtration is used (step 4) to remove a contaminant of M_r 60 000 from adenylate kinase (M_r 21 000). Sephadex G-75 fractionates in the M_r range 3000–70 000.

The success of the isolation procedure was judged[93] by the homogeneity of the product during ultracentrifugation and gel electrophoresis. Subsequent structural studies[92] confirmed the absence of any significant contaminating proteins.

2.8.2 Ribulosebisphosphate carboxylase from spinach

Ribulosebisphosphate carboxylase catalyses the CO_2-fixation step in photosynthesis:

$$\text{D-Ribulose-1,5-bisphosphate} + CO_2 \overset{Mg^{2+}}{\rightleftharpoons} 2\ \text{3-phospho-D-glycerate.}$$

The enzyme also catalyses an oxygenase reaction:

$$\text{D-Ribulose-1,5-bisphosphate} + O_2 \overset{Mg^{2+}}{\rightleftharpoons} \text{3-phospho-D-glycerate} + \text{2-phosphoglycollate,}$$

and is thus often referred to as ribulosebisphosphate carboxylase/oxygenase. It has been claimed that the enzyme, which occurs at high concentrations in chloroplasts, is the most abundant protein in the world.[94] The enzyme is composed of large and small subunits (see Chapter 3, Table 3.5) that are encoded by chloroplast and nuclear genes,respectively.[94,95]

Purification of the enzyme from spinach had been previously reported using procedures involving rather time-consuming chromatographic and sucrose density-gradient centrifugation steps.[96,97] The procedure devised by Hall and Tolbert[98] is considerably quicker and is summarized in Figure 2.10.

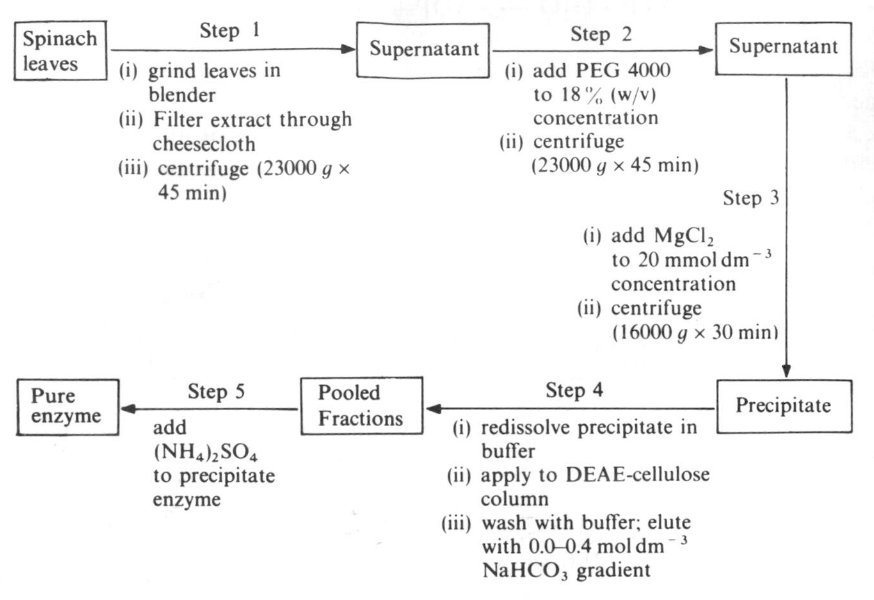

FIG. 2.10. Purification of ribulosebisphosphate carboxylase from spinach. All steps are carried out at 277 K (4 °C). PEG 4000 refers to polyethylene glycol of average M_r 4000.

In step 1, the spinach leaves are ground in buffer (50 mmol dm^{-3} *N*,*N*-bis(2-hydroxyethyl)glycine, 1 mmol dm^{-3} EDTA, and 10 mmol dm^{-3} 2-mercaptoethanol adjusted to pH 8.0 with KOH) containing 2 per cent (w/v) insoluble poly(vinylpolypyrrolidone) to adsorb oxidized phenolic substances (Section 2.5.2). Addition of poly(ethylene glycol) 4000 in step 2 precipitates nucleic acids, chlorophyll, and other pigments. Precipitation of ribulosebisphosphate carboxylase is achieved in step 3 by addition of Mg^{2+} to a final concentration of 20 mmol dm^{-3}. The precipitate is redissolved in grinding buffer and subjected to final purification on DEAE-cellulose (step 4). The poly(ethylene glycol) is not retained by the ion-exchanger and is washed through before the sodium bicarbonate gradient is applied. The final product is >95 per cent homogeneous on polyacrylamide gel electrophoresis and has a specific activity (1.4 μmol CO_2 fixed/mg protein/min) comparable with earlier preparations. On polyacrylamide gel electrophoresis in the presence of sodium dodecylsulphate, two bands are seen, corresponding to the two types of polypeptide chain (M_r values 56 000 and 14 000).

2.8.3 Adenosinetriphosphatase from beef-heart mitochondria

The mitochondrial adenosinetriphosphatase, which catalyses the hydrolysis of ATP,

$$ATP + H_2O \xrightarrow{Mg^{2+}} ADP + \text{orthophosphate},$$

occurs as 'knobs' on the inner (i.e. matrix) side of the inner mitochondrial membrane (see Chapter 8, Section 8.2.3). In oxidative phosphorylation, the reaction is driven in the direction of ATP synthesis by the proton motive force, and the enzyme (in association with various membrane components) is often referred to as the ATP synthetase complex.[99]

Beef heart is chosen as a source because this tissue is rich in mitochondria. Since the cells contain other adenosinetriphosphatases (e.g. that concerned with transport of Na^+ and K^+ across the plasma membrane) it is important to purify the mitochondria before extracting the required enzyme. In the procedure of Knowles and Penefsky[100] advantage is taken of the fact that exposure of mitochondria to sonic oscillations leads to the formation of submitochondrial particles in which the adenosinetriphosphatase points outwards into the medium (Fig. 2.11). The procedure is summarized in Figure (2.12).

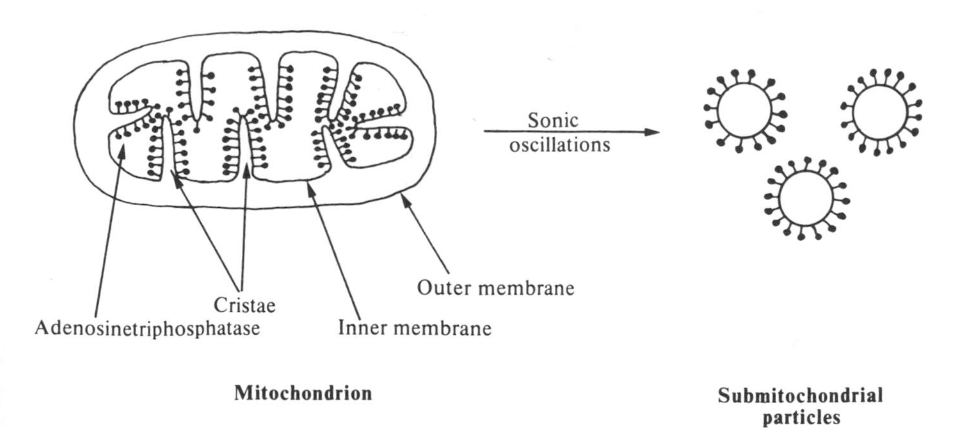

FIG. 2.11. The production of submitochondrial particles from a mitochondrion.

Particular points to note about this procedure are the following.

In step 1 an isotonic medium (0.25 mol dm^{-3} sucrose) is used in the extraction to prevent osmotic damage to the subcellular organelles.

In step 4 the adenosinetriphosphatase is detached from the submitochondrial particles by prolonged sonic oscillations in the presence of substrate (ATP). In

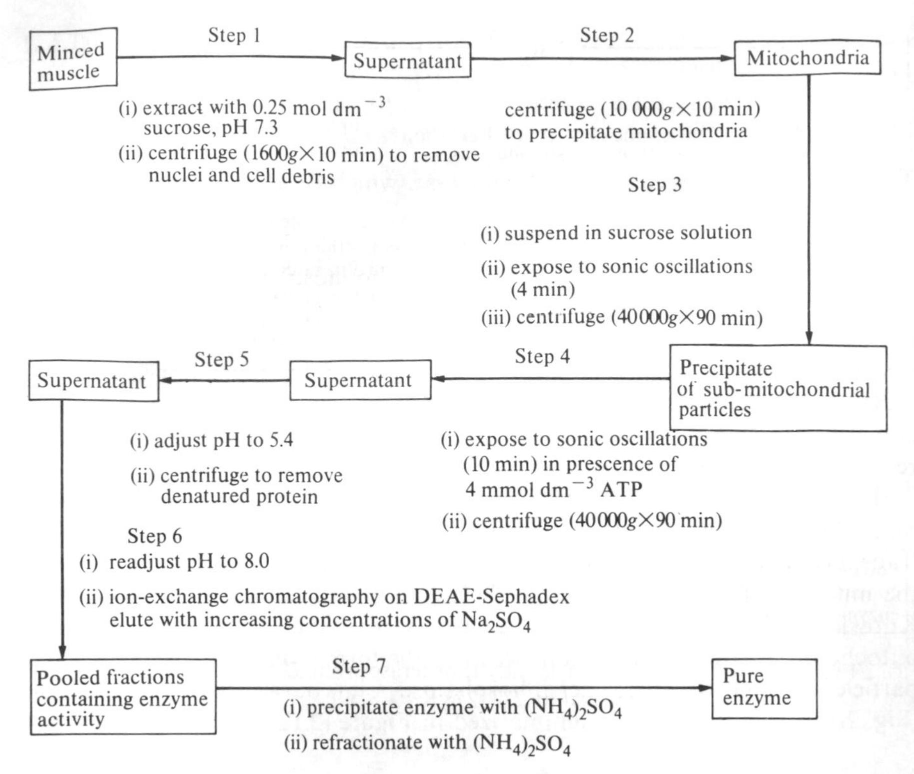

FIG. 2.12. Purification of mitochondrial adenosinetriphosphatase from beef heart.

the case of other membrane-bound enzymes alternative methods that can be used for extraction include the use of detergents or changes in ionic strength depending on the nature of the forces involved in the binding of the enzyme to the membrane (see Chapter 8, Section 8.4.1).

The purification factors in steps 4, 5, and 6 are 3.6-fold, 1.7-fold, and 3.8-fold, respectively, giving an overall purification factor from submitochondrial particles of approximately 23-fold. The final product appears homogeneous in the ultracentrifuge and on electrophoresis in the absence of denaturing agents. It should be noted that in the presence of sodium dodecylsulphate the enzyme dissociates into subunits of different size (see Chapter 3, Section 3.6.1.5) giving rise to multiple bands on sodium dodecylsulphate polyacrylamide gel electrophoresis.

2.8.4 RNA nucleotidyltransferase (RNA polymerase) from *E. coli*

RNA nucleotidyltransferase catalyses the DNA-dependent incorporation of the XMP moiety (X = nucleoside) of ATP, UTP, GTP, and CTP into high-M_r RNA

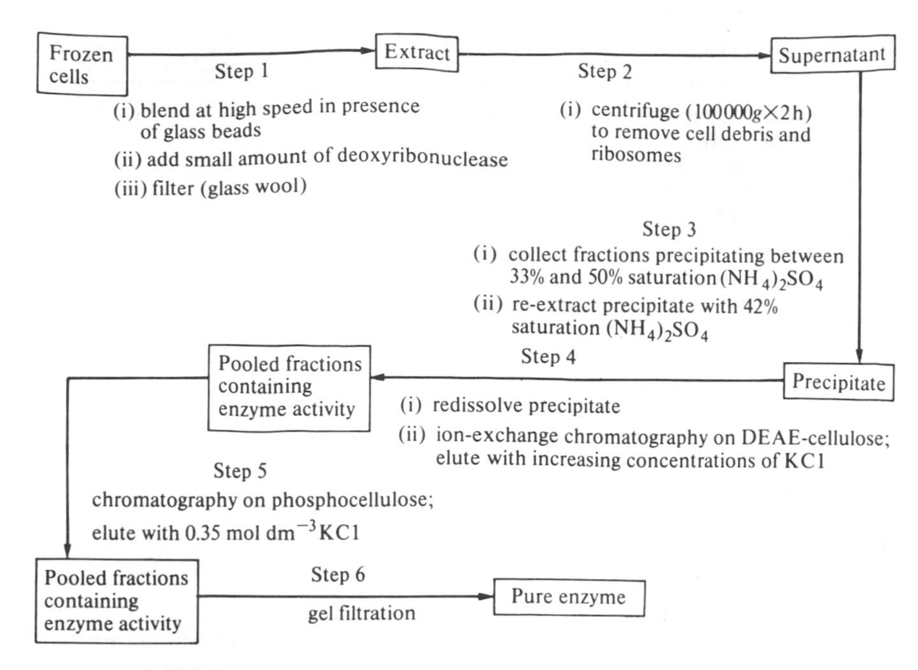

FIG. 2.13. Purification of RNA nucleotidyltransferase from *E. coli.* All steps are performed at 273–278 K (0–5 °C).

which is insoluble in acid. (For a more detailed discussion of the properties of the enzyme, see Chapter 7, Section 7.3.) The source of the enzyme in the procedure of Burgess[101] summarized in Fig. 2.13 is *E. coli* strain K-12; the cells are grown, harvested by centrifugation at 75 per cent of maximum growth, and frozen at 253 K (−20°C).

This procedure illustrates the harsh measures (step 1) needed to disrupt bacterial cell walls. It is possible that enzymatic methods could be used for this purpose (Section 2.5.2) but these are not easy to apply on a large scale. Because of the absence of subcellular organelles (except ribosomes) in bacteria, the extract contains significant amounts of DNA and is therefore highly viscous. Addition of deoxyribonuclease (step 1) reduces the viscosity of the medium, which is important in the later chromatographic steps. It also prevents formation of the high-M_r complex DNA-RNA-RNA nucleotidyltransferase, which would be sedimented along with ribosomes in step 2. In step 3 the oligonucleotides produced by the deoxyribonuclease treatment in step 1 are re-extracted into $(NH_4)_2SO_4$ at 42 per cent saturation. In step 6 gel filtration on a column of Biogel A −1.5 m (an agarose gel which fractionates in the M_r range 10 000–2 000 000) is used to remove a minor contaminant of higher M_r than the RNA nucleotidyltransferase. The overall yield of enzyme is 54 mg from 200 g of frozen cells corresponding to 56 per cent of the activity in the initial extract.

Purification factors for steps 2, 3, 4, 5, and 6 are 1.9-, 4.4-, 5.9-, 5.1-, and 1.1-fold, respectively, giving an overall purification of 260-fold. The homogeneity of the preparation is judged by ultracentrifugation (single peak) and by electrophoresis in 8 mol dm^{-3} urea (the four bands correspond to the correct amounts of the four subunits of different size). There is no detectable deoxyribonuclease, ribonuclease, or polynucleotide nucleotidyl transferase activity in the preparation.

2.8.5 Cytochrome *c* oxidase from beef heart

Cytochrome *c* oxidase catalyses the terminal step in the respiratory chain, in which electrons are donated to the oxidizing agent oxygen:

$$\underset{(\mathrm{Fe^{II}})}{4\ \text{Ferrocytochrome}\ c} + \mathrm{O_2} \rightleftharpoons \underset{(\mathrm{Fe^{III}})}{4\ \text{Ferricytochrome}\ c} + 2\mathrm{H_2O}.$$

The enzyme is firmly bound to the inner mitochondrial membrane (Fig. 2.11) and contains two molecules of haem (cytochrome *a* and cytochrome a_3), two atoms of copper, at least seven polypeptide chains as well as a substantial amount of phospholipid.

We shall consider the purification of enzyme from beef heart mitochondria as outlined in Fig. 2.14.[102,103] The enzyme can be prepared in both a phospholipid-depleted and a phospholipid-rich form.

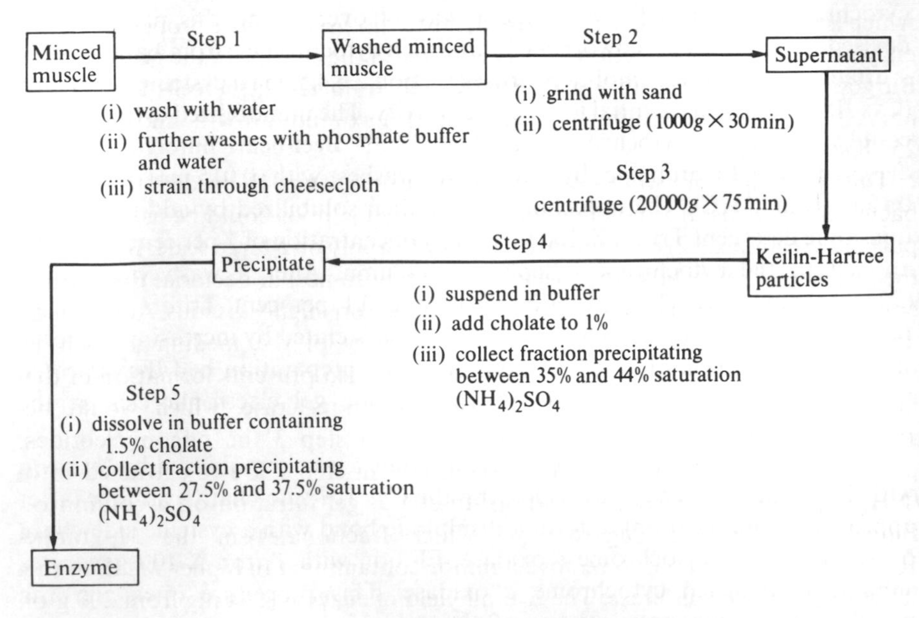

FIG. 2.14. Purification of cytochrome *c* oxidase from beef heart. All steps are performed at 273–278 K (0–5°C).

In this procedure the washes in step 1 are performed to remove non-cytochrome haemoproteins such as myoglobin. Grinding with sand (step 2) disrupts the mitochondria and the low-speed centrifugation is used to precipitate the sand, cell debris, and nuclei. Centrifugation at a higher speed (step 3) precipitates Keilin–Hartree particles, which consist of physically disintegrated mitochondrial membranes. (This preparation is named after two British biochemists who did pioneering work on cytochromes and their role in mitochondrial function in the 1920s and 1930s.) The subsequent steps (4 and 5) in the isolation of cytochrome *c* oxidase depend on extraction with a detergent (cholate) and fractionation by changes in ionic strength. The combined purification factor in these two steps is 11-fold. Cytochrome *c* oxidase prepared by this procedure is phospholipid-depleted (i.e. phospholipid constitutes <0.1 per cent of the material). By varying the concentrations of ammonium sulphate in the fractionation steps, it is possible to prepare a phospholipid-rich enzyme, in which phospholipid constitutes about 20 per cent of the material. The phospholipid-depleted enzyme is inactive; activity can be restored to the level of the phospholipid-rich form by addition of suitable mixtures of phospholipids.[103]

The isolation procedure for cytochrome *c* oxidase has been improved by the use of affinity chromatography. If cytochrome *c* is linked to CNBr-activated Sepharose (an agarose gel) (see Fig. 2.5), it is found that it does not in general function as a suitable affinity matrix for the purification of cytochrome *c* oxidase.[104] This failure is probably because the lysine side-chains through which the cytochrome *c* is linked to the Sepharose are involved in binding to cytochrome *c* oxidase.[105] An elegant way of overcoming this problem was devised by Azzi and his coworkers[105] in which cytochrome *c* from baker's yeast is attached to activated thiol-Sepharose (Section 2.6.4.5) via a cysteine side-chain (Cys 107) that is not essential for enzyme activity. The immobilized cytochrome *c* could now bind to cytochrome *c* oxidase.[105,106] Beef-heart mitochondria are depleted of cytochrome *c* by successive washes with 0.015 mol dm^{-3} and 0.15 mol dm^{-3} KCl. The mitochondria are then solubilized by addition of the non-ionic detergent Triton X-100 to a final concentration of 1 per cent (v/v), and applied to the cytochrome *c*-Sepharose column. After a wash with buffer (50 mmol dm^{-3} tris-HCl, pH 7.2, containing 0.1 per cent Triton X-100) to remove unbound protein, cytochrome *c* oxidase is eluted by increasing the ionic strength by addition of NaCl to the buffer. The preparation had the expected specific activity and pattern on polyacrylamide gel electrophoresis in the presence of sodium dodecylsulphate.[105,106]

In a further set of experiments,[107] cytochrome *c* was immobilized by reaction with CNBr-activated Sepharose, leaving the Cys 107 side-chain available (after appropriate modification) to form a disulphide bond with a cysteine side-chain on subunit III of cytochrome *c* oxidase. Elution with Triton X-100 generates subunit III-depleted cytochrome *c* oxidase. This procedure opens up the possibilities of systematic studies of the role of this subunit in the proton pumping activity of the enzyme.[108]

2.8.6 The *arom* multienzyme protein from *Neurospora*

This multienzyme protein* from *Neurospora crassa* contains five distinct enzyme activities, which catalyse adjacent steps in the biosynthesis of chorismic acid, the common precursor of the three aromatic amino acids tyrosine, phenylalanine, and tryptophan. Although purification of this protein had previously been reported, there was considerable doubt about the M_r of the protein and its subunit composition (see Chapter 3, Section 3.6). It was realized that these difficulties were probably due to proteolysis during the purification procedure and accordingly a new isolation procedure was developed[109] in which stringent precautions were taken to minimize the effects of proteases. These precautions appear to be especially necessary when purifying enzymes from fungi,[110] since they contain large amounts of proteases. The procedure (outlined in Fig. 2.15) starts from cells that have been harvested, dried, stored at 253 K (−20°C), and powdered in a blender.

Protease inhibitors are added at each step of this procedure. In the extraction step (step 1), phenylmethanesulphonyl fluoride (PMSF) and EDTA are included; the former irreversibly inactivates 'serine' proteases (see Chapter 5, Section 5.5.1.2), whereas the latter chelates metal ions, thereby inactivating bivalent-metal-ion-dependent aminopeptidases. Most of the intracellular proteases are removed by adsorption on DEAE-cellulose under the conditions used in step 2. Deoxyribonuclease is added in step 3 to hydrolyse the DNA in the extract released from the nuclei and thus reduce the viscosity of the solution for the later chromatographic steps. In step 4 the inhibitor benzamidine is added; this is an effective reversible inhibitor of proteases with trypsin-like specificity. It is added at this step because, unlike PMSF, it is not salted out of aqueous solutions by high concentrations of ammonium sulphate.

The final yield of enzyme is about 1 mg from 20 g powdered cells, representing a yield of some 25 per cent. Purification factors for steps 2, 4, 5, and 6 are 1.3-, 5.2-, 5.9-, and 15-fold, respectively, giving an overall purification of nearly 600-fold. The purified material appeared to be homogeneous on polyacrylamide gel electrophoresis in the presence of sodium dodecylsulphate and on centrifugation in a glycerol density gradient. There was no trace of species of low-M_r (which might be formed by proteolysis), although these could be detected in samples stored at 277 K (4°C) for 19 days, suggesting that traces of proteases were still present. The material could be stored satisfactorily at 253 K (−20°C) in the presence of 50 per cent glycerol.

Peptide mapping (see Chapter 3, Section 3.4.3) suggests that the *arom* multienzyme protein consists of two identical subunits each of M_r 165 000 and each possessing five distinct enzyme activities.[109] This protein is thus a multienzyme polypeptide (see the discussion in Chapter 7, Section 7.11.2), in which the polypeptide chain is presumably folded into a number of globular domains each possessing an active site. The linking portions of the polypeptide chain are probably the sites of attack by proteases.

* The definition of multienzyme proteins is discussed in Chapter 7, Section 7.2.

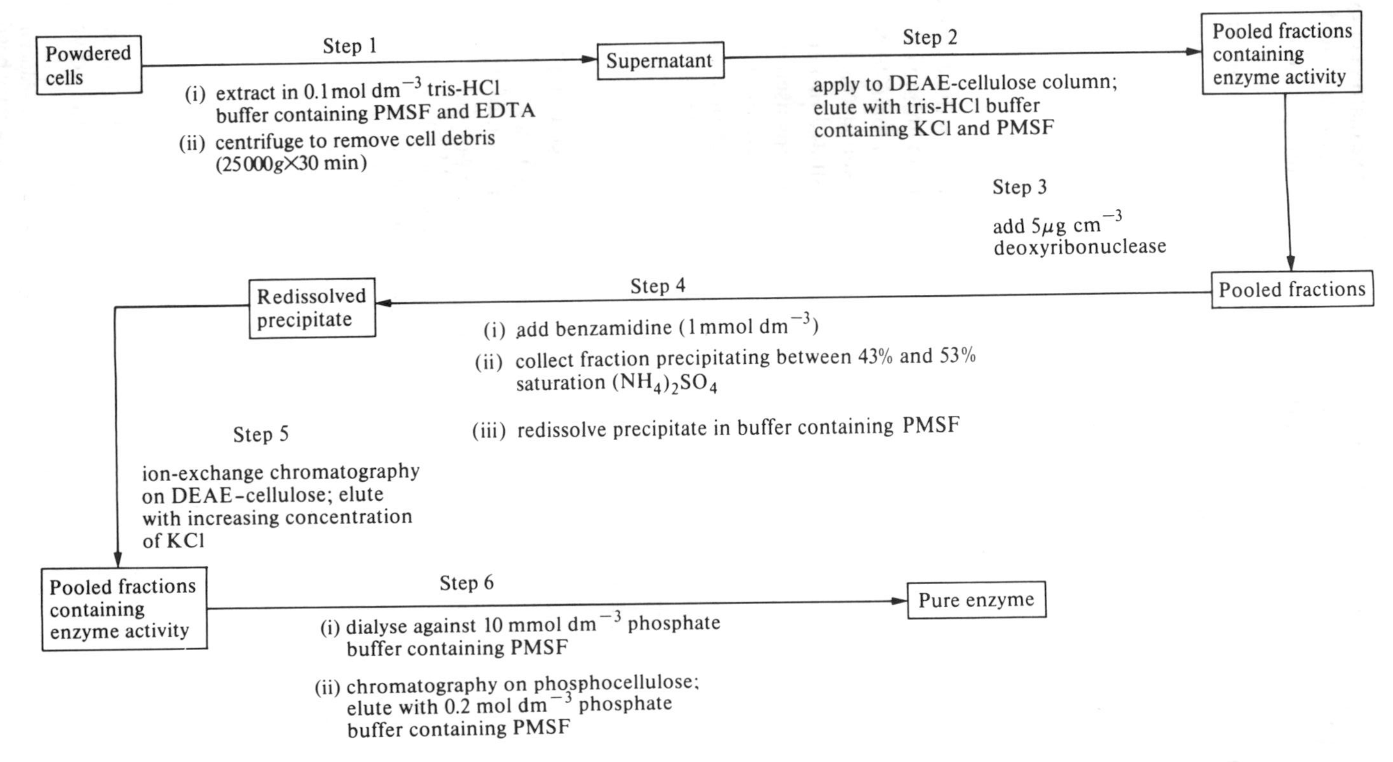

FIG. 2.15. Purification of *arom* multienzyme protein from *N. crassa*. All steps are performed at 273–278 K (0–5° C). PMSF is phenylmethanesulphonyl fluoride.

2.8.7 Glutathione reductase from *E. coli*

Glutathione reductase catalyses the NADPH-dependent reduction of oxidised glutathione* (GSSG):

$$GSSG + NADPH + H^+ \rightleftharpoons 2GSH + NADP^+.$$

Glutathione reductase belongs to the flavoprotein oxidoreductases, a family of enzymes (including dihydrolipoamide dehydrogenase (see Chapter 7, Section 7.7.1)) that possess a disulphide bond that is alternately oxidized and reduced as part of the catalytic mechanism. The enzyme had been previously prepared from *E. coli* using a complex two-stage affinity-chromatography procedure.[111] However, for more detailed investigations of the enzyme it was necessary to develop more a rapid procedure that could be scaled up. The preparation described by Scrutton *et al.*[112] made two important modifications. Firstly, by recombinant DNA techniques a strain of *E. coli* was constructed that produced 100–200 times more enzyme than the wild-type strain. (The fragment of the plasmid pGR containing the gene for the enzyme was inserted into the plasmid pKK 233-3 to form a new plasmid pKGR which was used to transform *E. coli* strain JM101). Secondly, dye-ligand chromatography (Section 2.6.4.3) was used to give a substantial degree of purification. The procedure is summarized in Fig. 2.16.

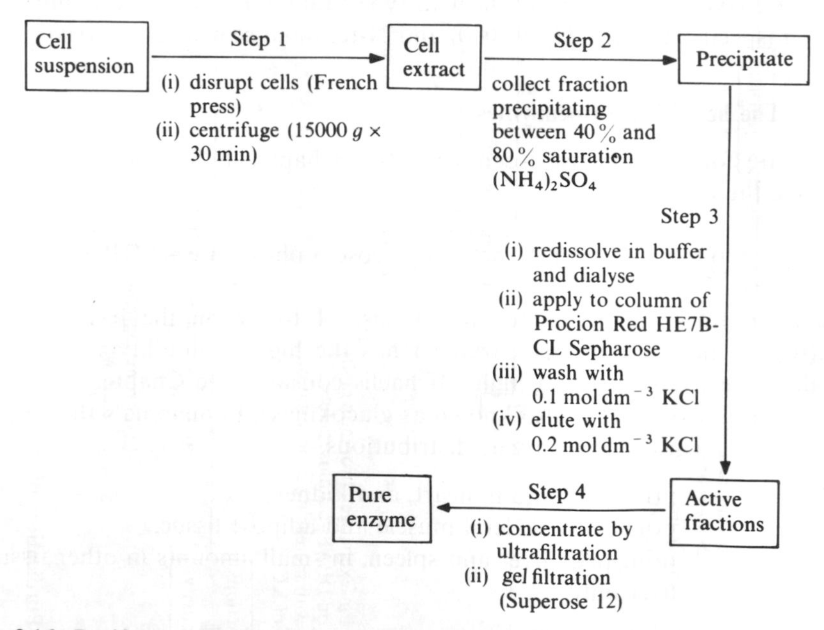

FIG. 2.16. Purification of glutathione reductase from *E. coli*. All steps are carried out at 277 K (4 °C).

* Glutathione (GSH) is a tripeptide γ–Glu–Cys–Gly that acts as an intracellular reducing agent. In its oxidized form (GSSG), two molecules of glutathione are linked by a disulphide bond.

The procedure uses a French press (Section 2.5.2) to disrupt the cell suspension followed by centrifugation to remove cell debris.

During the addition of ammonium sulphate (step 2) the pH is maintained at 7–7.5 by addition of 2 mol dm^{-3} K_2HPO_4 as necessary. The precipitate is redissolved in a low-ionic-strength buffer (5 mmol dm^{-3} potassium phosphate, pH 7.0, containing 1 mmol dm^{-3} EDTA and 1 mmol dm^{-3} 2-mercaptoethanol) and dialysed against three changes of this buffer, before being applied to a column of Procion Red HE-7B linked to cross-linked agarose. As mentioned previously (Section 2.6.4.3), Procion dyes can show considerable specificity for different classes of enzymes and this dye had been found in preliminary work to bind glutathione reductase strongly. After washing with buffer containing 0.1 mol dm^{-3} KCl to remove weakly bound proteins, glutathione reductase is eluted by buffer containing 0.2 mol dm^{-3} KCl (step 3). In a final step the enzyme is concentrated by ultrafiltration (Section 2.6.1.3) and subjected to gel filtration on Superose 12 (which fractionates in the M_r range 1000–300 000) under h.p.l.c. conditions (Section 2.6.6). The overall yield is 11.6 mg enzyme from 100 cm^3 of cell suspension, in turn derived from 10 dm^3 of initial cell culture. The purification factors for steps 2, 3, and 4 are 1.6-, 33.6- and 1.08-fold respectively, giving an overall purification of 58.3-fold. The enzyme is homogeneous on polyacrylamide gel electrophoresis in the presence of sodium dodecylsulphate and is identical with the enzyme prepared from the wild-type strain of *E. coli*[111] by a number of criteria (specific activity, M_r (97 000) and N-terminal sequence analysis).

2.8.8 The hexokinase isoenzymes

There are four hexokinase isoenzymes (see Chapter 1, Section 1.5.2) which catalyse the reaction

$$\text{D-Glucose} + \text{ATP} \overset{Mg^{2+}}{\rightleftharpoons} \text{D-glucose 6-phosphate} + \text{ADP}$$

The isoenzymes are designated hexokinases I to IV on the basis of their electrophoretic mobilities (isoenzyme I has the highest mobility towards the anode). Isoenzyme IV has a high Michaelis constant (see Chapter 4, Section 4.3.1.3) for glucose and is also known as glucokinase. In mammals the various isoenzymes possess distinct tissue distributions:

Type I	principally brain, heart, and kidney;
Type II	principally skeletal muscle and adipose tissue;
Type III	principally liver and spleen, in small amounts in other tissues;
Type IV	liver only.

Until the application of affinity chromatography techniques only isoenzyme I had been purified, e.g. from pig heart.[113] Isoenzyme IV was subsequently purified[114] from rat liver by a combination of ion-exchange chromatography and affinity chromatography using an immobilized derivative of *N*-acetylglucosamine (as shown in Fig. 2.17 with $n = 6$).

Although *N*-acetylglucosamine itself is a powerful competitive inhibitor (see Chapter 4, Section 4.3.2.2) of all four isoenzymes, it was found that isoenzymes I, II, and III were not retained by a column containing the derivative shown in Fig. 2.17 with $n=6$. A later investigation, however, showed that by varying the conditions isoenzyme II could be purified by this method.[115] These findings prompted a detailed study of the inhibitory action of the derivatives of *N*-acetylglucosamine in solution. The results of these studies suggested explanations for the behaviour of immobilized derivatives of *N*-acetylglucosamine and helped to indicate the modifications that were necessary for successful affinity chromatography of the hexokinase isoenzymes.[116]

The values of the dissociation constants for the various derivatives of *N*-acetylglucosamine from the enzyme showed a marked dependence on the length of the alkyl spacer arm (i.e. on *n* in Fig. 2.17). Table 2.4 shows the results for isoenzyme I; for the other isoenzymes the values of the dissociation constants are different reflecting small differences between the structures of the active sites of the isoenzymes.

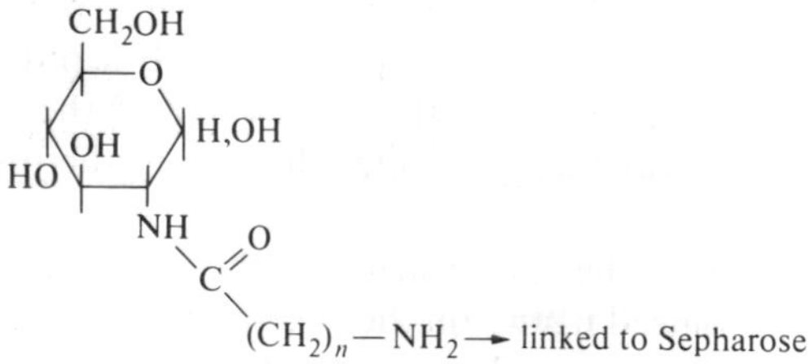

CH_2OH
O
OH
H,OH
HO
NH
C=O
$(CH_2)_n—NH_2$ → linked to Sepharose

FIG. 2.17. General structure of the immobilized derivative of *N*-acetylglucosamine used in affinity chromatography of hexokinase isoenzymes.

TABLE 2.4
Dissociation constants of various N-acetylglucosamine derivatives from hexokinase isoenzyme I

n	Dissociation constant (mmol dm^{-3})
3	14
4	5
6	3
8	1.3

The value of *n* refers to the side chain of derivatives depicted in Fig. 2.17.

The dissociation constant of the enzyme–inhibitor complex is derived from inhibition studies as described in Chapter 4, Section 4.3.2.2.

Since the concentration of immobilized ligand on the Sepharose gel is $\leqslant 10$ mmol dm^{-3}, it is possible to design an efficient chromatography system only if the dissociation constant of an enzyme–inhibitor complex is less than this value ($\leqslant 2$ mmol dm^{-3}).[116] Thus from Table 2.4 we can see that for isoenzyme I only the C_8 (i.e. $n=8$) derivative can provide the basis for an efficient chromatography system, as described in Section 2.8.8.1.

2.8.8.1 *Purification of isoenzyme I from rat kidney*[116]

Rat kidneys are homogenized in a triethanolamine-HCl buffer containing 0.2 mol dm^{-3} $MgCl_2$ and 50 mmol dm^{-3} glucose. After centrifugation the extract is partially purified by chromatography on DEAE-cellulose and then hydroxyapatite. The active fractions from these procedures are pooled, concentrated by ultrafiltration, and applied to an affinity column containing the C_8 ($n=8$) derivative at a ligand concentration of 5 mmol dm^{-3}. The column is washed with triethanolamine-HCl buffer containing 70 mmol dm^{-3} KCl to remove unwanted protein; isoenzyme I is then specifically eluted by inclusion of 0.2 mol dm^{-3} glucose in the buffer. Fractions containing activity are rechromatographed on the C_8 ($n=8$) affinity column to remove a minor contaminant.

The yield of isoenzyme I is about 65 per cent with a purification factor of 21 500-fold (both expressed relative to the crude extract); the material appeared homogeneous on polyacrylamide gel electrophoresis in the presence of sodium dodecylsulphate.

By varying the length of the spacer arm (i.e. the value of n in Fig. 2.17), the concentration of immobilized ligand on the gel, and the eluting conditions it has proved possible to devise procedures to isolate all four isoenzymes in a homogeneous form.[116] This detalied study illustrates the importance of a good understanding of the behaviour of ligands in solution for the design of affinity chromatography systems.

2.9 Conclusions from the examples of enzyme purification

The eight examples of enzyme purification discussed in Section 2.8 illustrate the range of problems faced in this type of work and some of the solutions available. It can be seen that many of the earlier procedures contain steps involving ammonium sulphate fractionation, ion-exchange chromatography, and gel filtration, whereas the more recent examples place greater emphasis on affinity methods involving the recognition of particular structural features. These affinity methods, together with developments in recombinant DNA technology,[5] have made it feasible to purify enzymes normally present in very small amounts, which would be difficult, if not impossible to purify by the earlier methods. The choice of methods to be used in any particular case depends on a number of factors[11] (Section 2.6.6). If *yield* is the primary objective, then the number of steps in a purification should be kept to a minimum, since losses inevitably occur at each stage. (Of course the number of steps has to be sufficient to give a homogeneous product.) If *speed* is the primary concern, it should be remembered that methods based on changes in solubility are in general more rapid than those involving column chromatography, e.g. ion-exchange chromatography or gel filtration, although recent developments in h.p.l.c. techniques have made such chromatographic procedures much less time-consuming (Section 2.6.6).

References

1. Dixon, M., Webb, E. C., Thorne, C. J. R., and Tipton, K. F., *Enzymes* (3rd edn). Longman, London (1979). See p. 40.
2. Hardie, D. G., and Cohen, P., *FEBS Lett.* **91**, 1 (1978).
3. Demain, A. L., *Methods Enzymol.* **22**, 86 (1971).
4. Roberts, T. M., and Lauer, G. D., *Methods Enzymol.* **68**, 473 (1979).
5. Old, R. W., and Primrose, S. B., *Principles of gene manipulation* (3rd edn). Blackwell, Oxford (1985).
6. Duncan, K., Lewendon, A., and Coggins, J. R., *FEBS Lett.* **165**, 121 (1984).
7. Marston, F. A. O., *Biochem. J.* **240**, 1 (1986).
8. Herring, P. J. (ed.), *Bioluminescence in action.* Academic Press, London (1978).
9. McElroy, W. D., and Seliger, H. H., *Scient. Am.* **207**, 76 (1962).
10. Newsholme, E. A., and Leech, A. R., *Biochemistry for the medical sciences.* Wiley, Chichester (1983). See p. 206.
11. Scopes, R. K., *Protein purification: principles and practice.* Springer, New York (1982).
12. Avis, P. J. G., in *Subcellular components* (2nd edn.) (Birnie, G. D., ed.), pp. 1–13. Butterworths, London (1972).
13. Freese, E., and Oosterwyk, J., *Biochemistry* **2**, 1212 (1963).
14. Repaske, R., *Biochim. biophys. Acta* **30**, 225 (1958).
15. Villaneuva, J. R., and Acha, I. G., in *Methods in microbiology* (Booth, C., ed.), vol. 4., p. 665. Academic Press, New York (1971).
16. Emerson, S., *Genetica* **34**, 162 (1963).
17. Scopes, R. K., *Protein purification: priniciples and practice.* Springer, New York (1982). See p. 27.
18. Birnie, G. D., and Rickwood, D. (eds.), *Centrifugal separations in molecular and cell biology.* Butterworths, London (1978).
19. Reiland, J., *Methods Enzymol.* **22**, 287 (1971).
20. Scopes, R. K., *Protein purification: principles and practice.* Springer, New York (1982). See p. 151ff.
21. McPhie, P., *Methods Enzymol.* **22**, 23 (1971).
22. Scopes, R. K., *Protein purification: principles and practice.* Springer, New York (1982). See p. 16.
23. Blatt, W. F., *Methods Enzymol.* **22**, 39 (1971).
24. Scopes, R. K., *Protein purification: principles and practice.* Springer, New York (1982). See p. 182.
25. Craig, L. C., *Methods Enzymol.* **11**, 870 (1967).
26. Himmelhoch, S. R., *Methods Enzymol.* **22**, 273 (1971).
27. Scopes, R. K., *Protein purification: principles and practice.* Springer, New York (1982). See Ch. 4.

28. Dixon, M., Webb, E. C., Thorne, C. J. R., and Tipton, K. F., *Enzymes* (3rd edn). Longman, London (1979). See p. 36.
29. Sluyterman, L. A. A. and Wijdenes, J., *J. Chromatogr.* **206**, 441 (1981).
30. Scopes, R. K., *Protein purification: principles and practice.* Springer, New York (1982). See p. 163ff.
31. Shuster, L., *Methods Enzymol.* **22**, 412 (1971).
32. Vesterberg, O., *Methods Enzymol.* **22**, 389 (1971).
33. Righetti, P. G., and Drysdale, J. W., in *Laboratory techniques in biochemistry and molecular biology* (Work, T. S. and Work, E., eds.), Vol. 5, p. 337. North-Holland, Amsterdam (1976).
34. Scopes, R. K., *Protein purification: principles and practice.* Springer, New York (1982). See p. 172ff.
35. Scopes, R. K., *Protein purification: principles and practice.* Springer, New York (1982). See p. 139.
36. Rosengren, J., Påhlman, S., Glad, M. and Hjertén, S., *Biochim. biophys. Acta* **412**, 51 (1975).
37. Scopes, R. K., *Protein purification: principles and practice.* Springer, New York (1982). See p. 40.
38. Scopes, R. K., *Protein purification: principles and practice.* Springer, New York (1982). See p. 43ff.
39. Ferdinand, W., *Biochem. J.* **92**, 578 (1964).
40. Lai, C. Y., and Horecker, B. L., *Essays Biochem.* **8**, 149 (1972).
41. Jakoby, W. B., *Methods Enzymol.* **22**, 248 (1971).
42. Kaufman, S., *Methods Enzymol.* **22**, 233 (1971).
43. Scopes, R. K., *Protein purification: principles and practice.* Springer, New York (1982). See p. 52ff.
44. Fried, M., and Chun, P. W., *Methods Enzymol.* **22**, 238 (1971).
45. Cuatrecasas, P., and Anfinsen, C. B., *Methods Enzymol.* **22**, 345 (1971).
46. Lowe, C. R., and Dean, P. D. G., *Affinity chromatography.* Wiley, London (1974).
47. Lowe, C. R., in *Laboratory techniques in biochemistry and molecular biology* (Work, T. S. and Work, E., eds.), Vol. 7, p. 267. North Holland, Amsterdam (1979).
48. *Methods Enzymol.* **34** (1974) (many articles).
49. *Affinity chromatography: a practical approach* (Dean, P. D. G., Johnson, W. S. and Middle, F. A., eds.) IRL Press, Oxford (1985).
50. Cuatrecasas, P., and Anfinsen, C. B., *Methods Enzymol.* **22**, 355 (1971).
51. Cuatrecasas, P., Wilchek, M., and Anfinsen, C. B., *Proc. natn. Acad. Sci. USA* **61**, 636 (1968).
52. Cuatrecasas, P., and Anfinsen, C. B., *A. Rev. Biochem.* **40**, 259 (1971).
53. Lowe, C. R., in *Laboratory techniques in biochemistry and molecular biology* (Work, T. S. and Work, E., eds.), Vol. 7, p. 323. North-Holland, Amsterdam (1979).
54. O'Carra, P., Barry, S., and Griffin, T., *Methods Enzymol.* **34**, 108 (1974).
55. Scopes, R. K., *Protein purification: principles and practice.* Springer, New York (1982). See p. 117.
56. Nichol, L. W., Ogston, A. G., Winzor, D. J., and Sawyer, W. H., *Biochem. J.* **143**, 435 (1974).
57. Andersson, L., Jörnvall, H., Åkeson, Å., and Mosbach, K., *Biochim. Biophys. Acta* **364**, 1 (1974).
58. Scopes, R. K., *Protein purification: priniciples and practice.* Springer, New York (1982). See p. 101ff.

59. Pogell, B. N., *Biochem. Biophys. Res. Commun.* **7**, 225 (1962).
60. Scopes, R. K., *Biochem. J.* **161**, 253 (1977).
61. Scopes, R. K., *Biochem. J.* **161**, 265 (1977).
62. Davies, J. R., and Scopes, R. K., *Analyt. Biochem.* **114**, 19 (1981).
63. Scopes, R. K., *Protein purification: principles and practice.* Springer, New York (1982). See p. 125ff.
64. Thompson, S. T., Cass, K. H., and Stellwagen, E., *Proc. natn. Acad. Sci. USA.* **72**, 669 (1975).
65. Stellwagen, E., *Acc. Chem. Res.* **10**, 92 (1977).
66. Biellmann, J.-F., Samana, J.-P., Bränden, C.-I., and Eklund, H., *Eur. J. Biochem.* **102**, 107 (1979).
67. Beissner, R. S., Quiocho, F. A., and Rudolph, F. B., *J. molec. Biol.* **134**, 847 (1979).
68. Edwards, R. A. and Woody, R. W., *Biochemistry* **18**, 5197 (1979).
69. Watson, D. H., Harvey, M. J., and Dean, P. D. G., *Biochem. J.* **173**, 591 (1978).
70. Hey, Y. and Dean, P. D. G., *Biochem. J.* **209**, 363 (1983).
71. Scopes, R. K., *Protein purification: principles and practice.* Springer, New York (1982). See p. 132ff.
72. Köhler, G. and Milstein, C., *Nature* **256**, 495 (1975).
73. Eisenbarth, G., *Analyt. Biochem.* **111**, 1 (1981).
74. Nishigaki, I., Ichinose, H., Tamai, K., and Nagatsu, T., *Biochem. J.* **252**, 331 (1988).
75. Brocklehurst, K., Carlsson, J., Kiersten, M. P. J., and Crook, E. M., *Biochem. J.* **133**, 573 (1973).
76. Hillson, D. A., and Freedman, R. B., *Biochem. J.* **191**, 373 (1980).
77. Dixon, M., Webb, E. C., Thorne, C. J. R., and Tipton, K. F., *Enzymes* (3rd edn). Longman, London (1979). See p. 30.
78. Scopes, R. K., *Protein purification: principles and practice.* Springer, New York (1982). See p. 60.
79. Dixon, M., Webb, E. C., Thorne, C. J. R., and Tipton, K. F. *Enzymes* (3rd edn). Longman, London (1979). See pp. 33–5.
80. Dixon, M., Webb, E. C., Thorne, C. J. R., and Tipton, K. F. *Enzymes* (3rd edn). Longman, London (1979). See p. 35.
81. Scopes, R. K., *Protein purification: principles and practice.* Springer, New York (1982). See p. 138.
82. Everse, J. and Stolzenbach, F. E., *Methods Enzymol.* **22**, 33 (1971).
83. Wrigley, C. W., *Methods Enzymol.* **22**, 559 (1971).
84. Hare, D. L., Stimpson, D. I. and Cann, J. R., *Archs. Biochem. Biophys.* **187**, 274 (1978).
85. Scopes, R. K., *Protein purification: principles and practice.* Springer, New York (1982). See p. 176.
86. Kédzy, F. J. and Kaiser, E. T., *Methods Enzymol.* **19**, 3 (1970).
87. Price, N. C., *Analyt. Biochem.* **73**, 447 (1976).
88. Baines, B. S., and Brocklehurst, K., *Biochem. J.* **173**, 345 (1978).
89. Fersht, A. R., and Kaethner, M. M., *Biochemistry* **15**, 818 (1976).
90. Pösö, H., and Pegg, A. E., *Biochim. biophys Acta* **747**, 209 (1983).
91. Fersht, A. R., *Enzyme structure and mechanism* (2nd edn). Freeman, New York (1985). See p. 143.
92. Heil, A., Müller, G., Noda, L., Pinder, T., Schirmer, H., Schirmer, I., and von Zabern, I., *Eur. J. Biochem.* **43**, 131 (1974).
93. Schirmer, I., Schirmer, R. H., Schulz, G. E., and Thuma, E., *FEBS Lett.* **10**, 333 (1970).

94. Ellis, R. J., *Trends Biochem. Sci.* **4**, 241 (1979).
95. Roy, H., and Cannon, S., *Trends Biochem. Sci.* **13**, 163 (1988).
96. Siegel, M. I., and Lane, M. D., *Methods Enzymol.* **42**, 472 (1975).
97. Andrews, T. J., Lorimer, G. H., and Tolbert, N. E., *Biochemistry* **12**, 11 (1973).
98. Hall, N. P., and Tolbert, N. E., *FEBS Lett.* **96**, 167 (1978).
99. Nicholls, D. G., *Bioenergetics.* Academic Press, London (1982).
100. Knowles, A. F., and Penefsky, H. S., *J. Biol. Chem.* **247**, 6617 (1972).
101. Burgess, R. R., *J. biol. Chem.* **244**, 6160 (1969).
102. King, T. E., *Methods Enzymol.* **10**, 202 (1967).
103. Yu, C.-A., Yu, L., and King, T. E., *J. biol. Chem.* **250**, 1383 (1975).
104. Weiss, H., Juchs, B., and Ziganke, B., *Methods Enzymol.* **53**, 98 (1978).
105. Bill, K., Casey, R. P., Broger, C., and Azzi, A., *FEBS Lett.* **120**, 248 (1980).
106. Azzi, A., Bill, K., and Broger, C., *Proc. natn. Acad. Sci. USA* **79**, 2447 (1982).
107. Bill, K. and Azzi, A., *Biochem. Biophys. Res. Commun.* **106**, 1203 (1982).
108. Brunori, M., Antonini, E., Malatesta, F., Sarti, P., and Wilson, M. T., *Eur. J. Biochem.* **169**, 1 (1987).
109. Lumsden, J., and Coggins, J. R., *Biochem. J.* **161**, 599 (1977).
110. Pringle, J. R., *Methods cell. Biol.* **12**, 149 (1975).
111. Mata, A. M., Pinto, M. C., and Lopez-Barea, J., *Z. Naturforsch. C: Biosci.* **39**, 908 (1984).
112. Scrutton, A. S., Berry, A., and Perham, R. N., *Biochem. J.* **245**, 875 (1987).
113. Easterby, J. S. and O'Brien, M. J., *Eur. J. Biochem.* **38**, 201 (1973).
114. Holroyde, M. J., Chesher, J. M. E., Trayer, I. P., and Walker, D. G., *Biochem. J.* **153**, 351 (1976).
115. Holroyde, M. J., and Trayer, I. P., *FEBS Lett.* **62**, 215 (1976).
116. Wright, C. L., Warsy, A. S., Holroyde, M. J., and Trayer, I. P., *Biochem. J.* **175**, 125 (1978).

3
The structure of enzymes

3.1 Introduction

In this chapter we shall consider various aspects of enzyme structure. The ultimate aim of an investigation is to establish the complete three-dimensional structure of an enzyme at atomic (or near atomic) resolution. This information provides a firm basis for understanding the properties of the enzyme, especially its catalytic activity (see Chapter 5).

The determination of this detailed structure is a daunting task for even the smallest enzyme and has been found to require several years of work. Assuming that a supply of purified enzyme is available, the principal stages of the work can be listed as:

(1) determination of relative molecular mass, M_r* (Section 3.2);
(2) determination of amino acid composition (Section 3.3);
(3) determination of primary structure (Section 3.4);
(4) determination of secondary and tertiary structure (Section 3.5);
(5) determination of quaternary structure (Section 3.6);

The term primary structure refers to the sequence of amino acids in a polypeptide chain; it contains only one-dimensional information and tells us little about the three-dimensional structure. The terms secondary and tertiary structure refer to different aspects of the three-dimensional structure; secondary structure refers to regular elements of structure such as helix, sheet, etc. (Section 3.5) in which interactions between regions of the enzyme in close proximity in the sequence are involved. Tertiary structure refers to the folding of a chain, by which portions of the molecule well separated in the sequence are brought into close contact with each other. The term quaternary structure refers to the arrangement of the individual subunits† in an enzyme consisting of more than one subunit.

We shall discuss the various aspects of structure in the order given above, but it should be emphasized that structural investigations do not necessarily proceed in this order. It is often found that information from one type of investigation can be useful in another line of work; thus the definitive M_r value is established once

* The term *relative molecular mass* (M_r) is now used in place of *molecular weight*. M_r is a dimensionless number and is the ratio of the molecular mass of a molecule to 1/12 the mass of one atom of ^{12}C. The latter value is known as a *dalton* (1.663×10^{-24} g). Molecular masses are often quoted in daltons or kilodaltons (abbreviated as Da or kDa, respectively).

† The term subunit refers to the smallest covalent unit. A subunit may consist of one polypeptide chain or of two or more chains linked by disulphide bonds. Thus the subunit of insulin consists of two chains; six of these subunits can aggregate to form a hexameric species.

the primary structure of the enzyme is known (Section 3.4.9.1) and the primary structure is needed to help in the elucidation of the three-dimensional structure by X-ray crystallography (Section 3.5.1). In each of the sections, the main experimental approaches involved will be indicated, together with a description of the principal results obtained and the application of these results.

3.2 The determination of M_r

Enzymes are macromolecules with M_r values ranging from about 10 000 to several million. Therefore methods for determining M_r such as mass spectrometry or freezing point depression, which are applicable to small molecules, are not suitable for enzymes. The majority of present-day determinations of M_r values of enzymes are performed by use of one or more of the following techniques:

(1) ultracentrifugation;
(2) gel filtration;
(3) sodium dodecylsulphate polyacrylamide gel electrophoresis.

Of these methods, (2) and (3) are 'semi-empirical' and comparisons are made with standard molecules of known M_r. In (1), however, the M_r can be calculated using equations derived from first principles.

Other techniques less commonly employed involve measurements of osmotic pressure and light scattering. The osmotic pressure is the pressure in excess of atmospheric pressure which must be applied to prevent movement of solvent from one solution to a second solution (containing a macromolecule) when the two solutions are separated by a membrane permeable to the solvent but not to the macromolecule. More details can be found in the book by Price and Dwek.[1]

The scattering of light by a solution arises because the electrons of solute molecules can be distorted (or *polarized*) by the fluctuating electromagnetic field of the radiation, leading to scattering of the radiation in different directions. From measurements of the intensity of scattering as a function of the angle of scattering the M_r of the macromolecule can be calculated.[2] If the wavelength of the radiation used is of the same order of magnitude as the dimensions of the macromolecule, we can obtain additional information on its shape.[3]

3.2.1 Ultracentrifugation

An ultracentrifuge is capable of generating intense centrifugal fields; in a typical machine a rotor speed of 65 000 r.p.m. corresponds to a field of the order of 300 000 times that of gravity. Under these conditions macromolecules will have a tendency to sediment, provided that the density of the macromolecule is greater than that of the solution (this generally holds for enzymes in aqueous solution). The ultracentrifuge can be used in two main ways to determine M_r values weights: sedimentation velocity and sedimentation equilibrium.

3.2.1.1 *Sedimentation velocity*

In this type of experiment the ultracentrifuge is operated at high speeds to generate centrifugal forces that are sufficiently intense to sediment the macromolecules. The sedimentation of an enzyme can be monitored by suitable optical means (including for instance changes in refractive index or absorption of 280 nm radiation)[4] and from these measurements the sedimentation coefficient, s, can be calculated. This coefficient gives a measure of the sedimentation velocity in a unit gravitational field. The sedimentation coefficient cannot by itself be used to calculate the M_r of the enzyme, since the rate of sedimentation will depend on other factors such as the shape of the macromolecule. However, if we have other information, such as the value of the diffusion coefficient (D) of the macromolecule, its partial specific volume ($\bar{v}$) and the density of the solution (ρ), the M_r can be calculated[5] from the formula:

$$M_r = \frac{RTs}{D(1 - \bar{v}\rho)},$$

where R is the gas constant, and T is the temperature.

The partial specific volume ($\bar{v}$) of a solute is defined as the volume change upon addition of 1 g of that solute to a large volume of solution, keeping all other parameters (temperature, pressure, etc.) constant. The value of $\bar{v}$ can be determined from very accurate measurements of the densities of solutions[6] or calculated from the amino-acid composition of the enzyme[7] by using the known values of the molar volumes of each increment of structure of the various amino-acids. The diffusion coefficient, D, can be measured in a separate experiment in which the ultracentrifuge is operated at low speeds so that no detectable sedimentation occurs.[4]

3.2.1.2 *Sedimentation equilibrium*

If the rotor speed is not great enough to cause complete sedimentation of the macromolecule, then after a while an equilibrium state is reached in which the tendency of the macromolecules to be sedimented is balanced by their tendency to diffuse from the region of high concentration at the bottom of the ultracentrifuge cell to the region of low concentration towards the meniscus of the solution. From measurements of the distribution of concentration, c, of the macromolecule as a function of distance, r, along the cell (i.e. distance from the axis of rotation) the M_r can be calculated provided that we also know $\bar{v}$ and ρ, as in the sedimentation velocity experiment (Section 3.2.1.1):

$$M_r = \frac{2RT}{(1 - \bar{v}\rho)\omega^2} \frac{\mathrm{d} \ln c}{\mathrm{d}r^2},$$

where R is the gas constant, T is the temperature, and ω is the angular velocity in rad s^{-1}. The value of $(\mathrm{d} \ln c/\mathrm{d}r^2)$ is obtained from the slope of the plot of $\ln c$ against r^2.

The advantage of the sedimentation equilibrium method is that a system is being studied at equilibrium and thus there is no dependence on the shape of the solute or the viscosity of the solution. Using a short column of solution (0.15–0.30 cm) the time required to reach equilibrium is quite short (4–8 hours). Under favourable circumstances M_r can be determined to within a few per cent of the true value (i.e. the value calculated subsequently from the primary structure). For further details of these methods, a number of articles can be consulted.[8-10] Provided that the appropriate values of $\bar{v}$ and ρ are used it is also possible to calculate the M_r of the enzyme under denaturing conditions (Section 3.7.1).

3.2.2 Gel filtration

As described in Chapter 2 (Section 2.6.1.2) certain cross-linked polymers such as Sephadex, Sephacryl and Bio-Gel are able to separate molecules according to size. Large molecules are unable to penetrate the pores of the cross-linked polymer and are eluted from a Sephadex column ahead of small molecules. For molecules of a similar shape (e.g. a series of globular proteins) there has been found to be a linear relationship between the logarithm of M_r and the distribution coefficient, K_d, over a certain range of M_r values.[11]

The value of K_d for a given solute is defined by the relationship

$$K_d = \frac{V_e - V_o}{V_i - V_o},$$

where V_e is the elution volume of the molecule of interest, V_0 is the elution volume of a molecule completely excluded by the column (e.g. dextran blue with an M_r of 2×10^6), and V_i is the elution volume of a small molecule (e.g. glucose) which is totally included by the column. Thus $K_d = 0$ for solutes which are totally excluded and $K_d = 1$ for solutes which are totally included.

For Sephadex G-200 the 'linear' range of M_r values is from about 40 000 to about 200 000. By reference to the standard curve, the M_r of the enzyme of interest can be determined (Fig. 3.1). Other types of cross-linked polymers can be used over different ranges of M_r values.

While the gel filtration method is extremely simple to use it must be emphasized that the M_r value obtained should only be used as a guide until more definitive evidence is available. It has been known for some time[11, 12] that glycoproteins, particularly those rich in carbohydrate, show anomalous behaviour on gel filtration, and this will also be the case for proteins that are not globular in shape. Thus, the heat-stable phosphatase inhibitor protein isolated from skeletal muscle (Chapter 6, Section 6.4.2.4) appears to have an M_r of 60 000 on gel filtration through Sephadex G-100, whereas the true M_r is close to 20 000.[13] Fortunately for our purposes it does appear that the vast majority of enzymes are globular proteins and are well behaved on gel filtration. The M_r derived by this technique seems to be accurate to within about 10 per cent.[11] It is also possible to measure M_r values under denaturing conditions (see Section

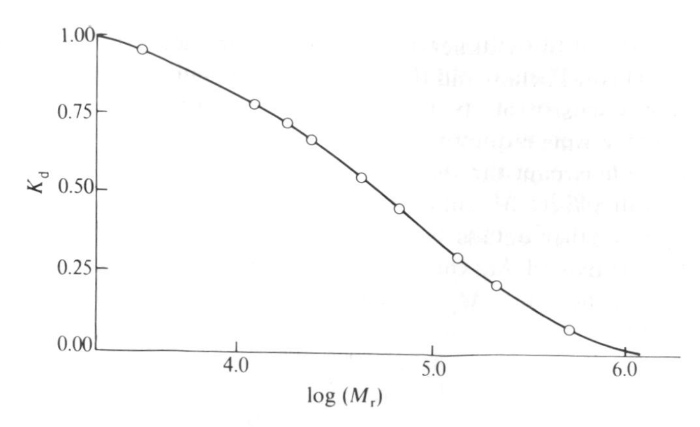

FIG. 3.1. Gel filtration of proteins on Sephadex G-200.[11] The proteins used (in order of increasing M_r were): glucagon, cytochrome *c*, myoglobin, chymotrypsinogen, ovalbumin, bovine serum albumin, lactate dehydrogenase, catalase, and β-D-galactosidase, K_d is the distribution coefficient for a solute and is defined in the text.

3.7.1) provided appropriate standard curves are constructed.[11, 14] Gel filtration can also be performed under high-performance liquid chromatography (h.p.l.c.) conditions using silica-based column support materials. The M_r can be determined in a much shorter time (ca. 30 min) than by conventional gel filtration.[15]

3.2.3 Sodium dodecylsulphate polyacrylamide gel electrophoresis

The mobility of a charged molecule in an electric field is normally a function of various factors such as the size and shape of the molecule and the charge it carries and it would therefore be expected that electrophoresis would not normally give any reliable estimates of M_r*. However, in the case of proteins we can overcome this difficulty by addition of the detergent sodium dodecylsulphate, SDS, which has the structure $CH_3(CH_2)_{11}OSO_3^- Na^+$. (SDS is also known as sodium lauryl sulphate). A reducing agent such as 2-mercaptoethanol is also added to break disulphide bonds (see Section 3.3.2). Addition of the detergent has two principal effects.[18]

1. Nearly all proteins bind SDS in a more or less constant ratio, 1.4 g SDS per gram of protein. Since the negative charge carried by the SDS overwhelms any charge carried by the protein, the protein-SDS complex has a constant charge/mass ratio.
2. The three-dimensional structure of the protein is lost and the protein-SDS complex is rod-shaped with a length proportional to the M_r of the protein.

* Hedrick and Smith[16] have shown that the M_r of a protein can be estimated by measuring its mobility as a function of acrylamide concentration. However, this method is only reliable if the standard proteins for calibration have the same shape, degree of hydration and partial specific volume as the unknown protein.[17]

Since the charge and hydrodynamic properties of the protein–SDS complex are both simple functions of the M_r, it is not particularly surprising that the mobility on electrophoresis is a function of M_r alone. The fact that larger molecules have lower mobilities means that the hydrodynamic effects (i.e. sieving) predominate over the charge effects. In practice it is found that a graph of the logarithm of the M_r against mobility is linear (Fig. 3.2) and the M_r of an unknown protein can be determined by reference to the standard line. Different ranges of M_r can be examined by the use of gels of different polyacrylamide concentration or by the use of gradient gels[19]; thus, 10 per cent gels give good separation in the range 10 000 to 70 000 and 5 per cent gels are satisfactory in the range 25 000 to 200 000. The accuracy of the M_r obtained is estimated to be 10 per cent or better.[18, 20]

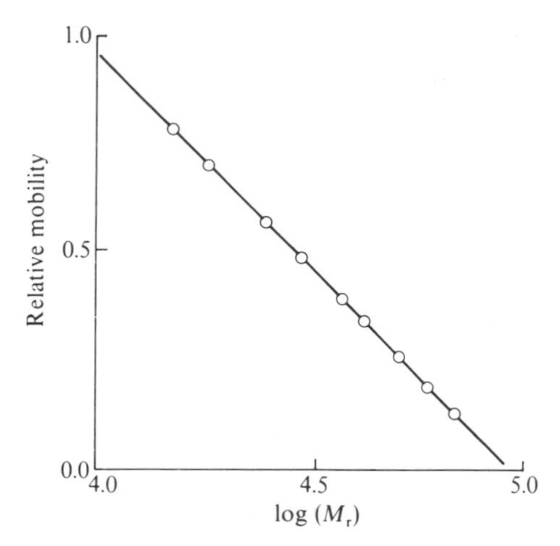

FIG. 3.2. Sodium dodecylsulphate polyacrylamide gel electrophoresis of proteins.[18] 10 per cent polyacrylamide gels were used and mobilities are expressed relative to bromophenol blue. The proteins used (in order of increasing subunit M_r) were: lysozyme, myoglobin, trypsin, carbonate dehydratase, glyceraldehyde-phosphate dehydrogenase, fructose-bisphosphate aldolase, fumarate hydratase, catalase, and bovine serum albumin.

The comments made in Section 3.2.2 about gel filtration also apply to SDS polyacrylamide gel electrophoresis; namely that the M_r obtained should be checked by other methods. There are cases known in which anomalous behaviour is observed on electrophoresis. Thus histones, which are highly positively charged, contribute significantly to the charge of the protein–SDS complex and give rise to rather low electrophoretic mobilities.[18] In addition, glycoproteins often show impaired binding of SDS and display low mobilities.[18] Additional complications can arise when proteins possess highly flexible regions (see the example of the lipoyl domains in the E2 component of the pyruvate dehydrogenase complex, Chapter 7, Sections 7.7.1 and 7.7.2).

Finally, we should note that addition of SDS dissociates a protein into its constituent subunits. If the subunit contains disulphide bonds, the addition of 2-mercaptoethanol ensures that these are broken, so that the molecular weight(s) we obtain by this method will be that those of the individual polypeptide chain(s).

3.2.4 Association–dissociation systems

In the previous Sections (3.2.1, 3.2.2, and 3.2.3) we have assumed that the enzyme in question has a unique M_r, i.e. that it is monodisperse. While this is a reasonable assumption for most purified enzymes, there are systems known that are capable of undergoing association and dissociation. A good example is provided by glutamate dehydrogenase from beef liver, where the fundamental unit of M_r 336 000 (itself made up of six polypeptide chains) can form an aggregate:

$$nA \rightleftharpoons A_n$$

where A is the hexamer of M_r 336 000 and A_n is the aggregate of M_r up to at least 2×10^6.

The position of equilibrium in this system depends on the concentration of enzyme with the aggregate form being favoured at high concentrations.[21] At any given concentration the M_r will contain contributions from the various species present and the measured value of this average M_r will depend on the technique employed. For instance, from osmotic pressure measurements we would determine a 'number average' M_r ($\bar{M}_n$) defined by

$$\bar{M}_n = \frac{\sum n_i m_i}{\sum n_i},$$

where n_i is the relative number of the ith species, and m_i is its relative molecular mass.

However, from sedimentation equilibrium or light-scattering measurements a 'weight average' M_r ($\bar{M}_w$) is determined.

$$\bar{M}_w = \frac{\sum n_i m_i^2}{\sum n_i m_i}.$$

Other examples of association-dissociation systems include enolase, in which a monomer-dimer equilibrium exists,[22] and acetyl-CoA carboxylase, in which extensive polymerization can occur to give large filamentous aggregates.[23]

It is possible to adapt some of the methods described to determine the M_r of the active unit of the enzyme when this may be in doubt, e.g. in an association–dissociation system. Cohen and Mire[24, 25] described the technique of *active enzyme centrifugation* in which enzyme at very low (catalytic) concentrations is sedimented through a solution of substrate. The rate of sedimentation down the ultracentrifuge cell is determined by monitoring the progress of conversion of substrate to product using absorption spectrophotometry or other convenient means of following the reaction. In addition, gel filtration can be

performed at low concentrations of enzyme if the measur
activity is used to monitor the position of elution of enzym
Using techniques such as these it has been possible to show,
active unit of beef-liver glutamate dehydrogenase is
336 000.[25, 26]

3.2.5 Uses of M_r information

The M_r of an enzyme is a fundamental piece of information because it allows us to convert the concentration of a solution from units of mass per volume (e.g. mg cm^{-3}) to units of molarity. This information can then be used in a whole variety of ways such as in considerations of composition (how many amino acids of a given type are present per molecule of enzyme?), catalytic activity (how many molecules of substrate are transformed by one molecule of enzyme per second?), and ligand binding (how many molecules of ligand are bound per molecule of enzyme?). Many of these applications will be described in the course of this and the succeeding three chapters.

Measurements of M_r made in the absence and presence of denaturing agents (see Section 3.7.1) will show whether or not the enzyme is composed of subunits and may indicate the number of such subunits (see Section 3.6.1.1). Alcohol dehydrogenase from yeast has an M_r of 140 000 on gel filtration, but on SDS-polyacrylamide gel electrophoresis the M_r is close to 35 000, indicating that the enzyme is a tetramer, composed of four subunits.

3.3 The determination of amino-acid composition

3.3.1 The amino acids

As mentioned in Chapter 1 (Section 1.2) virtually all enzymes are proteins and are thus made up of α-amino acids linked together. In some cases additional components such as metal ions, cofactors, or carbohydrate may be present. The composition of the protein part is defined in terms of the numbers of each of the 20 amino acids that occurs in each molecule of enzyme. These amino acids are shown in Table 3.1 in the form that is predominant at neutral pH, i.e. in the *zwitterion* form $^{+}NH_3$—CHR—CO_2^- rather than the uncharged form NH_2—CHR—CO_2H.* It must be pointed out that different ionized forms of amino acids from those shown in Table 3.1 are important in certain properties, e.g. in the reaction of cysteine with an electrophile such as iodoacetamide it is the —CH_2—S^- form of the cysteine side-chain that is the reactive species even though fewer than 10 per cent of the molecules are in this form at pH 7. In general, therefore, we shall, for convenience, depict amino acids as being *in their*

* Since the pK_a values of the carboxyl and amino groups attached to the α-carbon atom in the general formula are approximately 2 and 10 respectively, the 'zwitterion' form will predominate at neutral pH.

TABLE 3.1

The amino acids that are normally found in enzymes

α-Amino acids are of the general formula

$$^{+}NH_3-\underset{\displaystyle R}{\underset{|}{CH}}-CO_2^-$$

(except for proline). Different amino acids differ in the nature of the R group, or side-chain. One type of classification of the amino acids can be made on the basis of the chemical nature of the R group. The pK_a values of the carboxyl and amino groups attached to the α-carbon atom are approximately 2 and 10 respectively.

Amino acid	Abbreviations	Side-chain	pK_a of side-chain
Non-polar R group			
Alanine	Ala (A)	$-CH_3$	—
Valine	Val (V)	$-CH(CH_3)_2$	—
Leucine	Leu (L)	$-CH_2-CH(CH_3)_2$	—
Isoleucine	Ile (I)	$-CH(CH_3)-CH_2-CH_3$	—
Phenylalanine	Phe (F)	$-CH_2-C_6H_5$ (phenyl ring)	—
Tryptophan	Trp (W)	$-CH_2-$ (indole ring, N–H)	—
Methionine	Met (M)	$-CH_2-CH_2-S-CH_3$	—
Proline (see foot of table)			
Polar, uncharged R group			
Glycine	Gly (G)	$-H$	—
Serine	Ser (S)	$-CH_2-OH$	≈ 14.0
Threonine	Thr (T)	$-CH(OH)-CH_3$	≈ 14.0
Cysteine	Cys (C)	$-CH_2-SH$	≈ 8.3
Tyrosine	Tyr (Y)	$-CH_2-C_6H_4-OH$	≈ 10.1

TABLE 3.1 (*Contd.*)

Amino acid	Abbreviations	Side chain	pK_a of side chain
Asparagine	Asn (N)	$-CH_2-C(=O)NH_2$	
Glutamine	Gln (Q)	$-CH_2-CH_2-C(=O)NH_2$	
Acidic R group			
Aspartic acid	Asp (D)	$-CH_2-CO_2^-$	≈ 3.9
Glutamic acid	Glu (E)	$-CH_2-CH_2-CO_2^-$	≈ 4.3
Basic R group			
Lysine	Lys (K)	$-CH_2)_4-{}^+NH_3$	≈ 10.5
Arginine	Arg (R)	$-(CH_2)_3-NH-C(={}^+NH_2)NH_2$	≈ 12.5
Histidine	His (H)	$-CH_2-C=CH$ (ring: HN, CH=N)	≈ 6.0

Proline (abbreviated Pro or P) is an imino acid with a secondary nitrogen atom. The predominant form at pH 7 is

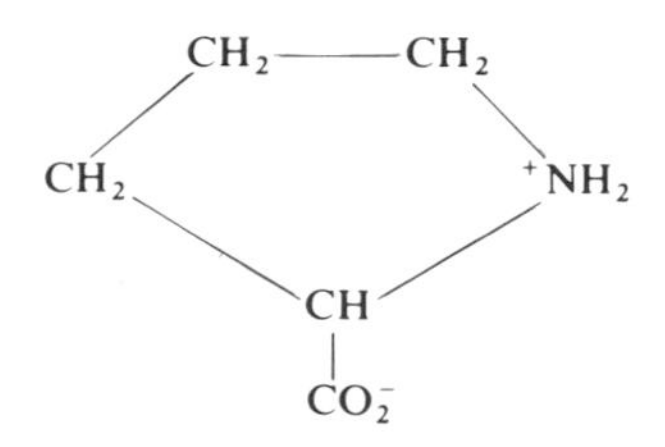

In this book we shall use the three-letter abbreviations for the amino acids. The one-letter code is frequently used in displays of amino-acid sequences to save space.

uncharged forms, and leave it to the reader to ascertain the predominant or most appropriate ionized form under a given set of circumstances.

Apart from glycine, all the amino acids have a chiral centre; indeed, isoleucine and threonine have two such centres. They are thus capable of existing in more than one enantiomeric form. It is found that in enzymes all the amino acids occur in the L-form in which the three-dimensional configuration about the α-carbon atom is as shown in Fig. 3.3(a).

D-Amino acids, which have the opposite configuration (see Fig. 3.3(b), do occur occasionally in some small peptides; thus D-phenylalanine occurs in the antibiotic gramicidin S. The mode of insertion of D-amino acids into such

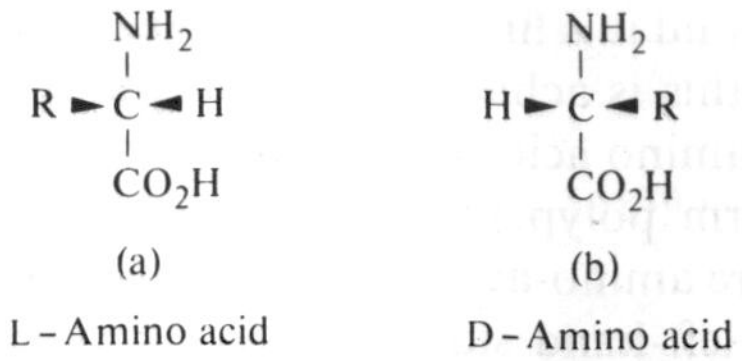

$$\begin{array}{cc} \begin{array}{c} NH_2 \\ | \\ R \blacktriangleright C \blacktriangleleft H \\ | \\ CO_2H \end{array} & \begin{array}{c} NH_2 \\ | \\ H \blacktriangleright C \blacktriangleleft R \\ | \\ CO_2H \end{array} \\ \text{(a)} & \text{(b)} \\ \text{L-Amino acid} & \text{D-Amino acid} \end{array}$$

FIG. 3.3. Structures of (a) an L-amino acid, and (b) a D-amino acid. In these diagrams, —R and —H project towards the reader; —NH_2 and —CO_2H project away from the reader.

systems is different,[27] however, from the normal mode of protein synthesis; a brief account of the latter is given in Chapter 9, Section 9.2.

3.3.2 Linkages between amino acids

The bond that joins amino acid units together to form a chain is known as an *amide* or *peptide bond.* We can view the bond as arising by the elimination of water as in the reaction:

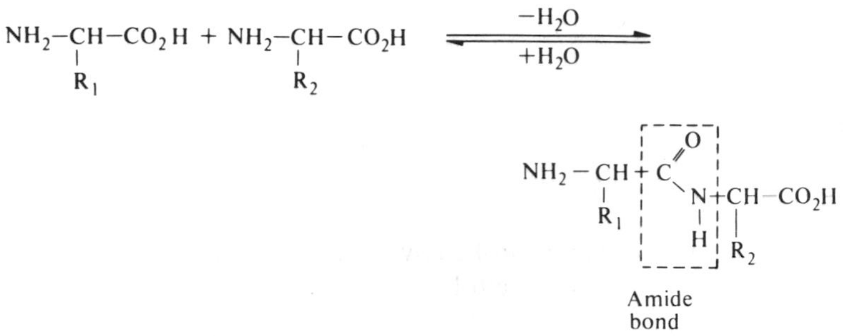

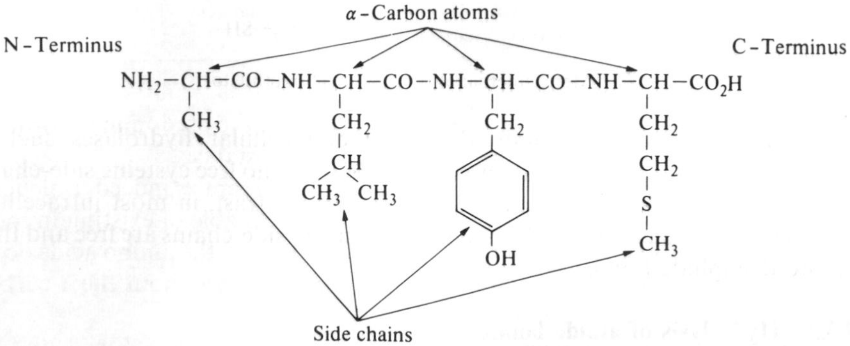

$$NH_2{-}\underset{R_1}{CH}{-}CO_2H + NH_2{-}\underset{R_2}{CH}{-}CO_2H \underset{+H_2O}{\overset{-H_2O}{\rightleftharpoons}} NH_2{-}\underset{R_1}{CH}{-}\overset{O}{\overset{\|}{C}}{-}\underset{H}{N}{-}\underset{R_2}{CH}{-}CO_2H$$

Amide bond

$$\underset{\text{N-Terminus}}{NH_2}{-}\underset{CH_3}{CH}{-}CO{-}NH{-}\underset{CH_2CH(CH_3)_2}{CH}{-}CO{-}NH{-}\underset{CH_2C_6H_4OH}{CH}{-}CO{-}NH{-}\underset{CH_2CH_2SCH_3}{CH}{-}\underset{\text{C-Terminus}}{CO_2H}$$

α-Carbon atoms

Side chains

FIG. 3.4. The structure of alanylleucyltyrosylmethionine. In naming the peptide, all amino acids except the C-terminal one are given the suffix -yl.[28]

In order to form this bond it is first necessary to activate the carboxyl group; the mechanism by which this is achieved is outlined in Chapter 9, Section 9.2.

When a number of amino acids are linked together in this way we build up a peptide chain. The term polypeptide or protein, is usually applied to chains consisting of 50 or more amino-acid units. By convention, the N-terminal amino acid is written at the left-hand side and the C-terminal one at the right. The amino-acid units are numbered from the N-terminal end. Figure 3.4 shows the tetrapeptide, alanylleucyltyrosylmethionine, which would be abbreviated as Ala-Leu-Tyr-Met or ALYM in the one letter code.[28]

Evidence for the presence of amide bonds in proteins comes from a variety of sources, e.g. u.v. and i.r. spectroscopy, the action of enzymes which are known to cleave amide bonds specifically and direct structure determinations using X-ray crystallography.

Apart from the amide bond, the only other type of covalent bond involved in linking amino acid units in enzymes is the disulphide (—S—S—) bond, which can be formed between two cysteine side-chains under oxidizing conditions:

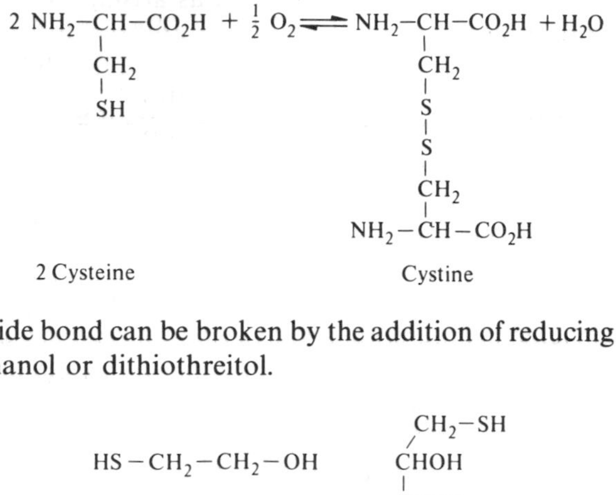

$$2\ NH_2{-}\underset{\underset{SH}{|}}{\underset{CH_2}{\underset{|}{CH}}}{-}CO_2H + \tfrac{1}{2}O_2 \rightleftharpoons NH_2{-}CH(CH_2{-}S{-}S{-}CH_2{-}CH(NH_2){-}CO_2H){-}CO_2H + H_2O$$

2 Cysteine — Cystine

The disulphide bond can be broken by the addition of reducing agents such as 2-mercaptoethanol or dithiothreitol.

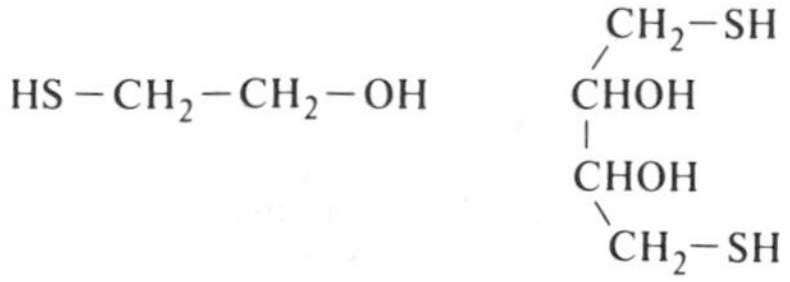

$$HS-CH_2-CH_2-OH \qquad HS-CH_2-CHOH-CHOH-CH_2-SH$$

2-Mercaptoethanol — Dithiothreitol

In some enzymes, especially the small extracellular hydrolases such as lysozyme, ribonuclease, and chymotrypsin, there are no free cysteine side-chains, i.e. all are paired to form disulphide bonds. By contrast, in most intracellular enzymes such as lactate dehydrogenase the cysteine side-chains are free and there are no disulphide bonds.

3.3.3 Hydrolysis of amide bonds

Hydrolysis of the amide bonds in proteins can be effected by a variety of reagents but the most commonly used conditions are 6 mol dm^{-3} HCl at 383 K (110°C) for 24 hours *in vacuo*. However, under these conditions some amino acids, e.g. tryptophan are partially or wholly destroyed (see Section 3.3.5). For determination of the tryptophan content of a protein, hydrolysis is carried out using *p*-toluene-sulphonic acid[29] or sodium hydroxide[30] in place of HCl. A single hydrolysis method claimed to be suitable for all amino acids (including cysteine and tryptophan) has been described.[31] It is based on alkylation of cysteine, followed by hydrolysis by methanesulphonic acid in the presence of tryptamine, which serves as an amino-acid protectant. In view of the problems associated with the various chemical methods, attention has been paid to the possibility of carrying out complete hydrolysis of proteins using proteolytic enzymes. As yet, however, no completely satisfactory procedure has been devised.[31, 32]

3.3.4 Analysis of amino-acid mixtures

After hydrolysis of a protein according to one of the procedures described in Section 3.3.3, the sample is dried and then analysed for the component amino acids. In the early days of protein analysis, the various amino acids were separated by complex crystallization procedures[33] and estimated gravimetrically or by microbiological assays.[34] These methods required many months of work and large amounts (up to several grams) of sample.

Following the development of chromatographic procedures, a great advance was made in 1958 with the development of an automated apparatus for performing amino-acid analysis;[35] with this apparatus it is possible to obtain the amino-acid composition of less than 1 mg of protein in a matter of a few hours. For their achievement and its application to the primary structure of ribonuclease, Stein and Moore were awarded the Nobel Prize for Chemistry in 1972. The amino-acid analyser consists essentially of an ion-exchange column, normally a sulphonated polystyrene resin. Samples to be analysed are dissolved in buffer at pH 2.2 and eluted from the column by defined volumes of citrate buffers of increasing pH, usually 3.25, 4.25, and 5.28, with the pH 5.28 buffer being of greater ionic strength than the first two. The order of elution of the amino acids (Asp, Thr, Ser, Glu, Pro, Gly, Ala, Val, Met, Ile, Leu, Tyr, Phe, Lys, His, Arg) can be seen to be broadly in line with expectations, i.e. the acidic amino acids are not retained by the negatively charged column, the neutral amino acids are retained to some extent, and the basic amino-acids are strongly retained. However, it is clear that other types of interactions between the amino acids and the column are involved, since the aromatic amino acids Tyr and Phe are eluted after the aliphatic amino acids such as Ala; this is presumably a consequence of interactions between the aromatic rings of these compounds with those of the column.[36]

As the amino acids are eluted from the column they are detected by reaction with ninhydrin (Fig. 3.5) which gives rise to a purple colour[37] (monitored at 570 nm). From the colour yield the amount of the amino acid can be evaluated and hence the composition of the protein determined. A more sensitive detection method involves the reaction of the amino acids with the reagents fluorescamine[38] or phthalaldehyde (Fig. 3.6),[39,40,41] which produce highly fluorescent compounds. With these reagents amino-acid analyses can be carried out successfully on as little as 1 μg of hydrolysed protein. Phthalaldehyde has some advantages over fluorescanine in terms of stability and sensitivity.[41]

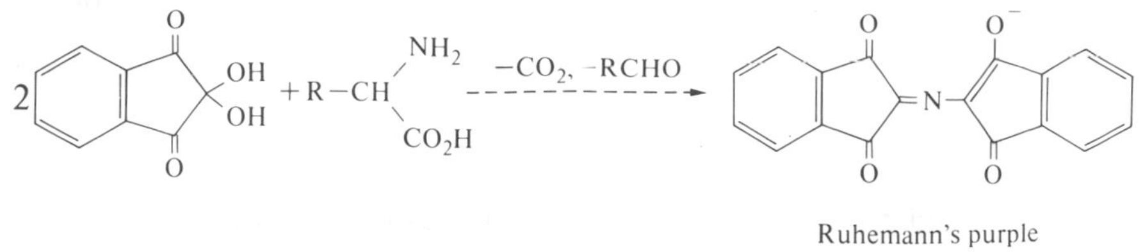

Ruhemann's purple

FIG. 3.5. The reaction of amino acids with ninhydrin. Proline does not give this reaction (Section 3.3.5.5).

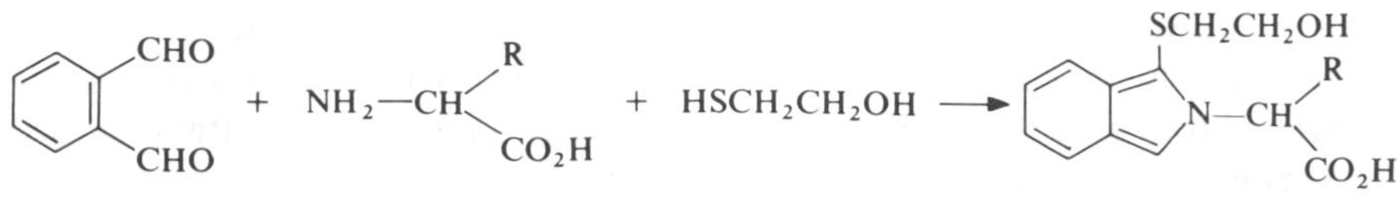

FIG. 3.6. The reaction of amino acids with phthalaldehyde in the presence of 2-mercaptoethanol to give the highly fluorescent 1-alkylthio-2-substituted isoindole product.

H.p.l.c. has been applied successfully to the analysis of amino-acid mixtures.[42] The mixture is reacted with reagents such as dimethylaminoazobenzenesulphonyl chloride or phthalaldehyde and the derivatives separated by reverse-phase h.p.l.c. using an acetonitrile gradient. The detection of the derivatives is by absorbance or fluorescence, respectively. Analyses can be completed in under 40 min.

3.3.5 Problems encountered in amino-acid analysis

As mentioned in Section 3.3.3 a number of amino acids are subject to destruction during acid hydrolysis of proteins and certain precautions must be taken to obtain reliable results. A full discussion of these precautions are described in a review[43] that also outlines procedures for the analysis of the carbohydrate components of glycoproteins, but we shall mention some of the more commonly encountered problems in Sections 3.3.5.1 to 3.3.5.5.

3.3.5.1 Hydroxyamino acids

The hydroxyamino acids, serine, threonine, and tyrosine, are partially destroyed during acid hydrolysis (up to about 10 per cent in 24 hours). In these cases, it is usual to perform hydrolyses for different lengths of time, e.g. 24, 48, and 72 hours, and to extrapolate the results of the analyses back to zero time to obtain the true content of these amino acids.

ion 69

·d amide bonds

›onds in proteins, especially those between the branched side-
and isoleucine are rather resistant to hydrolysis, e.g. the Ile–Ile
aved to the extent of about 50 per cent in 24 hours. In order to
values of the contents of these amino acids, extended periods of
hydrolysis (up to 120 hours) may be necessary.

3.3.5.3 Amide amino acids

The amide amino acids asparagine and glutamine are hydrolysed to yield aspartic acid and glutamic acid, respectively. Thus the content of glutamic acid determined in the amino-acid analyser will correspond to the sum of the contents of Glu plus Gln in the original protein. The decision as to whether a particular amino acid in the sequence of the protein is in fact Glu or Gln is normally made on the basis of the electrophoretic behaviour of peptide fragments (Section 3.4.7.2) or by the chromatographic properties of the phenylthiohydantoin derivatives following Edman degradation (Section 3.4.4.3). If there is a doubt about the identity, the amino acid is designated as Glx (Z), or Asx (B) in the case of a choice between aspartic acid and asparagine.

3.3.5.4 Sulphur-containing amino acids

Special problems arise in the determination of the sulphur-containing amino acids, cysteine, cystine, or methionine, since these are especially susceptible to oxidation even when care has been taken to remove oxygen prior to hydrolysis. These amino acids can be determined reasonably reliably as oxidized derivatives if the protein is oxidized with performic acid prior to hydrolysis.[44]

3.3.5.5 Proline

Proline, being an imino acid, cannot react with ninhydrin to yield the usual purple derivative (Fig. 3.5). The product formed is yellow and the amino-acid analyser is adapted to allow for monitoring at 440 nm to detect this derivative.

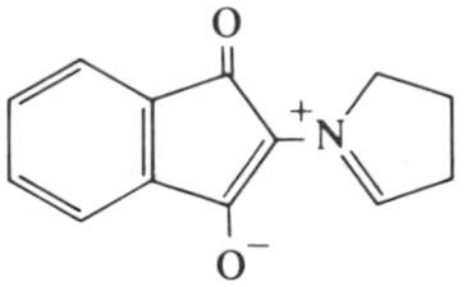

O
$\overset{+}{N}$
O^-

Product formed in reaction between proline and ninhydrin

For detection by reaction with phthalaldehyde (Section 3.3.4), proline must first be reacted with alkaline sodium hypochlorite solution.[41]

3.3.6 Uses of amino-acid composition data

Four particular uses of amino-acid data are described in Sections 3.3.6.1 to 3.3.6.4.

3.3.6.1 As a balance sheet

The composition of a protein can be used as a 'sum total' of amino acids to which the composition of derived peptide fragments (Section 3.4.2) must be related, since the total composition of the various fragments must equal the composition of the whole protein. This type of comparison could be used to show, for instance, that a particular fragment had not been accounted for during the purification of the fragments.

3.3.6.2 To explore similarities between enzymes

Composition data can be used to assess the degree of relatedness between enzymes, although much more detailed comparisons are possible when sequence data are available (Section 3.4.9.3). One method of assessing the relatedness has been described by Metzger[45] who defined a difference index (DI) as

$$\mathrm{DI} = 50 \times \left[\sum_i (\text{difference in mole fractions of amino acid}_i) \right].$$

DI equals zero if there is complete identity of composition and equals 100 if there are no amino acids in common between the two enzymes. The values of DI when human glyceraldehyde-phosphate dehydrogenase is compared with the enzyme isolated from rabbit, chicken, sturgeon, and *E. coli* are 2.9, 3.3, 5.4, and 7.5, respectively. These values indicate a high degree of relatedness between these enzymes; a conclusion that has been confirmed by the limited sequence data subsequently available. A somewhat similar method of comparison has been used to show that fructose-bisphosphate aldolase from yeast is related to the isoenzymes of fructose-bisphosphate aldolase from rabbit muscle, liver, and brain.[46] A fuller discussion of the various methods of comparison has been given by Cornish-Bowden,[47] who has drawn attention to the problems encountered when comparing the compositions of proteins of differing sizes. Composition data have been used to deduce the evolutionary relationships between organisms; although the information is less complete than that obtained from comparisons of amino-acid sequences (Section 3.4.9.5), it is much more readily obtained.[47]

3.3.6.3 As a guide to the properties of an enzyme

Many structural proteins have extremely simple amino-acid compositions, e.g. in the case of silk, glycine plus alanine plus serine account for 85 per cent of the total amino acids. This simplicity is correlated with the regular structure of these proteins, which is important for their function.

For enzymes, in general, no such regularities exist. Nevertheless, inspection of the amino-acid composition data can give a general idea of some of the properties of an enzyme. Thus, the protease pepsin has an overwhelming preponderance of acidic amino acids (Glu + Asp = 43) over basic amino acids (Lys + His + Arg = 4) in the enzyme isolated from the pig;[48] this can be related to the fact that the enzyme is active at the acid pH values (1–5) found in the stomach.

Hatch evaluated the ratio of polar to non-polar amino acids in a number of proteins and found that this ratio varied between about 1 and about 3. Proteins with a low ratio, i.e. a relatively high content of non-polar amino acids appeared to have a high content of helical structure and vice versa.[49] An extreme example of an enzyme with a very high content of non-polar amino acids (ratio = 0.6) is the C_{55}-isoprenoid alcohol phosphokinase from *Staphylococcus aureus.*[50] This extremely non-polar enzyme will preferentially dissolve in the butanol layer of a butanol–water system.

3.3.6.4 To calculate the partial specific volume of an enzyme

The amino-acid composition of a protein can be used to calculate[7] its partial specific volume, $\bar{v}$, which is used in determinations of molecular weight using the ultracentrifuge (see Section 3.2.1).

3.4 The determination of primary structure

3.4.1 The strategy underlying determination of primary structure

Up to a few years ago, determination of the primary structure of a protein or enzyme would have invariably proceeded according to the general scheme shown in Fig. 3.7.

The first protein sequence to be determined was that of insulin by Sanger in 1953;[51] since then the methods have been refined and developed to a considerable extent. Sequences are now known for well over a thousand proteins and the

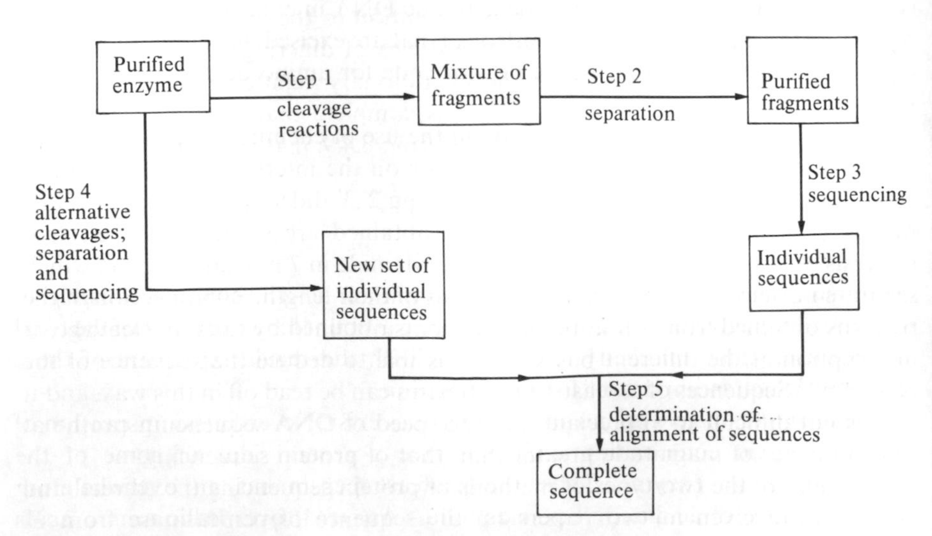

FIG. 3.7. Outline scheme for the determination of the amino acid sequence of a protein.

list is growing steadily. The largest polypeptide chains sequenced by the scheme outlined are those of phosphorylase from rabbit muscle, with 841 amino acids,[52] and β-D-galactosidase from *E. coli*, with 1021 amino acids.[53] By any standards these are remarkable achievements requiring many years of dedicated effort. We shall describe the various steps referred to in Fig. 3.7 in more detail in Sections 3.4.2 to 3.4.6.

Since the late 1970s, developments in the methods of 'direct' protein sequencing have been overtaken by those in the methods of sequencing bases in deoxyribonucleic acid (DNA). If the portion of DNA that encodes a particular enzyme or protein can be isolated and sequenced, the genetic code can be used to 'convert' the sequence of nucleotide bases into the amino acid sequence of the enzyme or protein (the 'indirect' sequencing method). Details of the methods involved in isolating and sequencing the DNA can be found in a number of books and reviews.[54–59] Isolation of suitable DNA fragments has been aided by the application of 'restriction endonucleases', enzymes that cleave DNA at a relatively small number of sites by recognizing certain sequences of bases[60]; over 400 such 'restriction enzymes' of different specificity are now known.[58] If an oligonucleotide whose sequence corresponds to a portion of the protein sequence is synthesized[61] and labelled with ^{32}P (to allow detection by autoradiography), it can be used to probe the mixture of fragments produced by the action of a restriction enzyme by the technique known as 'Southern blotting' (Fig. 3.8). In this technique the oligonucleotide probe hybridizes (forms hydrogen bonds) to the restriction fragment that contains the gene of interest. An alternative approach is to isolate the messenger RNA (mRNA) corresponding to the protein of interest. 'Reverse transcriptase' can then be used to produce the DNA that is complementary (known as cDNA) to that messenger. The latter technique has the advantage that the cDNA will correspond to the actual sequence of the protein, whereas the genomic DNA in eukaryotes may contain regions (intervening sequences or 'introns') that are excised during the formation of mature mRNA and that hence do not code for amino acids in the protein product.[58,59]

The DNA sequencing methods rely on the use of chemical reagents to bring about cleavage of DNA at given bases,[62] or on the interruption of enzymatic replication of DNA by DNA polymerase using 2′,3′-dideoxynucleotide triphosphates.[63] In both methods the products obtained are separated by electrophoresis on high-resolution gels (6–20% acrylamide in 7 $mol\,dm^{-3}$ urea), which separate nucleic-acid fragments on the basis of their length. By inspection of the patterns obtained from a number of fragments obtained by causing cleavage or interruption at the different bases, it is possible to deduce the sequence of the DNA.[62,63] Sequences of at least 100–200 bases can be read off in this way, and it is thus not difficult to appreciate that the speed of DNA sequencing can be at least an order of magnitude greater than that of protein sequencing.

Examples of the two types of methods of protein sequencing (i.e. 'direct' and 'indirect') are given in two papers on the sequence of penicillinase from *E. coli*.[64,65] The 'direct' method took about four years, whereas the nucleotide

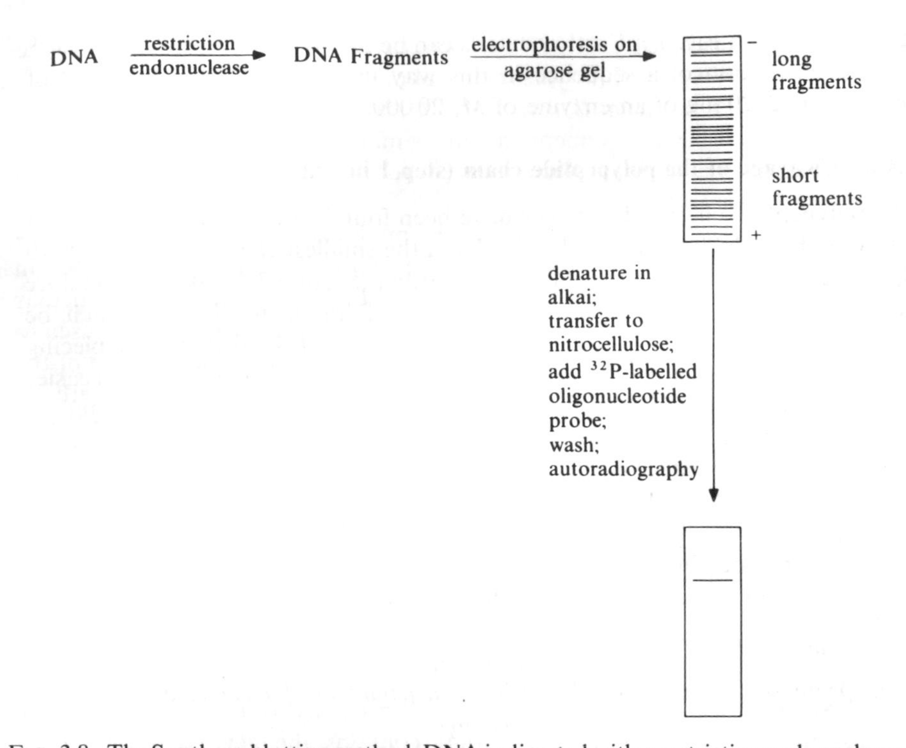

FIG. 3.8. The Southern blotting method. DNA is digested with a restriction endonuclease to give a mixture of fragments that can be separated by electrophoresis. After transfer to nitrocellulose ('blotting'), the fragment of interest can be located by adding a suitable oligonucleotide probe labelled at the 5′ end with ^{32}P (using polynucleotide 5′-hydroxyl-kinase). The probe hybridizes to the fragment of interest and can be detected by autoradiography. For further details, see references 58 and 59.

sequence of the DNA was deduced in about six months and was used to resolve some of the ambiguities encountered in the 'direct' method. The two types of method have been used in a complementary way in the determination of the sequence of phosphoglycerate kinase from baker's yeast (Section 3.4.8). Despite the impressive speed of the DNA sequencing methods it is still important to have some amino-acid sequence information about a protein, in order to identify the portion of DNA that is actually transcribed and subsequently translated. Also, knowledge of the amino-acid sequence of part of the protein can be used to design a suitable oligonucleotide probe[61] to locate the appropriate DNA fragment (see Fig. 3.8). In any particular case, if suitable quantities of purified protein are available, it may well be easier to sequence the protein directly. Finally, the DNA sequence cannot reveal any post-translational modifications e.g. N-terminal blocking (section 3.4.7.1), glycosylation, or formation of di-sulphide bonds; these can only be established by analysis of the protein itself.

In Sections 3.4.2 to 3.4.6 we shall discuss the various steps involved in determination of the primary structure of an enzyme according to the general scheme shown in Fig. 3.7. Further details can be found in the review by Allen.[66] The determination of a sequence in this way usually requires 0.1–1 μmol of enzyme, i.e. 2–20 mg of an enzyme of M_r 20 000.

3.4.2 Cleavage of the polypeptide chain (step 1 in Fig. 3.7)

The polypeptide chains of enzymes have been found to contain between about 100 and about 2500 amino-acid units. Even the smallest chains are too large to be sequenced directly from one end to the other (Section 3.4.4.4). It is therefore necessary to split the chain into a number of fragments that can each be sequenced before the sequence of the whole chain can be deduced by piecing together the sequences of the fragments. This piecing together will be much easier if we can cleave the polypeptide chain at a restricted number of points and so produce a limited number of well-defined fragments rather than cleave it randomly and so produce a collection of fragments of all sorts of lengths. Specific cleavages can be brought about by the use of certain proteases, since many of these possess high degrees of specificity; some of the more commonly used proteases are listed in Table 3.2. It is particularly important to ensure that pure preparations of proteases are used, since otherwise non-specific cleavages will

TABLE 3.2
Some proteases commonly employed in sequencing work for cleaving polypeptide chains

Protease	Nature of amino acid side-chains at cleavage site in polypeptide chain		
	amino ——— XY ↓ Z ——— carboxyl		
	X	Y	Z
Trypsin[67]	—	Basic side-chain (Lys, Arg)	—
Clostripain[68] (from *Clostridium histolyticum*)	—	Arg	—
Chymotrypsin[69]	—	Aromatic (Trp, Tyr, Phe) or hydrophobic (e.g. Leu, Met) side-chain	—
Elastase[70]	—	Small hydrophobic side-chain (e.g. Ala)	—
Papain[71, 72]	Phe (preference)	Preference for Lys over Ala	Leu or Trp (preference)
Thermolysin[73] (from *Bacillus thermoproteolyticus*)	—	—	Bulky hydrophobic side-chain (e.g. Val, Leu, Ile, Phe)
Endoprotease Glu-C[74] (from *Staphylococcus aureus* V8)	—	Glu or Glu, Asp	—
Pepsin[75]	—	Phe (preference)	Phe, Trp, Tyr (preference)

occur; in Chapter 5, Section 5.5.1.4 we describe the use of an affinity label to ensure that trypsin preparations are free from contamination by chymotrypsin. In order to obtain complete cleavage of the chain, e.g. on the C-terminal side of every lysine and arginine in the case of trypsin, the enzyme should be treated with a denaturing agent such as guanidine hydrochloride or urea (Section 3.7.1) so that the compact three-dimensional structure of the enzyme is destroyed and the susceptible bonds are exposed to the protease.[66]

The specificity of the proteolytic cleavages can sometimes be altered to advantage. For instance, lysine side-chains can be reacted with an anhydride such as maleic anhydride[76] or citraconic anhydride;[77] this prevents attack by trypsin at these points and limits cleavage to the C-terminal side of arginine side-chains, permitting the preparation of large fragments. Lysine side-chains can subsequently be regenerated by exposure to low pH (Fig. 3.9).

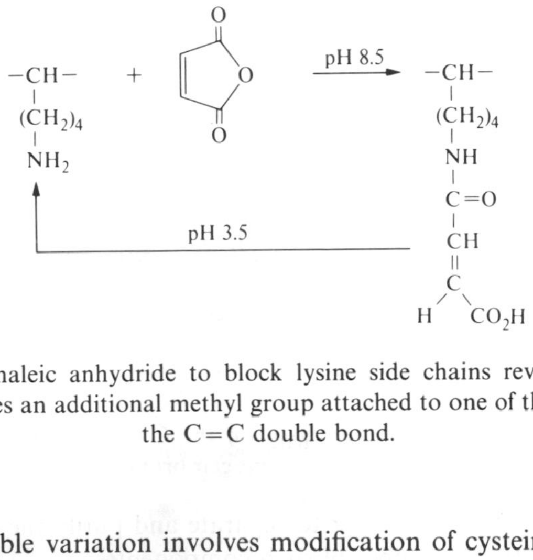

$$-CH(CH_2)_4NH_2- \;+\; \text{maleic anhydride} \xrightarrow{pH\ 8.5} -CH(CH_2)_4NH{-}C(=O){-}CH{=}CH{-}CO_2H-$$

pH 3.5

FIG. 3.9. Use of maleic anhydride to block lysine side chains reversibly. Citraconic anhydride possesses an additional methyl group attached to one of the carbon atoms in the C=C double bond.

Another possible variation involves modification of cysteine side-chains by reaction with 2-bromoethylamine to produce a side-chain analogous to that of lysine and which is in fact 'recognized' by trypsin[78] (Fig. 3.10). This modification reaction can generate additional cleavage sites and has sometimes proved useful in generating fragments from large polypeptides where successive lysine or arginine groups are well separated in the chain.[78]

Polypeptide chains can also be cleaved specifically by the use of certain chemical reagents. By far the most commonly used reagent is CNBr, cyanogen bromide, which under acidic conditions will cleave a polypeptide chain on the C-terminal side of methionine side chains[79] (Fig. 3.11).

This method of cleavage is important since methionine is a relatively uncommon amino acid in proteins; hence CNBr will generally cleave a protein into a smaller number of fragments than will be produced with attack by trypsin. For instance, in the case of adenylate kinase from pig muscle, cleavage with CNBr produces six fragments, whereas attack by trypsin would yield 32

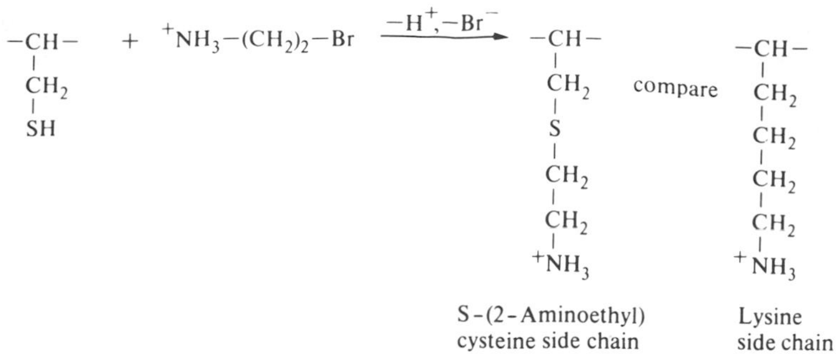

$-CH(CH_2SH)- + {}^{+}NH_3-(CH_2)_2-Br \xrightarrow{-H^+, -Br^-} -CH(CH_2-S-CH_2-CH_2-{}^{+}NH_3)-$ compare $-CH(CH_2-CH_2-CH_2-CH_2-{}^{+}NH_3)-$

S-(2-Aminoethyl) cysteine side chain — Lysine side chain

FIG. 3.10. Reaction of a cysteine side-chain with 2-bromoethylamine.

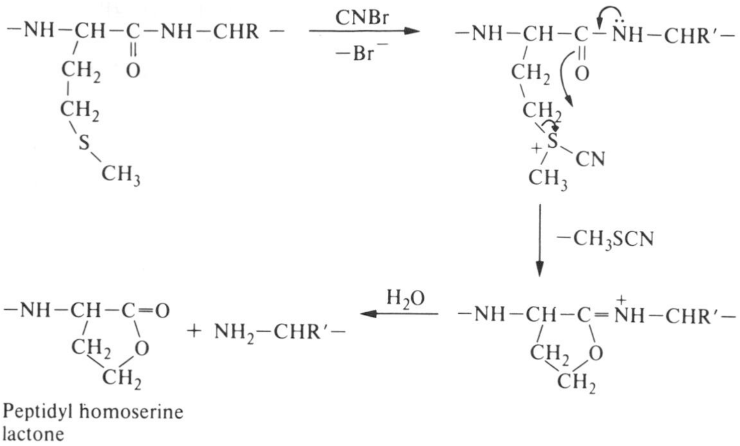

$-NH-CH(CH_2-CH_2-S-CH_3)-C(=O)-NH-CHR- \xrightarrow[-Br^-]{CNBr} -NH-CH(CH_2-CH_2-{}^{+}S(CH_3)-CN)-C(=O)-\ddot{N}H-CHR'-$

$\xrightarrow{-CH_3SCN} -NH-CH-C(=\overset{+}{N}H-CHR'-)$ (ring: CH_2-CH_2-O)

$\xrightarrow{H_2O} -NH-CH-C=O$ (ring: CH_2-CH_2-O) $+ NH_2-CHR'-$

Peptidyl homoserine lactone

FIG. 3.11. Cleavage of a polypeptide chain on the C-terminal side of methionine by reaction with cyanogen bromide.

fragments.[80] It is obviously easier to separate and purify the components of a mixture the smaller the number of such components.

Chemical methods for bringing about selective cleavage at other types of amino-acid side-chains including cysteine (e.g. by incubation at pH 9 after cyanylation[81]) and tryptophan (e.g. by incubation with *o*-iodosobenzoic acid under acid conditions,[82] see Section 3.4.8) have been described. For reviews of work in this area, the articles by Spande[83] and Allen[66] should be consulted.

3.4.3 Separation of the fragments (step 2 in Fig. 3.7)

The fragments produced by the various types of cleavage reactions described in Section 3.4.2 would be expected to differ in terms of their size, charge properties, chemical characteristics, etc. We could therefore utilize some of the methods described in Chapter 2 (Section 2.6) to separate and purify the various fragments present in the mixture. Methods that have been widely used include the following.

1. Separation on the basis of size by gel filtration. Sephadex G-50 would be most appropriate since it would separate species in the approximate M_r range 500 to 10 000. Peptide fragments that are eluted from a gel filtration column can be detected by measurements of absorption at 280 nm (which arises from the aromatic side chains of Tyr, Trp, and Phe) or at 215 nm (which arises from the amide bonds), or by measurements of other properties such as conductivity.[80]
2. Separation on the basis of charge, which reflects the numbers and types of ionizing side chains in the various fragments. Typically the mixture of fragments is subjected to paper electrophoresis in two dimensions at different pH values, e.g. at pH 6.5 followed by pH 3.5,[84] or to ion-exchange chromatography[66] (see Chapter 2, Section 2.6.2.1).
3. Separation by a combination of paper electrophoresis, say at pH 6.5, in one dimension, followed by paper chromatography in a second dimension. A suitable solvent system for the chromatography consists of a mixture of 1-butanol, acetic acid, water, and pyridine.[85] In this procedure the fragments are separated on the basis of charge and of chemical characteristics.

In methods (2) and (3) the peptide fragments can be located by reaction with ninhydrin.[86] These methods are suitable for small-scale work with a few milligrams of the mixture of fragments and are often used after a partial separation has been achieved on a larger scale by method (1). The separation pattern of peptide fragments obtained by methods (2) or (3) is referred to as the 'fingerprint' or 'peptide map' of the protein or enzyme and is widely used to look for changes in amino-acid sequences between related proteins.

Enzymes with a polypeptide chain even of modest size can give rise to a large number of fragments each of which must be obtained in a pure state to allow the sequence to be determined. In the early 1980s the separation of the peptide fragments represented the most time-consuming part of the entire sequencing operation. However, since that time, the separation problem has been greatly aided by the application of h.p.l.c. techniques. Reserve-phase h.p.l.c., in which the stationary phase consists of silica microspheres coated with long-chain (C_8 or C_{18}) alkyl groups and the mobile phase is a gradually increasing proportion of acetonitrile or propan-2-ol in aqueous solution, is capable of rapid, high resolution of peptide mixtures. Although the equipment involved is relatively expensive, h.p.l.c. is now the method of choice for peptide separation.[87] An example of the use of h.p.l.c. is given in Section 3.4.8.

3.4.4 Sequencing of the purified fragments (step 3 in Fig. 3.7)

Determination of the amino-acid sequence of the individual fragments will be discussed under a number of headings (Sections 3.4.4.1 to 3.4.4.4).

3.4.4.1 Determination of the N-terminal amino acid

The most widely used method involves reaction of the peptide with dansyl chloride (dimethylaminonaphthylsulphonyl chloride abbreviated DNS-Cl) under alkaline conditions, followed by acid hydrolysis to break the amide bonds (Fig. 3.12). The labelled N-terminal amino acid can be readily identified by electrophoresis or chromatography.[88] Dansyl chloride has now almost totally replaced the earlier N-terminal labelling reagent 1-fluoro-2,4-dinitrobenzene (FDNB), because dansylated amino acids are highly fluorescent and can be detected at amounts (less than 1 nmol) much lower than are required for the FDNB procedure.

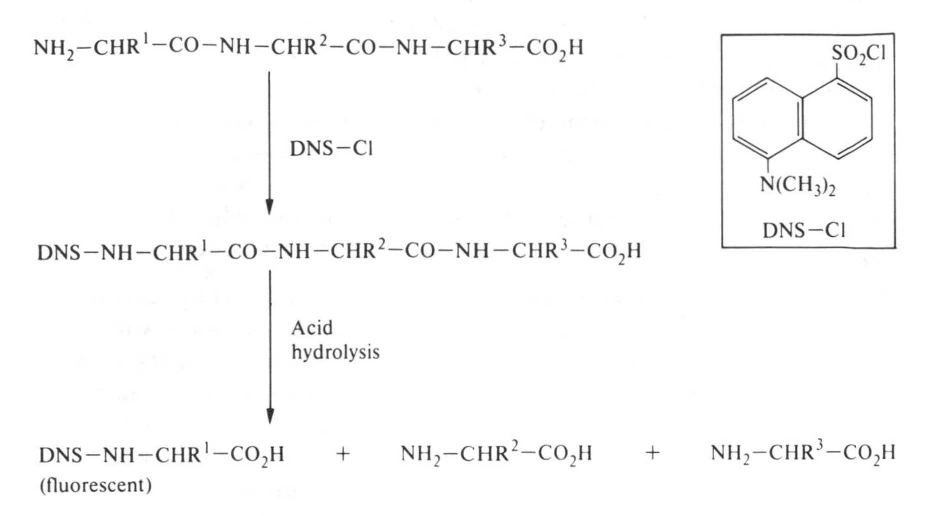

FIG. 3.12. The labelling of a tripeptide by reaction with dansyl chloride (DNS-Cl). The DNS-amino acid bond is stable to acid.

3.4.4.2 Determination of the C-terminal amino acid

Methods for determination of the C-terminal amino acid in peptides have not, in general, been as successful as those developed for the determination of the N-terminal amino acid. The most frequently used chemical method is that of hydrazinolysis, in which the peptide is reacted with hydrazine to convert all the amino acids except the C-terminal one to their hydrazides (Fig. 3.13). The unique C-terminal amino acid can be separated from the hydrazides by chromatographic procedures and identified using the amino-acid analyser (Section 3.3.4).

Schroeder[89] has outlined a procedure for hydrazinolysis of peptides that has overcome some of the more notable problems previously encountered, such as low yields and difficulties in separating and identifying the amino acid liberated.

An alternative method for the determination of the C-terminal amino acid involves the use of carboxypeptidase. This enzyme removes amino acids one at a time from the C-terminal end of a polypeptide chain (see Chapter 5, Section 5.5.1.7). The amino acid released can be identified with the aid of an amino-acid analyser. The application of this method and the various difficulties associated with it, including the fact that the rate of release of the C-terminal amino acid

$$NH_2{-}CHR^1{-}CO{-}NH{-}CHR^2{-}CO{-}NH{-}CHR^3{-}CO_2H$$

$$\downarrow N_2H_4$$

$$NH_2{-}CHR^1{-}CO{-}NH{-}NH_2 + NH_2{-}CHR^2{-}CO{-}NH{-}NH_2 + NH_2{-}CHR^3{-}CO_2H$$

(separated and identified)

FIG. 3.13. Hydrazinolysis of a tripeptide. The C-terminal amino acid is separated and identified.

depends on the nature of its side-chain, are discussed in reviews by Ambler[90] and Ward.[91]

3.4.4.3 Sequential degradation of the peptide chain

The determination of the N- and C-terminal amino acids in a peptide chain provides some useful information, but in order to obtain sequence data we require a method to degrade the peptide chain in a sequential manner under carefully controlled conditions. Such a method was developed by Edman[92] and is based on the reaction of the N-terminal amino acid with phenyl isothiocyanate (Fig. 3.14).*

The initial adduct, a phenylthiocarbamyl derivative, A, rearranges under anhydrous acidic conditions, cleaving the adjacent peptide bond and giving a heterocyclic anilinothiazolinone derivative, B. Derivative B is separated from the residual peptide by extraction with an organic solvent and converted to the more stable phenylthiohydantoin (PTH) derivative, C, before identification.[94] The Edman reaction permits the removal of amino acids one at a time from the N-terminus of a peptide chain. Although the conditions can be arranged so that the reactions involved take place with close to 100 per cent efficiency,[92] there is a gradual accumulation of chains of differing length resulting from incomplete removal of the N-terminal amino acid at each cycle. In practice, the useful upper limit is around 30–40 cycles of degradation.

Identification of the amino acid released in each cycle of Edman degradation can be achieved directly by identification of the PTH derivative, (C in Fig. 3.14), using chromatographic procedures.[95] An alternative, indirect, method uses the Edman reaction merely to remove the N-terminal amino acid in a controlled fashion; the new exposed N-terminal amino acid is identified by reaction with dansyl chloride as described in Section 3.4.4.1. This 'dansyl-Edman' procedure[96] is illustrated in Fig. 3.15; determination of the sequence of a decapeptide, for example, can be achieved with about 20 nmoles of sample, rather less than is required for the direct method.

* Methods for sequential degradation of a polypeptide chain from the C-terminus have not generally been successful. Reaction with acetic anhydride and thiocyanate can be used for the sequential removal of up to five amino acids from the C-terminus.[93]

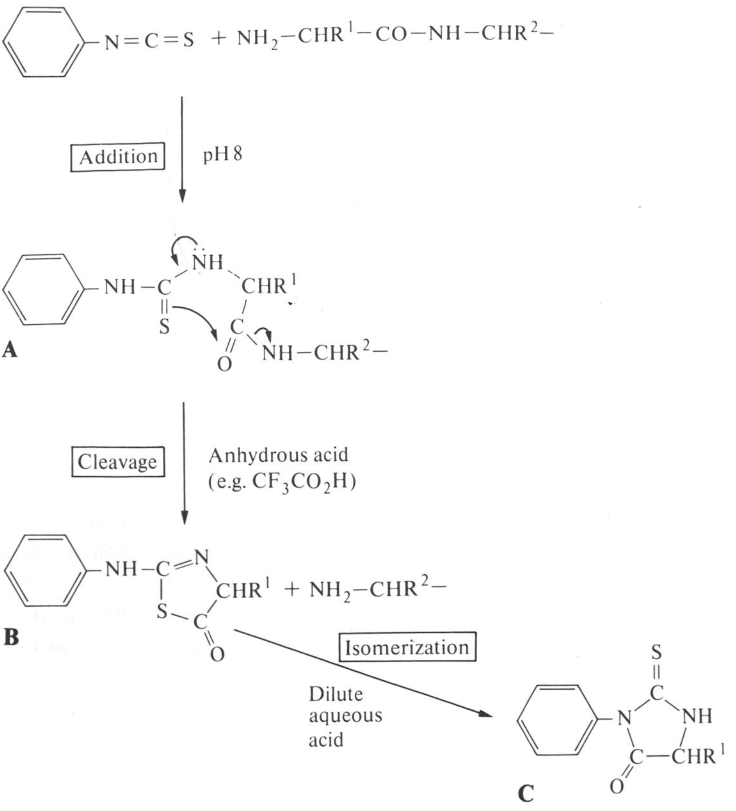

$C_6H_5-N=C=S + NH_2-CHR^1-CO-NH-CHR^2-$

Addition pH 8

A

$C_6H_5-NH-C(=S)-NH-CHR^1-C(=O)-NH-CHR^2-$

Cleavage Anhydrous acid (e.g. CF_3CO_2H)

B

$C_6H_5-NH-C=N-CHR^1-C(=O)-S$ (ring) $+ NH_2-CHR^2-$

Isomerization Dilute aqueous acid

C

FIG. 3.14. Sequential degradation of a peptide chain by reaction with phenyl isothiocyanate.

3.4.4.4 Recent developments in methods of sequencing

Two developments have helped to speed up the determination of sequences of peptides. These are (i) automated sequencing using the Edman degradation, and (ii) mass spectrometry. We shall discuss these in turn.

(i) Automated sequencing

In 1967, Edman and Begg described the design and application of an instrument, known as a sequenator, for determining automatically the amino-acid sequences of peptides and proteins.[92] The peptide is immobilized by being spread as a thin film in a spinning cylindrical glass cup and the necessary reagents and solvents are passed over the peptide in a programmed cycle. The PTH-amino acid formed after each cycle of the Edman degradation can be identified by thin-layer chromatography[92] or h.p.l.c.[97] By careful attention to the conditions of the reaction, an average yield of slightly better than 98 per cent was obtained at each

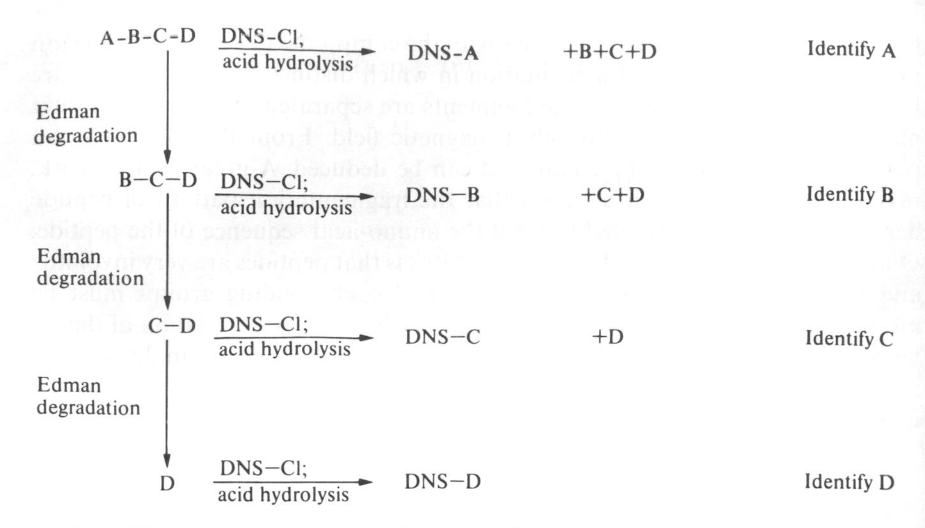

FIG. 3.15. The 'dansyl-Edman' procedure. Part of the sample at each stage is taken for N-terminal analysis using dansyl chloride; the remainder is subjected to Edman degradation.

cycle of a 60-residue degradation* of sperm whale myoglobin.[92] An alternative approach has been to immobilize the peptide by covalent attachment to a solid support, e.g. a modified acrylamide resin.[98] However, in this 'solid phase' approach lower yields are obtained at each cycle of degradation,[99] thus limiting the number of cycles that can be performed before unambiguous information is no longer obtained. A more recent alternative has been a 'gas phase' sequenator in which the reagents for the Edman degradation are delivered in a stream of argon or nitrogen and the protein or peptide is immobilized by adsorption onto a glass-fibre disc. The repetitive yields have been claimed to be better than with the 'liquid phase' instrument.[100]

Although there are many problems associated with the use of the sequenator,[97] for instance the corrosive nature of some of the reagents, less than 100 per cent yields at each cycle, internal peptide bond cleavage, and hydrolysis of the side-chains of glutamine and asparagine, there is no doubt that the instrument has made a substantial contribution to determination of protein sequences. At present, it would be unrealistic to suppose that the sequence of a polypeptide chain of, say 150 amino acids could be determined 'in one go'; the sequenator is best employed to determine the sequences of fragments (up to about 50 amino acids in length) produced by the cleavage reactions described in Section 3.4.2 (see the example of phosphoglycerate kinase in Section 3.4.8).

(ii) Mass spectrometry

In mass spectrometry a compound is converted, in the gas phase, to an ion by bombardment with a beam of electrons or with a gas such as methane that can

* If each cycle proceeds with a 98 per cent yield, the overall yield after 60 cycles is only 30 per cent.

take place in proton-transfer reactions with the compound of interest.[101] The ion produced then undergoes fragmentation in which distinct chemical entities are lost from the parent molecule; the fragments are separated on the basis of their mass, usually by passage through a magnetic field. From the fragmentation pattern the structure of the compound can be deduced. A great deal of work, reviewed by Priddle,[101] has shown that the fragmentation pattern of peptide derivatives can be interpreted to yield the amino-acid sequence of the peptide. One of the most difficult problems in this work is that peptides are very involatile and in order to apply the technique the hydrogen-bonding groups must be modified to increase the volatility. This is usually done by acetylation of the N-terminal amino group, followed by complete methylation of the amide nitrogen atoms and esterification of the C-terminal carboxyl group, achieved by reaction with methyl iodide under strongly basic anhydrous conditions[102] (Fig. 3.16).

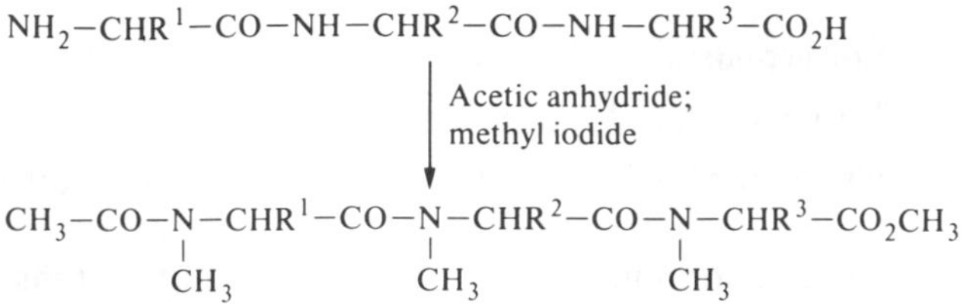

$$NH_2{-}CHR^1{-}CO{-}NH{-}CHR^2{-}CO{-}NH{-}CHR^3{-}CO_2H$$

$$\big\downarrow \text{ Acetic anhydride; methyl iodide}$$

$$CH_3{-}CO{-}\underset{\displaystyle CH_3}{\underset{|}{N}}{-}CHR^1{-}CO{-}\underset{\displaystyle CH_3}{\underset{|}{N}}{-}CHR^2{-}CO{-}\underset{\displaystyle CH_3}{\underset{|}{N}}{-}CHR^3{-}CO_2CH_3$$

FIG. 3.16. Modification of a peptide by acetylation and methylation in order to prepare a volatile derivative for mass spectrometry.

The resulting derivatives are sufficiently volatile if the number of amino acids in the peptide is about 12 or fewer[102] and are suitable for mass spectrometry because fragmentation is largely confined to the peptide bond.[101] The entire sequence of 162 amino acids of tetrahydrofolate dehydrogenase from *Lactobacillus casei* has been deduced by mass-spectrometric analysis of fragments produced by the action of various proteases on the enzyme.[103] The development of fast-atom bombardment (f.a.b.) mass spectrometry may overcome a number of the difficulties associated with conventional mass spectrometry.[104] In this technique there is no need for prior derivatization of the sample. The peptide is dissolved in glycerol and bombarded with high-energy atoms (Xe or Ar). The transfer of momentum from the atoms to the peptide leads to the production of gaseous ions. Peptides up to 20 amino acids in length have been analysed by this method.[104]

Although mass spectrometry must still be regarded as a highly specialized technique, it has one advantage over conventional sequencing procedures, namely that it is not necessary to work with purified peptides. This was shown in spectacular fashion when the enkephalin, a peptide that binds to opiate receptor sites in the brain, isolated from the pig, was shown to be a mixture of two pentapeptides of sequences[105]

Tyr-Gly-Gly-Phe-Met (a)

and Tyr-Gly-Gly-Phe-Leu (b)

The ratio of (a) to (b) is about 3 to 1. These structures were determined using only about 10 nmol (i.e. 5 μg) of sample.

3.4.5 Preparation and analysis of new fragments (step 4 in Fig. 3.7)

Once the peptide fragments have been purified and sequenced as described in Sections 3.4.2 to 3.4.4, the final task is to position them correctly so as to build up the overall sequence of the protein. This is usually done with the aid of a new set of fragments produced from the original protein by use of a different cleavage procedure (see Section 3.4.6). Thus, if trypsin were used to produce the first set of fragments, we might use chymotrypsin to produce the new set. These new fragments are purified and sequenced using the procedures already described (Sections 3.4.3 and 3.4.4).

3.4.6 Alignment of peptide sequences and determination of the overall sequence (step 5 in Fig. 3.7)

When the peptide sequences of the fragments produced by the various cleavage procedures are known, it is possible to position the peptides in the overall sequence by determining the regions of overlap of sequence. For instance, in the case of adenylate kinase from pig muscle,[80] two of the cyanogen bromide fragments CBc and CBd could be positioned with respect to each other by

Leu-Ser-Glu-Ile-Met	Glu-Lys-Gly-Glu . . . Met
CBc	CBd

the isolation of a peptide produced by the action of trypsin on the enzyme which had the sequence:

Met-Leu-Ser-Glu-Ile-Met-Glu-Lys.

Thus the overall sequence of this portion of the polypeptide chain is

Met-Leu-Ser-Glu-Ile-Met-Glu-Lys-Gly-Glu . . . Met

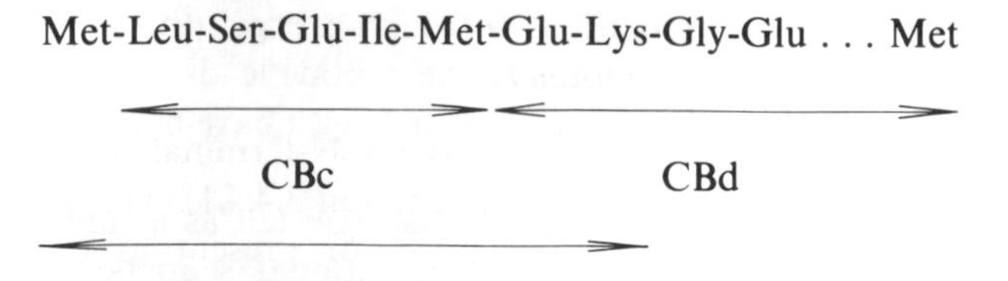

Tryptic peptide

This conclusion was confirmed by the analysis of fragments produced by the action of thermolysin on the enzyme.

The following considerations may give the reader some feel for the number of overlapping amino acids that are required to position the peptides unambiguously. If the amino-acid sequence of a protein is purely random, then the probability of occurrence of a given sequence of two amino acids, e.g. Asp-Gly, will be 1/400 (i.e. $1/20 \times 1/20$: there are 20 different amino acids found in proteins). The probability of occurrence of a given sequence of three amino acids, e.g. Asp-Gly-Leu, will be 1/8000. It is clear that a given sequence of three amino acids is unlikely to be repeated in a polypeptide chain of length 100 to 2500 amino acids and hence an overlap of this size should serve to position peptide fragments fairly unambiguously in the overall sequence.*

The positioning of fragments in the overall sequence of a protein is greatly helped if the sequence of a related protein is available. For instance, determination of the sequence of elastase was considerably speeded up because this enzyme shows a high degree of similarity to chymotrypsin, the sequence of which was already available.[106]

Recently it has proved possible to position fragments in the overall sequence by making use of the electron-density maps of proteins available from X-ray crystallography (see Section 3.4.9.3). Examples of the interplay of crystallography and sequencing are to be found in the studies on phosphoglycerate mutase[107] and hexokinase[108] from yeast.

3.4.7 Some problems encountered in the determination of amino-acid sequences

Using the procedures outlined in the previous sections (3.4.2 to 3.4.6) the sequence of a protein can usually be deduced, provided that sufficient care is taken in the purification and characterization of the various fragments. It is important to check that the composition of the enzyme deduced from the proposed sequence agrees with the composition determined by amino-acid analysis and that the sequence is in accordance with any DNA-derived sequence data available (Section 3.4.8). Any discrepancy between the various sets of data could indicate that a fragment has not been accounted for during the separation procedures. Some special problems arise in determinations of sequence; three of the most common problems are mentioned below.

3.4.7.1 Blocked N-terminal amino acid

Some proteins do not appear to possess an N-terminal amino acid when subjected to the dansyl chloride reaction (Section 3.4.4.1). This can be because the amino group is acetylated, formylated, or present as a pyroglutamyl moiety.[109] Examples of an acetylated N-terminal amino acid and an N-terminal pyroglutamyl group are provided by carbonate dehydratase from human erythrocytes[109] and peroxidase from horseradish[110] respectively. Myristic acid

* The numerical aspects of this argument must not be taken too seriously. The 20 amino acids do not occur with equal frequency and the sequence is almost certainly not random. Thus, in adenylate kinase, which has 194 amino acids, the sequence Leu-Lys occurs three times and the sequence Met-Glu occurs twice, i.e. much above the 'random' frequency.

($C_{13}H_{27}CO_2H$) is the N-terminal blocking group in the catalytic subunit of cAMP-dependent protein kinase[111] (see Chapter 6, Section 6.4.2.1).

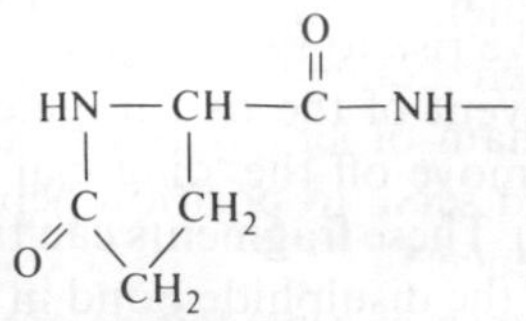

$$\begin{array}{l} \quad\quad\quad\quad\quad\quad \text{O} \\ \quad\quad\quad\quad\quad\quad \| \\ \text{HN}-\text{CH}-\text{C}-\text{NH}- \\ \ \ | \quad\quad\ \ | \\ \ \ \text{C} \quad\ \ \text{CH}_2 \\ \text{O}^{\not/} \ \diagdown \ \ \diagup \\ \quad\quad \text{CH}_2 \end{array}$$

The pyroglutamyl group in which an N-terminal glutamic acid has formed an internal amide bond.

Allen[66] has given a general discussion of the problems of structure determination in those cases where post-translational modification has occurred.

3.4.7.2 *Amide assignments*

In Section 3.3.5.3 it was mentioned that the 'amide' amino acids asparagine and glutamine are converted to aspartic acid and glutamic acid, respectively, during acid hydrolysis of a protein or peptide fragment. This conversion can cause difficulties in deciding whether a particular amino acid in the sequence is asparagine or aspartic acid (or glutamine or glutamic acid). A correct assignment can usually be made on the basis of the electrophoretic mobility at pH 6.5 of a fragment containing the amino acid in question[112] since the side-chain of aspartic acid (glutamic acid) will carry a negative charge at this pH, whereas the side-chain of asparagine (glutamine) will carry no charge. In order to prepare the fragment of the enzyme it is necessary to perform the hydrolysis under conditions in which the amide bond in the side-chain of asparagine or glutamine will not be hydrolysed, e.g. by the use of various proteases (see Table 3.2). Using this approach it was possible to show that amino acid 102 in chymotrypsin was aspartic acid rather than asparagine.[113] As we shall describe in Chapter 5 (Section 5.5.1.2), this aspartic acid side-chain plays a crucial role in the catalytic mechanism of the enzyme. Correct assignments can usually be made on the basis of Edman degradation, since thc phenylthiohydantoin derivatives (Section 3.4.4.3) can readily be distinguished. Reference to the nucleotide sequence (if available) would confirm these assignments.

3.4.7.3 *Location of disulphide bonds*

If a protein is found to contain disulphide bonds, formed between pairs of cysteine side-chains, it is generally necessary to break these bonds either by reduction (Section 3.3.2) followed by reaction with iodoacetate or by oxidation with performic acid (Section 3.3.5.4) in order to form derivatives that are stable during the procedures involved in sequence determination. However, when these disulphide bonds are broken, the information concerning which pairs of cysteine side-chains are linked is lost, and in order to determine the location of these linkages it is necessary to isolate fragments with the disulphide bonds intact. One very useful way of detecting fragments that have these bonds intact, i.e. that contain cystine, is the 'diagonal' electrophoresis technique developed by Brown and Hartley in their studies on the structure of chymotrypsinogen.[114] This technique is based on the fact that oxidation of cystine (R—S—S—R) by performic acid gives rise to two molecules of cysteic acid (R—SO_3H) which, being a strong acid, will carry a negative charge at pH 6.5. Thus a peptide fragment that contains cystine will after oxidation give rise to two fragments of greater mobility towards the anode (except in the event of the two fragments having identical amino-acid composition), and will move off the 'diagonal' on electrophoresis in two dimensions at pH 6.5 (Fig. 3.17). These fragments can then readily be isolated and sequenced and the position of the disulphide bond in the overall sequence can be determined. Recently it has been shown that analysis by fast-atom bombardment mass spectrometry of proteolytic fragments prior to and after reduction of the disulphide bonds can also be used to assign the positions of the disulphide bonds.[115]

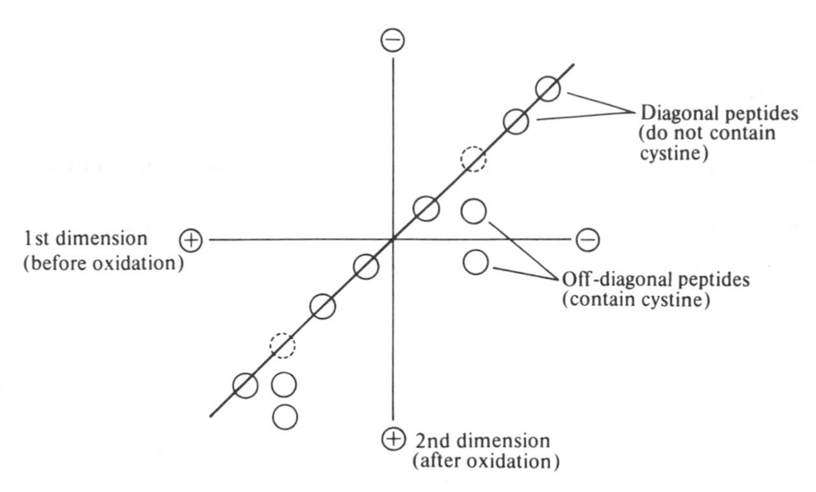

FIG. 3.17. The 'diagonal' electrophoresis method to locate cystine-containing peptides. The off-diagonal peptides would have moved to ◌ if no oxidation had been performed.

3.4.8 An example of sequence determination

The various methods described in Sections 3.4.2 to 3.4.7 have been used to determine the sequences of a large number of proteins. The example we have chosen to discuss is that of phosphoglycerate kinase from baker's yeast, since this work nicely illustrates the way that 'direct' (i.e. protein) and 'indirect' (via nucleic acid) sequencing techniques can complement each other.[116]

The sequence of yeast phosphoglycerate kinase (a monomer of M_r 44 500) was required to permit detailed interpretation of the electron-density map determined previously by X-ray crystallography.[117] Phosphoglycerate kinase was isolated by a procedure involving cell breakage using ammonia solution, ammonium sulphate fractionation, and affinity elution from CM-cellulose with 3-phosphoglycerate.[118, 119] Traces of a contaminating protein were removed by gel filtration on Sephadex G-75.

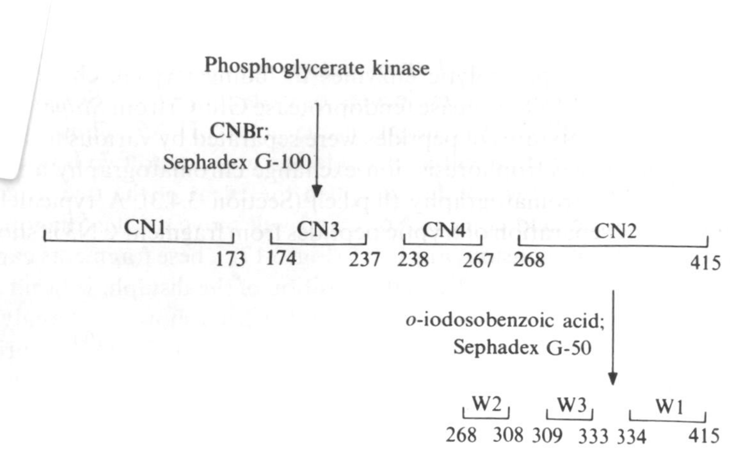

FIG. 3.18. The digestion of phosphoglycerate kinase to prepare fragments for sequencing. The numbering of fragments is described in the text.

The fragmentation steps involved in the 'direct' sequencing are illustrated in Fig. 3.18.

Amino-acid analysis revealed that the molecule possesses three methionine side-chains. Cleavage with CNBr (Section 3.4.2) yielded four fragments that could readily be separated by gel filtration on Sephadex G-100. The numbering of the fragments (CN1–4) is on the basis of their order of elution on gel filtration, CN1 being eluted first. The fractions pooled as 'CN1' were found to be heterogeneous on polyacrylamide gel electrophoresis in the presence of sodium dodecylsulphate; further purification was achieved by covalent chromatography on activated thiol-sepharose (see Chapter 2, Section 2.6.4.5), taking advantage of the fact that CN1 contained the single Cys in the molecule.

Fragments CN1, CN3, and CN4 were found by amino-acid analysis to contain homoserine, whereas CN2 did not. This placed CN2 as the C-terminal fragment. Fragments CN2, CN3, and CN4 showed free N-terminal amino groups by reaction with dansyl chloride (Section 3.4.4.1). CN1 did not show any free amino group and was therefore positioned at the N-terminal end of the molecule (see Fig. 3.18). (It had been previously established[120] by analysis of small peptides produced by proteolytic digestion that the N-terminus of yeast phosphoglycerate kinase was *N*-acetylserine.) The order of CN3 and CN4 was determined by nucleotide sequencing as described below.

The large CN2 fragment was further digested by treatment with *o*-iodosobenzoic acid; under the appropriate conditions, cleavage occurs specifically at the two Trp side-chains.[82] (Since Trp is abbreviated W in the one letter code (Table 3.1) the fragments are denoted W1-3). W1 was eluted first on gel filtration on Sephadex G-50.

The sequences of the various fragments in Fig. 3.18, i.e. CN2, CN3, CN4, W1, W2, and W3 were established by automated Edman degradation (Section 3.4.4.4) and by manual Edman sequencing of peptides derived by digestion of these fragments with various proteolytic enzymes including trypsin, chymotrypsin, thermolysin, pepsin, and V8 protease (endoprotease Glu-C) from *Staphylococcus aureus*. The resulting mixtures of peptides were separated by various techniques, such as high-voltage electrophoresis, ion-exchange chromatography and high-performance liquid chromatography (h.p.l.c.) (Section 3.4.3). A typical h.p.l.c. profile showing the separation of tryptic peptides from fragment CN3 is shown in Fig. 3.19.

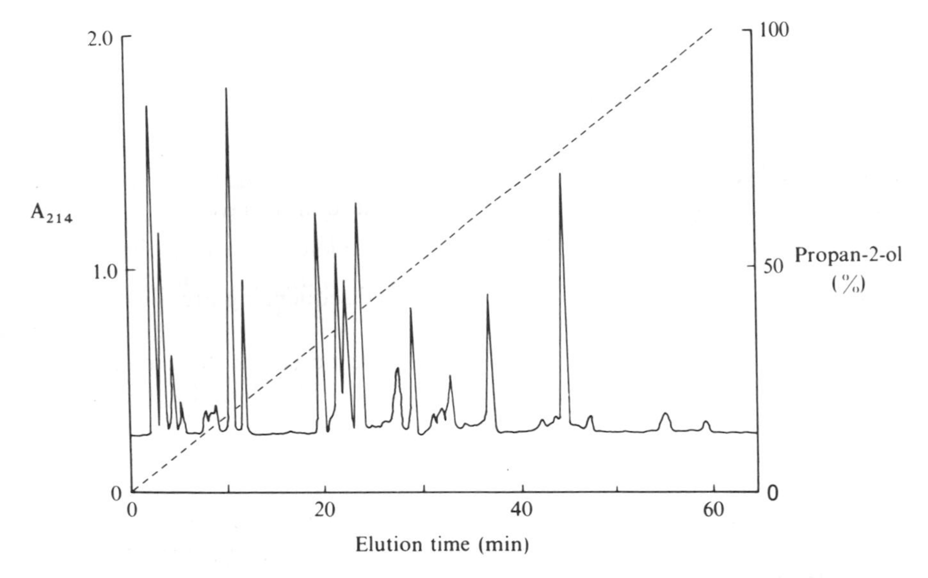

FIG. 3.19. H.p.l.c. separation of tryptic peptides from fragment CN3. Approximately 0.2 mg of the peptide mixture in 0.1 per cent trifluoroacetic acid was applied to a C_{18} reverse-phase column and eluted with a gradient of propan-2-ol as indicated. The elution of peptides was monitored at 214 nm at which the amide bond absorbs strongly.

The order of the fragments W1, W2, and W3 in CN2 was determined by automated Edman sequencing from the N-terminus of CN2 and by the isolation of overlapping peptides, e.g. peptide S29 from the V8 protease digest of CN2 (Fig. 3.20).

Amide assignments (Gln vs. Glu; Asn vs. Asp) were made from direct identification of the phenylthiohydantoin derivatives during Edman degradation (Section 3.4.4.3).

At this point, the sequences of the fragments CN3, CN2, and CN4 and the fact that CN4 was C-terminal had been established. The remaining parts of the protein sequence were established by complementary work being undertaken on the structure of the gene coding for phosphoglycerate kinase in yeast. It had been previously established[121] that this gene (the PGK gene) was located on a 2900-base (2.9-kb) fragment resulting from digestion of the vector pMA3 with the

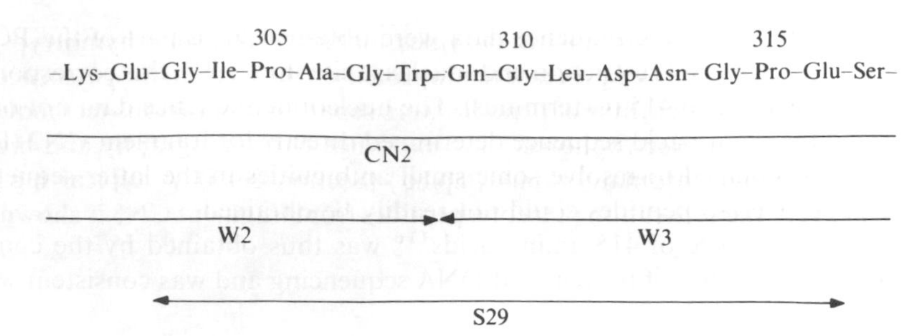

FIG. 3.20. An overlap peptide (S29) to establish the order of fragments W2 and W3 in the CN2 fragment.

restriction endonuclease Hind III*. By treatment of the pMA3 vector with the restriction endonuclease Sal I followed by an exonuclease (BAL 31) a series of overlapping gene fragments were prepared. After appropriate selection, these could be sequenced (see Fig. 3.21), to provide a complete sequence of the first 624 bases of the PGK gene (starting from ATG as the initiation codon).

By use of the genetic code, the DNA sequence could then be expressed as a protein sequence (amino acids 1 to 207).

The part of the protein sequence deduced by the DNA sequence covers the entire CN1 fragment as well as a portion of the CN3 fragment (see Fig. 3.18), thus showing the relative arrangement of these two fragments in the overall sequence.

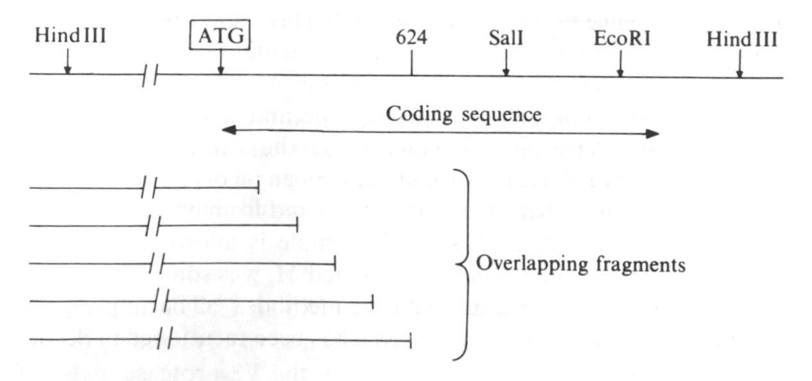

FIG. 3.21. General structure (not to scale) of the PGK gene in the pMA3 vector. The initiation codon (ATG) is shown together with the sites of cleavage by various restriction endonucleases (see reference 58 for details of these). The overlapping fragments were sequenced in the left←right direction, so that an unambiguous DNA sequence of the first 624 bases could be established.

* A complete understanding of the concepts involved in the DNA cloning and sequencing methods are not important for the purposes of this discussion. An excellent account of this area is given in reference 58.

A vector is a piece of DNA that carries genetic information and can replicate autonomously in an appropriate host. pMA3 is an artificially constructed 'shuttle vector' that can carry genetic information between different organisms, in this case *E. coli* and yeast.

In addition, nucleotide sequence data were obtained for a part of the PGK gene around the EcoRI restriction endonuclease site (Fig. 3.21); this corresponds to amino acids 281 to 415 (C-terminus). The nucleotide sequence data not only confirmed the amino-acid sequence determined directly for fragment CN2 (Fig. 3.18), but also helped to resolve some small ambiguities in the latter sequence where good overlap peptides could not readily be obtained.

The final sequence of 415 amino acids[116] was thus obtained by the complementary approaches of protein and DNA sequencing and was consistent with all the available data.

3.4.9 Uses of sequence information

The information contained in the primary structure of an enzyme can be used in a number of ways. As we have already noted (Section 3.4.1), even a small amount of sequence information can be useful for designing an oligonucleotide probe for the gene coding for the enzyme. However, the complete sequence can give much more information; five of the more important applications are discussed in Sections 3.4.9.1 to 3.4.9.5.

3.4.9.1 To calculate the M_r of an enzyme

There are a number of experimental methods (Section 3.2) for determining the M_r of an enzyme. However, the definitive M_r is only obtained when the covalent structure, i.e. sequence of the enzyme, is established, and this calculated M_r can be used as a check on the validity of the experimental methods. For instance in the case of β-D-galactosidase from *E. coli* the M_r of the subunit had been measured as 135 000, but when the sequence became available it was clear that this value was some 15 per cent too high and that the correct value was 116 300.[53] The discrepancy may have arisen from the use of an incorrect value of the partial specific volume of the enzyme (Section 3.2.1.1) and from errors in the interpretation of ultracentrifuge data. A second example is afforded by the work on pencillinase from *E. coli* in which the calculated M_r was some 20 per cent higher than that measured by physical chemical methods.[64] The importance of an accurate knowledge of the M_r of an enzyme has been mentioned in Section 3.2.5.

3.4.9.2 To locate a particular amino acid in an enzyme

The primary structure of an enzyme provides a framework within which the importance of particular amino-acid side-chains can be discussed. In Chapter 5 (Section 5.4.4) we shall describe how chemical modification of an amino-acid side-chain can be used to indicate the involvement of that side-chain in the catalytic mechanism of an enzyme. For instance, the reagent diisopropyl fluorophosphate modifies one serine side-chain in chymotrypsin leading to inactivation; analysis of the peptide fragment containing this modified amino acid shows that this serine is number 195 in the sequence (see Chapter 5, Section

5.5.1.2). It is essential to have this type of information in conjunction with three-dimensional structural data (Section 3.5) to formulate detailed proposals concerning the mechanism of action of an enzyme.

3.4.9.3 *To interpret data from X-ray crystallography*

X-ray crystallographic studies of an enzyme (Section 3.5) yield a three-dimensional map of the electron density in the molecule. In order to interpret this pattern of electron density in terms of the arrangement of atoms it is normally necessary to know the sequence of the enzyme so that the path of the chain of α-carbon atoms and amide bonds can be traced and the positions of the side-chains located.

The X-ray structure of phosphoglycerate mutase from yeast was determined in 1974[122] to a resolution of 0.35 nm. At this level of resolution some of the main structural features such as helices and sheets (Section 3.5.4) were clearly visible. However, it was not until sequence information on the enzyme became available that the higher-resolution electron-density maps could be interpreted in sufficient detail to allow a plausible mechanism of action to be proposed.[107] In this case the X-ray and sequence studies complemented each other nicely, since the electron density map helped to order the various peptide fragments in the overall sequence (Section 3.4.6).

3.4.9.4 *To predict the three-dimensional structure of an enzyme*

As described in Section 3.7.2 a number of experiments have shown that the primary structure contains the information necessary to specify the three-dimensional structure of an enzyme. We might therefore ask whether it is possible to use sequence data to predict the three-dimensional structure. The answer to this question is a moderately encouraging 'yes' provided that we set ourselves the more limited task of predicting which parts of the amino acid sequence might occur in regions of distinct secondary structure (helices, sheets, etc.). This type of approach is described in Section 3.7.3.

Sometimes it is possible from the sequence to guess at parts of an enzyme that may have a distinct functional role. In the case of phospholipase A_2, it is likely that the non-polar N-terminal portion of the polypeptide chain

Ala-Leu-Trp-Gln-Phe-

is involved in the interaction of the enzyme with the non-polar portions of phospholipid substrates. This conclusion is supported by the results of chemical modification of amino-acid side-chains.[123] It is well recognized that certain types of N-terminal sequences are involved in directing newly synthesized proteins to particular destinations, e.g. into the lumen of the endoplasmic reticulum prior to export[124] or to the various compartments of the mitochrondrion.[125] Stretches of non-polar amino acids of membrane-associated proteins have usually been assigned to membrane-spanning segments of the polypeptide chain[126] (see Chapter 8, Section 8.4). Finally, certain types of amino-acid sequence are associated with high degrees of flexibility, e.g. the Ala/Pro-rich sequences in the lipoyl domains of the E2 component of the pyruvate dehydrogenase complex (see Chapter 7, Section 7.7.2).

3.4.9.5 *To explore evolutionary relationships between enzymes*

Amino-acid sequences provide a wealth of data with which to make comparisons between enzymes.[127] One of the goals of such comparisons is to learn something about the processes involved in enzyme evolution.* A change in the primary structure of an enzyme usually arises from a change in the base sequence of the DNA coding for that enzyme. We shall mention only two of the various types of genetic modification that can occur: point mutations and gene duplication. A comprehensive review of this subject has been given by Smith.[129]

(i) A point mutation refers to the replacement of a single nucleotide in a triplet codon (a sequence of three bases in DNA required to specify which amino acid is incorporated into the sequence of a protein). Thus the codon AAG specifies lysine; replacement of the first base by G would give GAG which specifies glutamic acid. Comparison of the sequences of homologous enzymes, i.e. those that catalyse the same reaction, in different organisms shows that most, if not all, of the changes in sequence can be explained by point mutations.[129] With the recent developments in DNA sequencing (see Section 3.4.1), it is now also possible to examine DNA sequences in order to trace evolutionary events more directly.[130] The results confirm that the steady evolution of proteins results from the steady evolution of DNA. However, there are some additional points to be borne in mind when DNA sequences are compared that are not apparent from comparisons of protein sequences. (a) The degenerate nature of the genetic code means that changes in the third position of codons can occur at a greater rate than either the first or second positions without leading to a change in the amino acid incorporated. (b) The frequency of usage of synonymous codons (codons which specify the same amino acid) may be very different in different organisms.[131] (c) The final protein sequence is specified after the intervening sequences ('introns') have been removed during formation of mature mRNA; thus a range of proteins could be produced by appropriate combinations of 'exons' (expressed sequences).[132] We shall see in Chapter 5 (Section 5.4.5) that the technique of site-directed mutagenesis can be employed to introduce defined mutations at known sites in a protein.

Amino-acid substitutions do not appear to occur randomly along the sequence, since there is a high degree of identity around those amino acids that are known to be functionally important, e.g. those involved in the catalytic mechanism. In Chapter 5, Section 5.5.3.4 we shall see how comparisons of the sequences of fructose-bisphosphate aldolases have indicated that a histidine near

* Comparisons of three-dimensional structures of enzymes can also be used to examine these evolutionary processes (Section 3.5.3.3). It should be noted that a number of different amino-acid sequences can give rise to very similar three-dimensional arrangements, so in functional terms there may be much less restraint on amino-acid sequences.[128]

the C-terminus is probably not involved in the catalytic mechanism, contrary to an earlier proposal. Presumably the pattern of amino-acid substitutions reflects a combination of two processes: *mutation* which can occur randomly along the sequence and *survival* of the mutants which is not a random process. Replacement of one amino acid by another with different characteristics (e.g. the basic side-chain of lysine replaced by the acidic side-chain of glutamic acid) would produce a non-functional enzyme if the lysine side-chain were involved in the catalytic mechanism or in the maintenance of the structure of the enzyme. The operation of natural selection would serve to ensure that organisms with such non-functional enzymes did not survive.

(ii) When enzymes catalysing similar reactions within an organism are compared it is often possible to find evidence for the occurrence of gene duplication, in which two copies of a particular gene have been produced, followed by independent evolution of these two copies by processes such as point mutation. This type of duplication process would mean that an organism would be capable of producing enzymes of differing specificity allowing a greater variety of foodstuffs to be utilized and thereby enhancing the survival prospects of the organism. The high degree of sequence homology of various proteases of differing specificity, e.g. trypsin, chymotrypsin, elastase, and thrombin, especially around those amino acids known to be involved in the catalytic mechanism (Table 3.3) suggests very strongly that these enzymes evolved from some common precursor protease by a process of gene duplication, followed by subsequent independent evolution. Other examples of such 'divergent' evolution (i.e. from a presumed common precursor) are provided by the group of thiol proteases papain, ficin and bromelain, and by the two types of alcohol dehydrogenase in mammalian liver.[129]

On the other hand, comparisons between certain other enzymes suggests that a process of 'convergent' evolution may have occurred in which a similar type of

TABLE 3.3

Amino-acid sequences around the essential serine, histidine, and aspartic acid groups in various proteases[133]

Enzyme	Amino-acid sequences		
	Essential serine	Essential histidine	Essential aspartic acid
Trypsin (bovine)	Gly-Asp-SER-Gly-Gly	Ala-Ala-HIS-Cys-Tyr	Asn-Asn-ASP-Ile-Met
Chymotrypsin (bovine)	Gly-Asp-SER-Gly-Gly	Ala-Ala-HIS-Cys-Gly	Asp-Asn-ASP-Ile-Thr
Elastase (porcine)	Gly-Asp-SER-Gly-Gly	Ala-Ala-HIS-Cys-Val	Gly-Tyr-ASP-Ile-Ala
Thrombin (bovine)	Gly-Asp-SER-Gly-Gly	Ala-Ala-HIS-Cys-Leu	Asp-Arg-ASP-Ile-Ala
Subtilisin BPN′	Gly-Thr-SER-Met-Ala	Asn-Ser-HIS-Gly-Thr	Val-Ile-ASP-Ser-Gly

functional unit has been evolved starting from quite different precursor polypeptide chains. For example there is very little similarity in primary structure (Table 3.3) or in three-dimensional structure between the enzymes subtilisin, a protease isolated from *Bacillus subtilis*, and bovine chymotrypsin, suggesting that these two enzymes did not arise by a process of divergent evolution from a common precursor. However, both enzymes possess the 'charge relay system' of three precisely positioned side chains, Ser . . . His . . . Asp (see Chapter 5, Section 5.5.1.2) of amino acids well separated from each other in the sequence. This system is essential for the catalytic mechanism and its occurrence in the two quite different enzymes suggests that a process of 'convergent' evolution has occurred.[129] ('Convergent' evolution is sometimes referred to as 'parallel' evolution, since it is not clear that the enzymes are 'converging' towards one another with the passage of time!)

3.5 The determination of secondary and tertiary structure

Knowledge of the primary structure of an enzyme does not allow us to explain properties such as catalytic power and specificity. We must also consider how the polypeptide chain is folded up so that different parts of the chain are brought into close proximity with each other to create binding sites for substrates and unusual environments which facilitate catalysis (see Chapter 5, Section 5.3). The determination of the three-dimensional structure of an enzyme relies very heavily on X-ray crystallography since this is the only technique that has up to now been shown to be capable of giving structural information at atomic detail. We shall describe the technique in outline before considering the relationship between the structure in the crystal and that in solution. There are a number of reviews of X-ray crystallography that can be consulted for further details.[134, 135, 136, 137, 138]

3.5.1 X-ray crystallography

X-ray crystallography relies on the scattering of electromagnetic radiation of suitable wavelength by electrons belonging to the atoms in a molecule. In the case of a regularly arranged array of atoms, such as is present in a crystal lattice, we can have constructive or destructive interference between the scattered waves; only constructive interference will give rise to a detectable signal. X-rays (which are emitted when an electron falls to a lower orbital of an atom from a higher occupied orbital) provide suitable radiation to bring about these diffraction effects since their wavelength is comparable with the interatomic distances in a molecule.* Virtually all structures of proteins have been determined using the K_α radiation emitted by Cu; this is of wavelength 0.154 nm. Recently, increasing attention has been paid to the possibilities of using synchrotron radiation sources (see Chapter 5, Section 5.4.3). The intensities of these (highly specialized)

* This phenomenon is analogous to the diffraction of visible light by a diffraction grating. The spacing between the lines on the grating is of the same order as the wavelength of the incident light.

sources are much higher than those of conventional X-ray sources, thereby shortening the time required for data collection.[139]

The positions and intensities of diffracted rays (in which constructive interference has occurred) are measured using either photographic film, so that a pattern of spots is obtained, or a diffractometer that can measure intensities directly. From the pattern of spots on a photographic film it is possible to deduce certain features such as the overall symmetry of the crystal and the dimensions of the repeating unit, but additional information is required in order to determine the three-dimensional structure of the enzyme, as described below.

If we know the three-dimensional structure of a molecule, it is a relatively straightforward task to calculate the positions and intensities of the diffracted rays. Proceeding in the reverse direction from the diffraction pattern to the three-dimensional structure is difficult because we need to have information on the phases of the scattered X-rays, i.e. how the waves are arriving at the detector related in terms of numbers of wavelengths from a common origin. This information on phases is lost when only the positions and intensities of the diffracted rays are recorded. However, it is possible to solve the phase problem by the method of isomorphous replacement, in which a limited number of heavy atoms are introduced at selected sites in the molecule without distorting the crystal structure.[140] The electron-dense heavy atom scatters the X-rays more strongly than the lighter atoms (H, C, N, O, and S) of the enzyme and will add its scattering power to the diffracted rays. By measuring the changes in intensities of the diffracted rays it is possible to extract the required information on the phases; usually, at least two heavy-atom derivatives are required for this purpose. Heavy atoms can be introduced by chemical modification of amino-acid side-chains as in the following examples:

$$\mathrm{E{-}CH_2{-}SH + CH_3{-}Hg{-}NO_3 \longrightarrow E{-}CH_2{-}S{-}Hg{-}CH_3 + NO_3^- + H^+}$$

(Cysteine side chain)

$$\mathrm{E{-}CH_2{-}C_6H_4{-}OH + I_2 \longrightarrow E{-}CH_2{-}C_6H_3(I){-}OH + I^- + H^+}$$

(Tyrosine side chain)

Once the phases of the diffracted rays are known it is possible, by a process known as Fourier summation, to calculate the three-dimensional map of the electron density within the molecule.[141] By considering only a relatively limited number of spots in the diffraction pattern, i.e. those rays diffracted through the lowest angles, a map at low resolution (e.g. 0.6 nm) can be obtained. From this low-resolution map a good idea of the overall shape of the molecule can be gained. In order to obtain a map at the degree of resolution at which amino-acid side-chains can be resolved (0.3 nm or better) it is necessary to take into account the intensities of many more spots, perhaps several thousands, arising from rays diffracted through larger angles. Even with the advent of modern computing techniques, the analysis of diffraction data and calculation of electron-density maps is still a time-consuming process.

In order to interpret the electron-density map of an enzyme in terms of its atomic structure, it is necessary to know the amino-acid sequence of the molecule. We can then trace the path of the chain of α-carbon atoms and amide bonds and locate the various side-chains in the electron-density map. As mentioned in Section 3.4.9.3, it is not always necessary to have the complete sequence to hand, because the electron-density map can be used to order the various peptide fragments in the overall sequence. When the polypeptide chain has been fitted to the electron-density map, we can reasonably claim to have determined the three-dimensional structure of the enzyme.

When the basic requirements for X-ray crystallography are considered (Table 3.4), it is not difficult to see why the determination of the three-dimensional structure of an enzyme represents a major undertaking in terms of time and resources. Since the first high-resolution structure of a protein, myoglobin, was published in 1960[142] a number of technical and theoretical developments have aided the crystallographer and the structures of well over 200 proteins have now been determined to varying degrees of resolution. From this wealth of structural information a number of interesting conclusions have been drawn (see Section 3.5.3).

The structure of an enzyme deduced by X-ray crystallography is not entirely a static one, since atomic movements such as vibration and rotation about single bonds can still occur even though the packing of molecules in the crystal will

TABLE 3.4
The basic requirements for X-ray crystallography of enzymes

Requirement	Comments
Crystals of enzyme	Crystals must be of a suitable size, i.e. of dimensions approximately 0.5 mm or larger. They can be formed by gradual changes in the ionic strength or polarity of a solution of enzyme (see Chapter 2, Section 2.6.3). The crystals must also be reasonably stable in the X-ray beam during the course of the experiment
Isomorphous heavy-atom derivatives	The derivatives are necessary to provide information on phases. They can be prepared by chemical modification of amino-acid side-chains, e.g. with mercurials, or by soaking the crystals in a solution containing heavy-metal ions. Incorporation of the heavy atom must be shown not to distort the structure, i.e. to alter only the intensities not the positions of the spots in the diffraction pattern
Computing facilities	These facilities are required to calculate the electron-density map of the molecule from information on the position, intensities, and phases of the diffracted X-rays
Primary structure	The sequence of amino acids is necessary to interpret the electron-density map in terms of the arrangement of atoms, i.e. covalent structure, of the molecule

tend to dampen these motions.[143] A good example of atomic movements in a crystal is provided by trypsinogen, the inactive precursor of the protease trypsin (see Chapter 6, Section 6.2.1.1), of which the electron-density map has no density corresponding to four major segments of the chain: N-terminus to Gly 19, Gly 142–Pro 152, Gly 184–Gly 193, Gly 216–Asn 223.[144] It is concluded that these segments are waggling in a flexible manner or adopting a number of conformations in the crystal leading to a 'smearing out' of electron density. In the electron-density map of trypsin these segments are well defined, presumably because the motion is now restricted. Other examples of the loss of electron density are provided by the C-terminal regions of fructose-bisphosphate aldolase and tyrosyl-tRNA synthetase (see Chapter 5, Sections 5.5.3.1 and 5.5.5.4, respectively). As described in Section 3.5.2, the flexibility of an enzyme molecule is considerably greater in solution than in the crystal.

3.5.2 The structure of an enzyme in solution

Our knowledge of the structure of an enzyme in solution is much less exact than of its structure in the crystal. Several lines of evidence, such as measurements of fluorescence quenching, hydrogen-deuterium exchange, and nuclear magnetic resonance show that proteins in solution possess fairly flexible structures.[145, 146] It is thus more appropriate to think of an enzyme in solution as existing in a number of conformational states (related to each other by rotation about single bonds) of roughly equivalent energy. The structure in solution will represent a time average of these various conformational states of the enzyme.

There is no technique that can give information on the structures of enzymes in solution in the atomic detail provided by X-ray crystallographic studies of the solid state. However, recent advances in the application of nuclear magnetic resonance to solutions of small enzymes such as lysozyme have yielded an impressive amount of structural information.[147, 148] Optical rotatory dispersion and circular dichroism,[149] which rely on the interaction of polarized light with chiral (optically active) compounds can be used to measure the amount of regular secondary structure (Section 3.5.4) in an enzyme. Chemical modification of amino-acid side-chains (see Chapter 5, Section 5.4.4) can be used to divide side-chains of a particular type into those that are 'exposed', i.e. on the surface of the enzyme, and those which are 'buried', i.e. in the interior of the enzyme.[150] The overall conclusion from these and other studies is that the time average structure of an enzyme in solution is on the whole very similar to its structure in the solid state (crystal)[145, 151] and this finding provides the justification for using the crystallographic data to discuss the properties of the enzyme in solution. One observation that shows that the structures in the crystal and in solution must be reasonably similar is that certain enzymes, e.g. pancreatic ribonuclease and triosephosphate isomerase can display catalytic activity in the crystalline state. In general the rates of processes in the crystalline state are rather slower than those in solution, reflecting the slower diffusion of molecules through a crystal, whereas the values of equilibrium properties such as dissociation constants are similar in the two phases.[151]

3.5.3 The importance of knowing the three-dimensional structure of an enzyme

The advances in our detailed understanding of the properties of enzymes over the last twenty years or so can largely be ascribed to the increase in detailed structural information provided by X-ray crystallography. These advances in understanding are discussed in the next sections (3.5.3.1 to 3.5.3.3).

3.5.3.1 To test models of macromolecular structure

The three-dimensional structure provides the experimental data with which to test theoretical models of macromolecular structure. The relative contributions of the various types of forces involved in the maintenance of the overall structure (Section 3.5.5) can be evaluated by referring to the experimentally determined structure. We can also use the experimental structure to test the success of structure prediction methods (Section 3.7.3).

3.5.3.2 To propose a mechanism of catalysis

The structure of an enzyme provides the framework within which the catalytic power of that enzyme can be understood. It is possible (see Chapter 5) to propose chemically reasonable mechanisms for a number of enzymes by combining these structural data with results from other studies such as enzyme kinetics, which tell us about the dynamic aspects of catalysis.

3.5.3.3 To explore similarities between enzymes

In Section 3.4.9.5 we described how the comparisons of amino-acid sequences have helped to explore evolutionary relationships between enzymes. Useful comparisons have also been made between the three-dimensional structures of enzymes that catalyse related reactions. Thus, Rossmann and his coworkers have demonstrated that a common nucleotide-binding structure consisting of helix and sheet elements (Section 3.5.6.4) is present in NAD^+-requiring dehydrogenases and in kinases that bind ADP (which forms part of the NAD^+ molecule).[152, 153] This regularity in the three-dimensional structures would not have been deduced from comparisons of the amino-acid sequences. A more general method for making structural comparisons between enzymes has been described.[128, 154, 155]

3.5.4 Features of structures adopted by enzymes

In this section we shall describe the principal structural elements that have been found in proteins and then proceed to outline the principles that govern the three-dimensional structures adopted by enzymes. A later section (3.5.6) will look at the detailed structures of three particular enzymes: α-chymotrypsin, adenylate kinase, and glyceraldehyde-3-phosphate dehydrogenase.

3.5.4.1 The amide bond

The amide bond which provides the means of linking amino acids in the polypeptide chain (Section 3.3.2) is not adequately represented by the formula $-\overset{\overset{\displaystyle O}{\|}}{C}-NH-$, since this fails to show that there is in fact considerable resistance to rotation about the C—N bond. Estimates of the activation energy for this rotation are of the order of 80 kJ mol^{-1}.[156] It is therefore more appropriate to depict the amide bond as a planar unit with extensive delocalization of the lone pair of electrons on the nitrogen atom imparting a partial double-bond character to the C—N bond (Fig. 3.22).

The form of the amide bond unit shown in Fig. 3.22 in which the two α-carbon atoms are *trans* to one another, is generally more stable than the *cis* form by some 12 kJ mol^{-1} because of the steric crowding in the latter.[156] However, when the amide bond involves the amino acid proline, the difference in energy between *cis* and *trans* forms is much less pronounced and the *cis* form is found occasionally in proteins.[156] Rotation is allowed about the single bonds linking the α-carbon atoms to the carbonyl carbon atom and to the nitrogen atom of the amide bond. We can thus describe the structure of a dipeptide unit in terms of the dihedral angles (ϕ, ψ), i.e. the angles of rotation about these bonds (Fig. 3.23)

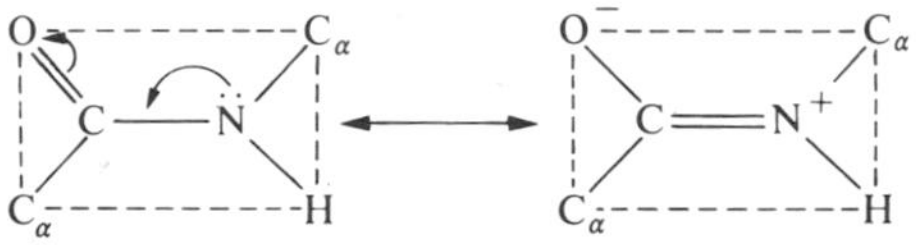

FIG. 3.22. The amide bond showing delocalization of electrons. The dotted lines represent the planar unit of two α-carbon atoms and the carbon, nitrogen, oxygen, and hydrogen atoms. The length of the amide C—N bond is 0.132 nm and the distance between the two α-carbon atoms in this trans configuration is approximately 0.38 nm.

Rotations about these bonds, i.e. changes in the angles ϕ and ψ, will change the distances between non-bonded atoms and thus the energy of the dipeptide unit will vary. (This is because when non-bonded atoms are brought into close proximity with each other, overlap of the electron clouds leads to repulsion and hence destabilization.) Calculations of the energy of the dipeptide unit as a function of the angles ϕ and ψ have been made,[157] and these show that there are certain well-defined conformations that are of lowest energy, i.e. of greatest stability. The most important of these regular secondary structures are the α-helix and β-sheet structures, which had already been proposed as the basic structural units of fibrous proteins;[158] thus, wool consists largely of α-helical structure and silk largely of β-sheet structure.

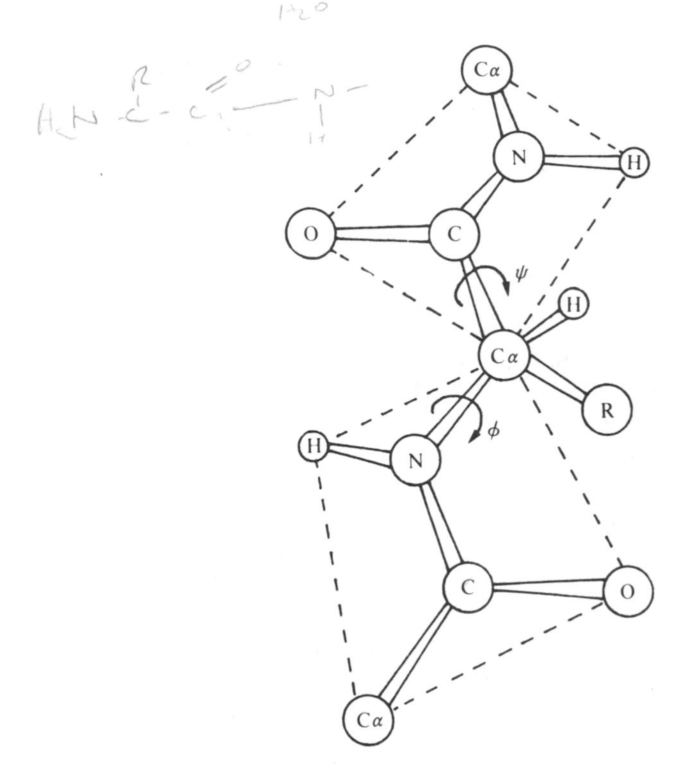

FIG. 3.23. A dipeptide unit of a polypeptide chain showing the planar amide units and the relevant angles of rotation about the bonds to the central α-carbon atom. $\phi = \psi = 180°$ corresponds to fully extended polypeptide chain.

3.5.4.2 The α-helix

The α-helix is depicted in Fig. 3.24(a). It consists of a right-handed helix* (i.e. the direction of twist from the N-terminus to the C-terminus is that of a right-handed corkscrew) in which hydrogen bonds (Section 3.5.5.1) are formed between the carbonyl group of the amide bond between amino acids n and $n+1$ and the amino group of the amide bond between amino acids $n+3$ and $n+4$ (Figs. 3.24(a) and 3.24(b)). Because of the restrictions on rotation around the bonds to the α-carbon atom in proline, this amino acid cannot be incorporated into an α-helix without seriously distorting it. Proline therefore acts as a helix-breaker.

Since there are approximately 3.6 amino acids per turn of the α-helix and a total of 13 atoms in the ring closed by formation of the hydrogen bond, the α-helix is sometimes referred to as the 3.6_{13} helix. The pitch of the α-helix is 0.54 nm

* For L-amino-acids (Section 3.3.1) the right-handed helix is far more stable than the left-handed helix because of the crowding of the side-chains in the latter. Only the right-handed helix is found in proteins.

(Fig. 3.24(a)) which corresponds to a translation of 0.15 nm per amino acid. Other types of helical structure are possible, such as the 3_{10} helix, but these are less stable than the α-helix because the hydrogen bonds are somewhat distorted. The 3_{10} helix occurs only rarely in enzymes;[159] short lengths are found in lysozyme and in carbonate dehydratase. The ways in which α-helices can pack together in protein structures have been discussed by Richardson.[160]

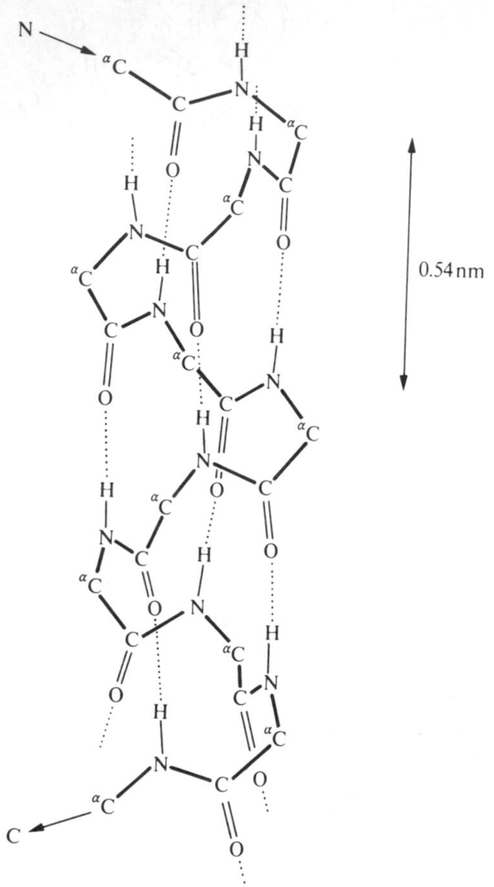

0.54 nm

(b)

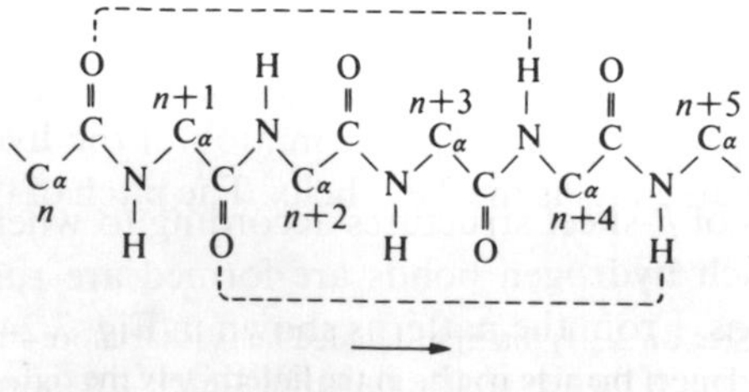

FIG. 3.24. (a, b)

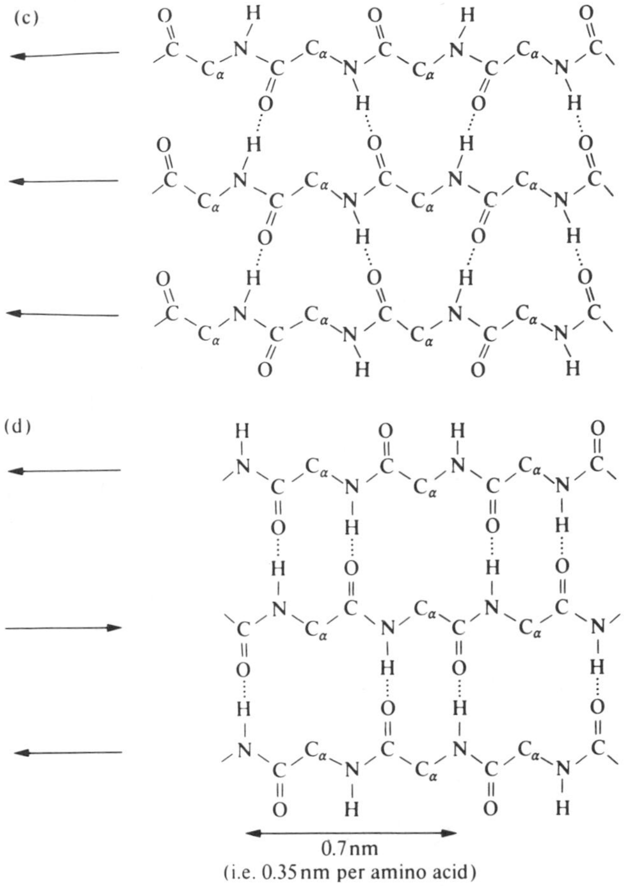

(c)

(d)

0.7 nm
(i.e. 0.35 nm per amino acid)

FIG. 3.24. Elements of regular structure found in proteins and enzymes, showing the arrangement of hydrogen bonds, which are depicted by dotted lines. The arrows show the direction of a polypeptide chain (N→C). (a) α-helix; the heavy lines trace the path of the polypeptide chain. (b) The hydrogen bonding arrangement in the α-helix more clearly. (c) and (d) parallel and antiparallel β-sheet structures, respectively. In three dimensions the sheets represented in (c) and (d) are in fact corrugated, with the side-chains of the amino acids projecting perpendicularly to the plane of the sheet. For the α-helix, parallel sheet and antiparallel sheet structures the values of (ϕ, ψ) (see Fig. 3.23) are approximately $(-57°, -47°)$, $(-119°, +113°)$ and $(-139°, +135°)$ respectively.

3.5.4.3 *The β-sheet*

There are two types of β-sheet structures according to whether the polypeptide chains between which hydrogen bonds are formed are running in parallel or antiparallel directions. From the patterns shown in Fig. 3.24(c) and 3.24(d) it can be seen that the hydrogen bonds in the parallel sheets are somewhat distorted: no such distortion occurs in the antiparallel sheet. Several enzymes contain extensive networks of sheet structure: parl of thc arrangement found in carboxypeptidase A[161] is illustrated in Fig. 3.25. As noted by Chothia,[162] the β-sheets in globular proteins show a right-handed twist (as viewed along the polypeptide chain) rather than being flat as in silk. This twist is seen clearly in Figs. 3.31(a) and 3.32(b).

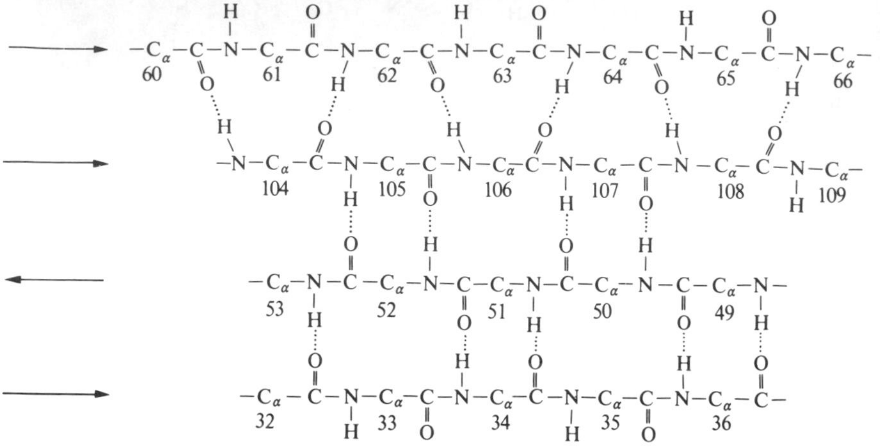

```
              H     O           H     O           H     O
              |     ‖           |     ‖           |     ‖
———→   -Cα-C-N-Cα-C-N-Cα-C-N-Cα-C-N-Cα-C-N-Cα-C-N-Cα-
       60  ‖  61   /  62 ‖   63   /  64  ‖  65   /  66
           O      H      O       H       O      H
           :      :      :       :       :      :
           H      O      H       O       H      O
           \      //     \       //      \      //
———→     -N-Cα-C-N-Cα-C-N-Cα-C-N-Cα-C-N -Cα- C-N-Cα-
          104    |  105 ‖  106   |  107 ‖  108   |  109
                 H      O        H      O        H
                 :      :        :      :
                 O      H        O      H
                 ‖      |        ‖      |
←———     -Cα-N-C -Cα-N-C-Cα-N-C-Cα- N-C-Cα-N-
          53 |    52   ‖ 51 |   50     ‖ 49 |
             H         O    H          O    H
             :         :    :          :    :
             O         H    O          H    O
             ‖         |    ‖          |    ‖
———→     -Cα - C-N-Cα-C-N-Cα-C-N-Cα-C-N-Cα-C-
          32     |  33 ‖  34   |  35 ‖  36
                 H     O       H     O
```

FIG. 3.25. Part of the β-sheet structure of carboxypeptidase A.[161] The directions of the strands are shown by arrows. Hydrogen bonds are represented by dotted lines.

3.5.4.4 *Other structural features*

Of the various other elements of regular structure in proteins, the most abundant is the β-turn. The β-turn is a sequence of four amino acids in a protein, in which the polypeptide chain folds back on itself by nearly 180° (Fig. 3.26(a)), thus giving the protein a globular rather than a linear shape.[163] A total of 11 types of β-turn have been identified which differ in terms of the angles of rotation (ϕ, ψ in Fig. 3.23) about the various single bonds involved.[163] A β-bulge occurs where an extra amino acid is inserted in one strand of a β-sheet, disrupting the hydrogen bonding[164] (see Fig. 3.26(b)). A β-hairpin is a portion of polypeptide chain that folds back on itself so that the two halves constitute two adjacent strands of an antiparallel sheet[165] (see Fig. 3.26(c)).

The importance of these structural features in proteins has been reviewed.[160, 166]

3.5.4.5 *Principles governing the structures adopted by enzymes*

Enzymes, and globular proteins in general, do not possess the simple regular structures that characterize fibrous proteins such as wool or silk. However, by examining the known three-dimensional structures, it is possible to formulate some general principles that appear to govern the structures adopted by enzymes;[156] these principles are summarized below.

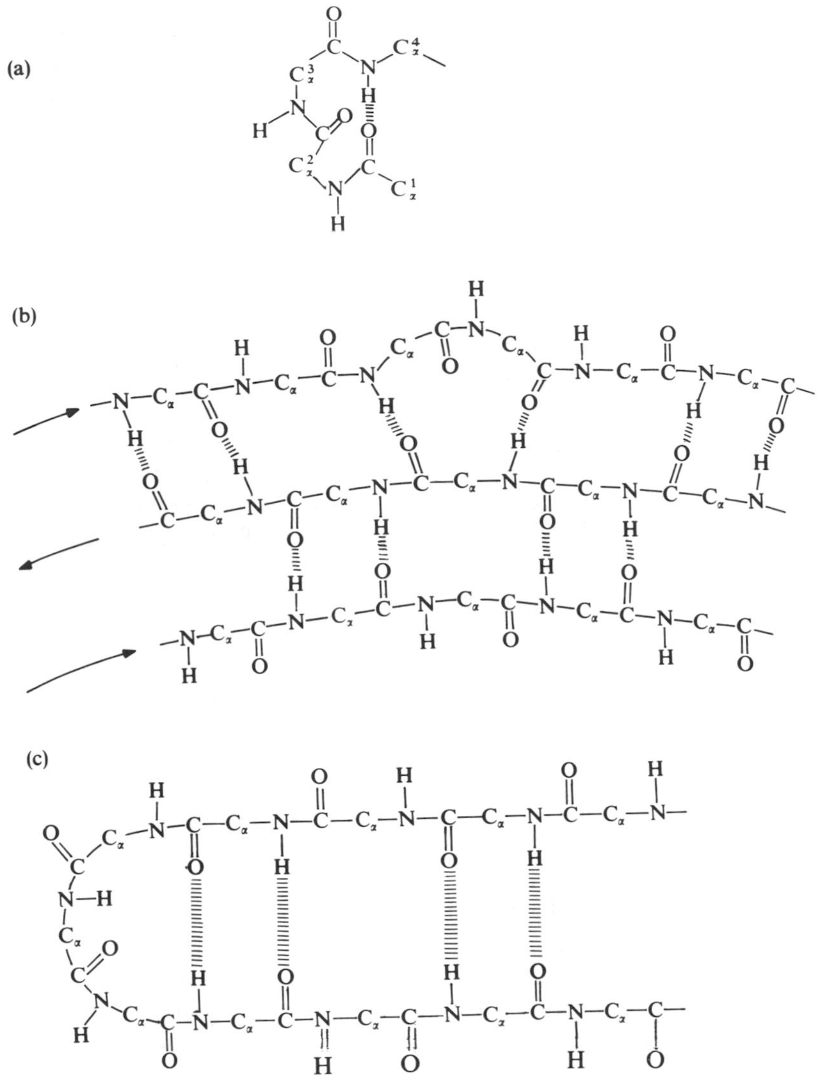

FIG. 3.26. Diagrammatic illustration of some structural features found in proteins. Dotted lines indicate hydrogen bonds. (a) A Type I β-turn;[163] (b) β-bulge;[164, 166] (c) β-hairpin (of the Class 3 three residue type[166]).

1. Enzymes are generally very closely packed globular structures with only a small number of internal cavities which are normally filled by water molecules.

2. The structural elements (amide bonds, helices, sheets, etc.) generally possess similar geometries to those observed in model compounds. However, some small deviations from the planarity of amide bonds have been noted in lysozyme, and α-helices can be distorted towards 3_{10} helices.

3. Non-polar amino-acid side-chains, e.g. those of Leu, Val, Phe, etc., are generally buried in the interior of the enzymes away from the solvent water. Ionized side-chains, e.g. those of Lys, Asp, etc. tend to be on the exterior of enzymes and able to interact with the solvent. Exceptions to these generalizations usually point to side-chains that have particular functions, e.g. the buried side-chain of aspartic acid 194 in chymotrypsin that is involved in a strong electrostatic interaction with the protonated α-amino group of isoleucine 16 (see Chapter 5, Section 5.5.1.2). Non-polar side-chains on the exterior of enzymes may well be involved in binding to a non-polar substrate or to a membrane.

4. Polar groups in the interior of enzymes are normally paired in hydrogen bonds. A very large percentage of the carbonyl and amino groups of amide bonds that are placed in the interior by the folding of the polypeptide chain are paired in hydrogen bond formation or form hydrogen bonds to internal water molecules and are thus rendered essentially non-polar.[167] While these internal hydrogen bonds may not be of great significance in terms of the energy of the folded structure, they are undoubtedly important in specifying the correct folded structure (Section 3.5.5.5).

5. Most larger enzyme molecules tend to exist in structural domains, i.e. globular units, connected by segments of polypeptide chain. Some good examples of this type of structure are found amongst the kinases and dehydrogenases,[135] which consist of two domains or 'lobes'. One of the domains is involved in binding the common substrate, i.e. NAD^+ for dehydrogenases; the other is termed the catalytic domain and will be different in the different enzymes. It has been suggested that movements of the domains with respect to each other may be an important part of the mechanism of catalysis in these enzymes.

3.5.4.6 Classification of enzyme structures

In the previous section (3.5.4.5) we described some of the basic principles that seem to govern the types of structures adopted by enzymes. A given enzyme will possess a certain amount of helix, sheet, and other elements of regular secondary structure but this does not tell us very much about the overall structure of the enzyme. Several attempts have been made to look for patterns in the distribution of structural elements in proteins: one of the best known classifications is the four-class system introduced by Levitt and Chothia.[168] The four classes of protein structure are as follows.

1. All-α-proteins which have only α-helix structure. Myoglobin and citrate synthase are examples of this type of protein.

2. All-β-proteins which have mainly β-sheet structure. Chymotrypsin and the constant and variable regions of immunoglobulins belong to this class.

3. α + β-proteins, which have α-helix and β-sheet structural segments that do not mix but are separated along the polypeptide chain. Thermolysin, lysozyme, and ribonuclease belong to this class.

4. α/β-proteins, which have mixed or alternating segments of α-helix and β-sheet structure. The kinases and dehydrogenases with their nucleotide-binding domains are good examples of this class, which also includes other enzymes of the glycolytic pathway such as triosephosphate isomerase and phosphoglycerate mutase.

Richardson[160] has presented a more detailed classification of protein structures based on patterns within domains rather than of whole protein molecules.

The occurrence of ordered assemblies of structural elements found in a number of proteins suggests that there are stable types of 'supersecondary structure'. These types of structures would be of importance in directing the way in which a polypeptide chain could fold to adopt its correct three-dimensional structure (Section 3.7.2).

3.5.5 The forces involved in stabilizing the folded structures of enzymes

The forces that maintain the folded three-dimensional structure of a globular protein such as an enzyme are weak, non-covalent interactions. Experiments of the type described in Section 3.7.2 have shown that in the case of enzymes that possess disulphide bonds, these bonds are formed by oxidation of pairs of cysteine side-chains only after the three-dimensional structure of an enzyme has been acquired. It would therefore seem that disulphide bonds are important in stabilizing a folded structure that has already formed rather than directing the acquisition of this structure. The bond energy of a disulphide bond (approximately 200 kJ mol^{-1}) is much greater than that of the non-covalent interactions we shall describe in Sections 3.5.5.1 to 3.5.5.4.

The non-covalent interactions can be discussed in four categories: hydrogen bonds, electrostatic forces, Van der Waals forces, and hydrophobic forces. The division between these categories is somewhat arbitrary, since all attractive forces between atoms ultimately arise from favourable interactions between electrons and nuclei.

3.5.5.1 Hydrogen bonds

The hydrogen bond arises from the attraction between a hydrogen atom attached to an electronegative atom (O and N are the most important examples in biological systems) and another electronegative atom. Figure 3.27 shows the arrangement that is important in α-helix and β-sheet structures.

The hydrogen bond is most stable when (i) the distance between the two electronegative atoms is within closely specified limits. In the case shown in Fig. 3.27 the O$\leftrightarrow$N distance is within the range 0.28–0.30 nm; and (ii) the arrangement of atoms e.g. O . . . H—N is linear.

A typical value for the energy of a hydrogen bond *in vacuo* is approximately 20 kJ mol^{-1}, but it is unlikely that the many internal hydrogen bonds in enzymes (Section 3.5.4.5) make any substantial contribution to the stabilization of the folded structure. This is because in the unfolded state the 'internal' polar groups

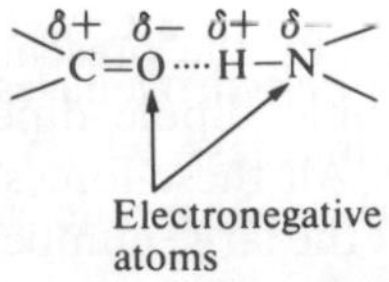

FIG. 3.27. The hydrogen bond involved in helix and sheet structures in enzymes and proteins.

could form alternative hydrogen bonds with the solvent water[169] and there is therefore little gain in stability when comparing the folded state with the unfolded. However, because of the strict geometrical requirements referred to above, hydrogen bonds are undoubtedly important in specifying the correct folded structure (Section 3.5.5.5).

3.5.5.2 *Electrostatic forces*

This category consists of the attractive forces that exist between charged groups, e.g. between $—\overset{+}{N}H_3$ (Lys) and $—COO^-$ (Asp); these interactions are also known as salt bridges.

The force between two charges is inversely proportional to the dielectric constant of the medium separating them, and the lack of detailed knowledge of the value of the dielectric constant in an enzyme makes any quantitative assessment of the contribution of electrostatic forces to the stability of the folded structure very difficult.* Although salt bridges play a crucial role in particular enzymes, e.g. in the activation of chymotrypsinogen (see Chapter 5, Section 5.5.1.2), it appears that in general the number of salt bridges is limited and hence their contribution to the stability of enzyme structures is probably rather small.

3.5.5.3 *Van der Waals forces*

Van der Waals forces are weak forces that occur when molecules or groups of atoms are in close contact with one another (less than about 0.4 nm apart in the case of atoms found in proteins†). Attractive forces result from favourable interactions between dipoles (separations of charge) that may be permanent or transient. Transient dipoles arise by the local fluctuations in electron density in

* Recently, progress has been made in determining the effective dielectric constant within proteins. The effects of changing charged amino-acid side-chains elsewhere in the protein on the pK_a of the active-site histidine side-chain (His 64) of subtilisin have been measured. Calculations suggest that the effective dielectric constant is between 30 and 60 (the values for water and benzene are 80 and 2.3, respectively). The relatively high value for the protein is thought to be due partly to the surrounding water and also to polar side-chains.[170, 171]

† When atoms are brought too close to each other, the electron clouds repel one another. The balance of attractive and repulsive forces leads to an optimum distance, the van der Waals distance, between the atoms. This is the sum of the van der Waals radii of the atoms concerned.

atoms and can also be induced by the presence of a neighbouring dipole. A variety of interactions are possible: dipole–dipole, dipole–induced dipole, and induced dipole–induced dipole. All these forces are weak, with energies of the order of 10 kJ mol^{-1} or less, but the large number of van der Waals contacts that occur in proteins may well mean that they make a significant contribution to the stability of protein structures.

3.5.5.4 *Hydrophobic forces*

It is well known that hydrocarbons, such as methane, benzene, etc., are much more soluble in liquid hydrocarbons or most other organic solvents than in water. In fact, water has an almost unique place as an extremely poor solvent for hydrocarbon solutes.[172] The probable reason for this lies in the fact that water consists of transient clusters of hydrogen bonded molecules; when a hydrocarbon is introduced into the aqueous medium there is a reorganization of the water molecules around the hydrocarbon that makes the system more ordered.[173] This increase in order, i.e. decrease in entropy, is unfavourable on thermodynamic grounds[174] and hence there is a tendency to force the hydrocarbon out of contact with the water and into the organic phase, so that this decrease in entropy does not occur.

In the case of proteins, these considerations dictate that the non-polar portions of the molecule, for example the side-chains of amino acids Val, Leu, Ile, Phe, Trp, should be buried in the interior of the molecule away from the solvent water. Calculations show that this burying of the non-polar side chains makes by far the major contribution to the stability of the folded structure compared with that of the unfolded structure.[175] For lysozyme this stabilization energy is calculated to be of the order of 1200 kJ mol^{-1}. *It should be emphasized that it is the burying of the non-polar side chains away from water that provides this stabilization, rather than interactions between the non-polar side chains themselves.*[176]

3.5.5.5 *Some conclusions about the forces involved in the maintenance of the folded structures of enzymes*

From the discussion in the previous section (3.5.5.4) it is evident that hydrophobic forces make the greatest contribution in energy terms to the stability of the folded state compared with the unfolded state. However, hydrophobic forces do not confer any geometrical specificity on interactions, since they arise essentially from exclusion from the aqueous phase,[176] and thus do not serve to specify any particular folded structure. In order to specify a structure, interactions such as hydrogen bonding, that possess geometrical requirements (Section 3.5.5.1) are of great importance. The overall stability of the folded structure compared with the unfolded structure is much smaller than the values of hydrophobic stabilization energy would suggest. This difference can largely be accounted for by the fact that it is unfavourable on entropy grounds to convert a highly flexible unfolded polypeptide chain to a more ordered folded chain.[175]

3.5.6 Examples of enzyme structures

In this section we shall describe the three-dimensional structures of the enzymes α-chymotrypsin, adenylate kinase, and glyceraldehyde-3-phosphate dehydrogenase. However, it is first necessary to mention some of the ways of depicting these structures.

3.5.6.1 Representations of enzyme structures

Considerable problems are posed by trying to depict the complex three-dimensional structures adopted by enzymes, especially when using two-dimensional representations. Undoubtedly the most informative representations are three-dimensional scale models built directly from the coordinates of the electron-density map. An example of such a model is shown in Fig. 3.28 which depicts the structure of adenylate kinase from pig muscle at a scale of 20 cm = 1 nm. This model uses wire parts that keep the bond angles and bond lengths of the planar amide bond units constant but allow rotation about the single bonds to the α-carbon atoms. Although features such as the deep cleft at the bottom right-hand side of the model are clearly visible, it is difficult to appreciate the details of the structure without having the model on hand. A similar limitation arises with 'space-filling' models in which each atom is represented by a sphere, sometimes flattened, of radius equal to its van der Waals radius. 'Space filling' models are very suitable for highlighting features of the surface of an enzyme, such as a cleft or depression, but are of little use in illustrating internal features such as the folding of the polypeptide chain.

One way of providing a more simplified representation of the structure would be to strip off the side-chains of the amino acids and merely show the folding of the polypeptide chain backbone; this approach is illustrated in Figs. 3.29(a) and 3.30 for the cases of α-chymotrypsin and adenylate kinase, respectively. Two limitations of this approach are: (i) the model does not show the nature of the side-chains in any particular region of the enzyme such as the active (catalytic) site, and (ii) it is not always easy to pick out elements of regular secondary structure. In order to meet the second limitation, it has become popular to show the folding of the polypeptide chain with particular emphasis on these elements of regular structure, helices being shown as cylinders and strands of sheet as arrows. Examples of this type of representation are shown in Figs. 3.31(a) and 3.32 for adenylate kinase and glyceraldehyde-3-phosphate dehydrogenase, respectively. Richardson has developed another approach in which α-helices are shown as spiral ribbons, the β-strands as thick arrows and the irregular structures as ropes.[160] Examples of this type of representation are shown in Fig. 3.31(b) and Fig. 5.23(b).These approaches have helped to focus attention on the 'super-secondary structures' (Section 3.5.4.6) found in a number of enzymes.

3.5.6.2 The structure of α-chymotrypsin from beef pancreas

α-Chymotrypsin consists of three polypeptide chains held together by five disulphide bonds: the origin of this three-chain structure is discussed in Chapter

FIG. 3.28. The structure of adenylate kinase shown as a three-dimensional scale model built with wire parts (scale 20 cm = 1 nm). (Reproduced with permission from *Principles of protein structure* by G. E. Schulz and R. H. Schirmer, p. 136. Copyright Springer, New York (1979).

5, Section 5.5.1.1. The chains run between amino acids 1–13 (chain A), 16–146 (chain B), and 149–245 (chain C). The five disulphide bonds are between cysteine side-chains at positions 1 and 122, 42 and 58, 136 and 201, 168 and 182, and 191 and 220.

The structure of the enzyme is represented in Fig. 3.29(a), which shows the positions of the α-carbon atoms.[177] The overall shape of the molecule is that of an ellipsoid with a maximum dimension of 5.1 nm. There is a shallow depression at the active site in which there are three side-chains, Ser 195, His 57, and Asp 102 of crucial importance in the catalytic mechanism (see Chapter 5, Section 5.5.1.2).

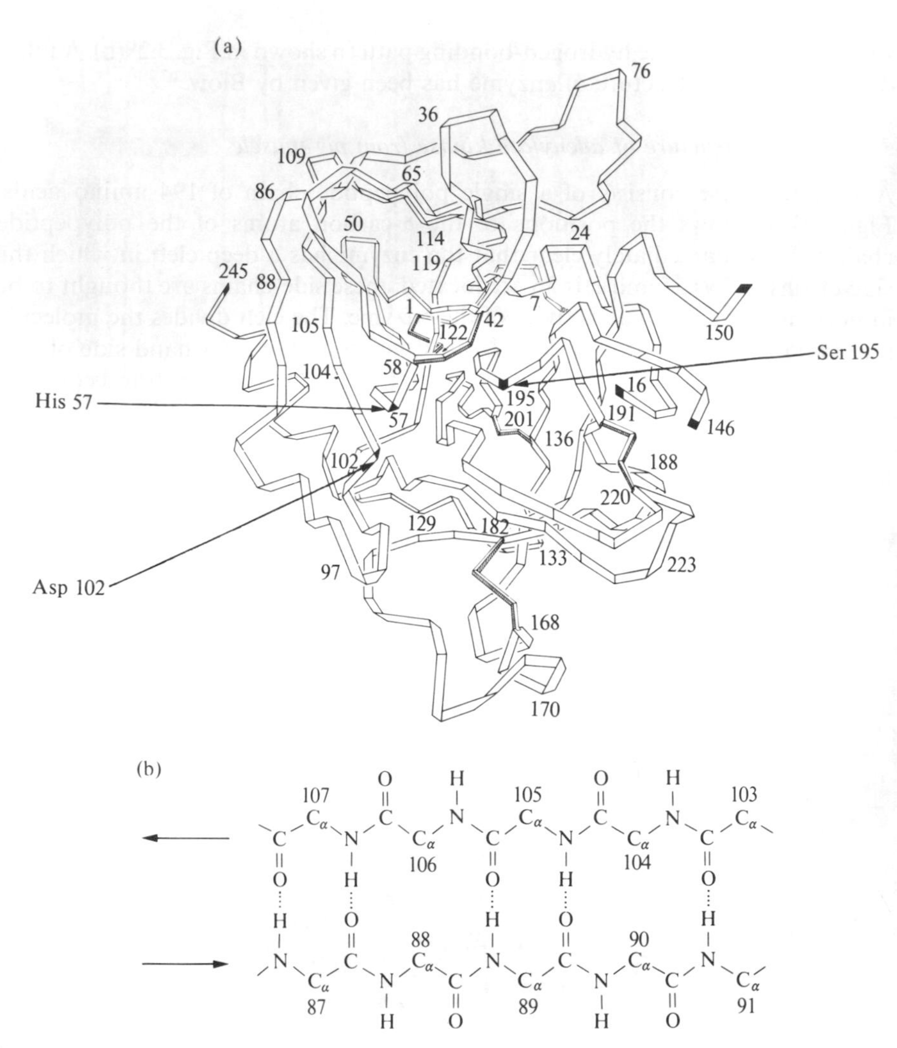

FIG. 3.29. The structure of α-chymotrypsin. (a) The positions of α-carbon atoms. (Reproduced with permission from *Enzymes*, 3rd edn (P. D. Boyer ed.) Vol. 3, p. 194. Copyright Academic Press, New York (1971).) (b) The arrangement of hydrogen bonds in part of the antiparallel sheet structure of the enzyme.

There are only two short stretches of helix between amino acids 164 and 173 and between amino acids 235 and 245. Both helices are somewhat distorted from an α-helix towards a 3_{10} helix. There is a considerable amount of antiparallel β-sheet structure in the enzyme and in fact the structure can be described as consisting of two folded units (amino acids 27–112 and 133–230) in each of which there are six strands of antiparallel sheet forming a highly distorted hydrogen-bonded 'cylinder'. While this point is difficult to see in Fig. 3.29(a), it should be relatively easy to appreciate that amino acids 87–91 and 103–107 form part of a sheet structure with the hydrogen-bonding pattern shown in Fig. 3.29(b). A fuller discussion of the structure of enzyme has been given by Blow.[177]

3.5.6.3 *The structure of adenylate kinase from pig muscle*

Adenylate kinase consists of a single polypeptide chain of 194 amino acids. Figure 3.30 shows the positions of the α-carbon atoms of the polypeptide chain.[178] It is immediately clear that the enzyme has a deep cleft in which the side-chains of Cys 25 and His 36 are located; these side-chains are thought to be important in the catalytic activity of the enzyme. The cleft divides the molecule into two distinct domains or 'lobes'; the domain on the right-hand side of Fig. 3.30 has three stretches of helix, whereas that on the left has five strands of sheet with interconnecting stretches of helix. The elements of regular secondary structure are more clearly represented in Fig. 3.31(a) and (b). Altogether a very high proportion (over two-thirds) of the amino acids in the molecule are in some type of regular secondary structure; this may account for the great stability of the enzyme at low pH (see Chapter 2, Section 2.8.1).

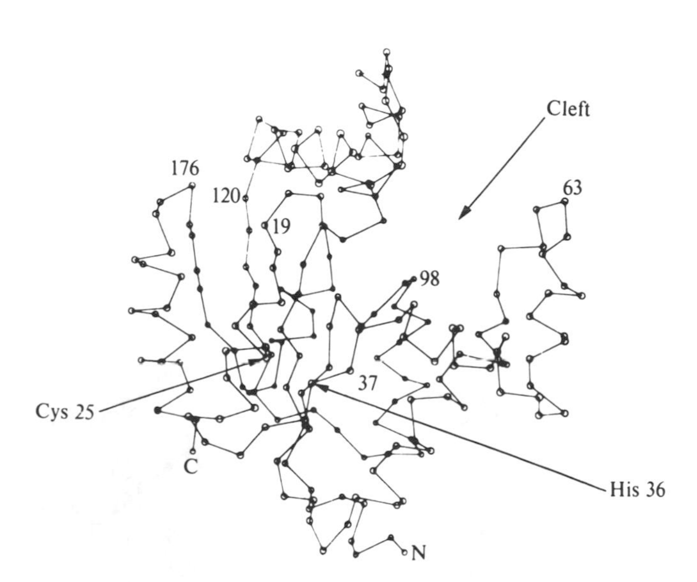

FIG. 3.30. Positions of the α-carbon atoms in adenylate kinase. The locations of Cys 25 and His 36, which are thought to be involved in the catalytic mechanism of the enzyme, are shown. (Reprinted by permission from reference 178, *Nature*, *Lond*. 250, 120–3. Copyright 1974, Macmillan Journals Limited.)

3.5.6.4 *The structure of glyceraldehyde-3-phosphate dehydrogenase from lobster muscle*

Glyceraldehyde-3-phosphate dehydrogenase is a tetramer, consisting of four subunits each of M_r 35 000. The dimensions of the tetrtramer are approximately

15 nm × 14 nm × 8 nm. Figure 3.32 shows the structure of a subunit in two distinct halves representing the two domains.[179] It is interesting that the dividing point between the domains is close to Cys 149, the side-chain of which is involved in the catalytic activity of the enzyme.

The NAD^+-binding domain which occupies the N-terminal part of the polypeptide chain (Fig. 3.32 (a)) is dominated by a structure consisting of strands of a twisted parallel β-sheet; the strands are labelled A–F from the N-terminus. Between these strands are stretches of helix, such as helix α_E between strands D

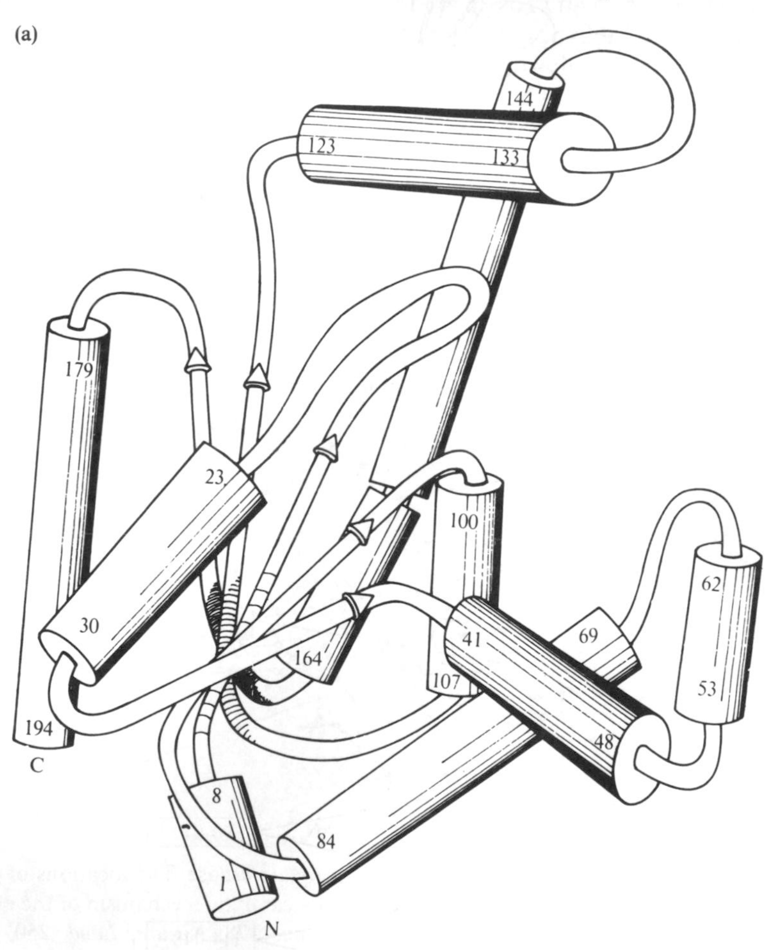

FIG. 3.31. Elements of regular structure in adenylate kinase. (a) The path of the polypeptide chain; helices are depicted as cylinders and strands of sheet as arrows. (b) The path of the polypeptide chain; helices are depicted as ribbons and strands of sheet as arrows. (c) The amino acids involved in these elements of structure. (Part (a) reproduced with permission from *Principles of protein structure* by G. E. Schulz and R. H. Schirmer, p. 136. Copyright Springer, New York (1979). Part (b) reproduced with permission from *Advances in Protein Chemistry* **34**, 167–339. Copyright, J. S. Richardson (1981).)

(b)

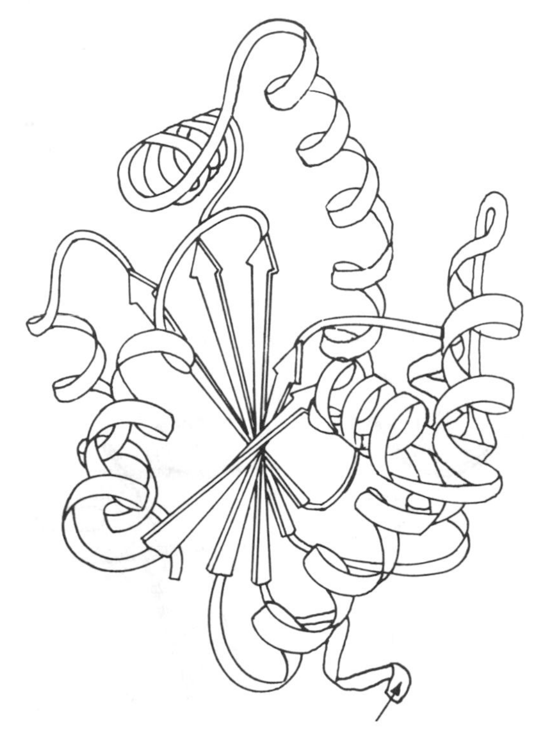

Adenylate kinase

(c)

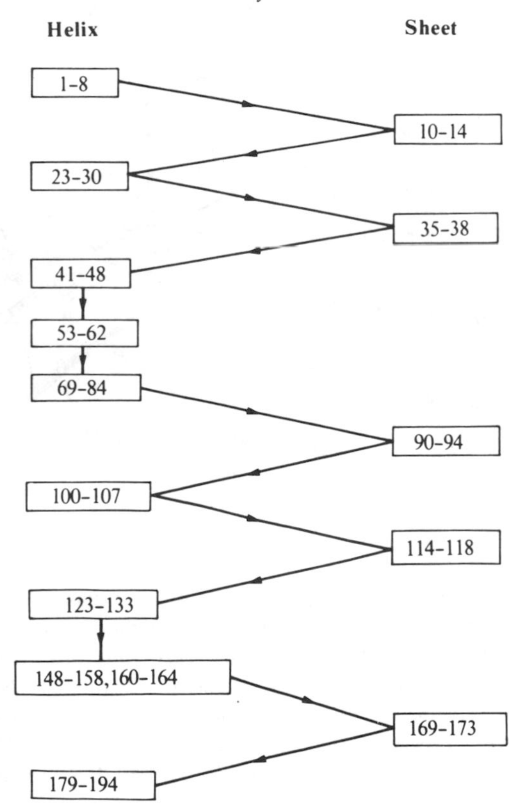

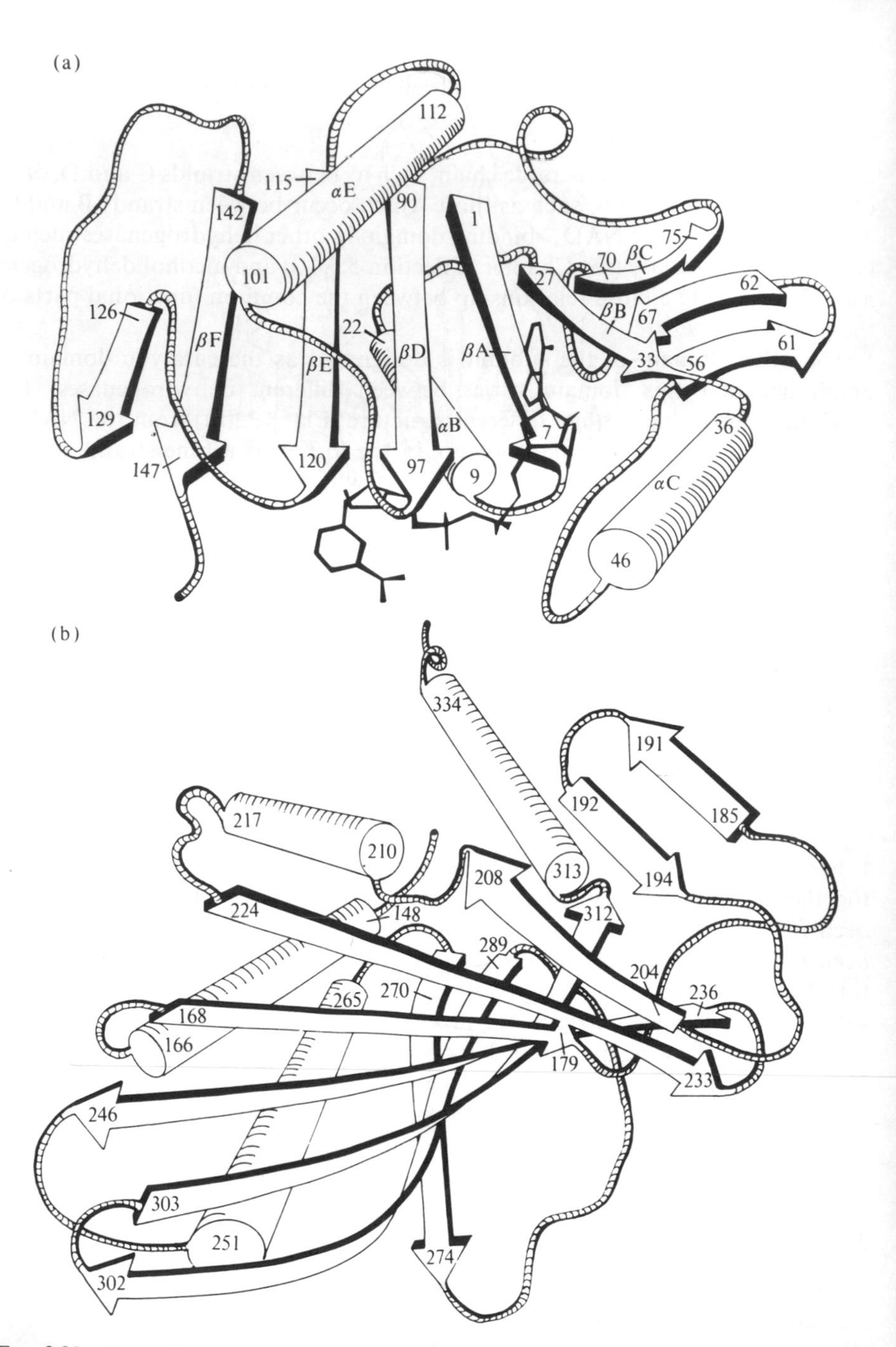

FIG. 3.32. The structure of a subunit of glyceraldehyde-3-phosphate dehydrogenase.[179] Helices are represented by cylinders and strands of sheet by arrows. (a) The NAD^+-binding domain; the adenine ring binds near β_A and the nicotinamide ring binds between β_D and β_E. (b) The catalytic domain. (Part (a) is reprinted with permission from *Journal of Molecular Biology* **90**, 25–49. Copyright Academic Press Inc., (London) (1974). Part (b) reprinted with permission from *Journal of Biological Chemistry* **250**, 9137–62. Copyright American Society of Biological Chemists Inc. (1975).)

and E, or of extended polypeptide chain, such as between strands C and D, or of other antiparallel strands, such as those which occur between strands B and C. There is a very similar NAD^+-binding domain in other dehydrogenases such as lactate dehydrogenase (see Chapter 5, Section 5.5.4.3) and alcohol dehydrogenase,[153] suggesting a close relationship between the common functional parts of these enzymes.

The other domain of the subunit is designated as the catalytic domain to emphasize that this domain varies between different dehydrogenases. The catalytic domain has a quite different structure (Fig. 3.32(b)) from the NAD^+-binding domain. The dominant feature in the former is a nine-strand, mainly antiparallel, sheet structure on one side of which are three associated helices(148–166, 210–217, and 251–265) with axes roughly parallel with the axes of the sheet strands. At the C-terminus of the polypeptide chain there is a stretch of helix that packs against the central β-sheet of the NAD^+-binding domain.

The binding site for NAD^+ is located near the interface between subunits and it may well be that some of the side-chains involved in forming the binding site of one subunit are supplied by a neighbouring subunit. This situation has also been found in other cases, e.g. triosephosphate isomerase (see Chapter 5, Section 5.5.2.2), citrate synthase[180] and glutamine synthetase.[181]

3.6 The determination of quaternary structure

It has been found that most large enzymes consist of a number of subunits* held together by non-covalent forces, i.e. they are *oligomeric*. In some ways the occurrence of multiple subunits in an enzyme can be considered analogous to the occurrence of folded globular domains (Section 3.5.4.5) along a single polypeptide chain. The term quaternary structure refers to the arrangement of the subunits in oligomeric enzymes (or proteins in general). In this section we shall deal with the following questions that can be asked about oligomeric enzymes. (i) How many subunits are there and of what type? (Section 3.6.1). (ii) How are the subunits arranged? (Section 3.6.2). (iii) What forces are involved in holding subunits together? (Section 3.6.3). (iv) What is the significance of multiple subunits in an enzyme? (Section 3.6.4).

3.6.1 Number and type of subunits

In this section we shall discuss the various experimental approaches to determining the number and type of subunits in an oligomeric enzyme.

3.6.1.1 Studies of M_r

An indication that an enzyme consists of multiple subunits is provided by results of studies of M_r (Section 3.2) performed in the absence and presence of denaturing agents such as guanidine hydrochloride (Section 3.7.1). In favourable

* The definition of the term subunit has been given in Section 3.1.

cases the number of subunits can be deduced directly from the M_r data. For instance, alcohol dehydrogenase from yeast has an M_r of about 145 000 as determined by ultracentrifugation. However, under denaturing conditions a value of 36 000 is obtained, suggesting that the enzyme consists of four subunits that are probably identical. In this type of work it is particularly important to make sure that traces of proteases are absent from the enzyme preparations, since such proteases may well be active under the denaturing conditions and cause extensive fragmentation of the unfolded polypeptide chains of the enzyme of interest. For instance, from the results of sodium dodecylsulphate polyacrylamide gel electrophoresis (Section 3.2.3) it was deduced that pyruvate kinase from yeast consisted of eight subunits each of M_r about 20 000.[182] However, when isolated under conditions chosen to inactivate traces of proteases, the enzyme was found to consist of four subunits, each of M_r about 55 000.[183] (The unit of M_r 20 000 presumably represented the predominant fragment produced by protease action.) In Chapter 7 (Sections 7.9 and 7.11.2) we shall refer to similar difficulties in the determination of the subunit structures of tryptophan synthase and of the *arom* multienzyme polypeptide protein.

3.6.1.2 *Cross-linking studies*

A second method for determining the number of subunits in an oligomeric enzyme relies on the use of cross-linking agents such as dimethylsuberimidate.[184] This compound reacts with pairs of lysine side-chains, which almost invariably occur on the surfaces of enzymes (Section 3.5.4.5) to form cross-linkages that are stable in the presence of denaturing agents (Fig. 3.33).

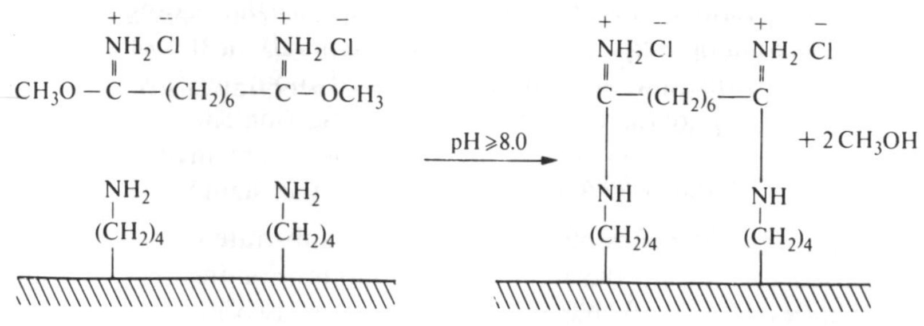

$$CH_3O-\overset{\overset{+\ \ -}{NH_2Cl}}{\overset{\|}{C}}-(CH_2)_6-\overset{\overset{+\ \ -}{NH_2Cl}}{\overset{\|}{C}}-OCH_3 \quad + \quad 2\ \text{Lys}-(CH_2)_4-NH_2 \xrightarrow{pH \geq 8.0} \text{Lys}-(CH_2)_4-NH-\overset{\overset{+\ \ -}{NH_2Cl}}{\overset{\|}{C}}-(CH_2)_6-\overset{\overset{+\ \ -}{NH_2Cl}}{\overset{\|}{C}}-NH-(CH_2)_4-\text{Lys} + 2CH_3OH$$

FIG. 3.33. Cross-linking of lysine side-chains by reaction with dimethylsuberimidate.

If we react an enzyme that consists of, say, four subunits with this cross-linking agent, a mixture of species will be formed in which different numbers of subunits are cross-linked (Fig. 3.34). The various species can be separated by sodium dodecylsulphate polyacrylamide gel electrophoresis (Section 3.2.3) giving a total of four bands.

The number of bands will give the number of subunits directly. It is clearly

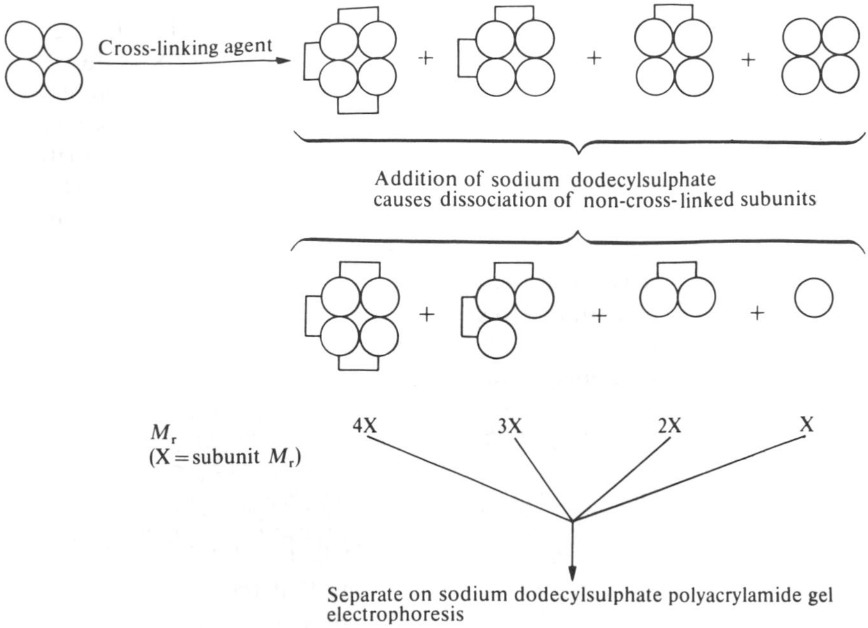

FIG. 3.34. The use of a cross-linking agent to determine the number of subunits in an oligomeric enzyme. For convenience the intrasubunit cross-linkages are not shown.

important in this work (i) to avoid complete intersubunit cross-linking, otherwise the mixture of species shown in Fig. 3.34 will not result and only one band corresponding to an M_r of 4X will be seen on electrophoresis; (ii) to avoid intermolecular cross-linking, otherwise large aggregates will be formed. These complications can be minimized by using low concentrations of cross-linking agent and of protein respectively.[185]

3.6.1.3 Studies of ligand binding

The number of binding sites on an enzyme for a substrate or other ligand can often be used to indicate the number of subunits. For instance, the alcohol dehydrogenases from yeast and liver bind four and two moles of NADH per mole of enzyme, respectively, in agreement with the tetrameric (four subunit) and dimeric (two subunit) structures of the respective enzymes. Problems can arise, however, when multiple binding sites in an oligomeric enzyme do not behave independently. Thus in the case of aspartate carbamoyltransferase from *E. coli*, studies of the binding of the inhibitor CTP suggested that the enzyme contained four regulatory subunits. It was subsequently established that there are in fact six such subunits[186, 187] and that the earlier results can be accounted for by the negative cooperativity (see Chapter 6, Section 6.2.2.2) that occurs between these binding sites.

3.6.1.4 *Studies of symmetry*

The type of symmetry possessed by an enzyme, deduced from X-ray crystallographic studies (Section 3.5.1), can sometimes be used to indicate the number of subunits in an enzyme or at least to exclude various proposed subunit structures. The observation of both a three-fold and a two-fold axis of symmetry in aspartate carbamoyltransferase from *E. coli*[187] excluded the previously proposed subunit structure of the enzyme (four catalytic plus four regulatory subunits). In conjunction with other results from studies of M_r and of the C-terminal amino acids of the subunits,[186] these data indicated that the enzyme consisted of six catalytic and six regulatory subunits.

3.6.1.5 *Identity of subunits*

If, under denaturing conditions, an enzyme gives rise to a number of species that differ in M_r, it can be concluded that the enzyme contains non-identical subunits.* In order to establish the relative numbers of each type of subunit, we can perform quantitative determinations of the N- and C-terminal amino acids of the various subunits (Sections 3.4.4.1 and 3.4.4.2). The subunit composition can then be confirmed by checking that the observed M_r of the intact enzyme equals the sum of the M_r values of the component parts, e.g. the M_r of aspartate carbamoyltransferase from *E. coli* (300 000) corresponds to the sum of the M_r of six catalytic (6 × 34 000) and six regulatory (6 × 17 000) subunits. In certain cases, however, it is not difficult to 'miss' a subunit of very different size from the other subunits: thus, in the case of phosphorylase kinase (see Chapter 6, Section 6.4.2.1) a small calcium-binding subunit was only detected a number of years after the subunit structure of the enzyme had been thought to be firmly established.

When subunits appear to be identical in terms of M_r (and/or terminal amino acids), it is desirable to check on their identity in terms of amino-acid sequence. A reasonably sound check can be made by determining the number of distinct fragments produced by the action of a specific protease such as trypsin. In each subunit the number of tryptic peptide fragments should equal the number of lysines plus the number of arginines plus one. If an enzyme consists of identical subunits, the number of tryptic peptides in the peptide map (Section 3.4.3) will equal this number (Lys + Arg + 1). On the other hand, if the subunits are not of identical sequence, the number of tryptic peptides will be greater than this. For instance lactate dehydrogenase from dogfish muscle had been shown to consist of four subunits each of M_r 36 000. Amino-acid composition data showed that there were 36 arginines per molecule (144 000) of enzyme, i.e. nine per subunit if the subunits were identical. Only nine distinct arginine-containing peptides could be isolated after digestion of lactate dehydrogenase by trypsin, indicating that the four subunits were identical in terms of sequence.[188] A similar approach

* This assumes that we are dealing with a pure preparation of enzyme and have taken steps to avoid problems arising from the action of traces of proteases.

was used to show the identity of the six subunits of glutamate dehydrogenase from beef liver.[189] It is sometimes possible to underestimate the number of distinct fragments produced by the action of a protease, since there may be peptides that differ only slightly in sequence from one another and are not sufficiently resolved in the standard separation procedures (Section 3.4.3). A further complication might arise if the enzyme possesses an insoluble 'core' that is resistant to attack by the protease. The ultimate proof of subunit identity can therefore only be obtained when the complete amino-acid sequences of the subunits have been determined.

An extensive compilation has been made of the subunit stoichiometries of oligomeric enzymes.[190] In the majority of such enzymes, subunits are identical with one another; some of the more notable examples of enzymes that contain non-identical subunits are listed in Table 3.5. It should also be noted that almost all oligomeric enzymes that consist of identical subunits contain even numbers of such subunits. There are, however, some well-documented examples of enzymes containing odd numbers of identical subunits, such as phospho-2-keto-3-deoxygluconate aldolase from *Pseudomonas putida*[191] and ornithine carbamoyltransferase from *Bacillus subtilis*,[192] each of which are trimeric enzymes with subunit M_r values of 24 000 and 47 000, respectively.

Finally, it should be remembered that some enzymes are capable of indefinite association. The examples of glutamate dehydrogenase[21] and acetyl-CoA carboxylase[23] have already been mentioned: a more comprehensive list of such systems has been given by Frieden.[193] In such systems it is inappropriate to speak of a definite quaternary structure except perhaps under very carefully defined conditions.

3.6.2 Arrangement of subunits

The arrangement of the subunits in an oligomeric enzyme can usually be deduced from the type of symmetry possessed by the molecules (this is found by X-ray crystallography; Section 3.5.1). In general, it is found that the arrangement of the subunits is such as to maximize the number of intersubunit contacts.[194] For a tetrameric enzyme such as lactate dehydrogenase the preferred geometrical arrangement will be tetrahedral, rather than square or linear (Fig. 3.35). In the case of a hexameric enzyme, the preferred arrangement will be octahedral. (For a more detailed discussion of the possible geometrical arrangements of oligomeric enzymes, references 194 and 195 should be consulted.)

The detailed arrangement of the subunits and the nature of the intersubunit contacts can only be decided when the complete three-dimensional structure of the enzyme has been obtained. However, in a system that has not been studied by X-ray crystallography it is possible to learn something about the arrangement of the subunits by the use of cross-linking agents. This type of approach showed that the subunits of the Ca^{2+}- and Mg^{2+}-activated adenosinetriphosphatase of the inner membrane of *E. coli* were probably arranged as shown in Fig. 3.36. In this work, a reversible cross-linking agent, dithiobis(succinimidyl propionate)

TABLE 3.5

Some enzymes composed of non-identical subunits

Enzyme	Source	Subunit composition	Comments
Lactose synthase	Bovine mammary tissue	$\alpha\beta$	
Haemoglobin[a]	Human red blood cells	$\alpha_2\beta_2$	
Tryptophan synthase (see Chapter 7, Section 7.9.1)	*E. coli*	$\alpha_2\beta_2$	
cAMP-dependent protein kinase (see Chapter 6, Section 6.4.2.1)	Rabbit skeletal muscle	$\alpha_2\beta_2$	α and β represent catalytic and regulatory subunits, respectively
Aspartate carbamoyltransferase	*E. coli*	$\alpha_6\beta_6$	α and β represent catalytic and regulatory subunits, respectively. The molecule is assembled as $(\alpha_3)_2(\beta_2)_3$
Ribulosebisphosphate carboxylase	Spinach	$\alpha_8\beta_8$	
RNA nucleotidyltransferase (see Chapter 7, Section 7.3)	*E. coli*	$\alpha_2\beta\beta'\sigma$	
Phosphorylase kinase (see Chapter 6, Section 6.4.2.1)	Rabbit skeletal muscle	$\alpha_4\beta_4\gamma_4\delta_4$	
Adenosinetriphosphatase (see Chapter 2, Section 2.8.3)	Beef-heart mitochondria	$\alpha_3\beta_3\gamma\delta\varepsilon$	

[a] Haemoglobin is included (although it is not an enzyme) since a great deal is known about the inter-subunit contacts in this protein (Section 3.6.3).

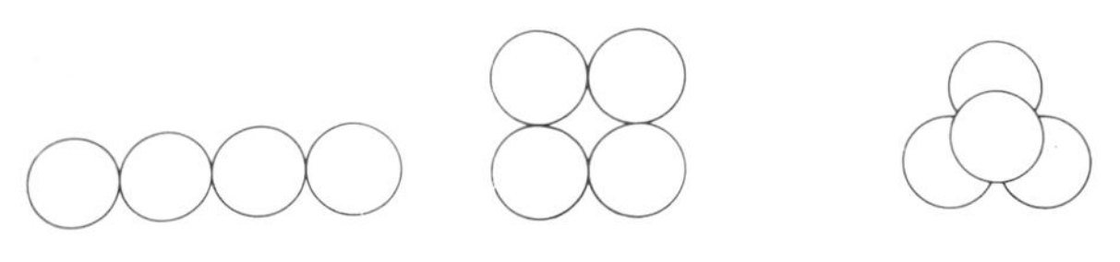

Linear (3 contacts) Square (4 contacts) Tetrahedral (6 contacts)

FIG. 3.35. Possible arrangements of subunits in a tetrameric protein, showing the number of intersubunit contacts.

Top view Side view

FIG. 3.36. Proposed arrangement of the Ca^{2+}- and Mg^{2+}-activated adenosinetriphosphatase from *E. coli*, as deduced by cross-linking studies.[227] The subunit stoichiometry of this enzyme is $\alpha_3\beta_3\gamma\delta\varepsilon$ (see Table 3.5).

was used.[227] This bifunctional ester results in the cross-linking of pairs of lysine side-chains by two propionyl groups that are joined by a disulphide bond. It is thus possible to identify the components of the cross-linked species by treatment with dithiothreitol (Section 3.3.2) to break the disulphide bond, and subsequent electrophoresis in the presence of sodium dodecylsulphate. The results showed that it was possible to cross-link α and β subunits to each other or to the γ subunit. However, no α–α or β–β cross-links could be formed. Other evidence showed that the smaller δ and ε subunits were attached only peripherally (Fig. 3.36).

3.6.3 Forces involved in the association between subunits

Oligomeric enzymes are usually dissociated into subunits by treatment with denaturing agents (Section 3.7.1) such as guanidine hydrochloride. The forces involved in the association of the subunits are thus of the weak, non-covalent type that are involved in stabilizing the folded structure of a polypeptide chain (Section 3.5.5), i.e. hydrogen bonds, electrostatic forces, van der Waals forces, and hydrophobic forces. There is only limited information available about the nature of inter-subunit contacts because the number of oligomeric enzymes whose structures have been solved by X-ray crystallography to the required resolution is still rather small. From the studies on systems such as haemoglobin, lactate dehydrogenase, and glyceraldehyde-3-phosphate dehydrogenase, the following generalization can be made:[196]

Number of non-polar contacts > number of hydrogen bonds
> number of electrostatic interactions.

The finding that there are large numbers of non-polar groups on the surfaces that are involved in interaction between subunits suggests that van der Waals forces and hydrophobic forces * provide a strong tendency for subunits to associate; the importance of hydrophobic forces is also suggested by a thermodynamic analysis of protein–protein interactions.[197] Fisher has suggested that there is a strong correlation between the tendency for subunits to associate and the content of amino acids with non-polar side-chains.[199] Proteins that contain more than about 30 per cent non-polar amino acids are unable to bury the non-polar side-chains within the interior of a folded polypeptide chain. The exposed non-polar groups will then tend to interact with others on other subunits giving rise to an oligomeric system.

As mentioned previously (Section 3.5.5.5), hydrophobic forces do not confer any geometrical specificity to an interaction. Specific interactions between subunits must therefore rely on forces such as hydrogen bonding that have distinct geometrical requirements. The importance of interactions between particular side-chains in different subunits is well illustrated by the various mutant haemoglobins. [195] In normal haemoglobin, the side-chain of tyrosine 35 on the β-chain forms a hydrogen bond with the side-chain of aspartic acid 126 on the α-chain. In haemoglobin Philly, the tyrosine is replaced by phenylalanine, which cannot form the hydrogen bond. This loss of a hydrogen bond leads to increased dissociation of haemoglobin Philly into subunits and is associated with clinical symptoms of mild anaemia. In Chapter 5 (Section 5.5.5.5) it is shown how the replacement of phenylalanine 164, which lies near the subunit interface in tyrosyl-tRNA synthetase, by aspartic acid leads to dissociation of the subunits at pH 7.78, at which pH, the aspartic acid carries a negative charge. At low pH, the side-chains are protonated and the subunits can associate.[200]

3.6.4 The significance of multiple subunits

There seem to be at least three reasons why the possession of multisubunit enzymes is advantageous to an organism. Firstly, the presence of multiple subunits confers additional possibilities of regulating the catalytic activity of an enzyme. This is discussed in detail in Chapter 6, Section 6.2.2.1. Secondly, the assembly of different types of subunits into a large complex permits variation in catalytic properties; this is well illustrated by the examples of tryptophan synthase and RNA nucleotidyl transferase discussed in Chapter 7, Sections 7.9 and 7.3 respectively. Another example of this type of system is lactose synthase (Table 3.5), which consists of two different subunits. One of these subunits (β) is lactalbumin, which is a protein that is devoid of any catalytic activity but able to alter the specificity of the other subunit (α), a galactosyltransferase, so that it can catalyse the synthesis of lactose rather than *N*-acetyllactosamine:[201]

* Hydrophobic forces are dominated by the entropy term (Section 3.5.5.4) and are thus weaker at lower temperatures. The dissociation of a number of enzymes such as adenosinetriphosphatase from mitochondria and phosphorylase b at low temperatures probably indicates that hydrophobic forces are involved in the association of subunits in these cases. Alternative explanations for such 'cold lability' have also been given.[198]

α subunit
UDPgalactose + *N*-acetylglucosamine ⇌ UDP + *N*-acetyllactosamine

αβ complex
UDPgalactose + D-glucose ⇌ UDP + lactose.

Lactalbumin thus acts as a specifier protein that can alter the specificity of the α subunit so that when necessary (e.g. during lactation) lactose can be synthesized in preference to *N*-acetyllactosamine that is used in the biosynthesis of glycoproteins.

Many oligomeric enzymes do not display these properties; thus, lactate dehydrogenase does not show regulatory characteristics. In such cases we must look for another reason for the occurrence of multiple subunits. A third reason may well be to confer stability on the enzyme. It is difficult to gain reliable data on this point because treatments that disrupt the bonds between subunits also tend to disrupt the folded structure of the subunits themselves, so that correctly folded isolated subunits are not easy to obtain.* The observation that the surfaces of subunits contain a number of non-polar side-chains (Section 3.6.3) suggests that correctly folded subunits, were they to occur, would, in general, be rather unstable.

3.7 The unfolding and refolding of enzymes

In Sections 3.5 and 3.6 we discussed the three-dimensional structures adopted by enzymes. It is now appropriate to consider how the three-dimensional structure can be lost (Section 3.7.1) and subsequently regained (Section 3.7.2). Section 3.7.3 deals with methods of predicting the three-dimensional structure of an enzyme.

3.7.1 Unfolding of enzymes

The compact folded form of an enzyme is generally thermodynamically more stable than the modified form by only a relatively small margin of the order of 15–60 kJ mol^{-1} [175] (Section 3.5.5.5). It is therefore not surprising that most enzymes can be relatively easily unfolded (and dissociated in the case of multisubunit enzymes) by a variety of conditions such as extremes of pH, heating, and addition of organic solvents, detergents, chaotropic agents,† or high concentrations of urea or guanidine hydrochloride. Loss of the three-dimensional structure of the folded enzyme is known as denaturation and agents that bring about denaturation are known as denaturing agents. Denaturation of an enzyme can be monitored by the loss of catalytic activity, by changes in

* Some progress has been made in this area using enzymes linked to insoluble supports, thereby permitting the preparation of matrix-bound subunits.[202] Other evidence for the existence of catalytically active subunits can be obtained from a study of the kinetics of regain of activity during refolding.[203]

† Chaotropic agents are ions of low charge density, e.g. trichloroacetate, that disorder water structure and facilitate the transfer of non-polar groups to the solvent water.

spectroscopic parameters such as the circular dichroism spectrum (Section 3.5.2), or by changes in other properties such as solubility.

The modes of action of denaturing agents have been extensively investigated.[204, 205] Heating will disrupt the folded structure of an enzyme by increasing the vibrational and rotational motions of atoms. A change in pH will affect the state of ionization of amino-acid side-chains that may well be involved in the maintenance of the folded structure. Addition of organic solvents or detergents would be expected to unfold an enzyme by interacting with the non-polar amino-acid side-chains previously buried in the interior of the folded structure. Careful measurements have shown that both urea and guanidine hydrochloride also act by increasing the solubility of the non-polar amino-acid side-chains[206, 207] while maintaining the hydrogen-bonding capacity of the aqueous solvent.[208] Guanidine hydrochloride has some advantages over urea as a denaturing agent, since solutions of the latter slowly decompose to yield cyanate.[209]

In most cases that have been investigated, the transition between folded and unfolded states seems to be relatively sharp, i.e. there is a small range of concentration of denaturing agent over which the enzyme changes more or less completely from one form to the other. Recent studies on the unfolding of myoglobin by guanidine hydrochloride[209] show that the number of binding sites for the denaturing agent increases as the protein unfolds; this provides a clue to the sharpness of the transition between folded and unfolded states. A two-state model, in which only the fully folded and fully unfolded states are considered, has generally provided a satisfactory description of the thermodynamic and some of the kinetic aspects of denaturation.[210]

3.7.2 Refolding of enzymes

Several years ago, Anfinsen performed an important experiment that showed that a denatured enzyme could regain its correct folded structure when the denaturing agent was removed.[211] This observation led to the conclusion that the primary structure of an enzyme contained the information required to specify its three-dimensional structure. (This conclusion is hardly surprising since the enzyme is synthesized *in vivo* as a polypeptide chain that folds up to assume its three-dimensional structure.) In outline, Anfinsen's experiment (Fig. 3.37) involved treating pancreatic ribonuclease with 2-mercaptoethanol to break the four disulphide bonds (Section 3.3.2) and then adding 8 mol dm^{-3} urea to produce inactive, unfolded, reduced enzyme. When the urea and reducing agent were removed by dialysis and the enzyme was oxidized to re-form the disulphide bonds, it was found that almost 100 per cent of the original enzyme activity was regained. Anfinsen argued that the complete regain of activity showed that the refolded enzyme possessed the correct disulphide bonds and therefore that the polypeptide chain had folded in such a way as to bring the correct pairs of cysteine side-chains into juxtaposition. There are 105 possible ways ($8!/2^4 \times 4!$) in which four disulphide bonds can be formed from eight cysteine side-chains,[212] so

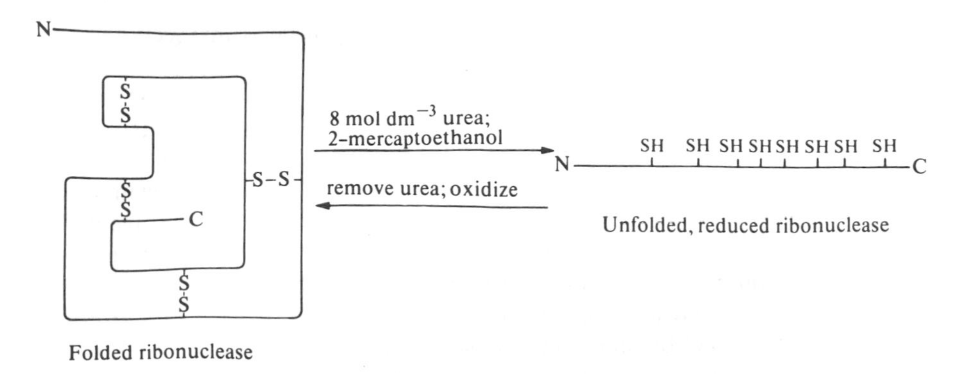

FIG. 3.37. The reduction and re-oxidation of pancreatic ribonuclease.

formation of the disulphide bonds on a random basis would lead to the regain of only about 1 per cent enzyme activity.

It has now become clear that the refolding of unfolded reduced, ribonuclease is a very complex process. Initially a number of distinct intermediates are observed in which different numbers of incorrect disulphide bonds are formed; these intermediates slowly rearrange to give the final, active product containing the correct disulphide bonds. For an account of this work, articles by Creighton[213, 214] should be consulted. Interestingly, there is an enzyme, protein disulphide-isomerase, that catalyses the conversion of those species containing incorrect disulphide bonds to that containing the correct bonds. It seems likely that this enzyme ensures that *in vivo* only a protein containing the correct disulphide bonds will be formed.[215, 216]

An important problem concerns the rate at which denatured proteins refold. It has been calculated[217] that it would take something like 10^{77} years for a polypeptide chain of 100 amino acids to 'sample' all the possible conformations; this time should be compared with the age of the earth, i.e. 4.6×10^9 years. Since proteins refold fairly quickly, on a time scale of seconds or minutes, it is clear that refolding does not involve a random search of all possible conformations until the most stable one is found. The current view[218] is that folding involves initially the formation of small local elements of structure that then act as nucleation centres to direct further folding of the polypeptide chain until the final folded state is reached. The elements of regular secondary structure such as α-helix or β-sheet (Section 3.5.4) are strong candidates for such nucleation centres. Attention has also been directed to the possibility that *cis* $\rightleftharpoons$ *trans* isomerization of prolines in the polypeptide chain (Section 3.5.4.1) may be a rate-determining step in the folding process,[219] and an enzyme (prolyl isomerase) catalysing this isomerization has been isolated.[220] Studies on the refolding of denatured proteins can give valuable insights into the way the polypeptide chain acquires its three-dimensional structure during biosynthesis. However, it is important to realize that the two processes cannot always be compared, because it appears that the growing polypeptide chain generally acquires recognizable folded structure while still attached to the ribosome. The complexities of the situation *in vivo* have been discussed.[221, 222]

3.7.3 Structure-prediction methods

The experiment outlined in Fig. 3.37 shows that the amino-acid sequence of a protein specifies its three-dimensional structure. For some time, biochemists have been investigating the possibility of predicting the three-dimensional structure from the sequence data. An *a priori* approach, in which the most stable conformation of a polypeptide chain is calculated by taking into account interactions between non-bonded atoms, is extremely laborious and useful results have been obtained only for short peptide chains. A greater degree of success has been obtained with semi-empirical approaches in which known three-dimensional protein structures are analysed to determine the probability that a certain amino acid, or sequence of amino acids, occurs in a particular element of regular structure on a more than random basis. With this information it is possible to draw up a set of rules to predict structural features in other proteins.[212] The Chou–Fasman rules[223] exemplify this approach. Amino acids are rated according to the likelihood that they occur in α-helices, β-sheets, or β-turns; for instance glutamic acid shows a strong preference for α-helices, a very low preference for β-sheets, and a middle to low preference for β-turns; proline and glycine have very low preferences for α-helices. The empirical rule for α-helix formation is that a cluster of four 'helix-forming' amino acids out of a sequence of six will nucleate a helix; the helical segment will extend in both directions until terminated by a sequence of four 'helix-breaking' amino acids. Analogous rules for the formation of β-sheet and β-turn structures have also been given.[223]

The Chou–Fasman and other prediction methods predict the elements of regular secondary structure (α-helix and β-sheet) with a success rate of about 60 per cent.[224] While this achievement is noteworthy, it is clear that we are still some considerable way from being able to predict the structure of a protein.* It has been suggested that combinations of separate prediction methods can give more reliable results.[226] The fact that the success rates of the prediction methods are only relatively modest implies that while short-range forces, extending over a few amino acids in the sequence, are important in specifying the final folded structure of a protein, the contribution of long-range forces between amino acids more remote in the sequence cannot be neglected.

3.8 Concluding remarks

The last two decades have witnessed an explosive growth in the amount of information concerning the structures of enzymes. This has been particularly

* An interesting example of the use of the structure-prediction methods occurred in the case of adenylate kinase. Several workers in the structure-prediction field were given the amino-acid sequence of the enzyme *before* the results of X-ray crystallography became available. A comparison of the actual structure with the various predicted structures was then made.[225]

true for the three-dimensional structures revealed by X-ray crystallography. We are now in a position to discuss, for instance, the mechanism of enzyme catalysis in much greater detail than was previously possible. Chapter 5 illustrates the substantial progress that has been made in this direction.

Probably the aspect of protein structure over which the greatest question mark remains is that of the refolding of enzymes (Section 3.7.2) and the relationship between the amino-acid sequence and the three-dimensional structure. The X-ray crystallographic studies on kinases and dehydrogenases have shown that functionally related enzymes contain common elements of three-dimensional structure, although this regularity is not readily apparent when comparisons are made between the amino-acid sequences of these enzymes. It is reasonable to expect that, as more data are accumulated, the underlying basis of such structural regularities and of the evolutionary relationships between functionally related enzymes will become clearer.

References

1. Price, N. C., and Dwek, R. A., *Principles and problems in physical chemistry for biochemists* (2nd edn). Clarendon Press, Oxford (1979). See Ch. 5.
2. Van Holde, K. E., *Physical biochemistry*. Prentice-Hall, Englewood Cliffs, New Jersey (1971). See Ch. 9.
3. Sund, H., Pilz, I., and Herbst, M., *Eur. J. Biochem.* **7**, 517 (1969).
4. Van Holde, K. E., *Physical biochemistry*. Prentice-Hall, Englewood Cliffs, New Jersey (1971). See Ch. 4.
5. Price, N. C., and Dwek, R. A., *Principles and problems in physical chemistry for Biochemists* (2nd edn). Clarendon Press, Oxford (1979). See p. 73.
6. Lee, J. C., and Timasheff, S. N., *Biochemistry* **13**, 257 (1974).
7. Cohn, E. J., and Edsall, J. T., in *Proteins, amino acids and peptides* (Cohn, E. J. and Edsall, J. T. eds.). Reinhold, New York (1943). See p. 370.
8. Creeth, J. M., and Pain, R. H., *Prog. Biophys. molec. Biol.* **17**, 217 (1967).
9. Aune, K. C., *Methods Enzymol.* **48**, 163 (1978).
10. Van Holde, K. E., *Physical biochemistry*. Prentice-Hall, Englewood Cliffs, New Jersey (1971). See Ch. 5.
11. Andrews, P., *Methods Biochem. Anal.* **18**, 1 (1970).
12. Alhadeff, J. A., *Biochem. J.* **173**, 315 (1978).
13. Nimmo, G. A., and Cohen, P., *Eur. J. Biochem.* **87**, 341 (1978).
14. Mann, K. G., and Fish, W. W., *Methods Enzymol.* **26**, 28 (1972).
15. Regnier, F. E., *Methods Enzymol.* **91**, 137 (1983).
16. Hedrick, J. L., and Smith, A. J., *Archs. Biochem. Biophys.* **126**, 155 (1968).
17. Hames, B. D., and Rickwood, D., *Gel electrophoresis of proteins: a practical approach*. IRL Press, Oxford (1981). See p. 14.
18. Weber, K., Pringle, J. R., and Osborn, M., *Methods Enzymol.* **26**, 3 (1972).
19. Hames, B. D., and Rickwood, D., *Gel. electrophoresis of proteins: a practical approach*. IRL Press, Oxford (1981). See page 71.
20. Laemmli, U. K. *Nature, Lond.* **227**, 680 (1970).
21. Reisler, E., Pouyet, J., and Eisenberg, H., *Biochemistry* **9**, 3095 (1970).
22. Wold, F., *Enzymes* (3rd edn) **5**, 499 (1971).

23. Lane, M. D., Moss, J., and Polakis, S. E., *Curr. Top. Cell. Reg.* **8**, 139 (1974).
24. Cohen, R., and Mire, M., *Eur. J. Biochem.* **23**, 267 (1971).
25. Cohen, R., and Mire, M., *Eur. J. Biochem.* **23**, 276 (1971).
26. Rogers, K. S., Hellerman, L., and Thompson, T. E., *J. biol. Chem.* **240**, 198 (1965).
27. Leland, S. G., and Zimmer, T.-L., *Essays Biochem.* **9**, 31 (1973).
28. IUPAC-IUB Joint Commission on Biochemical Nomenclature: Recommendations (1983). Reprinted in *Biochem. J.* **219**, 345 (1984).
29. Liu, T.-Y., and Chang, Y. H., *J. biol. Chem.* **246**, 2842 (1971).
30. Hugli, T. E., and Moore, S., *J. biol. Chem.* **247**, 2828 (1972).
31. Inglis, A. S., *Methods Enzymol.* **91**, 137 (1983).
32. Glazer, A. N., DeLange, R. J., and Sigman, D. S., in *Laboratory techniques in biochemistry and molecular biology* (Work, T. S. and Work, E. eds.) Vol. 4, p. 3. North-Holland, Amsterdam (1976). See p. 38.
33. Tristram, G. R., *Trends Biochem. Sci.* **4**, N31 (1979).
34. Velick, S. F., and Udenfriend, S., *J. biol. Chem.* **203**, 575 (1953).
35. Spackman, D. H., Stein, W. H., and Moore, S., *Analyt. Chem.* **30**, 1190 (1958).
36. Svasti, J., *Trends Biochem. Sci.* **5**, VIII (January 1980).
37. Bottom, C. B., Hanna, S. S., and Siehr, D. J., *Biochem. Educ.* **6**, 4 (1978).
38. Udenfriend, S., Stein, S., Böhlen, P., Dairman, W., Leimgruber, W., and Weigele, M., *Science, NY* **178**, 871 (1972).
39. Benson, J. R., and Hare, P. E., *Proc. natn. Acad. Sci. USA* **72**, 619 (1975).
40. Simons, S. S. Jr. and Johnson, D. F., *J. amer. Chem. Soc.* **98**, 7098 (1976).
41. Böhler, P., *Methods Enzymol.* **91**, 17 (1983).
42. Chang, J.-Y., Knecht, R., and Braun, D. G., *Methods Enzymol.* **91**, 41 (1983).
43. Glazer, A. N., DeLange, R. J., and Sigman, D. S., in *Laboratory techniques in biochemistry and molecular biology* (Work, T. S. and Work, E. eds.) Vol. 4, p. 3. North-Holland, Amsterdam (1976). See p. 13 ff.
44. Glazer, A. N., DeLange, R. J., and Sigman, D. S., in *Laboratory techniques in biochemistry and molecular biology* (Work, T. S. and Work, E. eds.) Vol. 4, p. 3. North-Holland, Amsterdam (1976). See p. 21.
45. Metzger, H., Shapiro, M. B., Mosimann, J. E., and Vinton, J. E., *Nature Lond.* **219**, 1166 (1968).
46. Harris, C. E., Kobes, R. D., Teller, D. C., and Rutter, W. J., *Biochemistry* **8**, 2442 (1969).
47. Cornish-Bowden, A., *Methods Enzymol.* **91**, 60 (1983).
48. Tang, J., Sepulveda, P., Marciniszyn, J. Jr, Chen, K. C. S., Huang, W.-Y., Tao, N., Liu, D., and Lanier, J. P., *Proc. natn. Acad. Sci. USA* **70**, 3437 (1973).
49. Hatch, F. T., *Nature, Lond.* **206**, 777 (1965).
50. Sandermann, J. Jr and Strominger, J. L., *Proc. natn. Acad. Sci. USA* **68**, 2441 (1971).
51. Sanger, F., and Thompson, E. O. P., *Biochem. J.* **53**, 366 (1953).
52. Titani, K., Koide, A., Hermann, J., Ericsson, L. H., Kumer, S., Wade, R. D., Walsh, K. A., Neurath, H., and Fischer, E. H., *Proc. natn. Acad. Sci. USA* **74**, 4762 (1977).
53. Fowler, A. V. and Zabin, I., *J. biol. Chem.* **253**, 5521 (1978).
54. *Methods Enzymol.* **68** (1979).
55. *Methods Enzymol.* **100** (1983).
56. *Methods Enzymol.* **101** (1983).
57. *Methods Enzymol.* **152** (1987).
58. Old, R. W., and Primrose, S. B., *Principles of gene manipulation* (3rd edn). Blackwell, Oxford (1985).

59. Adams, R. L. P., Knowler, J. T., and Leader, D. P., *The biochemistry of the nucleic acids* (10th edn). Chapman and Hall, London (1986).
60. Malcolm, A. D. B., *Biochem. Soc. Symp.* **44**, 1 (1979).
61. Itakura, K., Rossi, J. J., and Wallace, R. B., *A. Rev. Biochem.* **53**, 323 (1984).
62. Maxam, A. M., and Gilbert, W., *Proc. natn. Acad. Sci. USA* **74**, 560 (1977).
63. Sanger, F., Nicklen, S., and Coulson, A. R., *Proc. natn. Acad. Sci. USA* **74**, 5463 (1977).
64. Ambler, R. P., and Scott, G. K., *Proc. natn. Acad. Sci. USA* **75**, 3732 (1978).
65. Sutcliffe, J. G., *Proc. natn. Acad. Sci. USA* **75**, 3737 (1978).
66. Allen, G., in *Laboratory techniques in biochemistry and molecular biology* (Work, T. S. and Burdon, R. H. eds.) Vol. 9. North-Holland, Amsterdam (1981).
67. Keil, B., *Enzymes* (3rd edn.) **3**, 249 (1971). See p. 263.
68. Mitchell, W. M., and Harrington, W. F., *Enzymes* (3rd edn), **3**, 699 (1971). See p. 710.
69. Blow, D. M., *Enzymes* (3rd edn) **3**, 185 (1971). See p. 205.
70. Hartley, B. S., and Shotton, D. M., *Enzymes* (3rd edn) **3**, 323 (1971). See p. 333.
71. Glazer, A. N., and Smith, E. L., *Enzymes* (3rd edn) **3**, 501 (1971). See p. 519.
72. Alecio, M. R., Dann, M. L., and Lowe, G., *Biochem. J.* **141**, 495 (1974).
73. Matsubara, H., and Feder, J., *Enzymes* (3rd edn). **3**, 721 (1971). See p. 781.
74. Houmard, J., and Drapeau, G. R., *Proc. natn. Acad. Sci. USA* **69**, 3506 (1972).
75. Fruton, J. S., *Enzymes* (3rd edn) **3**, 119 (1971). See p. 140.
76. Butler, P. J. G., Harris, J. I., Hartley, B. S., and Leberman, R., *Biochem. J.* **103**, 78P (1967).
77. Dixon, H. B. F., and Perham, R. N., *Biochem. J.* **109**, 312 (1968).
78. Thomas, J. O., in *Companion to biochemistry* (Bull, A. T., Lagnado, J. R., Thomas, J. O., and Tipton, K. F., eds.) Vol. 1, p. 87. Longman, London (1974). See p. 101.
79. Gross, E., *Methods Enzymol.* **11,** 238 (1967).
80. Heil, A., Müller, G., Noda, L., Pinder, T., Schirmer, H., Schirmer, I., and von Zabern, I., *Eur. J. Biochem.* **43**, 131 (1974).
81. Stark, G. R., *Methods Enzymol.* **47**, 129 (1977).
82. Mahoney, W. C., Smith, P. K., and Hermodson, M. A., *Biochemistry* **20**, 443 (1981).
83. Spande, T. F., Witkop, B., Degani, Y., and Patchornik, A., *Adv. Protein Chem.* **24**, 97 (1970).
84. Weeds, A. G., and Hartley, B. S., *Biochem. J.* **107**, 531 (1968).
85. Waley, S. G., and Watson, J., *Biochem. J.* **55**, 328 (1953).
86. Bennett, J. C., *Methods Enzymol.* **11,** 330 (1967).
87. Waterfield, M. D., in *Practical protein chemistry: a handbook* (Darbre, A., ed.). Wiley, Chichester (1986). See Ch. 6.
88. Gray, W. R., *Methods Enzymol.* **25**, 121 (1972).
89. Schroeder, W. A., *Methods Enzymol.* **25**, 138 (1972).
90. Ambler, R. P., *Methods Enzymol.* **25**, 143 (1972).
91. Ward, C. W., in *Practical protein chemistry: a handbook* (Darbre, A., ed.). Wiley Chichester (1986). See Ch. 18.
92. Edman, P., and Begg, G., *Eur. J. Biochem.* **1**, 80 (1967).
93. Meuth, J. L., Harris, D. E., Dwulet, F. E., Crowl-Powers, M. L., and Gurd, F. R. N., *Biochemistry* **21**, 3750 (1982).
94. Niall, H. D., *Methods Enzymol.* **27**, 942 (1973).
95. Waterfield, M. D., Scrace, G., and Totty, N. in *Practical protein chemistry: a handbook* (Darbre, A., ed.). Wiley, Chichester (1986). See Ch. 13.
96. Gray, W. R., *Methods Enzymol.* **25**, 333 (1972).

97. Hunkapiller, M. W., and Hood, L. E., *Methods Enzymol.* **91**, 486 (1983).
98. Stark, G. R., *Adv. protein. Chem.* **24**, 261 (1970). See p. 294.
99. Stark, G. R., *Adv. protein Chem.* **24**, 261 (1970). See p. 299.
100. Hunkapiller, M. W., Hewick, R. M. Dreyer, W. J., and Hood, L. E., *Methods Enzymol.* **91**, 399 (1983).
101. Priddle, J. D., in *International review of biochemistry* (Offord, R. E., ed.), Vol. 24, p. 1. University Park Press, Baltimore (1979).
102. Morris, H. R., Williams, D. H., and Ambler, R. P., *Biochem. J.* **125**, 189 (1971).
103. Morris, H. R., *Phil. Trans. R. Soc.* **A293**, 39 (1979).
104. Williams, D. H., Bradley, C. V., Santikarn, S., and Bojesen, G., *Biochem. J.* **201**, 105 (1982).
105. Hughes, J., Smith, T. W., Kosterlitz, H. W., Fothergill, L. A., Morgan, B. A., and Morris, H. R., *Nature, Lond.* **258**, 577 (1975).
106. Shotton, D. M., and Hartley, B. S., *Nature, Lond.* **225**, 802 (1970).
107. Winn, S. I., Watson, H. C., Fothergill, L. A., and Harkins, R. N., *Biochem. Soc. Trans.* **5**, 657 (1977).
108. Anderson, C. M., Stenkamp, R. E., and Steitz, T. A., *J. molec. Biol.* **123**, 15 (1978).
109. Narita, K., in *Protein sequence determination* (Needleman, S. B., ed.), p. 25. Chapman and Hall, London; Springer, Berlin (1970). See p. 82.
110. Welinder, K. G., *Eur. J. Biochem.* **96**, 483 (1979).
111. Carr, S. A., Biemann, K., Shoji, S., Parmelee, D. C., and Titani, K., *Proc. natn. Acad. Sci. USA* **79**, 6128 (1982).
112. Offord, R. E., *Nature, Lond.* **211**, 591 (1966).
113. Blow, D. M., Birktoft, J. J., and Hartley, B. S., *Nature, Lond.* **221**, 337 (1969).
114. Brown, J. R., and Hartley, B. S., *Biochem. J.* **101**, 214 (1966).
115. Yazdanparast, R., Andrews, R. C., Smith, D. L., and Dixon, J. E., *J. biol. Chem.* **262**, 2507 (1987).
116. Perkins, R. E., Conroy, S. C., Dunbar, B., Fothergill, L. A., Tuite, M. F., Dobson, M. J., Kingsman, S. M., and Kingsman, A. J., *Biochem. J.* **211**, 199 (1983).
117. Bryant, T. N., Watson, H. C., and Wendell, P. J., *Nature, Lond.* **247**, 14 (1974).
118. Scopes, R. K., *Biochem. J.* **122**, 89 (1971).
119. Fifis, T., and Scopes, R. K., *Biochem. J.* **175**, 311 (1978).
120. Yoshida, A., *Analyt. Biochem.* **49**, 320 (1972).
121. Dobson, M. J., Tuite, M. F., Roberts, N. A., Kingsman, A. J., Kingsman, S. M., Perkins, R. E., Conroy, S. C., Dunbar, B., and Fothergill, L. A., *Nucleic Acids Res.* **10**, 2625 (1982).
122. Campbell, J. W., Watson, H. C., and Hodgson, G. I., *Nature, Lond.* **250**, 301 (1974).
123. Slotboom, A. J., and de Haas, G. H., *Biochemistry* **14**, 5394 (1975).
124. Blobel, G. and Dobberstein, B., *J. Cell Biol.* **67**, 835 (1972).
125. Hurt, E. C. and van Loon, A. P. G. M., *Trends Biochem. Sci.* **11**, 204 (1986).
126. Eisenberg, D., *A. Rev. Biochem.* **53**, 595 (1984).
127. Dayhoff, M. O., Barker, W. C., and Hunt, L. T., *Methods Enzymol.* **91**, 524 (1983).
128. Bajaj, M., and Blundell, T., *A. Rev. Biophys. Bioeng.* **13**, 453 (1984).
129. Smith, E. L., *Enzymes* (3rd edn) **1**, 267 (1970).
130. Diamond, J. M., *Nature, Lond.* **332**, 685 (1988).
131. Adams, R. L. P., Knowler, J. T., and Leader, D. P., *The biochemistry of the nucleic acids* (10th edn). Chapman and Hall, London (1986). See p. 437.
132. Adams, R. L. P., Knowler, J. T., and Leader, D. P., *The Biochemistry of the nucleic acids* (10th edn). Chapman and Hall, London (1986). See p. 289.

133. Markland, F. S. Jr and Smith, E. L., *Enzymes* (3rd edn) **3**, 561 (1971).
134. Blundell, T. L., and Johnson, L. N., *Protein crystallography*. Academic Press, New York (1976).
135. Blake, C. C. F., *Essays Biochem.* **11**, 37 (1975).
136. Eisenberg, D., *Enzymes* (3rd edn) **1**, 1 (1970).
137. *Methods Enzymol.*, **114** (1985).
138. *Methods Enzymol.*, **115** (1985).
139. Sweet, R. M., *Methods Enzymol.*, **114**, 20 (1985).
140. Green D. W., Ingram, V. M., and Perutz, M. F., *Proc. R. Soc.* **A225**, 287 (1954).
141. Blundell, T. L., and Johnson, L. N., *Protein crystallography*. Academic Press, New York (1976). See Ch. 12.
142. Kendrew, J. C., Dickerson, R. E., Strandberg, B. E., Hart, R. G., Davies, D. R., Phillips, D. C., and Shore, V. C., *Nature, Lond.* **185**, 422 (1960).
143. Artymiuk, P. J., Blake, C. C. F., Grace, D. E. P., Oatley, S. J., Phillips, D. C., and Sternberg, M. J. E., *Nature, Lond.* **280**, 563 (1979).
144. Huber, R., and Bode, W., *Acc. Chem. Res.* **11**, 114 (1978).
145. Williams, R. J. P., *Biol. Rev.* **54**, 389 (1979).
146. Mobility and migration of biological molecules. *Biochem. Soc. Symp.* **46** (1981).
147. Markley, J. L., and Ulrich, E. L., *A. Rev. Biophys. Bioeng.* **13**, 493 (1984).
148. Redfield, C., and Dobson, C. M., *Biochemistry* **27**, 122 (1988).
149. Campbell, I. D., and Dwek, R. A., *Biological spectroscop.* Benjamin/Cummings, Menlo Park, California (1984). See Ch. 10.
150. Kronman, M. J., and Robbins, F. M., in *Fine structure of proteins and nucleic acids* (Fasman, G. D. and Timasheff, S. N., eds.) Vol. 4, p. 271. Marcel Dekker, New York (1970).
151. Rupley, J. A., in *Structure and stability of biological molecules* (Timasheff, S. N. and Fasman, G. D., eds.) p. 291. Marcel Dekker, New York (1969).
152. Rossmann, M. G., Moras, D., and Olsen, K. W., *Nature, Lond.* **250**, 194 (1974).
153. Rossmann, M. G., Liljas, A., Bränden, C.-I., and Banaszak, L. J., *Enzymes* (3rd edn) **11**, 61 (1975).
154. Rossman, M. G., and Argos, P., *J. molec. Biol.* **111**, 75 (1976).
155. Matthews, B. W., and Rossmann, M. G., *Methods Enzymol.* **115**, 397 (1985).
156. Creighton, T. E., *Prog. Biophys. molec. Biol.* **33**, 231 (1978).
157. Ramachandran, G. N., and Sasisekharan, V., *Adv. protein Chem.* **23**, 283 (1968).
158. Dickerson, R. E., and Geis, I., *The structure and action of proteins*. Harper and Row, New York (1969). See Ch. 2.
159. Schulz, G. E., and Schirmer, R. H., *Principles of protein structure*. Springer, New York (1979). See Ch. 5.
160. Richardson, J. S., *Adv. Protein Chem.* **34**, 167 (1981).
161. Lipscomb, W. H., Reeke, G. N. Jr. Hartsuck, J. A., Quiocho, F. A., and Bethge, P. H., *Phil. Trans. R. Soc.* **B257**, 177 (1970).
162. Chothia, C., *J. molec. Biol.* **75**, 295 (1973).
163. Chou, P. Y., and Fasman, G. D., *J. molec. Biol.* **115**, 135 (1977).
164. Richardson, J. S., Getzoff, E. D., and Richardson, D. C., *Proc. natn. Acad. Sci. USA* **75**, 2574 (1978).
165. Milner-White, E. J., and Poet, R., *Biochem. J.* **240**, 289 (1986).
166. Milner-White, E. J., and Poet, R., *Trends Biochem. Sci.* **12**, 189 (1987).
167. Chothia, C., *Nature, Lond.* **248**, 338 (1974).
168. Levitt, M., and Chothia, C., *Nature, Lond.* **261**, 552 (1976).

169. Fersht, A. R., Shi, J.-P., Knill-Jones, J., Lowe, D. M., Wilkinson, A. J., Blow, D. M., Brick, P., Carter, P., Waye, M. M. Y., and Winter, G., *Nature, Lond.* **314**, 235 (1985).
170. Gilson, M. K., and Honig, B. H., *Nature, Lond.* **330**, 84 (1987).
171. Sternberg, M. J. E., Hayes, F. R. F., Russell, A. J., Thomas, P. G., and Fersht, A. R., *Nature, Lond.* **330**, 86 (1987).
172. Tanford, C., *Science, NY* **200**, 1012 (1978).
173. Gutfreund, H., and Knowles, J. R., *Essays Biochem.* **3**, 25 (1967).
174. Price, N. C., and Dwek, R. A., *Principles and problems in physical chemistry for biochemists* (2nd edn). Clarendon Press, Oxford (1979). See Ch. 2.
175. Creighton, T. E., *Prog. Biophys. molec. Biol.* **33**, 231 (1978). See p. 248.
176. Janin, J., and Chothia, C., *J. molec. Biol.* **100**, 197 (1976).
177. Blow, D. M., *Enzymes* (3rd edn) **3**, 185 (1971). See p. 194.
178. Schulz, G. E., Elzinga, M., Marx, F., and Schirmer, R. H., *Nature, Lond.* **250**, 120 (1974).
179. Buehner, M., Ford, G. C., Moras, D., Olsen, K. W., and Rossmann, M. G., *J. molec. Biol.* **90**, 25 (1974). See also *J. biol Chem.* **250**, 9137 (1975).
180. Remington, S. J., Wiegand, G., and Huber, R., *J. molec. Biol.* **158**, 111 (1982).
181. Almassy, R. J., Janson, C. A., Hamlin, R., Xuong, N.-H., and Eisenberg, D., *Nature, Lond.* **323**, 304 (1986).
182. Ashton, K. and Peacocke, A. R., *FEBS Lett.* **16**, 25 (1971).
183. Fell, D. A., Liddle, P. F., Peacocke, A. R., and Dwek, R. A., *Biochem. J.* **139**, 665 (1974).
184. Davies, G. E., and Stark, G. R., *Proc. natn. Acad. Sci. USA* **66**, 651 (1970).
185. Coggins, J. R., in *Theory and practice of affinity techniques* (Sundaram, P. V. and Eckstein, F., eds.) p. 89. Academic Press, London (1978).
186. Weber, K., *Nature, Lond.* **218**, 1116 (1968).
187. Wiley, D. C., and Lipscomb, W. N., *Nature, Lond.* **218**, 1119 (1968).
188. Allison, W. S., Admiraal, J., and Kaplan, N. O., *J. biol. Chem.* **244**, 4743 (1969).
189. Appella, E., and Tomkins, G. M., *J. molec. Biol.* **18**, 77 (1966).
190. Dixon, M., Webb, E. C., Thorne, C. J. R., and Tipton, K. F., *Enzymes* (3rd edn). Longman, London (1979). See pp. 550–67.
191. Wood, W. A., *Trends Biochem. Sci.* **2**, 223 (1977).
192. Simon, J.-P., and Stalon, V., *Europ. J. Biochem.* **88**, 287 (1978).
193. Frieden, C., *A. Rev. Biochem.* **40**, 653 (1971).
194. Schulz, G. E., and Schirmer, R. H., *Principles of protein structure*. Springer, New York (1979). See p. 100.
195. Klotz, I. M., Langerman, N. R., and Darnall, D. W., *A. Rev. Biochem.* **39**, 25 (1970).
196. Liljas, A., and Rossmann, M. G., *A. Rev. Biochem.* **43,** 475 (1974).
197. Chothia, C., and Janin, J., *Nature, Lond.* **256**, 705 (1975).
198. Bock, P. E., and Frieden, C., *Trends Biochem. Sci.* **3**, 100 (1978).
199. Fisher, H. F., *Proc. natn. Acad. Sci. USA* **51**, 1285 (1964).
200. Jones, D. H., McMillan, A. J., Fersht, A. R., and Winter, G., *Biochemistry* **24**, 5852 (1985).
201. Hall, L., and Campbell, P. N., *Essays Biochem.* **22**, 1 (1986).
202. Chan, W. W.-C., *Can. J. Biochem.* **54**, 521 (1976).
203. Grossman, S. H., Pyle, J., and Steiner, R. J., *Biochemistry* **20**, 6122 (1981).
204. Tanford, C., *Adv. protein Chem.* **23**, 121 (1968).
205. Tanford, C., *Adv. protein. Chem.* **24**, 1 (1970).
206. Nozaki, Y., and Tanford, C., *J. biol. Chem.* **238**, 4074 (1963).

207. Wetlaufer, D. B., Malik, S. K., Stoller, L., and Coffin, R. L., *J. amer. chem. Soc.* **86**, 508 (1964).
208. Roseman, M., and Jencks, W. P., *J. amer. chem. Soc.* **97**, 631 (1975).
209. Pace, C. N., and Vanderburg, K. E., *Biochemistry* **18**, 288 (1979).
210. Creighton, T. E., *Prog. Biophys. molec. Biol.* **33**, 231 (1978). See p. 249.
211. Anfinsen, C. B., *Harvey Lect.* **61**, 95 (1967).
212. Anfinsen, C. B. and Scheraga, H. A., *Adv. protein Chem.* **29**, 205 (1975).
213. Creighton, T. E., *J. molec. Biol.* **129**, 411 (1979).
214. Creighton, T. E., *Prog. Biophys. molec. Biol.* **33**, 231 (1978). See p. 283.
215. Freedman, R. B., *Trends Biochem. Sci.* **9**, 438 (1984).
216. Freedman, R. B., *Nature, Lond.* 329, 196 (1987).
217. Creighton, T. E., *Prog. Biophys. molec. Biol.* **33,** 231 (1978). See p. 255.
218. Jaenicke, R., *Prog. Biophys. molec. Biol.* **49**, 117 (1987). See p. 156.
219. Jaenicke, R., *Prog. Biophys. molec. Biol.* **49**, 117 (1987). See p. 161.
220. Lang, K., Schmid, F. X., and Fischer, G., *Nature, Lond.* **329**, 268 (1987).
221. Jaenicke, R., *Prog. Biophys. molec. Biol.* **49**, 117 (1987). See p. 120.
222. Tsou, C.-L., *Biochemistry* **27**, 1809 (1988).
223. Chou, P. Y. and Fasman, G. D., *A. Rev. Biochem.* **47**, 251 (1978).
224. Kabsch, W. and Sander, C., *FEBS Lett.* **155**, 179 (1983).
225. Chou, P. Y. and Fasman, G. D., *A. Rev. Biochem.* **47**, 251 (1978). See p. 263.
226. Sawyer, L., Fothergill-Gilmore, L. A., and Freemont, P. S., *Biochem. J.* **249**, 789 (1988).
227. Bragg, P. D. and Hou, C., *Arch. Biochem. Biophys.* **167**, 311 (1975).

4
An introduction to enzyme kinetics

4.1 Outline of the chapter

The subject of enzyme kinetics often generates considerable trepidation in biochemistry students. Many treatments of the topic make two assumptions: firstly, that the readers can confidently handle complex algebraic equations, and, secondly, that a detailed knowledge of kinetics is an indispensable part of a biochemist's training. In this chapter we do not make these assumptions but instead introduce some of the main ideas in enzyme kinetics that are useful in understanding the mechanism of action and control of isolated enzymes (see Chapters 5 and 6) and the role of enzymes in the cell (see Chapter 8). The emphasis will be on the information that can be gained from a study of enzyme kinetics.

The first part of the chapter (Section 4.2) describes some practical aspects of enzyme kinetics, answering the question 'How do we obtain kinetic data?' The next part (Section 4.3) deals with the analysis of the results, discussing the theoretical background, equations involved, and information obtained. In these sections we shall be dealing with 'steady-state kinetics', where an enzyme is present at very small molar concentrations (usually much less than 1 per cent) compared with the molar concentration(s) of substrate(s) acted upon. Under these conditions, the equations involved are *comparatively* straightforward and the data are usually collected over a time scale of minutes, so that 'conventional' methods of mixing and observation can be employed. In Section 4.4 we refer in outline to experiments in which enzyme and substrate(s) are present at comparable concentrations. Under these conditions it is usually necessary to use special techniques to ensure rapid mixing and observation. This type of experiment has given detailed insights into the various steps in an overall reaction, particularly those occurring in the early, 'pre-steady-state' period (see Fig. 4.2). In fact, these studies are often termed 'pre-steady-state kinetics'. As described in Chapter 8 (Section 8.5.4) there are many examples known where, in the cell, the concentration of an enzyme is comparable with that of its substrate and in such cases the study of pre-steady-state kinetics is particularly relevant.

We end this section by stating three of the reasons why it is important for the biochemistry student to have some knowledge of enzyme kinetics. Firstly, kinetics, *in conjunction with other techniques*, provides valuable information on the mechanism of action of an enzyme (see Chapter 5). Secondly, it can give an insight into the role of an enzyme under the conditions that exist in the cell and the response of an enzyme to changes in the concentrations of metabolites (see

Chapter 8). Thirdly, it can help to show how the activity of an enzyme can be controlled, which may provide a valuable pointer to mechanisms of regulation under physiological conditions (see Chapters 6 and 8).

4.2 How do we obtain kinetic data?

The aim of an experiment is to measure the rate of formation of product (or disappearance of substrate) under specified conditions. It is then possible to vary in turn certain parameters such as the concentration of substrate(s), pH, temperature, or concentration of modifying ligands, and collect data to analyse in terms of theoretical models.

The rate of a particular enzyme-catalysed reaction can often be measured in a number of ways, but there is normally one method that is more convenient than the others. This point can be illustrated by reference to the reaction catalysed by hexokinase:

$$\text{D-Glucose} + \text{ATP} \xrightarrow{Mg^{2+}} \text{D-glucose 6-phosphate} + \text{ADP}.$$

The rate of this reaction could be monitored by removing samples from the reaction mixture at known times after addition of enzyme, stopping the reaction quickly (e.g. by addition of acid to inactivate the enzyme) and measuring the amount of product formed. Ion-exchange chromatography (see Chapter 2, Section 2.6.2.1) would be a useful technique for separating the products from the substrates in this case. Clearly such a 'stop and sample' (or *discontinuous*) assay procedure involves possible sampling errors and considerable work in separation and estimation of the products.

A more convenient method would involve the *continuous* measurement of some property that changes during the course of the reaction. In the case of the reaction above, there is no convenient change in absorbance, for example, but we could bring about such a change if we *couple* the production of D-glucose 6-phosphate to the reduction of $NADP^+$ to NADPH using glucose-6-phosphate dehydrogenase:

$$\text{D-Glucose} + \text{ATP} \xrightarrow{Mg^{2+}} \text{D-Glucose 6-phosphate} + \text{ADP}$$

$$\downarrow \quad (NADP^+ \rightarrow NADPH)$$

$$\text{D-Glucono-}\delta\text{-lactone 6-phosphate}$$

$NADP^+$ does not absorb at 340 nm, whereas NADPH does, so it is possible to monitor the production of NADPH (and hence of D-glucose 6-phosphate) continuously. If we use a *coupled assay procedure* such as this we must add sufficient coupling enzyme and substrate(s) so that the D-glucose 6-phosphate formed in the first step is 'immediately' converted to D-glucose-δ-lactone 6-phosphate, i.e. so that the coupling reaction is not rate limiting. Several detailed analyses of the kinetics of coupled assay systems have indicated the conditions under which the true rate of the reaction of interest can be measured.[1, 2, 3, 4, 5] It is, of course, obvious that the coupling enzyme should be highly purified and certainly free from detectable quantities of the enzyme we are trying to assay.

In a number of cases it is not possible to monitor the reaction continuously (either directly on the reaction of interest or by using a coupled assay) so a 'stop and sample' method must be employed. For instance, in the reaction catalysed by ornithine decarboxylase,

$$\underset{\text{L-Ornithine}}{NH_2\text{—}(CH_2)_3\text{—}\underset{\underset{^{14}CO_2H}{|}}{\overset{\overset{NH_2}{|}}{CH}}} \longrightarrow {}^{14}CO_2 + \underset{\substack{\text{1,4-diaminobutane}\\\text{(putrescine)}}}{NH_2(CH_2)_4NH_2}$$

if L-[1-^{14}C]ornithine is used, the CO_2 liberated will be radioactive. This CO_2 can be trapped by a suitable base (e.g. ethanolamine dissolved in 2-methoxyethanol) and the radioactivity estimated by scintillation counting. Determination of the amount of CO_2 formed after various times of reaction allows the rate of the reaction to be calculated. Assay procedures involving radioactively labelled substrates are very sensitive and are of particular value when low concentrations of substrates are being used or when the amount of enzyme activity is low.

When the method of assay has been decided upon, it is important to take a number of precautions to obtain reliable data.[6, 7]

1. The substrates, buffers, etc. should be of as high a purity as possible, since contaminants may affect the activity of enzymes. For example, commerical preparations of NAD^+ sometimes contain inhibitors of dehydrogenases[8, 9] and it has been found that certain preparations of ATP contain trace amounts of vanadate (VO_4^{3-}) ions, which act as a powerful inhibitor of adenosinetriphosphatase (Na^+, K^+-activated).[10]
2. It must be ascertained that the enzyme preparation does not contain any compound (or other enzyme) that interferes with the assays. The possibility of non-enzyme-catalysed conversion of substrate to product should be tested by performing appropriate control experiments (e.g. using heat-inactivated enzyme).
3. The enzyme should be stable (i.e. not lose any significant amount of catalytic activity) during the time taken for assay. Breakdown of substrate (other than by the enzyme) should not occur.
4. Since the activity of an enzyme can be markedly affected by changes in pH, temperature, etc., it is important to ensure that these parameters are stabilized by use of buffers, thermostatted baths, etc.

5. It should be checked that, once the steady state has been achieved (Section 4.3.1.1), the measured rate of reaction is constant over the period of interest and is proportional to the amount of enzyme added.
6. The *initial* rate of reaction should be measured to avoid possible complications arising from product inhibition, occurrence of the reverse reaction and depletion of substrate.[11, 12]

4.3 How do we analyse kinetic data?

As in many other branches of experimental science, kinetic data are analysed in terms of theoretical models in order to test the correctness of the models and to deduce the values of constants in the equations derived from the models. In this section we shall describe for a number of situations the theoretical background to steady-state kinetics, then show how to treat data in terms of the resulting equations, and finally describe the significance of the results obtained.

4.3.1 One-substrate reactions

It is easiest to deal initially with reactions in which only one substrate is acted on by an enzyme. This would include reactions catalysed by enzymes in the following groups (see Chapter 1, Section 1.5.1): hydrolases (if H_2O is considered to be in a large excess), isomerases, and most lyases.

4.3.1.1 Theoretical background

We shall assume that catalysis occurs via rapid and reversible formation of a complex between enzyme, E, and substrate, S. (The part of the enzyme to which the substrate binds is known as the *active site* of the enzyme.) This complex then breaks down in a slow step to give the product, P, and regenerate enzyme (Scheme 1). In practice this scheme is likely to be an oversimplification (see Section 4.3.1.4).

$$\mathrm{E+S} \underset{k_{-1}}{\overset{k_1}{\rightleftharpoons}} \mathrm{ES}$$

$$\mathrm{ES} \overset{k_2}{\rightarrow} \mathrm{E+P}$$

Scheme 1. The conversion of substrate to product in an enzyme catalysed reaction. k_1, k_{-1}, k_2 represent the rate constants for the individual steps.

The equation describing the kinetics of this scheme can be derived by making one of two types of assumption as follows.

(i) The equilibrium assumption
Here we assume that the $\mathrm{E+S} \rightleftharpoons \mathrm{ES}$ equilibrium is only slightly disturbed by the breakdown of ES to give product. This is clearly a better assumption the lower the value of k_2 relative to k_{-1}.

The equilibrium constant, K, is defined by

$$K = \frac{[E][S]}{[ES]}, \tag{4.1}$$

where [E] and [S] are the concentrations of *free* enzyme and *free* substrate, respectively. However, since the total concentration of substrate is much greater than the total concentration of enzyme (see Section 4.1), essentially all the substrate is free and we can set $[S]_{free}$ equal to $[S]_{total}$.

At any concentration of S we can evaluate the fraction, F, of enzyme present as ES as follows:

$$F = \frac{[ES]}{[E]+[ES]}.$$

Now, from (4.1)

$$[ES] = \frac{[E][S]}{K},$$

$$\therefore \quad F = \frac{[E][S]}{K} \bigg/ \left([E] + \frac{[E][S]}{K}\right)$$

$$= \frac{[S]}{K+[S]}. \tag{4.2}$$

If the total concentration of enzyme is $[E]_0$, then $[ES] = F[E]_0$; thus, from (4.2)

$$[ES] = \frac{[E]_0[S]}{K+[S]}.$$

The rate of product formation, v, is given by

$$v = k_2[ES]$$

$$\therefore \quad v = \frac{k_2[E]_0[S]}{K+S} \tag{4.3}$$

Sometimes the equation (4.3) is written in the form

$$v = \frac{k_{cat}[E]_0[S]}{K+S}$$

where k_{cat} is a first-order rate constant equal to k_2 (k_{cat} is sometimes known as the *turnover number* of the enzyme). This formulation will be referred to later (see Section 4.3.1.3).

An equation of the type shown in (4.3) means that v will tend towards a maximum (or limiting) value as [S] increases (Fig. 4.1).

The maximum rate will be observed when all the enzyme is in the form of the ES complex. Let the maximum (limiting) rate = V_{max}; this will be equal to $k_2[E]_0$: then eqn (4.3) can be rewritten as:

$$v = \frac{V_{max}[S]}{K+[S]}. \tag{4.4}$$

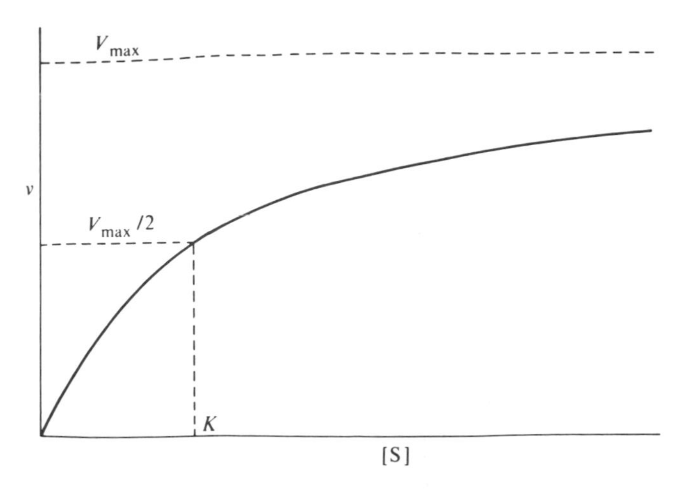

FIG. 4.1. Dependence of velocity (v) on substrate concentration ([S]) according to eqn (4.3).

We should note from eqn (4.4) and Fig. 4.1 that when [S] is small compared with K, the reaction is first order in [S], since $v = V_{max}[S]/K$. When [S] is large compared with K, the reaction is zero order in [S], since $v = V_{max}$. At intermediate values of [S], the reaction is of a fractional order in [S].

When $v = V_{max}/2$, $[S] = K$. Thus K corresponds to the concentration of substrate when the velocity is half-maximal. It is known as the Michaelis* constant, and is normally written as K_m. If we have used the equilibrium assumption, we can equate K_m with the dissociation constant of the ES complex (K in eqn (4.1)).

(ii) The steady-state assumption

In this approach, we abandon the assumption that the $E + S \rightleftharpoons ES$ equilibrium is not perturbed by the breakdown of ES. Instead, it is assumed that ES is in a 'steady state', i.e. that the concentration of ES remains constant because the rate of its formation equals the rate of its breakdown. If we were to examine the variation of [ES] with time in a typical experiment we would obtain a graph of the type shown in Fig. 4.2.

After an initial phase (the pre-steady-state period) the concentration of ES remains fairly constant and it is thus in order to apply the steady-state assumption to Scheme 1 to evaluate the fraction of enzyme in the form of the ES complex. (For the steady-state assumption to be valid the rate of change of [ES] must be small compared with the rate of change of [S] or [P]. In experiments in which the concentration of substrate is much greater than that of enzyme, the maximum value of [ES] and hence the rate of change of [ES], will be small.)

* Michaelis was one of the first workers to develop the mathematical analysis of enzyme kinetics along the lines indicated here. An equation of the type (4.4) is often referred to as the Michaelis–Menten equation.

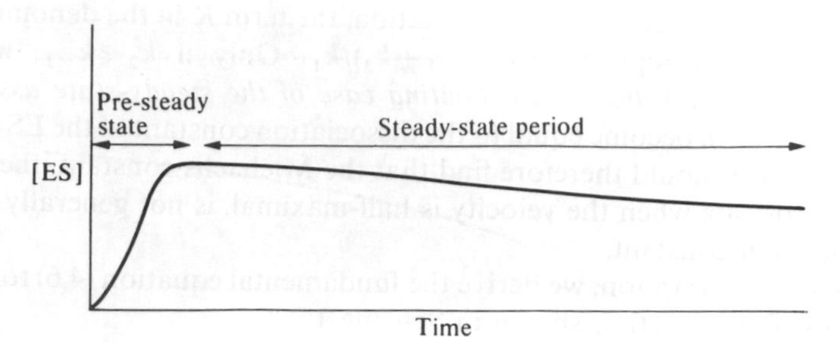

FIG. 4.2. The concentration of the ES complex as a function of time of reaction.

Now, according to the steady-state assumption, the rate of production of ES($=k_1$[E][S]) must equal the rate of its breakdown ($=k_{-1}$[ES]$+k_2$[ES])

$$\therefore \quad k_1[\mathrm{E}][\mathrm{S}]=k_{-1}[\mathrm{ES}]+k_2[\mathrm{ES}]$$

$$\therefore \quad [\mathrm{ES}]=\frac{k_1[\mathrm{E}][\mathrm{S}]}{k_{-1}+k_2}.$$

Proceeding as before,

$$F=\frac{[\mathrm{ES}]}{[\mathrm{E}]+[\mathrm{ES}]}$$

$$=\frac{[\mathrm{S}]}{\left[\dfrac{k_{-1}+k_2}{k_1}\right]+[\mathrm{S}]}.$$

If the total concentration of enzyme is $[\mathrm{E}]_0$, then

$$[\mathrm{ES}]=F[\mathrm{E}]_0$$

$$=\frac{[\mathrm{E}]_0[\mathrm{S}]}{\left[\dfrac{k_{-1}+k_2}{k_1}\right]+[\mathrm{S}]}$$

and the rate of product formation, v, is given by:

$$v=k_2[\mathrm{ES}]$$

$$=\frac{k_2[\mathrm{E}]_0[\mathrm{S}]}{\left[\dfrac{k_{-1}+k_2}{k_1}\right]+[\mathrm{S}]}.$$

Putting $k_2[\mathrm{E}]_0=V_{\mathrm{max}}$

$$v=\frac{V_{\mathrm{max}}[\mathrm{S}]}{\left[\dfrac{k_{-1}+k_2}{k_1}\right]+[\mathrm{S}]}. \qquad (4.5)^*$$

* This equation is sometimes referred to as the Briggs–Haldane equation because these workers (in 1925) first applied the steady-state assumption to enzyme kinetics.

Thus using the steady-state approximation, the term K in the denominator of eqn (4.4) has been replaced by $(k_{-1}+k_2)/k_1$. Only if $k_2 \ll k_{-1}$, *when the equilibrium assumption becomes a limiting case of the steady-state assumption,* does this latter term become equal to the dissociation constant of the ES complex $(K=k_{-1}/k_1)$. We should therefore find that the Michaelis constant, the concentration of substrate when the velocity is half-maximal, is not generally equal to this dissociation constant.

Using either assumption, we derive the fundamental equation (4.6) to describe the kinetics of the reaction shown in Scheme 1:

$$v=\frac{V_{max}[S]}{K_m+[S]}. \tag{4.6}$$

4.3.1.2 *Treatment of data*

It is very difficult to determine the limiting value of v (i.e. V_{max}) directly from a plot of v against [S] (Fig. 4.1) and therefore K_m cannot readily be determined in this way either. To overcome these diffculties, eqn (4.6) can be rearranged in a number of ways to give convenient graphical representations. Four of the best known rearranged forms are given below:

(i) The Lineweaver–Burk equation[13]

This equation is obtained by taking reciprocals of the two sides of eqn (4.6):

$$\frac{1}{v}=\frac{K_m}{[S]}\cdot\frac{1}{V_{max}}+\frac{1}{V_{max}}.$$

Thus a plot of $1/v$ against $1/[S]$ gives a straight line of slope K_m/V_{max} and with intercepts on the x and y axes of $-1/K_m$ and $1/V_{max}$, respectively (Fig. 4.3(a)).

(ii) The Eadie–Hofstee equation[14, 15]

Equation (4.6) is rearranged to give:

$$\frac{v}{[S]}=\frac{V_{max}}{K_m}-\frac{v}{K_m}.$$

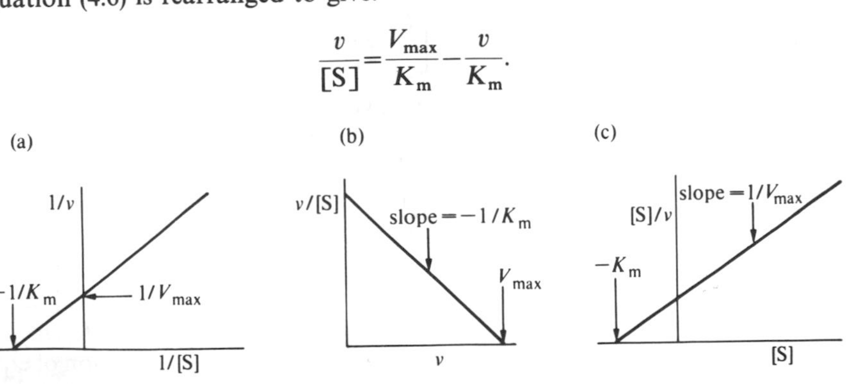

FIG. 4.3. Graphical representations of enzyme kinetic data according to the equations of (a) Lineweaver and Burk, (b) Eadie and Hofstee, and (c) Hanes. It should be remembered that the equation of a straight line is $y=mx+c$, where m equals the slope and c the intercept on the y-axis. The intercept on the x-axis equals $-c/m$.

Thus a plot of $v/[S]$ against v gives a straight line of slope $-1/K_m$ and with an x-axis intercept of V_{max} (Fig. 4.3(b)).

(iii) The Hanes equation[16]
Equation (4.6) is rearranged to give:

$$\frac{[S]}{v} = \frac{[S]}{V_{max}} + \frac{K_m}{V_{max}}.$$

Thus a plot of $[S]/v$ against $[S]$ is linear with a slope of $1/V_{max}$ and an x-axis intercept of $-K_m$ (Fig. 4.3(c)).

(iv) The direct linear plot[17, 18]
This method uses a rather different approach from those described in (i)–(iii). Equation (4.6) is arranged to give:

$$V_{max} = v + \frac{v}{[S]} \cdot K_m \qquad (4.7)$$

V_{max} and K_m are now treated as variables and v and $[S]$ as constants. If the first pair of observed values of v and $[S]$ (i.e. v_1 and $[S]_1$) are plotted as shown in Fig. 4.4(a) then reference to eqn (4.7) shows that the line connecting them describes pairs of values of V_{max} and K_m that are consistent with the observed values of v and $[S]$.

If we have a second set of values of v and $[S]$ (i.e. v_2 and $[S]_2$), a new line can be drawn (Fig. 4.4(b)). The point of intersection of the lines defines uniquely the values of V_{max} and K_m that satisfy the two sets of data. This procedure can be repeated for further data points (Fig. 4.4(b)). If there were no experimental error the various lines would all intersect at a common point. In practice a number of points of intersection are obtained (the maximum number of such points is $n(n-1)/2$, where n is the number of observations of v and $[S]$). The correct procedure is then to use the medians* of the values of V_{max} and K_m as the best-fit values.[17, 18, 19]

It is not possible to give simple answer to the question of which of the various methods of plotting data should be used. In any case it should be emphasized that the data should be as good as possible, since no graphical transformation allows sound conclusions to be drawn from poor data. The Lineweaver–Burk plot (Fig. 4.3(a)) is still the most commonly used and it has the advantage that the variables v and $[S]$ are plotted on separate axes. However, an analysis of the errors involved in the collection of the data (and hence in the determination of the parameters K_m and V_{max}) shows that there is a highly non-uniform distribution of errors over the range of values of $1/v$ and $1/[S]$ in the Lineweaver–Burk plot.[20] For this reason, the use of the Eadie–Hofstee and Hanes plots has been recommended, since in these plots the distribution of errors is more uniform.[20]

* The median is the middle value of a set of values arranged in order of magnitude, or the mean of the middle pair if the number of values is even.

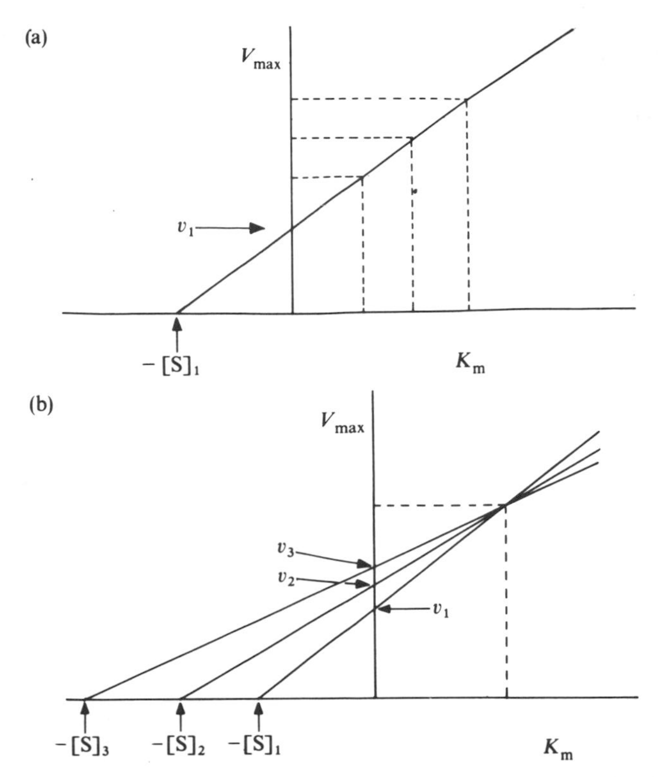

FIG. 4.4. The direct linear plot in which experimental values of v and [S] are plotted directly. (a) A straight line drawn through one set of experimental values; the dashed lines indicate pairs of values of V_{max} and K_m that are consistent with the experimental values. In (b) a number of sets of experimental values are plotted. The point of intersection of the various lines gives the values of V_{max} and K_m that uniquely satisfy the experimental data.

The direct linear plot has a number of advantages: (i) Values of v and [S] are plotted directly so that V_{max} and K_m can be determined without the need for calculations. (ii) It is statistically sound: the use of median values of V_{max} and K_m minimizes the influence of extreme values of v and [S] on these parameters.[19] However, the direct linear plot has some disadvantages. It is not very suitable for the display of data in multisubstrate reactions (Section 4.3.5) because of the resulting large number of lines. In addition, it is not easy to detect departures from the basic equation (4.6).

In recent years (especially with the greater availability of microcomputers) it has become more common to obtain the 'best fit' values of V_{max} and K_m by use of computer programs in which the data are fitted directly to eqn (4.6);[21, 22,23] a number of such programs are available commercially.

Finally, it should be mentioned that determinations of V_{max} and K_m can also be made from the integrated form of eqn (4.6). Substrate (or product) concentrations are monitored over a large fraction of the overall reaction, i.e. well beyond the 'initial rate' period. The values of V_{max} and K_m can be determined from a single progress curve that essentially represents a series of rate measurements at different values of substrate concentration.[12, 23, 24, 25] One way in which this is done is described in Appendix 4.1. A major disadvantage of the method is that the accuracy of the estimates of V_{max} and K_m can be very sensitive to errors in the estimates of the end-point of the reaction.[12] In addition, the progress curve can be used to calculate the true initial rates of reaction in those cases where there has been a significant decline in the rate during the time taken for mixing the assay components and obtaining readings.[12, 23, 25]

The units of K_m and V_{max} will be those in which [S] and v are measured. Thus K_m is expressed in units of concentration, i.e. mol dm^{-3}. The velocity, v, can be expressed in a number of ways, depending on the information available. For a purified enzyme, v is expressed as moles substrate consumed per unit time, per weight of enzyme. In the SI system this would be in terms of katal kg^{-1}, whereas in the older system it would be in terms of units mg^{-1} (see Chapter 2, Section 2.3.1). If the M_r of the enzyme is also known, we can calculate the molar concentration of enzyme active sites in the solution and hence evaluate k_{cat} ($= V_{max}/[E]_0$), which is also known as the *turnover number* of the enzyme (see Section 4.3.1.1).

4.3.1.3 *Significance of results*

The parameters K_m and V_{max} (k_{cat}) are of value (i) in characterizing the specificity of an enzyme for a particular substrate, (ii) in deciding between steady-state and equilibrium mechanisms, and (iii) in indicating the role of an enzyme in metabolism.

(i) The specificity of an enzyme for a substrate can be described as follows: From eqn (4.3) we have

$$v = k_{cat}\frac{[E]_0[S]}{K_m + [S]}.$$

Substituting $[E]_0 = [E] + [ES]$ and noting that $[ES] = [E][S]/K_m$ we have

$$v = \frac{k_{cat}}{K_m}[E][S]. \tag{4.8}$$

Thus k_{cat}/K_m is an apparent second-order rate constant that describes the rate in terms of the concentrations of the free enzyme and free substrate.

If there are two competing substrates S_1 and S_2 for the enzyme, it follows from eqn (4.8) that the rates of reaction are

$$v_{S_1} = \left(\frac{k_{cat}}{K_m}\right)_{S_1}[E][S_1]$$

$$v_{S_2} = \left(\frac{k_{cat}}{K_m}\right)_{S_2}[E][S_2].$$

The ratio of these rates of reaction is given by

$$\frac{v_{S_1}}{v_{S_2}}=\frac{(k_{cat}/K_m)_{S_1}}{(k_{cat}/K_m)_{S_2}}\cdot\frac{[S_1]}{[S_2]}.$$

Thus, at equal concentrations of S_1 and S_2 the relative rates of reaction of the two substrates are determined by the relative values of k_{cat}/K_m. The ratio k_{cat}/K_m can thus be used as a measure of the specificity of an enzyme for a substrate. For example, in the reaction catalysed by fumarate hydrotase, the ratios k_{cat}/K_m for fumarate, fluorofumarate, chlorofumarate and bromofumarate are 1.6×10^8, 9.8×10^7, 2.0×10^5 and 2.5×10^4 s^{-1} $(mol\,dm^{-3})^{-1}$, respectively.[26] There is thus high specificity shown towards fumarate and fluorofumarate; substitution by the larger halogens leads to a marked decline in the reactivity of the substrate.

(ii) The ratio of k_{cat}/K_m can be used to test for the applicability of the steady-state or equilibrium mechanisms (Section 4.3.1.1). Using the steady-state assumption, we note from eqns (4.5) and (4.6) that

$$K_m=\frac{k_{-1}+k_2}{k_1}.$$

Now, if $k_2>k_{-1}$ (which is the extreme form of the steady-state assumption), then $K_m=k_2/k_1$. Replacing k_2 by k_{cat}, it is clear that k_{cat}/K_m is equal to k_1, the rate constant for the association of enzyme with substrate. From fast-reaction studies (Section 4.4.3) it is known that the diffusion-controlled rate constant for association of enzyme with substrate is of the order of 10^9 $(mol\,dm^{-3})^{-1}\,s^{-1}$. Thus, if k_{cat}/K_m is of this order of magnitude, it can be concluded that the steady-state mechanism operates. This is the case for fumarate hydrotase, catalase, and triosephosphate isomerase, for which k_{cat}/K_m values are 1.6×10^8, 4×10^7 and 2.4×10^8 s^{-1} $(mol\,dm^{-3})^{-1}$, respectively.[27] However, if the value of k_{cat}/K_m is much lower than these values, it is reasonable to conclude that the equilibrium mechanism is more appropriate.

(iii) The role of an enzyme in metabolism may be judged by relating the K_m value to the prevailing concentration of substrate. This will be discussed further in Chapter 8 (Section 8.3.2) but two examples will serve to illustrate the ideas.

The isoenzyme IV of hexokinase (also known as glucokinase) is confined to the liver and has a high K_m for glucose (10 mmol dm^{-3}), whereas isoenzymes I–III have a much wider tissue distribution and a low K_m (40 μmol dm^{-3}). Thus, at the prevailing levels of blood glucose in a fasting subject (*ca.* 3 mmol dm^{-3}), isoenzymes I–III are working essentially at their maximum velocity, whereas isoenzyme IV is only working at about 25 per cent of its maximum velocity. After an intake of carbohydrate, the level of blood glucose rises to about 9.5 mmol dm^{-3} and the isoenzyme IV is now working at about half its maximum velocity (Fig. 4.5). Hence the liver isoenzyme (IV) can 'deal with' this extra glucose, converting it to D-glucose 6-phosphate, which is the first step in the process of storage as glycogen.

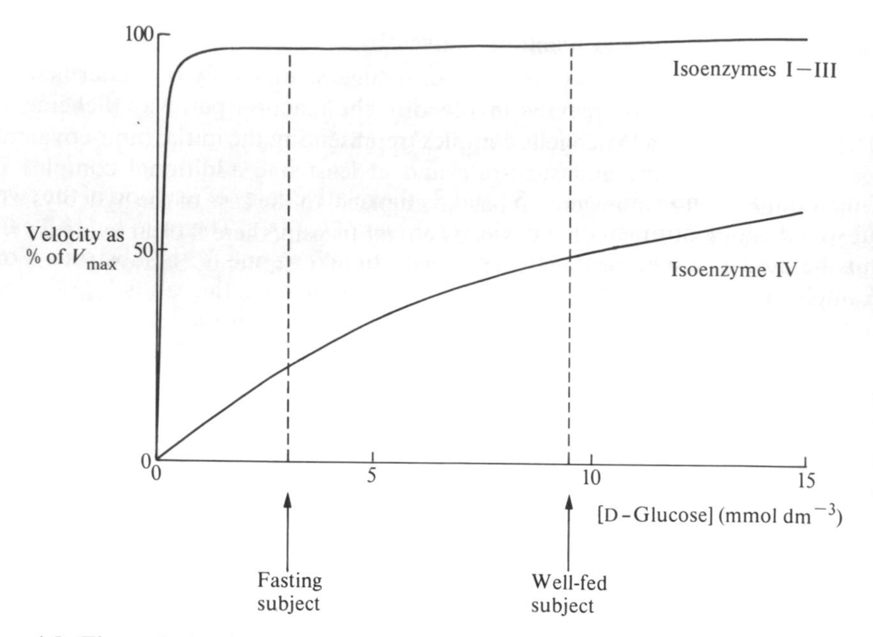

FIG. 4.5. The variation in velocity of the reactions catalysed by hexokinase isoenzymes with concentration of D-glucose.

The studies on ribulose bisphosphate carboxylase (see Chapter 2, Section 2.8.2), which catalyses a reaction involving assimilation of CO_2 in photosynthetic organisms, illustrate the need to exercise care in interpreting data on K_m and V_{max}. This enzyme is present in very large amounts in such organisms and is relatively easy to purify. However, early measurements showed that the activity of the purified enzyme was far too low to account for the observed rates of photosynthetic CO_2 fixation, and that, in particular, the K_m for CO_2 was about 50-fold higher than the apparent K_m measured in intact leaves or isolated chloroplasts. The solution to this problem is now thought[28] to lie in the fact that the enzyme is slowly activated by CO_2 and Mg^{2+}, and so considerable care has to be taken in performing and interpreting the assays of the purified enzyme. Activated enzyme has a K_m for CO_2 that is comparable with the apparent K_m in the intact system.

From this second example it is clear that it can be misleading to interpret the parameter K_m in anything other than purely operational terms (i.e. the concentration of substrate at which the velocity is half maximal). It is *sometimes* helpful to think of K_m as a crude measure of the affinity of an enzyme for its substrate (a high K_m implying weak affinity), but the comment made at the end of Section 4.3.1.1 must be borne in mind.

4.3.1.4 Some extensions to the simple model (Scheme 1)

It is worthwhile at this stage to mention some extensions to the relatively simple model (Scheme 1) we have considered up to now.

(i) More than one intermediate
In most cases it is probably unrealistic to suggest that only one intermediate (enzyme-containing complex) is involved in the reaction pathway (Scheme 1). There is likely to be a 'Michaelis complex' representing the initial (non-covalent) association of enzyme and substrate and at least one additional complex in which some bond rearrangement has taken place. In the case of the hydrolysis of ester and amide substrates catalysed by chymotrypsin, there is clear evidence for the involvement of an acyl-enzyme intermediate (Chapter 5, Section 5.4.1.5). If we extend Scheme 1 as follows:

$$\mathrm{E+S} \rightleftharpoons \mathrm{ES} \rightleftharpoons \mathrm{ES'} \rightarrow \mathrm{E+P},$$

then it can be shown by applying the steady-state assumption to the various intermediates in the pathway (ES and ES′) that the rate equation is of the same form as eqn (4.6) but that the interpretation of K_m in terms of the rate constants of the individual steps will be different from that in the simple case (eqn (4.5)).[29]

(ii) Substrate inhibition
Occasionally it is found that there is a decrease in v at high substrate concentrations. In the Lineweaver–Burk plot (Fig. 4.3(a)), for example, this would be manifested by an upward curvature at low values of 1/[S] as in the example of the hydrolysis of D-fructose 1,6-bisphosphate (FBP) catalysed by fructose-bisphosphatase, where the upward curvature is noted at FBP concentrations above 0.1 mmol dm^{-3} (Fig. 4.6).

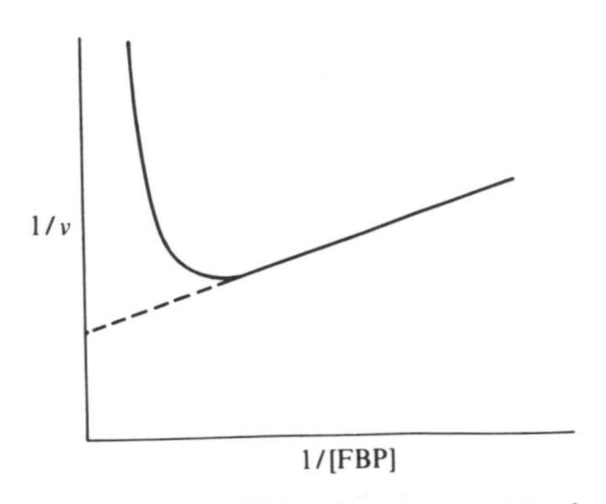

FIG. 4.6. Lineweaver–Burk plot for the reaction catalysed by fructose-bisphosphatase. The solid line represents the experimental data and shows that substrate inhibition occurs at high concentrations of FBP (above about 0.1 mmol dm^{-3}).

This phenomenon is known as substrate inhibition and is usually interpreted in terms of the existence of two types of substrate-binding site in the enzyme. Occupation of the first, high-affinity, type at low [FBP] leads to 'normal' kinetic behaviour (the linear part of Fig. 4.6). At high [FBP], the second, low-affinity, type of site becomes occupied and this is presumed to inhibit the catalytic reaction taking place at the first type of site. For a fuller discussion of this phenomenon, see reference 30.

(iii) Multiple active sites

Complex kinetics are often observed when an enzyme is composed of a number of subunits and possesses more than one active site. If interactions occur between these various sites non-linearity in the kinetic plots (Fig. 4.3) will be observed, characteristic of positive or negative cooperativity. These points are discussed in more detail in Chapter 6 (Section 6.2.2.2).

Other more detailed texts should be consulted for details of additional complications to the simple model (Scheme 1) such as inhibition by products[31, 32] and the occurrence of reverse reactions.[33, 34, 35]

4.3.2 Inhibition of one-substrate reactions

The study of the effects of inhibitors on enzyme-catalysed reactions is important not only to introduce various terms such as competitive inhibition, but also to give information on the active site of an enzyme (see Chapter 5, Section 5.4.1.3) and on inhibition which may be of possible physiological significance (see Chapter 6, Section 6.2.2.1). We shall confine our attention to *reversible inhibitors*, i.e. inhibitors which combine reversibly with an enzyme* rather than those which cause irreversible covalent modification. The use of the latter in the determination of enzyme mechanisms is discussed in Chapter 5, Section 5.4.4.

4.3.2.1 *Theoretical background*

One of the ways in which inhibition of enzyme-catalysed reactions can be discussed is in terms of a general scheme shown below (Scheme 2).

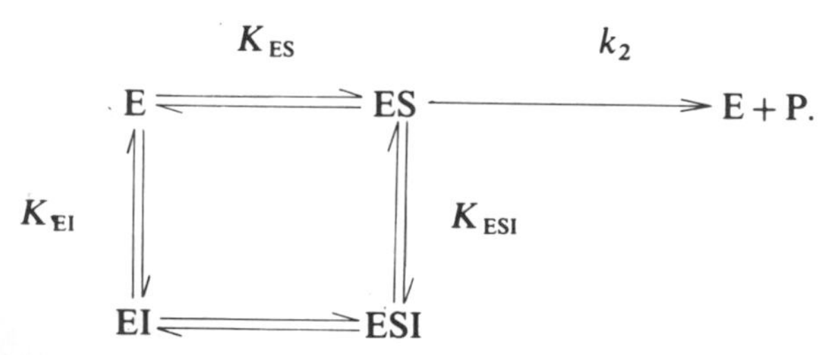

Scheme 2. A general scheme for inhibition of enzyme-catalysed reactions, K_{ES}, K_{ESI}, and K_{EI} represent dissociation constants. ESI is assumed to be inactive.

We shall assume that the enzyme-containing complexes are in equilibrium with each other, i.e. that the breakdown of ES to generate product does not significantly disturb the equilibrium. A general kinetic equation for this scheme

* The effect is reversible if it is decreased by lowering the concentration of the inhibitor (e.g. by dilution or dialysis). The distinction between reversible and irreversible inhibition is not absolute and may be difficult to make if the inhibitor binds very tightly to the enzyme and is released very slowly. In these circumstances the inhibitors are termed tight-binding inhibitors.[36]

can then be derived as described in Appendix 4.2 to this chapter:

$$v = V_{max} \frac{\frac{[S]}{K_{ES}}}{1 + \frac{[S]}{K_{ES}} + \frac{[I]}{K_{EI}} + \frac{[S][I]}{K_{ESI} \cdot K_{ES}}},$$

or in reciprocal form:

$$\frac{1}{v} = \frac{1}{V_{max}} \left[1 + \frac{[I]}{K_{ESI}} \right] + \frac{K_{ES}}{V_{max}} \left[1 + \frac{[I]}{K_{EI}} \right] \frac{1}{[S]}. \tag{4.9}$$

If we apply the steady-state approximation to Scheme 2, a complex kinetic expression is derived that is difficult to test experimentally[37] and will not be discussed further here.

The general equation (4.9) is simplified if certain assumptions are made about the magnitudes of the various dissociation constants. These *limiting cases* will be discussed in terms of the Lineweaver–Burk plots (Fig. 4.3(a)) of kinetic data, but we could, of course, use the other plots shown in Figs. 4.3 and 4.4. A more detailed classification of the various types of inhibition[36] and a description of the Dixon plot for the determination of inhibitor constants are given in Appendix 4.2.

4.3.2.2 Treatment of data

Case (i). Competitive inhibition

If we assume that $K_{ESI} = \infty$ (i.e. that the ES complex cannot combine with I nor the EI complex with S), then eqn (4.9) reduces to

$$\frac{1}{v} = \frac{1}{V_{max}} + \frac{K_{ES}}{V_{max}} \left[1 + \frac{[I]}{K_{EI}} \right] \frac{1}{[S]}$$

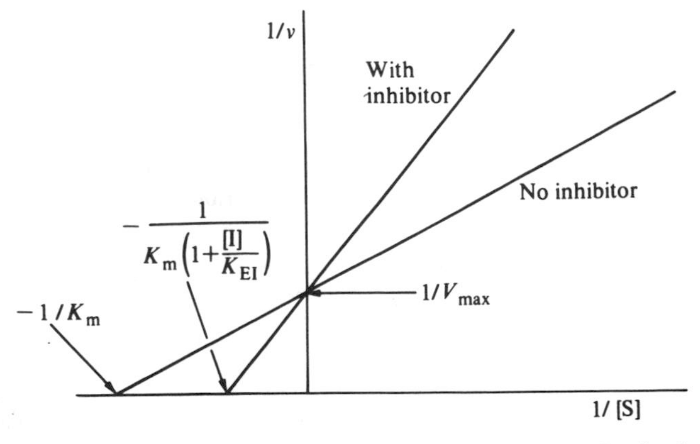

FIG. 4.7. The effect of a competitive inhibitor on a Lineweaver–Burk plot of enzyme kinetic data.

and the effect of the inhibitor on the Lineweaver–Burk plot is as shown in Fig. 4.7. V_{max} is unaffected but the apparent K_m is increased by a factor $(1+[I]/K_{EI})$. Effectively, the inhibitor pulls some of the enzyme over into the form of the EI complex. When the concentration of S is increased sufficiently, the effect on the velocity can be overcome (thus giving rise to the term 'competitive'). Many examples of competitive inhibition are found in one-substrate reactions; thus, carbamoylcholine is a competitive inhibitor with respect to acetylcholine in the reaction catalysed by acetylcholinesterase from bovine erythrocytes:

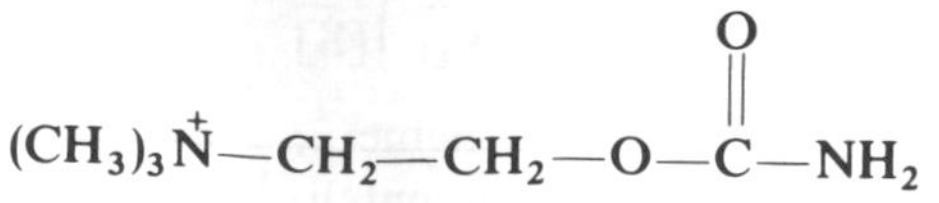

$$(CH_3)_3\overset{+}{N}—CH_2—CH_2—O—\overset{\overset{\large O}{\|}}{C}—NH_2 \qquad \text{carbamoylcholine}$$

$$(CH_3)_3\overset{+}{N}—CH_2—CH_2—O—\overset{\overset{\large O}{\|}}{C}—CH_3 \qquad \text{acetylcholine}$$

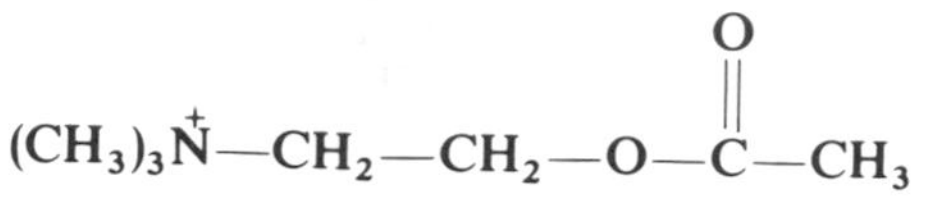

Although in this case the structural similarity between inhibitor and substrate makes it very likely that the two molecules bind to the same site on the enzyme, it is not always a justified conclusion that a competitive inhibitor binds to the active site.[38]

Case (ii). Non-competitive inhibition

If we assume that $K_{ESI}=K_{EI}$ (i.e. that the binding of S to the enzyme does not affect the binding of I), then eqn (4.9) reduces to

$$\frac{1}{v}=\frac{1}{V_{max}}\left[1+\frac{[I]}{K_{EI}}\right]+\frac{K_{ES}}{V_{max}}\left[1+\frac{[I]}{K_{EI}}\right]\frac{1}{[S]}$$

and the effect of the inhibitor on the Lineweaver–Burk plot is as shown in Fig. 4.8.

K_m remains unaffected, but V_{max} is descreased by a factor

$$1\bigg/\left(1+\frac{[I]}{K_{EI}}\right).^{*}$$

The inhibitor pulls both E and ES over into inactive forms (EI and ESI, respectively) but does not affect the distribution between them. Hence, increasing the concentration of S does not serve to overcome the effect of the inhibitor on the velocity. Examples of non-competitive inhibition in one substrate reactions are less common than those of competitive inhibition. In the case of fructose-bisphosphatase, AMP acts as a non-competitive inhibitor with respect to the substrate fructose 1,6-bisphosphate. There are many examples, however, of non-competitive inhibition in multisubstrate reactions (see Section 4.3.5.5).

* Reference to Appendix 4.2 shows that this situation is more correctly termed pure non-competitive inhibition.

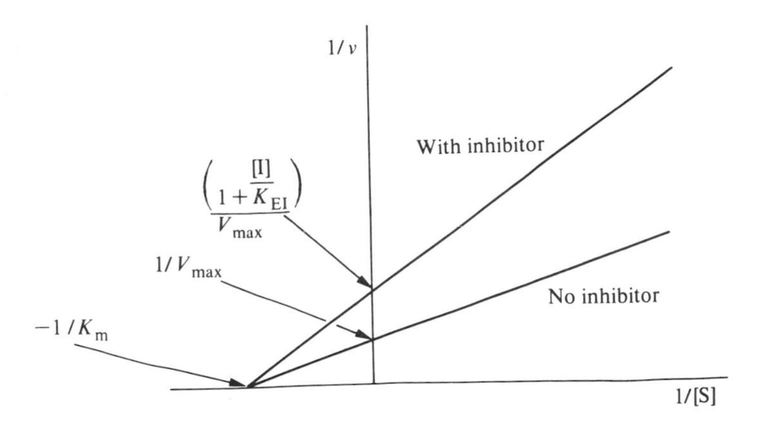

FIG. 4.8. The effect of a non-competitive inhibitor on a Lineweaver–Burk plot of enzyme kinetic data.

Case (iii). Uncompetitive inhibition

If we assume that $K_{EI} = \infty$ (i.e. that I cannot combine with E, but only with the ES complex) then eqn (4.9) reduces to

$$\frac{1}{v} = \frac{1}{V_{max}}\left[1 + \frac{[I]}{K_{ESI}}\right] + \frac{K_{ES}}{V_{max}}\frac{1}{[S]}$$

and the effect of the inhibitor on the Lineweaver–Burk plot is as shown in Fig. 4.9. Both K_m and V_{max} are affected by the inhibitor giving rise to parallel lines. There are very few cases indeed of uncompetitive inhibition in one-substrate reactions (an example is the inhibition of alkaline phosphatase from rat intestine by L-phenylalanine[39]) but more examples are found in multisubstrate reactions (e.g. *S*-adenosylmethionine behaves as an uncompetitive inhibitor towards ATP in the reaction catalysed by methionine adenosyltransferase from yeast.[40]

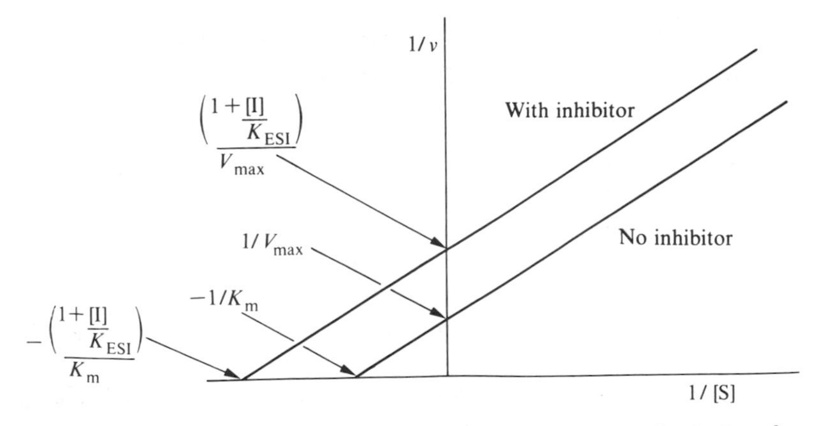

FIG. 4.9. The effect of an uncompetitive inhibitor on a Lineweaver–Burk plot of enzyme kinetic data.

4.3.2.3 Significance of results

We have introduced the terms competitive, non-competitive, and uncompetitive inhibitors by referring to the model depicted in Scheme 2 for one-substrate reactions. In our view it is better to define these types of inhibitors in terms of their effects on the parameters K_m and V_{max} (Figs. 4.7, 4.8, and 4.9) rather than by referring to relationships between the binding sites for S and I on the enzyme, because the latter are much more difficult to determine.[38] As we shall see later (Section 4.3.5.4), the terms competitive, non-competitive, and uncompetitive are extremely useful for classifying the effects of inhibitors in multisubstrate reactions, which can be of value in distinguishing between various types of reaction mechanism. In a number of cases (see Chapter 6, Section 6.2.2.1) the effects of an inhibitor do not fall into any of the three limiting cases depicted in Figs. 4.7, 4.8, and 4.9. In such cases the inhibition is usually referred to as being of a 'mixed' type and will require different equations and assumptions from those described in this section.[41] (See the discussion in Appendix 4.2)

4.3.3 The effect of changes in pH on enzyme-catalysed reactions

There are a number of distinct effects that a change in pH can have on enzyme-catalysed reactions, e.g. inactivation of the enzyme outside a certain pH range or a change in the ionization state of the substrate(s). A third possibility is that there could be a change in the equilibrium position if H^+ is involved in the reaction, e.g. in the reaction catalysed by creatine kinase:

$$\text{Creatine} + \text{MgATP}^{2-} \rightleftharpoons \text{phosphocreatine}^{2-} + \text{MgADP}^{-} + \text{H}^{+}.$$

In this case, increasing the pH will displace the equilibrium in favour of phosphocreatine synthesis.

However, the possibility of most interest to us is that there are changes in the ionization state of amino-acid side-chains that are essential for the catalytic activity of the enzyme. In this case we might hope to obtain information about the nature of these side-chains.

4.3.3.1 Theoretical background

The effects of pH on the velocity of an enzyme-catalysed reaction can be complex, since both K_m and V_{max} can be affected, and, in order to undertake a detailed analysis of the ionizations involved, values of both parameters should be obtained over a range of pH values. A number of excellent detailed treatments of the effects of pH on K_m and V_{max} have been given.[42, 43, 44, 45] We should note that it is usually easier to analyse changes in V_{max}, since this parameter generally reflects a single rate constant, whereas K_m is a function of several rate constants.*

* Analysis of a scheme in which the free enzyme and enzyme–substrate complex each possess two ionizing side-chains shows that changes in V_{max} depend on ionizations of the enzyme–substrate complex; changes in V_{max}/K_m depend on ionizations of the free enzyme, and changes in K_m depend on ionizations of both the free enzyme and the enzyme–substrate complex. For further details of these effects and the analysis of more complex schemes, references 42, 43, 44 and 45 should be consulted.

In order to see how pK_a values can be obtained from the experimental data, we can consider the simple case of a single ionizing side-chain in the enzyme. (Similar equations would result if we considered a single ionizing side-chain in the enzyme–substrate complex.)

$$EH^+ \rightleftharpoons E + H^+.$$

We shall assume that EH^+ is inactive and that E is the active form.

Now the acid dissociation constant, K_a, is given by

$$K_a = \frac{[E][H^+]}{[EH^+]},$$

$$\therefore \; [EH^+] = \frac{[E][H^+]}{K_a}.$$

The fraction, F, of enzyme in the unprotonated (active) form is given by

$$F = \frac{[E]}{[E]+[EH^+]} = \frac{K_a}{K_a + [H^+]}.$$

Let $(V_{max})_m$ equal the maximum rate when all the enzyme is in the unprotonated form. Then at any pH, the observed V_{max} is given by

$$V_{max} = (V_{max})_m \cdot F.$$

Thus

$$V_{max} = (V_{max})_m \cdot \frac{K_a}{K_a + [H^+]}. \tag{4.10}$$

If we have a second ionizing group in the enzyme, such as shown in the following scheme:

$$\underset{\text{inactive}}{\overset{HX \quad YH}{E}} \underset{+H^+}{\overset{-H^+}{\rightleftharpoons}} \underset{\text{active}}{\overset{HX \quad Y^-}{E}} \underset{+H^+}{\overset{-H^+}{\rightleftharpoons}} \underset{\text{inactive}}{\overset{X^- \quad Y^-}{E}}$$

with pK_{a_1} for the first step and pK_{a_2} for the second step,

the expression for V_{max} at any pH can be shown to be

$$V_{max} = \frac{[V_{max})_m}{1 + \frac{[H^+]}{K_{a_1}} + \frac{K_{a_2}}{[H^+]}}. \tag{4.11}$$

4.3.3.2 *Treatment of data*

Consider eqn (4.10). At pH values well below the pK_a (i.e. when $[H^+] \gg K_a$), then

$$V_{max} = (V_{max})_m \cdot \frac{K_a}{[H^+]}.$$

Taking logarithms:

$$\log_{10} V_{max} = \log_{10}(V_{max})_m - pK_a + pH,$$

so that a plot of $\log_{10} V_{max}$ against pH will be linear with a slope of 1. (In practice this will work well up to about 1.5 pH units below the pK_a.)

At pH values well above the pK_a (i.e. when $[H^+] \ll K_a$) then

$$V_{max} = (V_{max})_m,$$

i.e. there is no variation of V_{max} with pH.

The plot of $\log_{10} V_{max}$ against pH will therefore be of the form shown in Fig. 4.10. Extrapolation of the linear portions of the plot will give the value of pK_a at the point of intersection.

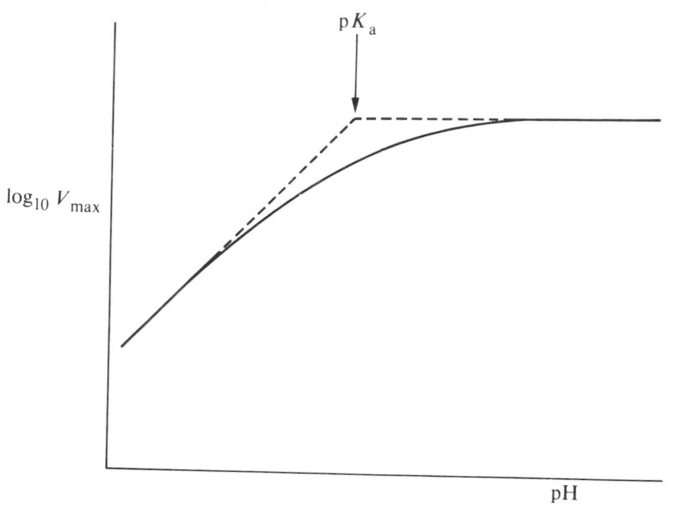

FIG. 4.10. The effect of pH on the V_{max} of an enzyme-catalysed reaction if only the unprotonated form of the enzyme is catalytically active. The solid curve represents the experimental data.

A similar analysis of eqn (4.11) shows that if two ionizing groups were involved, a plot of $\log_{10} V_{max}$ against pH would have three distinct regions of slope 1, 0, and -1 as shown in Fig. 4.11. These regions correspond to the cases where $[H^+] \gg K_{a_1}$, $K_{a_1} \ll [H^+] \gg K_{a_2}$ and $K_{a_2} \gg [H^+]$ respectively.

As shown in Fig. 4.11, the values of pK_{a_1} and pK_{a_2} can be obtained from the points of intersection of the extrapolated linear portions of the plot. However, if the two pK_a values are closer than about 1.5 pH units, the ionizations will not be independent and the pK_a values derived may need correction to obtain the true values.[46]

4.3.3.3 *Significance of results*

A study of the effect of changes in pH on the velocity of an enzyme-catalysed reaction generally leads to the conclusion that there is an 'optimum pH' for that

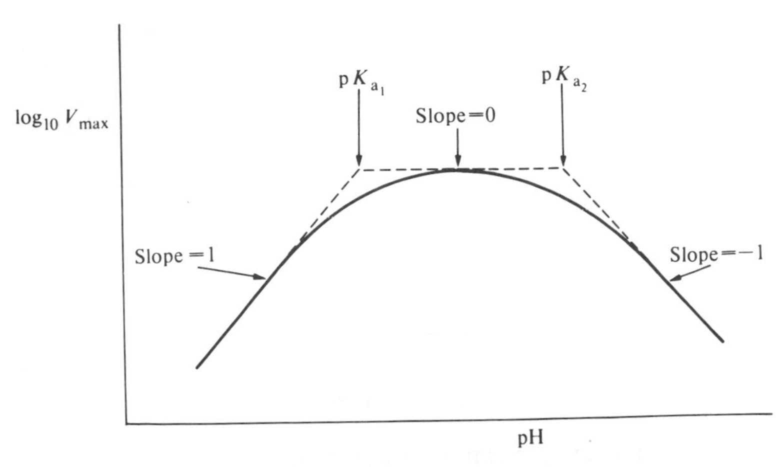

FIG. 4.11. The effect of pH on the V_{max} of an enzyme-catalysed reaction when two ionizing groups are involved. The solid curve represents the experimental data.

reaction. In view of the various types of effect of pH mentioned in Section 4.3.3, it is clear that such a term should be used with a certain degree of caution. There are cases known in which studies of enzyme kinetics at the optimum pH have given misleading results when considering the *in vivo* situation (see Chapter 8, Section 8.3.2).

It is often tempting to try to assign the measured pK_a values (Figs. 4.10 and 4.11) to particular types of amino-acid side-chain, using the values of pK_a for free amino acids (see Chapter 3, Section 3.3.1). However, there are considerable pitfalls in this procedure, since the environment of a side-chain in an enzyme can be very different from that of the free amino acid and this can cause a large shift in pK_a. Thus, in pepsin, one aspartic acid side-chain has a pK_a of about 1.0, some 3 pH units lower than that of free aspartic acid.[47] Nevertheless, in some cases it has been possible from pH-rate studies to implicate particular side-chains in the mechanisms of certain enzymes, e.g. histidine side-chains in pancreatic ribonuclease[48] and chymotrypsin[49] and carboxylic-acid side-chains in lysozyme.[50] As described in Chapter 5 (Section 5.4), these conclusions can be supported by the results of other experiments.

4.3.4 The effect of changes in temperature on enzyme-catalysed reactions

In general, the effects of changes in temperature on the rates of enzyme-catalysed reactions do not provide much useful information as far as the mechanism of catalysis is concerned.* However, these effects can be important in indicating

* A notable exception is the study of the chymotrypsin-catalysed hydrolysis of *N*-acetyl-L-tryptophanamide, in which, from the temperature dependence of the reaction rate and other data, an energy profile (see Chapter 5, Fig. 5.1) for the reaction could be constructed.[51]

structural changes in enzymes and in considering enzyme activity in poikilothermic organisms, in which intracellular temperatures vary considerably, so a short discussion of the effects will be given in this section. More detailed discussions can be found in references 52 and 53. The effects of temperature on the stability of enzymes have been mentioned in Chapter 3, Sections 3.6.3 and 3.7.1.

4.3.4.1 Theoretical background

From the transition-state theory of chemical reactions, we can derive an expression for the variation of the rate constant, k, with temperature. This is of the form of eqn (4.12), which is sometimes referred to as the Arrhenius expression (see also Chapter 5, Section 5.3).

$$k = A\mathrm{e}^{-E_a/RT} \tag{4.12}$$

where A is known as the pre-exponential factor,
R is the gas constant,
T is the (absolute) temperature, and
E_a is the activation energy for the reaction.

From eqn (4.12) it is clear that there is an exponential increase in reaction rate with temperature. We can introduce a quantity, known as the Q_{10}, which is the ratio (or quotient) of the reaction rate at $(T+10)$K compared with that at TK. Using eqn (4.12) it can be shown that at temperatures around 300 K (27°C), Q_{10} is approximately given by $\mathrm{e}^{E_a/75000}$. For many chemical reactions, the values of Q_{10} are in the range 2–4, corresponding to activation energies of about 50–100 kJ mol^{-1}.

Similar considerations apply in the case of enzyme-catalysed reactions, but the values of E_a (and hence of Q_{10}) are generally lower than the corresponding values for non-enzyme-catalysed reactions when a comparison can be made. For instance, in the hydrolysis of urea catalysed by acid, the value of E_a is 100 kJ mol^{-1}, whereas the same reaction catalysed by urease has a much lower E_a (42 kJ mol^{-1}).

Above a certain temperature an enzyme will tend to lose the compact three-dimensional structure that is required for catalytic activity. Incubation of many enzymes at temperatures above about 323 K (50°C) leads to a fairly rapid loss of catalytic activity.

4.3.4.2 Treatment of data

According to the Arrhenius expression (eqn (4.10)), a plot of ln(velocity) against $1/T$ gives a straight line of slope $-E_a/R$. Taking into account the loss of catalytic activity at high temperatures, we would expect the Arrhenius plot for an enzyme-catalysed reaction to resemble that shown in Fig. 4.12(a).*

* Cornish-Bowden[54] has pointed out that the analysis of Arrhenius plots is more satisfactory if the variations of K_m and V_{max} are analysed separately over a range of temperatures. At any temperature, the observed velocity will depend on both parameters.

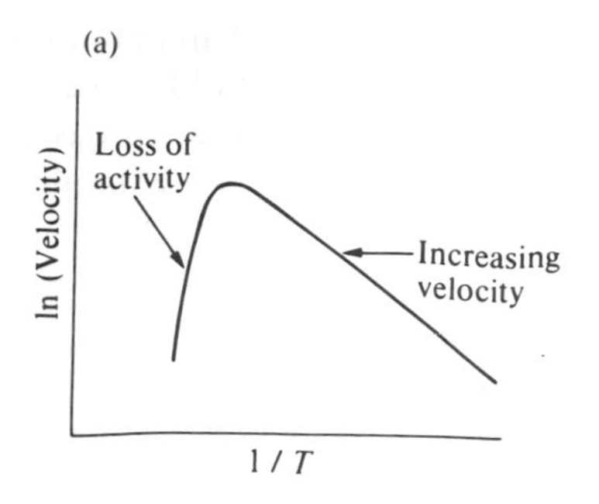

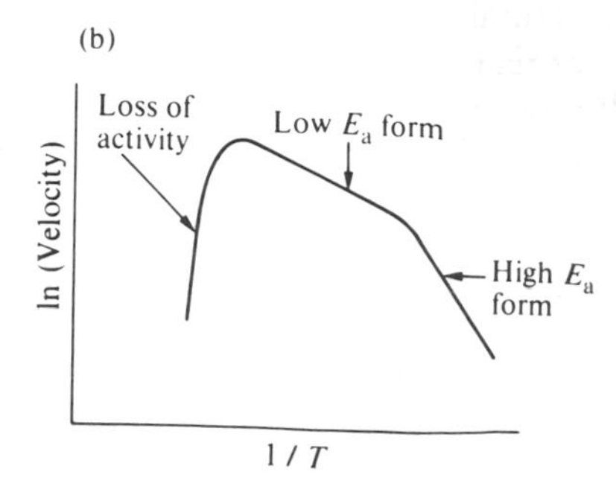

FIG. 4.12. Arrhenius plots for enzyme-catalysed reactions illustrating the loss of catalytic activity at high temperature. (a) Only one form of an enzyme involved; (b) two interconvertible forms with different activation energies involved.

Complications arise when an enzyme can arise in two (or more) interconvertible forms with different activation energies. There will then be a discontinuity in the Arrhenius plot around the temperature where the change over between the two forms becomes significant (Fig. 4.12(b)). An example of this type of behaviour is provided by the enzyme adenosinetriphosphatase (Na^+, K^+-activated) for which the transition probably arises from structural changes in the tightly bound phospholipid molecules associated with the enzyme.[55] A detailed discussion of the causes of discontinuous Arrhenius plots has been given.[56]

4.3.4.3 *Significance of results*

As mentioned in Section 4.3.4.1 the values of Q_{10} (and hence of E_a) for enzyme-catalysed reactions are generally lower than for reactions not subject to enzyme catalysis. For most enzymes in homoiothermic species (e.g. mammals) the value of Q_{10} is approximately 2. However, in species that have to adapt to cold conditions it is important that the value of Q_{10} is lower than this so that essential metabolic reactions are not slowed down too much. It is found, for instance, that many enzymes from intertidal species such as anemones and winkles have values of Q_{10} of about 1 (i.e. the rates do not change significantly with temperature).[57] It would be of considerable interest to know how these enzymes have evolved to this state. In a number of species of terrestrial insects there are major alterations in metabolism to coordinate the production of cryoprotectants such as glycerol and sorbitol. In such cases there appear to be differential effects on the kinetics and regulatory properties of the glycolytic enzymes.[58]

Although, as shown in Fig. 4.12, enzymes generally lose activity at temperatures above about 323 K (50°C), it should be noted that the enzymes from thermophilic bacteria (e.g. *Bacillus stearothermophilus*, which grows at temperatures up to 363 K (90°C)) are stable at these high temperatures. The cause of the exceptional stability of these enzymes is currently under investigation; in the case of glyceraldehyde-3-phosphate dehydrogenase (which contains four subunits) it has been suggested that the stabilization arises from additional inter-subunit ionic bonds and hydrophobic interactions.[59]

4.3.5 Two-substrate reactions

Up to now our discussion has been concerned with enzyme-catalysed reactions involving a single substrate and this has allowed us to introduce terms such as K_m, V_{max}, competitive inhibition, etc. This treatment would appear to be of limited value, since oxidoreductases, transferases, and ligases (Chapter 1, Section 1.5.1) catalyse reactions which by definition involve more than one substrate. Nevertheless, many of the concepts involved in one substrate kinetics can be useful in the analysis of the kinetics of these more complex reactions. In this section we shall explain the broad divisions of types of mechanism of two-substrate reactions and indicate how the necessary equations are derived. We shall then show how to use these equations in the analysis of experimental data, and conclude by describing how it is possible to distinguish between the various types of possible mechanism. More detailed treatments of these topics can be found in a number of textbooks.[60-64]

4.3.5.1 Theoretical background

We can divide two-substrate reactions into two main categories.

(i) Those involving a ternary complex (i.e. a complex containing enzyme and both substrates)

In these cases the reaction

$$E + A + B \rightarrow E + P + Q$$

proceeds via ternary complexes of the type EAB and EPQ:

$$E + A + B \rightarrow EAB \rightarrow EPQ \rightarrow E + P + Q.$$

This category can be further subdivided:

(i)(a) Those reactions in which the ternary complex is formed in an *ordered* manner, i.e. the second substrate (say, B) can bind to the enzyme only after A has already bound:

$$E + A \rightarrow EA$$

$$EA + B \rightarrow EAB \qquad \text{(but } E + B \not\rightarrow EB\text{)}.$$

(i)(b) Those reactions in which the ternary complex is formed in a *random* manner (i.e. either substrate can bind first):

$$E + A \rightarrow EA \qquad E + B \rightarrow EB$$

or

$$EA + B \rightarrow EAB \qquad EB + A \rightarrow EAB.$$

(ii) Those not involving a ternary complex

The most important class of reactions in this category proceed by *enzyme substitution* or *ping-pong* mechanisms, i.e. a modified form of the enzyme (E′) is formed together with the first product, before the second substrate is bound:

$$\mathrm{E}+\mathrm{A} \rightarrow \mathrm{E}'+\mathrm{P}$$

$$\mathrm{E}'+\mathrm{B} \rightarrow \mathrm{E}+\mathrm{Q}.$$

A second class of enzymes in this category operates via a Theorell–Chance mechanism (named after the original investigators) in which a ternary complex is presumably formed but its breakdown to yield the first product is very fast so that the ternary complex is kinetically insignificant. This type of mechanism has been shown to apply in the oxidation of ethanol and other primary alcohols by NAD^+ catalysed by alcohol dehydrogenase from horse liver.[65, 66]

An alternative way of representing enzyme-catalysed reactions has been proposed by Cleland.[67] The progress of the reaction is shown by a horizontal line (branched if necessary) with enzyme forms depicted below the line. Successive additions of substrates and release of products are depicted by vertical arrows. The rate constants can be indicated, if necessary adjacent to these arrows.

Thus, the random ternary complex mechanism (type (i)(b) above) would be

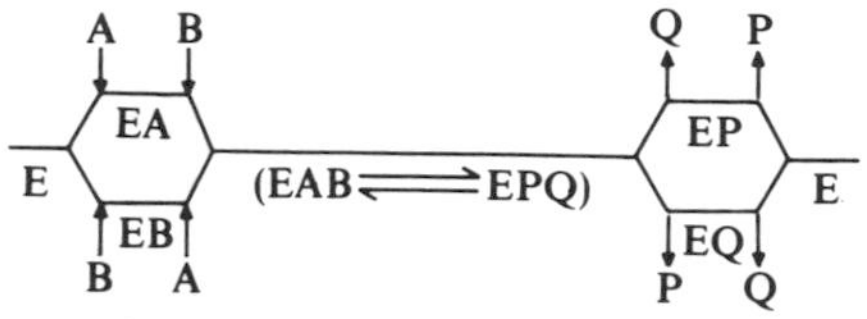

A B Q P
EA EP
E (EAB⇌EPQ) E
EB EQ
B A P Q

and the enzyme-substitution mechanism (type (ii) above) would be

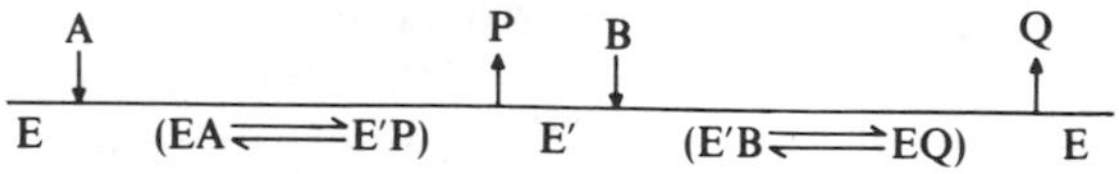

A P B Q
E (EA⇌E′P) E′ (E′B⇌EQ) E

4.3.5.2 *Derivation of equations for two-substrate reactions*

The derivation of equations to describe the kinetics of the various types of mechanism does not involve any fundamental principles in addition to those used for one-substrate reactions (Section 4.3.1.1), although the algebra is obviously more complex. (A good account of the derivations is given by Engel; in this book there is also an explanation of the King–Altman procedure, which simplifies the derivations considerably.[68, 69] Computer-assisted methods for deriving the rate equations have been described.[70]) The basic idea is to evaluate the concentrations of the various enzyme-containing complexes in terms of the total concentration of enzyme, under the stated conditions of substrate concentrations. In order to do this we use the steady-state assumption* in the same way as for one substrate reactions (Section 4.3.1.1). The velocity of the overall reaction is then equal to the concentration of the complex that precedes regeneration of free enzyme multiplied by the rate constant for the regeneration step, e.g. in the reaction scheme

$$E \rightleftharpoons EA \rightleftharpoons EAB \rightleftharpoons EPQ \rightarrow EP \xrightarrow{k_i} E+P$$

the rate of the overall reaction, v, equals k_i[EP]. This is because we need to regenerate free enzyme (by the k_i step above) in order to allow the reaction to continue.

The equations for v, the initial rate of the reaction, which result from these treatments are of the following forms.†

For the ternary complex mechanisms ((i)(a) and (i)(b) in Section 4.3.5.1):

$$v=\frac{V_{max}[A][B]}{K'_A K_B+K_B[A]+K_A[B]+[A][B]} \tag{4.13}$$

(an equation of this type is also derived for the Theorell–Chance mechanism).

For the enzyme substitution mechanisms ((ii) in Section 4.3.5.1):

$$v=\frac{V_{max}[A][B]}{K_B[A]+K_A[B]+[A][B]}. \tag{4.14}$$

4.3.5.3 *Significance of the parameters in the equations*

V_{max} in eqns (4.13) and (4.14) represents the maximum velocity at saturating levels of substrates A and B.

In a purely *practical* sense, the constants K_A and K_B in the eqns (4.13) and (4.14) represent the Michaelis constants (for substrates A and B, respectively) in the presence of saturating concentrations of the other substrate. This is readily shown; e.g. in eqn (4.13) if we divide the numerator and denominator by [B] we obtain

$$v=\frac{V_{max}[A]}{\frac{K'_A K_B}{[B]}+\frac{K_B[A]}{[B]}+K_A+[A]},$$

* If we apply the steady-state assumption to the case of the random order ternary complex mechanism (i)(b) in Section 4.3.5.1, the equation obtained is complex, with terms containing the square of the concentrations of substrate. In this case we can apply the equilibrium assumption (Section 4.3.1.1) and assume that E, EA, EB, and EAB are all in equilibrium with one another. The resulting equation (4.13) describes many reactions of this type and to that extent the equilibrium assumption is justified.

† There are a number of different formulations of these equations. Details of some of these are given in Appendix 4.3.

and when $[\mathrm{B}] \to \infty, \dfrac{1}{[\mathrm{B}]} \to 0$, so that

$$v = \frac{V_{max}[\mathrm{A}]}{K_\mathrm{A} + [\mathrm{A}]},$$

which is of the same form as eqn (4.6).

K'_A in eqn (4.13) does not have a simple practical meaning.

In terms of the *mechanisms* of the reactions, the constants K'_A, K_A, and K_B represent combinations of the rate constants of individual steps in the reaction (compare eqn (4.5) for one-substrate reactions). Their precise meanings vary according to the type of mechanism under discussion. In the case of the random-order ternary complex mechanism ((i)(b) in Section 4.3.5.1), K'_A, K_A, and K_B have simple meanings in terms of dissociation constants, because in this case we used the equilibrium assumption to derive eqn (4.13).

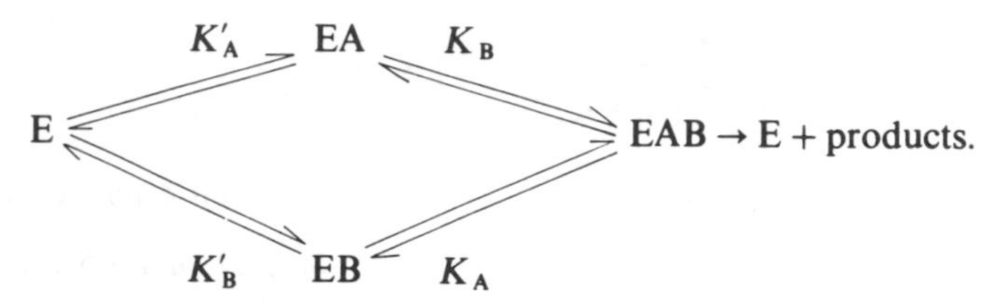

K'_A, K_A, and K_B represent the dissociation constants for EA, EAB to yield EB, and EAB to yield EA respectively. (Note that $K'_\mathrm{B} = K'_\mathrm{A} K_\mathrm{B} / K_\mathrm{A}$.)

4.3.5.4 *Treatment of data*

The values of the parameters V_{max}, K'_A, K_A, and K_B in eqn (4.13) can be derived from the experimental data by a computer fitting procedure[22] analogous to that mentioned for one-substrate reactions (Section 4.3.1.2). However, the values are more commonly obtained by graphical methods involving primary and secondary plots.

The velocity, v, is measured at various values of [A], keeping the concentration of B constant; this procedure is then repeated at other fixed values of [B].

By taking the inverse of eqn (4.13), i.e.

$$\frac{1}{v} = \left[1 + \frac{K_\mathrm{A}}{[\mathrm{A}]} + \frac{K_\mathrm{B}}{[\mathrm{B}]} + \frac{K'_\mathrm{A} K_\mathrm{B}}{[\mathrm{A}][\mathrm{B}]}\right] \frac{1}{V_{max}}, \tag{4.15}$$

we see that a *primary* plot of $1/v$ against $1/[\mathrm{A}]$ at a fixed value of [B] will be linear (Fig. 4.13) with

$$slope = \frac{1}{V_{max}} \left[K_\mathrm{A} + \frac{K'_\mathrm{A} K_\mathrm{B}}{[\mathrm{B}]}\right] \tag{4.16}$$

and

$$intercept\ (on\ y\text{-}axis) = \frac{1}{V_{max}} \left[1 + \frac{K_\mathrm{B}}{[\mathrm{B}]}\right]. \tag{4.17}$$

As [B] increases, both the slope and intercept will decrease.

(The lines in the primary plot (Fig. 4.13) intersect at a point that can be above, on, or below the *x*-axis, depending on the relative values of K'_A, K_A, and K_B. At any value of [B] an 'apparent' K_m for A can be derived from the intercept on the *x*-axis. From eqn (4.15) it can be shown that this K_m is $([B]K_A + K'_A K_B)/([B] + K_B)$; K_m will vary with [B], unless $K'_A = K_A$ when the point of intersection is on the *x*-axis.)

Secondary plots of the slopes and intercepts of the primary plot against 1/[B] can then be constructed (Fig. 4.14(a) and (b)).

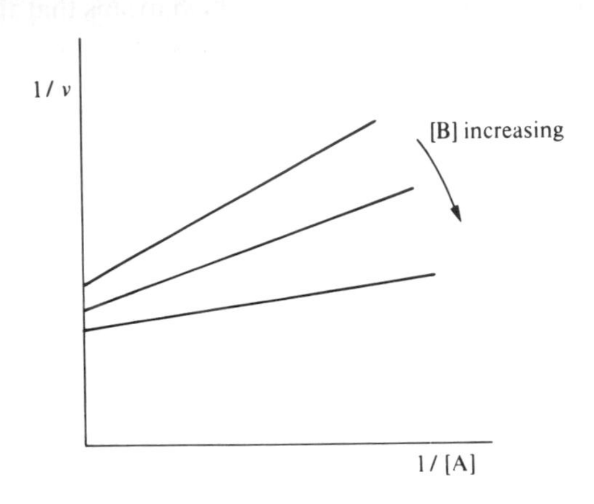

FIG. 4.13. Primary plot of enzyme kinetic data according to eqn (4.15).

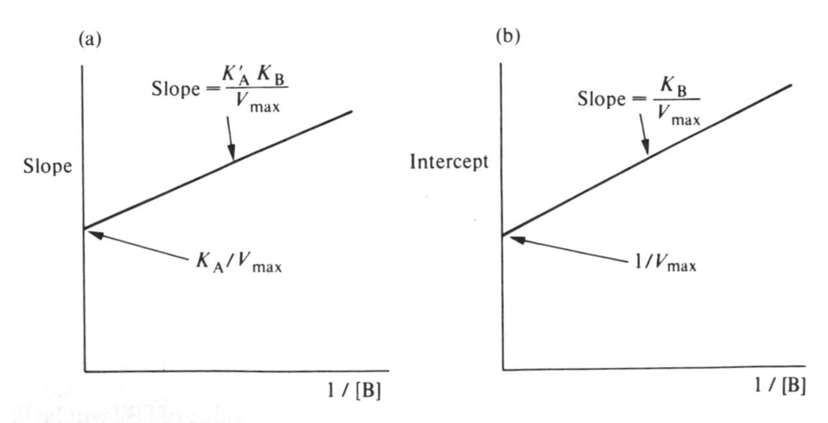

FIG. 4.14. Secondary plots of (a) slopes and (b) intercepts of primary plot (Fig. 4.13) against 1/[B].

By inspection of eqn (4.16) we see that a plot of the slope against 1/[B] (Fig. 4.14(a)) is linear with a *slope* of $K'_A K_B/V_{max}$ and an *intercept on the y-axis* of K_A/V_{max}.

From eqn (4.17) we see that a plot of the intercepts against 1/[B] (Fig. 4.14(b)) is linear with a *slope* of K_B/V_{max} and an *intercept on the y-axis* of $1/V_{max}$. Thus it is possible to determine in turn V_{max}, K_B, K_A, and K'_A from the slopes and intercepts of these secondary plots.

As mentioned in Section 4.3.5.3, the parameters K'_A, K_A, and K_B have a simple meaning in the random-order ternary complex mechanism ((i)(b) in Section 4.3.5.1). If it is known from other studies (described in Section 4.3.5.5) that this mechanism applies to a particular enzyme then we can draw some conclusions from the relative values of these parameters. Thus, for example, in the case of creatine kinase from rabbit muscle assayed in the direction of phosphocreatine synthesis, it is found that $K'_{MgATP} > K_{MgATP}$, which means that the binding of one substrate is enhanced by the binding of the other. (This is known as *substrate synergism.*) By contrast, in the oxidation of ethanol by NAD^+ catalysed by alcohol dehydrogenase from yeast, $K'_{NAD^+} = K_{NAD^+}$ and hence the binding of one substrate is independent of the binding of the other.

Turning now to eqn (4.14) for the enzyme substitution mechanism ((ii) in Section 4.3.5.1) we find that a primary plot of $1/v$ against $1/[A]$ at fixed values of [B] consists of a set of parallel lines (Fig. 4.15). This can be shown by taking the inverse of eqn (4.14):

$$\frac{1}{v} = \left[1 + \frac{K_A}{[A]} + \frac{K_B}{[B]}\right] \frac{1}{V_{max}}. \tag{4.18}$$

Hence, a plot of $1/v$ against $1/[A]$ has a *slope* of K_A/V_{max}; this slope is independent of $[B]$, resulting in a set of parallel lines. We can derive the parameters K_B and V_{max} (and hence K_A) by performing a *secondary* plot of the *y-axis intercepts* of the primary plot against $1/[B]$. This secondary plot will have a *slope* of K_B/V_{max} and a *y-axis intercept* of $1/V_{max}$.

4.3.5.5 *Significance of results: distinction between the various mechanisms for two-substrate reactions (Section 4.3.5.1)*

From the discussion in Section 4.3.5.4 it can be seen how an enzyme substitution mechanism can be distinguished from a ternary complex mechanism, since the former gives rise to a series of parallel lines in the primary plot (Fig. 4.15). It

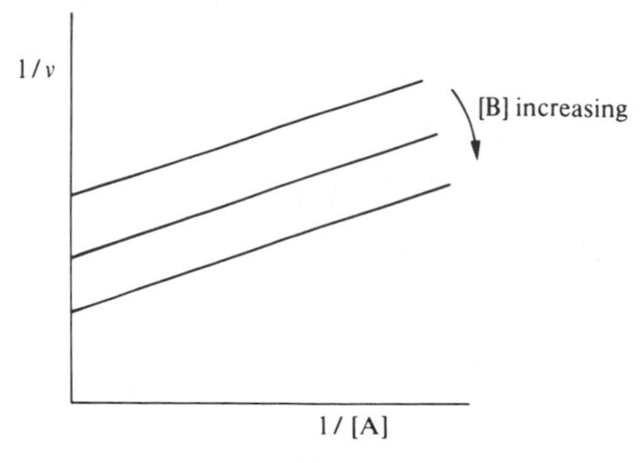

FIG. 4.15. Primary plot of enzyme kinetic data according to eqn (4.18) (the enzyme substitution mechanism).

should be emphasized that considerable care is needed to ensure that lines in a plot such as Fig. 4.15 are truly parallel, and it is probably best to make an estimate of the magnitude of the K'_A term in the general equation (4.13) and of the error of this estimate (K'_A should equal zero for the enzyme substitution mechanism). If the primary plot suggests that an enzyme substitution mechanism is operative in a particular case, confirmatory evidence can be obtained from:

(1) demonstration of partial reactions, i.e. conversion of A to P in the absence of B:

$$E + A \rightleftharpoons E' + P$$

(2) isolation of a modified form of the enzyme (E′ in the above scheme), often using a suitable radioactive label.

Thus, in a reaction catalysed by nucleoside diphosphate kinase from erythrocytes, an enzyme substitution mechanism is indicated by the kinetic data:

$$GTP + dGDP \rightleftharpoons GDP + dGTP.$$

In this case it is possible to isolate a modified radioactive enzyme, in which a histidine side-chain is phosphorylated,[71] after incubation of the enzyme with [γ-^{32}P]GTP (i.e. GTP labelled at the γ position of the triphosphate group).

$$E + GTP \rightleftharpoons E-Ⓟ + GDP,$$

which would be followed by

$$E-Ⓟ + dGDP \rightleftharpoons E + dGTP.$$

(Ⓟ represents a phosphoryl group.)

Other enzymes that operate via an enzyme substitution mechanism include the amino-transferases, in which the pyridoxal phosphate cofactor (see Chapter 1, Section 1.4) attached to the enzyme undergoes a covalent change to form E′, and phosphoglycerate mutase from yeast or mammalian sources*. The reaction catalysed by the latter enzyme would appear to be a one-substrate reaction in which-2-phospho-D-glycerate (2PG) is converted into 3-phospho-D-glycerate (3PG). However, in this case a primer molecule of 2,3-bisphospho-D-glycerate (BPG) is required. We thus have two partial reactions:

$$E + BPG \rightleftharpoons E-Ⓟ + 2PG,$$

$$E-Ⓟ + 3PG \rightleftharpoons E + BPG,$$

giving a net reaction

$$3PG \rightleftharpoons 2PG.$$

From the steady-state kinetic analysis given in Section 4.3.5.4 we are unable to distinguish between an ordered and a random ternary complex mechanism

* There are in fact two types of phosphoglycerate mutases. The enzyme from plant sources is independent of the 2,3-bisphospho-D-glycerate primer, whereas the enzyme from yeast or mammalian sources requires this primer. However, both types are now classified as isomerases (EC 5.4.2.1).

because both give rise to an expression of the form of eqn (4.13).* However, a distinction between the mechanisms can be made if other data are available, e.g. from the following types of experiment.

(i) Substrate-binding experiments
In the random-order ternary complex mechanism, each substrate, A and B, should bind separately to the enzyme, whereas in an ordered mechanism the second substrate (e.g. B) cannot bind in the absence of the first substrate, A. For instance, in the reaction catalysed by lactate dehydrogenase,

$$\text{L-Lactate} + \text{NAD}^+ \rightleftharpoons \text{pyruvate} + \text{NADH} + \text{H}^+,$$

it is found that NAD^+ binds tightly to the enzyme but there is no detectable binding of lactate.[73] This suggests that the reaction proceeds via an ordered mechanism with NAD^+ binding preceding that of lactate, a conclusion that is supported by other evidence (see Chapter 5, Section 5.5.4).

(ii) Product-inhibition patterns
The type of inhibition shown by the products P and Q towards the substrates A and B can be used to indicate a probable mechanism for the reaction (or exclude an alternative mechanism). The equations underlying these experiments are complex and full details can be found in more comprehensive textbooks.[74, 75] An example of the method is to distinguish between the ordered and random ternary complex mechanisms by studying the inhibition caused by product Q. In the random mechanism, Q behaves as a competitive inhibitor with respect to substrate B, whereas in the ordered mechanism, Q behaves as a mixed-type inhibitor with respect to B (i.e. the Lineweaver–Burk plots in the presence and absence of inhibitor intersect at a point to the left of the y-axis).

(iii) Isotope exchange at equilibrium[76]
In this type of experiment we study the enzyme-catalysed rate of exchange of isotope between substrate(s) and product(s) when the components of the reaction mixture are present at their equilibrium concentrations. For instance, in the reaction catalysed by malate dehydrogenase,

$$\text{L-Malate} + \text{NAD}^+ \rightleftharpoons \text{oxaloacetate} + \text{NADH} + \text{H}^+,$$

we set up a mixture containing substrates and products so that, at the pH in question, the ratio [oxaloacetate] [NADH]/[malate] [NAD^+] equals the previously measured equilibrium constant for the reaction. One of the substrates, say NAD^+, is isotopically labelled (e.g. with ^{14}C) and after addition of a catalytic amount of enzyme we study the rate at which this isotope is incorporated into NADH. (Note that there is no *net* conversion of NAD^+ to NADH, because the

* Although the Theorell–Chance mechanism (Section 4.3.5.1) also gives an expression of the form of eqn. (4.13), it can be distinguished from the ternary complex mechanisms by comparison of the magnitudes of the various parameters in the rate equation with the rates of the forward and reverse reactions (see reference 72).

system is at equilibrium; there will, however, be forward and backward reactions occurring so that *exchange* between substrate and product can occur.) It was found that as the concentrations of malate and oxaloacetate were raised (in a constant ratio so that the equilibrium state was maintained), the rate of isotope exchange between NAD^+ and NADH fell to zero, whereas the rate of exchange between malate and oxaloacetate increased to a plateau value.[77] This result indicates that the enzyme obeys an ordered mechanism of the type

$$E + NAD^+ \rightleftharpoons E^{NAD^+} \rightleftharpoons E^{NAD^+}_{malate} \rightleftharpoons E^{NADH}_{oxaloacetate} \rightleftharpoons E^{NADH} \rightleftharpoons E + NADH.$$

As we raise the concentrations of malate and oxaloacetate, we increase the rate at which the binary complexes (E^{NAD^+} and E^{NADH}) are converted to the ternary complexes. The enzyme thus becomes confined to the central 'box' shown in the above scheme and NAD^+ (or NADH) never dissociates to join the pool of free NAD^+ (or NADH). Since the enzyme is present only in very small amounts compared with the concentrations of substrates, this means that we will observe no exchange of isotope between free NAD^+ and free NADH. The results are not consistent with a random ternary complex mechanism.

As a result of experiments of the type described in (i), (ii), and (iii) it has been possible to assign a number of enzyme-catalysed reactions to the types of mechanism discussed in Section 4.3.5.1. The reactions catalysed by alcohol dehydrogenase (from yeast), creatine kinase, and phosphorylase appear to proceed via random ternary complex mechanisms, whereas the reactions catalysed by lactate dehydrogenase and malate dehydrogenase operate via ordered ternary complex mechanisms. A compilation of mechanisms has been made by Fromm.[78]

4.3.6 Reactions involving more than two substrates

A number of enzyme-catalysed reactions involve three substrates; some examples are shown below.

Glyceraldehyde 3-phosphate dehydrogenase

$$\text{D-Glyceraldehyde-3-phosphate} + NAD^+ + \text{orthophosphate} \rightleftharpoons \text{3-phospho-D-glyceroyl phosphate} + NADH$$

Glutamate dehydrogenase

$$\text{2-Oxoglutarate} + NH_3 + NAD(P)H \rightleftharpoons \text{L-Glutamate} + NAD(P)^+ + H_2O$$

Isocitrate dehydrogenase

$$\text{2-Oxoglutarate} + CO_2 + NADH \rightleftharpoons \textit{threo}\text{-D}_S\text{-isocitrate} + NAD^+$$

Tyrosyl-tRNA synthetase

$$\text{ATP} + \text{L-tyrosine} + \text{tRNA}^{\text{Tyr}} \rightleftharpoons \text{AMP} + \text{pyrophosphate} + \text{L-tyrosyl-tRNA}^{\text{Tyr}}$$

In these reactions there are a number of possible mechanisms, e.g. enzyme substitution or quaternary complex formation by ordered, random, or partly ordered mechanisms. The principles involved in the derivation of equations and the treatment of data are similar to those used in the analysis of two-substrate reactions (Section 4.3.5), although the algebra is considerably more complex[79] and the requirement for accurate data is more stringent. An example of the analysis of a three-substrate reaction is provided by the studies on glutamate dehydrogenase, in which it was shown that the reaction is most likely of the random type with 2-oxoglutarate, NH_4^+, and NADH binding in any order to form the quaternary complex.[80]

4.4 Pre-steady-state kinetics[81–83]

4.4.1 Background: the need for special techniques

From kinetic studies performed under steady-state conditions we can obtain only very limited information about the rates of individual steps in an enzyme-catalysed reaction. The maximum velocity (V_{max}) can be used to calculate the value of k_{cat}, or turnover number (see Section 4.3.1.1), according to the equation:

$$k_{cat} = \frac{V_{max}}{[\text{E}]_0}$$

where $[\text{E}]_0$ is the concentration of enzyme active sites.

Values of k_{cat} are known for many enzymes and range from about 10 to about $10^7\ \text{s}^{-1}$.[84] The value of k_{cat} gives an idea of the rate constant of the slowest step in an enzyme-catalysed reaction. In steady-state experiments we generally work with very low concentrations of enzyme (1 nmol dm^{-3} or less is not unusual) and under these conditions the reaction is slow enough to allow the use of 'conventional' methods (e.g. manual addition of enzyme to start the reaction and observation by a spectrophotometer). For instance if $[\text{E}]_0 = 1$ nmol dm^{-3} and $k_{cat} = 10^3\ \text{s}^{-1}$, V_{max} would equal 60 μmol dm^{-3} min^{-1}, which would be quite convenient to monitor spectrophotometrically.

However, if we wish to examine processes that occur on a time scale of less than about a few seconds (and these will include the steps other than the slowest step of the overall reaction), we need to employ faster methods of mixing and observation. The problem has been neatly solved by the introduction of 'stopped-flow' methods, which can allow observation of reactions that occur on a time scale of a few milliseconds. The basic elements of a 'stopped-flow' apparatus are shown in Fig. 4.16.

The drive barrier is pushed in, usually mechanically, and the contents of the two reactant syringes (e.g. enzyme and substrate(s)) are mixed. The flow of liquid

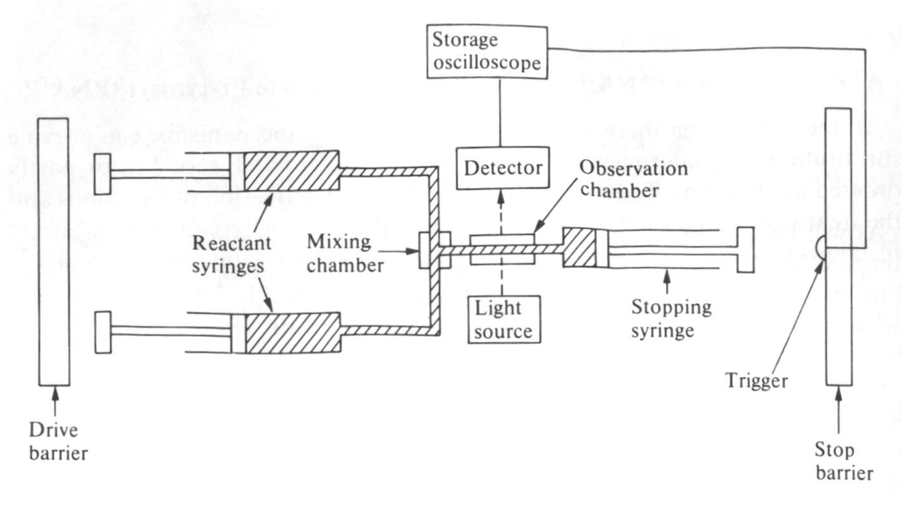

FIG. 4.16. A typical stopped-flow apparatus.

forces the piston of the stopping syringe out until it hits the stop barrier, when the flow stops. At this point the oscilloscope is triggered and the changes in absorbance or fluorescence of the liquid in the observation chamber can be recorded. The trace on the storage oscilloscope is usually photographed; the photograph can then be analysed at leisure. For this type of apparatus there is a 'dead' time, corresponding to the time interval between the start of the reaction by mixing and the stop of the flow of liquid, of about 1 ms.

4.4.2 Applications of the stopped-flow technique

The stopped-flow technique has been valuable in elucidating the details of enzyme-catalysed reactions. We shall describe two of the most important types of study; the determination of rate constants of individual steps in the reaction and the identification of transient species.

4.4.2.1 Determination of rate constants

An example of the determination of the rate constant of an individual step in a reaction is provided by the binding of NADH to lactate dehydrogenase (isoenzyme LDH-5) which was monitored by the increase in the fluorescence of NADH that occurs on binding.[85] Equal concentrations of enzyme active sites and substrate (8 μmol dm^{-3} of each) were mixed and the fluorescence increase was monitored over a period of about 16 ms.* By analysing the data in terms of equations developed for reversible reactions, the rate constants could be deduced

* For a second-order reaction in which the concentrations of reactants are equal, the half time is given by $1/k[A]$, where [A] is the initial concentration of the reactants and k is the rate constant. In this case (ignoring the reverse reaction) we can see that if $k = 6.3 \times 10^7$ (mol dm^{-3})$^{-1}$s^{-1} and $[A] = 8\mu$mol dm^{-3}, then $t_{1/2} \approx 2$ ms, which is within the time scale of the stopped-flow apparatus.

as:

$$\mathrm{E + NADH} \underset{k_{-1}}{\overset{k_1}{\rightleftharpoons}} \mathrm{E.NADH}$$

$$k_1 = 6.3 \times 10^7\ (\mathrm{mol\,dm^{-3}})^{-1}\,\mathrm{s^{-1}}$$

$$k_{-1} = 450\ \mathrm{s^{-1}}$$

As will be seen in Chapter 5 (Section 5.5.4.2), the determination of the rate constant for the dissociation step (k_{-1}) is of value in deciding the nature of the slow step of the overall reaction. Further work, using stopped-flow measurements of the fluorescence changes in a mutant form of lactate dehydrogenase, showed that the slow step of the overall reaction corresponds to the movement of a flexible loop region in the enzyme. This movement casues the loop to close over the active site, with consequent changes in the fluorescence of the tryptophan side-chain in the loop (see Chapter 5, Section 5.5.4.5).

4.4.2.2 Identification of transient species

In the stopped-flow technique, the concentration of enzyme is high enough to permit the detection and identification of enzyme-containing complexes in the reaction pathway, and to determine the rate constants of the steps that involve formation and decay of these species. Using absorption spectrophotometry, concentrations of the order of 10 μmol dm^{-3} can be readily detected; using fluorescence the limit is probably tenfold lower. An early experiment by Chance showed that the oxidation of malachite green by H_2O_2 catalysed by peroxidase involved an enzyme–H_2O_2 complex; this experiment provided the first direct evidence for the participation of an enzyme-substrate complex in an enzyme-catalysed reaction.[86] In Chapter 5 (Section 5.5.4.2) we shall see that the application of the stopped-flow technique to the reaction catalysed by lactate dehydrogenase (i.e. L-lactate + $NADH^+ \rightleftharpoons$ pyruvate + NADH + H^+) has demonstrated the transient production of enzyme-bound NADH. A detailed analysis of this reaction has allowed the rate constants of all the elementary steps in the reaction to be determined.[87, 88]

An extension of the stopped-flow technique has been described; this involves the use of a rapid scanning detecting device that can scan a spectrum over a 200 nm range in 1 ms.[89] By this means it is possible to record rapid changes in the absorption spectrum and hence identify transient species with more certainty. The technique has been used to show, for instance, that the NADH produced in the rapid phase of the reaction catalysed by alcohol dehydrogenase from horse liver (i.e. ethanol + $NAD^+ \rightleftharpoons$ acetaldehyde + NADH + H^+) is bound to the enzyme, because its absorption maximum (320 nm) is different from that of free NADH (340 nm).

4.4.3 Relaxation methods

Stopped-flow methods (Section 4.4.1) are extremely useful for the study of processes that occur over a period of several milliseconds. The lower limit is set by the time required to achieve uniform mixing of liquids and this appears to be of the order of 1 ms, or slightly less. We have seen in Section 4.4.2.1 that it is possible to use stopped-flow methods to study the reaction between NADH and lactate dehydrogenase provided that relatively low concentrations of substrate and enzyme are employed. However, if it is not possible to adopt this 'low concentration' procedure to bring the rate of a rapid reaction into a convenient time scale,* we can use the so-called 'relaxation' methods, which do not involve the mixing of reactants. In these methods, a system at equilibrium is perturbed by a sudden change in temperature or other parameter (pressure, pH, etc.) and then allowed to 'relax' to the new position of equilibrium. (By discharging a condenser between electrodes in the solution the temperature can be raised by 5–10 K in 1 μs. The position of equilibrium will change provided that the enthalpy change (ΔH°) for the reaction is not equal to zero). The rate of the 'relaxation' is monitored by appropriate spectroscopic (or other) techniques, and can be related theoretically to the rate constants for the forward and reverse reactions occurring at equilibrium. From measurements at a variety of concentrations of reactants, the rate constants can be evaluated. More details of these procedures are given in review articles.[81, 83, 90]

The temperature-jump method has been used, for instance, to show that the rate constant for the association of NADH with malate dehydrogenase ($k = 5 \times 10^8$ (mol dm^{-3})$^{-1}$ s^{-1}) is close to the limit calculated for diffusion control of association of enzyme with substrate ($k \approx 10^9$ (mol dm^{-3})$^{-1}$ s^{-1}), i.e. that almost every collision or encounter between enzyme and substrate leads to formation of a complex.

The method was of value in a study of the reaction catalysed by alkaline phosphatase from *E. coli*; this enzyme catalyses the hydrolysis of a variety of phosphate esters, R—O—Ⓟ, via formation and breakdown of a phosphoryl-enzyme intermediate. Because there is little dependence of the rate of reaction on the nature of the R group, it had been postulated that the hydrolysis of the phosphoryl enzyme was the common rate-limiting step for hydrolysis of the different phosphate esters. However, the rate of this step (k_4 in the scheme below) was measured by stopped-flow techniques and found to be faster than the rate of the overall reaction.[91] The nature of the slow step was deduced from experiments in which the rate of binding of a non-hydrolysable substrate analogue, 4-nitrobenzylphosphonate,

$$NO_2-C_6H_4-CH_2-PO_3^{2-}$$

to the enzyme was studied by the temperature-jump method.[92] The results indicated that a structural change in the enzyme occurred after binding the inhibitor and that the rate of this change corresponded to that of the slow step of

* This could be, for instance, because of the small size of an absorbance change at low concentrations or because of the order of the process. The half time of a first-order process does not depend on the concentration of the reactant.

the overall reaction (k_2 in the scheme below):

$$\mathrm{E + R{-}O{-}}\text{Ⓟ} \underset{}{\overset{k_1}{\rightleftharpoons}} \mathrm{E.R{-}O{-}}\text{Ⓟ} \overset{k_2}{\rightleftharpoons} \mathrm{E^*.R{-}O{-}}\text{Ⓟ} \overset{k_3}{\rightarrow} \mathrm{E{-}}\text{Ⓟ} \;(+\,\mathrm{ROH}) \overset{k_4}{\rightarrow} \mathrm{E} + \text{orthophosphate}$$

slow step (k_2)

4.5 Concluding remarks

In this chapter we have introduced the principles and equations that are required to give an understanding of the important features of enzyme kinetics. The application of both steady-state and pre-steady-state measurements is of considerable value in helping to deduce the sequence of enzyme-containing complexes in a reaction. We shall see in Chapter 5 how these data can be used in conjunction with other information to formulate a detailed picture of the mechanism of an enzyme-catalysed reaction in both kinetic and structural terms. As will be shown in Chapters 6 and 8, kinetic information helps in gaining an understanding of the control of enzyme activity and of the role of an enzyme in the metabolic processes in the cell.

References

1. McClure, W. R., *Biochemistry* **8**, 2782 (1969).
2. Storer, A. C., and Cornish-Bowden, A., *Biochem. J.* **141**, 205 (1974).
3. Rudolph, F. B., Baugher, B. W., and Beissner, R. S., *Methods Enzymol.* **63**, 22 (1979).
4. Easterby, J. S., *Biochem. J.* **219**, 843 (1984).
5. Yang, S-Y., and Schulz, H., *Biochemistry* **26**, 5579 (1987).
6. Allison, R. D. and Purich, D. L., *Methods Enzymol.* **63**, 3 (1979).
7. Tipton, K. F., in *Techniques in the life sciences: protein and enzyme biochemistry.* Elsevier/North-Holland, Amsterdam (1978). Volume B112. See page 1.
8. Dalziel, K., *J. biol. Chem.* **238**, 1538 (1963).
9. Winer, A. D., *J. biol. Chem.* **239**, PC 3598 (1964).
10. Cantley, L. C. Jr., Josephson, L., Warner, R., Yanagisawa, M., Lechene, C., and Guidotti, G., *J. biol. Chem.* **252**, 7421 (1977).
11. Cornish-Bowden, A., *Fundamentals of enzyme kinetics.* Butterworths, London (1979). See Ch. 3.
12. Wharton, C. W., *Biochem. Soc. Trans.* **11**, 817 (1983).
13. Lineweaver, H., and Burk, D., *J. amer. chem. Soc.* **56**, 658 (1934).
14. Eadie, G. S., *J. biol. Chem.* **146**, 85 (1942).
15. Hofstee, B. H. J., *J. biol. Chem.* **199**, 357 (1952).
16. Hanes, C. S., *Biochem. J.* **26**, 1406 (1932).
17. Eisenthal, R., and Cornish-Bowden, A., *Biochem. J.* **139**, 715 (1974).
18. Cornish-Bowden, A., and Eisenthal, R., *Biochem. J.* **139**, 721 (1974).
19. Cornish-Bowden, A., *Fundamentals of enzyme kinetics.* Butterworths, London (1979). See Ch. 10.

20. Cornish-Bowden, A., *Fundamentals of enzyme kinetics*. Butterworths, London (1979). See pp. 26–8.
21. Cleland, W. W., *Adv. Enzymol.* **29**, 1 (1967).
22. Cleland, W. W., *Methods Enzymol.* **63**, 103 (1979).
23. Atkins, G. L., and Nimmo, I. A., *Analyt. Biochem.* **104**, 1 (1980).
24. Orsi, B. A., and Tipton, K. F., *Methods Enzymol.* **63**, 159 (1979).
25. Moreno, J., *Biochem. Educ.* **13**, 64 (1985).
26. Teipel, J. W., Hass, G. M., and Hill, R. L., *J. biol. Chem.* **243**, 5684 (1968).
27. Fersht, A. R., *Enzyme structure and mechanism* (2nd edn). Freeman, New York (1985). See p. 103.
28. Jensen, R. G., and Bahr, J. T., *A. Rev. Plant Physiol.* **28**, 379 (1977).
29. Fersht, A. R., *Enzyme structure and mechanism* (2nd edn). Freeman, New York (1985). See p. 102.
30. Dixon, M., Webb, E. C., Thorne, C. J. R., and Tipton, K. F., *Enzymes* (3rd edn). Longman, London (1979). See p. 126.
31. Cornish-Bowden, A., *Fundamentals of enzyme kinetics*. Butterworths, London (1979). See p. 33.
32. Laidler, K. J., and Bunting, P. S., *The chemical kinetics of enzyme action* (2nd edn). Clarendon Press, Oxford (1973). See p. 170.
33. Cornish-Bowden, A., *Fundamentals of enzyme kinetics*. Butterworths, London (1979). See p. 30.
34. Laidler, K. J., and Bunting, P. S., *The chemical kinetics of enzyme action* (2nd edn). Clarendon Press, Oxford (1973). See p. 81.
35. Ferdinand, W., *The enzyme molecule*. Wiley, London (1976). See p. 141.
36. *Symbolism and terminology in enzyme kinetics*. Recommendations (1981) of the Nomenclature Committee of the International Union of Biochemistry. Reprinted in *Eur. J. Biochem.* **128**, 281 (1982).
37. Engel, P. C., *Enzyme kinetics* (2nd edn). Chapman and Hall, London (1981). See p. 34.
38. Price, N. C., *Trends Biochem. Sci.* **4**, N272 (1979).
39. Ghosh, N. K., and Fishman, W. H., *J. biol. Chem.* **241**, 2516 (1966).
40. Greene, R. C., *Biochemistry* **8**, 2255 (1969).
41. Todhunter, J. A., *Methods Enzymol.* **63**, 383 (1979).
42. Cornish-Bowden, A., *Fundamentals of enzyme kinetics*. Butterworths, London (1979). See pp. 130–41.
43. Dixon, M., Webb, E. C., Thorne, C. J. R., and Tipton, K. F., *Enzymes* (3rd edn). Longman, London (1979). See pp. 138–64.
44. Laidler, K. J., and Bunting, P. S., *The chemical kinetics of enzyme action* (2nd edn). Clarendon Press, Oxford (1973). See Ch. 5.
45. Tipton, K. F., and Dixon, H. B. F., *Methods Enzymol.* **63**, 183 (1979).
46. Engel, P. C., *Enzyme kinetics* (2nd edn). Chapman and Hall, London (1981). See p. 41.
47. Fersht, A. R., *Enzyme structure and mechanism* (2nd edn). Freeman, New York (1985). See p. 423.
48. Findlay, D., Mathias, A. P., and Rabin, B. R., *Biochem. J.* **85**, 139 (1962).
49. Hammond, B. R., and Gutfruend, H., *Biochem. J.* **61**, 187 (1955).
50. Parsons, S. M., and Raftery, M. A., *Biochemistry* **11**, 1623 (1972).
51. Bender, M. L., Kézdy, F. J., and Gunter, C. R., *J. amer. Chem. Soc.* **86**, 3714 (1964).
52. Dixon, M., Webb, E. C., Thorne, C. J. R., and Tipton, K. F., *Enzymes* (3rd edn). Longman, London (1979). See pp. 164–82.

53. Laidler, K. J., and Peterman, B. F., *Methods Enzymol.* **63**, 234 (1979).
54. Cornish-Bowden, A., *Fundamentals of enzyme kinetics.* Butterworth, London (1979). See Ch. 7.
55. Dahl, J. L., and Hokin, L. E., *A. Rev. Biochem.* **43**, 327 (1974). See p. 343.
56. Londesborough, J., *Eur. J. Biochem.* **105**, 211 (1980).
57. Hazel, J. R., and Prosser, C. L., *Physiol. Rev.* **54**, 620 (1974). See p. 621.
58. Storey, K. B., and Storey, J. M., *Trends Biochem. Sci.* **8**, 242 (1983).
59. Walker, J. E., Wonacott, A. J., and Harris, J. I., *Eur. J. Biochem.* **108**, 581 (1980).
60. Cornish-Bowden, A., *Fundamentals of enzyme kinetics.* Butterworths, London (1979). See Ch. 6.
61. Dixon, M., Webb, E. C., Thorne, C. J. R., and Tipton, K. F., *Enzymes* (3rd edn). Longman, London (1979). See pp. 82–119.
62. Laidler, K. J., and Bunting, P. S., *The chemical kinetics of enzyme action* (2nd edn). Clarendon Press, Oxford (1973). See Ch. 4.
63. Ferdinand, W., *The enzyme molecule.* Wiley, London (1976). See pp. 165–85.
64. Engel, P. C., *Enzyme kinetics* (2nd edn). Chapman and Hall, London (1981). See Ch. 5.
65. Theorell, H., and Chance, B., *Acta chem. scand.* **5**, 1127 (1951).
66. Dalziel, K., and Dickinson, F. M., *Biochem. J.* **100**, 34 (1966).
67: Cleland, W. W., *Enzymes* (3rd edn). **2**, 3 (1970). See p. 5.
68. Engel, P. C., *Enzyme kinetics* (2nd edn). Chapman and Hall, London (1981). See Ch. 6.
69. Huang, C. Y., *Methods Enzymol.* **63**, 54 (1979).
70. Fromm, H. J., *Methods Enzymol.* **63**, 84 (1979).
71. Parks, R. E. Jr. and Agarwal, R. P., *Enzymes* (3rd edn) **8**, 307 (1973). See p. 315.
72. Engel, P. C., *Enzyme kinetics* (2nd edn). Chapman and Hall, London (1981). See p. 59.
73. Holbrook, J. J., Liljas, A., Steindel, S. J., and Rossmann, M. G., *Enzymes* (3rd edn) **11**, 191 (1975). See p. 281.
74. Laidler, K. J., and Bunting, P., *The chemical kinetics of enzyme action* (2nd edn). Clarendon Press, Oxford (1973). See pp. 127–33.
75. Tipton, K. F., in *Companion to biochemistry* (Bull, A. T., Lagnado, J. R., Thomas, J. O., and Tipton, K. F., eds), Vol. 1, p. 277. Longman, London (1974). See p. 238.
76. Purich, D. L., and Allison, R. D., *Methods Enzymol.* **64**, 3 (1980).
77. Silverstein, E., and Sulebele, G., *Biochemistry* **8**, 2543 (1969).
78. Fromm, H. J., *Methods Enzymol.* **63**, 42 (1979).
79. Dalziel, K., *Biochem. J.* **114**, 547 (1969).
80. Engel, P. C. and Dalziel, K., *Biochem. J.* **118**, 409 (1970).
81. Halford, S. E., in *Companion to biochemistry* (Bull, A. T., Lagnado, J. R., Thomas, J. O., and Tipton, K. F., eds), Vol. 1, Ch. 5. Longman, London (1974).
82. Hammes, G. G., and Schimmel, P. R., *Enzymes* (3rd edn) **2**, 67 (1970).
83. Fersht, A. R., *Enzyme structure and mechanism* (2nd edn). Freeman, New York (1985). See Ch. 4.
84. Fersht, A. R., *Enzyme structure and mechanism* (2nd edn). Freeman, New York (1985). See p. 152.
85. Stinson, R. A., and Gutfreund, H., *Biochem. J.* **121**, 235 (1971).
86. Chance, B., *J. biol. Chem.* **151**, 553 (1943).
87. Südi, J., *Biochem. J.* **139**, 251 (1974).
88. Südi, J., *Biochem. J.* **139**, 261 (1974).
89. Holloway, M. R., and White, H. A., *Biochem. J.* **149**, 221 (1975).
90. Malcolm, A. D. B., *Prog. Biophys. molec. Biol.* **30**, 205 (1975).

91. Trentham, D. R. and Gutfreund, H., *Biochem, J.* **106**, 455 (1968).
92. Halford, S. E., Bennett, N. G., Trentham, D. R., and Gutfreund, H., *Biochem. J.* **114**, 243 (1969).

Appendix 4.1

The integrated form of the Michaelis–Menten equation[1]

Equation (4.6) states that

$$v=\frac{V_{max}[S]}{K_m+[S]}.$$

The velocity, v, can be expressed as $-d[S]/dt$, therefore

$$-\frac{d[S]}{dt}=\frac{V_{max}[S]}{K_m+[S]}.$$

Separating variables and integrating between limits at time zero when the substrate concentration $=[S]_0$ and time t where the substrate concentration $=[S]_t$:

$$\int_{[S]_0}^{[S]} d[S]+K_m\int_{[S]_0}^{[S]}\frac{d[S]}{[S]}=-V_{max}\int_0^t dt,$$

$$\text{therefore } [S]-[S]_0+K_m\cdot\ln\frac{[S]}{[S]_0}=-V_{max}\cdot t.$$

Rearranging

$$\frac{[S]_0-[S]}{t}=-K_m\cdot\frac{1}{t}\cdot\ln\frac{[S]_0}{[S]}+V_{max}.$$

Thus a plot of $([S]_0-[S])/t$ against $1/t\cdot\ln([S]_0/[S])$ (Fig. A4.1) gives a straight line of slope $-K_m$ and intercept V_{max}.

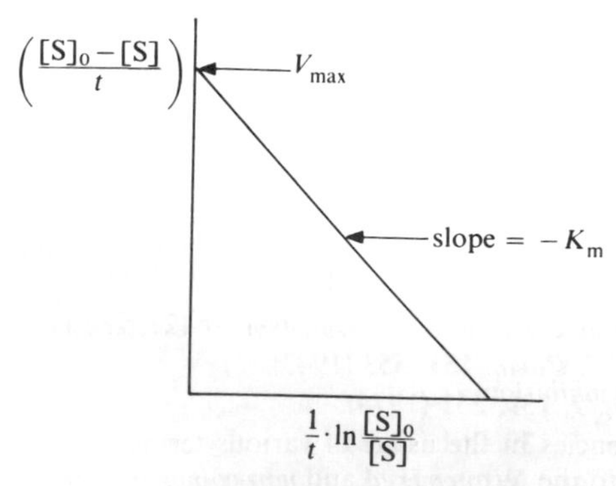

FIG. A4.1. Analysis of the progress curve of an enzyme-catalysed reaction using the integrated form of the Michaelis–Menten equation.

Appendix 4.2

The interaction of an enzyme with substrate (S) and inhibitor (I)

(*a*) *General considerations*

Consider the scheme below, where E, S, I represent enzyme, substrate, and inhibitor, respectively:

$$\begin{array}{ccccc} E & \underset{}{\overset{K_{ES}}{\rightleftharpoons}} & ES & \xrightarrow{k_2} & E+P \\ K_{EI}\updownarrows & & \updownarrows K_{ESI} & & \\ EI & \rightleftharpoons & ESI & & \end{array}$$

ESI is assumed to be inactive and K_{ES}, K_{EI}, and K_{ESI} represent dissociation constants.

Now

$$[ES]=\frac{[E][S]}{K_{ES}}, \quad [EI]=\frac{[E][I]}{K_{EI}},$$

and

$$[ESI]=\frac{[ES][I]}{K_{ESI}}=\frac{[E][S][I]}{K_{ESI}\cdot K_{ES}}.$$

The fraction (F) of enzyme in the form of the ES complex is given by

$$F=\frac{[ES]}{[E]+[ES]+[EI]+[ESI]}$$

$$=\frac{\dfrac{[S]}{K_{ES}}}{1+\dfrac{[S]}{K_{ES}}+\dfrac{[I]}{K_{EI}}+\dfrac{[S][I]}{K_{ESI}\cdot K_{ES}}}.$$

Now the observed velocity, v, is related to the maximum velocity, V_{max}, by

$$v=V_{max}\cdot F.$$

Hence,

$$v=V_{max}\frac{\dfrac{[S]}{K_{ES}}}{1+\dfrac{[S]}{K_{ES}}+\dfrac{[I]}{K_{EI}}+\dfrac{[S][I]}{K_{ESI}\cdot K_{ES}}}.$$

(*b*) *Classification of inhibition*

Because of inconsistencies in the usage of various terms relating to enzyme inhibition, particularly relating to the terms *mixed* and *non-competitive inhibition*, the International Union of Biochemistry has produced a number of recommendations.[2] The classification of types of inhibition should be in terms of the effects on V_{max} and V_{max}/K_m.

If the apparent value of V_{max}/K_m is decreased by the inhibitor, the inhibition is said to have a *competitive* component; if the inhibitor has no effect on the apparent value of V_{max}, the inhibition is said to be *competitive.*

If there is an effect on the apparent value of V_{max}, the inhibition has an *uncompetitive component*; if the inhibitor has no effect on the apparent value of V_{max}/K_m, the inhibition is said to be *uncompetitive.*

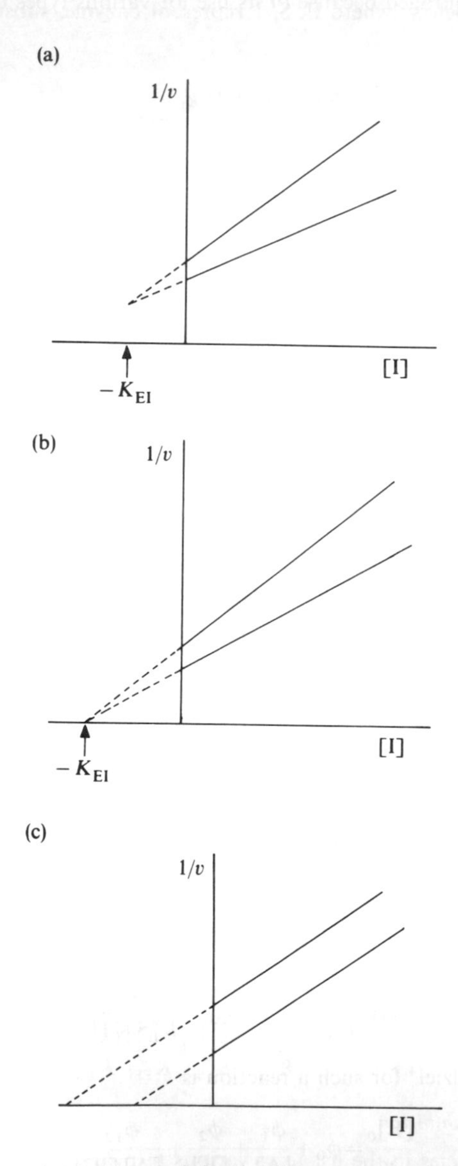

FIG. A.4.2. The Dixon plot for determining inhibitor constants: (a), (b) and (c) show the plots for the cases of competitive, pure non-competitive, and uncompetitive inhibition, respectively. Each line in the plots represents data obtained at a constant [S].

If both competitive and uncompetitive components are present, the inhibition is said to be *mixed*. The case in which the effects on V_{max}/K_m are greater than on V_{max} is called *predominantly competitive inhibition*; that in which the effects on V_{max} are greater than on V_{max}/K_m is called *predominantly uncompetitive inhibition*. Where the effects on V_{max}/K_m and V_{max} are the same (i.e. there is no change in K_m) this may be called *pure non-competitive inhibition* (see Section 4.3.2.2). However, usage of the term non-competitive inhibition is to be discouraged because of its use for various types of mixed inhibition.

(c) The Dixon plot[3]

The general equation (4.9) can be rearranged as

$$\frac{1}{v}=\frac{1}{V_{max}}+\frac{K_{ES}}{V_{max}}\cdot\frac{1}{[S]}+[I]\left(\frac{1}{V_{max}\cdot K_{ESI}}+\frac{1}{V_{max}}\cdot\frac{K_{ES}}{K_{EI}}\cdot\frac{1}{[S]}\right).$$

Thus, a plot of $1/v$ against [I] at constant [S] will be a straight line. If two such lines are drawn (from measurements at two different values of [S]), the values of K_{EI} can be found from the point of intersection (see Fig. A4.2). (This can be shown by setting the values of $1/v$ equal at two values of [S].) The plot is therefore particularly useful if v is measured at a large number of inhibitor concentrations but only a limited number of substrate concentrations (compare Figs. 4.7, 4.8 and 4.9). In the case of competitive inhibition, the lines intersect above the abscissa (Fig. A4.2(a)), whereas in the case of pure non-competitive inhibition, the point of intersection is on the abscissa (Fig. A4.2(b)). There is no point of intersection in the case of uncompetitive inhibition, since the lines are parallel (Fig. A4.2(c)). The Dixon plot cannot be used to deduce the value of K_{ESI}.

Appendix 4.3

Formulation of equations for two-substrate kinetics

The International Union of Biochemistry[2] proposes the following formulation for the initial rate, v, of a two-substrate reaction obeying Michaelis–Menten kinetics:

$$v=\frac{[E]_0}{\frac{1}{k_0}+\frac{1}{k_A[A]}+\frac{1}{k_B[B]}+\frac{1}{k_{AB}[A][B]}},$$

where $[E]_0$ is the total concentration of enzyme and the ks are individual rate constants.
In terms of eqn (4.13),

$$k_0=\frac{V_{max}}{[E]_0},\qquad k_A=\frac{k_0}{K_A},$$

$$k_B=\frac{k_0}{K_B}\quad\text{and}\quad k_{AB}=\frac{k_0}{K_B\cdot K'_A}.$$

The formulation of Dalziel[4] for such a reaction is

$$\frac{[E]_0}{v}=\phi_0+\frac{\phi_1}{[A]}+\frac{\phi_2}{[B]}+\frac{\phi_{12}}{[A][B]}$$

where ϕ_0, etc. are known as Dalziel coefficients. These coefficients allow the equation to be written more concisely, since they represent combinations of constants.

In terms of eqn (4.13),

$$\phi_0 = \frac{[E]_0}{V_{max}}, \quad \phi_1 = \frac{[E]_0 K_A}{V_{max}},$$

$$\phi_2 = \frac{[E]_0 K_B}{V_{max}} \quad \text{and} \quad \phi_{12} = \frac{[E]_0 K'_A K_B}{V_{max}}.$$

For further details of the various formulations of these equations, see references 5, 6, and 7.

References for Appendices

1. Moreno, J., *Biochem. Educ.* **13**, 64 (1985).
2. Symbolism and Terminology in Enzyme Kinetics. Recommendations (1981) of the Nomenclature Committee of the International Union of Biochemistry. Reprinted in *Europ. J. Biochem.* **128**, 281 (1982).
3. Dixon, M., *Biochem. J.* **55**, 170 (1953).
4. Dalziel, K., *Acta chem. scand.* **11**, 1706 (1957).
5. Tipton, K. F., in *Companion to biochemistry* (Bull, A. T., Lagnado, J. R ., Thomas, J. O., and Tipton, K. F., eds.), Vol. 1, p. 227. Longman, London (1974). See p. 247.
6. Rudolph, F. B. and Fromm, H. J., *Methods Enzymol.* **63**, 138 (1979).
7. Cornish-Bowden, A., *Fundamentals of enzyme kinetics.* Butterworths, London (1979). See Ch. 6.

5
The mechanism of enzyme action

5.1 Introduction

In this chapter, we shall consider the mechanism of action of enzymes and gain some insight into two of the remarkable properties of enzymes referred to in Chapter 1, Section 1.3; namely catalytic power and specificity. In Section 5.3 we introduce some of the basic principles of catalysis and show how the study of the mechanisms of reactions in organic chemistry can provide a framework in which to interpret catalysis by enzymes. The main experimental approaches available to determine the mechanism of enzyme action are described in Section 5.4. Section 5.5 shows how these different approaches have been combined to provide coherent pictures of the mechanism of five particular enzymes, namely chymotrypsin, triosephosphate isomerase, fructose-bisphosphate aldolase, lactate dehydrogenase, and tyrosyl-tRNA synthetase. In the concluding Section (5.6), some more general aspects of enzyme catalysis are discussed.

Enormous progress has been made in the understanding of enzyme mechanism in the last 20 years or so; the major reason for this has been the wealth of structural information that has become available from X-ray crystallographic studies (see Chapter 3, Section 3.5.1). However, although chemically plausible mechanisms have been proposed for a number of enzymes it is doubtful that we are yet in a position to account *in quantitative terms*, for the catalytic power shown by any one enzyme.

5.2 Definition of the mechanism of an enzyme-catalysed reaction

We can justifiably claim to have deduced the mechanism of an enzyme-catalysed reaction if we have determined (i) the *sequence* of the enzyme-containing complexes* as substrate(s) is (are) converted to product(s); (ii) the *rates* at which these complexes are interconverted; and (iii) the *structures* of these complexes. This definition implies that we require both kinetic and structural information and it is therefore not surprising that kinetic and X-ray crystallographic techniques have made major contributions to our understanding of enzyme mechanisms. Valuable information has also been gained from studies involving detection of intermediates (Section 5.4.2), chemical modification of amino-acid side-chains (Section 5.4.4) and, most recently, by site-directed mutagenesis (Section 5.4.5).

* This sequence constitutes the *elementary steps* of the overall reaction.

5.3 Background to catalysis

Studies of mechanisms in organic chemistry have highlighted various factors that contribute to the enhancement of the rate of reactions. These factors are described (Sections 5.3.1 to 5.3.5) so that their importance in the specific examples of enzyme catalysis discussed in Section 5.5 can be assessed. However, not all the factors are necessarily important in every enzyme-catalysed reaction.

Inspection of the 'energy profile' for a typical reaction (Fig. 5.1(a)) shows that in order to proceed from reactant(s) to product(s) an energy barrier ($\Delta G^{\ddagger}$) must be surmounted. The highest point of the energy profile is designated the *transition state* of the reaction. From the transition-state theory of reaction rates,[1] an equation can be derived that expresses the relationship between the rate constant, k', and the free energy of activation ($\Delta G^{\ddagger}$):

$$k' = \frac{kT}{h} e^{-\Delta G^{\ddagger}/RT} = \frac{kT}{h} e^{-\Delta H^{\ddagger}/RT} \cdot e^{\Delta S^{\ddagger}/R}$$

where k is the Boltzmann constant,
h is Planck's constant,
T is the temperature,
R is the gas constant,
and $\Delta H^{\ddagger}$ and $\Delta S^{\ddagger}$ are the enthalpy and entropy of activation, respectively.

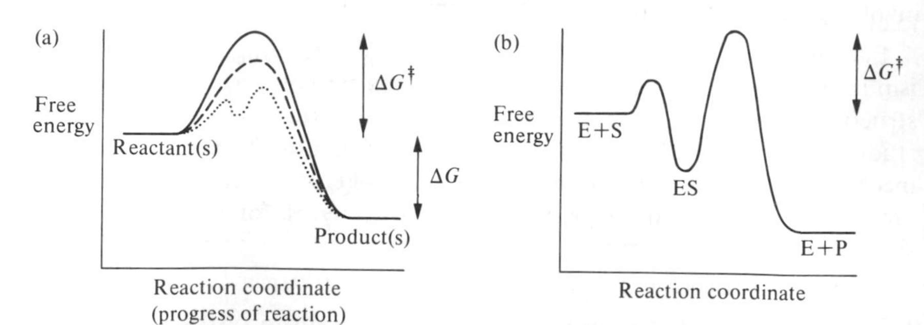

FIG. 5.1. Energy profiles for reactions. (a) Profiles for an uncatalysed reaction (solid line), a catalysed reaction (dashed line), and a catalysed reaction involving intermediate formation (dotted line). (b) Profile for an enzyme-catalysed reaction $E+S \leftrightarrows ES \rightarrow E + P$. Note that comparisons should not be made between the vertical scales in (a) and (b).

In order to speed up the rate of a reaction at constant temperature, the value of $\Delta G^{\ddagger}$ must be reduced. Therefore, the function of a catalyst is to provide an alternative reaction pathway so that a lowered energy barrier has to be surmounted (see, for example, the dashed line in Fig. 5.1(a)). One particular way in which a catalyst might function is via the rapid formation and decomposition of an *intermediate*, which is represented by the local minimum in the dotted line in Fig. 5.1(a). This is known as covalent catalysis. In all cases, the catalyst does not alter the magnitude of free-energy change (ΔG in Fig. 5.1(a)) and therefore does not cause a shift in the equilibrium between reactant(s) and product(s);* it merely increases the rate at which that equilibrium is attained.

These general considerations also apply to catalysis by enzymes; however, we should note that the energy profile corresponding to even the simplest type of enzyme mechanism is more complex (see Fig. 5.1(b)). The energy profiles of a number of enzyme catalysed reactions have been determined e.g. triosephosphate isomerase (Fig. 5.29), proline racemase[2] and tyrosyl-tRNA synthetase.[3]

Various factors leading to rate enhancements that may be relevant to enzyme-catalysed reactions are proximity and orientation effects, acid–base catalysis, covalent catalysis, strain or distortion, and changes in environment. These will now be discussed in turn.

5.3.1 Proximity and orientation effects

It seems intuitively obvious that an enzyme could increase the rate of a reaction involving more than one substrate by binding the substrates at adjacent sites and therefore bringing them into close proximity with each other. The reaction would then occur more readily than if it depended on chance encounters between the reacting molecules in solution. Many examples from organic chemistry suggest that reactions can be greatly accelerated when the reacting groups are combined within a single molecule. For example, the following data were obtained for the amine-catalysed hydrolysis of *p*-nitrophenyl esters[4] (a reaction involving nucleophilic attack by the amine of the carbonyl group of the ester):

Reacting groups separated

$N(CH_3)_3$ + $CH_3{-}C({=}O){-}O{-}C_6H_4{-}NO_2$

$k = 4.3\ (\text{mol dm}^{-3})^{-1}\ \text{min}^{-1}$

Reacting groups combined

$(CH_3)_2N{-}CH_2{-}CH_2{-}CH_2{-}C({=}O){-}O{-}C_6H_4{-}NO_2$

$k = 21500\ \text{min}^{-1}$

Difficulties arise in such situations when we try to establish the correct basis for comparison; thus, in the above example we are comparing the magnitudes of rate constants that are of different orders and hence of different dimensions. There could also be some differences between the detailed mechanisms of the two reactions. Despite these difficulties, Page and Jencks[5, 6] have given a detailed thermodynamic treatment of proximity effects in enzyme-catalysed reactions and their conclusions are mentioned at the end of this section after the importance of orientation effects has been discussed.

There is no doubt that the orientation of the reacting molecules with respect to each other can greatly influence the rate of reaction. Catalysis could thus be

* This is of course true for catalysis brought about by enzymes. However, if the concentration of enzyme is comparable with that of the substrates and products there can be an *apparent* shift in the equilibrium, see Chapter 8, Section 8.5.4.

achieved by ensuring that the reactants are in the correct orientation as they approach each other. Hydrolysis of the *p*-bromophenyl esters shown below is catalysed in each case by the neighbouring carboxylate group; a process which involves nucleophilic attack by the carboxylate on the carbonyl group of the ester. The presence of a bridge structure that almost totally restricts rotation about the single bond increases the rate of hydrolysis by some 230-fold:[7]

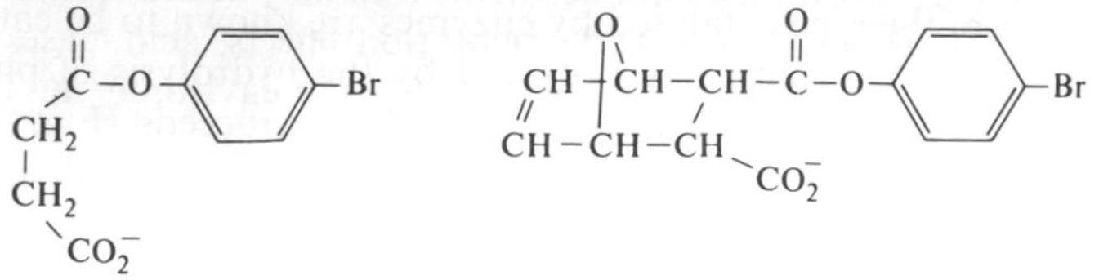

Relative rate of hydrolysis	1	230

A more spectacular example of rate enhancement is provided by the rates of lactone (internal ester) formation in compounds (I) and (II) below:[8]

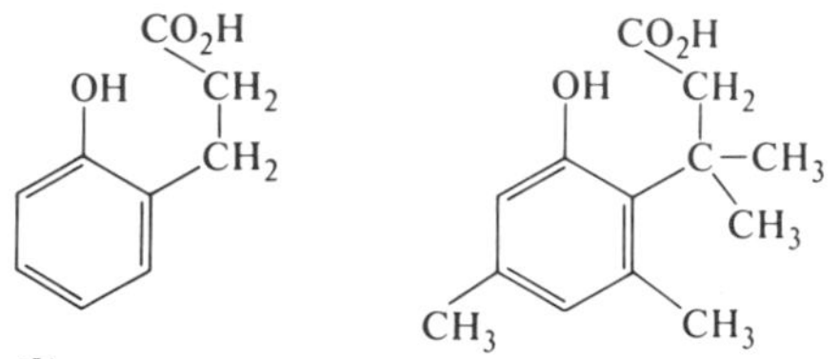

(I) (II)

The presence of the two methyl groups on the carbon atom adjacent to the aromatic ring in (II) increases the rate of lactone formation by a factor of 4×10^5.

In these examples the restrictions placed on rotation about single bonds by a bridge structure or by bulky substituents ensures that the preferred orientation of the reacting groups closely resembles that of the transition state of the reaction. A simple way of looking at this is to say that the molecule will spend a greater proportion of the time in the conformation that leads to reaction, i.e.

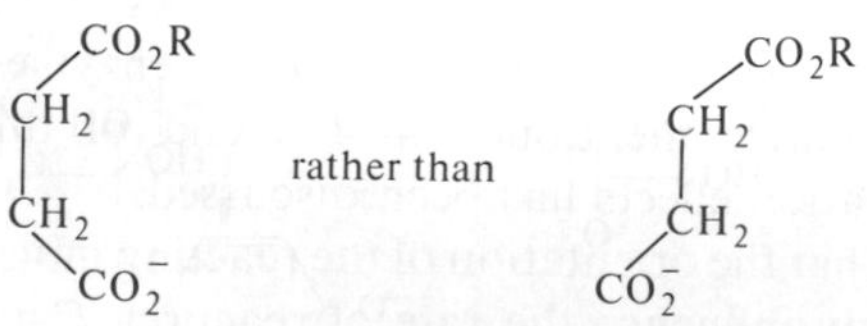

rather than

This will mean that less rotational entropy (fewer degrees of freedom) will be lost as the reaction proceeds towards the transition state. The smaller negative entropy of activation will lead to an increase in the rate of reaction.

Estimates of the effects of orientation and proximity on the entropy of activation, and hence on the rate of reaction, have been by Page and Jencks.[5,6] Their calculations suggest that each factor can contribute a rate enhancement of at least 10^4-fold for a bimolecular reaction, giving a possible total rate enhancement of at least 10^8-fold from these two factors combined.

5.3.2 Acid–base catalysis

Many reactions of the type catalysed by enzymes are known to be catalysed by acids and/or bases. An example is provided by the hydrolysis of phenyl-β-D-glucopyranoside[9] shown in Fig. 5.2. Acid catalysis proceeds (Fig. 5.2(a)) via protonation of the glycosidic oxygen followed by dissociation of phenol to yield a stabilized carbonium ion which is then attacked by water to form glucose. On the other hand, base catalysis proceeds (Fig. 5.2(b)) via abstraction of a proton from the C-2 OH group followed by nucleophilic attack of the —O^- group on C-1 to form an epoxide. The epoxide is then attacked by water to form glucose.

Since enzymes contain a number of amino-acid side-chains that are capable of acting as proton donors or acceptors (see Chapter 3, Section 3.3.1), it is reasonable to suppose that acid–base catalysis would be important in enzyme-catalysed reactions. In simple organic reactions, acid catalysis is divided into *specific acid catalysis* (where the rate expression contains contributions from H^+ only) and *general acid catalysis* (where the rate expression includes contributions from H^+ and from other potential proton donors in the solution, e.g.

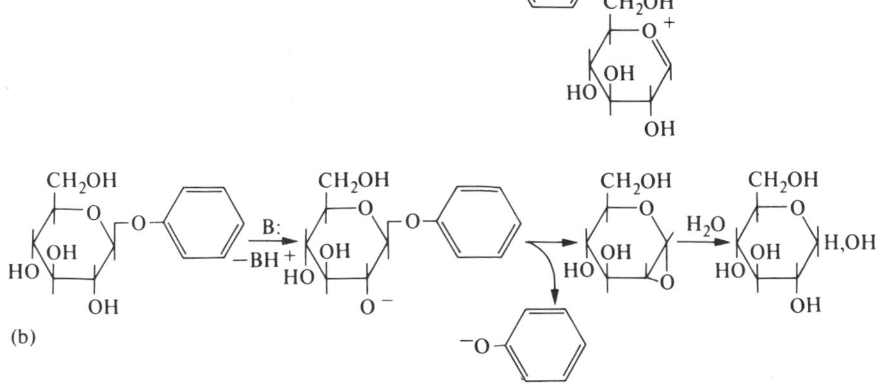

FIG. 5.2. Hydrolysis of phenyl-β-D-glucopyranoside subject to catalysis by acid (a) or base (b).

CH_3CO_2H). A similar division is made for base catalysis. As far as enzyme-catalysed reactions are concerned only *general* acid or base catalysis can occur, because enzymes do not possess any mechanism for concentrating H^+ or OH^-.

A good example of general acid catalysis occurs in the hydrolysis of oligosaccharides catalysed by lysozyme.[10] In this enzyme the side-chain of Glu 35 is in a highly non-polar environment, and this has the effect of raising its pK_a so that the carboxyl group remains unionized at pH values up to about 6 (the pK_a of the side-chain of free glutamic acid is 4.3). The carboxyl group donates a proton to the glycosidic oxygen of the substrate, facilitating cleavage of the C–O bond, and leading to the formation of a carbonium ion that is stabilized by the neighbouring ionized side-chain of Asp 52 (Fig. 5.3). The analogies with the acid-catalysed hydrolysis of the glucoside in Fig. 5.2(a) should be noted.

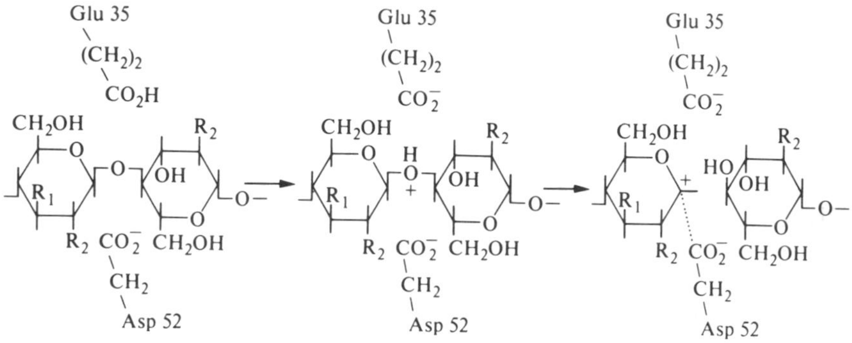

FIG. 5.3. Part of the proposed mechanism of action of lysozyme.[10] The process is completed by reaction of the carbonium ion with water, with overall retention of configuration at C–1.

Histidine side-chains often play important roles as acid–base catalysts in enzyme action (see, for example the cases of chymotrypsin, fructose-bisphosphate aldolase, and lactate dehydrogenase discussed in Section 5.5). This is because the histidine side-chain has a pK_a close to 7, so that near neutral pH there is a reasonable balance between the proton-donating (protonated) form and proton-accepting (deprotonated) forms.

It is conceivable that suitably positioned proton-donating and proton-accepting side-chains in an enzyme could act in a concerted fashion, so that electrons are both 'pulled' and 'pushed'. This type of effect has been demonstrated in a model system, namely the mutarotation of tetra-*O*-methylglucose (Fig. 5.4) but the importance of the effect in enzyme-catalysed reactions is less clear.[11]

A fuller discussion of acid–base catalysis and its importance in enzyme-catalysed reactions has been given in the monograph by Fersht.[12]

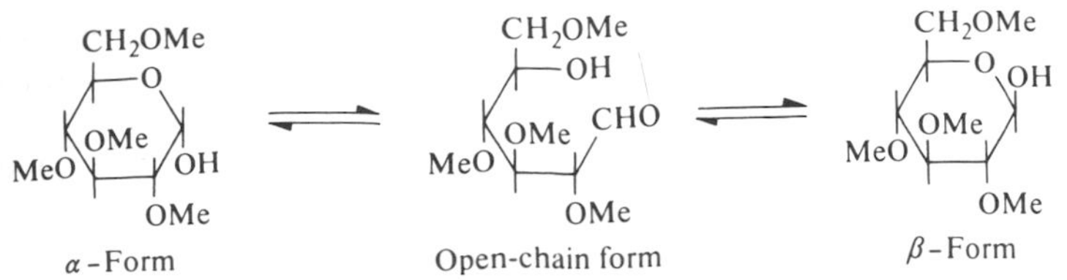

FIG. 5.4. The mutarotation of tetra-*O*-methyl glucose in which α- and β-forms are interconverted via the open-chain form. 2-Hydroxypyridine, in which acid and base groups are combined in one species, acts as a much more effective catalyst than a mixture of pyridine and phenol, indicating that concerted acid–base catalysis occurs.[11]

5.3.3 Covalent catalysis (intermediate formation)

It has been recognized for many years that reactions can be speeded up by the formation of intermediates, provided that such intermediates are both rapidly formed and rapidly broken down (see the dotted line in Fig. 5.1(a)). An example is provided by the decarboxylation of acetoacetate, which is catalysed by primary amines via the rapid formation and breakdown of a Schiff base or ketimine,[13] see Fig. 5.5.

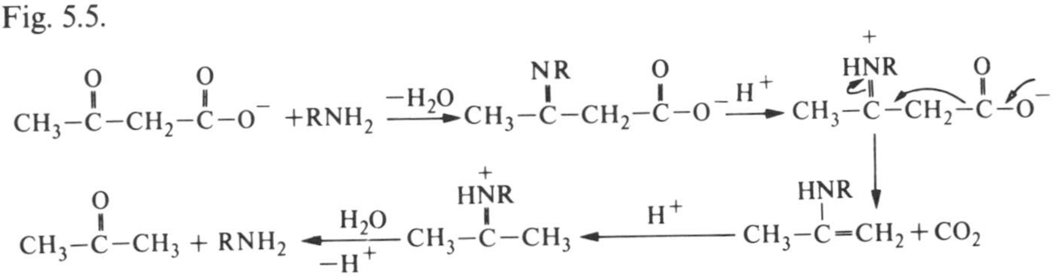

FIG. 5.5. The decarboxylation of acetoacetate catalysed by primary amines.

The advantage conferred by intermediate formation in this case is that the Schiff base is readily protonated, thus providing greater electron-withdrawing power to aid the decarboxylation than would be provided by the original carbonyl group.

Many of the examples of covalent catalysis in enzyme-catalysed reactions involve attack by a nucleophilic side-chain at an electron-deficient centre in the substrate; such attack is termed nucleophilic catalysis. Examples of nucleophilic catalysis are provided by chymotrypsin (Section 5.5.1) and fructose-bisphosphate aldolase (Section 5.5.3); in these cases the nucleophilic side-chains are those of serine and lysine, respectively. Other types of side-chain that have been found to participate in nucleophilic catalysis include cysteine (e.g. papain and glyceraldehyde-3-phosphate dehydrogenase) and histidine (e.g. phosphoglycerate mutase and nucleosidediphosphate kinase).

5.3.4 Strain or distortion

There is the possibility that a substrate may be distorted on binding to the appropriate enzyme. This would speed up the reaction if the distortion (or strain) lowered the free energy of activation by making the geometry and electronic structure of the substrate more closely resemble that of the postulated transition state.[14] Examples of this effect can be found in model systems, e.g. the very rapid hydrolysis of strained cyclic phosphate esters compared with the open-chain forms.[15] (Fig. 5.6).

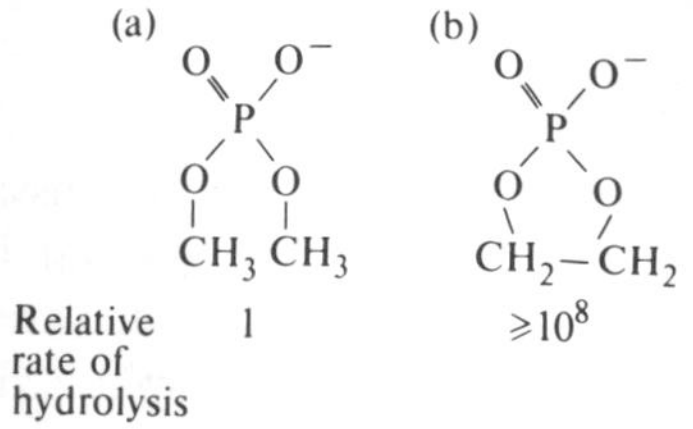

(a) $(CH_3O)_2PO_2^-$ (b) CH_2-CH_2 cyclic phosphate

Relative rate of hydrolysis: (a) 1; (b) $\geqslant 10^8$

FIG. 5.6. Structures of phosphate esters. The faster rate of hydrolysis of (b) is attributed to the strain in the molecule, which is relieved on hydrolysis.

As is often the case, the evidence is more difficult to interpret in enzyme-catalysed reactions. It has been proposed that the glucose ring at the point of cleavage of an oligosaccharide substrate is distorted on binding to lysozyme, so that the geometry approaches that in the postulated transition state, which involves a carbonium ion[10] (see Figs. 5.3 and 5.7); this should lead to the enhancement of the rate of reaction.* Another example of substrate distortion is provided by studies using i.r. spectroscopy that has shown that dihydroxyacetone phosphate is distorted on binding to triosephosphate isomerase.[17]

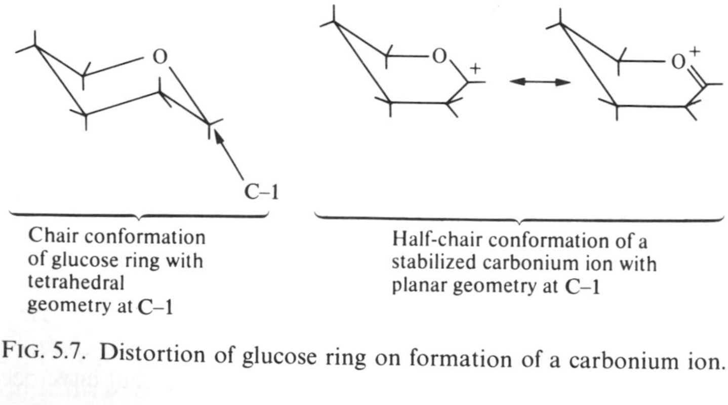

C–1

Chair conformation of glucose ring with tetrahedral geometry at C–1

Half-chair conformation of a stabilized carbonium ion with planar geometry at C–1

FIG. 5.7. Distortion of glucose ring on formation of a carbonium ion.

A different way of regarding strain is to think in terms of stabilization of the transition state of the reaction rather than destabilization of the substrate. In other words, it is envisaged that an enzyme is more suited to making favourable contacts with the transition state of the substrate than with the substrate in its

* Detailed calculations of the energy of the lysozyme–substrate complex suggest, however, that this distortion does not in fact contribute greatly to the rate enhancement[16] (see Section 5.3.5).

normal conformation. From this it would follow that a compound that resembles the transition state in geometry and electronic structure will bind more tightly to an enzyme than will the substrate. There has been considerable interest[18] in the search for such 'transition-state analogues', since these would help to confirm ideas about transition states of reactions. A transition state analogue for triosephosphate isomerase is described in Section 5.5.2.3.

5.3.5 Changes in environment

The rates of many organic reactions are highly sensitive to the nature of the solvents in which they occur. In particular, dipolar aprotic solvents such as dimethylsulphoxide and dimethylformamide, which are not capable of solvating anions, are extremely good solvents for nucleophilic displacement reactions.[19] The reaction below occurs over 12 000 times faster in dimethylsulphoxide than in water.

$$N_3^- + F\text{-}C_6H_4\text{-}NO_2 \longrightarrow F^- + N_3\text{-}C_6H_4\text{-}NO_2$$

X-ray crystallographic studies show that enzymes are capable of providing unusual environments for reactions. In the case of lysozyme, the cleft in the molecule that forms the substrate binding site is lined by a number of non-polar amino-acid side-chains that obviously provide an environment markedly different from that of the solvent water. In addition, within the active site there is an appropriately positioned negative charge, the ionized side-chain of Asp 52, which helps to stabilize the positive charge on the carbonium ion in the transition state of the reaction (Fig. 5.3). Calculations[16] suggest that this electrostatic stabilization, which would be strong in a medium of low dielectric constant, may be the most important factor in this case, contributing a rate enhancement of 3×10^6-fold.

5.3.6 Discussion of the factors likely to be involved in enzyme catalysis

In Sections 5.3.1 to 5.3.5 we have discussed various factors that are generally accepted to be important in bringing about rate enhancement in enzyme-catalysed reactions. The examples of enzyme mechanisms described in Section 5.5 will show how these various factors operate in particular situations. Additional factors may be involved in certain cases; for example, in a metalloenzyme the metal ion can display powerful electron-withdrawing ability thereby polarizing a chemical bond and making it more susceptible to nucleophilic attack. An instance of this type of effect is provided by carboxypeptidase, where a zinc ion serves to polarize the carbonyl group of an amide substrate (see Section 5.5.1.7).

Although the causes of rate enhancement are reasonably well understood in qualitative terms, we are still some way from a *quantitative* understanding of their importance in enzyme-catalysed reactions. However, progress is being made in this direction and recently it has been claimed that the 10^{16}-fold rate enhancement of DNA hydrolysis shown by staphylococcal nuclease can be accounted for by a combination of transition-state stabilization, metal-ion catalysis, general base catalysis, and catalysis by proximity of attacking water.[20]

It is possible to make practical applications of the knowledge we have concerning rate enhancements both in altering the catalytic properties of enzymes by site-directed mutagenesis (see Section 5.5.5.5) and in designing synthetic 'model' enzymes (or 'synzymes') that are capable of catalysing fairly simple types of reactions. One of the most successful of these attempts has involved the synthesis of an 'artificial' chymotrypsin in which a catalytic site containing imidazole, carboxyl, and hydroxyl groups (see Section 5.5.1.2) has been incorporated into cyclodextrin, which provides the binding site. The artificial enzyme had an M_r only about 5 per cent that of chymotrypsin but showed comparable kinetic features in the esterase reaction and enhanced stability at elevated temperatures and high pH.[21] Reviews of the design and applications of synthetic enzymes have been given by Wharton[22] and Breslow.[23]

5.4 Experimental approaches to the determination of enzyme mechanisms

Having discussed the theoretical basis of rate enhancements, we now describe the main experimental approaches that can be used to elucidate the mechanism of action of an enzyme. These approaches will be dealt with under five main headings: kinetic studies (Section 5.4.1), detection of intermediates (Section 5.4.2), X-ray crystallography (Section 5.4.3), chemical modification of amino-acid side-chains (Section 5.4.4), and site-directed mutagenesis (Section 5.4.5). These approaches are to be regarded as complementary; the information gained from one approach should be interpreted in the light of results from the others in order to build up the overall picture of the mechanism.

5.4.1 Kinetic studies

In Chapter 4, we described how the catalytic activity of an enzyme could be assayed and then proceeded to introduce some of the principal concepts and equations of enzyme kinetics. Our object in this section is to show how information about enzyme mechanisms can be gained from these studies. The types of information available from kinetic studies are summarized in Table 5.1 and the various techniques are discussed in turn (Sections 5.4.1.1 to 5.4.1.5).

5.4.1.1 Variation of substrate concentration(s)

Steady-state kinetic studies support the proposal that one-substrate reactions proceed via the formation and decay of one or more enzyme–substrate com-

TABLE 5.1
The types of information on enzyme mechanisms available from kinetic studies

Experiment	Information available
Variation of substrate concentration(s)	Sequence of complexes in reaction. Distinction between possible mechanisms
Variation of substrate structure	Structural features responsible for binding and catalytic activity. (Mapping the active site)
Reversible inhibition (see Chapter 4, Section 4.3.2)	Competitive inhibitors can help to define the active site
Variation of pH (see Chapter 4, Section 4.3.3)	The pK_a values of side-chains involved in catalytic activity. Assignment of the pK_a values to particular amino acid side-chains may be possible
Pre-steady-state kinetics (see Chapter 4. Section 4.4)	Detection of enzyme-containing complexes. Rates of elementary steps in the reaction

plexes (see Chapter 4, Section 4.3.1), but cannot by themselves give any indication of the sequence of such complexes. In the case of two-substrate reactions, steady-state kinetic studies can distinguish between ternary complex and enzyme substitution mechanisms (see Chapter 4, Section 4.3.5.1). With additional information from product inhibition, substrate binding, or isotope-exchange experiments, it is possible to distinguish between mechanisms in which a ternary complex is formed in an ordered or a random fashion (see Chapter 4, Section 4.3.5.5).

5.4.1.2 Variation of substrate structure

A good deal has been learnt about the general features of enzyme active sites by correlating the rates of reactions with the structures of the substrates. For instance, by comparing the rates of hydrolysis of a large number of amide derivatives of amino acids it has been demonstrated that chymotrypsin has a strong preference for substrates containing aromatic or bulky hydrophobic R groups, whereas elastase has a strong preference for substrates containing small hydrophobic R groups. Clearly the substrate-binding sites of these enzymes must contain features that account for the observed specificites (see Section 5.5.1.2).

```
                  O  site of hydrolysis
                  ‖ ↓
X—NH—CH—C—NHY
         |
         R
```

It is possible to develop this approach and map active sites in greater detail. For example exhaustive studies have been carried out on the specificity displayed by papain towards synthetic peptide substrates.[24] These studies showed that there are seven 'subsites' on the enzyme (see Fig. 5.8), and that subsite S_2 interacts specifically with an L-phenylalanine side-chain. Later work showed that subsite

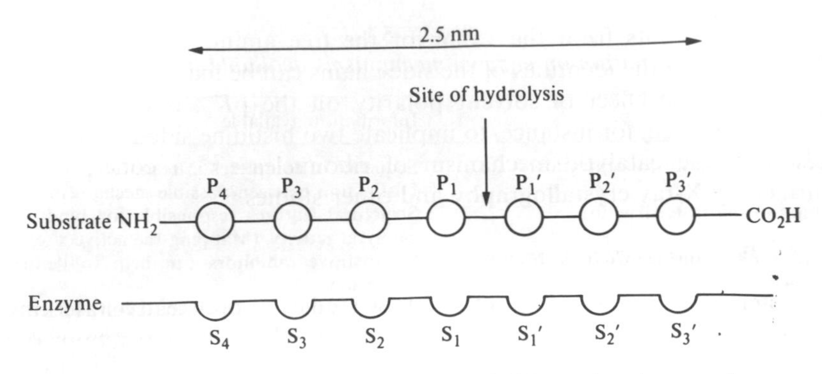

FIG. 5.8. The 'subsites' of papain as revealed by active-site mapping.

S'_1 is stereospecific for L-amino acids with a preference for the hydrophobic side-chains of leucine and tryptophan.[25] The explanation in structural terms for the precise interaction between enzyme and substrate in this case presents a formidable challenge to X-ray crystallographers; however, the probable locations of subsites S'_1, S_1, and S_2 in the enzyme have been deduced.[24, 26]

5.4.1.3 *Reversible inhibition*

The study of inhibition of enzyme-catalysed reactions (see Chapter 4, Section 4.3.2) can give information on the structures of active sites. One of the most likely explanations for competitive inhibition is that substrate and inhibitor bind to the same site on an enzyme. By comparison of the structures of the substrate and the competitive inhibitor it is possible to define the essential structural features of these molecules that are involved in their binding to the active site. A detailed study was carried out in the case of papain;[24] thus the tripeptide Ala-Phe-Arg acted as a powerful competitive inhibitor of the enzyme, presumably since it occuped the subsites S_3, S_2, and S_1 (Fig. 5.8) and could not therefore undergo hydrolysis. Competitive inhibitors are also of value in X-ray crystallographic work, where it is usually difficult to study the enzyme–substrate complex directly (Section 5.4.3).

5.4.1.4 *Variation of pH*

The catalytic activity of many enzymes is markedly dependent on pH. As described in Chapter 4, Section 4.3.3, there are a number of explanations for this phenomenon, but the one of most immediate concern here is the ionization of amino-acid side-chains that are involved in the catalytic mechanism. By suitable analysis of plot of reaction rate against pH,[27] it is usually possible to deduce the pK_a values of these ionizing side-chains and the identity of the side-chains can be guessed at by comparison with the pK_a values of side-chains of free amino acids (see Chapter 3, Section 3.3.1) or small peptides. However, it should be noted that the environment of a side-chain in an enzyme can shift the pK_a of the side-chain by up to four units from the value for the free amino acid. More definite deductions about the identities of the side-chains can be made if other information, such as the effect of solvent polarity on the pK_a, is available.* This approach was used, for instance, to implicate two histidine side-chains as being involved in the catalytic mechanism of ribonuclease,[28] a conclusion later confirmed by X-ray crystallography and other studies.[29]

5.4.1.5 Pre-steady-state kinetics

In Chapter 4, Section 4.4 we described how studies of pre-steady-state kinetics could be used to detect enzyme-containing complexes in a reaction and to determine the rates of formation and decay of such complexes. In these experiments, the concentration of enzyme more closely approaches that of the substrates than is the case with steady-state studies, and usually special techniques are required to achieve rapid mixing and rapid detection of the changes occurring.

A simple illustration of the use of pre-steady-state kinetics is afforded by the chymotrypsin-catalysed hydrolysis of *p*-nitrophenyl acetate. The production of *p*-nitrophenol (Fig. 5.9) shows a 'burst' phase[30] (the size of which corresponds to approximately 1 mol per mol enzyme) followed by a slower steady-state rate. This observation is consistent with the type of mechanism described in Section 5.5.1.3 for this enzyme, in which a fast step, corresponding to the

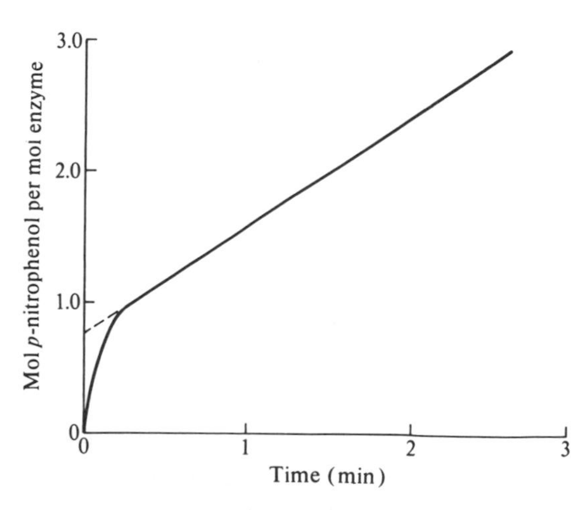

FIG. 5.9. The hydrolysis of *p*-nitrophenylacetate catalysed by chymotrypsin. The solid line represents the experimental data.

* If the dissociating group is neutral, e.g. $—CO_2H \rightleftharpoons —CO_2^- + H^+$, dissociation is accompanied by charge separation. A decrease in solvent polarity will discourage dissociation, i.e. raise the pK_a. However, if the dissociating group is cationic, e.g. $—ImH^+ \rightleftharpoons —Im + H^+$, there is no separation of charge and the pK_a will be much less sensitive to changes in solvent poarity. (Im = imidazole).

formation of acyl enzyme and release of *p*-nitrophenol, is followed by a slow step, corresponding to the rate of hydrolysis of the acyl enzyme to regenerate enzyme:

$$E{-}CH_2{-}OH + NO_2{-}C_6H_4{-}O{-}\overset{\overset{O}{\|}}{C}{-}CH_3 \xrightarrow{\text{fast}} E{-}CH_2{-}O{-}\overset{\overset{O}{\|}}{C}{-}CH_3 + NO_2{-}C_6H_4{-}OH$$

$$E{-}CH_2{-}O{-}\overset{\overset{O}{\|}}{C}{-}CH_3 + H_2O \xrightarrow{\text{slow}} E{-}CH_2{-}OH + CH_3CO_2H$$

This experiment can be performed without the need for specialized apparatus. Clearly, more information on the mechanism would be available from measurements of the early 'burst' phase of the reaction; for example we could measure the rate of formation of the acyl enzyme. The use of fast-reaction techniques is well illustrated by experiments carried out on the reaction catalysed by alcohol dehydrogenase from horse liver.[31] If the enzyme is mixed rapidly with saturating concentrations of the substrates ethanol and NAD^+, there is rapid production of 2 mol NADH per mol of the dimeric enzyme, followed by a slower steady-state rate (Fig. 5.10). If the production of NADH is monitored by fluorescence, when free and enzyme-bound NADH can be readily distinguished, it is found that the NADH produced in the rapid phase is enzyme-bound. These studies show that the release of NADH from the enzyme is the rate-determining (i.e. slowest) step in the overall reaction. Values of the rate constants of some of the individual steps in the reaction could be deduced from more detailed measurements.[31] A further

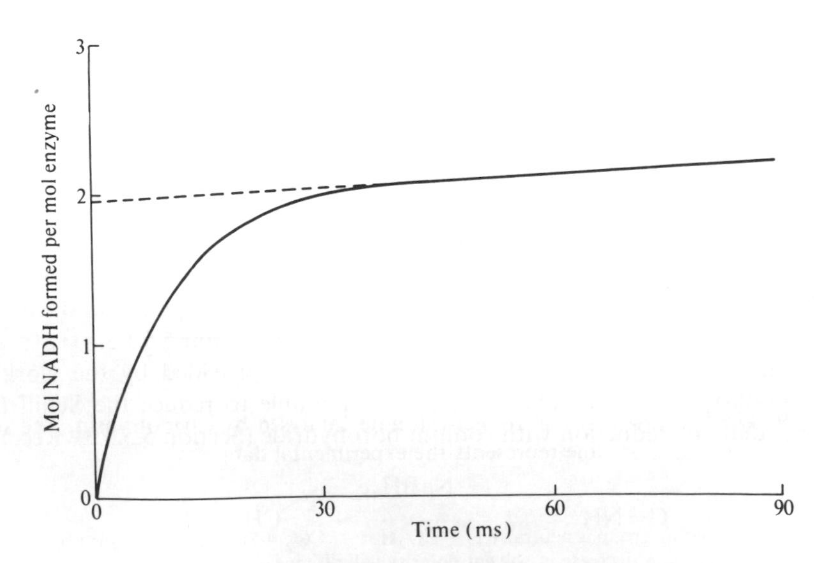

FIG. 5.10. Stopped-flow study of the reaction between ethanol and NAD^+ catalysed by alcohol dehydrogenase from horse liver (a dimeric enzyme).[31]

example of pre-steady-state kinetic studies is provided by the work on the reaction catalysed by lactate dehydrogenase (Section 5.5.4.2).

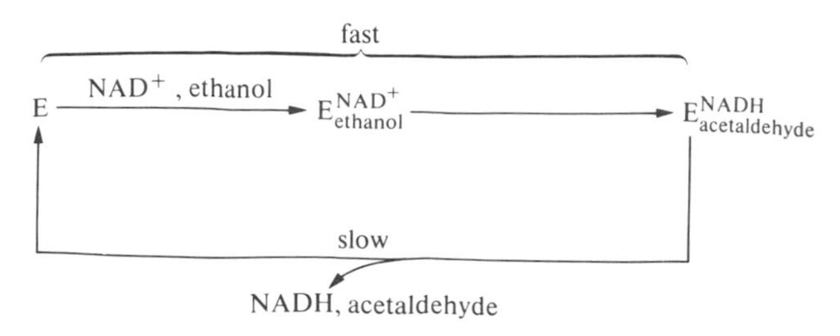

5.4.2 Detection of intermediates

One of the most direct methods for obtaining information about the pathway of a reaction is to detect any intermediates that may be involved in the reaction. In some cases an intermediate is sufficiently stable to be isolated and characterized (the stability will depend on the depth of the minimum in the energy profile shown in Fig. 5.1(a). In other cases, such as in the reaction catalysed by alcohol dehydrogenase from horse liver, it is possible to infer the existence of an intermediate by spectroscopic means. Whatever the method of detection, it must be shown that the rates of formation and decay of any presumed intermediate are consistent with the overall rate of the reaction. Thus, a postulated intermediate that is so stable that it breaks down more slowly than the overall rate of the reaction cannot be involved in that reaction.

We have already noted (Section 5.4.1.5) that pre-steady-state kinetic studies suggest that an acyl enzyme intermediate is involved in the chymotrypsin-catalysed hydrolysis of esters. The rate of breakdown of the acyl enzyme is very low at acid pH, so that the intermediate can be isolated (and crystallized) if the enzyme and ester substrate are mixed and the pH is then rapidly lowered. X-ray crystallographic work has shown that it is the side-chain of Ser 195 that becomes acylated.[32] As we shall see in Section 5.5.1.3, the chymotrypsin-catalysed hydrolysis of amide substrates is also thought to follow a pathway involving formation and breakdown of an acyl enzyme. However, in this case the breakdown step is faster than the formation step, so that the acyl enzyme does not accumulate. It is nevertheless possible to infer its participation in the overall reaction by 'trapping' experiments, as described in Section 5.5.1.3.

Another example of 'trapping' intermediates is provided by the work on fructose-bisphosphate aldolase, where it is possible to reduce the Schiff base intermediate by reduction with sodium borohydride (Section 5.5.3.3). Recently,

$$\gt C=\overset{+}{N}H- \xrightarrow{NaBH_4} \gt CH-\overset{+}{N}H_2-$$

Protonated Schiff base (unstable) | Protonated amine (stable) the possibility has been investigated that unstable intermediates can be stabilized at low temperatures, where the rate of breakdown will be decreased.[33, 34] In such experiments it is usually necessary to work in mixed-solvent systems such as dimethylsulphoxide–water to prevent freezing, although these systems can cause problems by altering reaction mechanisms. In some cases, the use of concentrated solutions of salts such as ammonium acetate may be preferable.[35] Considerable success has been achieved with this cryoenzymological approach; thus, in the case of ribonuclease, a series of enzyme-containing complexes on the reaction pathway has been studied by X-ray crystallography.[33] In studies of the elastase-catalysed hydrolysis of a *p*-nitroanilide substrate, the structure of a tetrahedral intermediate (Section 5.5.1.3) has been deduced.[34]

5.4.3 X-ray crystallographic studies

In Chapter 3 (Section 3.5) we saw that the use of X-ray crystallography has given a wealth of detail about the structures of enzymes. Clearly, we need to be able to locate the active site in an enzyme and then examine the mode of binding of substrate(s) in order to pinpoint the functional amino-acid side-chains that are involved in the catalytic mechanism. Up to fairly recently it has not proved generally possible to study the structure of the catalytically active complex of enzyme and substrate(s) directly, because the collection of diffraction data takes several hours, by which time conversion of substrate(s) to product(s) will have occurred. In the case of an enzyme that catalyses a reaction involving two or more substrates, it is possible to examine the complex formed between the enzyme and one of the substrates; see the example of lactate dehydrogenase in Section 5.5.4.3. A number of alternative approaches have been adopted to gain information on the structure of the catalytically active complex.

Firstly, it may be possible to examine the active complex in the case of a one-substrate reaction when the equilibrium lies very much to one side. Thus, in the case of triosephosphate isomerase, which catalyses the reaction below, it has been possible to examine the crystal structure of the enzyme-dihydroxyacetone phosphate complex (Section 5.5.2.2):

$$\text{Dihydroxyacetone phosphate} \rightleftharpoons \text{D-glyceraldehyde 3-phosphate.}$$

Secondly, an enzyme can be studied in the presence of a very poor substrate or a competitive inhibitor. These molecules are likely to bind to the active site in a manner similar to the substrate but will remain unchanged over the time course of the experiment. By inspection of a three-dimensional model of the enzyme it is then possible to work out how the normal substrate binds to the enzyme. This type of approach is illustrated in the cases of chymotrypsin (Section 5.5.1.2) and lactate dehydrogenase (Section 5.5.4.3).

Thirdly, it may be possible to examine unstable complexes at low temperatures, where the rate of decomposition of such complexes will be decreased. This type of approach has been mentioned in connection with work on ribonuclease and elastase (Section 5.4.2) and will undoubtedly be more widely applied in X-ray crystallographic work in the future.

Recently, there has been considerable interest in the application of synchrotron radiation (the radiation produced during the acceleration of charged particles) to the study of enzyme catalysis. The high intensity of the radiation compared to that of conventional X-ray sources means that the time required for data collection is dramatically reduced and in principle it is possible to obtain information on structural changes or catalytic steps that occur on the millisecond time scale.[36] The approach has been used to study the catalysis by phosphorylase *b* in the presence of AMP (see Chapter 6, Section 6.4.2.1) of the reaction between the poor substrate heptenitol and orthophosphate to yield heptulose 2-phosphate. The structure of the ternary complex (which accumulates because the rate of its conversion to product is slow) has been determined.[36]

The detailed information available from X-ray crystallography is very valuable in providing a framework in which to interpret the results of other experiments. Apart from locating the active site and providing information on the nature of side-chains likely to be involved in the catalytic mechanism, it is also possible to examine the extent of any structural (or conformational*) changes in the enzyme that accompany the binding of substrate(s). These conformational changes may well be of considerable importance in the catalytic process (Section 5.6). However, it should be remembered that even such a powerful technique as X-ray crystallography does have its limitations. Sometimes, attempts to prepare an enzyme–substrate complex by soaking crystals of enzyme in a solution of the substrate leads to cracking of the crystals, thus hampering further crystallographic analysis. The cracking is presumably a result of conformational changes within the enzyme that disrupt the crystal lattice. This type of problem was encountered in the attempt to study the complex between adenylate kinase and AMP.[37] In the studies on phosphorylase *a*, the problem posed by the cracking of the crystals on addition of substrates was overcome by prior treatment of the crystals with a cross-linking agent.[38] A second type of difficulty is presented by the rather artificial conditions under which crystals are studied. It is possible that in concentrated salt solutions, e.g. 3 $mol\,dm^{-3}$ ammonium sulphate, the interaction between an enzyme and its substrate may be rather different from that which occurs in solutions of lower ionic strength. Thus, there could be no binding at all or weaker binding to additional sites on the enzyme. Some of the difficulties involved in locating the active site of phosphorylase *a*[30] and adenylate kinase[40,41] have been discussed.

5.4.4 Chemical modification of amino-acid side-chains

The principle underlying the application of chemical modification techniques to the study of enzyme mechanisms is very simple. If an amino-acid side-chain involved in the catalytic activity is chemically modified, the enzyme will be inactivated. Provided that the identity of the modified side-chain can be

* Conformations of molecules are related to each other by rotations about single bonds. Conformational changes refer to interconversions between different conformations of an enzyme molecule.

established by standard structural techniques, e.g. isolating and sequencing a modified peptide (see Chapter 3, Section 3.4), then it should be possible to determine which side-chain is involved in the catalytic mechanism.

However, chemical modification techniques pose problems both of design and of interpretation (for reviews of the techniques see references 42–44). On the experimental side it is highly desirable that the modification be specific. Thus it should be checked that only one type of side-chain is modified, and further that only one of this type has reacted (e.g. only one of, say, 30 lysines in the enzyme). The requirement for specificity presents considerable problems when it is remembered that many amino-acid side-chains are nucleophilic, so that an electrophile such as 1-fluoro-2,4-dinitrobenzene could be attacked by a variety of side-chains, e.g. those of cysteine, lysine, histidine, tyrosine, etc. In Sections 5.4.4.1 to 5.4.4.3 we shall mention some approaches to improve the specificity of modification reactions; Section 5.4.4.4 will deal with problems encountered in the interpretation of chemical modification experiments.

5.4.4.1 Application of chemical principles

It is well known that mercury has a very strong affinity for sulphur (cf. the very low solubility product of HgS), so it would be expected that mercurial reagents should bring about highly specific modification of cysteine side-chains in enzymes:

$$E{-}CH_2{-}SH + Cl{-}Hg{-}C_6H_4{-}CO_2^- \longrightarrow E{-}CH_2{-}S{-}Hg{-}C_6H_4{-}CO_2^- + Cl^- + H^+$$

p-Chloromercuribenzoate

We would also expect that the activated aromatic ring in a tyrosine side-chain would be especially susceptible to electrophilic substitution for example by I_2 or by tetranitromethane:

$$E{-}CH_2{-}C_6H_4{-}OH + C(NO_2)_4 \longrightarrow E{-}CH_2{-}C_6H_3(NO_2){-}OH + H^+ + C(NO_2)_3^-$$

(It has been suggested[45] that this reaction takes place via initial formation of a charge-transfer complex between tetranitromethane and the phenolate ion, followed by electron transfer.)

It is also useful to take note of the different orders of reactivity observed for the side-chains of free amino acids with various reagents:

acylating reagents (e.g. iodoacetamide, iodoacetate)	Cys > Tyr > His > Lys
arylating reagents (e.g. 2,4,6-trinitrobenzene sulphonate, 1-fluoro-2,4-dinitrobenzene)	Cys > Lys > Tyr > His

and apply this information in enzyme-modification experiments. Thus agents such as 2,4,6-trinitrobenzenesulphonate should be used in pref acylating agents to modify lysine side-chains.

Using these principles it is possible to draw up a list of reagents that display reasonable specificity for modifying particular types of amino-acid side-chain (Table 5.2). However, the specificity of any modification reaction should always be checked, e.g. by analysis of the modified enzyme, since the reactivity of a side-chain can be considerably influenced by its local environment within an enzyme.

TABLE 5.2

Some chemical modification procedures that show reasonable specificity for amino-acid side-chains in enzymes

Side-chain	Reagent(s) used
Cysteine	Mercurials, e.g. *p*-chloromercuribenzoate. Disulphides, e.g. 5,5′-dithiobis-(2-nitrobenzoic acid). Iodoacetamide
Lysine	2,4,6-trinitrobenzenesulphonate. Pyridoxal phosphate (± reducing agent such as $NaBH_4$)
Histidine	Diethylpyrocarbonate Photo-oxidation
Arginine	Phenylglyoxal 2,3-Butanedione
Tyrosine	Tetranitromethane *N*-acetylimidazole Iodine
Tryptophan	*N*-bromosuccinimide
Aspartic acid or Glutamic acid	Water-soluble carbodiimide plus nucleophile, e.g. glycine methylester

Details of these procedures are given in reference 42.

Changes in the selectivity of certain reagents can be brought about by changing the pH. For instance, at pH values above about 7, the side-chain of cysteine ($pK_a \approx 8$) is extremely reactive towards iodoacetate because there is a significant proportion of the ionized form, $—CH_2—S^-$, which is the reactive nucleophile. At pH values below 6, the fraction of the ionized form of the cysteine side-chain is much less and the side-chain of methionine ($—CH_2—CH_2—S—CH_3$) is a more reactive nucleophile. By working at pH 5.6 it was possible to react a methionine side-chain with iodoacetate in isocitrate dehydrogenase from pig heart without any modification of cysteine side-chains.[47]

5.4.4.2 *Super-reactive side-chains*

Sometimes one particular amino-acid side-chain in an enzyme is especially reactive because of its unique environment. In such cases, specific modifications can be achieved in an 'accidental' manner. A good example of such a super-reactive group is provided by chymotrypsin, where it is found that diisopropylfluorophosphate modifies only the side-chain of Ser 195 and does not react with any of the other 27 serine side-chains in the enzyme or with free serine. The reason for the super-reactivity of the side-chain of Ser 195 has been found from

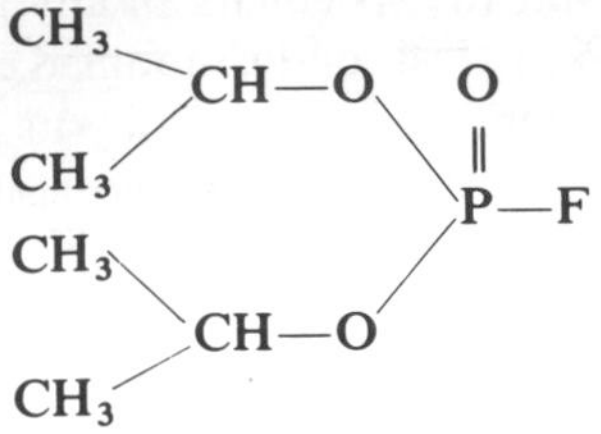

$$\begin{array}{l} (CH_3)_2CH{-}O \\ \qquad\qquad\qquad\quad P(=O){-}F \\ (CH_3)_2CH{-}O \end{array}$$

Diisopropylfluorophosphate

X-ray crystallographic studies (Section 5.5.1.2). Another example of a super-reactive group is provided by glutamate dehydrogenase from beef liver, where the side-chain of one lysine (Lys 126) out of a total of over 30 lysine side-chains in each subunit is especially reactive towards 2,4,6-trinitrobenzenesulphonate.[48] In chymotrypsin and glutamate dehydrogenase these super-reactive groups are involved in the catalytic mechanisms, but this is not always the case. For instance, the highly reactive thiol group in creatine kinase does not appear to be essential for the catalytic activityof the enzyme (Section 5.4.4.4).

5.4.4.3 *Affinity labelling*

One way of improving the specificity of a modifying reagent is to incorporate within it some structural feature that will 'direct' it to a certain site on the enzyme, such as the active site. The reactive part of the reagent will then react with an amino-acid side-chain in the vicinity of that site. In the case of triosephosphate isomerase, bromohydroxyacetone phosphate acts as an affinity label;[49] the resemblance between the affinity label and the substrate for the enzyme is shown below:

Substrate	*Affinity label*
CH_2OH	CH_2Br
|	|
$C=O$	$C=O$
|	|
$CH_2OPO_3^{2-}$	$CH_2OPO_3^{2-}$
Dihydroxyacetone phosphate	Bromohydroxyacetone phosphate

The affinity label binds to the active site of the enzyme and the labile Br atom (activated by the adjacent carbonyl group) can be displaced by a suitably positioned nucleophilic amino-acid side-chain (Section 5.5.2.4). Affinity labelling of chymotrypsin and of trypsin is described in Section 5.5.1.4.

The ideas underlying affinity labelling have been extended to the design of so-called 'suicide substrates' for enzymes.[50] When acted upon by the appropriate enzyme such a substrate is converted to a product that essentially irreversibly inactivates the enzyme, usually by covalent modification (see the scheme below)

$$\mathrm{E+X} \rightleftharpoons \mathrm{E.X} \longrightarrow \mathrm{E.X^*} \begin{cases} \xrightarrow{k_1} \mathrm{E-X^*} \\ \xrightarrow{k_2} \mathrm{E + X^*} \end{cases}$$

Reaction of an enzyme, E, with a suicide substrate, X. After binding to the enzyme, the substrate is converted by the enzyme to a reactive form (X*). The E.X* complex can either form E–X*, an irreversibly inactivated complex, or can dissociate to form E and X*. The balance between these two processes will depend on the relative values of the rate constants k_1 and k_2.

A number of suicide substrates (e.g. clavulanic acid and sulbactam) have been designed for penicillinase (the enzyme that confers reistance towards penicillin), and it is clear that this is an area of considerable pharmaceutical importance.[50]

5.4.4.4 *Interpretation of chemical modification experiments*

Considerable caution is required in interpreting the results of chemical modification experiments. If modification of a particular amino-acid side-chain leads to inactivation of an enzyme, it does not necessarily follow that the side-chain is directly involved in the catalytic mechanism. It could happen, for instance, that the reacting side-chain is some way from the active site, but that modification causes a conformational change in the enzyme leading to loss of activity (Fig. 5.11). There is obviously a whole range of possibilities for the positioning of

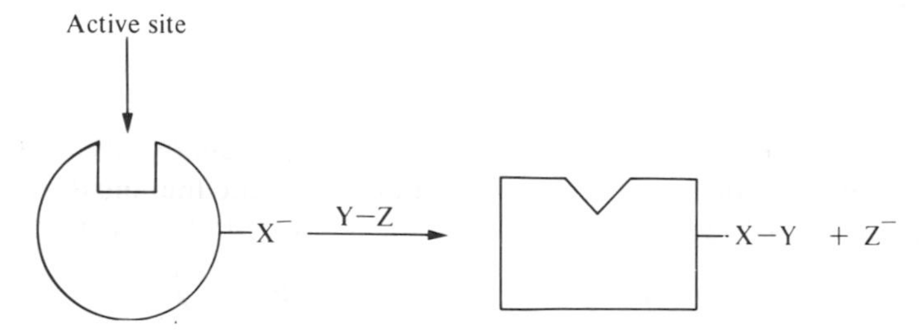

FIG. 5.11. Modification of a group, $—X^-$, outside the active site of an enzyme causes a conformational change leading to loss of activity.

such a side-chain relative to the active site. In many cases it may be possible to make an assessment of the extent of involvement of a side-chain in the catalytic mechanism by comparing the effects of modification by groups of different sizes. Thus, in the case of creatine kinase from rabbit muscle, modification of a single cysteine side-chain by iodoacetamide leads to complete loss of activity (a); however, the smaller perturbation (b) leads to only a 30 per cent loss of activity.[51]

(a) $\mathrm{E—CH_2—SH} \rightarrow \mathrm{E—CH_2—S—CH_2—CO—NH_2}$

(b) $\mathrm{E—CH_2—SH} \rightarrow \mathrm{E—CH_2—S—CN}$

The cysteine side-chain cannot therefore be directly involved in the catalytic mechanism. Side-chains that can be ascribed a secondary type of role may play a part in binding of substrates or in conformational changes in the enzyme that are involved in catalysis. Any modification of a side-chain that was directly involved in the catalytic mechanism, such as that of Ser 195 of chymotrypsin which takes part in the formation of the acyl enzyme, would lead to complete inactivation.

In a chemical modification experiment there are two criteria that must be fulfilled before it can be stated with any degree of certainty that a modified side-chain is at the active site of an enzyme.[52]

Firstly, there must be a stoichiometric relationship between the extent of inactivation and the extent of modification, so that complete inactivation corresponds to the modification of one side-chain per active site (Fig. 5.12(a)). Secondly the addition of substrate or competitive inhibitor must protect against inactivation (Fig. 5.12(b)), since the side-chain is no longer accessible to the modifying reagent. The need for proper analysis of such protection experiments has been emphasized.[53]

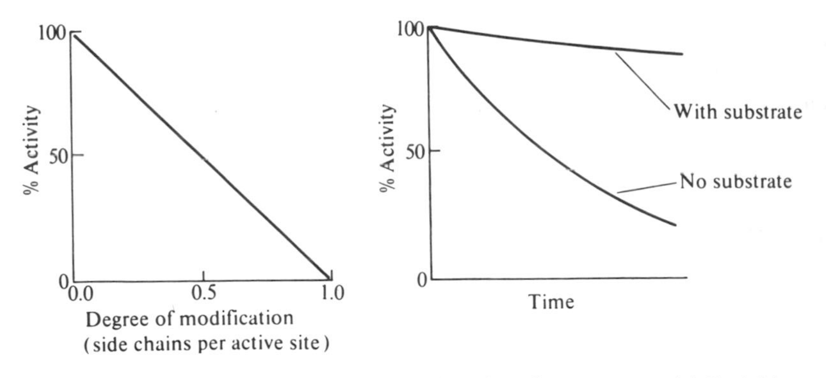

FIG. 5.12. Criteria for modification at the active site of an enzyme. (a) Stoichiometric relationship between extent of inactivation and extent of modification. (b) Protection against inactivation by addition of substrate (or competitive inhibitor).

These criteria should be regarded only as minimal and it is possible that they can be fulfilled even when modification is occurring at a side-chain remote from the active site; the example of the modification of the side-chain of Cys 165 in lactate dehydrogenase is mentioned in Section 5.5.4.4. Despite these problems, chemical modification techniques have been used very extensively to help elucidate details of enzyme mechanisms. We shall refer to chemical modification data in all examples of mechanisms discussed in Section 5.5.

5.4.5 Site-directed mutagenesis

Since the early 1980s the development of recombinant-DNA technology (see Chapter 11, Section 11.5.5) has led to the possibility of introducing amino-acid replacements at specified positions in a protein, provided that (a) the gene coding for the protein is available, and (b) there is a suitable vector for expression of the gene. The details of this technique, known as site-directed mutagenesis or oligonucleotide-directed mutagenesis are beyond the scope of this book, but can be found in a number of articles and reviews.[54–57]

A typical procedure is outlined in Fig. 5.13. The gene of interest is cloned into a vector such as the filamentous phage M13. The DNA of this phage, which is in the form of a covalent circle of single-stranded DNA (ssDNA), goes through a covalent double-stranded form (duplex) during replication in the *E. coli* host. One strand of the DNA is isolated. An oligonucleotide is synthesized that is complementary to the region of the gene to be mutated except for a single base (or double base) change that changes the codon from that (those) of the target amino acid(s) to that (those) of the desired replacement amino acid(s). (For example, the change 5′-TTG-3′ to 5′-TTC-3′ changes the codon from that for Gln to that for Glu.) This oligonucleotide is annealed to the DNA strand and serves as the primer for DNA synthesis by the Klenow fragment of DNA polymerase I (this fragment contains the 5′→3′ polymerase and 3′→5′ exonuclease activities, but not the 5′→3′ exonuclease activity of the intact enzyme). After joining (by DNA ligase) the ends of the new strand of DNA, the heteroduplex is used to transform a host (e.g. the JM strain of *E. coli*). Clones containing the mutant DNA can be selected by their ability to hybridize (form hydrogen bonds) to the

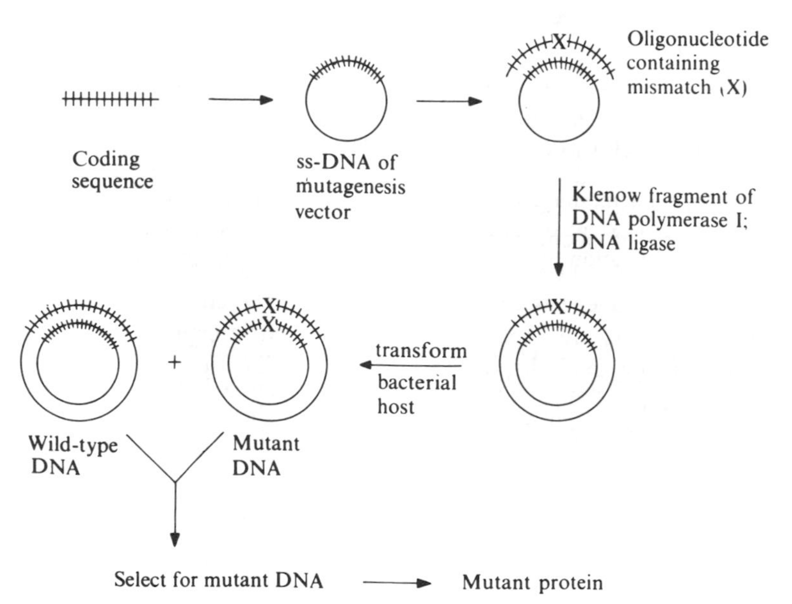

FIG. 5.13. Typical procedure for site-directed mutagenesis. The mismatch is depicted as (X) within a complementary (i.e. base-pairing) region. For further details, see references 54–57.

original oligonucleotide under more forcing conditions of temperature and ionic strength than clones containing the wild-type DNA. The mutant enzyme can then be produced in large quantities in the transformed cells.

A number of techniques are available to increase the frequency of mutants in the transformed cells. One of these involves the incorporation of sulphur-containing nucleotide analogues (phosphothioates) during the *in vitro* synthesis of DNA by the Klenow fragment, which then protects the mutant strand against 'nicking' by certain restriction endonucleases.[58, 59] This allows the amount of non-mutant strand (which is subject to 'nicking') to be reduced by digestion with exonuclease III.

As an alternative to the use of single-stranded templates such as M13, it is possible to use double-stranded vectors from plasmids in which the region requiring to be mutated is converted to a single-stranded form by 'nicking' with a restriction endonuclease followed by treatment with exonuclease III.[54, 60] The exposed portion can then be annealed with the oligonucleotide primer.

The various methods rely on the ready availability of oligonucleotides of defined sequence.[61] Oligonucleotides of 16–21 bases in length are generally used for single base mismatches; longer primers are used for multiple mismatches. The mismatch(es) is (are) best kept towards the middle of the oligonucleotide to protect it (them) from excision by the $3' \rightarrow 5'$ exonuclease activity of the Klenow fragment.

Site-directed mutagenesis is clearly a very powerful method for assessing the importance of particular amino-acid side-chains in an enzyme. It is necessary to perform some control experiments; thus the mutant gene should be sequenced to confirm that only the desired mutation(s) has (have) been introduced. In addition, it is desirable to check that the overall structural features of the enzyme have not been altered by the mutation, since otherwise it would be difficult to ensure that a change in the catalytic properties is solely a consequence of the replacement amino acid introduced. Spectroscopic techniques such as circular dichroism (see Chapter 3, Section 3.5.2) are useful in this regard, since the overall structural features of an enzyme can be ascertained quickly,[62] in contrast to the laborious process of structure determination by X-ray crystallography.

Site-directed mutagenesis requires the gene for the enzyme of interest to be available. The results are capable of more detailed interpretation if the X-ray structure of the enzyme is known. Chemical modification of amino-acid side-chains (see Section 5.4.4) is still widely used in cases where these conditions are not fulfilled and to gain an idea of the types of side-chains that may be important for the function of an enzyme. In addition, chemical modification can introduce functionalities into a protein other than those available in the range of amino-acid side-chains. Thus heavy atoms can be introduced for X-ray crystallography (See Chapter 3, Section 3.5.1) and spin labels or fluorescent groups can be introduced for spectroscopic studies.[63]

The impact of site-directed mutagenesis on studies of enzyme mechanisms is illustrated in most of the examples discussed in Section 5.5.

5.5 Examples of enzyme mechanisms

In this section we shall discuss current ideas about the mechanisms of action of five enzymes; chymotrypsin, triosephosphate isomerase, fructose-bisphosphate aldolase, lactate dehydrogenase and tyrosyl-tRNA synthetase. These enzymes catalyse different types of reactions, i.e. they belong to different classes in the classification scheme described in Chapter 1, Section 1.5.1. In each example it should become apparent how various experimental approaches have been used in a complementary fashion to formulate a reasonably detailed picture of the mechanism. Comments on the mechanisms of related enzymes will be made where appropriate.

5.5.1 Chymotrypsin

Chymotrypsin is a protein-hydrolysing enzyme. It is synthesized in the acinar cells of the pancreas as an inactive precursor (zymogen) known as chymotrypsinogen that is activated in the small intestine by the action of proteases (see Chapter 6, Section 6.2.1.1). The mechanism of chymotrypsin has been intensively studied[64] and is probably better understood than any other. We shall discuss in turn the information gained from studies of X-ray crystallography, site-directed mutagenesis, detection of intermediates, chemical modification, and kinetics before outlining the proposed mechanism of action.

5.5.1.1 Background information on the structure and action of the enzyme

Chymotrypsin has an M_r of about 25 000 and consists of three polypeptide chains held together by disulphide bonds. This unusual structure arises as a result of the activation of the zymogen by proteolytic cleavage (Fig. 5.14). The activation of chymotrypsinogen occurs in stages.[64] There is an initial cleavage, catalysed by trypsin, of the Arg 15-Ile 16 bond that yields a fully active, two-chain, enzyme known as π-chymotrypsin. Subsequent cleavage, catalysed by

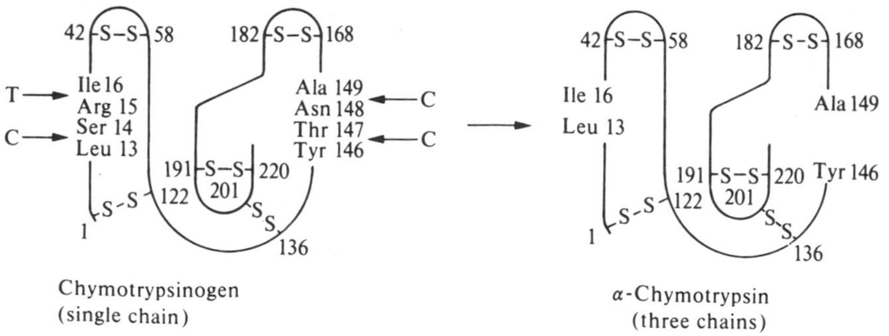

42 S–S 58 182 S–S 168
T → Ile 16
Arg 15
C → Ser 14
Leu 13
Ala 149 ← C
Asn 148
Thr 147 ← C
Tyr 146
191 S–S 220
122 201
1 S–S
S–S
136
Chymotrypsinogen
(single chain)

→

42 S–S 58 182 S–S 168
Ile 16
Leu 13
Ala 149
191 S–S 220 Tyr 146
122 201
1 S–S
S–S
136
α-Chymotrypsin
(three chains)

FIG. 5.14. Activation of chymotrypsinogen. T and C represent the actions of trypsin and chymotrypsin, respectively.

chymotrypsin then takes place to yield δ-chymotrypsin (by cleavage of the Leu 13-Ser 14 bond), and finally α-chymotrypsin (by cleavage of the Tyr 146-Thr 147 and Asn 148-Ala 149 bonds). α-Chymotrypsin is the most commonly studied form of the enzyme.

Chymotrypsin catalyses the hydrolysis not only of proteins but also of small substrates and in the latter case will act on both ester and amide bonds. The enzyme shows relatively little specificity when acting on ester substrates: a finding that can be correlated with the inherently higher reactivity of esters. In the case of amide bonds (which are intrinsically less easily hydrolysed) the only substrates that are acted on by chymotrypsin are those with large hydrophobic side-chains, principally those of Trp, Tyr, and Phe, and, to a lesser extent, those of Leu and Met. Specificity is shown for the nature of the R group attached to the carbon atom on the amino side of the amide bond. Despite the differences in specificity as far as amide and ester substrates are concerned, there are good grounds for believing that the mechanism of hydrolysis is the same for both types of substrates (Section 5.5.1.3).

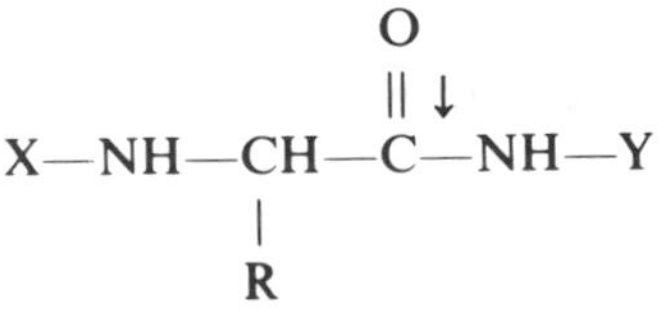

```
                O
                ‖ ↓
X—NH—CH—C—NH—Y
          |
          R
```

5.5.1.2 X-ray crystallographic studies

The structure of α-chymotrypsin has been discussed briefly in Chapter 3, Section 3.5.6.2, where it was mentioned that the molecule contains a significant amount of antiparallel β-sheet structure. There is a shallow depression at the active site, and also a distinct pocket that plays an important part in the binding of specific substrates.[65] Of special note is the finding that certain charged groups, e.g. the α-amino group of Ile 16* and the carboxylate side-chains of Asp 102 and Asp 194, are buried in the interior of the molecule.[65] The significance of these groups is discussed later in this section.

X-ray crystallographic studies have provided answers to a number of questions about chymotrypsin:

(1) Why is the side-chain of Ser 195 especially reactive?
(2) How do substrates bind to the enzyme?
(3) How do the differences in specificity between chymotrypsin and related proteases arise?
(4) How is the zymogen activated?

* The numbering of amino acids in chymotrypsin refers to the sequence of the intact polypeptide chain of chymotrypsinogen.

(i) Why is the side-chain of Ser 195 especially reactive?

Chymotrypsin is one of a number of enzymes (trypsin, elastase, and thrombin are other examples) that are classified as serine proteases, on the basis that modification of a single serine side-chain in each molecule by a reagent such as diisopropylfluorophosphate leads to inactivation. In chymotrypsin this reactive serine has been identified as Ser 195. The reason for the high reactivity of this side-chain (the other 27 serine side chains in chymotrypsin do not react with diisopropylfluorophosphate, nor does free serine) was unclear until X-ray crystallography showed that there was a precisely arranged group of three side-chains in the enzyme.[66] This arrangement allows the possibility of partial ionisation of the side-chain of Ser 195 by a 'charge relay' mechanism (Fig. 5.15).

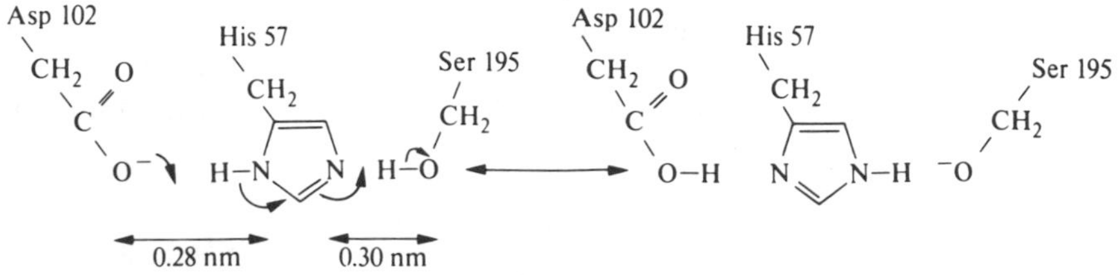

Asp 102 His 57 Ser 195 ⟷ Asp 102 His 57 Ser 195

0.28 nm 0.30 nm

FIG. 5.15. The charge-relay system in chymotrypsin.[66]

Normally ionization of serine side chains is insignificant because of the high pK_a (≈ 14) of the —CH_2—OH group. The partial negative charge on the oxygen atom increases the nucleophilicity of the serine side-chain enormously; this is exploited in the catalytic mechanism that proceeds via acyl enzyme formation. All other serine proteases so far examined possess this catalytic triad or charge-relay arrangement of Asp . . . His . . . Ser side-chains. The extensive similarities of amino-acid sequences around these amino acids in various serine proteases have already been noted (see Chapter 3, Section 3.4.9.5). The details of the charge-relay systems in these enzymes have been extensively investigated. Using n.m.r.[67, 68] and neutron diffraction[69] techniques it has been shown that His 57 is the ionizing group of pK_a around 7 and that the buried Asp 102 is of low pK_a. Site-directed mutagenesis has also been used to assess the importance of Asp 102 in the charge-relay system.[70] The mutant trypsin Asp 102 → Asn showed a k_{cat} value some 5000-fold lower than that of the native enzyme, showing that Asp 102 is important but not absolutely required for catalysis. Using affinity labels that modified different side-chains at the active site, it was shown that the Asp 102 → Asn mutation had considerably greater effects on the reactivity of Ser 195 than on that of His 57. Other studies using site-directed mutagenesis have shown that the three side-chains in the charge-relay system act synergistically to cause rate enhancement.[156]

(ii) How do substrates bind to the enzyme?

A number of crystallographic studies have been made of the binding of poor substrates or competitive inhibitors to the enzyme. The results of such studies[71] on the binding of the inhibitor *N*-formyl tryptophan (amides of which are good substrates for chymotrypsin) are shown in Fig. 5.16. The aromatic side-chain of the inhibitor binds in the pocket, which is lined with non-polar side-chains. The —NH— group forms a hydrogen bond with the peptide-chain carbonyl group of Ser 214 and the carboxyl group (and hence, by implication, the amide bond of a substrate) is close to the side-chain of Ser 195. The carbonyl group can also form hydrogen bonds with the peptide chain —NH— group of Ser 195 and/or Gly 193.

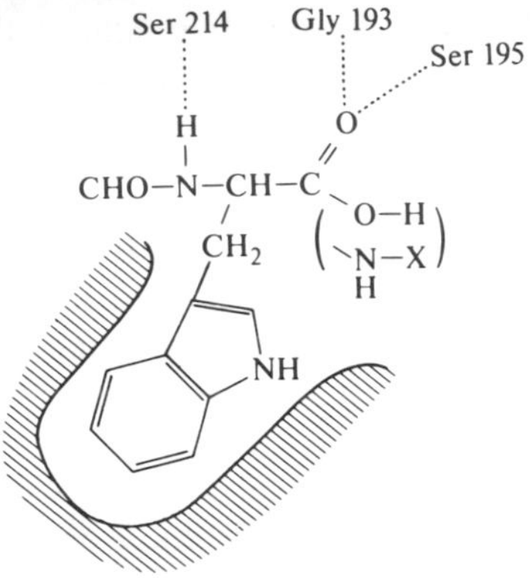

FIG. 5.16. Binding of the inhibitor *N*-formyl tryptophan to chymotrypsin. The structure of the substrate is indicated in brackets.

(iii) How do the differences in specificity between chymotrypsin and related proteases arise?

The serine proteases chymotrypsin, trypsin, and elastase possess very similar three-dimensional structures, but display quite different specificities for substrates. Chymotrypsin is specific for amides with aromatic or other large hydrophobic side-chains, trypsin is specific for amides with positively charged side-chains (Lys or Arg), and elastase has a somewhat broader specificity, showing a preference for amides with small hydrophobic side-chains, e.g. Ala.[72] The results of X-ray crystallography have shown that these differences in specificity can be correlated with differences in the substrate-binding pockets of the three enzymes (Fig. 5.17). In the case of chymotrypsin, the hydrophobic side-chain of the substrates makes a number of favourable contacts with the non-polar side-chains lining the pocket. At the bottom of the pocket is the uncharged side-chain of Ser 189. In trypsin, this serine is replaced by aspartic acid and the negatively charged side-chain is then exactly placed to bind electrostatically to the positively charged side-chain of a substrate. Elastase has a pocket very similar to that of chymotrypsin; however, access to the pocket is obstructed by the bulky side-chains of Val 216 and Thr 226 and hence bulky side-chains of potential substrates cannot enter the pocket.[72] Site-directed mutagenesis has been used to investigate the specificity of trypsin.[70] The side-chains of Gly 216 or

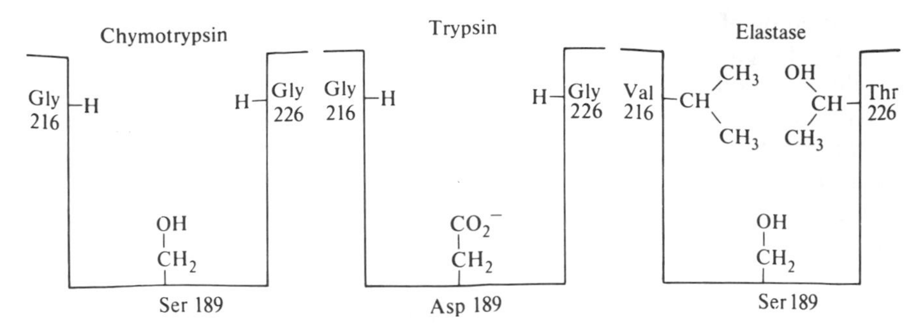

FIG. 5.17. Comparison of the substrate binding pockets of the serine proteases chymotrypsin, trypsin, and elastase.

Gly 226 or both have been replaced by Ala. The mutant enzymes show differing specificities towards lysine and arginine substrates; thus the mutation Gly 226 → Ala increases the preference of the enzyme for lysine rather than arginine.

(iv) How is the zymogen activated?

A comparison of the three-dimensional structures of chymotrypsinogen and chymotrypsin allows us to identify the differences between the two molecules and to explain the activation process in structural terms. One perhaps surprising finding is that the Asp . . . His . . . Ser charge-relay system is already present in the zymogen, with the three side-chains occupying very similar or identical positions to those found in the active enzyme.[73] The zymogen, however, is inactive* because the substrate-binding pocket is not properly formed, and thus a substrate cannot be positioned precisely to take advantage of the reactive serine side-chain. Formation of the substrate-binding pocket during the activation process proceeds as follows.[73] Cleavage of the Arg 15-Ile 16 bond by trypsin creates a new positive charge at the α-amino group of Ile 16. A strong electrostatic force between this positive charge and that of the side-chain of Asp 194 helps to move other parts of the molecule, such as the side-chains of Arg 145 and Met 192, and to form the substrate-binding pocket. The electrostatic interaction between Ile 16 and Asp 194 is strong because it occurs in a region of low dielectric constant in the interior of the enzyme (see Chapter 3, Section 3.5.5.2).

5.5.1.3 Detection of intermediates

We have already mentioned (Section 5.4.1.5) that the chymotrypsin-catalysed hydrolysis of esters such as *p*-nitrophenyl acetate proceeds via an acyl enzyme intermediate.

* Some careful work has shown that chymotrypsinogen does show very slight activity in catalysing the hydrolysis of esters; the activity is some 10^6-fold less than that of chymotrypsin. The mechanism of the zymogen-catalysed reaction appears to be similar to that of the active enzyme, proceeding via an acyl enzyme.[74]

$$\mathrm{E{-}CH_2{-}OH} \xrightarrow{\text{fast}} \mathrm{E{-}CH_2{-}O{-}\overset{\overset{\displaystyle O}{\|}}{C}{-}R}$$

(slow: $\mathrm{E{-}CH_2{-}O{-}CO{-}R} \rightarrow \mathrm{E{-}CH_2{-}OH}$)

The rate of the deacylation step is pH-dependent and can be slowed down to such an extent that at low pH the acyl enzyme can be isolated; the acyl group is linked to the side-chain of Ser 195.[32]

In the hydrolysis of amides, it would appear that the deacylation step is more rapid than formation of acyl enzyme; this can be correlated with the lower reactivity of amides compared with esters. There is, therefore, no accumulation of the acyl enzyme and its participation in the reaction can only be inferred indirectly by 'trapping' experiments. A suitable 'trapping' experiment involves conducting the enzyme-catalysed hydrolysis of a substrate in the presence of a high concentration of a nucleophile such as alaninamide or hydroxylamine that can compete with water in the breakdown of the acyl enzyme:

$$\mathrm{E{-}CH_2{-}O{-}\overset{\overset{\displaystyle O}{\|}}{C}{-}R} \xrightarrow{H_2O} \mathrm{E{-}CH_2{-}OH + R{-}C({=}O){-}OH} \quad \text{(a)}$$

$$\mathrm{E{-}CH_2{-}O{-}\overset{\overset{\displaystyle O}{\|}}{C}{-}R} \xrightarrow{NH_2CH(CH_3)CONH_2} \mathrm{E{-}CH_2{-}OH + R{-}C({=}O){-}NHCH(CH_3)CONH_2} \quad \text{(b)}$$

The ratio of products (a)/(b) was determined using an ester substrate for which the involvement of an acyl enzyme intermediate had been established directly. It was then found that, at the same concentration of nucleophile, an identical product ratio was observed using an amide substrate.[75] This provides very good evidence for the participation of an acyl enzyme intermediate in the hydrolysis of amides.

By analogy with hydrolytic reactions in organic chemistry, it has been presumed that the formation and decomposition of the acyl enzyme proceed via tetrahedral intermediates (Fig. 5.18).

Direct evidence for the existence of tetrahedral intermediates has been gathered from a number of experiments, mainly on enzymes related to chymotrypsin. Firstly, stopped-flow studies (Chapter 4, Section 4.4) on the elastase-catalysed hydrolysis of an amide substrate, a *p*-nitroanilide derivative of a tripeptide, have shown that the decomposition of the first tetrahedral intermediate to the acyl enzyme is the rate-limiting step of the overall hydrolytic reaction.[76] Secondly, X-ray crystallographic studies have been made of the very strong complexes formed by trypsin with small protein inhibitor molecules such

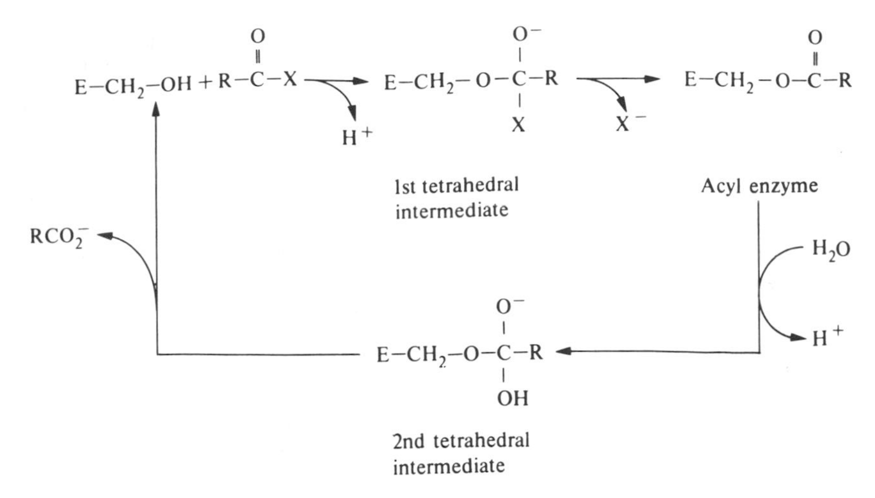

FIG. 5.18. The mechanism of action of chymotrypsin showing the tetrahedral intermediates involved in the formation and breakdown of acyl enzyme.

as bovine pancreatic trypsin inhibitor and soybean trypsin inhibitor.[77, 78] These studies have shown that the strong interactions arise at least in part from the fact that in these complexes the crucial amide bonds of the inhibitors are distorted so as to resemble a tetrahedral intermediate. Thirdly, as we have already mentioned (Section 5.4.2), it has proved possible by working at low temperatures to stabilize a tetrahedral intermediate in the elastase-catalysed hydrolysis of a *p*-nitroanilide substrate.[34]

5.5.1.4 Chemical modification of amino-acid side-chains

There have been a large number of chemical-modification studies on chymotrypsin, but we shall mention only some of those that have confirmed the roles of side-chains implicated by X-ray crystallographic work.

The importance of the side-chain of Ser 195 in the catalytic mechanism was originally deduced from chemical-modification studies using diisopropylfluorophosphate (Section 5.4.4.2). This side-chain can also be modified by phenylmethanesulphonyl fluoride which, being a solid, has certain advantages in handling over diisopropylfluorophosphate, which is a liquid.* Phenylmethanesulphonyl fluoride is widely used in enzyme-purification procedures to minimize degradation by proteases (see the isolation of the *arom* multienzyme protein discussed in Chapter 2, Section 2.8.6).

Some elegant affinity-labelling experiments (Section 5.4.4.3) have shown that the side-chain of His 57 is involved in the catalytic activity of the enzyme. The

* Diisopropylfluorophosphate is an extremely toxic substance, since it inactivates acetylcholinesterase by reaction with a highly reactive serine side-chain, thereby blocking the neurotransmission function of acetylcholine.

close structural relationship between the reagent *N-p*-toluenesulphonyl-L-phenylalanine chloromethylketone (TPCK) and the ester substrate is shown in Fig. 5.19. TPCK inactivates chymotrypsin and fulfils the usual criteria for modification at the active site (Section 5.4.4.4). The site of modification is the side-chain of His 57.[64] That the affinity label does indeed recognize a particular site in chymotrypsin is nicely demonstrated by the finding that TPCK does not inactivate trypsin, which possesses a different substrate-binding pocket (Fig. 5.16). Conversely, the reagent *N-p*-toluenesulphonyl-L-lysine chloromethylketone (TLCK) inactivates trypsin but has no effect on chymotrypsin.[79] The two reagents find considerable application in protein-sequencing studies (Chapter 3, Section 3.4.2), since in the preparation of peptide fragments it is essential to ensure that the chymotrypsin used is not contaminated by trypsin or vice versa. (Thus it is desirable to use TPCK-treated trypsin and TLCK-treated chymotrypsin.)

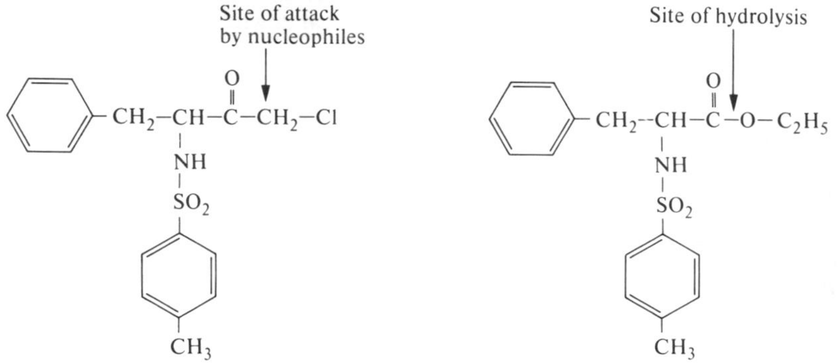

Site of attack by nucleophiles

$C_6H_5-CH_2-CH(NH-SO_2-C_6H_4-CH_3)-C(=O)-CH_2-Cl$

Affinity label (TPCK)

Site of hydrolysis

$C_6H_5-CH_2-CH(NH-SO_2-C_6H_4-CH_3)-C(=O)-O-C_2H_5$

Substrate

FIG. 5.19. The structure of the affinity label TPCK and its relationship to a substrate of chymotrypsin.

The importance of the α-amino group of Ile 16 in chymotrypsin is suggested by chemical modification studies.[64] It would normally be extremely difficult to modify this amino group selectively in view of the presence of the other amino groups in the molecule. This problem was overcome by a 'double-labelling' experiment as outlined in Fig. 5.20, in which chymotrypsinogen was first reacted with acetic anhydride (Ac_2O) to acetylate all the amino groups (α- and ε-) in the molecule. This acetylated chymotrypsinogen could then be cleaved by trypsin to yield a fully active acetylated δ-chymotrypsin which has a free α-amino group at Ile 16. On treatment with ^{14}C-labelled acetic anhydride it is found that activity is lost in proportion to the amount of radioactivity incorporated into the α-amino group of Ile 16, thus confirming the importance of this group in the activity of the enzyme.[64]

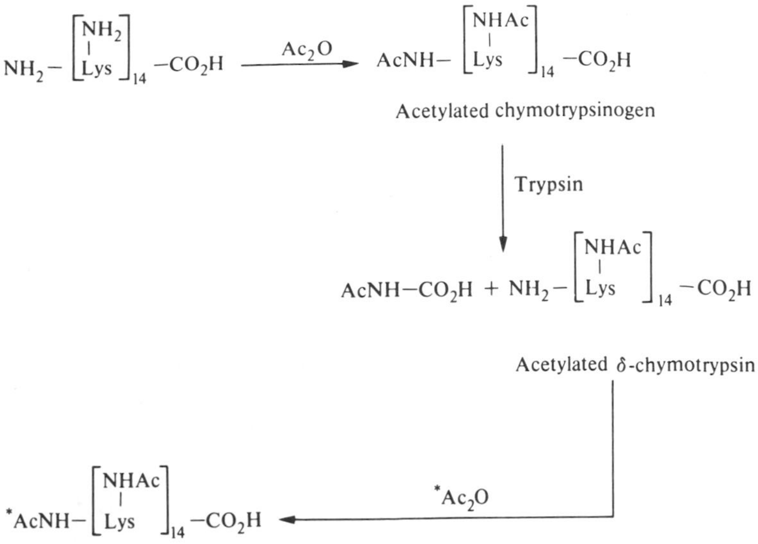

$$NH_2-\begin{bmatrix} NH_2 \\ | \\ Lys \end{bmatrix}_{14}-CO_2H \xrightarrow{Ac_2O} AcNH-\begin{bmatrix} NHAc \\ | \\ Lys \end{bmatrix}_{14}-CO_2H$$

Acetylated chymotrypsinogen

↓ Trypsin

$$AcNH-CO_2H + NH_2-\begin{bmatrix} NHAc \\ | \\ Lys \end{bmatrix}_{14}-CO_2H$$

Acetylated δ-chymotrypsin

$$\xrightarrow{{}^*Ac_2O} {}^*AcNH-\begin{bmatrix} NHAc \\ | \\ Lys \end{bmatrix}_{14}-CO_2H$$

FIG. 5.20. Acetylation of the α-amino group of Ile 16 in chymotrypsin by a double-labelling experiment. Ac_2O is acetic anhydride.

5.5.1.5 Kinetic studies

The kinetics of the reactions catalysed by chymotrypsin have been very intensively investigated. The results of these experiments are more easily interpreted if small synthetic ester or amide substrates are employed; many of these substrates (e.g. *p*-nitrophenyl esters or *p*-nitroanilide derivatives) possess convenient spectroscopic properties that allow the hydrolysis reactions to be monitored easily. The results of steady-state and pre-steady-state kinetic studies have allowed conclusions to be drawn regarding (i) the reaction pathway and the involvement of the acyl enzyme; (ii) the specificity of the enzyme and the nature of the substrate-binding site; and (iii) the magnitudes of the rate constants of a number of the individual steps in Fig. 5.18. For a summary of this work, reference 80 should be consulted.

5.5.1.6 The proposed catalytic mechanism of chymotrypsin

From the various types of experiment described in Section, 5.5.1.2 to 5.5.1.5, a mechanism for the chymotrypsin-catalysed hydrolysis of amide (and ester) substrates has been proposed.[66] The mechanism is outlined in Fig. 5.21, where for the sake of clarity details of the substrate-binding site and the movements of substrate and enzyme have been omitted. A fuller account of these aspects is given in reference 81. In the acyl enzyme, the charge-relay system is disrupted, since no proton is available to link the side-chains of His 57 and Ser 195. However, in the deacylation step, the $R'NH_2$ leaving group is replaced by H_2O and the charge-relay system can be re-established.

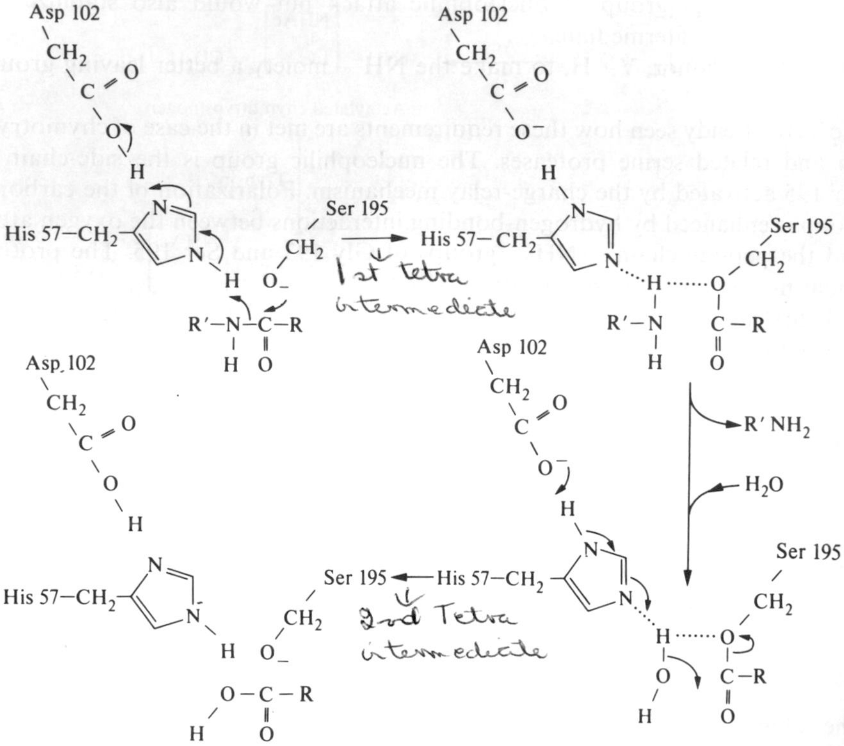

FIG. 5.21. The proposed mechanism of action of chymotrypsin on amide substrates.[66]

5.5.1.7 *The relationship between the mechanism of action of chymotrypsin and those of other proteases*

If we consider the hydrolysis of an amide bond, we can formulate the essential chemical requirements of a catalyst as the possession of the following.

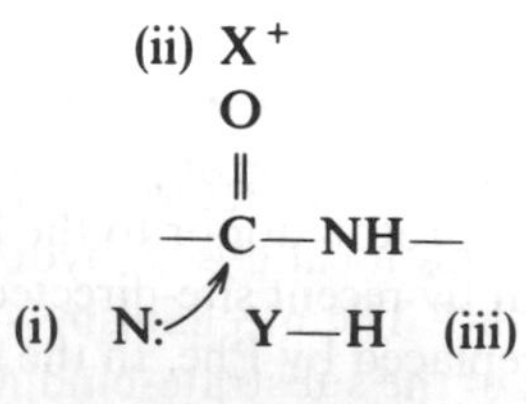

$$\begin{array}{c} \text{(ii) } X^+ \\ O \\ \| \\ -C-NH- \\ \text{(i) } N\!:\quad Y-H \text{ (iii)} \end{array}$$

(1) a nucleophilic group, N, to attack the carbonyl group, leading to the formation of a tetrahedral intermediate;
(2) some positively charged species, X^+, in the neighbourhood of the oxygen atom of the carbonyl group. This would not only increase the susceptibility of

the carbonyl group to nucleophilic attack but would also stabilize the tetrahedral intermediate;

(3) a proton donor, Y—H, to make the NH— moiety a better leaving group.

We have already seen how these requirements are met in the case of chymotrypsin and related serine proteases. The nucleophilic group is the side-chain of Ser 195 activated by the charge-relay mechanism. Polarization of the carbonyl group is enhanced by hydrogen-bonding interactions between the oxygen atom and the peptide chain —NH— groups of Gly 193 and Ser 195. The proton-donating species is the side-chain of His 57.

At first sight, other types of proteases possess very different mechanisms, but it is usually possible to interpret these mechanisms in terms of the requirements outlined above. For instance, in carboxypeptidase[81, 82] (Fig. 5.22), the nucleophilic group is the side-chain of Glu 270, or possibly a water molecule activated by this side-chain and the zinc ion acts as the polarizing influence on the carbonyl group.

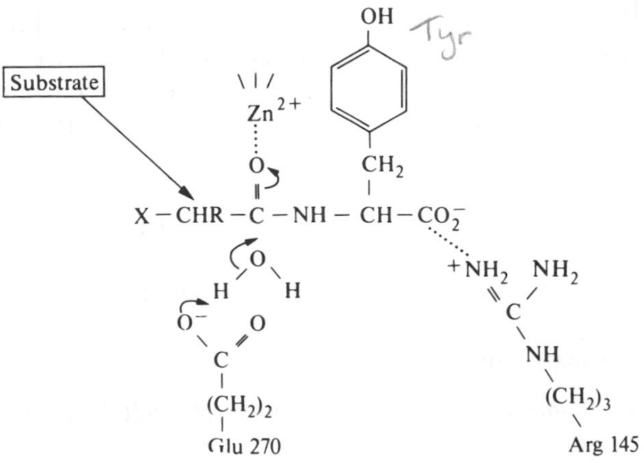

FIG. 5.22. Part of the proposed mechanism of action of carboxypeptidase.[82] An alternative mechanism involves direct attack of the side-chain of Glu 270 on the carbonyl group of the substrate.

It had been previously thought,[82] on the basis of chemical-modification and X-ray data, that the phenolic –OH group of the side-chain of a Tyrosine (probably Tyr 248) acted as a proton donor to the leaving group. However, this possibility has been ruled out by recent site-directed mutagenesis experiments[83] in which Tyr 248 has been replaced by Phe. In the mutant the k_{cat} is unchanged although the K_m for peptide and ester substrates is raised. These results suggest that Tyr 248 plays a role in the binding of substrates. Similar results have been found for the mutant in which Tyr 198 (also close to the active site) has been replaced by Phe and in the double mutant (Tyr 248 and Tyr 198 both replaced by Phe).[84]

In thiol proteases such as papain, the nucleophilic group is a cysteine side-chain that is partially ionized as a result of the presence of a neighbouring histidine side-chain. The polarization of the carbonyl group of the amide bond occurs by hydrogen bonding as in the case of chymotrypsin, and the proton donating group is a histidine side-chain.[81, 85–87] In acid proteases such as pepsin and penicillopepsin, the nucleophilic group appears to be a water molecule activated by an aspartic acid side-chain (Asp 32 in pepsin). The diad of Asp 32 and Asp 215 acts as a proton-donor to the –NH leaving group.[88]

From this discussion it can be appreciated that the various types of proteases have adopted different solutions to the problems associated with the hydrolysis of amide bonds, although all the different solutions can be rationalized in terms of chemical principles. The differences in specificity between the various proteases can be understood in terms of the details of the interactions between enzyme and substrate (see the examples of chymotrypsin, trypsin, and elastase in Section 5.5.1.2). The specificity displayed by carboxypeptidase is for the position of an amide bond in a polypeptide chain, since this enzyme removes the C-terminal amino acid (see Chapter 3, Section 3.4.4.2). By referring to the structure of the active site of the enzyme (Fig. 5.22) we can see that the positively charged side-chain of Arg 145 is in an appropriate position to bind to the C-terminal carboxylate group of the substrate. There are in fact two types of carboxypeptidases, A and B, which are very similar in overall structure. However, carboxypeptidase B, which is specific for the removal of C-terminal Lys or Arg has an appropriately positioned negatively charged side-chain (that of Asp 255) to bind the positively charged side-chains of the substrate.[89] Carboxypeptidase A will remove all C-terminal amino acids other than Lys or Arg from a polypeptide chain.

5.5.2 Triosephosphate isomerase

Triosephosphate isomerase catalyses the following reaction:

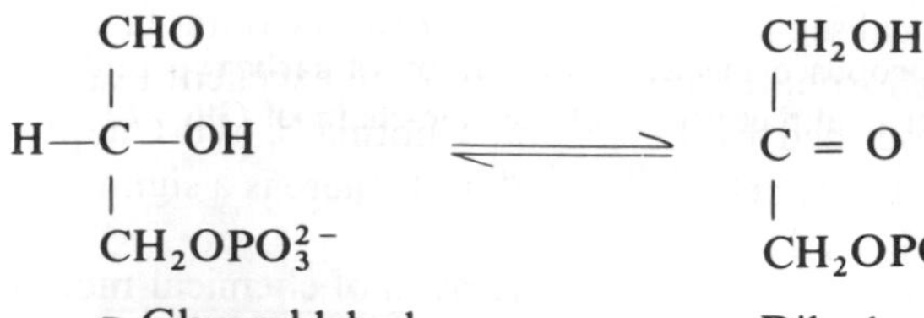

$$\begin{array}{c} CHO \\ | \\ H-C-OH \\ | \\ CH_2OPO_3^{2-} \end{array} \rightleftharpoons \begin{array}{c} CH_2OH \\ | \\ C=O \\ | \\ CH_2OPO_3^{2-} \end{array}$$

D-Glyceraldehyde 3-phosphate (G3P) ⇌ **Dihydroxyacetone phosphate (DHAP)**

The reaction is of great physiological importance in that it allows interconversion of the two three-carbon units produced by the cleavage of D-fructose 1,6-bisphosphate catalysed by fructose-bisphosphate aldolase (Section 5.5.3). D-Glyceraldehyde 3-phosphate is an intermediate in the glycolytic pathway leading from glucose to pyruvate. Dihydroxyacetone phosphate can be reduced to glycerol 1-phosphate, which acts as a precursor in the synthesis of various lipids.

The equilibrium of the triosephosphate isomerase-catalysed reaction is strongly in favour of dihydroxyacetone phosphate:

$$K = \frac{[\text{DHAP}]}{[\text{G3P}]} = 367^* \text{ at } 298 \text{ K } (25°\text{C}).$$

The enzyme has been intensively studied in recent years and a good understanding of its mechanism has been achieved by the use of X-ray crystallography, affinity labelling, and experiments involving the isotopic labelling of substrates. Attention will be focused on triosephosphate isomerase from chicken breast muscle, although it seems that the enzyme from a wide variety of other sources has very similar properties.

5.5.2.1 *Background information on the structure and action of the enzyme*

Triosephosphate isomerase has an M_r of 53 000 and is composed of two subunits of identical sequence. It is of special note that the enzyme is an extremely efficient catalyst, with a turnover number, expressed per active site, of about 250 000 min^{-1} in the direction of conversion of D-Glyceraldehyde 3-phosphate to dihydroxyacetone phosphate. As will be mentioned later (Section 5.5.2.5), there are good grounds for believing that triosephosphate isomerase is an almost perfectly evolved enzyme, i.e. that further evolution to produce a more efficient catalyst is not possible.[91]

5.5.2.2 *X-ray crystallographic studies*

Crystallographic studies have shown that each subunit of the enzyme is roughly spherical with a diameter of approximately 3.5 nm. There is a surprisingly regular manner of chain folding so that each subunit consists of an inner cylinder or 'barrel' of eight strands of parallel pleated sheet. The adjacent strands are linked mainly by helical segments, which thus form the outer face of each 'barrel' (Fig. 5.23)[92]. Triosephosphate isomerase is thus an excellent example of the α/β type of structure described by Levitt and Chothia[93] (see Chapter 3, Section 3.5.4.6). In the contact area between the subunits there is a significant number of amino acids with polar side-chains.

* Earlier values of K were quoted as being around 20. However, it is now recognized that both G3P and DHAP can exist as hydrated forms in aqueous solution.[90] These hydrated (diol) forms such as that shown here for G3P do not act as substrates for the enzyme. When the corrections for the amounts of true (unhydrated) substrate and product are made, K becomes much larger.

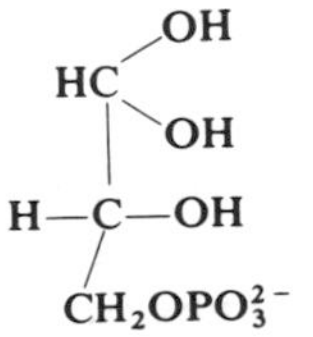

```
       OH
      /
    HC
     | \
     |  OH
     |
  H—C—OH
    /
  CH2OPO3^2-
```

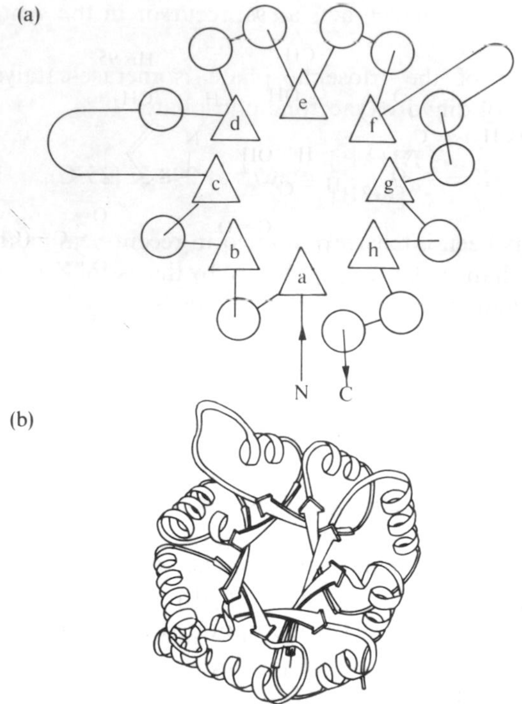

FIG. 5.23. Schematic representations of the chain folding in each subunit of triose-phosphate isomerase.[94] In (a), Circles and triangles represent helices and strands of sheet, respectively. In (b), ribbons and arrows represent helices and strands of sheet, respectively. Part (b) is reproduced with permission from *Advances in Protein Chemistry* **34**, 167–339. Copyright J. S. Richardson (1981).

One important aspect of the X-ray crystallographic work is that it is possible to study a true enzyme–substrate complex, because the equilibrium in this one-substrate reaction lies so much over to one side. An analysis of the complex of enzyme with dihydroxyacetone phosphate showed that the substrate is bound to each subunit in a 'pocket' that includes a few side-chains from the adjacent subunit, i.e. the active site seems to contain contributions from both subunits. The principal amino-acid side-chains that are within about 0.4 nm of the substrate are shown in Fig. 5.24.[92, 94]

The importance of subunit interactions in the catalytic properties of the enzyme has been emphasized by site-directed mutagenesis experiments in which side-chains near the subunit interface, e.g. Asn 14 and Asn 78, were mutated with consequent effects on k_{cat}.[95]

5.5.2.3 *Isotopic labelling of substrates*

The effects of isotopic substitution on the rate of a reaction can give valuable information about the elementary steps of that reaction (for reviews, see references 96 and 97. However, we shall confine our attention to experiments in

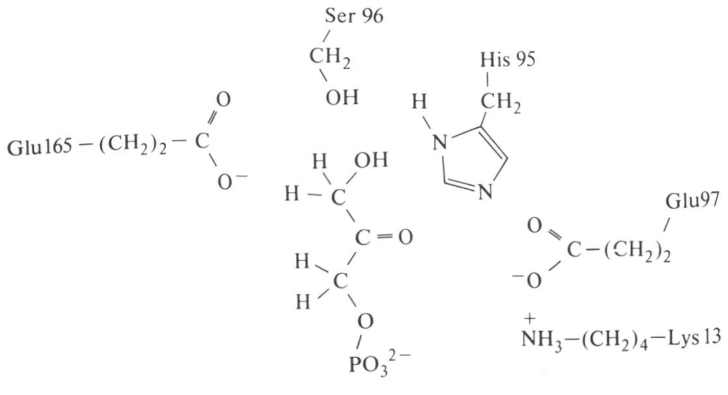

Ser 96 — CH_2 — OH; His 95 — CH_2 — imidazole (N–H, N); Glu165 — $(CH_2)_2$ — $C(=O)O^-$; substrate: H, OH, H–C, C=O, H, H, C, O, PO_3^{2-}; Glu97 — $(CH_2)_2$ — $C(=O)O^-$; $\overset{+}{N}H_3-(CH_2)_4-$Lys 13

FIG. 5.24. Side-chains within about 0.4 nm of the substrate in the triosephosphate isomerase—dihydroxyacetone phosphate complex.[94]

which isotopes are used as labels to follow the fate of a particular atom or group in a reaction.

Early work established that, in the presence of enzyme, up to one atom of tritium (^3H) can be incorporated into dihydroxyacetone phosphate from the solvent, ^3H_2O. This atom of ^3H is incorporated in a stereospecific manner.[98]

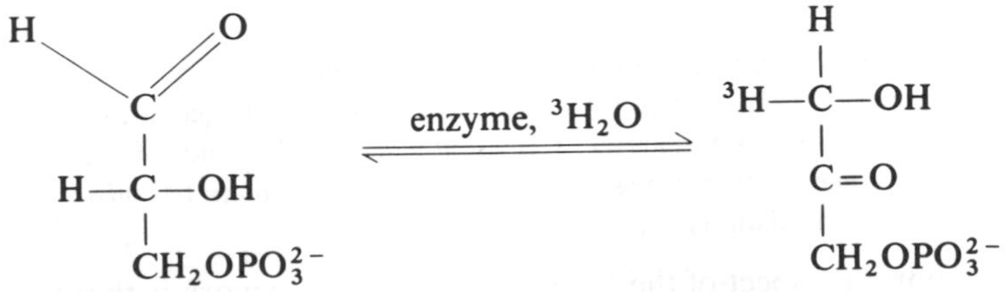

$$\mathrm{H{-}C(=O){-}CH(OH){-}CH_2OPO_3^{2-}} \underset{}{\overset{\text{enzyme, } ^3H_2O}{\rightleftharpoons}} \mathrm{H(^3H)C(OH){-}C(=O){-}CH_2OPO_3^{2-}}$$

This finding suggests that the reaction proceeds via abstraction of a proton from the substrate by a base (B), followed by an exchange with the solvent. On chemical grounds it seems reasonable that a *cis*-enediol intermediate is involved (Fig. 5.25). The extent of incorporation of ^3H into the product will depend on how the rate of exchange of BH^+ with the solvent compares with the rate at which the proton is returned to the *cis*-enediol to give the product. The *cis*-enediol mechanism implies that the proton (or ^3H) is returned to the same face of the diol from which it is abstracted; this is consistent with the observed stereochemistry of the reaction.

Some information on the nature of the base, B, was gained from an experiment in which specifically labelled dihydroxyacetone phosphate was converted in the presence of enzyme to D-glyceraldehyde 3-phosphate. The product was then 'trapped' by an essentially irreversible oxidation with NAD^+ catalysed by glyceraldehyde-3-phosphate dehydrogenase in the presence of arsenate.[91] It was found that there was a small amount (about 5 per cent) of transfer of isotope from C-1 to C-2 (Fig. 5.26). The fact that even a small amount of transfer of label is

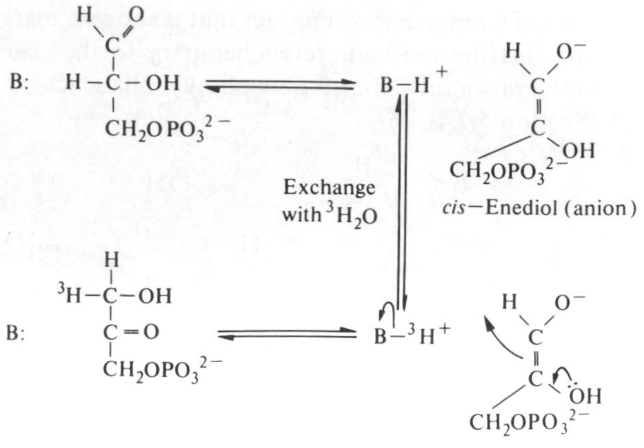

$$B{:}\ \ H-C(=O)-C(H)(OH)-CH_2OPO_3^{2-} \rightleftharpoons B-H^+\ \ H(O^-)C=C(OH)CH_2OPO_3^{2-}$$

cis-Enediol (anion)

Exchange with ^3H_2O

$$B{:}\ \ {}^3H-C(H)(OH)-C(=O)-CH_2OPO_3^{2-} \rightleftharpoons B-{}^3H^+\ \ H(O^-)C=C(OH)CH_2OPO_3^{2-}$$

FIG. 5.25. Incorporation of ^3H into dihydroxyacetone phosphate via a *cis*-enediol intermediate in the reaction catalysed by triosephosphate isomerase.

$${}^3H-C(H)(OH)-C(=O)-CH_2OPO_3^{2-} \xrightarrow{\text{Enzyme, } H_2O} \underset{(95\%)}{H-C(=O)-C(H)(OH)-CH_2OPO_3^{2-}} + \underset{(5\%)}{H-C(=O)-C({}^3H)(OH)-CH_2OPO_3^{2-}}$$

Trap with NAD^+, arsenate and glyceraldehyde-3-phosphate dehydrogenase

$$CO_2^- -C(H)(OH)-CH_2OPO_3^{2-} + CO_2^- -C({}^3H)(OH)-CH_2OPO_3^{2-}$$

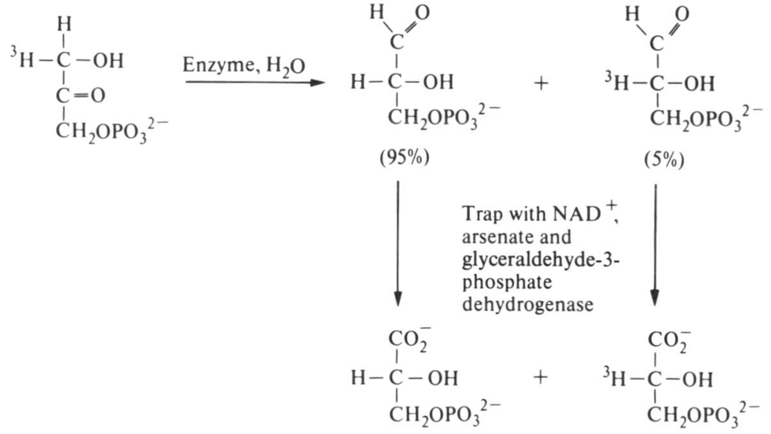

FIG. 5.26. Transfer of ^3H from C-1 to C-2 in the reaction catalysed by triosephosphate isomerase.

observed suggests that only a single base is likely to be involved in the abstraction and return of protons by the *cis*-enediol mechanism shown in Fig. 5.25. If there were more than one base involved, transfer of ^3H could only occur via an elaborate system of proton-exchange reactions within the enzyme. The amount of transfer is small because most of the ^3H on B^3H^+ has exchanged with the solvent (H_2O) to generate BH^+ before the proton is added back to give D-glyceraldehyde 3-phosphate (see Fig. 5.25).

Further support for the idea of a *cis*-enediol intermediate in the enzyme-catalysed reaction was obtained from studies of enzyme inhibition. 2-Phosphoglycollate is a very powerful competitive inhibitor of the enzyme, with a dissociation constant of 6 μmol dm^{-3}. The fact that it shows a marked similarity in terms of electron distribution and stereochemistry to the *cis*-enediol intermediate has led to the proposal[18] that 2-phosphoglycollate acts as a transition-state analogue* (Section 5.3.4).

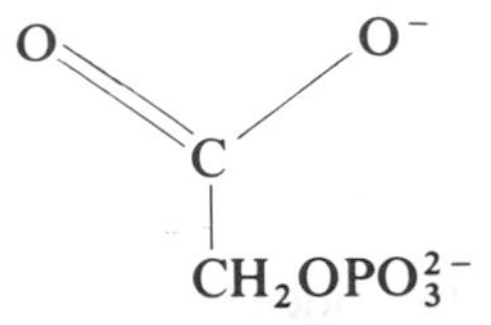

$$O{=}C(O^-)\text{—}CH_2OPO_3^{2-}$$

2-Phosphoglycollate

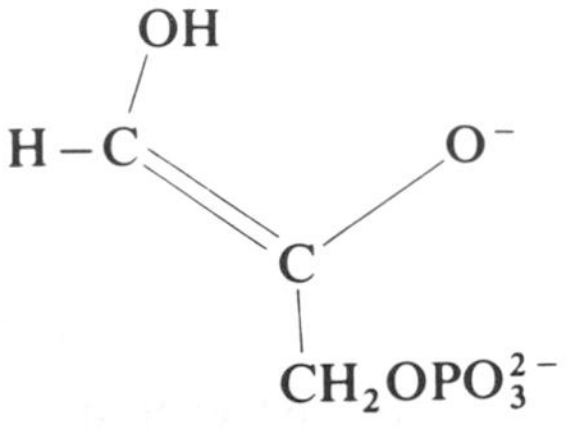

$$H\text{—}C(OH){=}C(O^-)\text{—}CH_2OPO_3^{2-}$$

cis-Enediol intermediate

5.5.2.4 *Affinity labelling*

Some elegant affinity-labelling experiments have helped to identify the base involved in the proton-transfer reactions (Fig. 5.25). Two different affinity labels have been used, bromohydroxyacetone phosphate (I),[49] and glycidol phosphate (II):[99]

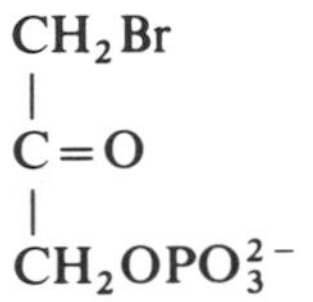

$$CH_2Br\text{—}C{=}O\text{—}CH_2OPO_3^{2-}$$

(I)

$$CH_2\text{—}O\text{—}CH \text{ (epoxide)}\text{—}CH_2OPO_3^{2-}$$

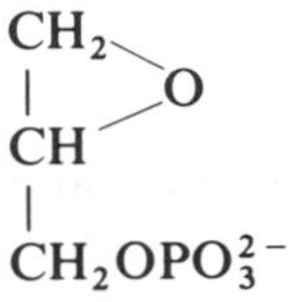

(II)

Both inhibitors conformed to the usual criteria for modification at the active site of the enzyme (Section 5.4.4.4), i.e. stoichiometric inactivation and protection by competitive inhibitors. Inhibitor I clearly resembles dihydroxyacetone phosphate and has an activated Br atom that can be displaced by a nucleophilic group. Inhibitor II resembles the *cis*-enediol intermediate because of the steric requirements of the epoxide ring; this ring can be opened following attack by a suitable nucleophile. (It should be mentioned that the experiments with inhibitor II were conducted with the enzyme isolated from rabbit skeletal muscle.)

In each case the modified, inactivated, enzyme was subjected to proteolytic digestion and the modified peptide isolated. By sequencing the peptides it was

* The value of the dissociation constant (6 μmol dm^{-3}) appears very low by comparison with the quoted K_m values for G3P (320 μmol dm^{-3}) and DHAP (620 μmol dm^{-3}), which give a rough measure of the affinity of the enzyme for these substrates. However, when the corrections are applied for the content of the hydrated forms of these compounds (see foot note on p. 217),[90] the K_m values for the true substrates become G3P (11 μmol dm^{-3}) and DHAP (340 μmol dm^{-3}). The value for G3P is not very different from the dissociation constant for the inhibitor, so there is some doubt about the designation of the inhibitor as a transition-state analogue.

shown that in each case the site of modification was the side-chain of Glu 165, that is also implicated by crystallographic studies as being at the active site (Section 5.5.2.2). In the case of enzyme modified by inhibitor I, it was noted that the attached label could subsequently migrate to the adjacent side-chain, that of Tyr 164; this migration could be prevented by reduction of the modified enzyme with sodium borohydride.[49] This example serves to emphasize that in some chemical-modification experiments there can be a possibility of migration of the modifying group, so that the site of attachment deduced by the normal methods of structural characterization (Chapter 3, Section 3.4) may not correspond to the initial site of modification where the observed effect, e.g. on enzyme activity, occurred.

The affinity-labelling experiments show that the side-chain of Glu 165 is indeed at the active site of the enzyme and the fact that the side-chain can act as a nucleophile makes it very likely that it acts as the general base involved in proton-transfer reactions (Fig. 5.25). The important role played by Glu 165 has been confirmed by site-directed mutagenesis experiments in which Glu 165 was replaced by Asp.[100] The mutant enzyme had a k_{cat} some 10^3-fold lower than that of the native enzyme, with an altered K_m. Determination of the rate of the individual steps in the catalytic process showed that the mutation had decreased the rates of the enolization steps (ES $\rightleftharpoons$ EZ and EZ $\rightleftharpoons$ EP in Fig. 5.29) rather than those of the binding steps (E+S $\rightleftharpoons$ ES and E+P $\rightleftharpoons$ EP).

5.5.2.5 *The mechanism of triosephosphate isomerase*

From the evidence discussed above, the outline mechanism of action of the enzyme is clear (Fig. 5.27).

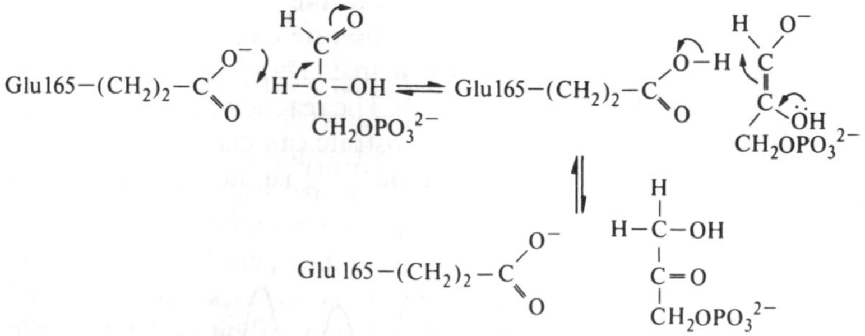

Glu165$-(CH_2)_2-$C(O^-)=O H$-$C(CHO)$-$OH $CH_2OPO_3^{2-}$ $\rightleftharpoons$ Glu165$-(CH_2)_2-$C(O$-$H)=O C(H)(O^-)=C$-$OH $CH_2OPO_3^{2-}$ $\rightleftharpoons$ Glu 165$-(CH_2)_2-$C(O^-)=O H$-$C(H)$-$OH C=O $CH_2OPO_3^{2-}$

FIG. 5.27. Outline of the mechanism of the reaction catalysed by triosephosphate isomerase. Details of the proton-transfer reactions in the second step have been omitted.

There are, however, a number of points that require clarification. Firstly, what (if any) is the role of the side-chain of His 95 that has been shown by X-ray crystallographic studies[92,94] to be close to the substrate-binding site (Fig. 5.24)? This side-chain might well play a part in transferring protons between the carbonyl and hydroxyl oxygen atoms. Secondly, what positively charged side-chain on the enzyme is involved in binding the phosphate group of the substrate? Thirdly, what is the nature of the electrophilic group on the enzyme that serves to polarize the carbonyl group of the substrate? The presence of such a group is not only desirable on chemical grounds, since it helps to make the hydrogen attached to the adjacent carbon atom more acidic, but is also indicated by the fact that dihydroxyacetone phosphate bound to the enzyme is much more readily reduced by hydride ion than is free dihydroxyacetone phosphate, implying polarization of the carbonyl group of the substrate by the enzyme[101] (Fig. 5.28). Additional evidence for the polarization of the carbonyl group of the substrate when it is bound to the enzyme has been obtained from i.r. spectroscopic studies that have provided direct evidence for distortion of the bound substrate.[17] The X-ray crystallographic studies show that the side-chains of Lys 13 and His 95 lie close to the carbonyl oxygen of dihydroxyacetone phosphate; these may well be involved in substrate polarization.[94]

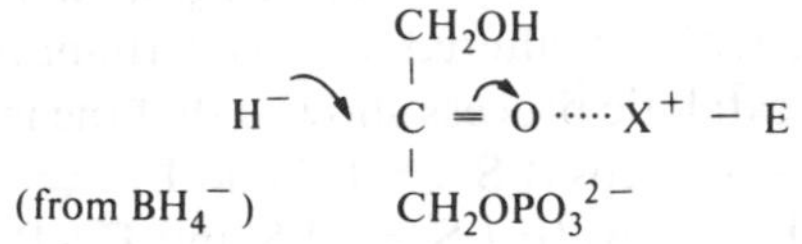

CH_2OH

H^- → C = O ····· X^+ − E

(from BH_4^-) $CH_2OPO_3^{2-}$

FIG. 5.28. Polarization of the carbonyl group of dihydroxyacetone phosphate bound to triosephosphate isomerase. This would facilitate the reduction of the substrate by borohydride.[101]

Very detailed measurements have been made of the rates of exchange of isotope between substrate and solvent during the enzyme-catalysed reaction. These results have been used to calculate the rate constants of the individual steps of the reaction and have enabled a free-energy profile of the reaction pathway to be constructed,[91,102] (Fig. 5.29). The reaction occurs very nearly at a rate that is limited by the rate at which substrate can encounter the enzyme by the normal process of diffusion in solution,[91,103] i.e. no further 'evolutionary

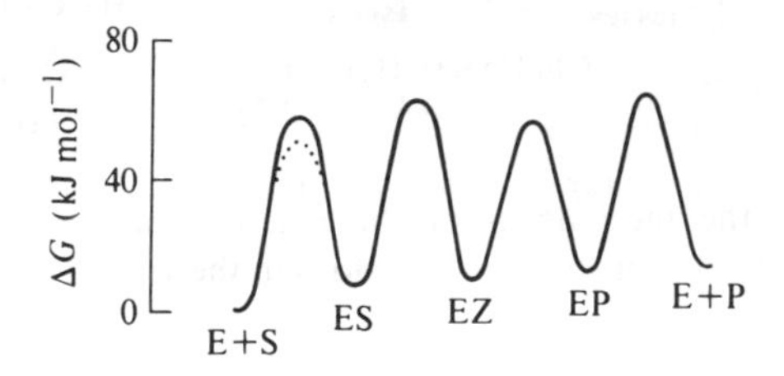

FIG. 5.29. Free-energy profile for the reaction catalysed by triosephosphate isomerase.[102] S is dihydroxyacetone phosphate, P is D-glyceraldehyde 3-phosphate, and Z is the *cis*-enediol intermediate. The dashed line between E + S and ES represents the diffusion-controlled encounter rate. Free energies are referred to a standard state of 40 μmol dm^{-3}, which is the concentration of dihydroxyacetone phosphate *in vivo*.

improvement' in catalytic power is possible. (It is, of course, possible that further evolution might lead to 'desirable' changes in the regulatory properties of the enzyme.)

The mechanism of the triosephosphate isomerase-catalysed reaction is probably analogous to those catalysed by a number of other aldose–ketose isomerases. For instance, the mechanism of glucosephosphate isomerase, which catalyses the interconversion of D-glucose 6-phosphate and D-fructose 6-phosphate, is also thought to proceed via a *cis*-enediol intermediate.[104]

5.5.3 Fructose-bisphosphate aldolase

Fructose-bisphosphate aldolase catalyses the following aldol condensation reaction:

$$\begin{array}{c} CH_2OPO_3^{2-} \\ | \\ C{=}O \\ | \\ CH_2OH \end{array} \quad + \begin{array}{c} CHO \\ | \\ R \end{array} \rightleftharpoons \begin{array}{c} CH_2OPO_3^{2-} \\ | \\ C{=}O \\ | \\ H{-}C{-}OH \\ | \\ HO{-}C{-}H \\ | \\ R \end{array}$$

Of particular interest is the case where R is $-CHOH-CH_2OPO_3^{2-}$ (i.e. where the aldehyde substrate is D-glyceraldehyde 3-phosphate) since this corresponds to the reaction in glycolysis in which the six-carbon unit D-fructose 1,6-bisphosphate (FBP) is cleaved to yield the two three-carbon units D-glyceraldehyde 3-phosphate and dihydroxyacetone phosphate. As befitting an enzyme that is involved in such an important pathway, fructose-bisphosphate aldolase occurs in nearly all plant and animal tissues and in most micro-organisms. The enzyme occurs in a variety of forms[105,106] (Fig. 5.30).

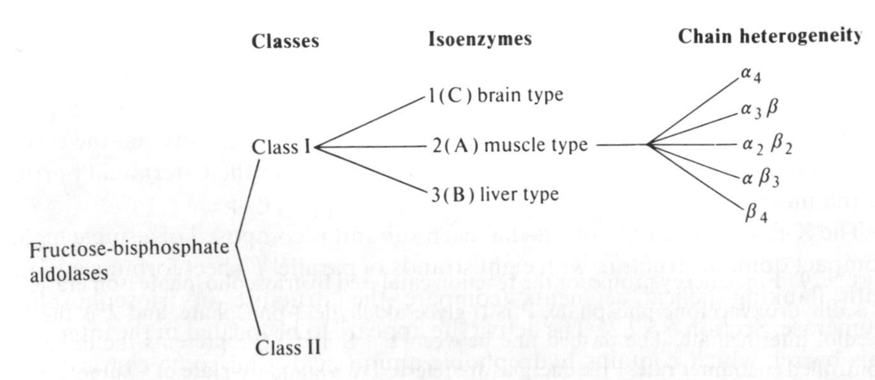

FIG. 5.30. The variety of forms in which fructose-bisphosphate aldolases occur.

There is a clear distinction between the fructose-bisphosphate aldolases from higher plants and animals (Class I), which possess a lysine side-chain at the active site, and those from fungi and bacteria (Class II), which possess a metal ion that is essential for activity. In animal tissues there are three distinct isoenzymes (see Chapter 1, Section 1.5.2). Isoenzyme 2 (also known as A or muscle-type isoenzyme) has a very much higher activity towards FBP than towards D-fructose 1-phosphate (FIP) (the ratio FBP/FIP = 50). By contrast, isoenzyme 3 (also known as B or liver-type isoenzyme) has a ratio of activities FBP/FIP almost equal to 1, and isoenzyme 1 (C or brain-type) has an intermediate ratio (FBP/FIP = 26). The ability of isoenzyme 3 to use FIP as a substrate can be correlated with the role of the liver in the metabolism of fructose. Fructose-bisphosphate aldolase from rabbit muscle consists of five distinct species[105] that can be separated by isoelectric focusing (see Chapter 2, Section 2.6.2.3); this is accounted for by the fact that there are five different ways in which a tetrameric protein can be assembled from two types of subunit (Fig. 5.30). The two types of subunit differ in that the α-type has a C-terminal sequence -Asn-His-Ala-Tyr, and the β-type a corresponding sequence -Asp-His-Ala-Tyr. The conversion $\alpha \rightarrow \beta$ occurs *in vivo*[105] and hence the different forms α_4, $\alpha_3\beta$, etc.) are not isoenzymes according to the definition given in Chapter 1, Section 1.5.2.

Most of the detailed work on fructose-bisphosphate aldolase has been carried out on the enzyme isolated from rabbit skeletal muscle. We shall discuss the mechanism of the rabbit enzyme, although reference will be made to the enzyme from yeast in Section 5.5.3.6. Despite the fact that the pure enzyme has been available in large quantities for a number of years, it is only recently that the X-ray structure of the rabbit muscle enzyme has been published.[107] Data concerning the mechanism of the enzyme have been obtained from structural studies (Section 5.5.3.1), kinetic studies (Section 5.5.3.2), experiments involving the detection of intermediates (Section 5.5.3.3) and the chemical modification of amino-acid side-chains (Section 5.5.3.4). From these studies a reasonably coherent picture of the mechanism can be gained, although further structural work is required to confirm and extend many of the proposals.

5.5.3.1 Background information on the structure of the enzyme

Fructose-bisphosphate aldolase is a tetrameric enzyme of M_r 160 000. The complete amino-acid sequences of a number of Class I aldolases are known.[108,109,110] These confirm[108] that the only difference between the α-type and β-type subunits is the substitution of Asp for Asn in the C-terminal portion of the molecule (Section 5.5.3).

The X-ray structure[107] shows that each subunit is composed of a single highly compact domain structure with eight strands of parallel β-sheet forming a barrel with flanking helical segments (compare the structure of triosephosphate isomerase, Section 5.5.2.2). The active site appears to be located in the interior of this barrel, which contains hydrophobic amino acids and some charged side-chains with acidic and basic groups arranged in an alternating fashion to allow charge neutralization. Amongst conserved side-chains that are located in this interior are Lys 229, Lys 107, Lys 146, and Arg 148. The C-terminal portion of the chain (residues 340–363) is not located clearly in the X-ray structure and is thus presumably highly mobile. It is thought that this portion corresponds to an 'arm-like' structure folded over the surface of the subunit to extend to the active site in the interior of the β-barrel.

5.5.3.2 *Kinetic studies*

By studying the rate of isotope exchange between ^3H_2O and dihydroxyacetone phosphate (DHAP) at equilibrium, and assessing how the rate was affected by changes in the concentration of D-glyceraldehyde 3-phosphate (G3P), it was shown[111] that the reaction is of the ordered ternary complex type (see Chapter 4, Section 4.3.5.1) in which DHAP binds to the enzyme first, followed by G3P:

$$E + DHAP \rightleftharpoons E^{DHAP} \underset{-G3P}{\overset{+G3P}{\rightleftharpoons}} E^{DHAP}_{G3P} \rightleftharpoons E^{FBP} \rightleftharpoons E + FBP.$$

5.5.3.3 *Detection of intermediates*

The nature of the E^{DHAP} complex was deduced from a 'trapping' experiment. Sodium borohydride was added to a mixture of the enzyme and radioactively labelled (either by ^{14}C or ^{32}P) DHAP; this resulted in incorporation of radioactivity into the enzyme and loss of catalytic activity. Analysis of the labelled enzyme showed that a Schiff base had been formed between the carbonyl group of DHAP and the ε-amino group of a lysine side-chain; this could be reduced by borohydride[105] (Fig. 5.31).

Additional evidence in support of the involvement of a Schiff base is provided by the observation that there is an enzyme-catalysed exchange of ^{18}O between

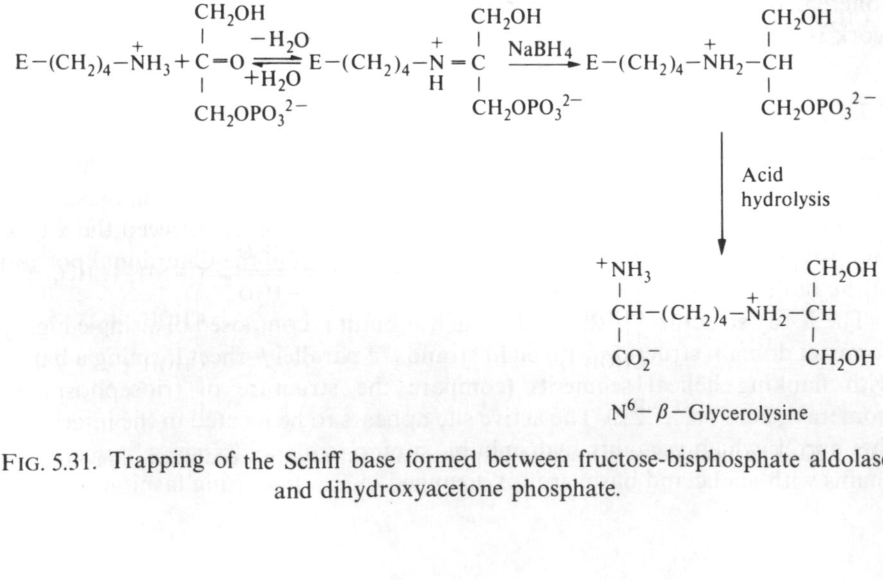

$$E{-}(CH_2)_4{-}\overset{+}{N}H_3 + \underset{CH_2OPO_3^{2-}}{\overset{CH_2OH}{C}}{=}O \underset{+H_2O}{\overset{-H_2O}{\rightleftharpoons}} E{-}(CH_2)_4{-}\underset{H}{\overset{+}{N}}{=}\underset{CH_2OPO_3^{2-}}{\overset{CH_2OH}{C}} \xrightarrow{NaBH_4} E{-}(CH_2)_4{-}\overset{+}{N}H_2{-}\underset{CH_2OPO_3^{2-}}{\overset{CH_2OH}{CH}}$$

$$\xrightarrow{\text{Acid hydrolysis}} \underset{CO_2^-}{\overset{^+NH_3}{CH}}{-}(CH_2)_4{-}\overset{+}{N}H_2{-}\underset{CH_2OH}{\overset{CH_2OH}{CH}}$$

N^6-β-Glycerolysine

FIG. 5.31. Trapping of the Schiff base formed between fructose-bisphosphate aldolase and dihydroxyacetone phosphate.

the carbonyl oxygen of DHAP and water.[112] The side-chain involved in Schiff base formation has been identified as that of Lys 229.[107,108] There is considerable homology between the structures of active site peptides from a variety of Class I fructose-bisphosphate aldolases.[105]

Formation of a Schiff base that is readily protonated is a useful way of generating an electron-withdrawing centre to aid the catalytic process (Section 5.3.3).

$$>C=\overset{+}{N}H \text{ is more electron withdrawing than } >C=O.$$

In the E^{DHAP} complex, this electrophilic centre serves to make a proton on an adjacent carbon atom more acidic. The carbanion can then attack the carbonyl group of the aldehyde substrate to yield the product (Fig. 5.32). This mechanism is analogous to that observed in a model system, namely the dealdolization of diacetone alcohol that is catalysed by primary and secondary amines via the formation and breakdown of a Schiff base intermediate.[113]

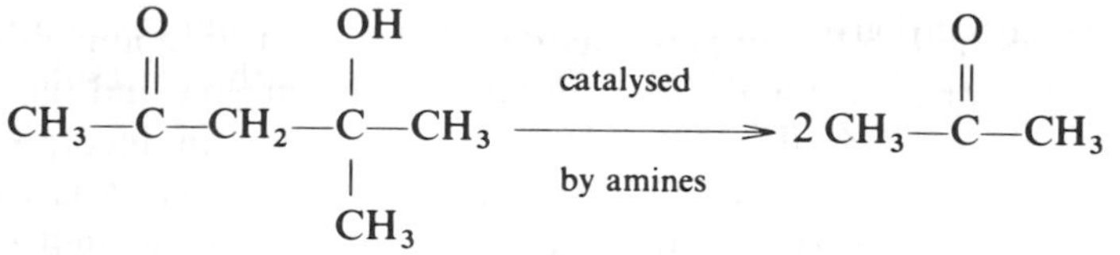

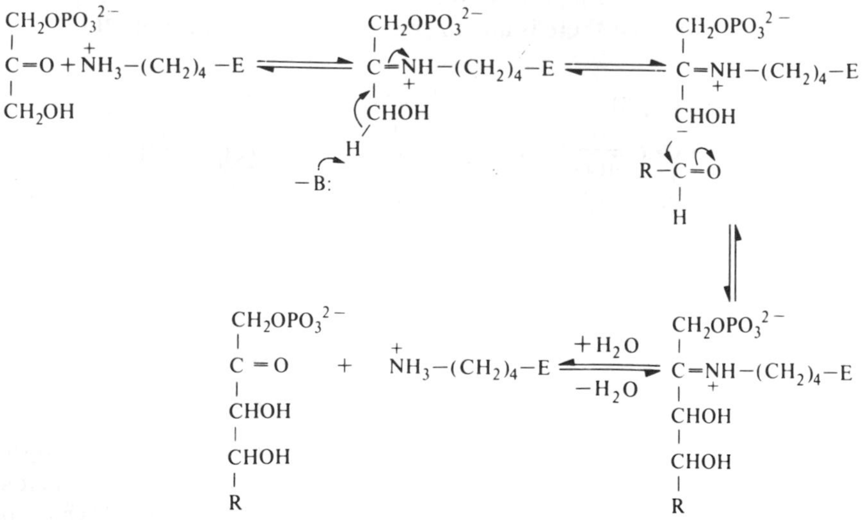

$$CH_3-\overset{O}{\overset{\|}{C}}-CH_2-\underset{CH_3}{\overset{OH}{C}}-CH_3 \xrightarrow[\text{by amines}]{\text{catalysed}} 2\,CH_3-\overset{O}{\overset{\|}{C}}-CH_3$$

$$\begin{matrix} CH_2OPO_3^{2-} \\ | \\ C{=}O + \overset{+}{N}H_3-(CH_2)_4-E \\ | \\ CH_2OH \end{matrix} \rightleftharpoons \begin{matrix} CH_2OPO_3^{2-} \\ | \\ C{=}\overset{+}{N}H-(CH_2)_4-E \\ | \\ CHOH \\ H \\ -B: \end{matrix} \rightleftharpoons \begin{matrix} CH_2OPO_3^{2-} \\ | \\ C{=}\overset{+}{N}H-(CH_2)_4-E \\ | \\ \bar{C}HOH \\ R-\underset{H}{C}{=}O \end{matrix}$$

$$\begin{matrix} CH_2OPO_3^{2-} \\ | \\ C=O \\ | \\ CHOH \\ | \\ CHOH \\ | \\ R \end{matrix} + \overset{+}{N}H_3-(CH_2)_4-E \underset{-H_2O}{\overset{+H_2O}{\rightleftharpoons}} \begin{matrix} CH_2OPO_3^{2-} \\ | \\ C{=}\overset{+}{N}H-(CH_2)_4-E \\ | \\ CHOH \\ | \\ CHOH \\ | \\ R \end{matrix}$$

FIG. 5.32. The mechanism of fructose-bisphosphate aldolase proceeding via a protonated Schiff base.

5.5.3.4 *Chemical modification of amino-acid side-chains*

Numerous chemical-modification experiments have been performed in order to clarify the roles played by various amino-acid side-chains in the mechanism of the enzyme. We shall concentrate on the modifications of lysine and histidine side-chains, as these appear to have provided the most clear cut results.

(i) Lysine

We have already mentioned (Section 5.5.3.3) that the side-chain of Lys 229 is involved in Schiff base formation in the E^{DHAP} complex. Other experiments have suggested that there is a second lysine side-chain at the active site of the enzyme.[114] Pyridoxal phosphate inactivates the enzyme by formation of a Schiff base (Fig. 5.33).

That the side-chain of Lys 229 is not involved in this reaction is shown by the fact that the reduced enzyme–pyridoxal phosphate derivative ((I) in Fig. 5.33) can still form a Schiff base with DHAP.[114] The side-chain modified by pyridoxal phosphate has been identified[108] as that of Lys 107. It has been proposed that this lysine side-chain is involved in electrostatic interactions with the 6-phosphate group of D-fructose 1,6-bisphosphate, and that this binding helps to position the carbonyl group at C-2 so that it can form a Schiff base with the side-chain of Lys 229. This suggestion is in accordance with the results of X-ray crystallography[107] (Section 5.5.3.1).

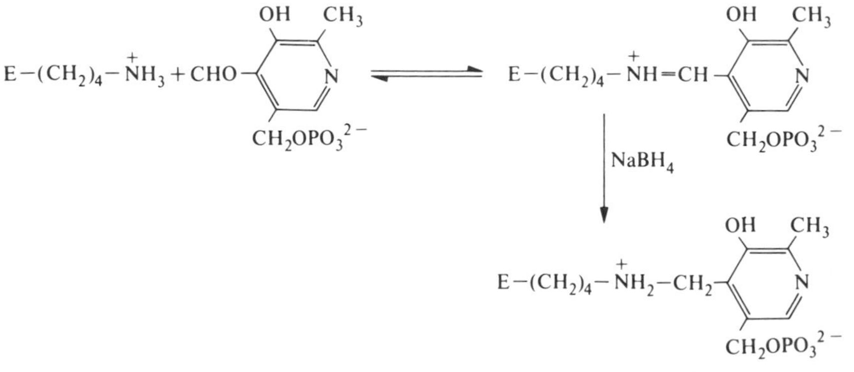

$$E{-}(CH_2)_4{-}\overset{+}{N}H_3 + CHO{-}[\text{pyridine ring: OH, } CH_3, N, CH_2OPO_3^{2-}] \rightleftharpoons E{-}(CH_2)_4{-}\overset{+}{N}H{=}CH{-}[\text{pyridine ring: OH, } CH_3, N, CH_2OPO_3^{2-}]$$

$$\xrightarrow{NaBH_4} E{-}(CH_2)_4{-}\overset{+}{N}H_2{-}CH_2{-}[\text{pyridine ring: OH, } CH_3, N, CH_2OPO_3^{2-}] \quad \text{(I)}$$

FIG. 5.33. Formation of a Schiff base by reaction of fructose-bisphosphate aldolase with pyridoxal phosphate.

(ii) Histidine

The involvement of histidine side-chains in the catalytic mechanism of fructose-bisphosphate aldolase was suggested by some early experiments in which the enzyme was photo-oxidized in the presence of a dye, rose Bengal.[115] This treatment led to inactivation of the enzyme, but the results were rather difficult to interpret because a number of histidine side-chains were modified. A more clear-cut chemical modification experiment involved the reagent *N*-bromoacetyl-ethanolamine phosphate, which inactivates the enzyme by alkylation of one histidine side-chain per subunit;[116] this has been identified as the side-chain of His 361.[108] It was thought that this histidine could well play a role in the proton-transfer reactions in the E^{DHAP} complex (see Fig. 5.32), but this suggestion is now doubtful since it appears that His 361 is not conserved in all Class I aldolases.[109,110]

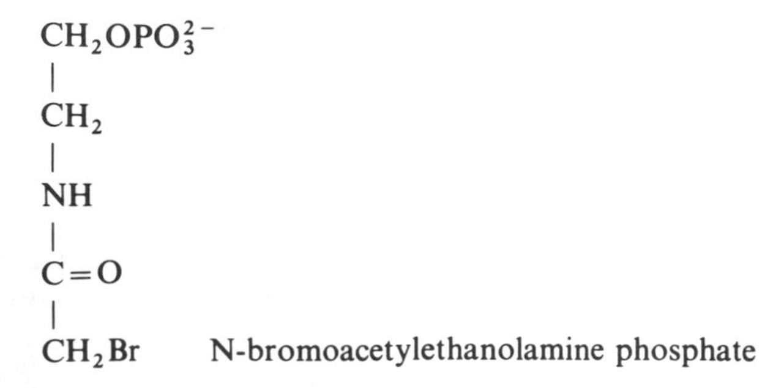

5.5.3.5 *An outline mechanism for fructose-bisphosphate aldolase*

Taking the various pieces of evidence in Sections 5.5.3.2 to 5.5.3.4 we can formulate an outline mechanism for the catalytic process (Fig. 5.34). DHAP binds to the enzyme to form a Schiff base, the carbanion then attacks G3P to form enzyme-bound FBP.

This mechanism should be regarded only as a working hypothesis for the following reasons:

1. The identity of the positively charged group that binds to the 1-phosphate group of the substrate (Fig. 5.34) is not known, although Arg 148 may fulfil this role.[107,117]
2. Chemical-modification experiments have suggested that cysteine and tyrosine side-chains may be involved in the catalytic mechanism,[105] but the roles of these side chains are not clear.
3. Detailed structures of enzyme–substrate or enzyme–inhibitor complexes are required in order to clarify the identities and roles of some of the side-chains involved.
4. The role played by amino acids in the mobile C-terminal portion of the polypeptide chain needs to be defined more clearly. Of the C-terminal 22 amino acids, only four are conserved in the Class I aldolases sequenced to date.[107,110] It is possible that this region is important in determining the activity and/or specificity of the various aldolases and their isoenzymes towards the different substrates acted on (Section 5.5.3). The identity of a base thought to be involved in proton transfer reactions remains unresolved (Section 5.5.3.4).

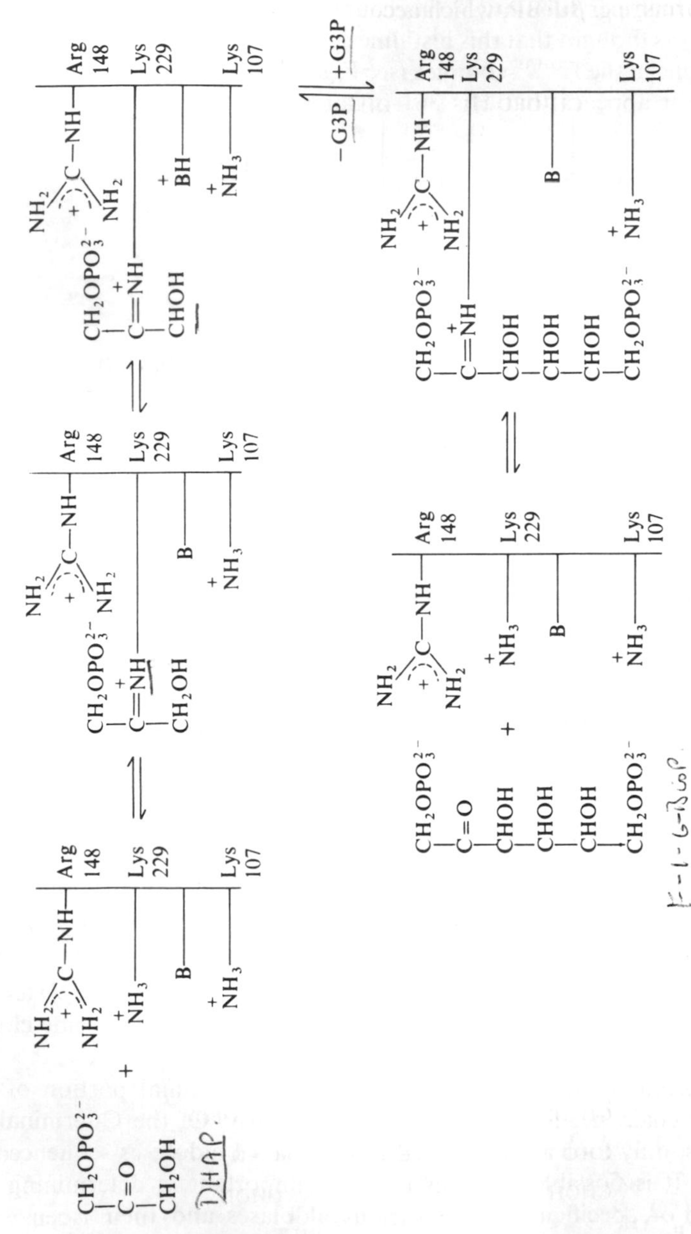

FIG. 5.34. The outline mechanism of fructose-bisphosphate aldolase. For the sake of clarity, details of the side-chains have been omitted. G3P is D-glyceraldehyde 3-phosphate. B represents a base though to be involved in proton-transfer reactions (Section 5.5.3.3).

5. The keto form of D-fructose 1,6-bisphosphate, which forms the Schiff base with the side-chain of Lys 229 (Fig. 5.34), constitutes only a small fraction (<2 per cent) of the total FBP in solution.[118] Most of the FBP is present in the furanose forms, e.g. β-FBP, which accounts for 90 per cent of the total FBP.

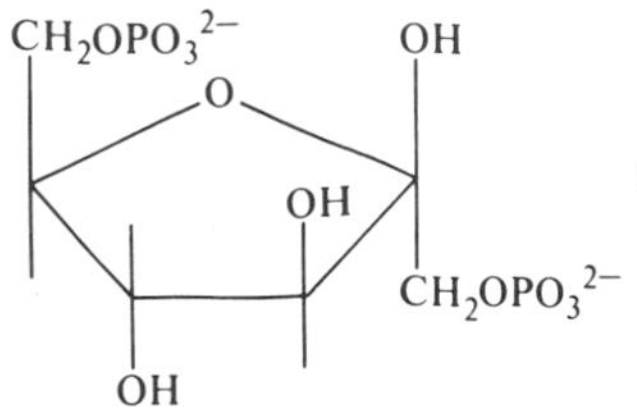

$CH_2OPO_3^{2-}$ OH O OH β-FBP $CH_2OPO_3^{2-}$ OH

Does FBP bind as a furanose form and then undergo ring opening to yield the keto form, or does only the keto form bind?

5.5.3.6 *The mechanism of Class II fructose-bisphosphate aldolases*

We have so far confined the discussion to fructose-bisphosphate aldolase isolated from rabbit skeletal muscle, but it can be assumed that the mechanisms of other Class I enzymes are similar, especially in view of the homology of sequence around the active-site lysine (Section 5.5.3.3). In Class II fructose-bisphosphate aldolases there is no such lysine side-chain but instead a metal ion that is required for activity. Thus, for instance, the enzyme from yeast is a dimer containing one zinc ion per subunit.[105] The metal ion could serve to polarize the carbonyl group of DHAP (Fig. 5.35) in much the same way as the protonated Schiff base in Class I enzymes. This polarization of the carbonyl group could be similar to that already described in the case of carboxypeptidase (Fig. 5.22). However, studies using i.r. spectroscopy[119] failed to reveal evidence for polarization of the carbonyl group of FBP or DHAP on binding to the yeast enzyme, although the carbonyl group of G3P appeared to be polarized substantially on binding. There is some evidence that carboxylate side-chains (those of Glu or Asp) can play a role in proton transfer reactions in Class II fructose-bisphosphate aldolases.[105]

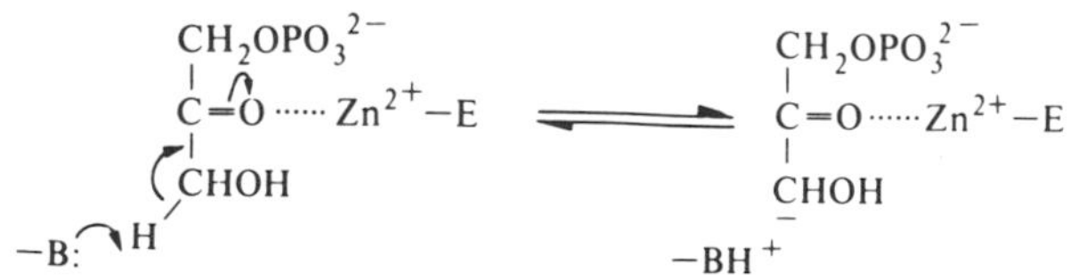

$CH_2OPO_3^{2-}$ C=O ······ Zn^{2+}—E CHOH —B: H ⇌ $CH_2OPO_3^{2-}$ C=O ······ Zn^{2+}—E CHOH —BH^+

FIG. 5.35. Polarization of the carbonyl group of dihydroxyacetone phosphate in Class II fructose-bisphosphate aldolases.

5.5.4 Lactate dehydrogenase

The enzyme lactate dehydrogenase catalyses the reaction:

$$\underset{\text{L-Lactate}}{H{-}\overset{\displaystyle CH_3}{\underset{\displaystyle CO_2^-}{\overset{|}{\underset{|}{C}}}}{-}OH} + NAD^+ \rightleftharpoons \underset{\text{Pyruvate}}{\overset{\displaystyle CH_3}{\underset{\displaystyle CO_2^-}{\overset{|}{\underset{|}{C}}}}{=}O} + NADH + H^+$$

In the reverse direction, this reaction represents the last step in the process of anaerobic glycolysis and provides a means of regeneration of NAD^+ required for the reaction catalysed by glyceraldehyde-3-phosphate dehydrogenase. The enzyme is present in the cytosol in sufficiently high concentrations that the reaction is close to equilibrium; this is discussed further in Chapter 8, Sections 8.3.1.4 and 8.5.4. Lactate dehydrogenase has been isolated from many sources, but most of the detailed work relating to the mechanism has been performed on the enzymes isolated from dogfish and pig (for a comprehensive review of this work, see reference 120). We shall discuss the results of kinetic studies (Section 5.5.4.2), X-ray crystallographic work (Section 5.5.4.3), experiments involving chemical modification of amino-acid side-chains (Section 5.5.4.4), and site-directed mutagenesis (Section 5.5.4.5). In Section 5.5.4.6 we shall discuss the mechanism of lactate dehydrogenase in relation to that of other dehydrogenases.

5.5.4.1 Background information on the structure of the enzyme

Lactate dehydrogenase is a tetramer of M_r 140 000. The enzyme provides a good example of the occurrence of isoenzymes (see Chapter 1, Section 1.5.2); in most tissues there are five forms of the enzyme, which can be separated by electrophoresis. The different forms arise from the five possible ways of assembling a tetramer from two types of subunit (α_4, $\alpha_3\beta$, $\alpha_2\beta_2$, $\alpha\beta_3$, and β_4). It is found that LDH-1 predominates in heart muscle, and is often referred to as the H_4 form, whereas LDH-5 predominates in skeletal muscle, and is often referred to as the M_4 form. Various other isoenzymes (e.g. the C, E, and F forms) are also known,[120] but they are much less widely distributed than the heart and muscle forms and will not be discussed further here.

5.5.4.2 Kinetic studies

Detailed steady-state kinetic studies on lactate dehydrogenase have shown that the enzyme follows an ordered ternary complex mechanism (see Chapter 4, Section 4.3.5.1) with NAD^+ (or NADH) binding preceding that of lactate (or pyruvate). This is shown in the following scheme, where the proton involved in the reaction has been omitted:

$$E + NAD^+ \rightleftharpoons E^{NAD+} \underset{-\text{lactate}}{\overset{+\text{lactate}}{\rightleftharpoons}} E^{NAD+}_{\text{lactate}} \rightleftharpoons$$

$$E^{NADH}_{\text{pyruvate}} \underset{+\text{pyruvate}}{\overset{-\text{pyruvate}}{\rightleftharpoons}} E^{NADH} \rightleftharpoons E + NADH.$$

The ordered ternary complex mechanism was deduced from studies of product inhibition and of substrate binding. For instance, the enzyme will not bind lactate or pyruvate in the absence of the dinucleotide substrate.[121]

Steady-state kinetic studies have also shown that oxamate (Fig. 5.36(a)) acts as a competitive inhibitor with respect to pyruvate (Fig. 5.36(b)) (note that the two molecules are isoelectronic) and that oxalate (Fig. 5.36(c)) is competitive with respect to lactate.[122] These inhibitors are very likely to bind to the active site of the enzyme and have found considerable value as substrate analogues in X-ray crystallographic studies on lactate dehydrogenase, allowing conclusions to be drawn regarding the structure of the catalytically active complex.

(a)	(b)	(c)	(d)
NH_2–C(=O)–CO_2^-	CH_3–C(=O)–CO_2^-	CO_2^-–CO_2^-	CH_3–CH(OH)–CO_2^-
Oxamate	Pyruvate	Oxalate	Lactate

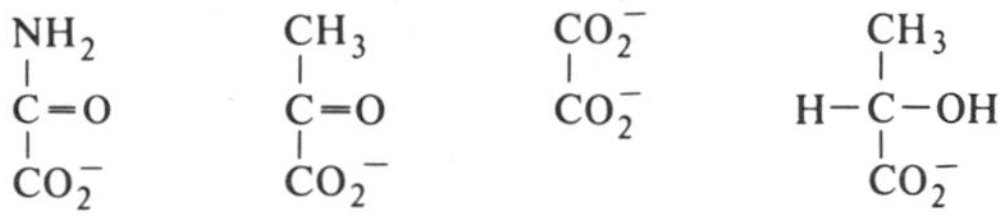

FIG. 5.36. Structures of some substrates and inhibitors of lactate dehydrogenase.

A more detailed analysis of the rates of the individual steps of the reaction has been undertaken using stopped-flow techniques (see Chapter 4, Section 4.4). The enzyme (LDH-1 from pig was used in these experiments) was mixed rapidly with saturating concentrations of the substrates NAD^+ and lactate, and production of NADH was monitored spectrophotometrically[123] (Fig. 5.37).

There was an initial burst of NADH production within the 'dead time' of the instrument, followed by a slow steady-state production of NADH. The size of the burst was equal to 1 mol NADH per mol of active sites. From these data, it can be concluded that all the steps in the reaction up to formation of enzyme-bound NADH are very rapid. Similar results were obtained with LDH-5 from pig,[124] but the size of the burst was only equal to 1 mol NADH per mol active sites at pH values of 8.0 or above.* Since NADH binds very tightly to the enzyme it might be thought that the dissociation of the E^{NADH} complex would be the slowest (rate-determining) step of the overall reaction. However, this is not the case, since the rate constant for the dissociation (measured in a separate experiment as 450 s^{-1})

* At lower pH values, the size of the burst was less; e.g. at pH 6.8 it was 0.5 mol per mol active sites.

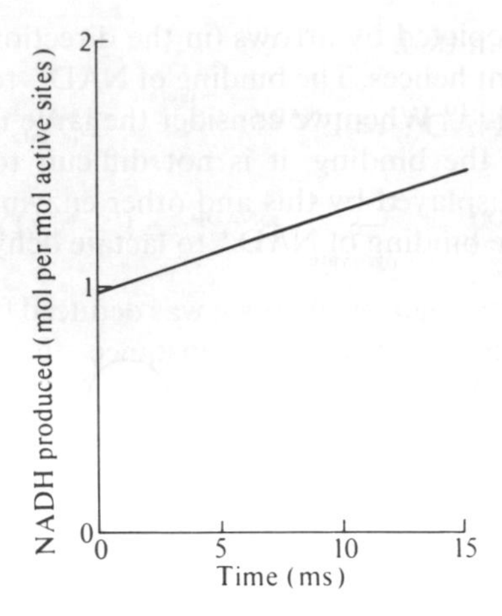

FIG. 5.37. Production of NADH in the reaction catalysed by lactate dehydrogenase.[123]

is considerably greater than the rate constant for oxidation of lactate in the steady state (80 s^{-1}). [124] It was therefore concluded that there is an extra step in the overall reaction that occurs after the interconversion of the ternary complexes that produces enzyme-bound NADH, but which precedes the dissociation of NADH. This extra step could be a slow conformational change in the E^{NADH} or $E^{NADH}_{pyruvate}$ complex[124] (see Section 5.5.4.5). The use of detailed pre-steady-state kinetic measurements has allowed the rate constants of all the individual steps in the reaction catalysed by LDH-1 from pig to be determined.[125,126]

5.5.4.3 *X-ray crystallographic studies*

Crystallographic and amino-acid sequence studies have been performed on LDH-5 from dogfish muscle.[120] There is a considerable amount of secondary structure (see Chapter 3, Section 3.5.4) in the molecule; approximately 40 per cent of the amino acids occur in helices and a further 23 per cent in various forms of β-structure.

Of particular interest is the occurrence of supersecondary structures in the molecule (see Chapter 3, Section 3.5.4.6). In the N-terminal half of the molecule there is a six-stranded parallel sheet, and in the C-terminal half there are two three-stranded antiparallel sheet structures.[120] The six-stranded sheet is involved in the binding of NAD^+ and seems to be a general feature of all the dehydrogenases whose structures have been examined so far, and has led to the development of ideas[127] about the evolutionary relationships between these and other enzymes such as kinases that act on mononucleotide substrates (e.g. ATP).* The six-stranded sheet[128] is shown schematically in Fig. 5.38, where the

* The tertiary structures of lactate dehydrogenase and malate dehydrogenase are very similar; indeed, by changing three amino acids in the active-site region of the enzyme it is possible to change the specificity of lactate dehydrogenase to that of malate dehydrogenase (see Section 5.5.4.5).

strands of β-sheet are depicted by arrows (in the direction N→C) and the thin connecting lines represent helices. The binding of NAD^+ to the enzyme has been examined in great detail.[120] When we consider the large number of amino-acid side-chains involved in the binding, it is not difficult to appreciate how the remarkable specificity displayed by this and other enzymes arises. Some of the important features of the binding of NAD^+ to lactate dehydrogenase are shown in Fig. 5.39.

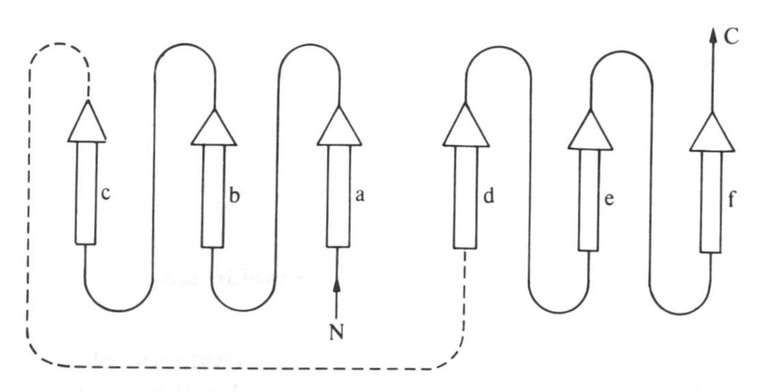

FIG. 5.38. The six-stranded sheet in lactate dehydrogenase. The arrows represent the strands of sheet and the connecting lines represent helices. (Between strands c and d there is a non-helical stretch of peptide chain.) The nicotinamide ring of NAD^+ binds between strands d and e and the adenine ring between strands a and b.[128]

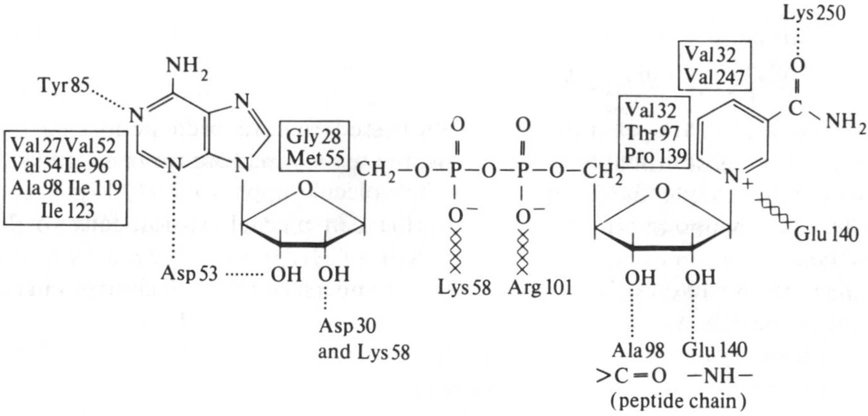

FIG. 5.39. Some of the interactions involved in the binding of NAD^+ to lactate dehydrogenase.[120] The dotted lines represent hydrogen bonds, the crosses electrostatic interactions, and the boxes hydrophobic interactions.

The adenine ring binds in a hydrophobic pocket provided by side-chains of Val, Ile, and Ala. Hydrogen bonds to the side-chains of Tyr 85 and Asp 53 (Fig. 5.39) impart specificity to the orientation of the bound adenine ring. Both ribose rings are involved in hydrogen bonds, as shown in Fig. 5.39; the specific interaction with the 2′-OH group of the adenine ribose helps to explain why $NADP^+$, which has a phosphate group at this position, is not a substrate for the enzyme.*

As far as the nicotinamide ring is concerned, we should note that there is a very important hydrogen-bonding interaction between the carbonyl oxygen of the side-chain and the side-chain of Lys 250. This interaction serves to orientate the nicotinamide ring in such a way as to expose the 'A'-side of the ring to the other substrate, lactate, thereby accounting for the 'A'-side specificity of the enzyme (Fig. 5.40). In the complex formed between NAD^+ and glyceraldehyde-3-phosphate dehydrogenase, there is a specific hydrogen bond between the carbonyl oxygen of the side-chain of the nicotinamide ring and the side-chain of Asn 313 that serves to expose the 'B'-side of the ring to the other substrate, accounting for the 'B'-side specificity of this enzyme (Fig. 5.40).[129, 130]

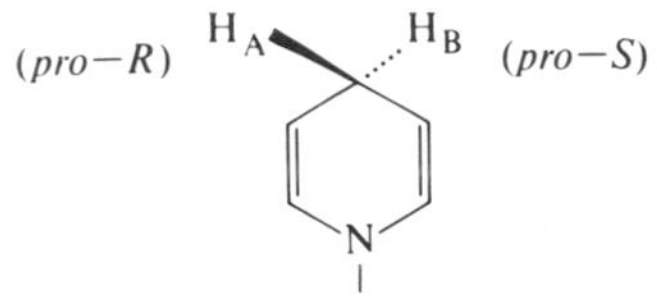

(*pro*–*R*) H_A H_B (*pro*–*S*)

N

FIG. 5.40. The specificity of NADH-linked dehydrogenases. 'A'-side dehydrogenases (e.g. malate, lactate and alcohol dehydrogenases) select the H_A hydrogen from NADH. 'B'-side dehydrogenases (e.g. glutamate and glyceraldehyde-3-phosphate dehydrogenases) select the H_B hydrogen from NADH. (See also Fig. 1.2.)

The structure of the catalytically active complex of lactate dehydrogenase has been deduced from an examination of the structures of complexes containing inhibitors: $E^{NAD^+}_{pyruvate}$, $E^{NAD^+}_{oxalate}$, and $E^{NADH}_{oxamate}$ and of the structure of a complex containing an NAD-pyruvate adduct.[131] These structures indicate that the site for lactate is most probably located between the nicotinamide ring and the side-chain of His 195, as shown in Fig. 5.41.

The side-chains of Arg 109 and Arg 171 can bind electrostatically to the carboxylate group of lactate. The side-chain of His 195 can form a hydrogen bond to the hydroxyl group of lactate, permitting a movement of electrons in the catalytic process as shown in Fig. 5.41.

Conformational changes in the enzyme appear to be of crucial importance in the formation of the catalytically active complex.[120] These changes are particularly marked on binding the second substrate to the binary E^{NAD^+} complex. There is a flexible loop (between amino acids 98 to 120), which in the absence of substrates extends into the solvent, but in the ternary complex moves down to enclose the substrates and thereby exclude water from the active site of the enzyme (see Section 5.5.4.5). For instance, the side-chain of Arg 109 moves 1.4 nm to be close to the substrate (Fig. 5.41). A smaller (0.1–0.2 nm[131]) but

* In general, a dehydrogenase acts on *either* NAD^+ *or* $NADP^+$. Glutamate dehydrogenase is almost unique in that it can act on both (see Chapter 8, Section 8.3.1.4).

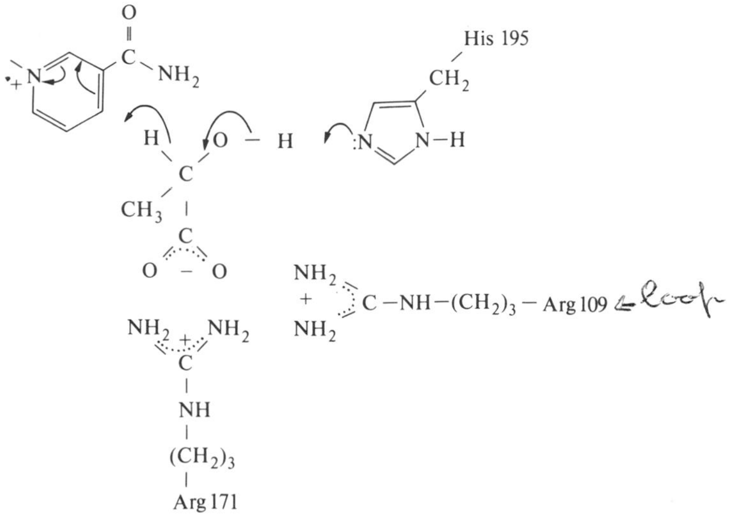

FIG. 5.41. Movement of electrons in the reaction between NAD^+ and lactate catalysed by lactate dehydrogenase.

highly important change occurs so as to bring the side-chain of His 195 into contact with the substrate and permit the movement of electrons as shown (Fig. 5.41).

5.5.4.4 *Chemical modification of amino-acid side-chains*

In this section we shall discuss experiments involving the modification of histidine, arginine, and cysteine side-chains. The results of these studies have complemented the information provided by X-ray crystallography.

(i) Histidine

The idea that there is a histidine side-chain at the active site of lactate dehydrogenase was suggested by the results of experiments using the affinity labels 3-bromoacetylpyridine and bromopyruvate.[120] (The latter acts as an affinity label in the presence of NAD^+.) In each case it was shown that the side-chain of His 195 was modified.

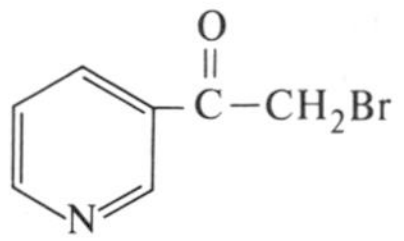

3-Bromoacetylpyridine

$Br-CH_2-C(=O)-CO_2^-$

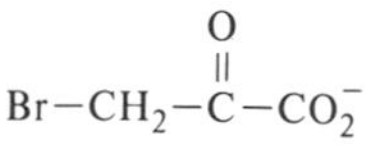

Bromopyruvate

It is also possible to modify the side-chain of His 195 selectively using diethylpyrocarbonate

$$(C_2H_5—O—\overset{\overset{\displaystyle O}{\|}}{C}—O—\overset{\overset{\displaystyle O}{\|}}{C}—OC_2H_5)$$

because of the much higher reactivity of this side-chain compared with the other histidine side-chains in the enzyme.[132] In all three cases modification led to inactivation of the enzyme, in keeping with the proposed role of the side-chain of His 195 (Fig. 5.41).

(ii) Arginine

Phenylglyoxal, which is fairly specific for modification of arginine side-chains (Table 5.2), inactivates lactate dehydrogenase. The findings that only one arginine side-chain per subunit was modified and that formation of the ternary complex provided protection against inactivation suggested that there is an arginine side-chain at the active site of the enzyme.[120] The X-ray crystallographic data (Fig. 5.41) shows that two arginine side-chains may be involved in the binding of substrate, so it would be of interest to know which one is modified by phenylglyoxal.

(iii) Cysteine

The experiments designed to show whether a cysteine side-chain is involved in the catalytic activity of lactate dehydrogenase provide good examples of the care needed in interpreting chemical-modification data. Reaction of the enzyme with reagents such as substituted maleimides led to inactivation[120] as a result of modification of a single cysteine side-chain per subunit (later identified as that of Cys 165[133]). The observations that various combinations of substrates afforded protection against inactivation, and that the amino-acid sequence around this cysteine was somewhat conserved in a variety of dehydrogenases, strengthened the conviction that this side-chain was indeed involved in the catalytic activity. However, X-ray crystallographic studies subsequently showed that the side-chain of Cys 165 was, in fact, some distance (1 nm) from the substrate-binding site.[120] The inactivation caused by this chemical modification is now interpreted in terms of a secondary, blocking, effect whereby a bulky reagent attached to the side-chain of Cys 165 would prevent the movement of the side-chain of His 195 towards the substrate and hence prevent formation of the catalytically active complex (Fig. 5.41). The X-ray crystallographic work has shown that in a ternary complex (e.g. $E^{NAD^+}_{pyruvate}$) the side-chain of Cys 165 would be inaccessible to modifying reagents; this observation would explain the protection against inactivation noted above. In the light of these comments, it would be expected that, if the side-chain of Cys 165 were modified with a small reagent, at least partial activity would be retained. This expectation has been confirmed in an experiment in which LDH-1 from pig was reacted with methanethiolsulphonate.[134] The modified enzyme,

$$E—CH_2—SH + CH_3—S—SO_2—CH_3 \rightarrow E—CH_2—S—S—CH_3 + CH_3SO_2^- + H^+,$$

in which the side-chain of Cys 165 was reacted, retained full activity but showed some differences from unmodified enzyme in terms of its affinity for substrates.

5.5.4.5 Site-directed mutagenesis experiments

Two recent sets of experiments on lactate dehydrogenase have contributed to our understanding of the roles played by amino acids in the flexible loop region of the enzyme (Section 5.5.4.3). These experiments have been undertaken on the enzyme from the thermophilic bacterium *Bacillus stearothermophilus* for which the appropriate gene had been isolated. The amino-acid sequence of this enzyme and its catalytic properties (when measured in the presence of the activator FBP) are similar to those of eukaryotic lactate dehydrogenase, indicating that the overall structures of the enzymes are likely to be similar.

In the first experiment,[135] Arg 109 in the flexible loop was replaced by Gln, thereby changing a positively charged side-chain into a neutral one. The mutation had no effect on the binding of NADH to the enzyme but decreased k_{cat} by some 400-fold. The results suggested that the side-chain of Arg 109 enhances the polarization of the carbonyl group of pyruvate (Fig. 5.41) and stabilizes the transition state of the reaction.

In a second experiment,[136] the rate of the movement of the loop region that is associated with formation of the catalytically active complex was measured. For this experiment Gly 106 in the loop was replaced by Trp and the other Trp side-chains (Trp 80, Trp 150 and Trp 203) were replaced by Tyr. (It should be noted that in proteins, Trp side-chains generally show much more intense fluorescence than Tyr side-chains because of the quenching of the excited state of the latter.[63]) The quadruple mutant enzyme thus has effectively a single fluorescent side-chain (Trp 106) within the loop region. The k_{cat} value for the mutant enzyme was approximately 60 per cent of that of the native enzyme and the K_m for pyruvate was unchanged. On mixing the mutant E^{NADH} complex with oxamate to form the ternary complex (Section 5.5.4.3), it was found that the fluorescence of Trp 106 decreased, reflecting movement of the loop. By stopped-flow measurements (see Chapter 4, Section 4.4.1) the rate of this movement was found to be 125 s^{-1}, i.e. very similar to the overall turnover rate of the enzyme as measured by k_{cat}. It is thus clear that the rate-determining step in the overall reaction corresponds to the movement of the loop to close over the active site (Section 5.5.4.3).

In a different type of experiment, the specificity of lactate dehydrogenase has been changed to that of malate dehydrogenase.[158] To take account of the increased size and additional negative charge of the new substrate, the following changes were introduced into the lactate dehydrogenase from *Bacillus stearothermophilus*: (i) the volume of the active site was increased (Thr 246→Gly); (ii) an acidic side-chain was replaced by a neutral one (Asp 197→Asn); (iii) a basic side-chain was introduced (Gln 102→Arg). These changes led to a new 'malate' dehydrogenase that catalyses the reduction of oxaloacetate by NADH 500 times faster than that of pyruvate. (The corresponding ratio for the original lactate dehydrogenase is 0.001.) This experiment confirms that the overall structures of the two enzymes are sufficiently similar that relatively small changes at the active site can cause marked changes in specificity.[158]

5.5.4.6 *The mechanism of lactate dehydrogenase and some other dehydrogenases*

The various approaches outlined in Sections 5.5.4.2 to 5.5.4.5 have given a fairly detailed insight into the mechanism of the reaction catalysed by lactate dehydrogenase. The structure of the active complex and the movement of electrons within it have been described (Fig. 5.41) and to this must be added details of the binding sites (e.g. Fig. 5.39) and of the rates of the various elementary steps in the reaction.

The mechanisms of a number of other dehydrogenases (e.g. alcohol dehydrogenase, malate dehydrogenase, and glyceraldehyde-3-phosphate dehydrogenase) are also known, at least in outline, and certain generalizations can be made.[81] All dehydrogenases contain a recognizable nucleotide binding domain in each subunit (see Section 5.5.4.3), with a second (catalytic) domain that can be variable in structure. We have already noted that the mode of binding of the nicotinamide ring of NAD^+ to the various enzymes can explain the specificity of hydrogen transfer (Fig. 5.40). In all the dehydrogenases mentioned, except alcohol dehydrogenase, it appears that a histidine side-chain acts as a general base catalyst by abstracting a proton from the —O—H group of the substrate,* thereby facilitating transfer of the hydride ion to NAD^+ (see Fig. 5.41). In the case of alcohol dehydrogenase, a zinc ion is involved in the catalytic activity and it is suggested that this zinc ion, or an ionized water molecule (OH^-) attached to the zinc ion, could play the part of the basic catalyst. It will be of great interest to see how the chemical problems involved in oxidation–reduction reactions have been solved in the cases of other dehydrogenases.

5.5.5 Tyrosyl-tRNA synthetase

The aminoacyl-tRNA synthetases catalyse formation of aminoacyl-tRNA derivatives according to the equation (tRNA ≡ transfer RNA):

$$\text{Amino acid} + \text{tRNA} + \text{ATP} \rightleftharpoons \text{aminoacyl}-\text{tRNA} + \text{AMP} + \text{pyrophosphate.}$$

The amino acid is linked to the adenosine at the 3′ terminus of its appropriate tRNA. In some cases the attachment is to the 2′ hydroxyl, in others to the 3′ hydroxyl, and in others to either.[137] Aminoacyl derivatives are delivered to the ribosomes, the sites of protein synthesis, in the order specified by the sequence of bases on messenger RNA (mRNA) (see Chapter 9, Section 9.2). A sequence of three bases on the mRNA (*a codon*) forms specific hydrogen bonds with a sequence of three bases on the tRNA (*an anti-codon*). The fidelity of protein synthesis depends on the linking of the correct tRNA to the correct amino acid,

* In the case of glyceraldehyde-3-phosphate dehydrogenase, the —O—H group is that of a thiohemiacetal attached to the side-chain of Cys 149. The thiohemiacetal is formed by attack of this side-chain on the carbonyl group of the aldehyde substrate.

and this in turn depends on the specificity of the interactions between a given aminoacyl-tRNA synthetase and the appropriate tRNA and amino acid. The aminoacyl-tRNA synthetases (one for each of the twenty amino acids found in proteins) have EC numbers 6.1.1. 1–7, 9–12, 14–22. Tyrosyl-tRNA synthetase has the EC number 6.1.1.1.

In this section we shall discuss the general aspects of the structures of aminoacyl-tRNA synthetases (Section 5.5.5.1) and the details of the reaction mechanism revealed by kinetic studies (Section 5.5.5.2). The problems posed by the need for recognition of the correct substrates are discussed in Section 5.5.5.3. In Section 5.5.5.4, the X-ray structure of tyrosyl-tRNA synthetase from *Bacillus stearothermophilus* is described. Finally, in Section 5.5.5.5, we shall examine some of the elegant experiments involving site-directed mutagenesis (Section 5.4.5) that have greatly extended our knowledge of the catalytic mechanism of the enzyme and the roles played by various amino acids.

5.5.5.1 *Structures of aminoacyl-tRNA synthetases*

The aminoacyl-tRNA synthetases for each amino acid have been purified from a variety of prokaryotic and eukaryotic sources.[137–139] There is a wide variety of structural types including α, α_2, α_4 and $\alpha_2\beta_2$ with total M_r values ranging from about 50 000 to 300 000. The variety of subunit structures of enzymes catalysing related reactions seems surprising; however, this diversity is also found in the sequences of the enzymes. In general, the subunit M_r of a given enzyme is larger in eukaryotes (yeast) than in prokaryotes (*E. coli*) and this is usually associated with an N-terminal extension (50–300 amino acids) in the former.[139] In the portions of sequence that can be aligned there is only moderate homology (30–50 per cent) between the corresponding enzymes from the two sources.[139] There is some evidence for internal repeats in the sequences of the larger monomeric aminoacyl-tRNA synthetases from prokaryotes.[137] In eukaryotes, nine of the enzymes form a complex of high M_r ($\approx 10^6$), although the significance of this complex is not established.[138] Detailed X-ray structural information is so far available on only one enzyme, namely tyrosyl-tRNA synthetase from *Bacillus stearothermophilus* (see Section 5.5.5.4).

5.5.5.2 *The reaction mechanism—kinetic studies*

A number of experiments have shown that the overall reaction catalysed by aminoacyl-tRNA synthetase is the sum of two reactions with an enzyme-bound aminoacyladenylate intermediate:[137]

$$\text{E} + \text{amino acid} + \text{ATP} \rightleftharpoons \text{E. aminoacyl-AMP} + \text{pyrophosphate}$$

$$\text{E. aminoacyl-AMP} + \text{tRNA} \rightleftharpoons \text{aminoacyl-tRNA} + \text{AMP} + \text{E.}$$

The enzyme-bound intermediate can readily be isolated; for instance, by mixing isoleucyl-tRNA synthetase with [^{3}H]isoleucine and ATP in the presence of necessary cofactors, such as Mg^{2+}, followed by gel filtration on Sephadex

G-50.[140] In a separate experiment it was shown that isoleucine was rapidly transferred from this intermediate to give an enzyme-bound product (isoleucyl-tRNA), that then slowly dissociated from the enzyme.

Since ATP, amino acid, and pyrophosphate can each bind to the enzyme, it has been concluded that the first reaction is of a random-order ternary-complex type[141] (see Chapter 4, Section 4.3.5.1). The occurrence of the first reaction has also been demonstrated by the observation of an enzyme-catalysed ATP $\rightleftharpoons$ pyrophosphate exchange reaction that requires amino acid, but not tRNA.[137, 142] In these experiments [^{32}P]pyrophosphate is added, the ATP is isolated at various intervals, and the extent of ^{32}P incorporation is determined. The rate of the exchange reaction can be measured and compared with the rate of the overall reaction (i.e. synthesis of aminoacyl-tRNA). In the majority of cases the rate of the first reaction is 10–100 times the rate of the overall reaction,[137] thus implying that the second reaction (aminoacyl transfer or product release) is rate determining. However, in other cases, e.g. tyrosyl-tRNA synthetase from *E. coli*, the rate constants for the two reactions are very similar and thus both contribute to the rate-determining step.[142] In some cases, additional complexities arise; thus, the dimeric tyrosyl-tRNA synthetase from *Bacillus stearothermophilus* shows 'half-of-the-sites' reactivity (see Chapter 6, Section 6.2.2.2), with only one molecule of tyrosyl-AMP formed and one molecule tRNA bound per molecule of enzyme. In this case, the enzyme-bound tyrosyl-AMP intermediate can bind additional tyrosine and ATP and undergo a conformational change.[143, 144]

5.5.5.3 *Recognition of substrates*

As mentioned in Section 5.5.5, it is crucial for the aminoacyl-tRNA synthetases to recognize their substrates correctly, and much experimental work has been devoted to an analysis of the mechanisms by which this recognition is achieved.

Because of their small size, the recognition of amino acids poses particular problems.[145] For instance, the factor by which isoleucine could be favoured over valine by isoleucyl-tRNA synthetase can be estimated as 20–40 fold (taking into account the binding energy available due to the extra methylene group in isoleucine and the relative intracellular concentrations of the two amino acids). However, the actual rate of misincorporation of valine for isoleucine is very low (1 part in 3000).[146] Experimentally, it has been found that isoleucyl-tRNA synthetase can activate valine, i.e. catalyse the formation of an enzyme-bound valyl-AMP complex in the absence of tRNA.[145] However, the valine cannot be subsequently transferred to tRNA; indeed, addition of $tRNA^{ile}$ to the complex leads to hydrolysis of the valyl-AMP to give valine and AMP. This type of 'editing' or 'proof-reading' mechanism has been found in a number of aminoacyl-tRNA synthetases and involves the operation of a second (editing) active site in the enzyme.[145] In some of these cases there is evidence for the transient formation of the 'wrong' aminoacyl-tRNA (e.g. threonyl-$tRNA^{val}$ formed in a reaction catalysed by valyl-tRNA synthetase), but such species are rapidly destroyed before dissociation from the enzyme.[145]

Some aminoacyl-tRNA synthetases do not have an editing function; thus cysteinyl-tRNA and tyrosyl-tRNA synthetases bind the correct amino acids much more tightly than possible competitors.[145] The side-chains responsible for the binding of the correct amino acid by tyrosyl-tRNA synthetase from *Bacillus stearothermophilus* have been identified by X-ray crystallography and site-directed mutagenesis (see Section 5.5.5.5).

The recognition of the correct tRNA would appear to present fewer problems, because of the larger size of this substrate. At present, there is no X-ray structure of a synthetase-tRNA complex available, so most of the conclusions regarding the interaction have been reached by indirect means, including the effects of modification of tRNA bases on binding and the use of photochemical cross-linking to establish which parts of tRNA are in contact with the enzyme.[137, 139] The overall conclusions appear to be that the main points of contact involve the 'inside' of the L-shaped tRNA molecule (Fig. 5.42) including the acceptor stem,[157] and that in some cases (though not all) the anti-codon seems to be involved in the interaction.[137, 139] The specificity must arise within these regions and the complementary parts of the synthetase. There is evidence that at least part of the recognition process involves a conformational change in the enzyme after the correct tRNA is bound.[137] The location of amino acids in tyrosyl-tRNA synthetase responsible for the binding of $tRNA^{tyr}$ has been examined by site-directed mutagenesis (Section 5.5.5.5).

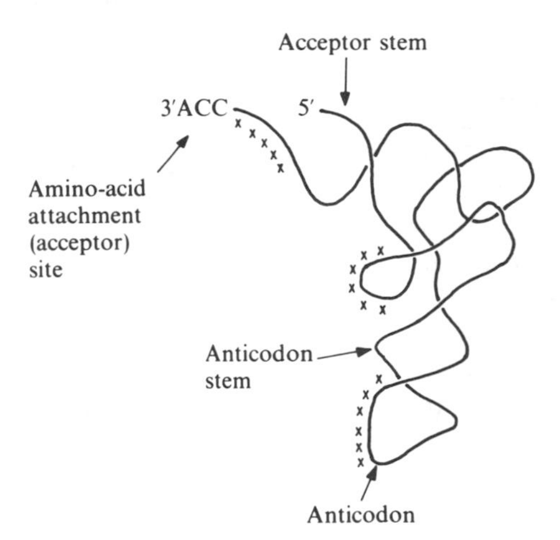

FIG. 5.42. General structure of a tRNA molecule. Regions marked × × × × indicate portions of the molecule identified as forming contacts with aminoacyl-tRNA synthetase.[137]

5.5.5.4 *The X-ray structure of tyrosyl-tRNA synthetase from* Bacillus stearothermophilus

The X-ray structure of one subunit of tyrosyl-tRNA synthetase from *Bacillus stearothermophilus* is shown in Fig. 5.43.[147] The enzyme contains 419 amino acids in each of the two subunits; however, only the N-terminal 319 amino acids can be distinguished in the electron-density map; the C-terminal 100 amino acids are presumably too disordered to be discerned (see Section 5.5.5.5). In the centre of each subunit is a 6-stranded β-sheet structure and there are five long (>1.2 nm) segments of α-helix with several shorter stretches. The location of the substrate binding site is shown in Fig. 5.43. A number of complexes have been examined by X-ray crystallography, namely with AMP, ATP, tyrosine, and tyrosyl-AMP (the last-named complex could be studied because of its high stability).[144] The amino-acid side-chains that interact with tyrosyl-AMP are described in Section 5.5.5.5.

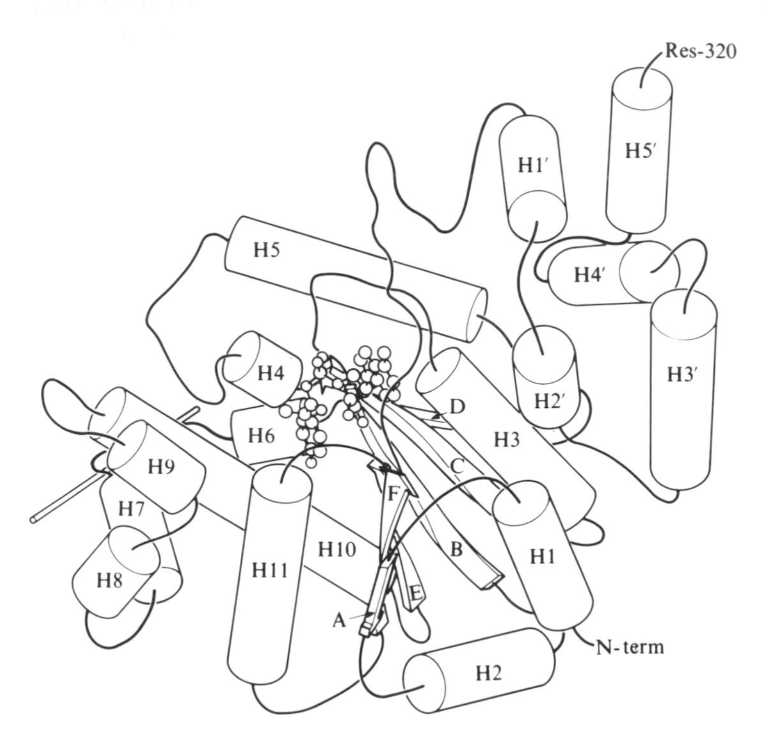

FIG. 5.43. Polypeptide chain conformation of residues 1–320 of the tyrosyl-tRNA synthase monomer. The binding site for tyrosyl-AMP and the molecular 2-fold axis are shown. The strands of β-sheet are labelled A to F. The α-helices run between the following amino acids:

H1, 2-10; H2, 19–27; H3, 49–60; H4, 71–74; H5, 91–105; H6, 124–128; H7, 132–139; H8, 145–149; H9, 152–156; H10, 164–183; H11, 195–208; H1′, 248–256; H2′, 263–269; H3′, 275–287; H4′, 293–306; H5′, 309–318.

(The figure was kindly supplied by Dr P. Brick and Prof. D.M. Blow, Imperial College, London, UK.)

5.5.5.5 *Studies by site-directed mutagenesis*

The powerful technique of site-directed mutagenesis (Section 5.4.5) has been employed by Fersht and his colleagues to undertake a series of very elegant experiments in which the roles of different parts of the tyrosyl-tRNA synthetase from *Bacillus stearothermophilus* have been explored. The design and interpretation of these experiments have drawn heavily on the detailed structural information available from X-ray crystallography. Much of the site-directed mutagenesis work has been summarized in a recent review.[144]

(i) Hydrogen bonds and specificity

The X-ray crystallographic work indicated that a number of hydrogen bonds were formed between the enzyme and tyrosyl-AMP (Fig. 5.44).

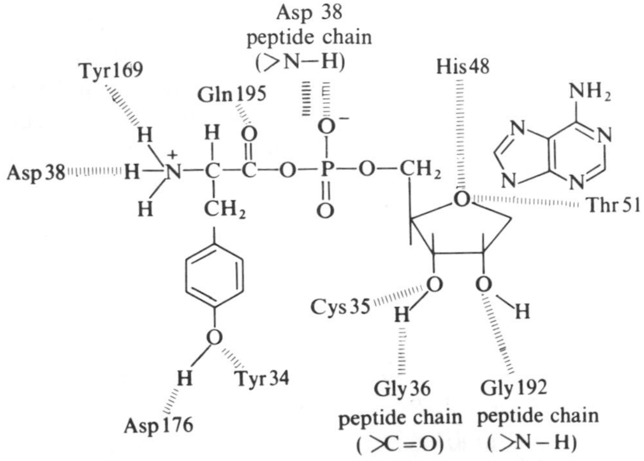

FIG. 5.44. Hydrogen bonds (||||||||||||) involved in the interaction between tyrosyl-AMP and tyrosyl-tRNA synthetase.[148]

The side-chains implicated have been systematically replaced by side-chains incapable of forming hydrogen bonds (e.g. Cys 35→Gly, Tyr 34→Phe) and the effects on the kinetics of the activation of tyrosine by ATP have been studied.[148] The results show that each of the contacts identified by X-ray crystallography is important and that other side-chains play a role (e.g. Asp 78 is involved in an interaction with the α-amino group of the tyrosyl-AMP). The side-chains responsible for the specificity of the enzyme for tyrosine as opposed to phenylalanine are those of Tyr 34 and Asp 176[144] (see Fig. 5.44). In the ribose-binding site the side-chains of Cys 35, Thr 51 and His 48 are involved in specific hydrogen-bonding interactions. Cysteine 35 had been implicated by chemical modification experiments as being involved in the catalytic activity and this residue is conserved in other prokaryotic tyrosyl-tRNA synthetases;[139] however, replacement by Gly or Ser did not lead to complete loss of activity (k_{cat} was reduced to 30 per cent of the value for wild-type enzyme) and thus Cys 35 is not an essential side-chain.[148]

The results of site-directed mutagenesis experiments showed that different types of hydrogen bonds made different contributions to the binding energy.[148] Thus, mutation of an uncharged side-chain (Tyr 169) that forms a hydrogen bond to a charged group on the substrate (the α-amino group) weakens the binding by 15.5 kJ mol^{-1}, whereas mutation of a side-chain (Tyr 34) that forms a hydrogen bond to an uncharged group (the phenolic OH group of tyrosyl-AMP) weakens the binding by only 2.2 kJ mol^{-1}.

In Fig. 5.44, the side-chain of Thr 51 is seen to form a hydrogen bond with the ribose of tyrosyl-AMP. However, this bond is thought to be weak, because of unfavourable geometry. The side-chain of Thr 51 could form a stronger hydrogen bond with water, thereby promoting dissociation of the enzyme–tyrosyl-AMP complex. It is interesting to note that Thr 51 is the only amino acid in the active site that is not conserved in the tyrosyl-tRNA synthetase from other prokaryotic sources, e.g. *E. coli* (Pro) and *Bacillus caldotenax* (Ala).

The Thr 51 of the *Bacillus stearothermophilus* enzyme has been replaced by other side-chains, e.g. Pro, Ala, that cannot make the hydrogen bond to ribose.[149] As measured by the ratio k_{cat}/K_m (see Chapter 4, Section 4.3.1.3) the activities of the Thr 51→Pro and Thr 51→Ala mutants are improved over that of the native enzyme by factors of 50 and 2, respectively. This type of experiment indicates that it is possible to change, and even improve, the catalytic properties of enzymes in a rational manner.

(i) Roles of the N- and C-terminal regions of the enzyme

A deletion mutant of tyrosyl-tRNA synthetase has been prepared consisting of the N-terminal 319 amino acids.[150] This mutant was found to be indistinguishable from the native enzyme as far as the activation of tyrosine (to form tyrosyl-AMP) was concerned, but was unable to bind to or aminoacylate tRNA. The conclusion that the C-terminal region of the polypeptide chain is involved in binding tRNA seems to be general for synthetases.[139] Some of the amino-acid side-chains of tyrosyl-tRNA synthetase involved in binding tRNA have been identified by site-directed mutagenesis.[151] In this work, heterodimers were constructed containing one full-length polypeptide chain (419 amino acids) and one truncated chain (319 amino acids). Positively charged side-chains were replaced by neutral ones (Arg, His → Gln; Lys → Asn). The interactions with tRNA were found to extend across the two subunits with sets of positively charged side-chains interacting with the acceptor and anti-codon stems (Fig. 5.42), clamping the tRNA in a fixed orientation. The flexible–CCA (3′ terminus) could then be aligned at the active site by additional contacts.[151]

In a further experiment, the nature of the interactions between the subunits of the enzyme was examined. Phe 164 lies on the symmetry axis at the subunit interface and the side-chains from the two subunits form hydrophobic contacts. On replacement of this Phe by Asp, the enzyme was found to dissociate at pH

7.78 because of the introduction of negative charges in close proximity. At lower pH, the Asp side-chains are protonated and the subunits can reassociate.[152]

5.6 Concluding comments on enzyme mechanisms

In this chapter we have shown that it is now possible to propose fairly detailed mechanisms for a number of enzyme-catalysed reactions. Clearly, the detailed structural information available from X-ray crystallography has had a profound influence on the progress made in this area in recent years, but it can be seen that the other techniques including site-directed mutagenesis have had a useful role to play. Many of the factors suggested by model studies (Section 5.3) as leading to rate enhancement, such as acid–base catalysis and intermediate-formation, are seen to occur in the mechanisms of the enzymes discussed in Section 5.5. It is therefore reasonable to state that enzyme catalysis does not involve any new principles in addition to those involved in non-enzyme-catalysed reactions; enzymes merely achieve highly effective combinations of the various established factors leading to rate enhancement. However, such a statement cannot be made with certainty until the observed rate enhancements in a number of enzyme-catalysed reactions can be accounted for quantitatively which, as yet, is not the case.

In the case of lactate dehydrogenase (Section 5.5.4) it is clear that binding of substrates is accompanied by large conformational changes in the enzyme. Such structural changes have been shown to occur in a number of enzymes (e.g. carboxypeptidase, lysozyme, and hexokinase) by X-ray crystallographic or other less-direct spectroscopic methods.[153] These observations provide support for Koshland's 'induced-fit' hypothesis,[154] which proposed that binding of substrate to enzyme induces a structural change in the enzyme so that the important amino-acid side-chains are brought into the correct spatial relationship for catalysis to occur. This 'induced fit' was viewed as playing a crucial role in enzyme catalysis and specificity (i.e. different substrates would be distinguished from one another by the extents to which they would induce the active conformation of the enzyme). More recently, however, the quantitative significance of 'induced fit' has been called into question,[155] and it has been proposed that specificity arises because of the differences in binding of the transition states of different substrates to the enzyme, i.e. the transition states of better substrates bind more effectively, see Section 5.3.4. It is difficult to believe that there will not be considerable debate on this topic in the future.

References

1. Price, N. C. and Dwek, R. A., *Principles and problems in physical chemisry for biochemists* (2nd edn). Clarendon Press, Oxford (1979). See p. 154.
2. Albery, W. J. and Knowles, J. R., *Biochemistry* **25**, 2572 (1986).
3. Fersht, A. R., Leatherbarrow, R. J., and Wells, T. N. C., *Phil. Trans. R. Soc. Lond.* **A317**, 305 (1986).

4. Bruice, T. C. and Benkovic, S. J., *J. amer. chem. Soc.* **85**, 1, (1963).
5. Page, M. I. and Jencks, W. P., *Proc. natn. Acad. Sci. USA* **68**, 1678 (1971).
6. Jencks, W. P. and Page, M. I., *Biochem. Biophys. Res. Commun.* **57**, 887 (1974).
7. Bruice, T. C. and Pandit, U.K., *Proc. natn. Acad. Sci. USA* **46**, 402 (1960).
8. Caswell, M. and Schmir, G. L., *J. amer. chem. Soc.* **102**, 4815 (1980).
9. Dahlquist, F. W., Rand-Meir, T., and Raftery, M. A., *Proc. natn. Acad. Sci. USA* **61**, 1194 (1968).
10. Imoto, T., Johnson, L. N., North, A. C. T., Phillips, D. C., and Rupley, J. A., *Enzymes* (3rd edn) **7**, 665 (1972).
11. Swain, C. G. and Brown, J. F. Jr., *J. amer. chem. Soc.* **74**, 2538 (1952).
12. Fersht, A. R., *Enzyme structure and mechanism* (2nd edn). Freeman, New York (1985). See Ch. 2.
13. Westheimer, F. H., *Proc. chem. Soc.* 253 (1963).
14. Jencks, W. P., *Adv. Enzymol.* **43**, 219 (1975). See p. 362.
15. Jencks, W. P., *Catalysis in chemistry and enzymology*. McGraw-Hill, New York (1969). See p. 305.
16. Warshel, A. and Levitt, M., *J. molec. Biol.* **103**, 227 (1976).
17. Belasco, J. G. and Knowles, J. R., *Biochemistry* **19**, 472 (1980).
18. Wolfenden, R., *Acc. Chem. Res.* **5**, 10 (1972). See also *A. Rev. biophys. Bioeng.* **5**, 271 (1976).
19. Parker, A. J., *Q. Rev. chem. Soc.* **16**, 163 (1962).
20. Serpersu, E. H., Shortle, D., and Mildvan, A. S., *Biochemistry* **26**, 1289 (1987).
21. D'Souza, V. T. and Bender, M. L., *Acc. Chem. Res.* **20**, 146 (1987).
22. Wharton, C. W., *Int. J. biol. Macromolec.* **1**, 3 (1979).
23. Breslow, R., *Adv. Enzymol.* **58**, 1 (1986).
24. Berger, A. and Schechter, I., *Phil. Trans. R. Soc.* **B257**, 249 (1970).
25. Alecio, M. R., Dann, M. L., and Lowe, G., *Biochem. J.* **141**, 495 (1974).
26. Lowe, G. and Yuthavong, Y., *Biochem. J.* **124**, 107 (1971).
27. Engel,P. C., *Enzyme kinetics* (2nd edn). Chapman and Hall, London (1981). See p. 38.
28. Findlay, D., Mathias, A. P., and Rabin, B. R., *Biochem. J.* **85**, 139 (1962).
29. Roberts, G. C. K., Dennis, E. A., Meadows, D. H., Cohen, J. S., and Jardetzky, O., *Proc. natn. Acad. Sci. USA* **62**, 1151 (1969).
30. Hartley, B. S. and Kilby, B. A., *Biochem. J.* **56**, 288 (1954).
31. Shore, J. D. and Gutfreund, H., *Biochemistry* **9**, 4655 (1970).
32. Henderson, R., *J. molec. Biol.* **54**, 341 (1970).
33. Douzou, P. and Petsko, G. A., *Adv. Protein Chem.* **36**, 246 (1984).
34. Fink, A. L. and Petsko, G. A., *Adv. Enzymol.* **52**, 177 (1981).
35. Cartwright, S. J. and Waley, S. G., *Biochemistry* **26**, 5329 (1987).
36. Hajdu, J., Acharya, K. R., Stuart, D. I., Barford, D., and Johnson, L. N., *Trends Biochem. Sci.* **13**, 104 (1988).
37. Pai, E. F., Sachsenheimer, W., Schirmer, R. H., and Schulz, G., *J. molec. Biol.* **114**, 37 (1977).
38. Madsen, N. B., Kasvinsky, P. J., and Fletterick, R. J., *J. biol. Chem.* **253**, 9097 (1978).
39. Sygusch, J., Madsen, N. B., Kasvinsky, P. J., and Fletterick, R. J., *Proc. natn. Acad. Sci. USA* **74**, 4757 (1977).
40. Fry, D. C., Kuby, S. A., and Mildvan, A. S., *Biochemistry* **24**, 4680 (1985).
41. Fry, D. C., Kuby, S. A., and Mildvan, A. S., *Biochemistry* **26**, 1645 (1987).
42. Means, G. E. and Feeney, R. E., *Chemical modification of proteins*. Holden-Day, San Francisco (1971).

43. Thomas, J. O., in *Companion to biochemistry* (Bull, A. T., Lagnado, J. R., Thomas, J. O., and Tipton, K. F., eds.), Vol. 1. Longman, London (1974). See Ch. 2.
44. Cohen, L. A., *Enzymes* (3rd edn) **1**, 147 (1970).
45. Bruice, T. C., Gregory, M. J., and Walters, S. L., *J. amer. chem. Soc.* **90**, 1612 (1968).
46. Cohen, L. A., *Enzymes* (3rd edn) **1**, 147 (1970). See p. 162.
47. Colman, R. F., *J. biol. Chem.* **243**, 2454 (1968).
48. Smith, E. L., Austen, B. M., Blumenthal, K. M., and Nyc, J. F., *Enzymes* (3rd edn) **11**, 293 (1975).
49. De La Mare, S., Coulson, A. F. W., Knowles, J. R., Priddle, J. D., and Offord, R. E., *Biochem. J.* **129**, 321 (1972).
50. Walsh, C. T., *Trends Biochem. Sci.* **8**, 254 (1983).
51. Der Terrossian, E. and Kassab, R., *Eur. J. Biochem.* **70**, 623 (1976).
52. Singer, S. J., *Adv. protein Chem.* **22**, 1 (1967).
53. Rakitzis, E. T., *Biochem. J.* **217**, 341 (1984).
54. Rossi, J. and Zoller, M., *Protein engineering*. Alan R. Liss, Inc., New York (1987). See Ch. 4.
55. Zoller, M. J. and Smith, J., *Nucl. Acids Res.* **10**, 6487 (1982).
56. Winter, G., Fersht, A. R., Wilkinson, A. J., Zoller, M., and Smith, M. *Nature, Lond.* **299**, 756 (1982).
57. Fersht, A. R., *Enzyme structure and mechanism* (2nd edn). Freeman, New York (1985). See Ch. 14.
58. Taylor, J. W., Schmidt, W., Cosstick, R., Okruszek, A., and Eckstein, F., *Nucl. Acids Res.* **13**, 8749 (1985).
59. Taylor, J. W., Ott, J., and Eckstein, F., *J. Nucl. Acids Res.* **13**, 8765 (1985).
60. Dalbadie-McFarland, G., Cohen, L. W., Riggs, A. D., Morin, C., Itakura, K., and Richards, J. H., *Proc. natn. Acad. Sci. USA* **79**, 6409 (1982).
61. Itakura, K., *A. Rev. Biochem.* **53**, 323 (1984).
62. Campbell, I. D. and Dwek, R. A., *Biological spectroscopy*. Benjamin/Cummings, Menlo Park, California (1984). See Ch. 10.
63. Campbell, I. D. and Dwek, R. A., *Biological spectroscopy*. Benjamin/Cummings, Menlo Park, California (1984). See Chs. 5 and 7.
64. Hess, G. P., *Enzymes* (3rd edn) **3**, 213 (1971).
65. Blow, D. M., *Enzymes* (3rd edn) **3**, 185 (1971).
66. Blow, D. M., Birktoft, J. J., and Hartley, B. S., *Nature, Lond.* **221**, 337 (1969).
67. Bachovchin, W. W., Kaiser, R., Richards, J. H., and Roberts, J. D., *Proc. natn. Acad. Sci. USA* **78**, 7323 (1981).
68. Porubcan, M. A., Westler, W. M., Ibañez, I. B., and Markley, J. L., *Biochemistry* **18**, 4108 (1979).
69. Kossiakoff, A. A. and Spencer, S. A., *Biochemistry* **20**, 6462 (1981).
70. Rutter, W. J., Gardell, S. J., Roczniak, S., Hilvert, D., Sprang, S., Fletterick, R. J., and Craik, C. S., *Protein engineering*. Alan R. Liss, Inc., New York (1987). See Ch. 23.
71. Steitz, T. A., Henderson, R., and Blow, D. M., *J. molec. Biol.* **46**, 337 (1969).
72. Hartley, B. S. and Shotton, D. M., *Enzymes* (3rd edn) **3**, 323 (1971).
73. Kraut, J., *Enzymes* (3rd edn) **3**, 165 (1971).
74. Gertler, A., Walsh, K. A., and Neurath, H., *Biochemistry* **13**, 1302 (1974).
75. Fastrez, J. and Fersht, A. R., *Biochemistry* **12**, 2025 (1973).
76. Hunkapiller, M. W., Forgac, M. D., and Richards, J. H., *Biochemistry* **15**, 5581 (1976).

77. Rühlmann, A., Kukla, D., Schwager, P., Bartels, K., and Huber, R., *J. molec. Biol.* **77**, 417 (1973).
78. Blow, D. M., Janin, J., and Sweet, R. M., *Nature, Lond.* **249**, 54 (1974).
79. Keil, B., *Enzymes* (3rd edn) **3**, 249 (1971).
80. Fersht, A. R., *Enzyme structure and mechanism* (2nd edn). Freeman, New York (1985). See Ch. 7.
81. Fersht, A. R., *Enzyme structure and mechanism* (2nd edn). Freeman, New York (1985). See Ch. 15.
82. Quiocho, F. A. and Lipscomb, W. N., *Adv. Protein Chem.* **25**, 1 (1971).
83. Gardell, S. J., Craik, C. S., Hilvert, D., Urdea, M. S., and Rutter, W. J., *Nature, Lond.* **317**, 551 (1985).
84. Gardell, S. J., Hilvert, D., Barnett, J., Kaiser, E. T., and Rutter, W. J., *J. biol. Chem.* **262**, 576 (1987).
85. Drenth, J., Jansonius, J. N., Koekoek, R., and Wolthers, B. G., *Adv. Protein Chem.* **25**, 79 (1971).
86. Angelides, K. J. and Fink, A. L., *Biochemistry* **18**, 2355 (1979).
87. Polgar, L. and Halasz, P., *Biochem. J.* **207**, 1 (1982).
88. Polgar, L., FEBS Lett. **219**, 1 (1987).
89. Schmid, M. F. and Herriott, J. R., *J. molec. Biol.* **103**, 175 (1976).
90. Reynolds, S. J., Yates, D. W., and Pogson, C. I., *Biochem. J.* **122**, 285 (1971).
91. Knowles, J. R. and Albery, W. J., *Acc. chem. Res.* **10**, 105 (1977).
92. Banner, D. W., Bloomer, A. C., Petsko, G. A., Phillips, D. C., Pogson, C. I., Wilson, I. A., Corran, P. H., Furth, A. J., Milman, J. D., Offord, R. E., Priddle, J. D., and Waley, S. G., *Nature, Lond.* **255**, 609 (1975).
93. Levitt, M. and Chothia, C., *Nature, Lond.* **261**, 552 (1976).
94. Phillips, D. C., Rivers, P. S., Sternberg, M. J. E., Thornton, J. M., and Wilson, I. A., *Biochem. Soc. Trans.* **5**, 642 (1977).
95. Casal, J. I., Ahern, T. J., Davenport, R. C., Petsko, G. A., and Klibanov, A. M., *Biochemistry* **26**, 1258 (1987).
96. Fersht, A. R., *Enzyme structure and mechanism* (2nd edn). Freeman, New York (1985). See pp. 91–5.
97. Laidler, K. J. and Bunting, P. S., *The chemical kinetics of enzyme action* (2nd edn). Clarendon Press, Oxford (1973). See Ch. 8.
98. Rieder, S. V. and Rose, I. A., *J. biol. Chem.* **234**, 1007 (1959).
99. Miller, J. C. and Waley, S. G., *Biochem. J.* **123**, 163 (1971).
100. Raines, R. T., Sutton, E. L., Straus, D. R., Gilbert, W., and Knowles, J. R., *Biochemistry* **25**, 7142 (1986).
101. Webb, M. R. and Knowles, J. R., *Biochem. J.* **141**, 589 (1974).
102. Albery, W. J. and Knowles, J. R., *Biochemistry* **15**, 5627 (1976).
103. Blacklow, S. C., Raines, R. T., Lim, W. A., Zamore, P. D., and Knowles, J. R., *Biochemistry* **27**, 1158 (1988).
104. Rose, I. A., *Adv. Enzymol.* **43**, 491 (1975).
105. Lai, C. Y. and Horecker, B. L., *Essays Biochem.* **8**, 149 (1972).
106. Lebherz, H. G. and Rutter, W. J., *Biochemistry* **8**, 109 (1969).
107. Sygusch, J., Beaudry, D., and Allaire, M., *Proc. natn. Acad. Sci. USA* **84**, 7846 (1987).
108. Lai, C. Y., *Archs. Biochem. Biophys.* **166**, 358 (1975).
109. Kelley, P. M. and Tolan, D. R., *Plant Physiol.* **82**, 1076 (1986).
110. Sawyer, L., Fothergill-Gilmore, L. A. and Freemont, P. S., *Biochem. J.* **249**, 789 (1988).
111. Rose, I. A., O'Connell, E. L., and Mehler, A. H., *J. biol. Chem.* **240**, 1758 (1965).

112. Model, P., Ponticorvo, L., and Rittenberg, D., *Biochemistry* **7**, 1339 (1968).
113. Westheimer, F. H. and Cohen, H., *J. Am. chem. Soc.* **60**, 90 (1938).
114. Shapiro, S., Enser, M., Pugh, E., and Horecker, B. L., *Archs Biochem. Biophys.* **128**, 554 (1968).
115. Hoffee, P., Lai, C. Y., Pugh, E. L., and Horecker, B. L., *Proc. natn. Acad. Sci. USA* **57**, 107 (1967).
116. Hartman, F. C. and Welch, M. H., *Biochem. Biophys. Res. Commun.* **57**, 85 (1974).
117. Hartman, F. C. and Brown, J. P., *J. biol. Chem.* **251**, 3057 (1976).
118. Gray, G. R., *Biochemistry* **10**, 4705 (1971).
119. Belasco, J. G. and Knowles, J. R., *Biochemistry* **22**, 122 (1983).
120. Holbrook, J. J., Liljas, A., Steindel, S. J., and Rossmann, M. G., *Enzymes* (3rd edn). **11**, 191 (1975).
121. Takenaka, Y. and Schwert, G. W., *J. biol. Chem.* **223**, 157 (1956).
122. Novoa, W. B., Winer, A. D., Glaid, A. J., and Schwert, G. W., *J. biol. Chem.* **234**, 1143 (1959).
123. Heck, H.d'A., McMurray, C. H., and Gutfreund, H., *Biochem. J.* **108**, 793 (1968).
124. Stinson, R. A. and Gutfreund, H., *Biochem. J.* **121**, 235 (1971).
125. Südi, J., *Biochem. J.* **139**, 251 (1974).
126. Südi, J., *Biochem. J.* **139**, 261 (1974).
127. Rossman, M. G., Liljas, A., Bränden, C.-I., and Banaszek, L. J., *Enzymes* (3rd edn) **11**, 61 (1975).
128. Blake, C. C. F., *Essays Biochem.* **11**, 37 (1975). See p. 72.
129. Biesecker, G., Harris, J. I., Thierry, J. S., Walker, J. E., and Wonacott, A. J., *Nature, Lond.* **266**, 328 (1977).
130. Moras, D., Olsen, K. W., Sabesan, M. N., Buehner, M., Ford, G. C., and Rossmann, M. G., *J. biol. Chem.* **250**, 9137 (1975).
131. Adams, M. J., Buehner, M., Chandrasekhar, K., Ford, G. C., Hackert, M. L., Liljas, A., Rossmann, M. G., Smiley, I. E., Allison, W. S., Everse, J., Kaplan, N. O., and Taylor, S. S., *Proc. natn. Acad. Sci. USA* **70**, 1968 (1973).
132. Holbrook, J. J. and Ingram, V. A., *Biochem. J.* **131**, 729 (1973).
133. Taylor, S. S., Oxley, S. S., Allison, W. S., and Kaplan, N. O., *Proc. natn. Acad. Sci. USA* **70**, 1790 (1973).
134. Bloxham, D. P., Sharma, R. P., and Wilton, D. C., *Biochem. J.* **177**, 769 (1979).
135. Clarke, A. R., Wigley, D. B., Chia, W. N., Barstow, D., Atkinson, T., and Holbrook, J. J., *Nature, Lond* **324**, 699 (1986).
136. Waldman, A. D. B., Hart, K. W., Clarke, A. R., Wigley, D. B., Barstow, D. A., Atkinson, T., Chia, W. N., and Holbrook, J. J., *Biochem. Biophys. Res. Commun.* **150**, 752 (1988).
137. Schimmel, P. R. and Söll, D., *A. Rev. Biochem.* **48**, 601 (1979).
138. Dang, C. V. and Dang, C. V., *Biochem. J.* **239**, 249 (1986).
139. Schimmel, P., *A. Rev. Biochem.* **56**, 125 (1987).
140. Eldred, E. E. and Schimmel, P. R., *Biochemistry* **11**, 17 (1972).
141, Söll, D. and Schimmel, P. R., *Enzymes* (3rd edn.) **10**, 489 (1974).
142. Fersht, A. R. and Kaethner, M. M., *Biochemistry* **15**, 818 (1976).
143. Fersht, A. R., Mulvey, R. S., and Koch, G. L. E., *Biochemistry* **14**, 13 (1975).
144. Fersht, A. R., *Biochemistry* **26**, 8031 (1987).
145. Fersht, A. R., *Enzyme structure and mechanism* (2nd edn). Freeman, New York (1985). See Ch. 13.
146. Loftfield, R. B. and Vanderjagt, M. A., *Biochem. J.* **128**, 1353 (1972).
147. Blow, D. M. and Brick, P., *Biological macromolecules and assemblies* (Jurnak, F. and

McPherson, A., eds), Vol. 2. p. 442. Wiley, New York (1985).

148. Fersht, A. R., Shi, J.-P., Knill-Jones, J., Lowe, D. M., Wilkinson, A. J., Blow, D. M., Brick, P., Carter, P., Waye, M. M. Y., and Winter, G., *Nature, Lond.* **314**, 235 (1985).
149. Wilkinson, A. J., Fersht, A. R., Blow, D. M., Carter, P., and Winter, G., *Nature, Lond.* **307**, 187 (1984).
150. Waye, M. M. Y., Winter, G., Wilkinson, A. J., and Fersht, A. R., *EMBO J.* **2**, 1827 (1983).
151. Bedouelle, H. and Winter, G., *Nature, Lond.* **320**, 371 (1986).
152. Jones, D. H., McMillan, A. J., Fersht, A. R., and Winter, G., *Biochemistry* **24**, 5852 (1985).
153. Citri, N., *Adv. Enzymol.* **37**, 397 (1973).
154. Koshland, D. E. Jr., *Proc. natn. Acad. Sci. USA* **44**, 98 (1958).
155. Fersht, A. R., *Enzyme structure and mechanism* (2nd edn). Freeman, New York (1985). See pages 331 and 342.
156. Carter, P. and Wells, J. A., *Nature, Lond.* **332**, 564 (1988).
157. Hou, Y.-M. and Schimmel, P., *Nature, Lond.* **333**, 140 (1988).
158. Clarke, A. R., Smith, C. J., Hart, K. W., Wilks, H. M., Chia, W. N., Lee, T. V., Birktoft, J. J., Banaszak, L. J., Barstow, D. A., Atkinson, T., and Holbrook, J. J., *Biochem. Biophys. Res. Commun.* **148**, 15 (1987).

6
The control of enzyme activity

6.1 Introduction

In this chapter we shall consider the third of the remarkable properties of enzymes mentioned in Chapter 1, Section 1.3, namely the control of enzyme activity. It should be self-evident why enzyme activity needs to be controlled in living organisms, since if no such control existed all metabolic processes would tend towards states in which they were at equilibrium with their surroundings. Thus, for instance, storage of glycogen as an energy reserve in muscle would be impossible, since there is an enzyme (phosphorylase) catalysing the breakdown of glycogen into glucose units, which would then be acted upon to yield ultimately carbon dioxide and water.* Clearly, the existence of glycogen as an energy store implies that the activity of phosphorylase can be controlled in such a way as to allow the store to be called on as the situation demands. We shall see later (Section 6.4.2) that there are a number of elaborate mechanisms to effect this control in muscle, thereby allowing the rate of glycogen breakdown to vary by several orders of magnitude in different circumstances.

Regulation of enzyme activities can be brought about in two ways. Firstly, the amount of an enzyme can be altered as a result of changes in the rate of enzyme synthesis or degradation (or both). This type of regulation is suitable for long-term regulation, e.g. in response to changes in nutrients, and is discussed in more detail in Chapter 9. Secondly, the activities of enzymes already present in a cell or organism can be altered, permitting a rapid response to changes in conditions. This chapter will be confined to this second means of regulation.

We shall first outline in Section 6.2 some of the mechanisms that have been shown to be important in controlling the activity of single enzymes, and then proceed in Sections 6.3 and 6.4 to consider the control of a sequence of enzyme-catalysed reactions that constitute a metabolic pathway. In the cell, there are usually a number of interconnected metabolic pathways and the control of these presents more complex features (see Chapter 8, Section 8.3).

6.2 Control of the activities of single enzymes

The principal mechanisms that exist for control of enzyme activity can be broadly classified into two categories: (i) those mechanisms that involve a change in the covalent structure of an enzyme, and (ii) those mechanisms that involve conformational changes in an enzyme caused by the reversible binding of

* This would be the case in an oxidizing atmosphere, i.e. under aerobic conditions. Under anaerobic conditions, glucose would be converted to pyruvate and thence lactate (see Fig. 6.20).

'regulator' molecules. These categories are discussed in Sections 6.2.1 and 6.2.2, respectively.

There are various other mechanisms by which enzyme activity could be controlled, including the following.

(i) Specific inhibitor macromolecules

The best studied cases of such macromolecules act as inhibitors of proteases. We have already mentioned how bovine pancreatic trypsin inhibitor and soybean trypsin inhibitor bind at the active site of trypsin by mimicking the structure of the tetrahedral intermediate that occurs in the enzyme-catalysed reaction (see Chapter 5, Section 5.5.1.3).* The function of this type of inhibitor molecule will be mentioned in Section 6.2.1.1.

In most other cases the precise details of the binding of inhibitor to enzyme are less clear. However, the interactions can be of great physiological significance, as is illustrated by the role of the phosphatase inhibitor in the regulation of glycogen metabolism (Section 6.4.2.4). A compilation of proteins that act as inhibitors of intracellular enzymes has been made. It has been suggested that these proteins be termed 'regulatory subunits' rather than 'antizymes'.[3]

(ii) Availability of substrate or cofactor

The intracellular concentrations of the substrates for many enzymes are significantly lower than the values of the Michaelis constant (K_m). For example the concentration of D-glyceraldehyde 3-phosphate in muscle (approximately 3 μmol dm^{-3}) is well below the K_m of this substrate for glyceraldehyde-3-phosphate dehydrogenase (approximately 70 μmol dm^{-3}).[4] Under these conditions the rate of the enzyme-catalysed reaction will depend on the concentration of substrate (see Chapter 4, Fig. 4.1). If the prevailing concentration of substrate is much greater than the K_m, changes in the concentration of substrate will have little effect on the rate of reaction. This would be the case for fructose-bisphosphate aldolase in mouse brain, where the concentration of fructose-bisphosphate (approximately 200 μmol dm^{-3}) is well above its K_m (12 μmol dm^{-3}).[5]

When data on the intracellular concentrations of substrates are being used to decide whether a certain enzyme is essentially saturated, it is important to bear in mind that cells usually contain a number of distinct organelles, such as nuclei, mitochondria, lysosomes, etc. These organelles are bounded by membranes which often display selective permeability and thus enable an uneven distribution of a metabolite to exist within the various compartments of a cell. Many methods of determining the concentrations of metabolites give only an average intracellular concentration, rather than the concentration in a particular compartment where the relevant enzyme may be located. The consequences of this compartmentation within cells are discussed more fully in Chapter 8, Section 8.3.

* Some other examples of protease inhibitors include α_1-antichymotrypsin, α_1-antiprotease, and α_2-macroglobulin, which are all found in blood plasma.[1] There are also several protease inhibitors of bacterial origin, e.g. leupeptin and antipain, which are of relatively low M_r.[2]

(iii) Product inhibition

The product of an enzyme-catalysed reaction may act as an inhibitor of that reaction, so that under circumstances in which product accumulates, the enzyme is inhibited and the rate of product formation is decreased. (See for example the inhibition of hexokinase caused by D-glucose 6-phosphate discussed in Section 6.4.1.2.)

(iv) Non-enzyme-catalysed reactions

A fourth possibility that has as yet been little explored is that in certain cases the rate of non-enzyme-catalysed reactions may be rate limiting. Attention has been focused on the fact that several enzymes involved in carbohydrate metabolism display anomeric specificity, i.e. they act on only one anomer of a sugar.[6] 6-Phosphofructokinase acts on the β-anomer of D-fructose 6-phosphate (F6P), whereas fructose-bisphosphatase acts on the α-anomer of D-fructose 1,6-bisphosphate (FBP) (see Fig. 6.1).

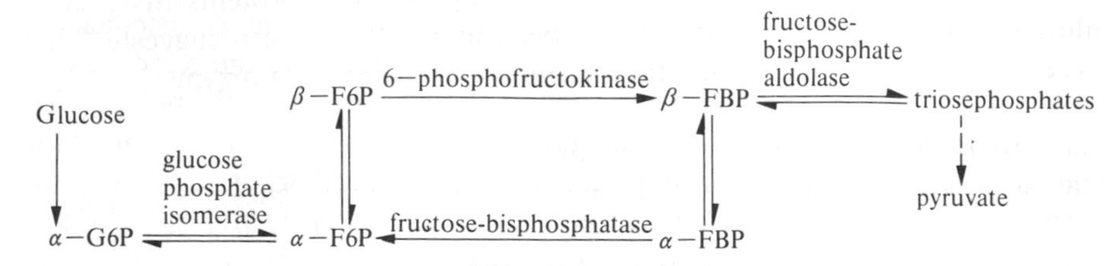

FIG. 6.1. Part of the pathways of glycolysis and gluconeogenesis, illustrating anomeric specificities of certain enzymes.[6] G6P, F6P, and FBP represent D-glucose 6-phosphate, D-fructose 6-phosphate, and D-fructose 1,6-bisphosphate, respectively.

The rates of anomerization (i.e. interconversion of α- and β-forms) of F6P and FBP are known. Calculations suggest that the non-enzyme-catalysed anomerization α-F6P $\rightarrow$ β-F6P, would not be rate limiting in glycolysis (glucose breakdown), but that the anomerization β-FBP $\rightarrow$ α-FBP might be so in gluconeogenesis (glucose synthesis).[6] These types of non-enzyme-catalysed reactions might well be rate limiting in other metabolic processes, especially when there are relatively high concentrations of the relevant enzymes, as is often the case in cells (see Chapter 8, Section 8.5).

Having briefly described these additional mechanisms of control, we shall now turn our attention to the principal means of regulating the activity of single enzymes.

6.2.1 Control of activity by changes in the covalent structures of enzymes

This category of control can be subdivided into those cases in which the change is essentially irreversible (Section 6.2.1.1) and those cases in which the change can be reversed provided that a source of energy, such as ATP, is available (Section 6.2.1.2).

6.2.1.1 Control of activity by essentially irreversible changes in covalent structure

It has been recognized for many years that a number of proteases are synthesized in the form of inactive precursors or zymogens, which can then be activated by the action of other proteases. The activation of chymotrypsinogen was discussed in detail in Chapter 5, Section 5.5.1.2. Some examples of enzymes that are activated in this way are given in Table 6.1. The first four enzymes mentioned in the table are synthesized as zymogens in the acinar cells of the pancreas and stored in zymogen granules. The release of the zymogens from these granules into the duodenum is under the control of hormones, primarily cholecystokinin-pancreozymin and secretin. Digestion of proteins requires the coordinated action of these various enzymes with their different specificities (see Chapter 5, Section 5.5.1.2); coordination is achieved by the common factor that the zymogens are all activated by the action of trypsin (Fig. 6.2.). In each case, the peptide bond cleaved during the activation of the zymogen lies on the C-terminal side of a lysine or arginine side-chain, in accordance with the known specificity of trypsin. For example, the activation of trypsinogen involves the removal of a hexapeptide Val-Asp-Asp-Asp-Asp-Lys from the N-terminus of the polypeptide chain. In order to account for the role of trypsin in initiating the activation of the zymogens, we must consider the involvement of another protease, entero-peptidase (previously referred to as enterokinase). Enteropeptidase, which is synthesized in the brush border of the epithelial cells of the small intestine, is an enzyme of high specificity catalysing the activation of trypsinogen as it enters the duodenum. The trypsin produced then catalyses the activation of the zymogens (Fig. 6.2). The scheme illustrates a number of important control features. It is essential that there is no premature activation of the zymogens, since this would

TABLE 6.1
Some enzymes activated by proteolytic action

Enzyme	Precursor	Function
Trypsin	Trypsinogen	Pancreatic secretions (see Section 6.2.1.1)
Chymotrypsin	Chymotrypsinogen	
Elastase	Proelastase	
Carboxypeptidase	Procarboxypeptidase	
Phospholipase A_2[7]	Prophospholipase A_2	Pancreatic secretion
Pepsin	Pepsinogen	Secreted into gastric juice: most active in pH range 1–5
Thrombin	Prothrombin	Part of the blood coagulation system (see Section 6.2.1.1)
Clr̄	Clr	Part of the first component of the complement system (see Section 6.2.1.1)
Chitin synthase[8]	Zymogen	Involved in the formation of the septum during budding and cell division in yeast

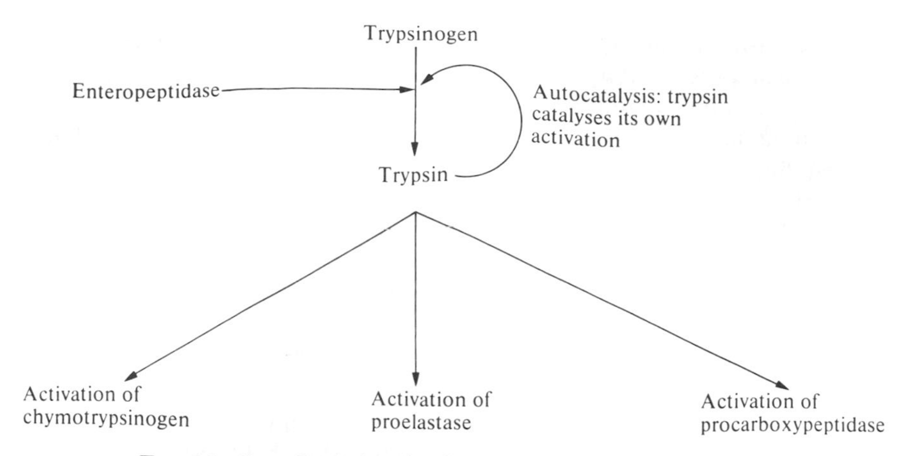

FIG. 6.2. Coordinated activation of the pancreatic proteases.

result in considerable damage to the pancreas.* In order to achieve this, the initial trigger provided by enteropeptidase is located at a site distinct from the site of production of the zymogens. The amount of active proteases produced is controlled by the amount of zymogens secreted into the duodenum, which is in turn under hormonal control. Eventually, the proteases digest themselves into amino acids and small peptides, which are reabsorbed from the intestinal tract and used for protein synthesis. The average daily production of the short-lived hydrolytic enzymes by the human pancreas is of the order of 10 g.[9]

An additional mechanism to prevent premature activation of trypsinogen is provided by the presence of a trypsin inhibitor protein in the pancreatic secretion. This protein, which is present to the extent of about 2 per cent of the protein content of the secretion (very much less in molar terms than the concentration of trypsinogen) binds very tightly to any active trypsin that might be present in the zymogen granules, and thus ensures that the activation of trypsinogen occurs only at the physiologically appropriate time and place, i.e. when it encounters enteropeptidase in the small intestine. The control of the production of enteropeptidase is less well understood and remains a problem for future work.

The scheme shown in Fig. 6.2 illustrates the principle of *signal amplification*, which will provide a recurrent theme in this Chapter. A small amount of enteropeptidase can produce rapidly a larger amount of trypsin, since one molecule of enzyme can act on many molecules of substrate in a short time. The trypsin, in turn, gives rise to an even larger quantity of active proteases.

This feature of amplification is also well illustrated by the systems involved in blood coagulation and complement activation. In the blood coagulation systems,[10] a series of enzymes and other protein factors act on each other in a

* In the condition known as acute pancreatitis, which can be lethal in severe cases, there is premature activation of the precursors of proteases and lipases synthesized in the pancreas.

sequential fashion, known as a *cascade*, eventually leading to the activation of the zymogen prothrombin to yield thrombin. Thrombin then catalyses the hydrolysis of a limited number of Arg-Gly bonds in fibrinogen to produce fibrin, which spontaneously forms an insoluble clot, strengthened by cross-linking between fibrin molecules. There are two pathways of activation of blood coagulation; extrinsic, which is triggered by a lipoprotein released from injured tissue, and intrinsic, which is triggered by contact of one of the factors with a surface such as the fibrous protein collagen that surrounds the damaged blood vessel. In both pathways, the sequence of enzyme-catalysed reactions provides a large amplification of the initial signal to ensure that sufficient fibrin is formed rapidly at the site of injury to restrict blood loss.

The complement system consists of a series of blood plasma proteins that cause damage to, and eventually lysis of, an invading organism, such as a bacterium.[11] The trigger for the system is usually provided by the aggregation of antibody molecules bound to the surface of the invading organism. This aggregation activates the components of the complement system by a sequence of activations of zymogens, leading to the rupture of the membrane of the invading organism.

In the various cases discussed, it is clear that the sequences of enzyme-catalysed reactions provide a *rapid, amplified response* to a small initial signal. Once the zymogens have been activated and the resulting proteases have performed their particular functions, the active enzymes are degraded rather than being converted back to zymogens. It is obviously important that the systems are 'switched off' at an appropriate time, so that for instance fibrin is formed only at the site of an injury and not in the entire blood system. The 'switching off' aspect has so far received less attention than the 'switching on'.

6.2.1.2 *Control of activity by reversible changes in covalent structure*

It is now clear that a large number of enzymes exist in two forms (of different catalytic properties), which can be interconverted by the action of other enzymes. The classic example of this type of enzyme is phosphorylase, which catalyses the reaction:

$$(\text{Glycogen})_n + \text{orthophosphate} \rightleftharpoons (\text{glycogen})_{n-1} + \alpha\text{-D-glucose 1-phosphate}$$

The enzyme exists in two forms: phosphorylase *b*, which requires AMP (or a few other ligands) for activity, and phosphorylase *a*, which is active in the absence of AMP, although AMP does activate it to a small extent. The only difference in covalent structure between the *a* and *b* forms of phosphorylase is that the side-chain of Ser 14 is phosphorylated in the *a* form but not phosphorylated in the *b* form.[12] Comparison of the three-dimensional structures of the two forms shows that the conformations of the first 19 amino acids from the N-terminus are markedly different. This segment is flexible in the *b* form, but well ordered in the *a* form, with the phosphorylated side-chain of Ser 14 interacting with the positively charged side-chain of Arg 69.[12]

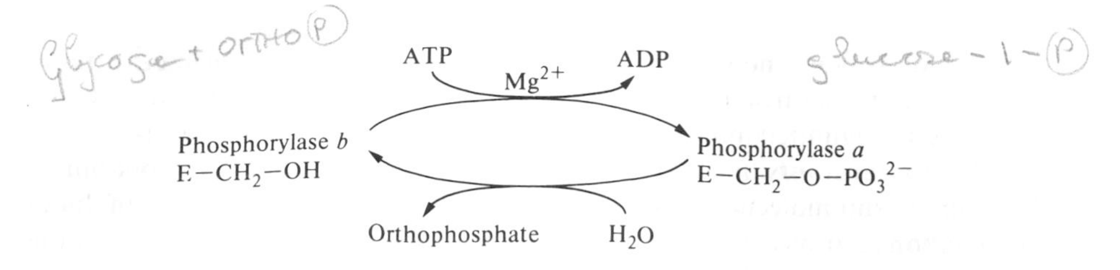

In the direction $b \rightarrow a$, the reaction requires ATP and Mg^{2+} and is catalysed by phosphorylase kinase; in the direction $a \rightarrow b$, the reaction is catalysed by phosphorylase phosphatase and liberates orthophosphate. If the reaction proceeds in a cyclical fashion, the net result would be the hydrolysis of ATP to ADP and orthophosphate.

A review published in 1986[13] lists over 140 enzymes and other proteins whose biological activity can be controlled by reversible covalent modification; some of these are listed in Table 6.2. (An indication of the growth of interest in this topic

TABLE 6.2
Some enzymes and proteins whose activity can be controlled by reversible covalent modification

Enzyme or protein	Modification	Biological function
Phosphorylase	Phosphorylation	Glycogen metabolism (see Section 6.4.2)
Glycogen synthase	Phosphorylation	
Phosphorylase kinase	Phosphorylation	
Phosphatase inhibitor protein	Phosphorylation	
Fructose 2,6-bisphosphatase/ 6-phosphofructo-2-kinase	Phosphorylation	Regulation of glycolysis (see Section 6.4.1.1)
Pyruvate dehydrogenase complex (mammalian)	Phosphorylation	Entry of pyruvate into tricarboxylic-acid cycle (see Chapter 7, Section 7.7.4)
Branched chain 2-oxoacid dehydrogenase complex	Phosphorylation	Breakdown of leucine, isoleucine and valine
Acetyl-CoA carboxylase	Phosphorylation	Synthesis of fatty acids
Troponin-I	Phosphorylation	Muscular contraction
Myosin light chain	Phosphorylation	
Glutamine synthetase (mammalian)	ADP-ribosylation	glutamine acts as N donor in a wide range of biosynthetic reactions
Glutamine synthetase (*E. coli*)	Adenylylation	
RNA nucleotidyltransferase (*E. coli*)	ADP-ribosylation	On infection by T4 phage, an Arg side chain in the α-subunit of the enzyme becomes modified. This shuts off transcription of the host genes
G-protein	ADP-ribosylation	G-proteins act as transducing agents relaying the effect of hormone binding to the activation of adenylate cyclase (see Chapter 8, Section 8.4.5)

For further details, see reference 13.

can be judged from the fact that only 32 such examples were listed in a 1978 review[14].) The roles of a number of these enzymes and proteins will be discussed in connection with the regulation of glycogen metabolism (Section 6.4.2). The majority of the reversible covalent modifications involve a phosphorylation—dephosphorylation cycle, usually at a serine side-chain, but at a threonine side-chain in the case of the phosphatase-inhibitor protein (Section 6.4.2.4). In the case of glutamine synthetase, a tyrosine side-chain (Tyr 397) in each of the 12 subunits of the enzyme can be adenylylated.[15]

$$\text{E}-\text{CH}_2-\text{C}_6\text{H}_4-\text{OH} + \text{ATP} \xrightarrow{\text{Mg}^{2+}} \text{E}-\text{CH}_2-\text{C}_6\text{H}_4-\text{O}-\text{AMP} + \text{Pyrophosphate}$$

In addition to those listed in Table 6.2, other types of modification include methylation, acetylation, tyrosinolation and sulphation.[13] The significance of reversible covalent modification of enzymes has been discussed in a number of reviews.[14, 16, 17] Two of the more noteworthy aspects of these systems are the following.

1. Because the conversion of an enzyme from one form to another is enzyme catalysed, there can be a *rapid change* in the amount of active enzyme present, and a *large amplification* of an initial signal. The degrees of amplification that can be achieved in various circumstances are discussed in references 13 and 16.
2. Reversible modification permits much more controlled responses to different metabolic circumstances than is the case with irreversible covalent modification. In the former case, a system can be viewed as being 'poised' for response; in this state there is continual activation and inactivation of the enzyme.

Such 'poising' obviously consumes a certain amount of energy, usually in the form of ATP, but some calculations suggest that this represents only a small fraction of the total cellular turnover of ATP.[13, 18] However, this conclusion has been challenged by Goldbeter and Koshland,[19] who maintain that a significant fraction of the total energy expenditure is required for the large number of reactions that involve reversible covalent modification of proteins. Whatever the actual fraction, however, it is clear that this is the price that the organism pays in order to support its sophisticated control mechanism. Under the influence of an appropriate stimulus, a system can rapidly be activated to produce almost exclusively the active form of the required enzyme. When the stimulus is removed, the system can be converted back to its resting state (almost exclusively the inactive form of the enzyme) and it is also possible to generate a whole range of intermediate levels of response. By contrast, in the cases of irreversible covalent modification, such as blood coagulation and complement activation, there is a rapid amplified response to some emergency (e.g. injury or infection) but, after use, the whole cascade systems must be replenished, since the steps in the cascades operate in one direction only.

6.2.2 Control of activity by ligand-induced conformational changes in enzymes

We shall first describe (Section 6.2.2.1) some of the observations that led to the formulation of ideas about the role of conformational changes in controlling the activities of enzymes, and then proceed (Section 6.2.2.2) to outline some of the theoretical models that have been used to describe such control systems. A final section (6.2.2.3) will discuss the biological significance of this type of control.

*6.2.2.1 Experimental observations on regulatory enzymes**

One of the earliest studies that demonstrated the importance of ligands in controlling the activity of enzymes was that of Cori and her coworkers[20] who showed that AMP was required to bring about the breakdown of glycogen in muscle (as a result of the activation of phosphorylase *b*; see Section 6.4.2.1).

However, many of the ideas about the role of conformational changes in controlling the activities of enzymes were developed as a result of work on biosynthetic pathways in microorganisms. In the mid 1950s, for example, it was found that threonine dehydratase, the first enzyme in the pathway of isoleucine biosynthesis in *E. coli*, was strongly inhibited by the end product, isoleucine, which bears only a limited structural resemblance to the substrate or the product of the reaction[21] (Fig. 6.3). Only the first enzyme in the pathway shows this inhibition.

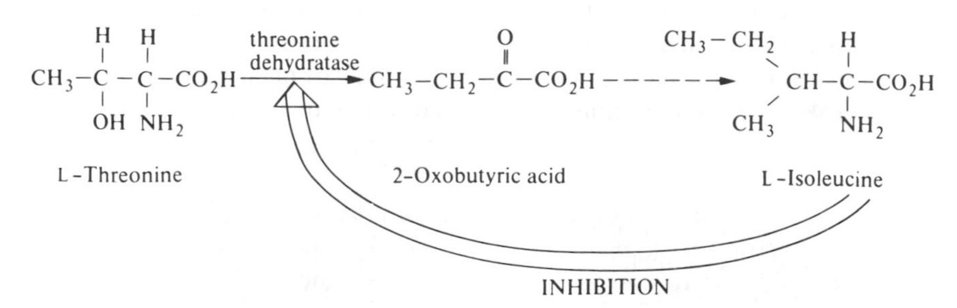

FIG. 6.3. Inhibition of threonine dehydratase, the first enzyme in the pathway of isoleucine biosynthesis in *E. coli*.

A more striking example was provided by aspartate carbamoyltransferase, the enzyme catalysing the first step in pyrimidine biosynthesis in *E. coli*, which is inhibited by the end products of the pathway, especially CTP.[22] It was also found that the purine nucleotide ATP could activate aspartate carbamoyltran-

* In one sense most, if not all, enzymes are regulatory, since their activities can be changed to some extent by, for example, changes in the concentration of substrates. etc. However, in this context, regulatory enzymes are those whose activity is controlled by conformational changes that accompany the binding of ligands. This definition includes those enzymes that exhibit sigmoid kinetics (see Fig. 6.4).

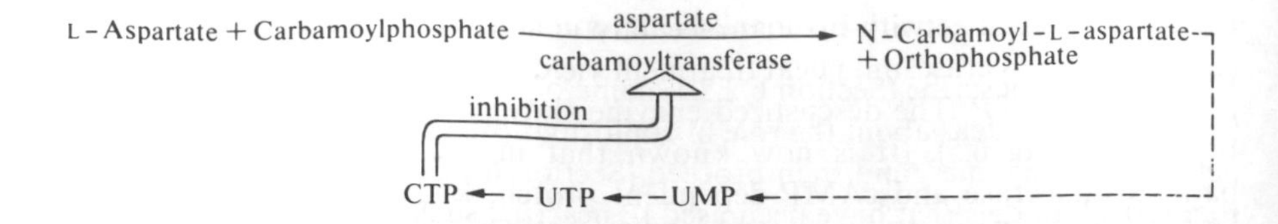

sferase, thus providing a mechanism for achieving a balance between the production of pyrimidine and purine nucleotides, which is required for the synthesis of nucleic acids.

By the early 1960s a number of examples had been described of this type of regulation, in which only the first enzyme in a pathway was subject to *feedback inhibition* by the end product of the pathway. In an incisive review[23] of these systems, Monod, Changeux, and Jacob noted several common features.

1. The ligands that were involved in regulation of enzyme activity (referred to as *regulator molecules* or *effectors*) were usually structurally distinct from the substrates or products of the relevant enzyme-catalysed reactions. It was therefore unlikely that an effector would bind at the active site of an enzyme. In most cases the inhibition caused by effectors was of a mixed type, i.e. did not conform to any of the three limiting cases of competitive, non-competitive, or uncompetitive inhibition (see Chapter 4, Section 4.3.2). This is also consistent with the proposal that the effector did not bind at the active site of the enzyme.

2. Many of the enzymes whose activity was controlled in this way did not show the normal type of kinetic behaviour, i.e. the plots of velocity against substrate concentration were of a sigmoidal (Fig. 6.4) rather than a hyperbolic (see Chapter 4, Fig. 4.1) nature. The sigmoidal shape implies that over a certain range of substrate concentration the velocity is more sensitive to substrate concentration than would be the case for an enzyme that showed normal kinetic behaviour.

3. It was often possible to distinguish between the binding of effector and the binding of substrate to a regulatory enzyme. Various physical or chemical treatments could lead to desensitization of the enzyme, i.e. loss of response to

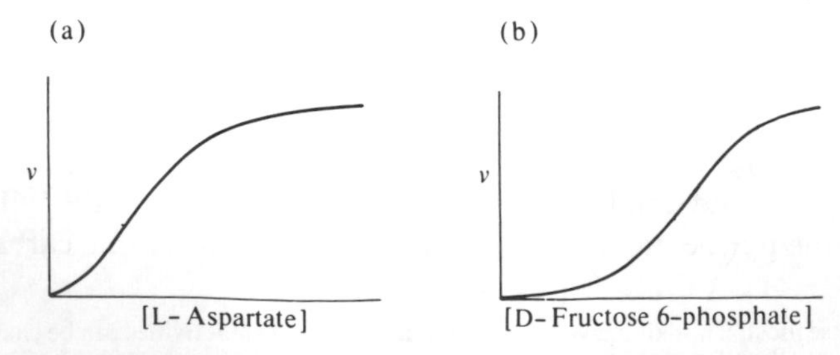

FIG. 6.4. Sigmoidal plots of velocity against substrate concentration for (a) aspartate carbamoyltransferase from *E. coli*, (b) 6-phosphofructokinase from rabbit skeletal muscle (assayed at high concentrations of ATP; see Fig. 6.21).

regulator molecules with no loss of catalytic activity. In the case of aspartate carbamoyltransferase, mild heat treatment yielded an active enzyme that was not inhibited by CTP. The desenstized enzyme showed normal, hyperbolic kinetic behaviour (Fig. 6.5). (It is now known that in the cases of aspartate carbamoyltransferase and cAMP-dependent protein kinase the binding sites for effectors and substrates are actually on different subunits, although this is not the case in most other regulatory enzymes.)

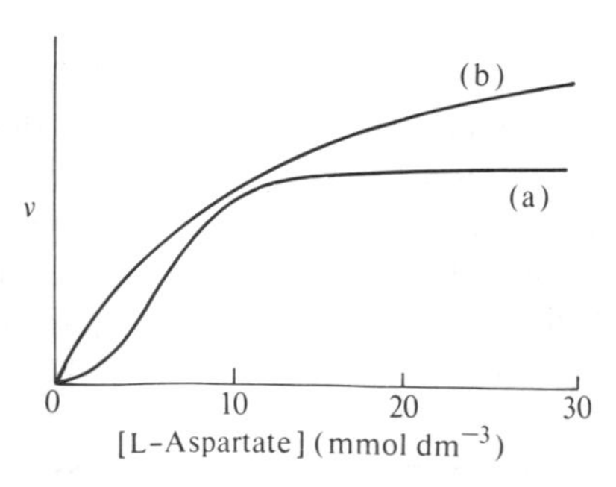

FIG. 6.5. Kinetic behaviour of aspartate carbamoyltransferase from *E. coli*. Curve (a) was obtained with enzyme assayed in the absence of CTP. Curve (b) was obtained with desensitized enzyme assayed in the absence or presence of CTP (0.2 mmol dm^{-3}).

4. In general, the regulatory enzymes were found to be oligomeric, i.e. they consisted of multiple subunits held together by non-covalent forces as described in Chapter 3, Section 3.6.

Consideration of these and other properties of regulatory enzymes led Monod, Changeux, and Jacob[23] to propose that in each case the regulator molecule (or effector) bound to a site distinct from the active site, so that the relationship between regulator molecule and substrate was an indirect or *allosteric* one (the term allosteric is derived from the Greek meaning 'different solid'). The binding of the regulator molecule to the regulator site induces a reversible conformational change in the enzyme that causes an alteration of the structure of the active site and a consequent change in the kinetic properties of the enzyme (Fig. 6.6).

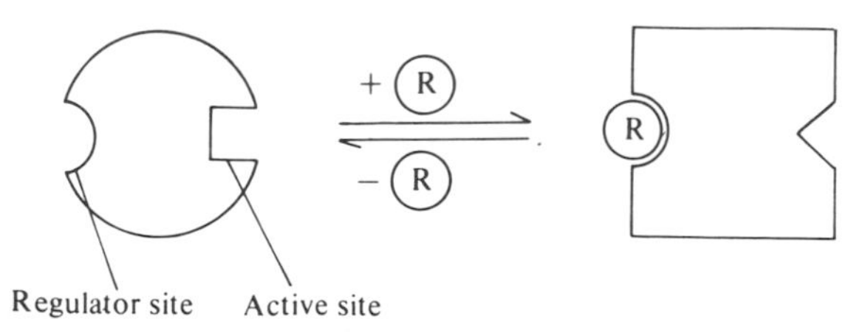

FIG. 6.6. Alteration of the structure of the active site of an enzyme by reversible binding of a regulator molecule R to a regulator site.

Although the point was not discussed at length in the 1963 review,[23] the sigmoidal nature of the kinetic plots (Fig. 6.4) was presumed to have its origin in conformational changes that were relayed between the subunits of multisubunit regulatory enzymes, i.e. the events at one active site could influence events at another. This subject will be discussed later when we consider the Monod–Wyman–Changeux model (Section 6.2.2.2).

6.2.2.2 *Theoretical models which account for the behaviour of regulatory enzymes*

A number of theoretical models have been proposed to try to account quantitatively for the behaviour of regulatory enzymes. In these models the major emphasis is placed on the interactions between the active sites on different subunits that can lead to sigmoidal kinetic plots of the type shown in Fig. 6.4. The models can then be extended to include the effects of regulator molecules on the kinetics. We shall describe four different models, those due to (i) Hill; (ii) Adair; (iii) Monod, Wyman, and Changeux; and (iv) Koshland, Némethy, and Filmer. Models (i) and (ii) are purely mathematical and make little reference to the behaviour of the enzyme or protein concerned, whereas models (iii) and (iv) make substantial reference to the behaviour of the enzyme or protein.

There are a number of other models that could account for sigmoidal kinetics, e.g. that of Rabin,[24] which involves slow conformational changes in an enzyme–substrate complex, and that of Ferdinand,[25] which applies to two-substrate reactions proceeding via random formation of a ternary complex (see Chapter 4, Section 4.3.5.1). The behaviour of glucokinase is discussed by Cornish-Bowden.[26]

Before considering these models it should be remembered that deviations from hyperbolic saturation curves had been detected in the early years of this century, when a study was made of the binding of oxygen to haemoglobin (Hb) and to myoglobin (Mb) (Fig. 6.7). It was later shown that myoglobin consisted of a single subunit with one oxygen-binding site, whereas haemoglobin consisted of four subunits each with an oxygen-binding site. The saturation curve for

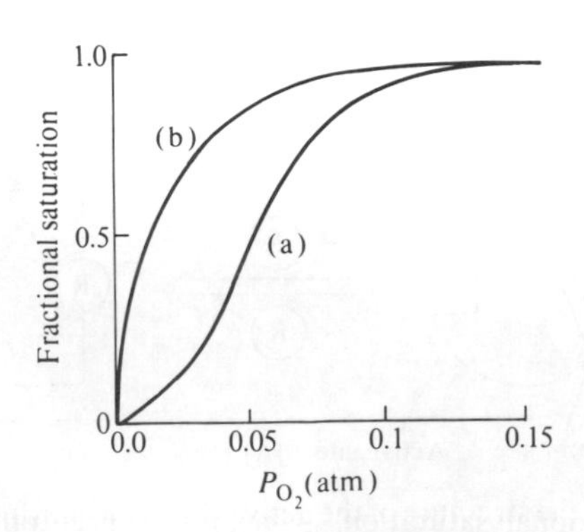

FIG. 6.7. Oxygen saturation curves for (a) haemoglobin and (b) myoglobin.

haemoglobin indicates that the binding of oxygen is *cooperative*, i.e. that binding of the first molecule of oxygen facilitates the binding of subsequent molecules of oxygen, thus giving rise to the sigmoid curve.

The hyperbolic saturation curve for myoglobin (note the analogy with the hyperbolic kinetic plot shown in Chapter 4, Fig. 4.1) can be accounted for as follows.

Consider the binding of a ligand, A, to a protein, P:

$$\mathrm{P} + \mathrm{A} \rightleftharpoons \mathrm{PA}.$$

The dissociation constant, K, is given by

$$K = \frac{[\mathrm{P}][\mathrm{A}]}{[\mathrm{PA}]},$$

$$\therefore \quad [\mathrm{PA}] = \frac{[\mathrm{P}][\mathrm{A}]}{K}. \qquad (6.1)$$

The fractional saturation, Y(i.e. the concentration of sites on the protein that are actually occupied, divided by the total concentration of such sites) is given by

$$Y = \frac{[\mathrm{PA}]}{[\mathrm{P}] + [\mathrm{PA}]}.$$

Substituting from (6.1),

$$Y = \frac{[\mathrm{A}]}{K + [\mathrm{A}]}. \qquad (6.2)$$

The plot of Y against [A] is shown in Fig. 6.8.

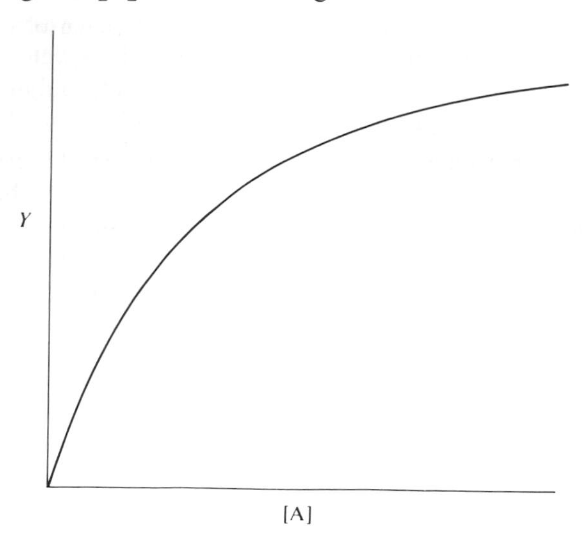

FIG. 6.8. Plot of fractional saturation, Y, against concentration of free ligand, [A], according to eqn (6.2).

(i) The Hill equation

In 1910, Hill[27] suggested the use of an equation that resembles (6.2) but has the concentration of free ligand, [A], raised to the power h:

$$Y = \frac{[A]^h}{K + [A]^h}. \tag{6.3}$$

An equation of this type can give rise to a sigmoidal plot of Y against [A] (see Fig. 6.9 for the case where $h = 3$), similar to the observed behaviour of the haemoglobin–oxygen system (Fig. 6.7). This suggests that there may be some physical basis for the assumption underlying the derivation of the equation.

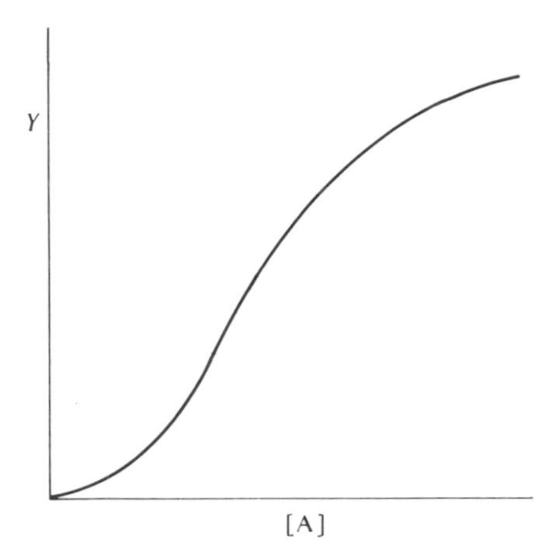

FIG. 6.9. Plot of fractional saturation, Y, against concentration of free ligand, [A], according to eqn (6.3) when $h = 3$.

An equation of the type given in (6.3) can be derived by assuming that a protein, P, has n binding sites for a ligand, A, and that the binding of A is *extremely cooperative*. This means that, as soon as one site on P is occupied, the remaining sites are immediately occupied, so that essentially we can consider an equilibrium between only two species P and PA_n, with no significant contribution from intermediate complexes (e.g. PA_{n-1}):

$$P + nA \rightleftharpoons PA_n,$$

$$K = \frac{[P][A]^n}{[PA_n]}. \tag{6.4}$$

The fractional saturation, Y, is given by

$$Y = \frac{[PA_n]}{[P] + [PA_n]}.$$

Substituting from (6.4), this becomes

$$Y = \frac{[A]^n}{K + [A]^n} \qquad \text{(which is of the same form as eqn (6.3))}$$

In eqn (6.3), h is known as the Hill coefficient. The value of h can be derived from experimental measurements of Y as a function of [A]. Rearranging eqn (6.3):

$$\frac{Y}{1-Y} = \frac{[A]^h}{K},$$

$$\therefore \quad \log\left[\frac{Y}{1-Y}\right] = h \log[A] - \log K. \qquad (6.5)$$

Thus a plot of $\log[Y/(1-Y)]$ against $\log[A]$ is linear with a slope equal to h (Fig. 6.10).

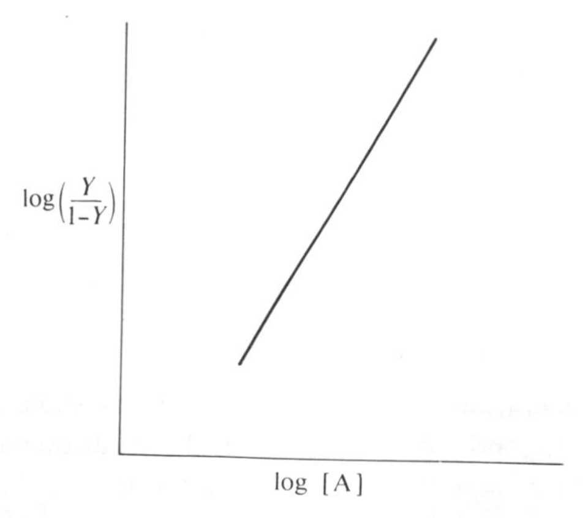

FIG. 6.10. Plot of $\log(Y/1-Y)$ against $\log[A]$ according to eqn (6.5). The slope of the line gives the value of h, the Hill coefficient.

For the binding of oxygen to haemoglobin, a straight line is obtained in this type of plot over most of the range of saturation of binding sites (at very low and very high degrees of saturation, deviations from linearity are observed because of a breakdown of the assumption underlying the derivation of eqn (6.3)). From the slope of the linear part of the plot, a value of $h = 2.6$ can be derived. The value of h can be taken as a measure of the strength of the cooperativity in ligand binding. If h is found to be equal to the number of binding sites, n, (four in the case of haemoglobin), this would imply extreme cooperativity in ligand binding, i.e. that intermediate species of the type PA_{n-1} make no significant contribution. In most cases, h is found to be less than the number of binding sites and is not necessarily integral. A value of h greater than 1 but less than the number of binding sites implies that the binding of ligand is cooperative (but not extremely cooperative). When h is equal to 1, there is no cooperativity in ligand binding and the saturation curve is hyperbolic (see eqn (6.2) and Fig. 6.8). If h is found to be less than 1, the binding could be described as *negatively cooperative*, i.e. the binding of the first ligand molecule discourages the binding of subsequent molecules of ligand.

An equation analogous to the Hill equation can be used to analyse enzyme kinetic data. In this case a plot of $\log[v/(V_{max}-v)]$ against log[substrate] gives a straight line, the slope of which equals h.

In summary, it can be stated that although the assumption underlying the derivation of eqn (6.3), i.e. that there is extreme cooperativity in ligand binding, is highly questionable in most cases, because intermediate complexes do make a significant contribution, the Hill coefficient, h, is nevertheless widely used as an empirical measure of the strength of cooperativity of ligand binding.

(ii) The Adair equation

In 1952, Adair[28] suggested that a more realistic description of the binding of oxygen to haemoglobin would include the various intermediate complexes. There would therefore be four equilibria, each with a characteristic dissociation constant:

$$\begin{array}{rcll} \mathrm{Hb}+\mathrm{O_2} & \rightleftharpoons & \mathrm{HbO_2} & K_1 \\ \mathrm{HbO_2}+\mathrm{O_2} & \rightleftharpoons & \mathrm{Hb(O_2)_2} & K_2 \\ \mathrm{Hb(O_2)_2}+\mathrm{O_2} & \rightleftharpoons & \mathrm{Hb(O_2)_3} & K_3 \\ \mathrm{Hb(O_2)_3}+\mathrm{O_2} & \rightleftharpoons & \mathrm{Hb(O_2)_4} & K_4 \end{array}$$

The fractional saturation of sites (Y) can be expressed in terms of the concentration of free ligand ([A]) and the four dissociation constants (eqn (6.6)). For a derivation of this equation, see Appendix 6.1.

$$Y=\frac{\dfrac{[\mathrm{A}]}{K_1}+\dfrac{2[\mathrm{A}]^2}{K_1K_2}+\dfrac{3[\mathrm{A}]^3}{K_1K_2K_3}+\dfrac{4[\mathrm{A}]^4}{K_1K_2K_3K_4}}{4\left[1+\dfrac{[\mathrm{A}]}{K_1}+\dfrac{[\mathrm{A}]^2}{K_1K_2}+\dfrac{[\mathrm{A}]^3}{K_1K_2K_3}+\dfrac{[\mathrm{A}]^4}{K_1K_2K_3K_4}\right]}. \tag{6.6}$$

From the experimental values of Y as a function of [A], it is possible to deduce the 'best fit' values of the four dissociation constants by curve-fitting procedures. The relative values of these dissociation constants (strictly speaking, of the *intrinsic* dissociation constants*) give a clue to the cooperativity of ligand

* In considering the binding of ligand to more than one site on a protein, statistical factors have to be taken into account. The dissociation constants (K_1, etc.) are related to intrinsic dissociation constants (K'_1, etc.) for a protein containing four sites, as follows:

$$K_1=K'_1/4, \quad K_2=2K'_2/3, \quad K_3=3K'_3/2, \quad K_4=4K'_4 \quad \text{(see Appendix 6.1).}$$

If the *intrinsic* dissociation constants followed a sequence $K'_1>K'_2>K'_3>K'_4$, the binding would be said to show (positive) cooperativity at each stage.

binding. For a further discussion of the Adair equation references 29 and 30 should be consulted.

(iii) The model of Monod, Wyman, and Changeux

In 1965, Monod, Wyman, and Changeux[31] proposed a model to account for the behaviour of allosteric proteins and enzymes, i.e. those that showed allosteric, or indirect, interactions between ligand-binding sites.* Starting from the observation that most allosteric proteins are oligomeric, that is they consist of more than one subunit, the model makes four principal statements about the symmetry and the conformation of an allosteric protein.

(a) The subunits occupy equivalent positions, so that within the oligomer, there is at least one axis of symmetry. (It now appears that the dimeric enzyme hexokinase from yeast may be an exception to this statement.[32])
(b) The conformation of each subunit depends on its interaction with the other subunits.
(c) There are two conformational states available to the oligomer. These are designated R (relaxed) and T (tense) and they differ in affinity for a given ligand.
(d) When the conformation of the protein changes from R to T (or *vice versa*), the symmetry of the oligomer is conserved.

If the conformations of the subunits in the T and R states are denoted by circles and squares, respectively, then statements (c) and (d) would mean that for a dimeric protein there is an equilibrium:

○○ ⇌ □□
T R

There would be no 'hybrid' species (□○), since in such species the symmetry of the oligomer would be lost. In other words, all the subunits change conformation in a 'concerted' fashion—this gives rise to the alternative name 'concerted' for this model.

The various equilibria involved in binding of a ligand, A, to a dimeric protein are illustrated in Fig. 6.11.

Monod, Wyman, and Changeux[31] defined two parameters ($\bar{Y}$ and $\bar{R}$) to describe the state of the allosteric system. $\bar{Y}$ is the fractional saturation of ligand sites on the protein, and $\bar{R}$ is the fraction of protein that is in the R state. The values of both $\bar{Y}$ and $\bar{R}$ are in the range from 0 to 1.

Using the equilibrium expressions, $\bar{Y}$ and $\bar{R}$ were evaluated in terms of the number of ligand-binding sites, n, the concentration of free ligand, α,† and two

* Monod, Wyman, and Changeux[31] introduced terms to define the two classes of allosteric effects. *Homotropic* effects are interactions between identical ligands, e.g. between substrate molecules bound to different subunits in an enzyme. *Heterotropic* effects are interactions between different ligands, e.g. between substrate and regulator molecule.

† For convenience, the concentration of free ligand [A] is divided by the dissociation constant of the ligand from the R state, K_R, to give a dimensionless ratio, α.

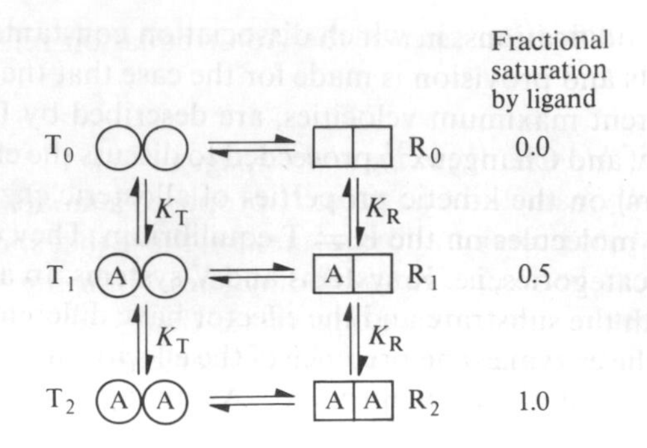

FIG. 6.11. Equilibria involved in binding of a ligand, A, to a dimeric protein according to the model of Monod, Wyman, and Changeux.[31] K_T and K_R are the intrinsic dissociation constants for ligand from the T and R states, respectively. Circles and squares represent the conformations of subunits in the T and R states, respectively.

constants L and c that could be adjusted to give the best fit to the experimental data. L is the ratio of the concentration of protein in the T state to that in the R state in the absence of ligand, i.e. $L=[T_0]/[R_0]$ in Fig. 6.11, and c is the ratio of the intrinsic dissociation constants for the ligand from the R and T states, respectively, i.e. $c=K_R/K_T$ in Fig. 6.11. The expressions derived[31] were

$$\bar{Y}=\frac{Lc\alpha(1+c\alpha)^{n-1}+\alpha(1+\alpha)^{n-1}}{L(1+c\alpha)^n+(1+\alpha)^n}, \tag{6.7}$$

$$\bar{R}=\frac{(1+\alpha)^n}{L(1+c\alpha)^n+(1+\alpha)^n}. \tag{6.8}$$

It is important to appreciate that the successive intrinsic dissociation constants for the T state are the same $(=K_T)$ and that the successive intrinsic dissociation constants for the R state are the same $(=K_R)$. *According to the model of Monod, Wyman, and Changeux, cooperativity in ligand binding arises because the protein is initially predominantly in one state (say T), but the ligand binds preferentially to the other state (R). As more ligand is added to the system, the protein is gradually swung over into the tighter-binding R state.* For the haemoglobin–oxygen system, the Monod, Wyman, and Changeux model gives a good fit to the experimental saturation curve with $L=9050$ and $c=0.014$,[31] i.e. the protein is initially more than 99.9 per cent in the T state, but oxygen binds over 70 times more tightly to the R state. An analysis of eqn (6.7) shows that cooperativity in ligand binding will be more pronounced when L is large and c is small.

Up to this point we have considered the Monod, Wyman, and Changeux model in terms of ligand saturation functions, which are readily applicable to a binding protein such as haemoglobin. However, in order to analyse kinetic data for an enzyme, where the observed velocity reflects both a Michaelis constant and a maximum velocity, some modifications to the basic equations are necessary. These modifications, in which dissociation constants are replaced by Michaelis constants and provision is made for the case that the two forms of the enzyme have different maximum velocities, are described by Dalziel.[33]

Monod, Wyman, and Changeux[31] proceeded to discuss the effects of regulator molecules (effectors) on the kinetic properties of allosteric enzymes in terms of the effects of these molecules on the R $\rightleftharpoons$ T equilibrium. They divided allosteric enzymes into two categories, i.e. K systems and V systems. In a K system, it was envisaged that both the substrate and the effector have different affinities for the T and R states of the enzyme. The presence of the effector modifies the affinity of the enzyme for the substrate, i.e. it affects K_m. Assuming that the substrate binds more tightly to the R state, it would follow that an effector that binds more tightly to the T state would displace the R $\rightleftharpoons$ T equilibrium towards the T state, thereby raising the K_m and decreasing the velocity at a given concentration of substrate. By contrast, an effector that bound more tightly to the R state would increase the velocity at a given concentration of substrate. Using these arguments, Monod, Wyman, and Changeux[31] were able to account for the behaviour of a K system such as aspartate carbamoyltransferase in the presence of activators (e.g. ATP) and inhibitors (e.g. CTP) (Fig. 6.12). (It should be noted that Monod, Wyman, and Changeux treated quantitatively only the case of exclusive binding, in which, for example, an inhibitor binds *only* to the T state. The more general case of non-exclusive binding, in which an inhibitor binds *preferentially* to the T state, is discussed by Rubin and Changeux[34]).

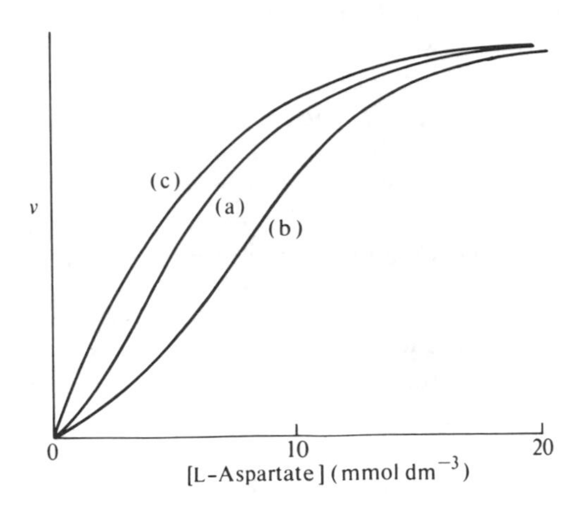

FIG. 6.12. Kinetic behaviour of aspartate carbamoyltransferase from *E. coli*. Curve (a) was obtained in the absence of effectors, curve (b) in the presence of 0.5 mmol dm^{-3} CTP, and curve (c) in the presence of 2 mmol dm^{-3} ATP.

In a V system, the substrate has the same affinity for the R and T states of the enzyme. However, the two states differ in their catalytic activities, so that the effector, by displacing the R $\rightleftharpoons$ T equilibrium, affects V_{max} rather than K_m. An example of a V system is phosphorylase *b*, which is activated by AMP.

(iv) The model of Koshland, Némethy, and Filmer

The model proposed in 1966 by Koshland, Némethy, and Filmer[35] to account for the ligand-binding properties of oligomeric proteins and enzymes can be seen as an extension of Koshland's 'induced-fit' hypothesis (see Chapter 5, Section 5.6) and as an attempt to define relationships between the various constants in the Adair equation (eqn (6.6)). In the 'induced-fit' hypothesis, it is postulated that the binding of substrate to an enzyme induces the appropriate conformation of the enzyme to allow catalysis to occur. Similarly, the binding of a ligand to one subunit in an oligomeric protein changes the conformation of that subunit, thereby altering the interaction of that subunit with its neighbours. In diagramatic terms the binding of a ligand A to a dimeric protein can be represented by the following equilibria:

○○ + A ⇌ ○[A]

○[A] + A ⇌ [A][A]

Features of the model are as follows.

(a) In the absence of ligand, the protein exists in one conformational state rather than as an equilibrium mixture of two conformational states as in the model of Monod, Wyman, and Changeux.

(b) The conformational change in the protein is *sequential*, i.e. the subunits change conformation sequentially as ligands bind, rather than in a concerted fashion as in the model of Monod, Wyman, and Changeux.

(c) The interactions between the subunits can be of a positive or a negative type, so that the binding of the second (and later) molecule of ligand can be more favourable (positive cooperativity) or less favourable (negative cooperativity) than the binding of the first molecule. By contrast, the model of Monod, Wyman, and Changeux allows only positive cooperativity in ligand binding, since the binding of ligand brings about a concerted transition of all the subunits to the form that has higher affinity for the ligand.

An expression for the fractional saturation of sites by ligand ($\bar{Y}$) is derived by considering the various equilibria with their characteristic dissociation constants, as was done for the Adair equation (eqn (6.6)). The model of Koshland, Némethy, and Filmer can then be used to derive relationships between the various dissociation constants in the Adair equation. Each dissociation constant is considered[35] to have three components:

(a) a dissociation constant characterizing the binding of ligand by the more favoured subunit conformation (□ in this scheme);

(b) an equilibrium constant for the conformational change of the subunit (i.e. ○ ⇌ □). This constant will be related to the difference in free energy between the two conformations;

(c) one or more interaction constants that depend on the degree of stability of the complex between the subunits relative to the stability of the isolated subunits. These constants depend on the geometrical arrangement of the oligomer, since this will determine the number and strength of the interactions between the subunits. For a tetrameric protein we could have the following arrangements:

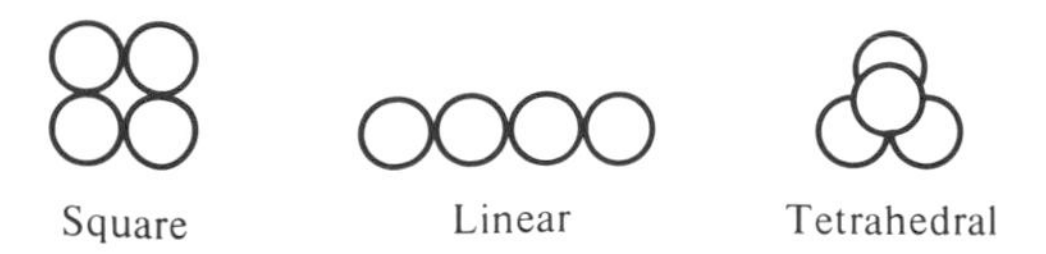

and the form of the interaction constants would be different in each case.

Assuming a particular arrangement of subunits in the oligomer, an equation can be derived for the $\bar{Y}$ in terms of the free ligand concentration and the various constants referred to above. The equations derived from the model of Koshland, Némethy, and Filmer are more complex than those derived from the model of Monod, Wyman, and Changeux and elaborate curve-fitting procedures are usually necessary to derive the values of the various constants that give the best fit to the experimental data. For detailed accounts of the model of Koshland, Némethy, and Filmer, references 30 and 35–37 should be consulted.

(v) Relationships between the model of Monod, Wyman, and Changeux and the model of Koshland, Némethy, and Filmer

The concerted model of Monod, Wyman, and Changeux and the sequential model of Koshland, Némethy, and Filmer can be viewed as limiting cases of a more general model involving all possible conformational and liganded states of an oligomeric protein[38] (Fig. 6.13). These more general models have been analysed (see, for instance, references 39 and 40), but the resulting equations are extremely complex and for our purposes it is more useful to discuss only the limiting cases (i.e. the concerted and the sequential models).

A great deal of effort has been expended in attempts to show which model provides a more appropriate description of the behaviour of a particular protein or enzyme. It is true that in a number of cases, such as the binding of oxygen to haemoglobin, both models can give a reasonably satisfactory fit to the experimental data. However, there are at least three types of observation in which a distinction between the models can be made.

(a) Negative cooperativity in ligand binding. As mentioned in (iv) above the model of Monod, Wyman, and Changeux cannot account for negative cooperativity in ligand binding, whereas the model of Koshland, Némethy, Filmer can account for either positive or negative cooperativity. Thus, the observation of negative cooperativity in ligand binding can be used to exclude the use of the concerted model in a description of that particular system. Of the known examples of negative cooperativity,[41] the best-investigated is that of the binding of NAD^+ to the tetrameric enzyme glyceraldehyde-3-phosphate dehydrogenase from rabbit muscle, where the successive dissociation constants have been

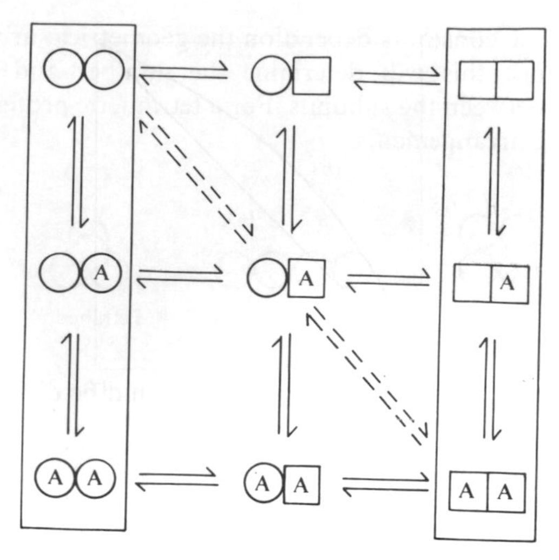

FIG. 6.13. A general model for the equilibria involved in the binding of a ligand, A, to a dimeric protein. The two vertical columns outlined show the species considered in the model of Monod, Wyman, and Changeux.[31] The dashed lines show the species considered in the model of Koshland, Némethy, and Filmer.[35]

determined as:

$$K_1 = 10^{-11}\ \mathrm{mol\,dm^{-3}},\quad K_2 = 10^{-9}\ \mathrm{mol\,dm^{-3}},\quad K_3 = 10^{-6}\ \mathrm{mol\,dm^{-3}},$$
$$K_4 = 2\times 10^{-5}\ \mathrm{mol\,dm^{-3}}.$$

It should be pointed out that it is very difficult to distinguish between negative cooperativity and non-identical binding sites. Thus, if in a dimeric protein there were one tight-binding site and one weak-binding site for a ligand, we would appear to observe negative cooperativity in the binding of the ligand. Presumably, to prove that the binding showed negative cooperativity we would need to establish that the sites were identical in the absence of ligands—a not inconsiderable task.

In its most extreme form, negative cooperativity can be manifested as 'half-of-the-sites reactivity', where, for instance, only half of the active sites of an oligomeric enzyme are able to catalyse the reaction at the same time.[41, 42] Tyrosyl-tRNA synthetase from *Bacillus stearothermophilus* shows this property (see Chapter 5, Section 5.5.5.2).

(b) Measurements of conformational changes and ligand binding. The conformational changes in enzymes and proteins that occur on binding of ligands can be monitored by a variety of means such as spectroscopic probes, changes in sedimentation coefficient, or changes in the reactivity of amino-acid side-chains.[43] Using such methods, the value of $\bar{R}$ (i.e. the fraction of protein in the R state) can be evaluated as increasing concentrations of ligand are added to the protein. The values of $\bar{Y}$ (i.e. the fraction of sites saturated) can be measured by

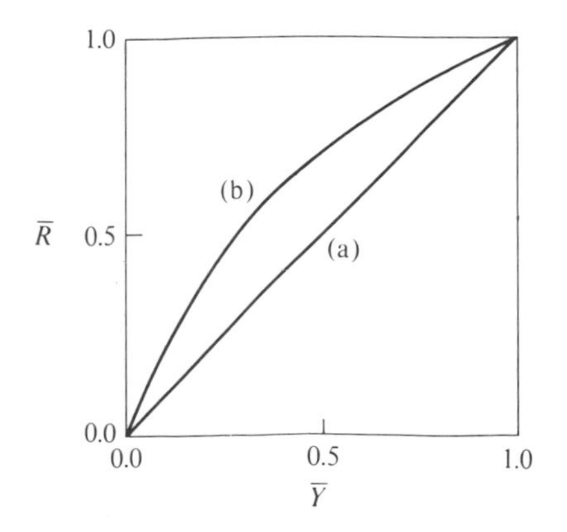

FIG. 6.14. Plots of $\bar{R}$ (the fraction of protein in the R state) against $\bar{Y}$ (the fraction of sites saturated). (a) Plot for a protein that conforms to the sequential model of Koshland, Némethy, and Filmer. (b) Typical plot for a protein conforming to the model of Monod, Wyman, and Changeux.

direct-binding studies.[44] The shape of the plot of $\bar{R}$ against $\bar{Y}$ can be used to distinguish between the concerted and sequential models (Fig. 6.14).

In the sequential model of Koshland, Némethy, and Filmer the conformational change in a subunit occurs only when a ligand binds to that subunit, and hence we would expect the plot of $\bar{R}$ against $\bar{Y}$ to be linear* (see (a) in Fig. 6.14). However, in the concerted model of Monod, Wyman, and Changeux, we would expect a non-linear plot of $\bar{R}$ against $\bar{Y}$ (see, for example, (b) in Fig. 6.14) because of the concerted nature of the $R \rightleftharpoons T$ transition. By examining the experimental data for $\bar{R}$ and $\bar{Y}$, it has been possible to classify the binding of AMP to phosphorylase *a* as being of the concerted type[45] and the binding of NAD^+ to glyceraldehyde-3-phosphate dehydrogenase from rabbit muscle as being of the sequential type.[46]

(c) Demonstration of isomerization reactions. The binding of NAD^+ to glyceraldehyde-3-phosphate dehydrogenase from yeast has been studied in great detail by the stopped-flow[47] and temperature-jump[48] methods (see Chapter 4, Section 4.4). The results were found to be in accord with predictions of the concerted model, but not with those of the sequential model, since the latter model cannot account for the observed isomerization reaction that occurs in the absence of ligand. (This isomerization reaction is denoted by the $T_0 \rightleftharpoons R_0$ equilibrium in the concerted model, see Fig. 6.11.)

(vi) Usefulness of the models

The various models described in (i)–(iv) can give only a limited insight into the nature of subunit interactions and cooperativity in ligand binding in regulatory

* It should be noted that a linear plot would only be expected in the simplest form of the sequential model, in which the ligand-induced conformational change is restricted to one subunit. If the conformational change extended across a subunit interface, deviations from linearity would be expected.

enzymes. It is necessary to have X-ray crystallographic data to understand the structural basis for these properties. At present, structural information is available on ligand-induced conformational changes in a number of enzymes, including phosphorylase *a*,[12] lactate dehydrogenase (see Chapter 5, Section 5.5.4), and glyceraldehyde-3-phosphate dehydrogenase,[49] but the best understood example is the haemoglobin–oxygen system. The elegant studies of Perutz[50] have shown how the binding of oxygen at the haem group on one subunit of deoxyhaemoglobin leads to a small movement (≈ 0.07 nm) of the iron into the plane of the haem; this movement triggers a series of structural alterations in the protein that can be sensed at the other haem groups in the molecule that are between 2.5 and 3.7 nm distant.

In spite of these qualifying comments, it remains true that the models, especially the concerted model of Monod, Wyman, and Changeux and the sequential model of Koshland, Némethy, and Filmer, have provided a relatively simple conceptual basis for understanding the cooperative binding of ligands to proteins. The models have served to focus attention on regulatory enzymes and have stimulated a large number of elegant and detailed studies on these enzymes. Some of the experimental approaches used to study regulatory enzymes are summarized in Table 6.3.

TABLE 6.3
Some methods of study of regulatory enzymes

Type of information sought	Methods of study
Is cooperative behaviour observed in enzyme kinetic studies?	Perform assays of enzyme activity over a wide range of substrate concentrations. Examine the kinetic plots (see Chapter 4, Fig. 4.3) for non-linearity. Use the Hill plot (Fig. 6.10) to give measure of cooperativity
Is cooperative behaviour observed in ligand binding?	Use direct-binding methods,[44] e.g. equilibrium dialysis or ultracentrifugation (these methods separate free ligand from bound ligand). Test for cooperativity using, e.g. the Hill plot (Fig. 6.10) or the Scatchard plot (Fig. 6.15)
Do regulator molecules bind to a site distinct from the active site?	Test effects of regulator molecules on the kinetic plots. Are these effects consistent with binding to distinct sites? (see for example Chapter 4, Section 4.3.2.3)
	A distinction between substrate and effector sites can be made if the enzyme can be desensitized to the effects of the regulator molecule while retaining activity (Section 6.2.2.1)
Do conformational changes occur on binding of ligand?	X-ray crystallographic data gives the most detailed information, but less-direct methods, involving spectroscopic probes or measurements of the reactivity of amino-acid side-chains, can also be useful[43]
Does the enzyme consist of subunits?	The number, type, and arrangement of subunits in an enzyme can be determined from determinations of M_r, symmetry and cross-linking patterns. These methods are discussed in Chapter 3, Section 3.6

6.2.2.3 *The significance of allosteric and cooperative behaviour in enzymes*

Allosteric interactions between regulator molecules and substrates provide a versatile means of regulation, since they allow the activity of an enzyme to be controlled by changes in the concentrations of species other than the substrate and product of that enzyme-catalysed reaction. We have already mentioned this point in connection with the feedback inhibition of early enzymes in biosynthetic pathways (Section 6.2.2.1) and shall return to it in Section 6.3.1.

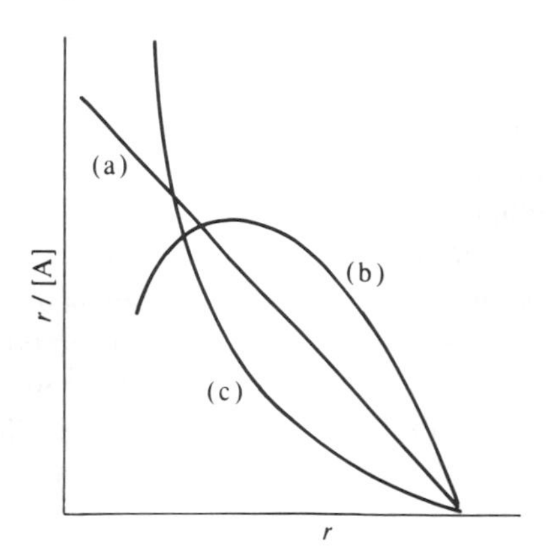

FIG. 6.15. Scatchard plots of ligand-binding data (see Appendix 6.1): r represents the number of molecules of A bound per molecule of P and [A] equals the concentration of free ligand. Case (a) corresponds to binding sites that are equivalent and independent; case (b) corresponds to positive cooperativity between ligand binding sites; case (c) corresponds to negative cooperativity between ligand binding sites (or possibly non-identical binding sites).

Enzymes that show cooperativity in kinetics have significantly different responses to changes in the concentrations of substrates from those that exhibit hyperbolic kinetics. The difference in response can be expressed in quantitative terms by calculating the change in substrate concentration required to increase the velocity from 10 per cent of V_{max} to 90 per cent of V_{max} as a function of the Hill coefficient, h, in eqn (6.9), which is analogous to the Hill equation (eqn 6.3):

$$v = \frac{V_{max}[S]^h}{K_m + [S]^h} \qquad (6.9)$$

Value of h in eqn (6.9)	Required change in [S] to increase velocity from 10 per cent of V_{max} to 90 per cent of V_{max}
1	81-fold
2	9-fold
3	4.33-fold
4	3-fold
0.5	6561-fold

For an enzyme displaying hyperbolic kinetics ($h = 1$ in eqn (6.9)) an 81-fold change in substrate concentration is required; this change is reduced to just over 4-fold for an enzyme showing cooperativity in kinetics with a Hill coefficient of 3. The enhanced response could well be important in enabling an enzyme to adjust to changes in conditions, but it should be stressed that large variations in the concentrations of substrates are unlikely to occur in cells. Additional means of amplifying the response of enzymes to changes in conditions are discussed in Section 6.3.2 in connection with the regulation of metabolic pathways.

It is also evident that an enzyme that shows negatively cooperative kinetics ($h < 1$ in eqn (6.9)) has a diminished response to changes in substrate concentration (over certain ranges of concentration) compared with an enzyme that displays hyperbolic kinetics. Negative cooperativity could thus serve to insulate an enzyme from the effects of changes in substrate concentration.[41] In at least one case (the binding of NAD^+ to glyceraldehyde-3-phosphate dehydrogenase from yeast) it appears that an enzyme can show a mixture of positive and negative cooperativity in ligand binding,[41, 51] permitting further variations in the response of the enzyme to changes in substrate concentration.

The best understood example of cooperativity in ligand binding is the haemoglobin–oxygen system. In this case the physiological significance of the sigmoidal saturation curve (Fig. 6.7) is clear.[52] Haemoglobin is almost fully saturated with oxygen in the lungs, where the partial pressure of oxygen is about 0.13 atm, and yet unloads the oxygen in the tissue capillaries, where the partial pressure of oxygen is about 0.01 atm, much more completely than if the binding curve were hyperbolic with the four binding sites acting independently. The equilibrium between oxyhaemoglobin and deoxyhaemoglobin is also influenced by other factors, including the concentration of 2,3-bisphospho-D-glycerate, which acts as an allosteric regulator.[52] The significance of cooperative and allosteric behaviour in regulatory enzymes is not so well understood, although it can be appreciated in qualitative terms (see the example of 6-phosphofructokinase in Section 6.4.1.1).

A final word should be added about the rate at which ligand-induced conformational changes occur in regulatory enzymes. Several lines of evidence have indicated that proteins in solution can have highly flexible structures (see Chapter 3, Section 3.5.2). The time scale of conformational changes in enzymes appears to be generally in the range 10^{-2} to 10^{-4} s,[53] which is roughly equivalent to the time scale of catalytic events (see Chapter 4, Section 4.4). However, a number of cases of slow conformational changes in enzymes are now known.[30, 53–55] An example is the inhibition of hexokinase isoenzyme II from ascites tumour cells by D-glucose 6-phosphate, which has a half-time of over 10 s.[56] The physiological function of slow conformational changes is not well understood; some possibilities have been discussed.[54, 57]

6.3 Control of metabolic pathways

In Section 6.2 we were largely concerned with the more important ways in which the activities of single enzymes can be controlled. We now turn to considering the regulation of metabolic pathways that consist of sequences of enzyme-catalysed reactions. It is convenient to distinguish between two types of control that can be exerted on metabolism.[58] *Intrinsic* (or *internal*) *control* refers to the regulation of enzyme activity by the concentrations of metabolites, e.g. substrates, products, end-products of pathways, or adenine nucleotides (see the example of 6-phosphofructokinase in Section 6.4.1.1). This type of control is predominant in unicellular organisms. In multicellular organisms, the metabolism within any given type of cell must be related to the requirements of the whole organism. In such cases an extra level of control, *extrinsic* (or *external*) *control* is brought about by means such as hormones or nervous stimulation (see Section 6.4.2). These signals generally act via *second messengers* to affect the activities of target enzymes. Examples of second messengers include 3′:5′-cyclic AMP (cAMP), Ca^{2+}, inositol 1,4,5-trisphosphate, and diacylglycerol (Fig. 6.16). An excellent discussion of these regulatory mechanisms is given in reference 59.

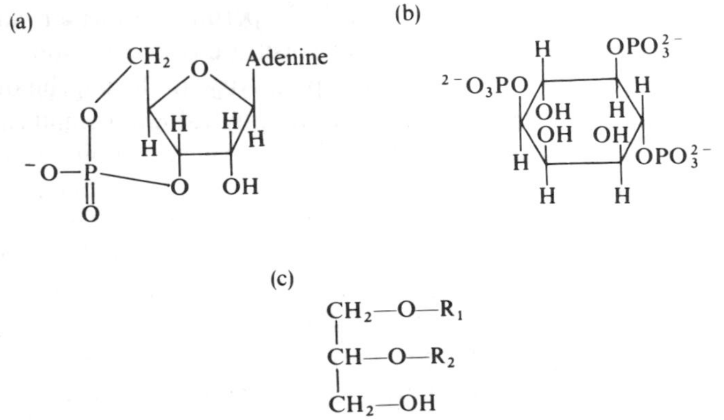

(a)

CH₂ O Adenine; O; H H; H H; $^-$O—P—O OH; O

(b)

H OPO_3^{2-}; $^{2-}O_3PO$ H; OH H; OH OH; H OPO_3^{2-}; H H

(c)

CH_2—O—R_1
|
CH—O—R_2
|
CH_2—OH

FIG. 6.16. Structures of some 'second messengers': (a) 3′:5′ cyclic AMP (cAMP); (b) inositol 1,4,5-trisphosphate; (c) diacylglycerol, where R_1 and R_2 represent fatty-acid chains, commonly stearic and arachidonic acids, respectively. Details of the action of these and other messengers can be found in reference 59.

6.3.1 General considerations

We shall consider a metabolic pathway in which A is converted to F in a sequence of steps catalysed by enzymes E_1 to E_5:

$$A \xrightarrow{E_1} B \xrightarrow{E_2} C \xrightarrow{E_3} D \xrightarrow{E_4} E \xrightarrow{E_5} F.$$

It would be most economical to regulate the flux through the pathway, i.e. the rate at which F is formed from A, by regulating the catalytic activity of an early enzyme in the pathway such as E_1. This would then avoid the wasteful accumulation of intermediates. If E_1 were subject to feedback inhibition (Section 6.2.2.1) by F, then if F accumulates or is supplied exogenously, the production of

F would be shut off. On the other hand, if the concentration of F falls, the production of F would be resumed.

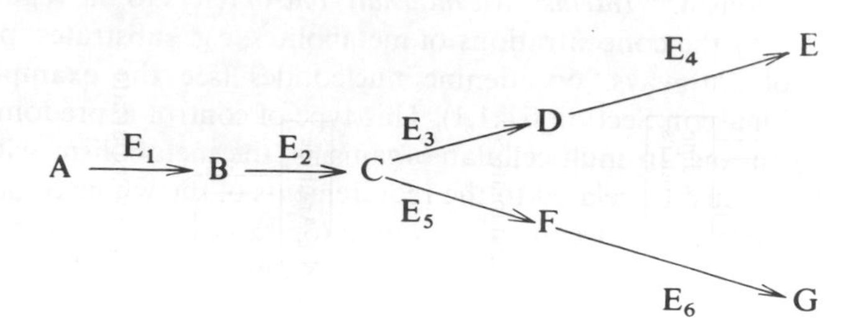

It might also be expected that if a pathway were branched, there would be a control point near the branch point. If an intermediate C is required for the synthesis of two products E and G, then it would seem reasonable that there should be some control mechanism to allow the rate of production of C to be altered to deal with various situations, for example, one in which E might be in excess but G not in excess. Such control could be achieved by a variety of interlocking controls near the branch point (for a review see reference 60). A good example is provided by the enzyme carbamoyl-phosphate synthase in *E. coli.* This enzyme catalyses the reaction:

$$\text{Glutamine} + 2\text{ATP} + \text{bicarbonate} \rightarrow \text{carbamoyl phosphate} + 2\text{ADP} + \text{orthophosphate} + \text{glutamate.}$$

Carbamoyl phosphate is used in the production of pyrimidines and of arginine. As shown in Fig. 6.17, there is a variety of control mechanisms to ensure that an appropriate rate of production of carbamoyl phosphate is maintained in a variety of circumstances. Carbamoyl-phosphate synthase is subject to strong feedback inhibition by pyrimidine nucleotides but not by arginine. Thus, when pyrimidine nucleotides are in excess, the synthesis of carbamoyl phosphate is inhibited and may become too slow to support adequate synthesis of arginine. Under these circumstances, ornithine will accumulate and, at sufficiently high concentrations, will antagonize the effects of pyrimidine nucleotides and thus restore the activity of carbamoyl-phosphate synthase. When this happens the carbamoyl phosphate produced will be used exclusively for the production of arginine since the enzyme catalysing the production of *N*-carbamoyl-L-aspartate (aspartate carbamoyltransferase) is feedback-inhibited by the high concentrations of pyrimidine nucleotides. If arginine is present in excess, the enzyme catalysing the production of *N*-acetyl-L-glutamate (amino-acid acetyltransferase) will be feedback-inhibited (Fig. 6.17).

Other elaborate control mechanisms that have been demonstrated to exist include sequential inhibition, cumulative inhibition, and enzyme multiplicity.[60] An example of enzyme multiplicity is found in the biosynthesis of the aromatic amino acids in *E. coli*[61]. The first step in the pathway, i.e. the production of 7-phospho-2-keto-3-deoxy-D-*arabino*-heptonate from phosphoenolpyruvate

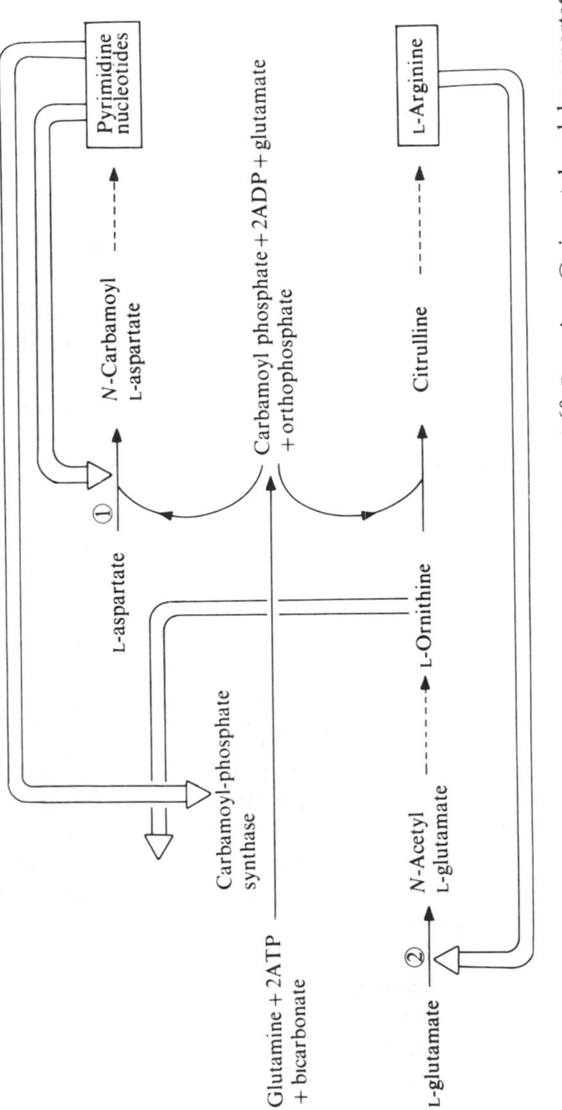

FIG. 6.17. Control of carbamoyl phosphate synthesis in *E. coli*.[60] Reaction ① is catalysed by aspartate carbamoyltransferase. Reaction ② is catalysed by amino-acid acetyltransferase.

and D-erythrose 4-phosphate is catalysed by three distinct enzymes, each of which is subject to control by a different end product. The first enzyme is subject to feedback inhibition by phenylalanine, the second to feedback inhibition by tyrosine, and synthesis of the third enzyme is repressed by tryptophan.

A final point that should be made is that within a cell there could be compartmentation of metabolites and enzymes. Our proposed mechanisms of control must therefore take into account any membrane-permeability barriers that may exist (see Chapter 8, Sections 8.2 and 8.3).

6.3.2 Amplification of signals

One of the features of a metabolic pathway (when compared with a single enzyme) is that considerable amplification of an initial signal is possible. The initial signal is usually a change in the concentration of a substrate or other ligand. Two particular mechanisms are considered to be important in signal amplification: these are substrate cycles (Section 6.3.2.1) and interconvertible enzyme cycles (Section 6.3.2.2).

6.3.2.1 Substrate cycles

Consider the following segment of a metabolic pathway:

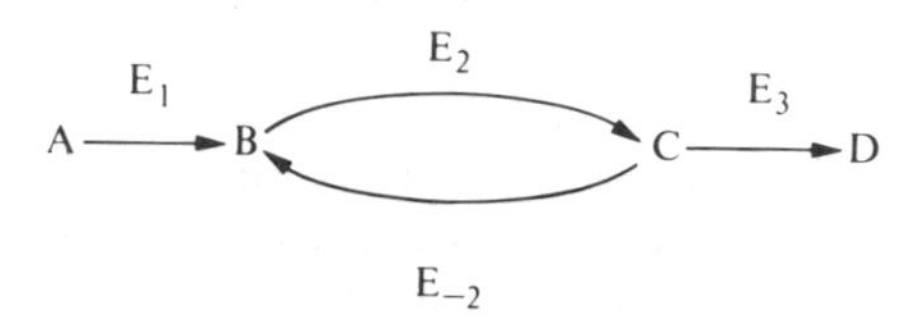

in which there are separate enzymes to catalyse the conversion of B to C (E_2) and of C to B (E_{-2}). If E_2 and E_{-2} are active simultaneously, a *substrate cycle* will occur in which B and C are interconverted. An example of a substrate cycle occurs in the glycolytic pathway (Fig. 6.18).[62] Reaction (1) is catalysed by

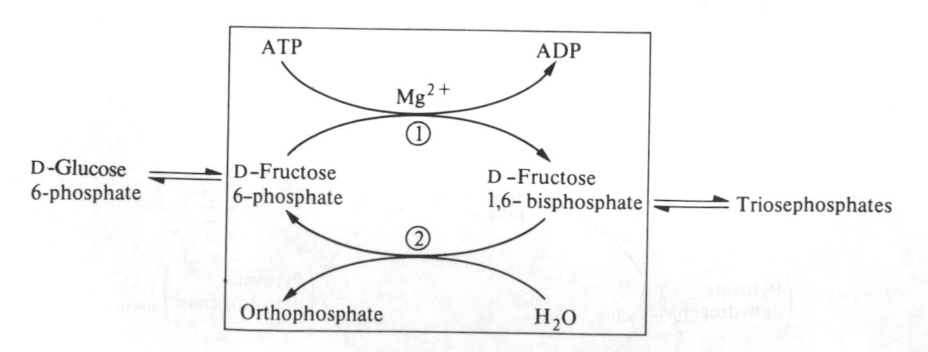

FIG. 6.18. The substrate cycle between D-fructose 6-phosphate and D-fructose 1,6-bisphosphate. Reaction (1) is catalysed by 6-phosphofructokinase and reaction (2) by fructose-bisphosphatase. Note the anomeric specificities of these enzymes (see Fig. 6.1).

6-phosphofructokinase, reaction (2) by fructose-bisphosphatase. If both enzymes are active simultaneously, the net result is the hydrolysis of ATP to yield ADP and orthophosphate. The importance of the substrate cycle in this case is that the activities of the two enzymes involved can be separately controlled (Section 6.4.1.1) and hence the net flux through this step, i.e. the rate of reaction (1) minus the rate of reaction (2), can be controlled much more precisely than if only a single enzyme were involved. The activities of the two enzymes are controlled in a reciprocal fashion, e.g. AMP activates 6-phosphofructokinase but inhibits fructose-bisphosphatase, so the effect of small changes in the concentrations of AMP on the net flux can be greatly amplified (Section 6.4.1.1).

A substrate cycle may also operate between D-glucose and D-glucose 6-phosphate in the liver.[63, 64]

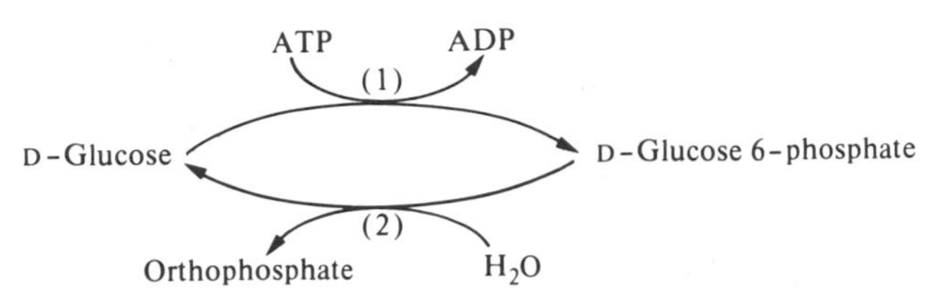

Reaction (1) is catalysed by hexokinase type IV (glucokinase) and reaction (2) by glucose 6-phosphatase. There is evidence that this substrate cycle may be important in the regulation of glucose metabolism in the liver.[63, 64]

Substrate cycles confer on the cell the possibility of more sensitive control of metabolic pathways. The price paid for this is consumption of energy in terms of ATP hydrolysis (Section 6.2.1.2). We shall discuss the evidence that substrate cycling does occur *in vivo* in Section 6.4.1.1.

6.3.2.2 *Interconvertible enzyme cycles*

In Table 6.2 we listed some examples of enzymes whose activity can be controlled by reversible covalent modification. An interconvertible enzyme cycle can provide a large amplification of an initial signal provided that the activities of the enzymes catalysing the interconversions are carefully controlled.[13, 18]

For example, the activity of the pyruvate dehydrogenase multienzyme complex from mammalian sources can be altered by phosphorylation (see Chapter 7, Section 7.7.4):

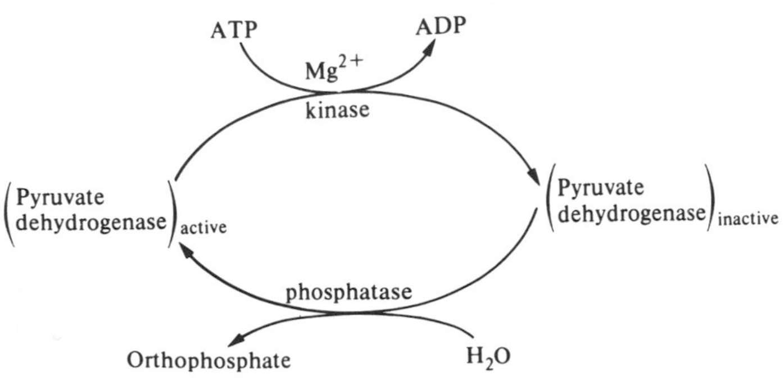

The kinase is active when the ratios of [NADH]/[NAD$^+$] and/or [acetyl-CoA]/[CoA] are high, so that when the concentration of NADH and/or acetyl-CoA is high, pyruvate dehydrogenase will be inactivated and the entry of pyruvate into the tricarboxylic acid cycle will be blocked.[14, 65] Some reciprocal effects are observed on the phosphatase; for example, the phosphatase is inhibited by a high ratio of [NADH]/[NAD$^+$], so that the balance between the active and inactive forms of pyruvate dehydrogenase can be very sensitive to changes in the concentrations of these metabolites.[14]

In the case of phosphorylase, the conversion of the inactive *b* form to the active *a* form is catalysed by phosphorylase kinase, which is itself subject to enzyme-catalysed activation and inactivation. A sequence of events can be traced back to the binding of a hormone to a receptor on the cell surface (Fig. 6.19).

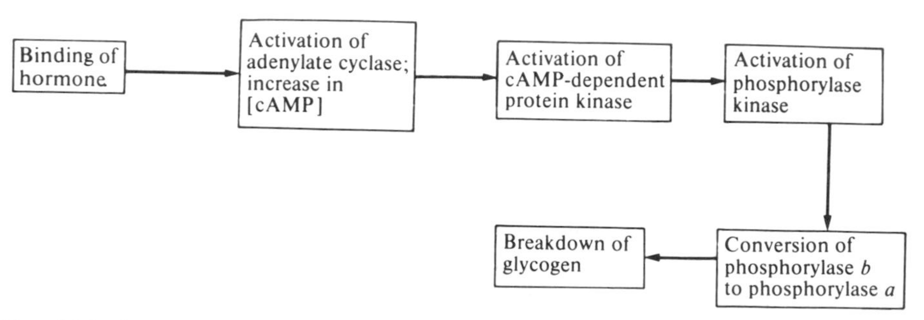

FIG. 6.19. Sequence of events involved in initiation of the breakdown of glycogen following binding of a hormone to a receptor on the cell surface.

The enzyme-catalysed activation of phosphorylase and of phosphorylase kinase can be reversed by the action of phosphatases that are themselves subject to regulation. There is also an elaborate mechanism to ensure that when phosphorylase is activated, the enzyme catalysing the synthesis of glycogen (glycogen synthase) is inactivated (Section 6.4.2.5).

For the present we should note that the sequence shown in Fig. 6.19 provides an excellent example of amplification. Experimentally it has been shown that there is only a small change in the concentration of cAMP on administration of a hormone such as adrenalin;[66] however, each molecule of protein kinase that is activated can rapidly activate many molecules of phosphorylase kinase and hence a great number of molecules of phosphorylase, thus leading to rapid breakdown of glycogen. It has been estimated that 50 per cent of the phosphorylase *b* could be converted to the active (*a*) form with only a 1 per cent increase in the concentration of cAMP,[67] the process occurring within about 2 s of the administration of hormone.[68]

6.3.3 Formulation of theories for the control of metabolic pathways

The development of a theory for the control of a particular metabolic pathway usually involves the following steps:

(1) detection of the major control points in the pathway;
(2) a study of the properties of the control enzymes;
(3) formulation of an initial theory;
(4) testing the correctness of the theory by obtaining and assessing experimental data (preferably from *in vivo* experiments); making any necessary modifications to the original theory.

We shall discuss these various steps in turn (Sections 6.3.3.1 to 6.3.3.4) and then illustrate them by discussing the present-day theories for the control of glycolysis (Section 6.4.1) and of glycogen metabolism (Section 6.4.2) in muscle.

6.3.3.1 Detection of major control points

Up to about 15 years ago it was general practice to divide the enzymes catalysing the various steps in a metabolic pathway into controlling and non-controlling types. Thus, the activity of a controlling enzyme would be changed by the regulatory signal, e.g. a change in the concentration of some metabolite, whereas the activity of a non-controlling enzyme would only change as a consequence of a change in the activity of a controlling enzyme. However, more recently it has become appreciated that the activities of all the enzymes in a pathway may have some influence on the overall flux through the pathway. In order to define metabolic control more precisely, and to provide experimentally testable models of control, a number of detailed mathematical analyses have been undertaken.[69–72] One of the best-known is that due to Kacser and Burns.[69,70] Each enzyme in the pathway is assigned a sensitivity coefficient, Z, which is a measure of the effect of a change in the activity of that enzyme on the flux through the pathway.* (The symbol Z has now been replaced by C^{F}_{E}, the flux control coefficient[73].) An enzyme with a high value of Z will be more important in controlling the flux through a metabolic pathway than one with a low value of Z.†

The different formulations and concepts of the various mathematical models of metabolic control continue to give rise to considerable debate.[75–78] However, the various formulations agree that in a metabolic pathway under given conditions, one (or very few) enzymes will dominate the regulation of the

* Mathematically, the sensitivity coefficient (Z_1) for an enzyme E_1 is defined as

$$Z_1 = \frac{\frac{dF}{F}}{\frac{dE_1}{E_1}} = \frac{d\ln F}{d\ln E_1},$$

where F is the flux through the pathway and E_1 is the catalytic activity of enzyme E_1

† In the glycolytic pathway, hexokinase and 6-phosphofructokinase have large values of Z (0.7 and 0.3, respectively, in erythrocytes; 0.47 and 0.53, respectively, in *S. cerevisiae*[74]). The other enzymes in the pathway have very low Z values and thus make a negligible contribution to the control of the flux.

pathway, though under different conditions the controlling reaction(s) may change. For our present purposes, therefore, we can effectively speak of major controlling and minor controlling enzymes. A major controlling enzyme should catalyse the slowest (or rate-limiting) step in the pathway. A reaction that is not at equilibrium *in vivo* is likely to be a rate-limiting step, since the reason that equilibrium is not achieved is presumably that there is insufficient active enzyme present to allow equilibration to occur. The initial identification of controlling enzymes therefore involves a search for those enzymes that (i) are present at low activities and (ii) catalyse reactions that are not at equilibrium. Additional methods involve searches for cross-over points and mathematical modelling techniques (see (iii) and (iv) below, respectively).

(i) The activity of each enzyme in a metabolic pathway is measured *in vitro* by performing an appropriate assay (see Chapter 4, Section 4.2) in a homogenate of the tissue concerned. It is important to measure the maximum velocity of each enzyme under conditions of pH, ionic strength, etc. that are similar to those existing in the cell. This may involve a partial purification of the homogenate to remove an inhibitor, for example. The results of the assays usually show that a reasonably clear distinction can be drawn between enzymes present in the cell at 'high activities' and those present at 'low activities'. In the glycolytic pathway (Fig. 6.20) hexokinase and 6-phosphofructokinase are 'low-activity' enzymes, whereas triosephosphate isomerase and pyruvate kinase are 'high-activity' enzymes.[5, 79]

(ii) In order to decide whether a given reaction in a pathway is at equilibrium *in vivo*, we must be able to measure the intracellular concentrations of the particular metabolites involved. The equilibrium constant for the reaction under conditions of pH, ionic strength, etc. that are similar to those that are thought to obtain in the cell can be determined by separate experiments using the isolated enzyme.

The usual method for determining the concentrations of metabolites in a given tissue involves the technique of 'freeze clamping'.[62] A sample of tissue is very rapidly frozen by compression between aluminium plates that are cooled to the temperature of liquid nitrogen, i.e. 77 K (−196°C). This rapid freezing effectively stops any processes that might tend to change the concentrations of metabolites. Before analysis of the tissue, it is essential to inhibit any enzyme-catalysed reactions; this is achieved by treating the frozen tissue with frozen perchloric acid, which will inactivate enzymes as the preparation is allowed to thaw. After centrifugation to remove protein and other precipitated material, the supernatant is neutralized and analysed for the metabolites of interest by appropriate chemical or enzymatic methods (see Chapter 10, Section 10.7). The content of a metabolite, determined as micromoles per gram of tissue can be expressed as a concentration, provided that the intracellular water content is known*

* The total water content of a tissue can be determined from the decrease in weight after drying. The contribution of extracellular water to this total can be assessed by measuring the concentration of some added compound (e.g. sorbitol) which is known not to penetrate the cells. The balance is the intracellular water.[80]

Results obtained on a rat heart perfused with glucose were:[81]

Metabolite	Intracellular concentration
ATP	11.5 mmol dm^{-3}
ADP	1.3 mmol dm^{-3}
D-Fructose 6-phosphate (F6P)	0.09 mmol dm^{-3}
D-Fructose 1,6-bisphosphate (FBP)	0.02 mmol dm^{-3}
AMP	0.17 mmol dm^{-3}

Thus, for the reaction catalysed by 6-phosphofructokinase,

$$\text{F6P} + \text{ATP} \overset{Mg^{2+}}{\rightleftarrows} \text{FBP} + \text{ADP},$$

the ratio of intracellular concentrations, known as the mass action ratio, Γ, is given by

$$\Gamma = \frac{[\text{FBP}][\text{ADP}]}{[\text{F6P}][\text{ATP}]}$$

$$= 0.025.$$

Since the equilibrium constant for this reaction is about 1200, it is clear that this reaction is far removed from equilibrium *in vivo*.

By contrast, for the reaction catalysed by adenylate kinase,

$$\text{ATP} + \text{AMP} \overset{Mg^{2+}}{\rightleftharpoons} 2\text{ADP},$$

the mass action ratio is 0.85 which is close to the equilibrium constant for this reaction (0.44). Therefore, adenylate kinase catalyses a reaction that is close to equilibrium *in vivo*. Rolleston[82] suggests that if the ratio of Γ/K (K is the equilibrium constant) for a reaction is less than 0.05, the reaction can be considered as being non-equilibrium. More recently, ^{31}P nuclear magnetic resonance (n.m.r.) has been demonstrated to have a great deal of potential for measurement of the intracellular concentrations of phosphorus-containing metabolites such as ATP, ADP, AMP, phosphocreatine, etc.[83–86] It is possible to measure the concentration of these metabolites in a sample of tissue, an organ, or even a whole organism inserted into the wide-bore magnet of the n.m.r. spectrometer, and to detect changes in the concentrations caused by treatments such as anoxia, or addition of metabolites or hormones. The results obtained have confirmed the general conclusions reached using the freeze-clamping method, but it is evident that the non-destructive n.m.r. technique will allow a more detailed insight into the balance of metabolites in cells than was previously possible.

One difficulty in this work is that only the total concentrations of metabolites are measured, and no account is taken of the possible effects of compartmentation within cells (see Chapter 8, Section 8.2). It appears, for instance, that a high proportion of the ADP in muscle is bound to the protein actin[87] and so the concentration of free ADP is likely to be much lower than the measured total concentration. For the purposes of calculating mass action ratios, the concentrations of free metabolites should be used. We should therefore make deductions about whether particular reactions are at equilibrium *in vivo* with a certain degree of caution.

(iii) A third method of locating control points in a metabolic pathway is based on 'cross-over' experiments.[82, 88] In this type of experiment, a system is perturbed, e.g. by addition of some inhibitor or activator or by anoxia, and the changes in the concentrations of the various metabolites in a metabolic pathway are measured. The simple 'cross-over' theorem states that 'following a perturbation of a metabolic steady state, the variations in the concentrations of the metabolites before and after a controlling enzyme have different signs'. For instance, in the case of mouse brain, it was found that anoxia causes a decrease in the concentrations of D-glucose 6-phosphate and D-fructose 6-phosphate, but an increase in the concentrations of D-fructose, 1,6-bisphosphate, dihydroxyacetone phosphate, and D-glyceraldehyde 3-phosphate.[89] There is thus a 'cross-over' between D-fructose 6-phosphate and D-fructose 1,6-bisphosphate and the perturbation (anoxia) has clearly activated 6-phosphofructokinase so that the concentrations of its substrates are lowered and of its products raised. This clearly indicates that 6-phosphofructokinase is a controlling enzyme whose activity can respond to the perturbation. For a detailed appraisal of the 'cross-over' theorem, reference 88 should be consulted.

(iv) As an additional indication of the points of control in a metabolic pathway, we can make use of mathematical-modelling techniques. In the glycolytic pathway (Fig. 6.20), the various reactions are divided into equilibrium and non-equilibrium categories (see (ii) above).[71] For each equilibrium enzyme (e.g. fructose-bisphosphate aldolase and triosephosphate isomerase), an equilibrium constant is defined, and for each non-equilibrium enzyme (e.g. hexokinase and 6-phosphofructokinase) the rate can be expressed in terms of the concentration of substrates and the appropriate maximum velocity (taking into account any product inhibition, etc.). It is then possible to write down expressions for the concentrations of metabolites and the flux through the pathway and to solve the resulting equations for various assumed conditions, such as a steady state.[71] This type of approach predicts the steady-state concentrations of glycolytic intermediates in erythrocytes reasonably well, and also shows the importance of the enzymes hexokinase and 6-phosphofructokinase in the control of the pathway.[90] A fuller description of mathematical-modelling techniques is beyond the scope of this text and the reader should consult references 70 and 91–93.

6.3.3.2 *Properties of the controlling enzymes*

From the various approaches described in Section 6.3.3.1, it is possible to locate the probable sites of control of a metabolic pathway. The next step is to study the catalytic properties of the controlling enzymes, especially those properties that might be involved in regulatory mechanisms. Of particular interest are the following questions.

1. What type of kinetic behaviour is observed (hyperbolic, sigmoidal, etc.)? How does the activity of the enzyme vary with changes in the concentration of substrate, particularly in the range of substrate (and enzyme) concentrations that occur *in vivo*?
2. Is the activity of the enzyme affected by any regulator molecules at concentrations that are known to occur in the cell? What is the nature of the regulation (allosteric, competitive inhibition, etc.)?
3. Is the activity of the enzyme subject to regulation by changes in its covalent structure, e.g. by phosphorylation or dephosphorylation? How are the enzymes catalysing these changes regulated?

6.3.3.3 *Formulation of an initial theory for the control of a metabolic pathway*

From the studies of the controlling enzymes (Section 6.3.3.2), it should be possible to decide which means of regulation is most likely to be involved in the control of a metabolic pathway. On the basis of the initial theory for the controls it is possible to make predictions that can be then tested.

6.3.3.4 *Test of the correctness of a theory for the control of a metabolic pathway*

In order to test the correctness of a theory, it is necessary to gather data about the activities of enzymes, the concentrations of substrates and effectors, and the state of covalent modification, etc. in a variety of metabolic circumstances. It is important to determine these parameters under conditions that reflect the *in vivo* situation as closely as possible; non-destructive techniques such as ^{31}P nuclear magnetic resonance[83–86] (Section 6.3.3.1) have great potential in such studies. If it emerges that, for instance, the variations in the concentrations of substrates and/or effectors are too small to account for the observed changes in enzyme activity or in the flux through the pathway, then the initial theory must be modified or discarded altogether and a new theory formulated and tested.

6.4 Examples of control of metabolic pathways

We shall now discuss how present-day ideas about the control of metabolic pathways have been developed by considering two particular pathways, namely glycolysis (Section 6.4.1) and glycogen metabolism (Section 6.4.2). These pathways illustrate most of the points made in Section 6.3.3. The experiments described refer mainly to the regulation of the pathways in muscle that have been particularly well studied because muscle is a highly specialized tissue in which the fluxes through the pathways vary dramatically in different metabolic circumstances. These metabolic pathways in muscle must therefore be under a high degree of control.

6.4.1 Regulation of glycolysis

The glycolytic pathway is the sequence of reactions in which glucose is converted to pyruvate, and is shown in Fig. 6.20. This pathway appears to be of almost universal importance in energy production.

Using the various approaches described in Section 6.3.3, it has been shown that the major controlling points for the pathway are the steps catalysed by 6-phosphofructokinase and by hexokinase,* since these enzymes are present at low activities and catalyse reactions which, *in vivo*, are far removed from equilibrium.[62, 79] It should also be mentioned that the reaction catalysed by pyruvate kinase is not at equilibrium *in vivo*, but since the activity of this enzyme is relatively high in muscle, its role in the control of the pathway is somewhat open to question.[62, 79]

The importance of 6-phosphofructokinase and hexokinase as controlling enzymes in the glycolytic pathway in a variety of tissues has also been confirmed by 'cross-over' experiments[89] and by mathematical-modelling studies.[74, 90]

6.4.1.1 *Regulation of the activity of 6-phosphofructokinase*

The kinetic behaviour *in vitro* of 6-phosphofructokinase is highly complex.[95–97] As shown schematically in Fig. 6.21, there are mutual interactions between the two substrates D-fructose 6-phosphate (F6P) and ATP. At low concentrations (in the physiological range) of F6P, it is found that high concentrations of ATP (in the physiological range) cause inhibition (Fig. 6.21(a)). This inhibition is due to binding of ATP to a site distinct from the active site (see Chapter 4, Section 4.3.1.4) as is shown by the fact that the enzyme can be desensitized (Section 6.2.2.1) to ATP inhibition by photo-oxidation in the presence of methylene blue.[98] The inhibition at high concentrations of ATP is accentuated by citrate but can be counteracted by various metabolites including AMP (see Fig. 6.21(a)), ADP, orthophosphate, and D-fructose, 1,6-bisphosphate. These latter compounds tend to accumulate when ATP is degraded, for example during anoxia. At high concentrations (in the physiological range) of ATP, the plot of velocity against concentration of F6P is markedly sigmoidal with a very low velocity at low concentrations of F6P (Fig. 6.21(b)).

An initial theory, based on the data shown in Fig. 6.21(a), is that the concentration of ATP controls the activity of 6-phosphofructokinase and hence the flux through the glycolytic pathway. Thus, when the concentration of ATP is high, as in resting muscle, 6-phosphofructokinase is inhibited and hence the rate of production of ATP falls. (Note that for each molecule of glucose converted to two molecules of pyruvate there is a net production of two molecules of ATP.) If this theory is correct, it would be expected that significant changes in the concentration of ATP would occur *in vivo* to account for the changes in the

* There is an additional point of control, namely the entry of glucose into the cell. This transport process is far removed from equilibrium and is regulated by insulin; an increase in the circulatory levels of this hormone will increase the rate of entry of glucose into the cell.[79, 94]

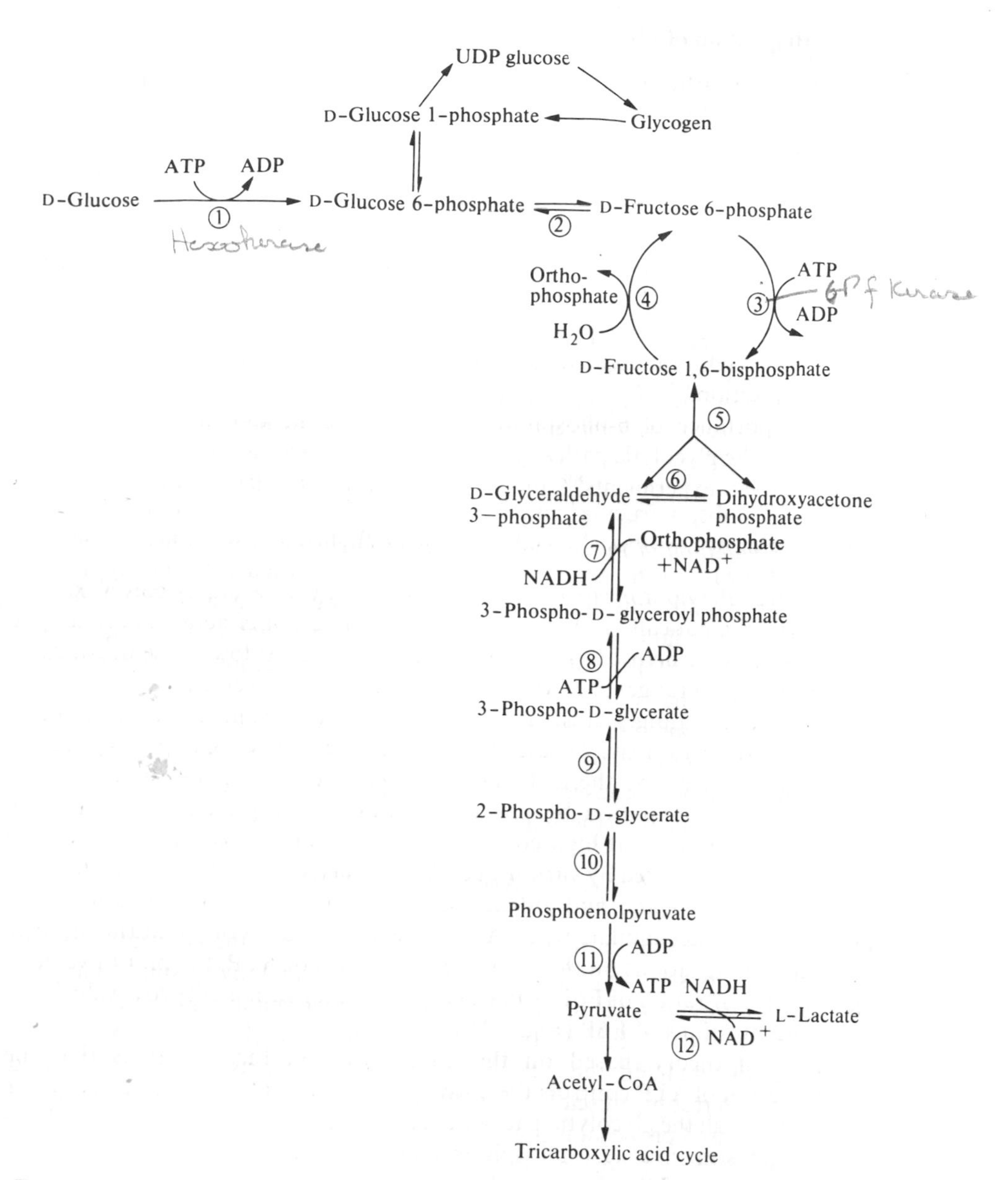

FIG. 6.20. The glycolytic pathway in muscle. The various steps are catalysed by the following enzymes: (1), hexokinase; (2), glucose-6-phosphate isomerase; (3), 6-phosphofructokinase; (4), fructose-bisphosphatase; (5), fructose-bisphosphate aldolase; (6), triosephosphate isomerase; (7), glyceraldehyde-3-phosphate dehydrogenase; (8), phosphoglycerate kinase; (9), phosphoglycerate mutase; (10), enolase; (11), pyruvate kinase; (12), lactate dehydrogenase. The magnesium ions associated with the kinase-catalysed reactions have been omitted.

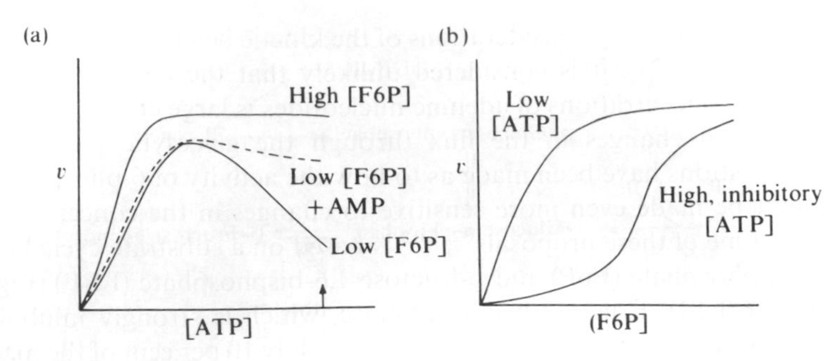

FIG. 6.21. Schematic representation of the kinetic behaviour of 6-phosphofructokinase.[95, 97] (a) Variation of velocity with concentration of ATP (the arrow on the abscissa indicates the prevailing physiological concentration of ATP). (b) Variation of velocity with concentration of F6P (D-fructose 6-phosphate).

activity of 6-phosphofructokinase (Fig. 6.21(a)). However, measurements on the flight muscles of blowflies showed that during flight there was only a small (10 per cent) decrease in the concentration of ATP compared with resting muscle, even though the flux through the glycolytic pathway had increased 100-fold.[99] Reference to the data shown in Fig. 6.21(a)), shows that this change in the concentration of ATP is nowhere near large enough to account for the change in the flux through the pathway.

It has been proposed that the effects of ATP on the activity of 6-phosphofructokinase can be amplified by other metabolites, especially ADP and AMP, which relieve the inhibition caused by high concentrations of ATP. The adenine nucleotides are linked in the reaction catalysed by adenylate kinase,

$$\mathrm{ATP} + \mathrm{AMP} \overset{Mg^{2+}}{\rightleftharpoons} 2\mathrm{ADP},$$

which is at equilibrium in muscle (Section 6.3.3.1). Since the concentrations of nucleotides in muscle are [ATP] > [ADP], [AMP], a relatively small change in the concentration of ATP brings about much larger changes in the concentrations of ADP and AMP.[62, 97] In the blowfly flight muscle, for instance, it was found that, during flight, the concentration of AMP increased by about 2.5-fold compared with resting muscle[99] (cf. a 10 per cent decrease in the concentration of ATP). The revised proposal is therefore that the activity of 6-phosphofructokinase is controlled by a combination of changes in the concentrations of ATP, AMP, ADP,* and other metabolites such as orthophosphate.

* The dependence of the activity of 6-phosphofructokinase on the concentrations of the adenine nucleotides has been explored by Atkinson[100, 101] who has introduced the term 'energy charge'. This is defined as 'one half of the phosphate anhydride bonds per adenosine' and is equal to $([\mathrm{ATP}] + \frac{1}{2}[\mathrm{ADP}])/([\mathrm{ATP}] + [\mathrm{ADP}] + [\mathrm{AMP}])$. The 'energy charge' varies from 0, when all the adenosine in a cell is present as AMP, to 1, when all the adenosine is present as ATP. Enzymes that are energy utilizing, such as ATP citrate (*pro-3S*)-lyase, have a different dependence on energy change from those that are energy-regenerating, such as 6-phosphofructokinase.

However, from detailed considerations of the kinetic behaviour of 6-phosphofructokinase (Fig. 6.21), it is considered unlikely that the combination of the changes in the concentrations of adenine nucleotides is large enough to account for the observed changes in the flux through the glycolytic pathway.[62, 97] Various suggestions have been made as to how the activity of 6-phosphofructokinase might be made even more sensitive to changes in the concentration of metabolites. One of these proposals[62, 102] is based on a substrate cycle between D-fructose 6-phosphate (F6P) and D-fructose 1,6-bisphosphate (FBP) (Fig. 6.18 and Section 6.3.2.1). Fructose-bisphosphatase, which is strongly inhibited by AMP, is found in significant amounts (approximately 10 per cent of the maximal activity of 6-phosphofructokinase) in various types of muscle where no significant gluconeogenesis occurs. The proposal has therefore been made that in resting muscle, in which the concentration of AMP is low, the low rate of conversion of F6P to FBP catalysed by 6-phosphofructokinase is opposed by conversion of FBP to F6P catalysed by fructose-bisphosphatase, giving a very low net flux through this step in the glycolytic pathway. However, when the muscle is made to do work, the concentration of ATP falls and that of AMP rises; 6-phosphofructokinase is now activated (i.e. no longer inhibited) and fructose-bisphosphatase is inhibited, giving a large net flux through this step. Calculations[102] suggest that the observed changes in net flux can be accounted for by changes in the concentration of AMP that are of the same order of magnitude as those detected by freeze-clamping measurements.[99] (Thus, a 250-fold change in net flux could be accounted for by a 2.5-fold change in the concentration of AMP.[102]) The hypothesis of substrate cycling accounts not only for the sensitivity of net flux to changes in the concentration of AMP, but also for the presence and distribution of fructose-bisphosphatase in various types of muscle. Fructose-bisphosphatase is not present in muscles such as heart muscle, where the energy demands are fairly constant and hence the need for regulation of glycolysis is much less than in the case of skeletal muscle. Lardy and his coworkers have also obtained direct evidence for substrate cycling in various tissues such as bumble bee flight muscle[103] and rat liver[104] using isotopically labelled phosphorylated monosaccharides (see Fig. 6.22).

However, the assumptions underlying the interpretation of the experimental data in these experiments have been questioned, particularly in the case of rat liver.[64] Also, mathematical modelling of the situation in rat liver suggests that under physiologically realistic conditions, when the compartmentation and the states of protonation of adenine nucleotides are taken into account, the rate of substrate cycling is negligible.[106]

Since about 1980, a good deal of attention has been focused on fructose 2,6-bisphosphate (F26BP) as a regulator of the activity of 6-phosphofructokinase.[107–109] This regulation appears to be of particular importance in the liver; indeed, F26BP is the most potent known activator of liver 6-phosphofructokinase, with effects being shown in the 0.1 $\mu mol\,dm^{-3}$ concentration range. The activation of 6-phosphofructokinase is synergistic with AMP. F26BP is also a powerful inhibitor of liver fructose-bisphophatase, again acting synergistically

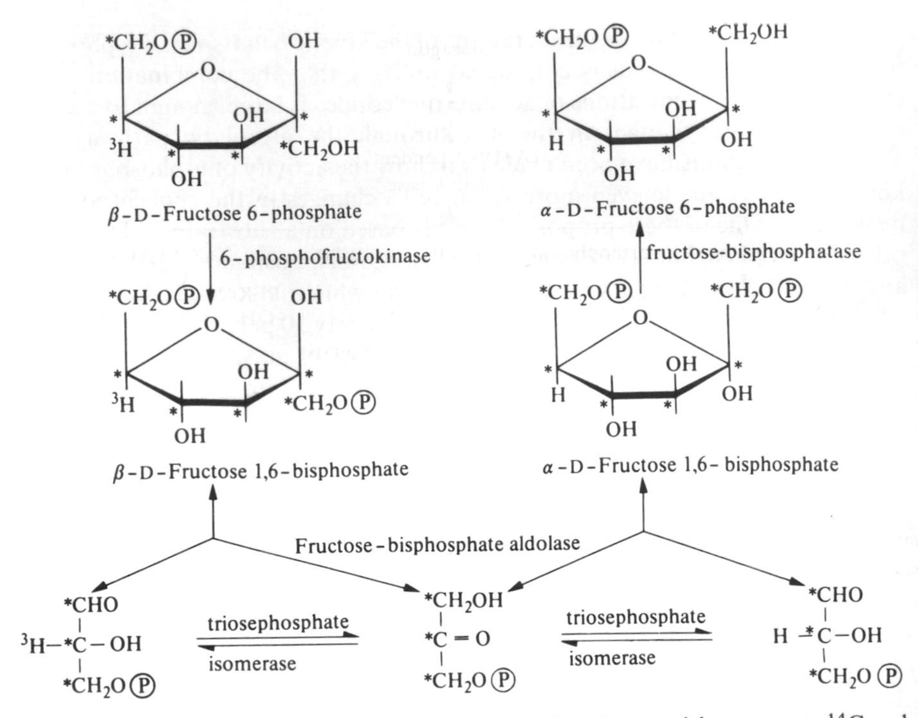

FIG. 6.22. System used to demonstrate substrate cycling. An asterisk represents ^{14}C and (P) a phosphoryl group. On addition of [5-^3H, U-^{14}C]F6P to the system, the action of 6-phosphofructokinase and fructose-bisphosphate aldolase produces isotopically labelled D-glyceraldehyde 3-phosphate and dihydroxyacetone phosphate. The action of triose-phosphate isomerase allows ^3H at C-2 of D-glyceraldehyde 3-phosphate to be exchanged for H from solvent water (see Chapter 5, Section 5.5.2.3). The action of fructose-bisphosphate aldolase and fructose-bisphosphatase then generates F6P in which ^3H is lost. If the remaining F6P is isolated at various times after the addition of [5-^3H, U-^{14}C]F6P and analysed for the content of ^3H and ^{14}C, the ratio of $^3H/^{14}C$ will fall if substrate cycling occurs. From the rate of decrease of this ratio, the rate of cycling can be deduced.[105]

with AMP. The synthesis and degradation of F26BP are catalysed by a kinase (6-phosphofructo-2-kinase) and phosphatase (fructose-2,6-bisphosphatase) whose activities are subject to control by reversible phosphorylation (Fig. 6.23). The latter kinase and phosphatase activities are located on an enzyme that consists of two identical polypeptide chains each of M_r 55 000.[107]

All the available evidence shows that, in liver, F26BP plays a major part in the hormonal and nutritional control of 6-phosphofructokinase and fructose-bisphosphatase activity and in determining the balance between glycolysis and gluconeogenesis.[109] Thus, the glycolytic flux in liver can be correlated with the level of F26BP but not with the levels of ATP, AMP, or citrate. The presence of F26BP is the signal that the level of glucose is high and that gluconeogenesis can

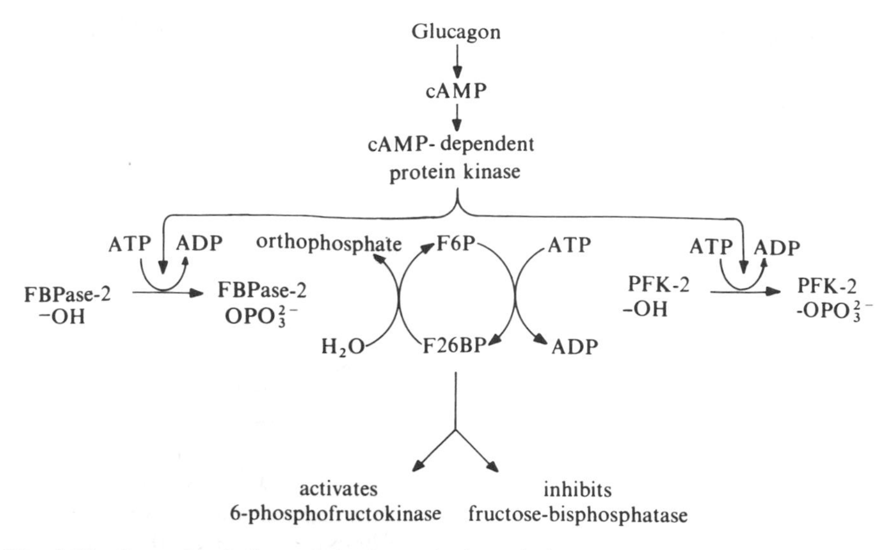

FIG. 6.23. Control of the synthesis and degradation of fructose 2,6-bisphosphate (F26BP).[106] In this scheme, FBPase-2 and PFK-2 represent fructose-2,6-bisphosphatase and 6-phosphofructo-2-kinase, respectively.

be switched off. The presence of glucagon, however, causes the level of F26BP to fall (see Fig. 6.23) and hence gluconeogenesis is favoured.

The importance of F26BP in the regulation of glycolysis in other tissues is less firmly established.[109–111] The concentration of F26BP in muscle is lower than that in the liver and this, coupled with the smaller effect of F26BP on muscle 6-phosphofructokinase compared with the liver enzyme, means that F26BP is unlikely to have a major effect on the glycolytic flux in muscle.

Among other factors that may be important in the regulation of the activity of 6-phosphofructokinase in muscle are the following.

1. 6-phosphofructokinase and fructose-bisphosphatase show different anomeric specificities (see Figs. 6.1 and 6.22), acting on the β-anomer of F6P and the α-anomer of FBP, and being activated by the α-anomer of FBP and the β-anomer of F6P, respectively. These specificities, taken together with the slow, spontaneous $\alpha \rightarrow \beta$ anomerizations of F6P and FBP (rate constants approximately 1 s^{-1}, reference 6) create non-equilibrium intracellular distributions of the α- and β-anomers of F6P and FBP. These non-equilibrium distributions could then amplify the effects of AMP and other metabolites on the enzymes.[112]
2. The calcium-binding protein calmodulin has been shown to interact with 6-phosphofructokinase from muscle.[113, 114] Calmodulin acts as a Ca^{2+}-dependent inhibitor of the enzyme and appears to compete with ATP for binding to the regulatory site of the enzyme.

3. A number of reports have shown that 6-phosphofructokinase from muscle can be reversibly phosphorylated.[115] However, the phosphorylation appears to cause only minor changes in the regulatory properties of the enzyme.

It should be noted that the factors (1)–(3) have been studied *in vitro* and their possible importance *in vivo* has not been clarified.

In summary, it can be stated that, although the importance of the step D-fructose 6-phosphate→D-fructose 1,6-bisphosphate as a control point in glycolysis is beyond dispute, the exact mechanism by which this step is regulated is still open to question and experimentation.

6.4.1.2 Regulation of the activity of hexokinase

Up to now we have concentrated attention on the regulation exerted at the D-fructose 6-phosphate→D-fructose 1,6-bisphosphate step in glycolysis. However, it is obvious that when 6-phosphofructokinase is inhibited, there would be an accumulation of the phosphorylated hexoses D-glucose 6-phosphate and D-fructose 6-phosphate unless some means existed to control the activity of hexokinase that catalyses the conversion of D-glucose to D-glucose 6-phosphate at the expense of ATP (see Fig. 6.20). It is found that hexokinases from a variety of mammalian tissues are inhibited by D-glucose 6-phosphate;* the inhibition is non-competitive (see Chapter 4, Section 4.3.2.2) with respect to D-glucose.[56] The inhibition of hexokinase would thus serve to regulate the rate of production of D-glucose 6-phosphate and hence D-fructose 6-phosphate.

There is, however, an additional factor to be borne in mind. As indicated in Fig. 6.20, hexokinase occurs near a branch point in the metabolism of monosaccharides, since D-glucose 6-phosphate is not only used to furnish D-fructose 6-phosphate for the glycolytic pathway but can also be used to produce D-glucose 1-phosphate via the action of phosphoglucomutase for the synthesis of glycogen. Under conditions in which the concentration of D-glucose is high, e.g. after an intake of carbohydrate in the food, it is desirable to build up reserves of glycogen. This would not be possible if hexokinase were inhibited by high prevailing concentrations of D-glucose 6-phosphate. The hormone insulin, secreted in response to high circulating concentrations of D-glucose, not only facilitates the transport of D-glucose into the cell, but also stimulates the synthesis of glycogen (Section 6.4.2.5). Thus when D-glucose concentrations are high, the excess D-glucose can be channelled via D-glucose 6-phosphate, D-glucose 1-phosphate, and UDP glucose to glycogen (Fig. 6.20). The dual control system afforded by regulation of the activities of hexokinase† and of 6-

* Hexokinase isoenzyme IV (also known as glucokinase) is not inhibited by D-glucose 6-phosphate. This isoenzyme, located in the liver, has a high K_m for D-glucose (see Chapter 4, Section 4.3.1.3). Thus phosphorylation of D-glucose can proceed in the liver at high prevailing concentrations of D-glucose 6-phosphate.

† As mentioned in Section 6.2.2.3, the inhibition of hexokinase isoenzyme II by D-glucose 6-phosphate from ascites tumour cells is rather slow.[56] This property allows the glycolytic intermediates D-glucose 6-phosphate and D-fructose 6-phosphate to rise to a new high level after addition of D-glucose to these cells before the hexokinase is inhibited.

phosphofructokinase that occur before and after the branch point, respectively, thus allows the cell to respond to a variety of metabolic situations and provides more flexible regulation of the glycolytic pathway than could be achieved with a single control point.

6.4.2 Regulation of glycogen metabolism

The second example of control of a metabolic pathway that we shall discuss is the regulation of glycogen metabolism in muscle. Glycogen is a polysaccharide made up of D-glucopyranose units linked by α-1,4 glycosidic bonds with branches formed by 1,6 bonds about every 12–18 groups (Fig. 6.24), i.e. there is no unique molecular structure for glycogen. The M_r of glycogen is in the range from about 10^6 to 10^8 (consisting of between 6000 and 600 000 D-glucose units). In a variety of organisms, e.g. *E. coli*[116] and rat liver,[117] a protein molecule is involved as a 'primer' backbone in the biosynthesis of glycogen.

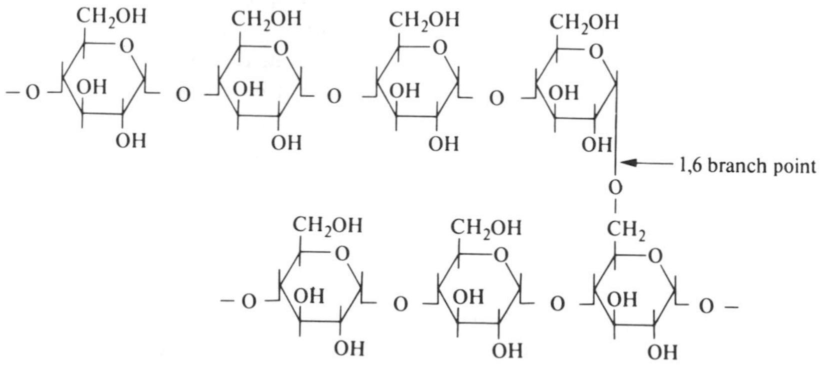

FIG. 6.24. The structure of glycogen showing the α-1,4 glycosidic bonds linking D-glucopyranose units and a 1,6 branch point.

The reactions involved in glycogen metabolism are relatively few in number (see Fig. 6.25) but the study of these reactions has revealed some very complex control mechanisms.

In both muscle and liver, glycogen has the function of being a reserve of D-glucose units. Liver glycogen can be used to help replenish D-glucose in the blood between meals and during exercise. In muscle, however, the enzyme glucose 6-phosphatase is lacking and hence the D-glucose 1-phosphate produced by the breakdown of glycogen is metabolized via the glycolytic pathway to provide ATP that acts as the fuel for muscular contraction. The amount of ATP that can be generated from the muscle glycogen is sufficient to sustain contraction for only a short time (of the order of one minute). Prolonged exercise requires the utilization of other sources of energy, e.g. triglycerides.[120]

Using the criteria for identification of controlling enzymes (Section 6.3.3.1), it has been established that the control points for glycogen breakdown and

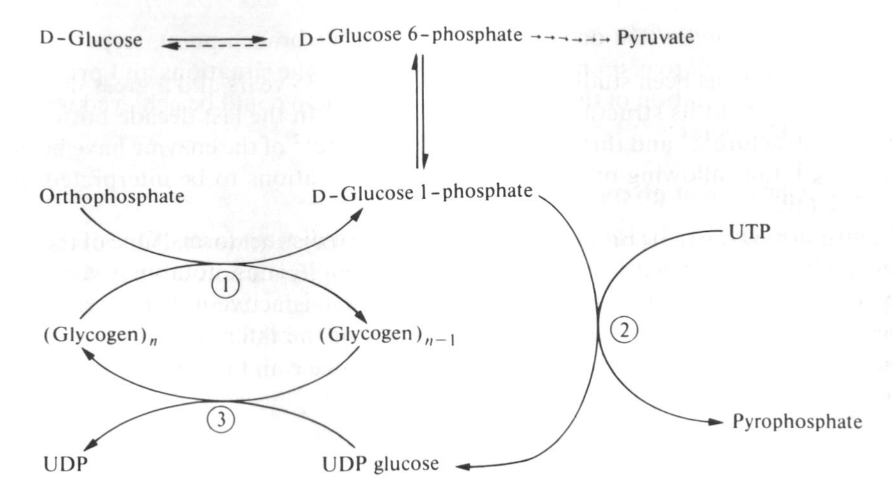

FIG. 6.25. Reactions involved in glycogen metabolism. The dashed line represents a reaction that occurs in liver but not in muscle. Reaction (1) is catalysed by phosphorylase plus debranching enzyme (which possesses two activities: amylo-1,6-glucosidase and 4-α-D-glucanotransferase[118]). Reaction (2) is catalysed by glucose 1-phosphate uridylyltransferase. Reaction (3) is catalysed by glycogen synthase plus branching enzyme (1,4-α-glucan branching enzyme[119]).

synthesis are the reactions catalysed by phosphorylase and glycogen synthase respectively.[62, 97, 121]

The main features of phosphorylase that implicate it as a controlling enzyme are that it is present at low activity (similar to the activity of 6-phosphofructokinase) and that it catalyses a reaction that, *in vivo*, is not at equilibrium.* A 'cross-over' type of experiment showed that when the breakdown of glycogen is stimulated by adrenalin, the catalytic activity of phosphorylase increases even though the concentration of its substrate decreases, i.e. phosphorylase is the controlling enzyme.[121] In the case of glycogen synthase, the main evidence that it is a controlling enzyme comes from the finding that it is present at low activity. 'Cross-over' experiments have helped to confirm this conclusion.[121]

* The equilibrium constant for the reaction catalysed by phosphorylase is given by

$$K = \frac{[(\text{Glycogen})_{n-1}]}{[(\text{Glycogen})_n]} \times \frac{[\text{G1P}]}{[\text{Orthophosphate}]},$$

but since it is impossible to determine the first term because of the variety of sizes of glycogen molecules, we write $K = [\text{G1P}]/[\text{orthophosphate}]$. At equilibrium this ratio is 0.3, but in muscle the ratio is approximately 100-fold lower. This is because in muscle the concentration of orthophosphate is high, and the concentration of G1P is low (the equilibrium between G1P and G6P lies 95 per cent to the side of G6P). Hence, in the muscle phosphorylase acts to break down glycogen. Glycogen synthesis proceeds via a different route, as shown in Fig. 6.25. Patients suffering from McArdle's disease lack phosphorylase but their ability to synthesise glycogen is not impaired.[122]

6.4.2.1 *Regulation of the activity of phosphorylase*

Phosphorylase has been studied intensively for over 35 years and a great deal is now known about its structure and function.[123, 124] In the last decade both the primary structure[125] and three-dimensional structure[12] of the enzyme have been elucidated, thus allowing many of the earlier observations to be interpreted in more detail.

Phosphorylase can be isolated from muscle in two distinct forms. One of these, phosphorylase *b*, is inactive in the absence of certain ligands, notably AMP (see Section 6.2.2.1), whereas the other, phosphorylase *a*, is active in the absence of these added ligands. Further work established that the interconversions of the two forms of phosphorylase were catalysed by a kinase and a phosphatase that could be isolated from muscle extracts.

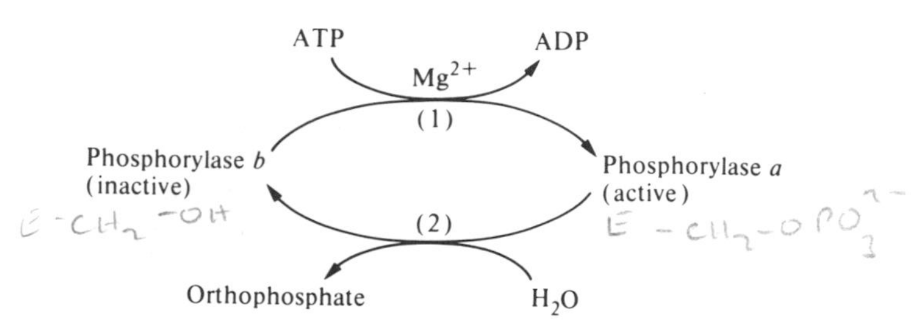

(1) is catalysed by phosphorylase kinase
(2) is catalysed by phosphorylase phosphatase

As mentioned earlier (Section 6.2.1.2) the conversion of phosphorylase *b* to phosphorylase *a* involves phosphorylation of a single side-chain (Ser 14) in each subunit. The conversion is accompanied by a change in the quaternary structure of the enzyme (the *b* form is a dimer, the *a* form a tetramer) but this change is not an integral part of the activation process.[126]

Thus there appear to be two major means of regulating the activity of phosphorylase.

1. By ligand-induced conformational changes in the enzyme. AMP (or IMP) will activate phosphorylase *b*; this effect is counteracted by certain other ligands, notably ATP, ADP, UDPglucose and D-glucose 6-phosphate.[124] Hence, under conditions in which the concentration of AMP is high and the concentrations of ATP and D-glucose 6-phosphate, in particular, are low, phosphorylase *b* will be active. This provides a 'crude' control mechanism for the enzyme.
2. By a change in the covalent structure of the enzyme, i.e. conversion of phosphorylase *b* to phosphorylase *a*, catalysed by phosphorylase kinase. Phosphorylase kinase is itself subject to regulation and can be activated in two main ways: by phosphorylation catalysed by a cAMP*-dependent protein kinase and by Ca^{2+}-mediated stimulation.

* cAMP is used as an abbreviation for 3′:5′-cyclic AMP (see Fig. 6.16).

The sequence of events in which phosphorylase kinase is activated by cAMP-dependent protein kinase is shown in Fig. 6.26.

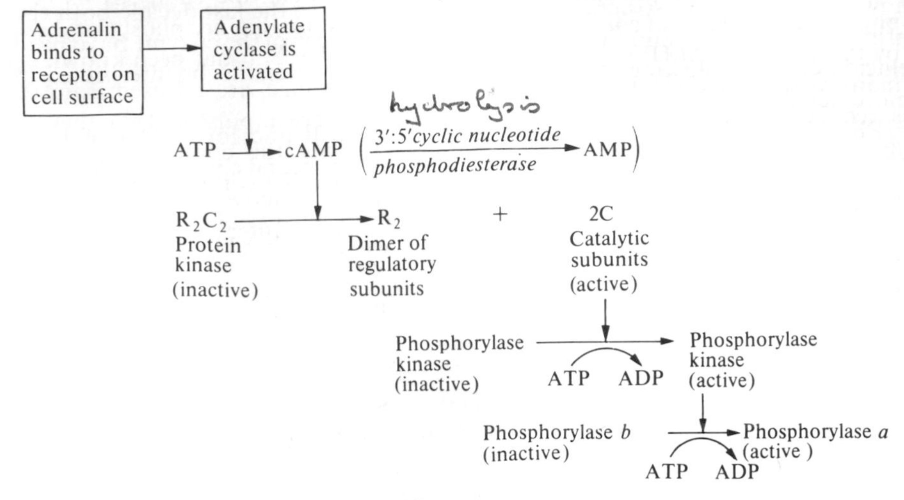

FIG. 6.26. The cascade mechanism for hormone-triggered activation of phosphorylase.

Binding of the hormone to the cell surface activates adenylate cyclase, which is a membrane-bound enzyme[127] (see Chapter 8, Section 8.4.5). The cAMP produced binds to the protein kinase,[128] causing dissociation of the inactive R_2C_2 complex to yield active catalytic subunits and a dimer of regulatory subunits. The active catalytic subunits catalyse the activation of phosphorylase kinase and hence of phosphorylase. This cascade mechanism allows for considerable amplification of the initial signal (Section 6.3.2.2). The administration of adrenalin to various tissues causes only small increases (less than fivefold) in the concentration of cAMP,[121] and indeed in the case of rabbit gracilis muscle it has been shown that the administration of 4 pmol of isoproterenol, an analogue of adrenalin, did not change the concentration of cAMP within experimental error, although the amount of phosphorylase *a* increased threefold.[129]

HO, HO–C_6H_3–C(OH)(H)–CH_2–N(H)–CH_3

Adrenalin

HO, HO–C_6H_3–C(OH)(H)–CH_2–N(H)–CH(CH_3)$_2$

Isoproterenol

Ca^{2+} ions in the concentration range of 0.1 to 1 μmol dm^{-3} can activate phosphorylase kinase in the presence of Mg^{2+} and ATP. This activation occurs by a process of autophosphorylation in which the enzyme catalyses its own phosphorylation,[130] at sites that are distinct from those that are phosphorylated by the action of cAMP-dependent protein kinase.[131,132] It has been known for many years that Ca^{2+} ions, in this concentration range, are released from the sarcoplasmic reticulum in muscle in response to a nerve impulse and can then act as a trigger for muscular contraction by allowing the components of the contractile apparatus (actin and myosin) to interact. Thus there is a clear link between the stimulation of muscular contraction, which requires ATP as a fuel, and activation of glycogen breakdown, which can furnish ATP via the glycolytic pathway (Fig. 6.20).

The mechanisms by which Ca^{2+} ions may affect the activity of phosphorylase kinase have been elucidated.[132] Phosphorylase kinase was originally thought to consist of three types of subunits with the quaternary structure $(\alpha\beta\gamma)_4$[133]. (The α and β subunits of M_r 145 000 and 128 000, respectively, are subject to phosphorylation and the γ subunit of M_r 45 000 contains the active site of the enzyme.) However, subsequent work[134] showed that the enzyme contains a fourth type of subunit (δ) of M_r 16 000 that is identical with the Ca^{2+}-binding protein, calmodulin, previously detected in a variety of other tissues such as brain and erythrocytes.[135] Activation of phosphorylase kinase by Ca^{2+} can proceed via Ca^{2+} binding to the δ subunit, which triggers a conformational change in the γ subunit converting it to an active form. A further degree of stimulation can be achieved by the binding of troponin-C (the Ca^{2+}-binding protein involved in the regulation of muscular contraction). Troponin-C interacts only with the β-subunit of phosphorylase kinase and only in the presence of Ca^{2+} ions.[132] Since the concentration of troponin-C in muscle is much higher than that of calmodulin, it may well be that activation of phosphorylase kinase via the binding of troponin-C is important in the synchronization of glycogen breakdown with muscular contraction.[132,136]

In summary, we have seen how the conversion of phosphorylase *b* to phosphorylase *a* can be brought about by the action of hormones such as adrenalin (acting via cAMP and cAMP-dependent protein kinase) or by a nerve impulse (acting via Ca^{2+} ions), thus providing a versatile amplified response to a variety of initial signals. It is now becoming increasingly recognized that the hormone-stimulated and nerve-impulse-stimulated pathways are interwoven.[132,137,138] For instance, Ca^{2+} ions acting via calmodulin are known to affect the activities of adenylate cyclase and 3′:5′-cyclic-nucleotide phosphodiesterase, which catalyse the synthesis and hydrolysis respectively, of cAMP. The Ca^{2+}-calmodulin system is also responsible for activating a multi-protein kinase of broad specificity that phosphorylates a number of proteins including glycogen synthase and tyrosine hydroxylase (which is involved in the synthesis of catecholamines).[137] In addition, cAMP-dependent protein kinase can catalyse the phosphorylation of troponin-I which may serve to regulate the interaction of actin and myosin and affect the force of muscular contraction.[139]

A number of experimental tests must be made in order to assess whether the various mechanisms for regulation of the activity of phosphorylase are operative *in vivo*. Two particular questions that need to be answered are: what are the concentrations *in vivo* of the various ligands such as AMP that affect the activity of phosphorylase *b*, and can the various events of the cascade mechanism depicted in Fig. 6.26 be demonstrated to occur *in vivo*? We shall deal with these questions in Sections 6.4.2.2 and 6.4.2.3, respectively.

6.4.2.2 *Concentrations of ligands* in vivo

The concentrations of a number of metabolites in muscle have been determined by the freeze-clamping technique (Section 6.3.3.1). These studies have shown that the concentration of AMP in resting muscle (approximately 100 $\mu mol\ dm^{-3}$) is greater than the concentration required to cause 50 per cent of the maximal activation of phosphorylase *b* (approximately 50 $\mu mol\ dm^{-3}$). The negligible activity of phosphorylase in resting muscle therefore implies that *in vivo* the effect of AMP is counteracted by high concentrations of ligands such as ATP, ADP, and D-glucose 6-phosphate; a conclusion confirmed by measurements of the concentrations of these ligands in resting muscle.[140] Ligand-induced activation of phosphorylase *b* can therefore occur *in vivo* only when the concentrations of activating ligands rise and/or those of counteracting ligands fall. Such a situation has been shown to arise in the perfused rat heart under anaerobic conditions,[141] and perhaps more convincingly in the muscles of I-strain mice that are deficient in phosphorylase kinase and cannot therefore activate phosphorylase by the $b \rightarrow a$ conversion.[142] A detailed study of the concentrations of various ligands in the muscles of I-strain mice before and after electrical stimulation of the muscle has led to the conclusion that, in this case, the increase in the activity of phosphorylase is largely due to a rise in the concentration of IMP rather than of AMP.[143]

Finally, mention should be made of the fact that *in vivo* a substantial proportion of the phosphorylase in muscle is probably bound to glycogen in the form of the 'glycogen particle' (see Chapter 7, Section 7.12). In this particle, the properties of the enzyme and the detailed effects of the various ligands may well be different from those observed in solution (see reference 124 for a discussion of this).

6.4.2.3 *Demonstration of the steps in the cascade mechanism (Fig. 6.26)*

The various steps in the cascade mechanism for the activation of phosphorylase have all been shown to occur *in vivo* in response to the administration of adrenalin.

(i) Conversion of phosphorylase b *to phosphorylase* a

The phosphorylation of phosphorylase *b* has been demonstrated *in vivo* in rabbit skeletal muscle.[144] On administration of adrenalin, up to 0.8 equivalents of ^{32}P were incorporated into phosphorylase *b* from ATP labelled at the γ-phosphoryl group with ^{32}P. In order to prevent the hydrolysis of phosphorylase *a* during the isolation procedure, it was necessary to homogenize the tissue in the presence of fluoride ions, which act as powerful inhibitors of phosphorylase phosphatase.

*(ii) Phosphorylation of phosphorylase kinase**

Careful work has shown that both the α- and the β-subunits of rabbit muscle phosphorylase kinase are phosphorylated in response to an intravenous injection of adrenalin.[145] In these experiments it was again necessary to homogenize the tissue in the presence of fluoride ions to prevent phosphatase action. By separation of the radioactively labelled peptides it was demonstrated that the sites of phosphorylation were the same as those that had previously been shown to be phosphorylated by the action of cAMP-dependent protein kinase *in vitro*.

(iii) Activation of cAMP-dependent protein kinase

It has proved more difficult to demonstrate convincingly that the cAMP-dependent protein kinase is activated *in vivo* in response to adrenalin. In theory it is possible to seek evidence for the cAMP-induced dissociation of the enzyme (Fig. 6.26) or for the action of the kinase. Both approaches involve a number of technical problems;[146] for instance in the second approach, account must be taken of the action of other protein kinases that are known to occur in tissues and that are not cAMP-dependent. Reasonably clear-cut demonstrations of the activation of cAMP-dependent protein kinase have been afforded by the work on phosphorylase kinase ((ii) above) and by the observation that, in rat liver, histone H1 could be phosphorylated in response to hormones at the site that is labelled by cAMP-dependent protein kinase *in vitro*.[147] A further discussion of the problems involved in this work are given in reference 146.

(iv) Increase in concentration of cAMP

We have already mentioned (Section 6.4.2.1) that administration of adrenalin and other hormones causes an increase in the concentration of cAMP, although the changes are relatively small. The reason that the activation of adenylate cyclase does not lead to large increases in the concentration of cAMP is that cAMP is hydrolysed to AMP in a reaction catalysed by 3′:5′-cyclic-nucleotide phosphodiesterase (Fig. 6.26). The K_m for cAMP is higher than its prevailing concentration, so that as the concentration of cAMP is raised by activation of adenylate cyclase, the activity of the phosphodiesterase is also raised.[121]† It is therefore essential that the relatively small changes in the concentration of

* The phosphorylation of α- and β-subunits increases the activity of phosphorylase kinase some 15–20 fold, but this activity is still completely dependent on Ca^{2+}.[132] At the concentration of Ca^{2+} in resting muscle (0.1 μmol dm^{-3}), the activity is likely to be very low. The ability of adrenalin to bring about the conversion of phosphorylase *b* to phosphorylase *a* is thought to be due in large part to the inhibition of protein phosphatase-1 (see Section 6.4.2.4).[132]

† There may be as many as 6 or 7 phosphodiesterases in the cell with differing K_m values for cAMP, but the one generally believed to play a major role in controlling the level of cAMP has a high K_m.

cAMP lead to an amplified physiological response; this amplification is achieved by the system of interconvertible enzymes shown in Fig. 6.26.

Thus, in conclusion it can be seen that the principal events in the proposed cascade system (Fig. 6.26) have been demonstrated to occur not only *in vitro*, but also *in vivo*. Before ending this discussion of the regulation of glycogen metabolism, two further aspects should be mentioned: firstly, the regulation of the enzymes that reverse the actions of the enzymes in the cascade system (Fig. 6.26), and secondly, the regulation of glycogen synthase, the controlling enzyme in the synthesis of glycogen (Fig. 6.25). These aspects are discussed in Sections 6.4.2.4 and 6.4.2.5, respectively.

6.4.2.4 How is phosphorylase switched off?

It appears inevitable that an elaborate control system of the type shown in Fig. 6.26 should possess mechanisms able to switch off phosphorylase and so prevent unnecessary depletion of glycogen reserves. As described in Section 6.4.2.3, cAMP is hydrolysed to AMP in a reaction catalysed by 3′:5′-cyclic-nucleotide phosphodiesterase. The action of the kinases (cAMP-dependent protein kinase, Ca^{2+}-calmodulin multiprotein kinase, and phosphorylase kinase) could be reversed by the action of one or more phosphatases (Section 6.4.2.1), thereby allowing much finer control of enzyme activity (Section 6.3.2.2) at the expense of consumption of a small amount of energy. These phosphatases have now been characterized, and appear to be of two main types.[137,148] Type 1 phosphatases dephosphorylate the β-subunit of phosphorylase kinase preferentially, whereas the type 2 enzymes act preferentially on the α-subunit of this enzyme. The type 1 enzymes each possess the same catalytic subunit of M_r 37 000, whereas there are three distinct type 2 enzymes (termed 2A, 2B, and 2C) each with a different catalytic subunit. In addition, one or more small inhibitor proteins are known that inhibit type 1 phosphatases, but not type 2. Inhibitor-1 for instance is a remarkably stable* protein of M_r 20 000 that is present in muscle at concentrations (1.5 μmol dm^{-3}) comparable with the concentrations of type 1 phosphatase, phosphorylase kinase, and glycogen synthase. To make the picture even more intriguing, it has been found that inhibitor-1 inhibits the type 1 enzyme only after it has itself been phosphorylated at a threonine side-chain by the action of cAMP-dependent protein kinase[149] (Fig. 6.27). This phosphorylation has been shown to occur *in vivo* in response to adrenalin,[150] thus permitting coordinated control of kinase and phosphatase in response to the hormone.

Phosphatases 1, 2A, and 2C account for the vast majority of protein phosphatase activity in muscle and act on a variety of enzymes involved in major metabolic pathways including glycogen metabolism, glycolysis, fatty-acid synthesis, and amino-acid breakdown.[137] In skeletal muscle, type 1 phosphatase appears to be associated largely with glycogen particles (see Chapter 7,

* The purification of this protein involves heating partially purified muscle extracts at 363 K (90 °C) to precipitate unwanted proteins.

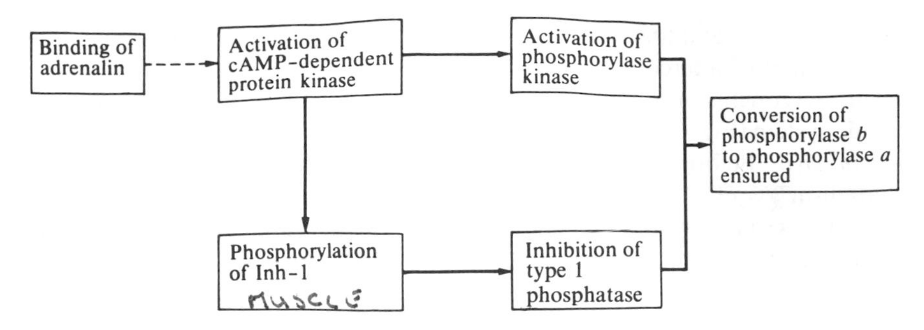

FIG. 6.27. Coordinated control of the activities of phosphorylase kinase and phosphorylase phosphatase following binding of adrenalin to the cell surface.

Section 7.12) via a glycogen-binding subunit of M_r 103 000. Type 2A and 2C phosphatases are confined to the cytosol.

Type 2B phosphatase is a Ca^{2+}-calmodulin-stimulated enzyme consisting of a catalytic subunit and a Ca^{2+}-binding subunit. It is of much narrower specificity than types 1, 2A, and 2C phosphatases. One of its most effective substrates is the phosphorylated form of inhibitor-1; thus, type 2B phosphatase could play a role in modulating the activity of type 1 phosphatase.[137]

6.4.2.5 *The regulation of glycogen synthase*

As mentioned in Section 6.4.2, glycogen synthase is the controlling enzyme in the synthesis of glycogen. Extensive work has shown that glycogen synthase is subject to a variety of controlling mechanisms that nicely complement those that exist for phosphorylase, thus permitting a coordinated control of glycogen metabolism.

Glycogen synthase can be converted from an active form (known as *a*, or I because its activity is **I**ndependent of D-glucose 6-phosphate) to an inactive form (known as *b* or D, because its activity is **D**ependent on D-glucose 6-phosphate) by the action of cAMP-dependent protein kinase.[151] Thus, as phosphorylase is switched on, glycogen synthase is switched off (Fig. 6.28).

The control of glycogen synthase is becoming better understood as more work is being performed on the purified enzyme (a tetramer of subunit M_r 86 000).[152] Two significant points emerge from recent work.

1. Glycogen synthase (like phosphorylase kinase; Section 6.4.2.3) can be phosphorylated at a number of sites (Fig. 6.29). It appears that there are at least nine distinct sites that can be phosphorylated *in vitro* (seven of which have been shown to be phosphorylated *in vivo*[152]). The cAMP-dependent protein kinase phosphorylates two serine side-chains near the C-terminus and a serine seven amino acids from the N-terminus; the latter can also be phosphorylated by phosphorylase kinase and by the Ca^{2+}-calmodulin multiprotein kinase (Section 6.4.2.1). At least three other kinases capable of phosphorylating glycogen synthase have been reported.[137,152] The various modifications produce forms of

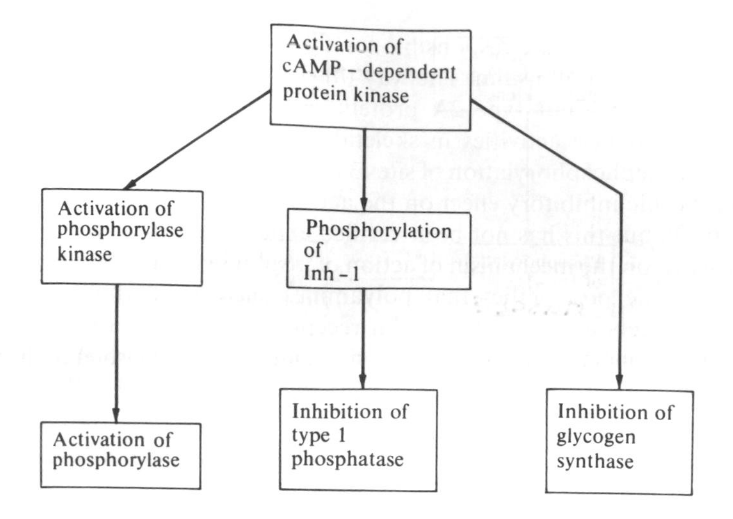

FIG. 6.28. Multiple effects of activation of cAMP-dependent protein kinase.

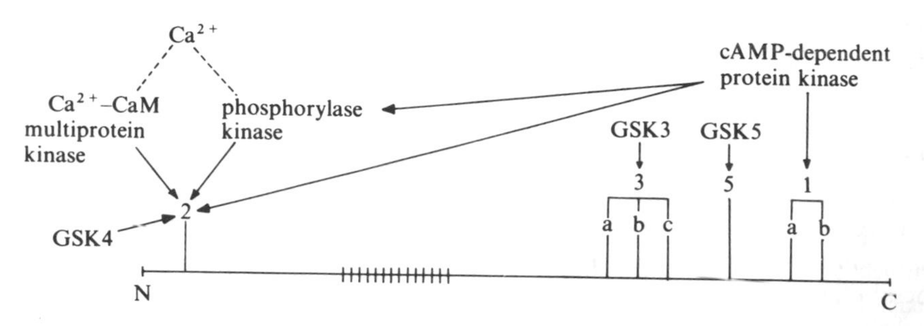

FIG. 6.29. The sites of phosphorylation of glycogen synthase.[137] GSK3, 4, and 5 are glycogen synthase kinases 3, 4, and 5. CaM is calmodulin. The polypeptide chain is not shown to scale.

glycogen synthase with distinctly different properties, thereby increasing the versatility of the response to signals such as cAMP and Ca^{2+}.

2. It has been known for many years[121] that insulin stimulates the activity of glycogen synthase, thus allowing D-glucose to be converted to glycogen under conditions in which glycolysis is inhibited (Section 6.4.1.2). The mechanism by which insulin exerts its effect is still not properly understood. The hormone has little effect on the concentration of cAMP or on the activity of cAMP-dependent protein kinase. Studies of the state of phosphorylation of the seven serine side-chains known to be phosphorylated in glycogen synthase *in vivo* have given some clues to the action of insulin.[153] In rabbits treated with propanolol (which blocks the action of adrenalin) plus insulin, there is a significant decrease in the degree of phosphorylation of sites 3a + 3b + 3c (Fig. 6.29) but no change in the degree of phosphorylation of the other sites (1a, 1b, 2, and 5). Dephosphorylation of sites

3a, 3b, and 3c is therefore responsible for the activation of glycogen synthase upon administration of insulin. The way this is brought about is not yet clear* since neither type 1 nor type 2A protein phosphatases, the major glycogen synthase phosphatase activities in skeletal muscle (Section 6.4.2.4), can bring about specific dephosphorylation of sites 3a, 3b, and 3c. It is possible that insulin has some specific inhibitory effect on the activity of glycogen synthase kinase 3 (see Fig. 6.29), but this has not been verified experimentally. There is currently intensive work on the mechanism of action of insulin in stimulating the synthesis of glycogen. The possibilities that polyamines such as spermine may act as second messengers and that the insulin receptor may act as a tyrosine kinase (catalysing the phosphorylation of tyrosine side-chains in proteins) are being explored.[137]

6.4.2.6 *Concluding remarks on the regulation of glycogen metabolism*

The preceding Sections (6.4.2.1 to 6.4.2.5) have illustrated the variety of mechanisms by which glycogen metabolism can be regulated, e.g. ligand-induced conformational changes in enzymes, hormonal action, and nerve impulses acting via Ca^{2+} ions. These mechanisms allow a variety of responses in different metabolic circumstances. It might appear that the recent work mentioned has only served to confuse the issue, since it has now been found that a bewildering variety of enzymes and other proteins are involved in regulation. However, some of the recent work has demonstrated that considerable economy is exercised on the part of an organism, since some of the enzymes and proteins have multiple effects. For instance, cAMP-dependent protein kinase and Ca^{2+}-calmodulin multiprotein kinase have effects on enzymes involved in a number of metabolic pathways† such as glycolysis, glycogen and lipid metabolism, and catecholamine synthesis.[137, 148] Protein kinases thus appear to play a central role in tramsmitting the effects of intracellular signals such as cAMP and Ca^{2+} to the target enzymes, which then bring about the metabolic changes in the cell. It also appears that the various phosphatases each have multiple activities (Section 6.4.2.4).

The regulation of glycogen metabolism is still under very active investigation and it is reasonable to expect that new features of the system will emerge; nevertheless, the account given here should give some impression of the present-day ideas on the subject.

6.5 Concluding remarks

In this chapter we have discussed the various means by which the activities of enzymes are controlled. The studies on single enzymes (Section 6.2) have

* Despite earlier reports to the contrary:[154] it now seems clear that insulin does not lead to any significant degree of dephosphorylation of inhibitor-1, which would activate type 1 protein phosphatase.

† The amino-acid sequences around the phosphorylated side-chain in the various substrates for cAMP-dependent protein kinase contain two (or more) basic amino acids on the N-terminal side and a hydrophobic amino acid on the C-terminal side.[137]

provided the conceptual basis on which theories of the control of metabolic pathways have been formulated (Section 6.3). At present, we are in the position that the control of a number of single enzymes can be understood in quantitative terms, and some progress has been made towards a quantitative understanding of the control of metabolic pathways. However, when the cellular level of organization is considered (see Chapter 8), the situation becomes even more complicated, since a number of new factors such as compartmentation are introduced. Much of the future work will be directed towards a detailed understanding of these new factors.

References

1. Heimburger, N., in *Proteases and biological control* (Reich, E., Rifkin, D. B., and Shaw, E., eds.), pp. 367–386. Cold Spring Harbor Laboratory (1975).
2. Aoyagi, T. and Umezawa, H., in reference 1, pp. 429–54.
3. Small, C. and Traut, T., *Trends Biochem. Sci.* **9**, 49 (1984).
4. Oguchi, M., Gerth, E., and Park, J. H., *J. biol. Chem.* **248**, 5571(1973).
5. Lowry, O. H. and Passonneau, J. V., *J. biol. Chem.* **239**, 31 (1964).
6. Benkovic, S. J. and Schray, K. J., *Adv. Enzymol.* **44**, 139 (1976).
7. Dijkstra, B. W., Drenth, J., and Kalk, K. H., *Nature, Lond.* **289**, 604 (1981).
8. Cabib, E., *Trends Biochem. Sci.* **1**, 275 (1976).
9. Banks, P., Bartley, W., and Birt, L. M., *The biochemistry of the tissues* (2nd edn). Wiley, London (1976). See p. 159.
10. Furie, B. and Furie, B. C., *Cell*, **53**, 505 (1988).
11. Reid, K. B. M., *Essays Biochem.* **22**, 27 (1986).
12. Fletterick, R. J. and Madsen, N. B., *A. Rev. Biochem.* **49**, 31 (1980).
13. Shacter, E., Chock, P. B., Rhee, S. G., and Stadtman, E. R., *Enzymes* (3rd edn.) **17**, 21 (1986).
14. Stadtman, E. R. and Chock, P. B., *Curr. Top. cell. Reg.* **13**, 53 (1978).
15. Almassy, R. J., Janson, C. A., Hamlin, R., Xuong, N.-H., and Eisenberg, D., *Nature, London.* **323**, 304 (1986).
16. Chock, P. B. and Stadtman, E. R., *Methods Enzymol.* **64**, 297 (1980).
17. *Curr. Top. cell. Reg.* **27** (1985).
18. Chock, P. B., Stadtman, E., Jurgensen, S. R., and Rhee, S. G., *Curr. Top. Cell. Reg.* **27**, 3 (1985).
19. Goldbeter, A. and Koshland, D. E. Jr., *J. biol. Chem.* **262**, 4460 (1987).
20. Cori, G. T., Colowick, S. P., and Cori, C. F., *J. biol. chem.* **123**, 381 (1938).
21. Umbarger, H. E. and Brown, B., *J. biol. Chem.* **233**, 415 (1958).
22. Gerhart, J. C. and Pardee, A. B., *J. biol. Chem.* **237**, 891 (1962).
23. Monod, J., Changeux, J.-P., and Jacob, F., *J. molec. Biol.* **6**, 306 (1963).
24. Rabin, B. R., *Biochem. J.* **102**, 22C (1967).
25. Ferdinand, W., *Biochem. J.* **98**, 278 (1966).
26. Cornish-Bowden, A., *Fundamentals of enzyme kinetics*. Butterworths, London (1979). See p. 172.
27. Hill, A. V., *J. Physiol., Lond.* **40**, iv (1910).
28. Adair, G. S., *J. biol. Chem.* **63**, 529 (1925).
29. Cornish-Bowden, A., *Fundamentals of enzyme kinetics*. Butterworths, London (1979). See pp. 154–9.

30. Ricard, J. and Cornish-Bowden, A., *Eur. J. Biochem.* **166**, 255 (1987).
31. Monod, J., Wyman, J., and Changeux, J.-P., *J. molec. Biol.* **12**, 88 (1965).
32. Steitz, T. A., Fletterick, R. J., Anderson, W. F., and Anderson, C. M., *J. molec. Biol.* **104**, 197 (1976).
33. Dalziel, K. *FEBS Lett.* **1**, 346 (1968).
34. Rubin, M. M. and Changeux, J.-P., *J. molec. Biol.* **21**, 265 (1966).
35. Koshland, D. E. Jr., Némethy, G., and Filmer, D., *Biochemistry* **5**, 365 (1966).
36. Cornish-Bowden, A., *Fundamentals of enzyme kinetics.* Butterworths, London (1979). See pp. 166–71.
37. Tipton, K. F., in *Companion to biochemistry* (Bull, A. T., Lagnado, J. R., Thomas, J. O., and Tipton, K. F., eds.), Vol. 2, p. 327. Longman, London (1979).
38. Eigen, M., *Q. Rev. Biophys.* **1**, 3 (1968).
39. Herzfeld, J. and Stanley, H. E., *J. molec. Biol.* **82**, 231 (1974).
40. Whitehead, E., *Prog. Biophys. molec. Biol.* **21**, 323 (1970).
41. Levitzki, A. and Koshland, D. E. Jr., *Curr. Top. cell. Reg.* **10**, 1 (1976).
42. Fersht, A. R., *Enzyme structure and mechanism* (2nd edn). Freeman, New York (1985). See p. 271.
43. Citri, N., *Adv. Enzymol.* **37**, 397 (1973).
44. Price, N. C. and Dwek, R. A., *Principles and problems in physical chemistry for biochemists* (2nd edn). Clarendon Press, Oxford (1979). See Ch. 4.
45. Griffiths, J. R., Price, N. C., and Radda, G. K., *Biochim. biophys. Acta* **358**, 275 (1974).
46. Fuller Noel, J. K. and Schumaker, V. N., *J. molec. Biol.* **68**, 523 (1972).
47. Kirschner, K., *J. molec. Biol.* **58**, 51 (1971).
48. Kirschner, K., Gallego, E., Schuster, I., and Goodall, D., *J. molec. Biol.* **58**, 29 (1971).
49. Biesecker, G., Harris, J. I., Thierry, J. C., Walker, J. E., and Wonacott, A. J., *Nature, Lond.* **266**, 328 (1977).
50. Perutz, M. F., *Nature, Lond.* **228**, 726·(1970).
51. Cornish-Bowden, A., and Koshland, D. E. Jr., *J. molec. Biol.* **95**, 201 (1975).
52. Stryer, L., *Biochemistry* (3rd edn). Freeman, New York (1988). See Ch. 7.
53. Citri, N., *Adv. Enzymol.* **37**, 397 (1973). See p. 557.
54. Frieden, C., *A. Rev. Biochem.* **48**, 471 (1979).
55. Bisswanger, H., *J. biol. Chem.* **259**, 2457 (1984).
56. Colowick, S. P., *Enzymes* (3rd edn). **9**, 1 (1973). See p. 45.
57. Neet, K. E. and Ainslie, G. R. Jr., *Methods Enzymol.* **64**, 192 (1980).
58. Martin, B. R., *Metabolic regulation: a molecular approach.* Blackwell, Oxford (1987). See p. 5.
59. Martin, B. R., *Metabolic regulation: a molecular approach.* Blackwell, Oxford (1987). See Chs. 2–12.
60. Stadtman, E. R., *Enzymes* (3rd edn). **1**, 397 (1970).
61. Pittard, J. and Gibson, F., *Curr. Top. cell. Reg.* **2**, 29 (1970).
62. Newsholme, E. A. and Leech, A. R., *Biochemistry for the medical sciences.* Wiley, Chichester (1983). See Ch. 7.
63. Newsholme, E. A. and Leech, A. R., *Biochemistry for the medical sciences.* Wiley, Chichester (1983). See Ch. 11.
64. Katz, J. and Rognstad, R., *Curr. Top. cell. Reg.* **10**, 237 (1976).
65. Kerbey, A. L., Randle, P. J., Cooper, R. H., Whitehouse, S., Pask, H. T., and Denton, R. M., *Biochem. J.* **154**, 327 (1976).
66. Newsholme, E. A. and Start, C., *Regulation in metabolism.* Wiley, London (1973). See p. 168.

67. Nimmo, H. G. and Cohen, P., *Adv. cyclic Nucleotide Res.* **8**, 145 (1977). See p. 193.
68. Cohen, P. and Antoniw, J. F., *FEBS Lett.* **34**, 43 (1973).
69. Kacser, H. and Burns, J. A., *Symp. Soc. exp. Biol.* **32**, 65 (1973).
70. Kacser, H. and Burns, J. A., *Biochem. Soc. Trans.* **7**, 1149 (1979).
71. Rapoport, T. A., Heinrich, R., Jacobasch, G., and Rapoport, S., *Eur. J. Biochem.* **42**, 107 (1974).
72. Crabtree, B. and Newsholme, E. A., *Curr. Top. cell. Reg.* **25**, 21 (1985).
73. Burns, J. A., Cornish-Bowden, A., Groen, A. K., Heinrich, R., Kacser, H., Porteous, J. W., Rapoport, S. M., Rapoport, T. A., Stucki, J. W., Tager, J. M., Wanders, R. J. A., and Westerhoff, H. V., *Trends Biochem. Sci.* **10**, 16 (1985).
74. Fell, D. A., *Trends Biochem. Sci.* **9**, 515 (1984).
75. Crabtree, B. and Newsholme, E. A., *Trends Biochem. Sci.* **12**, 4 (1987).
76. Kacser, H. and Porteous, J. W., *Trends Biochem. Sci.* **12**, 5 (1987).
77. Canela, E. I. and Franco, R., *Trends Biochem. Sci.* **12**, 218 (1987).
78. Savageau, M. A., *Trends Biochem. Sci.* **12**, 219 (1987).
79. Martin, B. R., *Metabolic regulation: a molecular approach.* Blackwell, Oxford (1987), See Ch. 1.
80. Newsholme, E. A. and Start, C., *Regulation in metabolism.* Wiley, London (1973). See pp. 142–4.
81. Williamson, J. R., *J. biol. Chem.* **240**, 2308 (1965).
82. Rolleston, F. S., *Curr. Top. cell. Reg.* **5**, 47 (1972).
83. Seeley, P. J., Sehr, P. A., Gadian, D. G., Garlick, P. B., and Radda, G. K., in *NMR in biology* (Dwek, R. A., Campbell, I. D., Richards, R. E., and Williams, R. J. P., eds.), pp. 247–75. Academic Press, London (1977).
84. Avison, M. J., Hetherington, H. P. and Shulman, R. G., *A. Rev. Biophys. Biophys. Chem.* **15**, 377 (1986).
85. Moore, G. R., Ratcliffe, R. G. and Williams, R. J. P., *Essays Biochem.* **19**, 142 (1983).
86. Radda, G. K., *Science* **233**, 640 (1986).
87. Seraydarian, K., Mommaerts, W. F. H. M., and Wallner, A., *Biochim. biophys. Acta* **65**, 443 (1962).
88. Heinrich, R. and Rapoport, T. A., *Eur. J. Biochem.* **42**, 97 (1974).
89. Lowry, O. H., Passonneau, J. V., Hasselberger, F. X., and Schultz, D. W., *J. biol. Chem.* **239**, 18 (1964).
90. Rapoport, T. A., Heinrich, R., and Rapoport, S. M., *Biochem. J.* **154**, 449 (1976).
91. Garfinkel, D., Garfinkel, L., Pring, M., Green, S. B., and Chance, B., *A. Rev. Biochem.* **39**, 473 (1970).
92. McMinn, C. L. and Ottaway, J. H., *J. theor. Biol.* **56**, 57 (1976).
93. Ottaway, J. H., *Biochem. Soc. Trans.* **7**, 1161 (1979).
94. Newsholme, E. A. and Leech, A. R., *Biochemistry for the medical sciences.* Wiley, Chichester (1983). See Ch. 15.
95. Stadtman, E. R., *Adv. Enzymol.* **28**, 41 (1966).
96. Bloxham, D. P. and Lardy, H. A., *Enzymes* (3rd edn) **8**, 239 (1973).
97. Martin, B. R., *Metabolic regulation: a molecular approach.* Blackwell, Oxford (1987). See Ch. 14.
98. Ahlfors, C. E. and Mansour, T. E., *J. biol. Chem.* **244**, 1247 (1969).
99. Sacktor, B. and Hurlbut, E. C., *J. biol. Chem.* **241**, 632 (1966).
100. Atkinson, D. E., *Biochemistry* **7**, 4030 (1968).
101. Shen, L. C., Fall, L., Walton, G. M., and Atkinson, D. E., *Biochemistry* **7**, 4041 (1968).
102. Newsholme, E. A. and Crabtree, B., *FEBS Lett.* **7**, 195 (1970).

103. Clark, M. G., Bloxham, D. P., Holland, P. C., and Lardy, H. A., *Biochem. J.* **134**, 589 (1973).
104. Clark, M. G., Bloxham, D. P., Holland, P. C., and Lardy, H. A., *J. biol. Chem.* **249**, 279 (1974).
105. Bloxham, D. P., Clark, M. G., Holland, P. C., and Lardy, H. A., *Biochem. J.* **134**, 581 (1973).
106. Garfinkel, L., Kohn, M. C., and Garfinkel, D., *Eur. J. Biochem.* **96**, 183 (1979).
107. Hers, H.-G., *Biochem. Soc. Trans.* **12**, 729 (1984).
108. Hers, H.-G. and van Schaftingen, E., *Biochem. J.* **206**, 1 (1982).
109. van Schaftingen, E., *Adv. Enzymol.* **59**, 315 (1987).
110. Vora, S., Oskam, R., and Staal, G. E. J., *Biochem. J.* **229**, 333 (1985).
111. Hue, L. and Rider, M. J., *Biochem. J.* **245**, 313 (1987).
112. Koerner, T. A. W. Jr., Voll, R. J., and Younathan, E. S., *FEBS Lett.* **84**, 207 (1977).
113. Mayr, G. W., *Eur. J. Biochem.* **143**, 513 (1984).
114. Mayr,. G. W., *Eur. J. Biochem.* **143**, 521 (1984).
115. Pilkis, S. J., Claus, T. H., Kountz, P. D., and El-Maghrabi, M. R., *Enzymes* (3rd edn) **18**, 3 (1987).
116. Barengo, R., Flawia, M., and Krisman, C. R., *FEBS Lett.* **53**, 274 (1975).
117. Krisman, C. R. and Barengo, R., *Eur. J. Biochem.* **52**, 117 (1975).
118. Taylor, C., Cox, A. J., Kernohan, J. C., and Cohen, P., *Eur. J. Biochem.* **51**, 105 (1975).
119. Caudwell, F. B. and Cohen, P., *Eur. J. Biochem.* **109**, 391 (1980).
120. Newsholme, E. A. and Leech, A. R., *Biochemistry for the medical sciences*. Wiley, Chichester (1983). See Ch. 9.
121. Newsholme, E. A. and Start, C., *Regulation in metabolism*. Wiley, London (1973). See Ch. 4.
122. Banks, P., Bartley, W., and Birt, L. M., *The biochemistry of the tissues* (2nd edn). Wiley, London (1976). See p. 208.
123. Graves, D. J. and Wang, J. H., *Enzymes* (3rd edn), **7**, 435 (1972).
124. Busby, S. J. W. and Radda, G. K., *Curr. Top. cell. Reg.* **10**, 89 (1976).
125. Titani, K., Koide, A., Hermann, J., Ericsson, L. H., Kumar, S., Wade, R. D., Walsh, K. A., Neurath, H., and Fischer, E. H., *Proc. natn. Acad. Sci. USA* **74**, 4762 (1977).
126. Birkett, D. J., Radda, G. K., and Salmon, A. G., *FEBS Lett.* **11**, 295 (1970).
127. Helmreich, A. J. M., Zenner, H. P., and Pfeuffer, T., *Curr. Top. cell. Reg.* **10**, 41 (1976).
128. Cohen, P., *Curr. Top. cell. Reg.* **14**, 117 (1978).
129. Nimmo, H. G. and Cohen, P., *Adv. cyclic Nucleotide Res.* **8**, 145 (1977). See p. 192.
130. Nimmo, H. G. and Cohen, P., *Adv. cyclic Nucleotide Res.* **8**, 145 (1977). See p. 183.
131. Nimmo, H. G. and Cohen, P., *Adv. cyclic Nucleotide Res.* **8**, 145 (1977). See p. 185.
132. Cohen, P., *Biochem. Soc. Trans.* **15**, 999 (1987).
133. Cohen, P., *Eur. J. Biochem.* **34**, 1 (1973).
134. Cohen, P., Burchell, A., Foulkes, J. G., Cohen, P. T. W., Vanaman, T. C., and Nairn, A. C., *FEBS Lett.* **92**, 287 (1978).
135. Means, A. R. and Dedman, J. R., *Nature, Lond.* **285**, 73 (1980).
136. Cohen, P., *Eur. J. Biochem.* **111**, 563 (1980).
137. Cohen, P., *Eur. J. Biochem.* **151**, 439 (1985).
138. Cohen, P., *Nature, London* **296**, 613 (1982).
139. Perry, S. V., *Biochem. Soc. Trans.* **7**, 593 (1979).
140. Morgan, H. E. and Parmeggiani, A., *J. biol. Chem.* **239**, 2440 (1964).
141. Morgan, H. E. and Parmeggiani, A., *J. biol. Chem.* **239**, 2335 (1964).
142. Lyon, J. B. Jr. and Porter, J., *J. biol. Chem.* **238**, 1 (1963).

143. Rahim, Z. H. A., Perrett, D., Lutaya, G., and Griffiths, J. R., *Biochem. J.* **186**, 331 (1980).
144. Mayer, S. E. and Krebs, E. G., *J. biol. Chem.* **245**, 3153 (1970).
145. Yeaman, S. J. and Cohen, P., *Eur. J. Biochem.* **51**, 93 (1975).
146. Nimmo, H. G. and Cohen, P., *Adv. cyclic Nucleotide Res.* **8**, 145 (1977). See p. 173.
147. Langan, T. A., *Proc. natn. Acad. Sci. USA* **64**, 1276 (1969).
148. Cohen, P., *Curr. Top. cell. Reg.* **27**, 23 (1985).
149. Nimmo, G. A. and Cohen, P., *Eur. J. Biochem.* **87**, 353 (1978).
150. Foulkes, J. G. and Cohen, P., *Eur. J. Biochem.* **97**, 251 (1979).
151. Soderling, T. R., Hickenbottom, J. P., Reimann, E. M., Hunkeler, F. L., Walsh, D. A., and Krebs, E. G., *J. biol. Chem.* **245**, 6317 (1970).
152. Cohen, P., *Enzymes* (3rd edn) **17**, 461 (1986).
153. Parker, P. J., Caudwell, F. B., and Cohen, P., *Eur. J. Biochem.* **130**, 227 (1983).
154. Foulkes, J. G., Jefferson, L. S. and Cohen, P., *FEBS Letters* **112**, 21 (1980).

Appendix 6.1

The Adair equation and the Scatchard equation for a protein containing multiple ligand-binding sites

Consider a system in which one molecule of protein, P, can bind up to n molecules of ligand, A.

	Sites occupied	Sites unoccupied
$P+A \rightleftharpoons PA$	1	$n-1$
$PA+A \rightleftharpoons PA_2$	2	$n-2$
$PA_2+A \rightleftharpoons PA_3$	3	$n-3$
$PA_{n-1}+A \rightleftharpoons PA_n$	n	0

Let the successive dissociation constants for PA, PA_2 . . . be K_1, K_2, etc.

$$\text{i.e.}\quad K_1=\frac{[P][A]}{[PA]},\qquad K_2=\frac{[PA][A]}{[PA_2]},\quad \text{etc.}$$

A.6.1. The Adair equation

The fractional saturation of ligand sites (Y) is given by

$$Y=\frac{\text{total concentration of A bound}}{\text{total concentration of sites for A}}$$

$$=\frac{[PA]+2[PA_2]+3[PA_3]\ldots *}{n([P]+[PA]+[PA_2]+[PA_3]\ldots)}$$

$$=\frac{\dfrac{[P][A]}{K_1}+\dfrac{2[P][A]^2}{K_1K_2}+\dfrac{3[P][A]^3}{K_1K_2K_3}+\cdots}{n\left[[P]+\dfrac{[P][A]}{K_1}+\dfrac{[P][A]^2}{K_1K_2}+\dfrac{[P][A]^3}{K_1K_2K_3}+\cdots\right]},$$

* Note the factor n in the denominator. This arises because each molecule of P has n sites for A. The factors 1, 2, 3, etc. in the numerator arise because each molecule of PA_2 contains two molecules of A, and so on.

$$\therefore \quad Y = \frac{\dfrac{[\mathrm{A}]}{K_1} + \dfrac{2[\mathrm{A}]^2}{K_1K_2} + \dfrac{3[\mathrm{A}]^3}{K_1K_2K_3} + \cdots}{n\left[1 + \dfrac{[\mathrm{A}]}{K_1} + \dfrac{[\mathrm{A}]^2}{K_1K_2} + \dfrac{[\mathrm{A}]^3}{K_1K_2K_3} + \cdots\right]}. \qquad \text{(A.6.1)}$$

This is the Adair equation.

For the special case of a tetrameric protein with four sites for A, $n = 4$. Then eqn (A.6.1) becomes

$$Y = \frac{\dfrac{[\mathrm{A}]}{\mathrm{K}_1} + \dfrac{2[\mathrm{A}]^2}{K_1K_2} + \dfrac{3[\mathrm{A}]^3}{K_1K_2K_3} + \dfrac{4[\mathrm{A}]^4}{K_1K_2K_3K_4}}{4\left[1 + \dfrac{[\mathrm{A}]}{K_1} + \dfrac{[\mathrm{A}]^2}{K_1K_2} + \dfrac{[\mathrm{A}]^3}{K_1K_2K_3} + \dfrac{[\mathrm{A}]^4}{K_1K_2K_3K_4}\right]}. \qquad \text{(A.6.2)}$$

This is the Adair equation for a tetramer (see eqn (6.6)).

A.6.2 The Scatchard equation

We introduce a new parameter, r, which is the average number of molecules of A bound per molecule of P.

By definition, $r = nY$ where n equals the number of ligand-binding sites and Y equals the fractional saturation of ligand-binding sites. From eqn (A.6.1),

$$r = \frac{\dfrac{[\mathrm{A}]}{K_1} + \dfrac{2[\mathrm{A}]^2}{K_1K_2} + \dfrac{3[\mathrm{A}]^3}{K_1K_2K_3} + \cdots}{1 + \dfrac{[\mathrm{A}]}{K_1} + \dfrac{[\mathrm{A}]^2}{K_1K_2} + \dfrac{[\mathrm{A}]^3}{K_1K_2K_3} + \cdots}.$$

In order to simplify this quite general expression, we have to derive a relationship between the successive K_s. Now if it is assumed that the sites are independent and equivalent (i.e. that the free energy of interaction of the ligand with each site is the same), then the K_s are related to each other by statistical factors, i.e. A can dissociate from the PA_2 complex in two ways, but A can associate with the $(n-1)$ vacant sites in the PA complex in $(n-1)$ ways. The general relationship is that the ith dissociation constant (K_i) is given by

$$K_i = \left[\frac{i}{n-i+1}\right]K,$$

where K is an intrinsic dissociation constant (i.e. one that takes into account these statistical factors).*

Thus $\qquad K_1 = \dfrac{K}{n}, \qquad K_2 = \dfrac{2K}{n-1}, \qquad K_3 = \dfrac{3K}{n-2}, \quad$ etc.

Our expression for r now becomes

$$r = \frac{[\mathrm{A}]\dfrac{(n)}{K} + \dfrac{2[\mathrm{A}]^2(n)(n-1)}{2K^2} + \dfrac{3[\mathrm{A}]^2(n)(n-1)(n-2)}{(2)(3)K^3} + \cdots}{1 + \dfrac{[\mathrm{A}](n)}{K} + \dfrac{[\mathrm{A}]^2(n)(n-1)}{2K^2} + \dfrac{[\mathrm{A}]^3(n)(n-1)(n-2)}{(2)(3)K^3} + \cdots}$$

* K is actually the geometric mean of all the dissociation constants, i.e. $K = (K_1K_2K_3 \ldots K_n)^{1/n}$.

$$=\frac{\dfrac{[\mathrm{A}](n)}{K}\left[1+\dfrac{[\mathrm{A}](n-1)}{K}+\dfrac{[\mathrm{A}]^2(n-1)(n-2)}{2K^2}+\cdots\right]}{1+\dfrac{[\mathrm{A}](n)}{K}+\dfrac{[\mathrm{A}]^2(n)(n-1)}{2K^2}+\dfrac{[\mathrm{A}]^3(n)(n-1)(n-2)}{6K^3}+\cdots}.$$

The expressions in the numerator and denominator in this equation are both binomial expansions. Thus the expression can be simplified to give

$$r=\frac{\dfrac{[\mathrm{A}](n)}{K}\left[1+\dfrac{[\mathrm{A}]}{K}\right]^{n-1}}{\left[1+\dfrac{[\mathrm{A}]}{K}\right]^{n}}$$

$$=\frac{\dfrac{[\mathrm{A}](n)}{K}}{\left[1+\dfrac{[\mathrm{A}]}{K}\right]}$$

$$\therefore\quad r=\frac{n[\mathrm{A}]}{K+[\mathrm{A}]}.$$

This equation can be rearranged to give the Scatchard equation:

$$\frac{r}{[\mathrm{A}]}=\frac{n}{K}-\frac{r}{K}. \qquad \text{(A.6.3)}$$

Thus, a plot of $r/[\mathrm{A}]$ against r gives a straight line of slope $-1/K$ and intercept on the x-axis of n, *provided that the sites are equivalent and independent* (see Fig. 6.15).

7
Enzymes in organized systems

7.1 Introduction

In the previous chapters we have discussed results obtained from the study of highly purified enzymes. By examining single purified enzymes it is possible to study the structure (Chapter 3), the kinetics (Chapter 4), and mechanism of action (Chapter 5) of a single enzyme in great detail and without interference from competing reactions, and this is essential for an understanding of the molecular basis of enzyme catalysis. However, in the intact cell, enzymes do not act as separated, isolated catalysts in dilute aqueous buffers with kinetics approximating to those described by the Michaelis and Menten equation (see Chapter 4, eqn 4.4). In many subcellular compartments the protein concentrations are high (e.g. in the mitochondrial matrix, 500 mg protein cm^{-3}),[1] and many enzymes compete with one another for substrates and effectors and they may be physically associated with other enzymes to varying extents. In the next two chapters we consider how enzymes may behave in the intact cell, firstly by examining the types of organized enzyme systems that exist in the cell, and secondly by examining the nature of the environment in which enzymes operate *in vivo* and how this differs from the conditions under which most enzyme assays are performed *in vitro*.

7.2 Organized enzyme systems

The organization of enzymes can be seen as a progression of increasing complexity extending from those enzymes that exist as single separate polypeptide chains, through to oligomeric enzymes, which may show allosteric interactions between the subunits, and ultimately to multienzyme proteins.* In the previous four chapters, we have already considered enzymes having single polypeptide chains and oligomeric enzymes and thus in the next sections we examine the following types of organization: (i) enzymes which although catalysing a single reaction, also interact in a specific manner with a macromolecular template and therefore require a more complex structure; (ii) multienzyme proteins with well-defined structures which have more than one distinct catalytic activity; and (iii) loose associations of enzymes in which the stoichiometry may be variable. RNA nucleotidyltransferase (DNA-dependent RNA polymerase) is the best example of the first type, since, in addition to catalysing nucleotide

* There is not yet a standard nomenclature for multienzymes. We use here that proposed by Karlson and Dixon.[2] Multienzyme proteins include all proteins with multiple catalytic domains, and they may be subdivided into those that are covalently linked (multienzyme polypeptides) and those that are non-covalently linked (multienzyme complexes).

transfers, the enzyme has to recognize specific regions on the DNA template in order to initiate and to terminate the synthesis of particular species of RNA. It is discussed in Section 7.3. There are several examples of the second type, e.g. pyruvate dehydrogenase (Section 7.7), tryptophan synthase (Section 7.9), and fatty-acid synthase (Section 7.11.2), in which more than one step is catalysed by the system concerned. The third type is represented by the glycogen particle (Section 7.12).

7.3 RNA nucleotidyltransferase from *E. coli*

RNA nucleotidyltransferase from *E. coli* represents an intermediate stage in the complexity of its organization between that of a typical allosteric enzyme and a large multienzyme protein. It comprises four different polypeptide chains, each having distinct functions in the complex but at the same time each helping to coordinate the synthesis of RNA on a DNA template. The RNA nucleotidyltransferase from *E. coli* is discussed specifically here for the following reasons: (i) it is the most highly purified and extensively studied RNA nucleotidyltransferase, (ii) all the evidence available suggests that it is responsible for the synthesis of all the RNA (mRNA, rRNA, and tRNA) present in uninfected *E. coli* cells, in contrast to the situation in eukaryotes where there is a multiplicity of RNA nucleotidyltransferases serving different functions. The enzyme catalyses the complete transcription cycle (Fig. 7.1; see also Chapter 9, Section 9.2 for an explanation of the process of transcription), which includes the steps of (i) recognition of the specific promoter regions on the genome, (ii) transcribing the appropriate gene or group of genes controlled by a particular operator, and (iii) terminating transcription at the appropriate region of the genome. The RNA nucleotidyltransferase is thus concerned with the synthesis of a large number of different mRNAs, each subject to different controls, and also with the synthesis of tRNA and rRNA. (For reviews on transcription and RNA nucleotidyltransferases, see references 3, 4, 5, and 6).

7.3.1 Structure of RNA nucleotidyltransferase

The enzyme from *E. coli* has been purified in several different laboratories and shown to be composed of four different polypeptide chains, $\alpha(M_r\ 36\,512)$, $\beta(M_r\ 150\,619)$, $\beta'(M_r\ 155\,162)$ and $\sigma(M_r\ 70\,236)$ and has the composition $\alpha_2\beta\beta'\sigma$. The genes for the polypeptides have been cloned and sequenced. The enzyme contains two Zn atoms, which are not removed by dialysis against chelating agents, although they are released when the enzyme is dissociated. Their role is at present unknown. Two models have been proposed for the quaternary structure (see Section 7.3.4). Genetic evidence from *E. coli* mutants suggests that these are the only proteins involved in transcription, but an additional protein, *nus* A protein is implicated in elongation.[7] The σ subunit can dissociate from the holoenzyme ($\alpha_2\beta\beta'\sigma$) to yield a core enzyme ($\alpha_2\beta\beta'$). Both the holoenzyme and the core enzyme will catalyse RNA synthesis on a DNA template *in vitro* (see

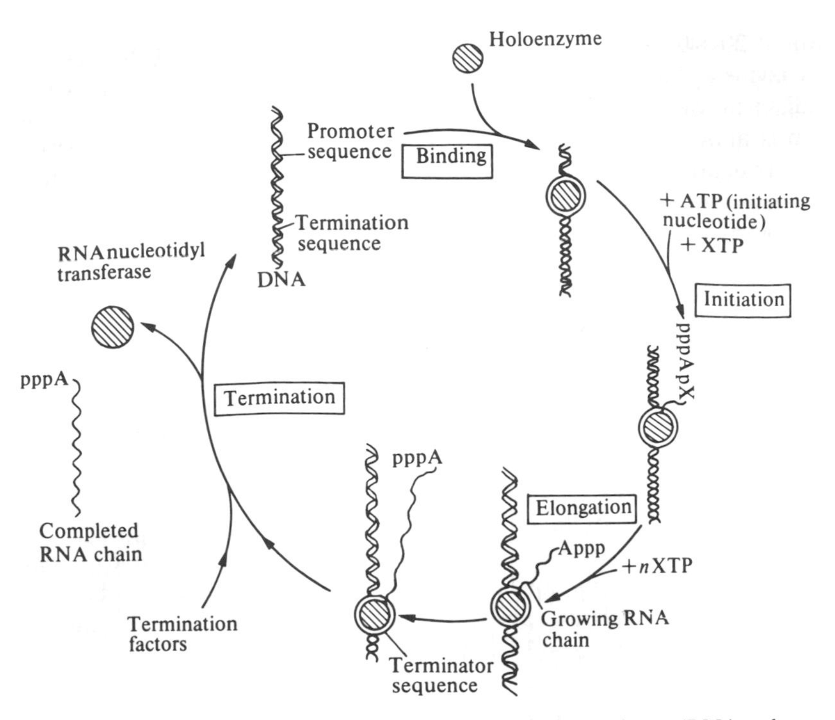

FIG. 7.1. The cycle of events catalysed by RNA nucleotidyltransferase (RNA polymerase) from *E. coli*. The release of the σ subunit following initiation is not shown.

Fig. 7.1 under binding, initiation, and elongation), but whereas the holoenzyme only initiates RNA synthesis on one strand of the DNA, commencing at the promoter sites, the core enzyme catalyses transcription on both strands starting at random positions. During normal transcription by the holoenzyme, the σ subunit is released soon after initiation. The σ subunit clearly has some role in normal initiation. The *nus* A protein is able to bind to the core enzyme in a similar fashion to the σ protein.[7]

7.3.2 Binding of RNA nucleotidyltransferase to DNA and initiation of transcription

We discuss next the present knowledge concerning the overall reaction mechanism and then subsequently the possible roles played by the individual subunits. The process of transcription is initiated by recognition of the promoter region on the DNA template (for a review of promoters see references 8, 9). It is estimated that the complete chromosome of *E. coli* contains in the region of 2000 promoters,[9] associated with different operons. The dissociation constants for the interaction between the core enzyme and DNA and between the holoenzyme and DNA are 0.5×10^{-11} mol dm^{-3} and 10^{-7} mol dm^{-3}, respectively. The σ subunit thus reduces the *general* binding affinity of RNA nucleotidyltransferase by about 20 000 times, and it is thought that this is important in enabling the holoenzyme to locate a particular region of DNA, namely the promoter site(s). The dissociation constant for the interaction of the holoenzyme with a promoter region is about 10^{-14} mol dm^{-3}. There is thus very tight binding between a promoter region and the holoenzyme and this has enabled the binary complex to be isolated on nitrocellulose filters. The different promoters will compete for RNA nucleotidyltransferase molecules and their effectiveness in competition will determine their frequency of usage. This is generally referred to as promoter strength, strong promoters being most effective.

The promoter regions are identified by nuclease digestion of the binary complex between DNA and the RNA nucleotidyltransferase, since the latter protects the promoter region. Over 100 promoter regions in *E. coli* have been sequenced[9] and by comparing these a consensus sequence can be formulated. There are two regions around positions −35 and −10 that are highly conserved. (The numbering indicates the number of nucleotides before the position at which transcription is initiated). In addition to the highly conserved regions (shown in large capitals), there are moderately conserved regions (small capitals) and weakly conserved regions (lower case letters). The contact points between the RNA nucleotidyltransferase and promoters are heavily concentrated in the consensus regions around −35 and −10.

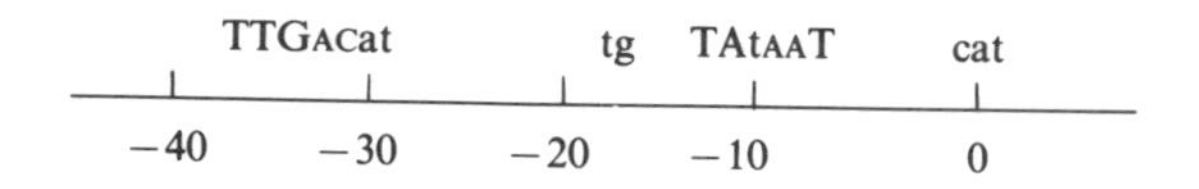

The positioning of the holoenzyme at the start sequence for transcription is accurate to within ±one nucleotide e.g. at the *lac* promoter in *E. coli*, transcription is initiated with the ribonucleotide sequence pppApApU, etc. or pppGpApApU, etc. Once the first phosphodiester bond has formed, the ternary complex (holoenzyme: DNA: oligoribonucleotide) is more stable and does not readily dissociate, although the σ subunit dissociates[10] from the holoenzyme after the first few phosphodiester bonds have formed, leaving the core enzyme: DNA: oligoribonucleotide complex to continue the process of elongation. Although certain prokaryotes have been shown to produce more than one σ factor, e.g. *Bacillus subtilis* uses at least five σ factors at different stages of its life cycle,[11, 12,134] for a long period only one σ factor had been isolated from uninfected *E. coli*. However, a second factor has more recently been discovered[13] as a result of studying proteins synthesized in response to heat shock.

An important question is how the holoenzyme locates the promoters within the double-stranded DNA of the chromosome. It is possible by using the values of estimated diffusion constants for DNA nucleotidyltransferase and the promoter to calculate the maximum rate at which the RNA nucleotidyltransferase could locate a promoter by a diffusion-controlled process.[14] However, when the calculated maximum rate is compared with experimentally determined rates for the strongest promoters, the latter are approximately 100-fold greater. In a detailed discussion Von Hippel *et al.*[14] suggest that the holoenzyme locates non-specific sites on the DNA initially, thus providing an 'expanded target', and then translocates along the DNA to the promoter region.

7.3.3 Elongation and termination of transcription

Elongation of the nascent RNA chain is catalysed by the core enzyme and the process continues, new nucleotides being added in a sequence complementary to that of one strand of the DNA. The mechanism is considered to be as shown in Fig. 7.2. It has been estimated that errors in transcription are between 1 in 2400 and 1 in 42 000 depending on conditions. The process of termination involves three steps: (i) termination in growth of the polynucleotide chain, (ii) release of nascent RNA, and (iii) release of the core enzyme from the DNA template. The mechanism of termination is not fully understood but is partly dependent on the nucleotide sequence and partly on the presence of a protein factor ρ.[15] The termination sequences of a number of genes have been examined and they show a number of common features, namely, a GC-rich region followed by several uracil residues and terminating with one or two adenine residues (Table 7.1); these terminate independently of protein factor ρ. The ρ factor is an oligomeric protein, the monomers having an M_r 46 000. It has been shown from *in vitro* experiments to promote termination and release of the RNA chain but not of the core enzyme. The termination sites that require ρ for termination appear to contain fewer conserved sequences.

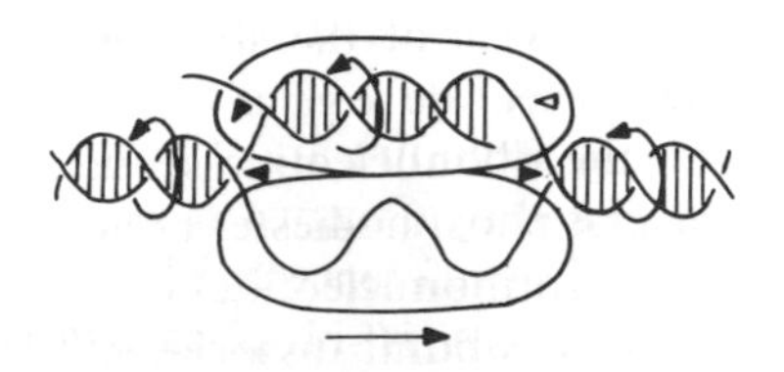

FIG. 7.2. Proposed mechanism of transcription. A limited region of the DNA is unwound and a DNA:RNA hybrid of about 12 base pairs is formed during transcription. The axis of the DNA template rotates relative to that of the enzyme. Unwinding of the DNA:RNA hybrid occurs during the rotation. The model is supported by photochemical cross-linking, chemical modification, and nuclease protection experiments.[14]

TABLE 7.1
Termination sequences in RNA recognized by RNA nucleotidyltransferase

λ 6S RNA	... A A C G G U U U C G G G G A U U U U U U A—3′—OH
λ 4S RNA	... U U G C C G C C G G G C G U U U U U U A—3′—OH
φ 80 M_3 RNA	... C C G C U U C A A G G G C U U U U U A—3′—OH
(λ and φ 80 M_3 are bacteriophages)	

Thus, the holoenzyme is necessary for correct initiation of RNA synthesis, after which the σ factor is no longer necessary, and the additional proteins such as *nus* A and ρ may be required for elongation and termination. In the next section we consider the roles of the subunits in the process of transcription.

7.3.4 The roles of the subunits of RNA nucleotidyltransferase

The roles of the individual subunits of RNA nucleotidyltransferase have been studied in the following ways: (i) examining the properties of isolated subunits, (ii) partial and complete reconstitution of the holo- and core enzymes, and (iii) from a study of mutations affecting the subunits.[16] The individual subunits have all been isolated. The α subunit is present in solution as α_2, it carries a binding site for the β subunit and is involved in the specificity of interaction with a gene, possibly by binding to the promoter region. However, in spite of the fact that its complete amino-acid sequence is known, it is the subunit about which there is least evidence for the function. The β' subunit is the most basic of the subunits; it is able to bind directly to DNA and is undoubtedly involved in template binding. The β subunit participates in binding the substrates, is involved in initiation and elongation, and it also carries the catalytic site for phosphodiester bond formation. The substrate-binding site was detected by affinity labelling (see Chapter 5, Section 5.4.4.3) using uridine triphosphate analogues. RNA nucleotidyltransferase from *E. coli* is strongly inhibited by the antibiotics rifamycin and streptolydigin and this is due to interaction with the β subunit. Proof of this is that resistant strains of *E. coli* have a modified β subunit, and that if RNA nucleotidyltransferase is rcconstituted using the β subunit from a resistant strain and the other subunits from a sensitive strain, then the resultant enzyme is insensitive to the antibiotics.

The details of the quaternary structure are still not completely clear. Alternative models have been proposed by Coggins *et al*,[17] and by Hillel and Wu[18] and they are discussed in detail in reference 5. The model of Coggins *et al.* is based on cross-linking studies using *N,N'*-bis-(2-carboximidoethyl)-tartaramide dimethyl ester dihydrochloride (see Chapter 3, Section 3.6.12) and that of Hillel and Wu[18] on low-angle neutron scattering and cross-linking with methyl 4-mercaptobutyrimidate.

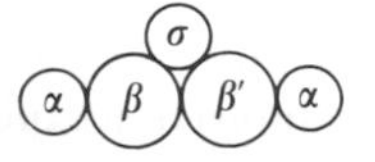

Model of Coggins et al.[17]

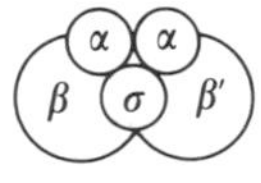

Model of Hillel and Wu[18]

The main point of difference between the two is whether the α-subunits are separated.

Partially reconstituted complexes have been isolated containing $\alpha_2\beta$ and $\beta'\sigma$. The $\alpha_2\beta$ does not bind DNA but it does bind rifamycin. The existence of $\alpha_2\beta$ suggests that β contains at least one strong binding site for α. Similarly the existence of $\beta'\sigma$ suggests that β' contains the binding site for σ. The holoenzyme has been reconstituted *in vitro* by the following pathway and a similar route is thought to operate *in vivo*:[16]

$$2\alpha \rightleftharpoons \alpha_2$$
$$\alpha_2 + \beta \rightleftharpoons \alpha_2\beta$$
$$\alpha_2\beta + \beta' \rightleftharpoons \alpha_2\beta\beta'$$
$$\alpha_2\beta\beta' + \sigma \rightleftharpoons \alpha_2\beta\beta'\sigma$$

This pathway is consistent with the arrangement of the subunits in the holoenzyme as determined by Hillel and Wu[18] and by Stöckel *et al.*[19] However, it is possible that, although the α subunits associate during the first stage of assembly, a subsequent rearrangement may occur. That this is likely is suggested by the observation that the initial $\alpha_2\beta\beta'\sigma$ complex formed is inactive, and thus may undergo rearrangement before becoming active.

7.3.5 RNA nucleotidyltransferases from bacteriophages

RNA nucleotidyltransferase from *E. coli* has to recognize a variety of promoter sites on the chromosome and transcription from these sites is subject to a variety of different controls and it is perhaps partly for this reason that the enzyme is a multisubunit protein. By contrast, the bacteriophage RNA nucleotidyltransferases have only to recognize a few promoters and this is reflected in a simple enzyme structure. For example, the RNA nucleotidyltransferases coded for by genes on the bacteriophages T_3 and T_7 comprise single polypeptide chains of M_r 110 000. These RNA nucleotidyltransferases are highly specific in their recognition of DNA. The T_7 enzyme recognizes only poorly the promoters of T_3 and vice versa. Thus, the complexity of any given RNA nucleotidyltransferases is partly a function of the range of species of RNA it is required to synthesize.

7.4 The occurrence and isolation of multienzyme proteins

In a metabolic pathway a number of enzymes catalyse a sequence of reactions linked so that the product of one enzyme-catalysed reaction becomes the substrate for the next enzyme. Such a sequence may be represented as

$$A \xrightarrow{E_1} B \xrightarrow{E_2} C \xrightarrow{E_3} D \xrightarrow{E_4} E \xrightarrow{E_5} F$$

in which A, B, C, D, and E are substrates for the enzymes E_1, E_2, E_3, E_4, and E_5. Both the efficiency of these reactions, measured as the yield of the end product F from the starting substrate A, and the overall rate of conversion of A to F will depend partly on the extent of coordination occurring between the five enzymes. For some metabolic pathways it has been shown that certain of the enzymes are physically associated with one another to form multienzyme proteins, e.g. the fatty-acid synthase complex (Section 7.11.2) and enzymes of the tryptophan biosynthetic pathway (Sections 7.9 and 7.11.3).

The multienzymes that have been most studied are probably amongst the more stable ones. There may be many others in which the associations are weak and are thus more difficult to detect as multienzymes. The isolation and characterization of a multienzyme protein usually presents more technical difficulties than isolation of a single enzyme. This is well illustrated by two examples discussed in more detail later in the chapter, namely, the pyruvate dehydrogenase complex (Section 7.7) and the yeast fatty-acid synthase complex (Section 7.11.2). In the case of the pyruvate dehydrogenase complex from *E. coli*, although it is clear that the multienzyme complex contains three different enzyme activities, the precise number of each of the polypeptide chains present in the complex is still debated. This may be because (i) different isolation procedures yield complexes of slightly different composition, (ii) some dissociation may occur during isolation, or (iii) in the intact cell complexes of slightly variable composition do exist. In a multienzyme complex having a large number of subunits, e.g. 50–100, a small percentage error in the determination of the M_r of the complex will significantly affect the proposed number of subunits.

Yeast fatty-acid synthase posed a different type of problem. It proved difficult to separate the polypeptide chains corresponding to the seven catalytic activities of the fatty-acid synthase. It was then shown that the whole complex comprises two multifunctional polypeptide chains (see Section 7.11.2.1) and that the smaller fragments observed in earlier experiments were the result of limited proteolysis during isolation.

Evidence for the existence of a multienzyme complex usually appears when a component enzyme of the complex is being isolated and it is found to copurify with another enzyme from the same metabolic pathway. If the ratio of the enzyme activities remains constant during the isolation procedure, then this is further evidence for the existence of a definite multienzyme complex, e.g. carbamoyl phosphate synthase, aspartate carbamoyltransferase, dihydro-orotase (see Section 7.10).

Two enzyme activities that were initially shown to copurify from rat liver and were subsequently found to be two catalytic activities associated with a single polypeptide chain are fructose 2,6-bisphosphatase and 6-phosphofructo-2-kinase. The enzyme catalysing both these activities has been purified from a number of tissues and it has been shown that the two activities reside in separate parts of the polypeptide chain.[20, 21] The importance of this enzyme is discussed in Chapter 6, Section 6.4.1.1. Another example of phosphatase and kinase activities residing in the same polypeptide chain is that of isocitrate dehydrogenase kinase and isocitrate dehydrogenase phosphatase. Both activities copurify and are coded for by a single gene.[22, 23]

7.5 Phylogenetic distribution of multienzyme proteins

Although only a limited number of multienzyme complexes have so far been studied, it does seem that in general multienzyme complexes and multienzyme polypeptides are both more common and more complex in eukaryotes than in prokaryotes; this is borne out by the examples of pyruvate dehydrogenase, fatty-acid synthase, and carbamoyl phosphate synthase, aspartate carbamoyltransferase, dihydro-orotase and the enzymes of the tryptophan biosynthetic pathway discussed in this chapter. A typical prokaryote cell, e.g. *E. coli*, has a very much smaller volume than a typical eukaryote cell, e.g. liver parenchyma (1 μm^3 as compared to 6000 μm^3). Even making allowance for the intracellular compartmentation in eukaryotes brought about by the various intracellular membranes, the volume in which free diffusion of substrates and enzymes can occur is much greater than in a prokaryote cell. Thus, if the intracellular concentrations of substrates and catalytic centers are similar in prokaryotes and eukaryotes, then the diffusion time is more likely to be rate limiting in the latter. The comparative biochemical information on carbamoyl phosphate synthase, aspartate carbamoyltransferase, dihydro-orotase, the tryptophan biosynthetic pathway, and fatty-acid synthase suggests that evolution has resulted in multienzyme polypeptides and more tightly associated multienzyme complexes. In addition, genetic evidence suggests that this is accompanied by a clustering or even fusion of the genes coding for these proteins.

7.6 Properties of multienzyme proteins

What are the advantages of enzymes being physically associated with one another rather than occurring as separate entities within cells? The answers may be found by comparing the properties of multienzyme proteins with those of separated enzymes. Firstly, the transit time, i.e. the time required for a product of one enzyme to diffuse to the catalytic site of a second enzyme will be less in a physically associated multienzyme protein than when the enzymes are not associated. Whether the transit time becomes the rate-limiting step in a metabolic pathway will depend on the catalytic activities of the enzymes. Two instances in which the transit times may become important are (a) when an intermediate has a short half life and (b) when an intermediate has a high M_r and thus would diffuse more slowly. There is as yet no well-documented example of the first situation. Carbamoyl phosphate probably has a short half life in the cytosol, since there are enzymes capable of rapidly degrading it. The association of carbamoyl phosphate synthase with aspartate carbamoyl transferase, which channels carbamoyl phosphate within the multienzyme protein, probably prevents its degradation.[24] An example of the second situation is that of the pyruvate dehydrogenase system, in which the second step in the reaction sequence involves the two enzyme-bound intermediates hydroxyethyl-thiamin-diphosphate-pyruvate dehydrogenase and dihydrolipoamide acetyltransferase.

Secondly, in addition to the transit time being reduced in a multienzyme protein, so also would be the transient time, that is the time required to change from one steady-state rate to another. Thus in a system with two enzymes

$$A \xrightarrow{E_1} B \xrightarrow{E_2} C$$

in a steady-state in a cell, if the concentration of A is increased then the time required to reach the new steady-state rate (transient time) will be much less in the case of a multienzyme protein.[25]

A third possibility afforded by multienzyme proteins is that of channelling or compartmentation. Once an intermediate has been formed in a multienzyme protein, it is not generally available to be acted upon by other enzymes outside the complex. Examples of this occur with carbamoyl phosphate formed in the multienzyme complex for pyrimidine biosynthesis in *Neurospora crassa* (see Section 7.10) that is not available for the biosynthesis of arginine, and also with ammonia generated by glutaminase in liver mitochondria, which is preferentially used by carbamoyl phosphate synthase.[27]

There is also some evidence for coordinate effects of a single ligand on a whole multienzyme protein, e.g. in the *arom* complex discussed later (Section 7.11.3), where the substrate of the first enzyme is able to activate all five catalytic activities of the multienzyme protein.

The examples of multienzymes discussed in the subsequent sections of this chapter have been chosen to illustrate the various properties of multienzyme proteins that have been described in this section. The pyruvate dehydrogenase, oxoglutarate dehydrogenase, and tryptophan synthase systems are the multi-enzyme complexes about which most structural information is available. They also illustrate how the intermediate substrates remain attached to the multi-enzyme complexes. Fatty acid synthase and the *arom* complex are examples of multienzyme polypeptides. The multienzyme complex of carbamoyl phosphate synthase and aspartate carbamoyltransferase is used to illustrate the importance of channelling an intermediate (carbamoyl phosphate) into one of two competing pathways. The 'glycogen particle' is much less well defined than the other multienzyme proteins and presumably represents a looser association between glycogen and the glycolytic enzymes.

7.7 The pyruvate dehydrogenase system

The pyruvate dehydrogenase multienzyme complex is one of the most thoroughly investigated multienzyme proteins. The complex catalyses the conversion of pyruvate to acetyl-CoA, which is the step connecting the glycolytic sequence to the tricarboxylic acid cycle and is an important control point in intermediary metabolism. The reactions catalysed by the complex, which comprises three distinct enzymes, are decarboxylation, dehydrogenation, and the formation of a thioester. Five cofactors are necessary for the complete reaction sequence, as shown below.

$$E_1\text{-TDP} + CH_3COCOO^- \rightleftharpoons E_1\text{-TDP—CHOHCH}_3 + CO_2 \quad (1)$$

$$E_1\text{-TDP—CHOHCH}_3 + E_2\text{-lip}\langle^{S}_{S}| \rightleftharpoons E_1\text{-TDP} + E_2\text{-lip}\langle^{SH}_{SCOCH_3} \quad (2)$$

$$E_2\text{-lip}\langle^{SH}_{SCOCH_3} + CoASH \rightleftharpoons E_2\text{-lip}\langle^{SH}_{SH} + CH_3COSCoA \quad (3)$$

$$E_2\text{-lip}\langle^{SH}_{SH} + E_3\text{-FAD} \rightleftharpoons E_2\text{-lip}\langle^{S}_{S}| + E_3\text{-FAD reduced*} \quad (4)$$

$$E_3\text{-FAD reduced} + NAD^+ \rightleftharpoons E_3\text{-FAD} + NADH + H^+ \quad (5)$$

E_1 = pyruvate decarboxylase† TDP = thiamin diphosphate

E_2 = dihydrolipoamide acetyl transferase $\text{lip}\langle^{S}_{S}|$ = oxidized lipoamide

E_3 = dihydrolipoamide dehydrogenase $\text{lip}\langle^{SH}_{SH}$ = reduced lipoamide

The complex that has been most extensively investigated is that from *E. coli*, although it has also been purified from a number of other bacteria[28–30] and mammalian sources.[31,32] The complex isolated from *E. coli* is discussed in Sections 7.7.1 to 7.7.3. The regulation of the complexes from mammalian sources differs from that of *E. coli* and these differences are discussed in Section 7.7.4.

7.7.1 Structure and isolation of the pyruvate dehydrogenase multienzyme complex from *E. coli*

Although it is clear that the multienzyme complex contains three distinct enzymes, pyruvate decarboxylase (E_1), dihydrolipoamide acetyltransferase (E_2), and dihydrolipoamide dehydrogenase (E_3), there is still uncertainty concerning the number of each of these enzymes in the complex. There is reasonable agreement about the M_r of the separate polypeptide chains at least of E_1 and E_3, and the M_r values of E_1 and E_2 have been inferred from the DNA sequence.[33,34] However, there is less agreement about the M_r of the intact complex (Table 7.2).

* E_3-FAD reduced. The precise structure of the reduced flavoprotein is not yet clear. There are two sulphydryl groups in close proximity to the FAD on E_3. It seems probable that a semiquinone intermediate FADH and the sulphydryl groups are both involved in the hydrogen transfer.[32]

† The Enzyme Commission recommended name for E_1 is pyruvate dehydrogenase (lipoamide). However, many workers in the field use the name pyruvate decarboxylase for E_1; this is both more informative and less confusing and is therefore adopted here.

Estimates of the M_r of the complex have ranged since 1970 as follows: 3×10^6 (1970),[35] 3.8×10^6 (1972),[36] 4.6×10^6 (1972),[37] 6.1×10^6 (1979)[38] and 5.36×10^6 (1985).[39] The values of M_r have been determined by different methods and using different preparations of the complex. It is difficult to determine the M_r of a large complex with sufficient accuracy to estimate its precise subunit composition. It is also possible either that some of the subunits may become detached during the isolation procedure or even that some variation or heterogeneity in the composition of the complex exists *in vivo*. Heterogeneity of composition is suggested when the M_r is determined by active enzyme centrifugation[40] (for details of this method, see Chapter 3, Section 3.2.4). Although knowledge of the precise subunit composition is necessary before the three-dimensional structure of the complex can be known with certainty, it is less important for understanding the mechanism of action of the multienzyme complex and will not be discussed further (for further details of the structure see references 31, 38 and 39).

TABLE 7.2
M_r values of the subunits of the pyruvate dehydrogenase multienzyme complex from E. coli

Pyruvate decarboxylase (E_1)	96 000–100 000
Dihydrolipoamide acetyltransferase (E_2)	60 500–91 000
Dihydrolipoamide dehydrogenase (E_3)	56 000
Pyruvate dehydrogenase multienzyme complex	3×10^6–6.1×10^6

References 40, 41, 42, and 43. For review, see reference 39.

The pyruvate dehydrogenase complex is isolated from *E. coli* by the steps indicated in Fig. 7.3. Precautions such as the inclusion of protease inhibitors have to be taken to prevent proteolysis occurring during isolation. The isolated multienzyme complex can be dissociated into its constituent subunits by use of 4 mol dm^{-3} urea, high pH, and calcium phosphate gel (Fig. 7.3). The multienzyme complex may also be reconstituted from the subunits and full enzyme activity restored. It is clear from both the dissociation and reconstitution studies that E_2 has binding domains for both E_1 and E_3, but that E_1 and E_3 do not bind in the absence of E_2. The separated acetyltransferase (E_2) has octahedral symmetry and this is consistent with the composition of the multienzyme complex having a total of 24 E_2 polypeptide chains and is also in agreement with the majority of M_r determinations. This arrangement of E_2 occurs not only in the *E. coli* pyruvate dehydrogenase complex, but also in the 2-oxoglutarate dehydrogenase complexes from *E. coli* and mammalian sources, and also in branched-chain oxo-acid dehydrogenases.[46] The 24 chains are thought to be arranged as shown in Fig. 7.4(a). The E_1 and E_3 enzymes are then considered to be associated with the E_2 chains. The stoichiometry favoured by most workers is

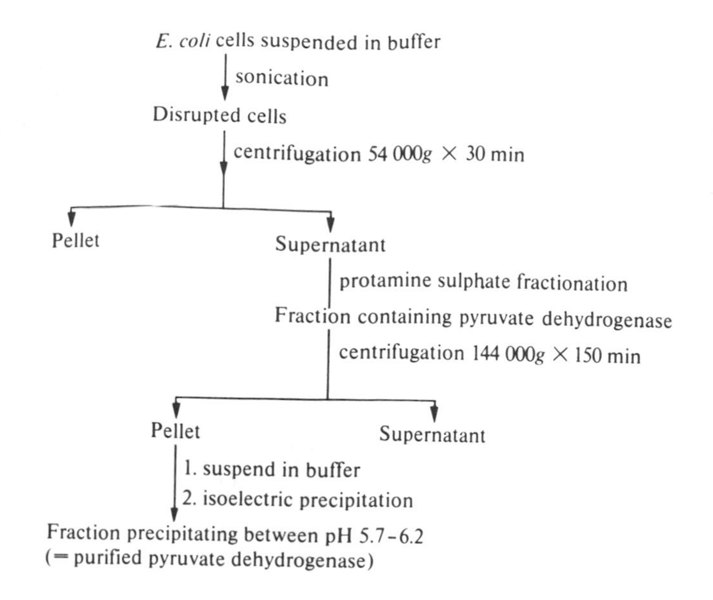

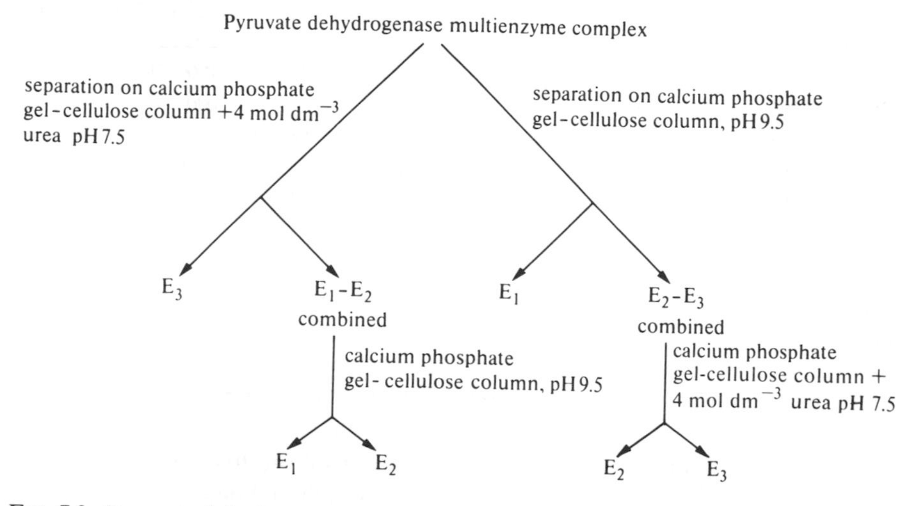

FIG. 7.3. Pyruvate dehydrogenase complex: isolation and separation of the subunits (see references 44 and 45).

24:24:12 for $E_1:E_2:E_3$[39], but Bates *et al.*[47] proposed a stoichiometry of 48:24:24.

Three of the cofactors thiamin diphosphate, lipoate, and FAD are firmly bound to the decarboxylase, the acetyltransferase, and the dehydrogenase, respectively, and do not dissociate during the reactions. Lipoate is covalently linked through an amide bond to a lysine residue on the acetyltransferase.

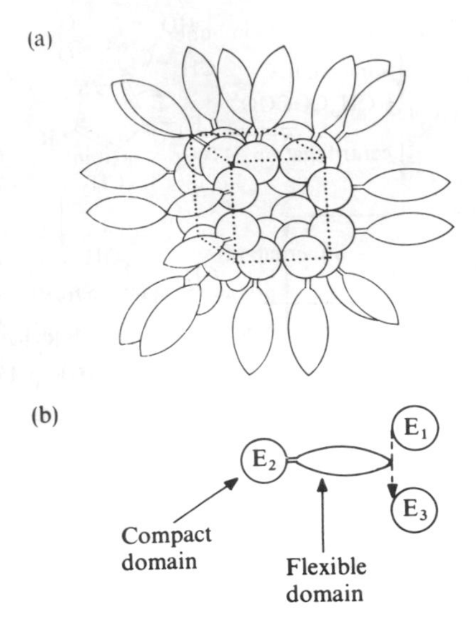

FIG. 7.4. Structure of pyruvate dehydrogenase multienzyme complex. (a) The probable arrangement of dihydrolipoamide acetyltransferase (E_2) forming the core of the pyruvate dehydrogenase multienzyme complex. Each E_2 polypeptide is thought to be made up of two domains, a central core containing the region of the active site and a flexible domain protruding out from the core. The latter contains the lipoate residues. The 24 E_2 polypeptide chains are arranged in the form of a cube, (a), and the E_1 and E_3 enzymes form a larger cube encasing the E_2 cube such that the flexible domain of E_2 can move between E_1 and E_3, as shown in (b). For details, see references 31 and 46.

$$\underbrace{\overset{\displaystyle CH_2}{CH_2\quad CH}\!-\!(CH_2)_4\!-\!CO}_{\text{Lipoate}}\;\underbrace{NH(CH_2)_4\!-\!\overset{\displaystyle NH\text{---}}{\underset{\displaystyle CO\text{---}}{C}}\!-\!H}_{\text{Lysyl group}}$$

(ring: CH_2–CH_2–CH closed by S—S)

← 1.4 nm →

Evidence shows that three lipoate residues are attached to each acetyltransferase molecule,[41,48] and the DNA sequence[49] shows three homologous regions in the N-terminal half of E_2, where the three lipoates are attached.[50] This arrangement seems to occur only in *E. coli* and *Azotobacter vinelandii*. All other pyruvate dehydrogenases so far studied have only one lipoate for each E_2.[46] Guest *et al.*[49] have shown that it is possible to delete one or two of the regions of the DNA coding for the lipoate attachments sites, and the E_2 that is expressed still shows full activity. Thus, the roles of the second and third lipoates are unclear.

7.7.2 Mechanism

The initial decarboxylation occurs when pyruvate combines with TDP on the decarboxylase probably by the mechanism shown below.

$$\text{(thiazolium ring of TDP)} + CH_3COCOO^- \rightleftharpoons \text{(adduct)} \rightarrow \text{(hydroxyethyl-TDP)} + CO_2$$

where $R_1 = CH_3$–(pyrimidine, NH_2)–CH_2– and $R_2 = {}^-O{-}P(O)(O^-){-}O{-}P(O)(O^-){-}OCH_2CH_2{-}$

The second step involves the oxidative transfer of the hydroxyethyl group bound to E_1 on to the disulphide group of lipoamide on E_2, forming a thioester (reaction 2 in Section 7.7). For this step the two enzymes have to be in close contact. The acetyl group so formed is then transferred to CoA in the third step. The fourth and fifth steps are concerned with the regeneration of oxidized lipoamide by transfer of 2H from dihydrolipoamide on E_2 via FAD of the dehydrogenase (E_3) to NAD^+. The fourth step also involves the interaction between two enzyme-bound intermediates of E_2 and E_3 (reaction 4, Section 7.7).

It has been suggested[31] that the transfer of the acetyl group and of the two hydrogen atoms from dihydrolipoamide involves the flexible arm of lysyllipoamide on E_2 swinging between E_1 and E_3. This arm would be approximately 1.4 nm and therefore could swing through a diameter of 2.8 nm. Estimates have been made[51,52] of the distances separating E_1 and E_3 to test this hypothesis. By using an analogue of TDP, thiochrome diphosphate, bound at E_1, it is possible to estimate the distance to the FAD on the active site of E_3 by fluorescence energy transfer. The distance appears to be about 4.5 nm, which is too great for the flexible lysyllipoamide arm alone to be the explanation. However, the polypeptide chain of E_2 adjacent to the lysyllipoamide arm is also highly flexible, being rich in alanine, proline, and charged amino-acid residues;[50] this could increase the range of the arm.[136].

A question that arises with a multienzyme complex of this nature in which there are several copies of each of the enzymes is whether, in a given catalytic cycle, one molecule of pyruvate is oxidatively decarboxylated by a single specific cluster of E_1–E_2–E_3 within the complex. Several experiments have been aimed at answering this question. In the multienzyme complex, pyruvate decarboxylase (E_1) catalyses the rate-limiting step. This has been demonstrated by using an active-site-directed inhibitor of E_1, thiamin thiazolone diphosphate, which inhibits pyruvate dehydrogenase activity in parallel with the decrease in the rate of conversion of pyruvate to acetyl-CoA.[42] In contrast, the acetyltransferase (E_2) can be partially inhibited by reacting with *N*-ethylmaleimide before any reduction in the rate of conversion of pyruvate to acetyl-CoA occurs. Also, it is possible to remove nearly all the lipoyl moieties using the enzyme lipoamidase, without significant loss of activity.[53]

It is possible to reconstitute a pyruvate dehydrogenase complex that is deficient in pyruvate decarboxylase (E_1) and to use this to demonstrate that pyruvate decarboxylase (E_1) is capable of transferring hydroxyethyl groups to several acetyltransferases.[43] This is done by allowing reaction to occur in the absence of CoA and measuring the extent to which the acetyltransferase (E_2) becomes acetylated. From these results it seems probable that there is a communicating network of acetyltransferases (E_2) in which the acetyl groups may be transferred from one E_2 to another, thus enabling a transfer of acetyl groups and the associated reduction of lipoamide residues from an E_1 site to a distant E_3 site within the complex.[54] The possible advantages of this arrangement are discussed in reference 55.

7.7.3. Control of the pyruvate dehydrogenase system in *E. coli*

The pyruvate dehydrogenase complex occupies a key position linking the glycolytic pathway to the tricarboxylic acid cycle. There are a number of factors that control its activity in different organisms. The decarboxylation step is effectively irreversible under physiological conditions and thus controls the entry of pyruvate into the tricarboxylic acid cycle. Pyruvate dehydrogenase activity is inhibited by two of the products of its activity, namely, acetyl-CoA, which is a competitive inhibitor with respect to CoA, and NADH, which is competitive with respect to NAD^+ (Fig. 7.5). The combination of acetyl-CoA and NADH may also inhibit by acetylation of the bound lipoamide. The activity of the pyruvate dehydrogenase complex is stimulated by nucleoside monophosphates and inhibited by GTP. The latter effect is reversed by GDP.

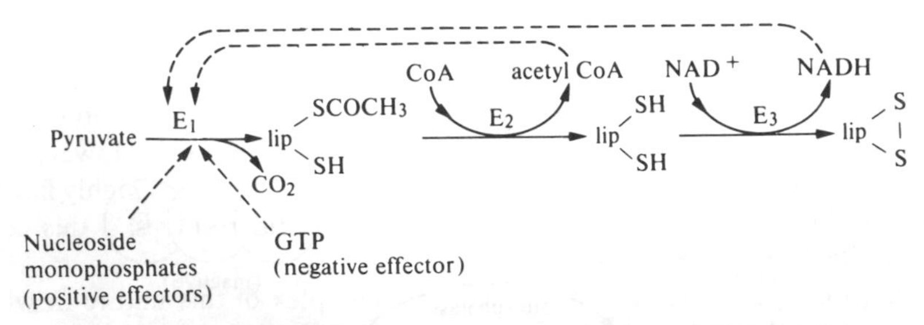

FIG. 7.5. Regulation of the pyruvate dehydrogenase complex in *E. coli*.

7.7.4 Organization and control of the pyruvate dehydrogenase system in mammalian tissues

The pyruvate dehydrogenase multienzyme complexes isolated from a number of mammalian tissues,[31] including pig heart, bovine kidney, and bovine heart, have been found to be larger than that of *E. coli*, and their regulation is subject to additional controls to those described in Section 7.7.3. The basic organization of the three components is similar, although the mammalian complexes show icosahedral rather than octahedral symmetry. The pyruvate decarboxylase (E_1) comprises two polypeptide chains α and β. There are 60 copies of each in the multienzyme complex.[46] It appears that in Gram-negative bacteria, e.g. *E. coli*, *Azotobacter vinelandii*, an octahedral arrangement occurs, whereas in Gram-positive bacteria, e.g. *Bacillus stearothermophilus*, fungi such as *Neurospora* and *Saccharomyces*, and higher eukaryotes, an icosahedral arrangement occurs.[56] The latter would mean that the components E_1, E_2, and E_3 are more likely to be present in multiples of 20, e.g. 60 rather than in multiples of eight, e.g. 24, which are expected for an octahedral arrangement.

The pyruvate dehydrogenase complexes from mammalian and avian sources, and from *Neurospora*, interact with two additional enzymes,[46] namely, [pyruvate dehydrogenase (lipoamide)] kinase and [pyruvate dehydrogenase (lipoamide)] phosphatase; the former is tightly bound to the complex, whereas the latter is very easily dissociated from the complex. The kinase is bound to the acetyltransferase (E_2) and sub-complexes containing E_2-kinase have been isolated. The substrate of the kinase is pyruvate-decarboxylase (E_1), which becomes phosphorylated on one of its serine* side-chains when the intracellular ratio of [ATP]/[ADP] is high; this results in inactivation of the multienzyme complex, specifically inhibiting the decarboxylation step (Fig. 7.6). The phosphatase catalyses the dephosphorylation of E_1, which converts the inactive multienzyme complex back into the active form (further details of this mechanism are given in references 57 and 58). Pyruvate inhibits the phosphorylation step, thus maintaining the complex in an active state. The ratio of phosphorylated to dephosphorylated pyruvate dehydrogenase multienzyme complex has been shown to vary with the physiological state of the mammal, e.g. whether it is well

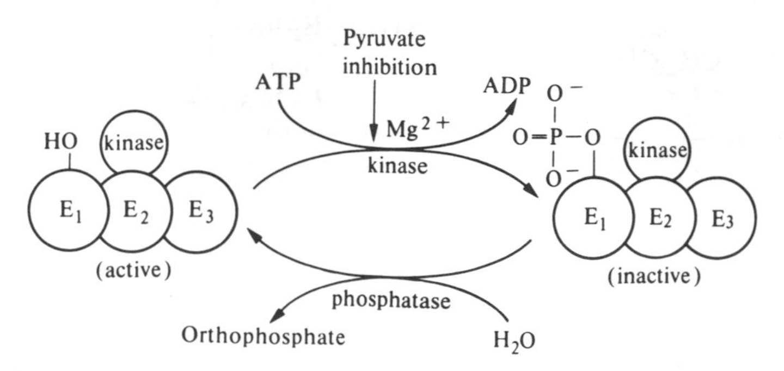

FIG. 7.6. Regulation of mammalian pyruvate dehydrogenase complexes by phosphorylation and dephosphorylation.

* There are three serine side-chains on E_1, which may become phosphorylated in succession, but phosphorylation of the first inactivates the complex. The significance of the three phosphorylation steps is not yet clear,[59–61] although it has been suggested that phosphorylation of the second and third site makes reactivation by the phosphatase slower.[62]

fed or starved (see also Chapter 6, Section 6.3.2.2). The pyruvate dehydrogenase multienzyme complexes from the lower eukaryotes *Neurospora crassa* and *Saccharomyces lactis* are also subject to similar covalent modification.[62]

In eukaryotes the pyruvate dehydrogenase multienzyme complex resides in the matrix within the inner mitochondrial membrane. The three enzyme components are coded for by nuclear genes. It has been shown that E_1, E_2, and E_3 are synthesized as cytosolic precursors each having a transit peptide to direct it into the mitochondrial matrix, where the peptide is removed by proteolysis and the multienzyme complex is assembled.[63]

7.8 Oxoglutarate dehydrogenase and branched chain 2-oxo acid dehydrogenase* multienzyme complexes

The oxoglutarate dehydrogenase multienzyme complex is not discussed here in detail because, in many respects, it is both structurally and functionally similar to that of pyruvate dehydrogenase. The complex has been isolated from *E. coli* and from mammalian sources. It consists of three distinct enzymes, oxoglutarate decarboxylase (E_1), dihydrolipoamide succinyltransferase (E_2), and dihydrolipoamide dehydrogenase (E_3) catalysing the conversion of 2-oxoglutarate to succinyl CoA, an integral step in the tricarboxylic acid cycle.

$$E_1\text{-TDP} + {}^-OOC.CO.CH_2.CH_2COO^- \rightleftharpoons E_1\text{-TDP—CHOH.CH_2.CH_2.COO^-} + CO_2 \qquad 1$$

$$E_1\text{-TDP—CHOH.CH_2.CH_2.COO^-} + E_2\text{-lip}\left\langle \begin{matrix} S \\ | \\ S \end{matrix} \right. \rightleftharpoons E_1\text{-TDP} + E_2\left\langle \begin{matrix} SH \\ SCO.CH_2.CH_2.COO^- \end{matrix} \right. \qquad 2$$

$$E_2\text{-lip}\left\langle \begin{matrix} SH \\ SCO.CH_2.CH_2.COO^- \end{matrix} \right. + CoASH \rightleftharpoons E_2\text{-lip}\left\langle \begin{matrix} SH \\ SH \end{matrix} \right. + {}^-OOC.CH_2.CH_2.COSCoA \qquad 3$$

$$E_2\text{-lip}\left\langle \begin{matrix} SH \\ SH \end{matrix} \right. + E_3\text{-FAD} \rightleftharpoons E_2\text{-lip}\left\langle \begin{matrix} S \\ | \\ S \end{matrix} \right. + E_2\text{-FAD reduced} \qquad 4$$

$$E_3\text{-FAD reduced} + NAD^+ \rightleftharpoons E_3\text{-FAD} + NADH + H^+ \qquad 5$$

TDP = thiamin diphosphate

$\text{lip}\left\langle \begin{matrix} S \\ | \\ S \end{matrix} \right.$ = oxidised lipoamide

$\text{lip}\left\langle \begin{matrix} SH \\ SH \end{matrix} \right.$ = reduced lipoamide

* These enzymes are sometimes referred to as α-ketoglutarate dehydrogenase and branched chain keto-acid dehydrogenase.

The arrangement of the enzymes within the multienzyme complex is similar to that in *E. coli* pyruvate dehydrogenase, based on octahedral symmetry.[46] The dihydrolipoamide succinyltransferase (E_2) binds both the decarboxylase (E_1) and the dehydrogenase (E_3). The amino-acid sequence of E_2 has been inferred from its DNA sequence.[64] Preparations of the isolated E_2 have a tetrad appearance similar to that of dihydrolipoamide acetyltransferase when examined by electron microscopy. The E_2 also comprises a compact and a flexible domain comparable with that of dihydrolipoamide acetyltransferase.[65] Dihydrolipoamide dehydrogenase isolated from pyruvate dehydrogenase and oxoglutarate dehydrogenase multienzyme complexes can be exchanged with one another in reconstitution experiments.

Like the pyruvate dehydrogenase complex, oxoglutarate dehydrogenase activity is regulated by the ratio of the concentrations of substrates and products ([succinyl CoA]/[CoA] and [NADH]/[NAD^+]), the products inhibiting activity. Activity is stimulated by AMP, which appears to promote tighter binding of the substrate, 2-oxoglutarate, and the cofactor TDP. ATP competes with AMP in reversing the stimulation. In the mammalian oxoglutarate dehydrogenase complexes there is no evidence for covalent modification by kinases and phosphatases. The branched chain 2-oxo acid dehydrogenase also has a similar structure.[132] It occurs inside the inner mitochondrial membrane and catalyses the overall reaction:

$$\begin{matrix}R_1 \\ R_2\end{matrix}\!\!>\!\text{CHCO}\cdot\text{COO}^- + \text{NAD}^+ + \text{CoASH} \rightleftharpoons \begin{matrix}R_1 \\ R_2\end{matrix}\!\!>\!\text{CHCOSCoA} + \text{NADH} + \text{H}^+ + \text{CO}_2$$

Its importance is in the oxidative decarboxylation of oxo-acids formed by transamination or deamination of valine, isoleucine and leucine. The multienzyme complex has E_1, E_2, and E_3 enzyme components and is also subject to regulation by a kinase and a phosphatase.[66]

7.9 The tryptophan synthase multienzyme complex from *E. coli*

The tryptophan synthase system is much smaller than either the pyruvate or oxoglutarate dehydrogenase systems, having an M_r of approximately 150 000. It is a striking example in which the interaction of two subunits greatly modifies their respective catalytic activities and this aspect is emphasized in the subsequent sections. The tryptophan synthase multienzyme complex catalyses the final steps in the biosynthesis of tryptophan (Fig. 7.7). The tryptophan synthase most studied is that from *E. coli* but it has also been investigated in a number of other bacteria, in fungi and in plants. Mammals lack tryptophan synthase and thus require a dietary source of tryptophan.

Tryptophan synthase from *E. coli* is coded for by two structural genes,* which, together with three other structural genes, form an operon† coding for the

* Structural genes are those genes that code for structural and enzyme proteins, as opposed to regulatory genes, which are concerned with the control of transcription or translation.

† An operon is a cluster of genes that are regulated as a unit and that code for metabolically related enzymes.

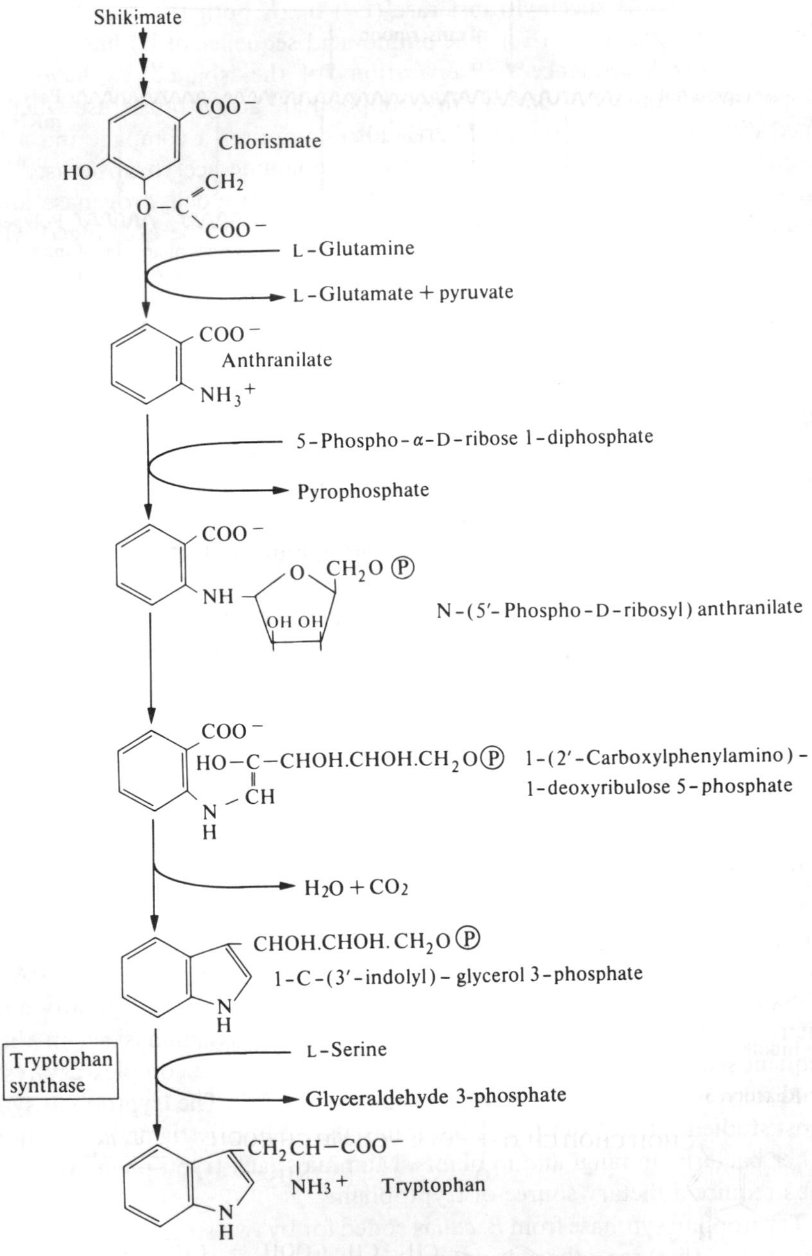

Shikimate

COO^-
Chorismate
HO
CH_2
O–C
COO^-

L-Glutamine
L-Glutamate + pyruvate

COO^-
Anthranilate
NH_3^+

5-Phospho-α-D-ribose 1-diphosphate
Pyrophosphate

COO^-
O CH_2O Ⓟ
NH
OH OH
N-(5′-Phospho-D-ribosyl) anthranilate

COO^-
HO–C–CHOH.CHOH.CH_2OⓅ
CH
N
H
1-(2′-Carboxylphenylamino)-
1-deoxyribulose 5-phosphate

$H_2O + CO_2$

CHOH.CHOH. CH_2O Ⓟ
1-C-(3′-indolyl)-glycerol 3-phosphate
N
H

Tryptophan synthase

L-Serine
Glyceraldehyde 3-phosphate

$CH_2CH–COO^-$
NH_3^+
Tryptophan
N
H

FIG. 7.7. The pathway for biosynthesis of tryptophan. Ⓟ = phosphoryl group.

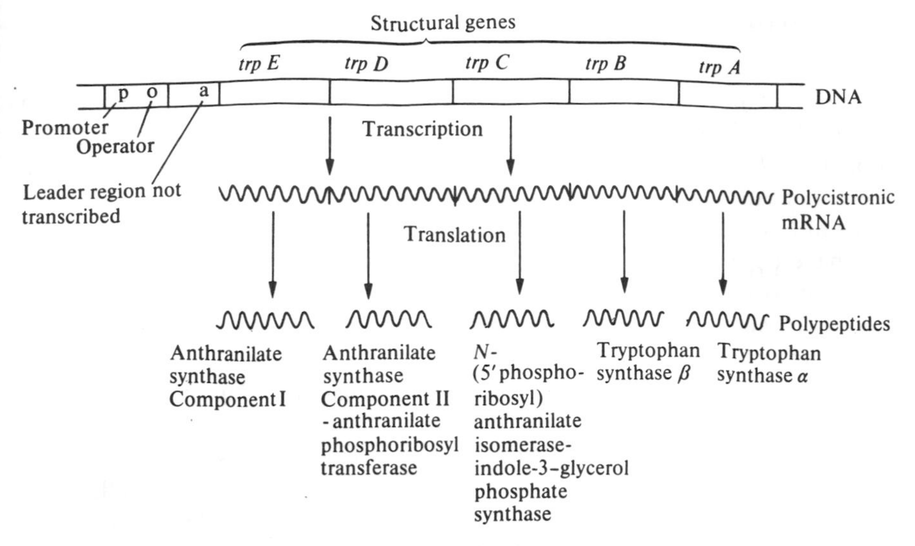

FIG. 7.8. The arrangement of the genes which constitute the tryptophan operon in *E. coli* (see reference 67).

enzymes required for the synthesis of tryptophan from chorismate (Fig. 7.8). It comprises two different polypeptide chains α and β and has the structure $\alpha_2\beta_2$. The separate polypeptide chains catalyse reactions 1 and 2, whereas the complex catalyses reaction 3, but will also catalyse reactions 1 and 2.

Reaction 1

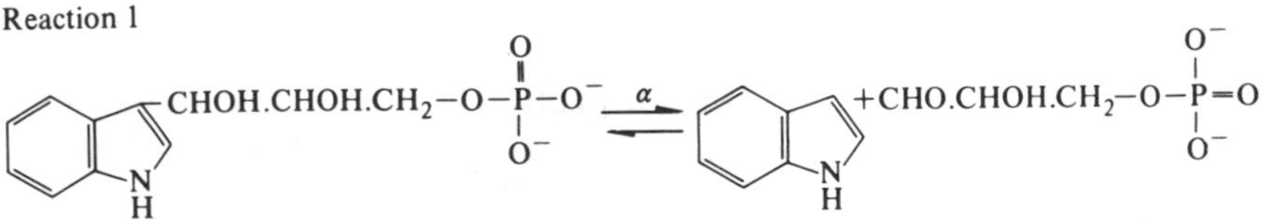

$$\text{Indole-CHOH.CHOH.CH}_2\text{-O-P(=O)(O}^-\text{)-O}^- \underset{}{\overset{\alpha}{\rightleftharpoons}} \text{Indole} + \text{CHO.CHOH.CH}_2\text{-O-P(=O)(O}^-\text{)O}^-$$

Indoleglycerol phosphate — Indole — Glyceraldehyde 3-phosphate

Reaction 2

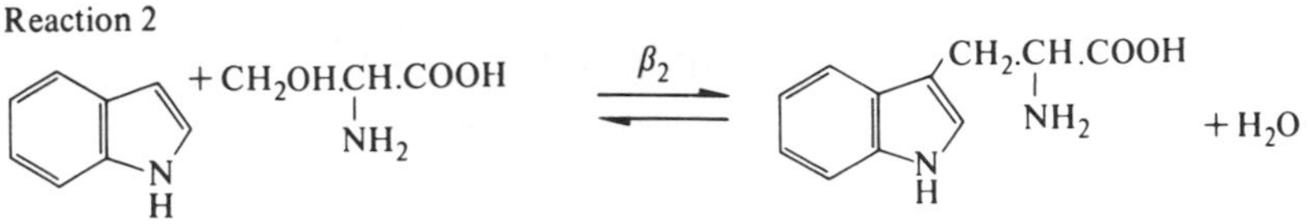

$$\text{Indole} + \text{CH}_2\text{OH.CH(NH}_2\text{).COOH} \overset{\beta_2}{\rightleftharpoons} \text{Indole-CH}_2\text{.CH(NH}_2\text{).COOH} + \text{H}_2\text{O}$$

Indole — L-Serine — L-Tryptophan

Reaction 3

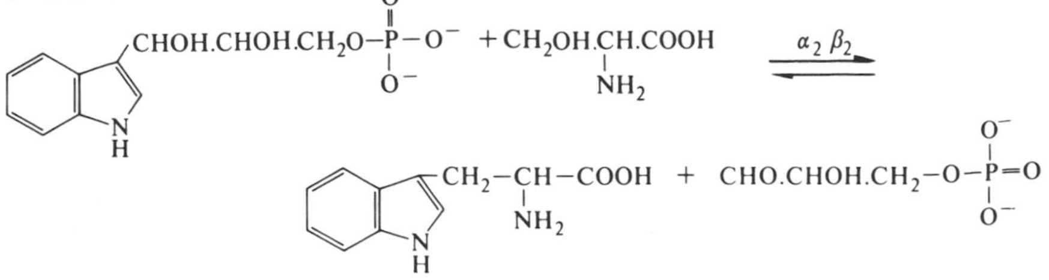

$$\text{Indole-CHOH.CHOH.CH}_2\text{O-P(=O)(O}^-\text{)-O}^- + \text{CH}_2\text{OH.CH(NH}_2\text{).COOH} \overset{\alpha_2\beta_2}{\rightleftharpoons}$$

$$\text{Indole-CH}_2\text{-CH(NH}_2\text{)-COOH} + \text{CHO.CHOH.CH}_2\text{-O-P(=O)(O}^-\text{)O}^-$$

7.9.1 Structure

The enzyme from *E. coli* can readily be dissociated into separate α and β_2 proteins, but more drastic conditions are required to separate β_2 into β (Fig. 7.9); this results in loss of activity. The activity is, however, restored upon reassociation. Each β polypeptide chain binds one molecule of pyridoxal phosphate. The complete sequences of the α subunit isolated from *E. coli*, *Salmonella typhimurium*, and *Enterobacter aerogenes* are known. Preliminary X-ray crystallographic data have been obtained for both subunits. Although the tryptophan synthase system in plants is also a two-component heteropolymer, the enzymes from *Euglena*, *Neurospora crassa*, and *Saccharomyces cerevisiae* are homopolymers (Table 7.3). The enzymes from *Neurospora crassa* and *Saccharomyces cerevisiae* were initially thought to be composed of separate α and β subunits but this was due to proteolysis during isolation and both have been shown to be homopolymers.[68–70] The homopolymeric nature of these enzymes is also supported by the genetic evidence of a single gene locus. This is believed to be an example where, during the course of evolution, gene fusion has occurred. The DNA sequence of the genes in both *E. coli* and *Saccharomyces cerevisiae* have been determined and they show a high degree of similarity.[71]

$$\underset{\substack{M_r \\ 146\,000}}{\alpha_2\beta_2} \underset{}{\overset{\text{Reversible dissociation promoted by removal of pyridoxal phosphate*}}{\rightleftharpoons}} \underset{\substack{M_r \\ 2\times 29\,000}}{2\alpha} + \beta_2 \xrightarrow{\text{Reversibly dissociated by urea or guanidine hydrochloride}} \underset{\substack{M_r \\ 2\times 44\,000}}{2\beta\ (\text{inactive})}$$

*Apparent association constant K_A (-pyridoxal phosphate) = 4×10^6 (mol dm^{-3})$^{-1}$ and K_A (+ pyridoxal phosphate + serine) = 2.6×10^9 (mol dm^{-3})$^{-1}$

FIG. 7.9. Dissociation of trytophan synthase (see reference 73).

TABLE 7.3
Structures of tryptophan synthases from different species

Source	Structure	M_r of complex	M_r of subunits	Comments
E. coli	$\alpha_2\beta_2$	146 000	2 × 29 000 2 × 44 000	Ready dissociation to $2\alpha + \beta_2$
Neurospora crassa	A_2	150 000	2 × 75 000	
Saccharomyces cerevisiae	A_2	154 000–167 000	2 × 80 000	
Euglena	A_2?	325 000	2 × 160 000	Complex contains four catalytic activities for E_9–E_{12} on Fig. 7.20

References 68–70 and 72.

7.9.2 Mechanism

The studies on the mechanism refer to the enzyme system from *E. coli*. The multienzyme complex, in addition to catalysing the complete reaction (reaction 3), will also catalyse the formation of indole (reaction 1) and the conversion of indole to tryptophan (reaction 2). The separate α and β_2 subunits will catalyse reactions 1 and 2, but at only 1/30th and 1/100th the rate catalysed by $\alpha_2\beta_2$. For this higher rate of reaction the α and β proteins have to be in physical contact. (They cannot, for example, be separated by a dialysis membrane.) Detailed kinetic studies, both rapid and steady state, have been made on reactions 1 and 2 catalysed by α and β_2, respectively, and these have been compared with similar studies using $\alpha_2\beta_2$.[74, 75] The general conclusion is that the mechanism is unchanged whether the single enzyme or the multienzyme complex is used, e.g. reaction 1 when studied in the direction of indole glycerol phosphate formation proceeds via an ordered ternary complex mechanism (see Chapter 4, Section 4.3.5.1) in which glyceraldehyde 3-phosphate binds first whether it is catalysed by α or by $\alpha_2\beta_2$. Although the mechanism appears the same, the rate constants are appreciably changed when a subunit is used compared to the multienzyme complex.

These increases in the rates of the partial reactions (reactions 1 and 2) that occur when either α or β_2 are combined may also be brought about by combining one defective protein with a catalytically active one. It is possible to obtain defective α or β_2 proteins either from suitable mutants of *E. coli* or by chemical modification of catalytically active subunits. These defective proteins can then be used to assemble hybrid $\alpha_2\beta_2$ proteins. The results of such experiments are summarized in Table 7.4. It can be seen that a subunit does not have to be catalytically active in order to effect a rate enhancement in the other reaction. It seems that so long as it can combine with the other subunit it will be effective.

TABLE 7.4
The catalytic activity of $\alpha_2\beta_2$ hybrids of tryptophan synthase formed by combining normal and defective subunits

Holoenzyme	Reaction 1	Reaction 2	Reaction 3
$\alpha_2\beta_2^d$	Rate enhanced with respect to α	Inactive	Inactive
$\alpha_2^d\beta_2$	Inactive	Rate enhanced with respect to β_2	Inactive
$\beta\beta^d$	Inactive	Half activity with respect to β_2	Inactive
$\alpha_2\beta\beta^d$	Active	Active	Full activity ($=\alpha_2\beta_2$)

Superscript 'd' indicates the defective subunit, i.e. one having no catalytic activity either isolated from a mutant, or produced by chemical modification.
References 76–78.

In addition to the increased rates of the partial reactions that occur when α and β_2 subunits are combined, another change occurs. Although the physiological role of the β_2 subunits is to catalyse reaction 2 it has been shown also to catalyse the following pyridoxal phosphate dependent reactions:

$$\text{L-serine} \overset{\beta_2}{\rightleftharpoons} \text{pyruvate} + NH_4^+.$$

$$\text{2-Mercaptoethanol} + \text{L-serine} \overset{\beta_2}{\rightleftharpoons} \text{S-hydroxyethyl-L-cysteine} + H_2O.$$

$$\text{2-Mercaptoethanol} + \text{L-serine} + \text{pyridoxal phosphate} \overset{\beta_2}{\rightleftharpoons}$$
$$S\text{-pyruvylmercaptoethanol} + \text{pyridoxamine phosphate} + H_2O.$$

When the subunits are combined to form $\alpha_2\beta_2$ not only is the catalytic efficiency of the partial reactions improved, but also the non-physiological side-reactions of the β_2 subunit are suppressed. Also, if the complete reaction (reaction 3) is carried out in the presence of [^{14}C]indole, it is found that no radioactivity is present in the tryptophan formed. This suggests that, when the complete reaction occurs, either that indole is not an intermediate or that indole formed in the first reaction remains enzyme bound, i.e. the indole is effectively channelled to the second catalytic site without equilibrating with indole in the surrounding medium. The kinctics have been studied in detail[79] and a number of schemes[75] have been suggested to explain how this might occur. At present the precise mechanism is not clear, especially since the three-dimensional structure of the complex, and hence the location of the active sites, has not been determined. A model that is consistent with kinetic and other evidence and that at present seems most favoured is shown in Fig. 7.10.

It is clear that the multienzyme complex must have two distinct active sites, since (i) the α and β_2 subunits are able to catalyse the partial reactions, and (ii) in *Neurospora*, where the enzyme is a single polypeptide chain, there is evidence for two separate sites. It is assumed that in the *E. coli* complex the two active sites must be closely juxtaposed so that the intermediate may be channelled from one catalytic centre to another. The distance of separation of the two catalytic sites has been estimated as 2 nm by measurement of fluorescence energy transfer between pyridoxal phosphate and fluorescent ligands attached to the α and β subunits.[82] The rate of the step in which indole is released in the partial reaction (reaction 1) becomes insignificant in the presence of L-serine. Thus the combination of the two subunits is assumed to induce certain conformational changes and also to bring the catalytic sites into close proximity; this increases the catalytic efficiency of the normal physiological reaction and also eliminates non-physiological side-reactions. For further details of the structure and mechanism, see references 68 and 73, and for the evolution of tryptophan biosynthetic enzymes, see reference 83.

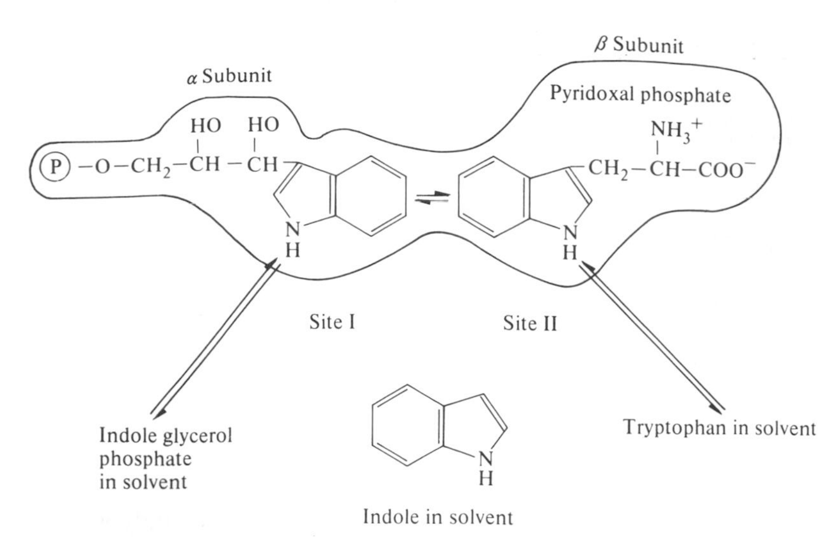

FIG. 7.10. Possible arrangement of the active sites of the α and β subunits of tryptophan synthase. Indole, present in the solvent medium, does not become incorporated into tryptophan when tryptophan synthase from *E. coli* is used to catalyse reaction 3 (see references 73, 80 and 81). Ⓟ = phosphoryl group.

7.10 Carbamoyl phosphate synthase and the associated enzymes of the pyrimidine and arginine biosynthetic pathways in *E. coli*, fungi, and mammalian cells

Carbamoyl phosphate is the metabolic precursor of both pyrimidines and arginine, and carbamoyl phosphate synthase catalyses a step that is at a branch point in the metabolic pathways (Fig. 7.11; and see also Chapter 6, Section 6.3.1). The carbamoyl moiety may become transferred either to aspartate or to ornithine. In certain tissues and organisms, carbamoyl phosphate synthase has been shown to be physically associated with aspartate carbamoyltransferase and in some cases also dihydro-orotase.

The multienzymes involved illustrate two important features, namely, (i) the channelling of a metabolite in a multienzyme protein and (ii) the evolutionary changes in the organization of a multienzyme protein.[84] We compare the organization of the pathways for the biosynthesis of pyrimidines and arginine in *E. coli*, *Saccharomyces cerevisiae*, *Neurospora crassa*, and mammalian systems. The main features of the systems in *E. coli*, *S. cerevisiae*, and *N. crassa* are illustrated in Fig. 7.12.

7.10.1 *E. coli*

It can be seen that in *E. coli* there is a single carbamoyl phosphate synthase that catalyses the formation of carbamoyl phosphate for both arginine and pyrimidine biosynthesis. Mutations in the carbamoyl phosphate synthase gene cause

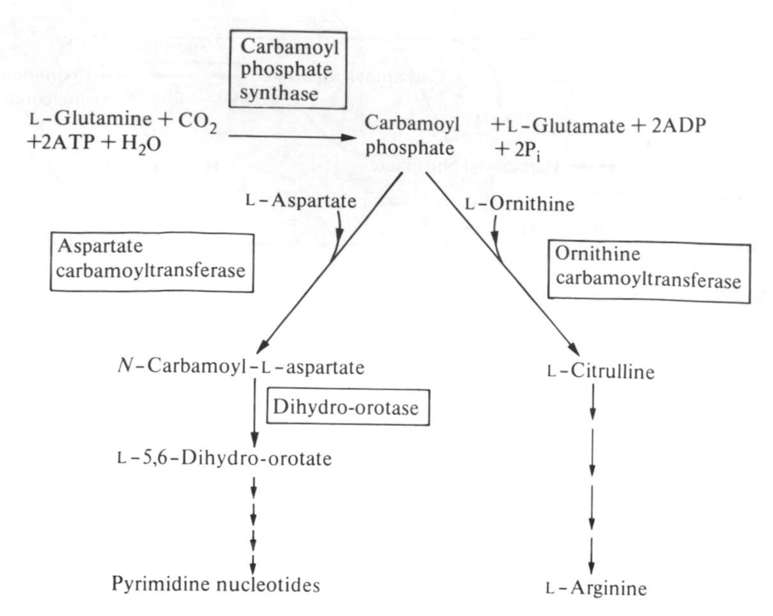

FIG. 7.11. The position of carbamoyl phosphate synthase on the metabolic pathways for arginine and pyrimidine biosynthesis. P_i = orthophosphate.

auxotrophy for both arginine and pyrimidines (i.e. both arginine and pyrimidine have to be supplied in the nutrient medium in order that growth can occur). Regulation of the balance in the supply of carbamoyl phosphate between pyrimidine and arginine biosynthesis is by the activator ornithine and the inhibitor UMP[85] (Fig. 7.12) and is more fully described in Chapter 6, Section 6.3.1. There is no evidence for any physical association of the enzymes in *E. coli.*

7.10.2 *Saccharomyces cerevisiae* (yeast)

In contrast to the situation in *E. coli*, the two fungi (*Saccharomyces*[86] and *Neurospora*[87, 88]) each have two distinct carbamoyl phosphate synthases, although there are differences between the two in the extent to which channelling of carbamoyl phosphate occurs. In *Saccharomyces* there are two distinct carbamoyl phosphate synthases (abbreviated to CPS_{Py} and CPS_{Arg}) and one of these, CPS_{Py}, is physically associated with aspartate carbamoyltransferase, but carbamoyl phosphate produced by either CPS_{Arg} or CPS_{Py} may be utilized in either biosynthetic pathway. The evidence for this is that mutants lacking either CPS_{Arg} or CPS_{Py} will grow on a minimal medium (i.e. without arginine or pyrimidine supplementation). However, mutants lacking CPS_{Py} will not grow on arginine-supplemented media and mutants lacking CPS_{Arg} will not grow on uracil-supplemented media. This is as expected from the feedback-inhibition patterns (Fig. 7.12), e.g. uracil inhibits CPS_{Py} and thus in a mutant lacking CPS_{Arg} no carbamoyl phosphate can be formed and growth is inhibited.

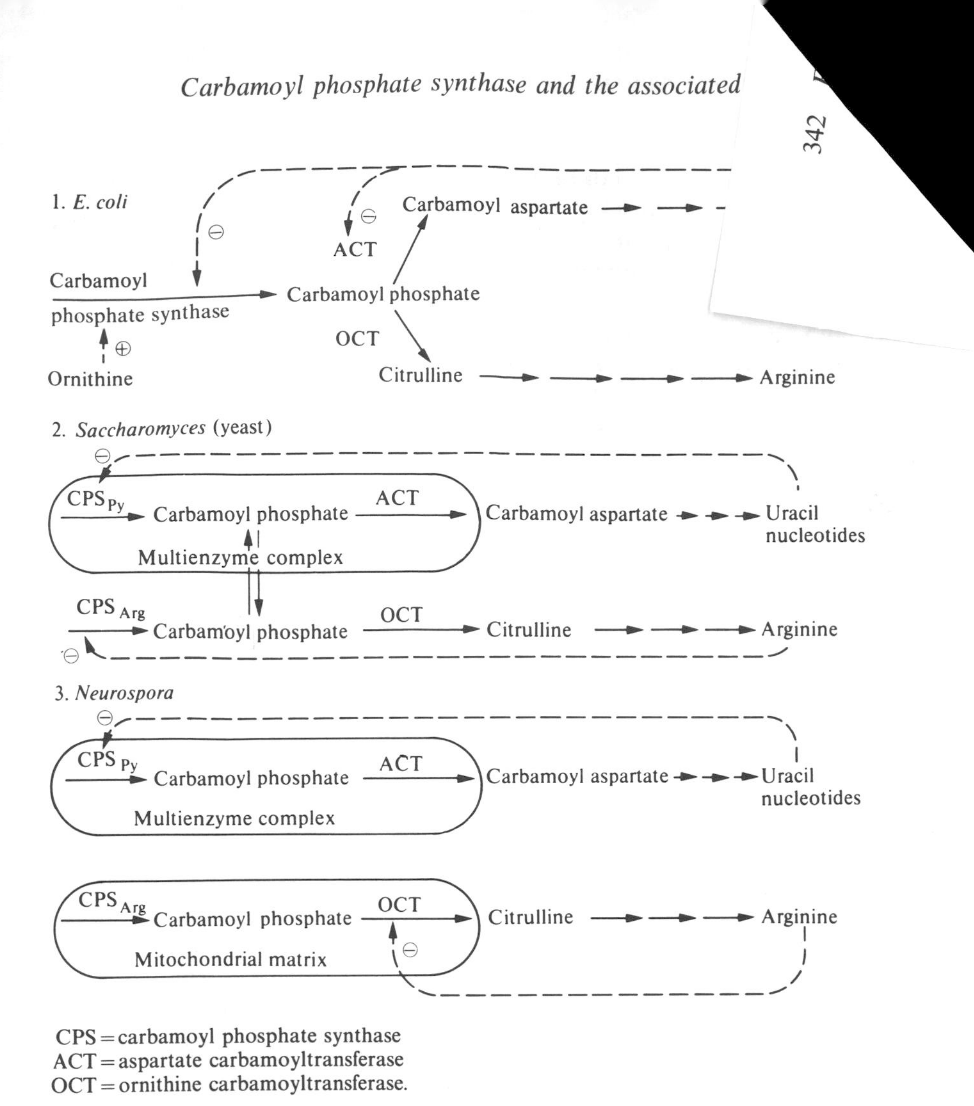

FIG. 7.12. Comparison of arginine and pyrimidine biosynthesis in *E. coli*, *Saccharomyces*, and *Neurospora*. ⊕ = activation, ⊖ = inhibition.

The fact that mutants lacking either CPS_{Arg} or CPS_{Py} will grow on a minimal medium shows that carbamoyl phosphate formed may be exchanged between the pathways. However, the growth rate of these mutants on minimal medium is appreciably slower than that of the wild type, which suggests there is not a completely free exchange of carbamoyl phosphate between the pathways, i.e. some channelling occurs.

7.10.3 *Neurospora*

In *Neurospora*[87] there is efficient channelling of carbamoyl phosphate synthesized by the two enzymes. The channelling is achieved by means of a bifunctional enzyme protein containing CPS_{Py} and aspartate carbamoyltransferase and by different subcellular locations of the enzymes (Fig. 7.12). CPS_{Arg} and ornithine carbamoyltransferase are located in the mitochondria, whereas the CPS_{Py}: aspartate carbamoyltransferase multienzyme protein is present in the cytosol. Consistent with this is the finding that mutants lacking CPS_{Py} or CPS_{Arg} are both auxotrophs (i.e. they require supplementation with uracil or arginine, respectively, for growth to occur). Further evidence is that carbamoyl phosphate formed by CPS_{Py} *in vitro* does not exchange with added radioactively labelled carbamoyl phosphate. CPS_{Arg} in *Neurospora* is not feedback-inhibited by arginine as is the corresponding enzyme in *Saccharomyces*.[89]

7.10.4 Mammalian tissues

In the cytosol of mammalian tissues there is a multienzyme protein having three enzyme activities, namely, carbamoyl phosphate synthase, aspartate carbamoyltransferase, and dihydro-orotase (E_1, E_2, and E_3; Fig. 7.13).

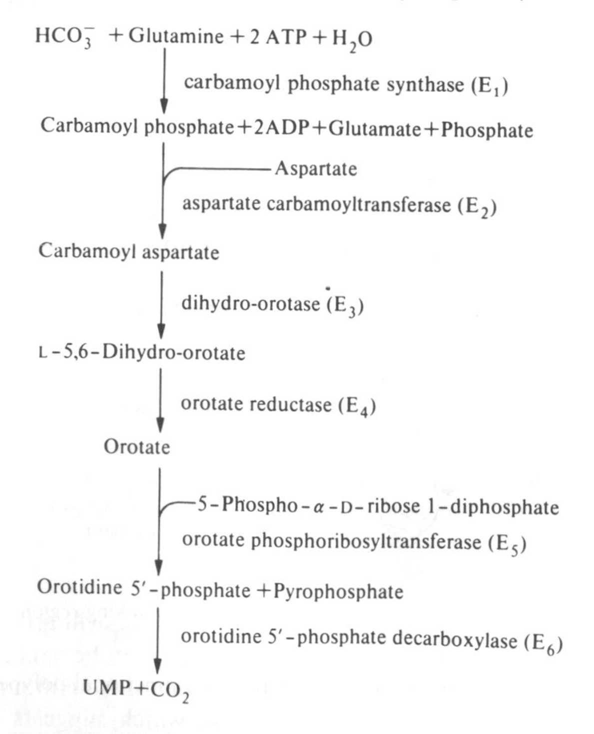

FIG. 7.13. Biosynthetic pathway for uridylate synthesis.

This complex has been studied in *Drosophila melanogaster*, bull frog eggs, rat liver, mouse spleen, and various cultured mammalian cells.[24, 90] It has been purified to homogeneity in mouse ascites hepatoma cells; the native complex has an M_r of about 800 000 and, after denaturation with sodium dodecylsulphate, a single size of polypeptide chain having an M_r of 200 000 is obtained.[91–94]

A method that has been used increasingly to study the separate domains of multienzyme polypeptides is limited proteolysis. In general the domains are compact but are separated by flexible regions and these latter are more sensitive to proteolysis (see also Sections 7.11.2.1 and 7.11.3). The CAD multienzyme polypeptide (carbamoyl phosphate synthase II, aspartate carbamoyltransferase, dihydro-orotase) has been studied in this way.[135] Using elastase and trypsin it is possible to remove the N-terminal domain having dihydro-orotase (DHO) activity, and the C-terminal domain having aspartate carbamoyl transferase (ATC), leaving a centre polypeptide with carbamoyl phosphate synthase II (CPSII) activity. (Carbamoyl phosphate synthase I is the other enzyme present in mitochondria functioning in the biosynthesis of citrulline.) This is shown diagramatically below.

	Dihydro-orotase	↓	Carbamoyl phosphate synthase	↓	Aspartate carbamoyltransferase	
N	$M_r = 43\,000$		$M_r = 150\,000$		$M_r = 40\,000$	C

There are two steps catalysed by CPSII, a glutaminase activity (1), followed by formation of carbamoyl phosphate (2);

$$\text{(1) glutamine} + H_2O \rightleftharpoons \text{glutamate} + NH_3$$

$$\text{(2) } NH_3 + HCO_3^- + 2ATP \rightleftharpoons H_2N\text{–}CO \cdot OPO_3^{2-} + 2ADP + P_1.$$

A model of the general structure of this multienzyme polypeptide is shown in Fig. 7.14, showing the exposed interdomain regions. It is found that when MgATP (substrate) and MgUTP (a negative effector) are bound to the enzyme, the rates of proteolysis by elastase and trypsin are retarded and this is believed to occur because a conformational change decreases the exposure of the interdomain regions.

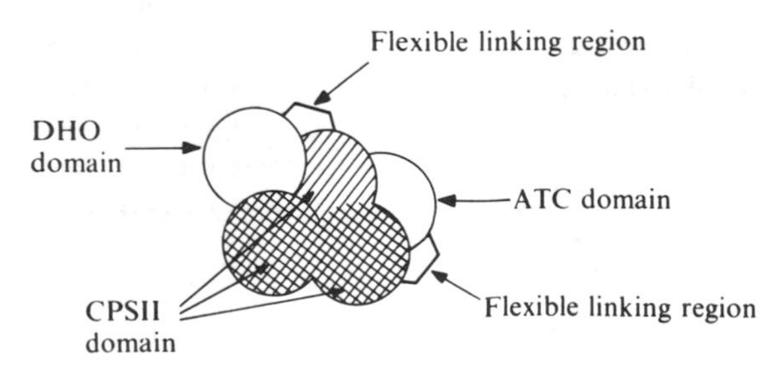

FIG. 7.14. A model of the domain structure of the multifunctional polypeptide CAD.[135]

It is interesting to compare the organization of the structural genes for the biosynthesis of UMP from carbon dioxide and glutamine or ammonia in different species. There are altogether six enzyme-catalysed reactions in the pathway, as shown in Fig. 7.13. In *E. coli* there are six gene loci corresponding to the six structural genes in the pathway. In *Saccharomyces* and *Neurospora* there are only five structural genes, since a single gene codes for the multienzyme polypeptide that contains E_1 and E_2 activities. In mammalian systems there appear to be only three structural genes coding for E_1–E_2–E_3, E_4, and E_5–E_6.[94] There is some evidence that E_5–E_6 in mammals exists as a multienzyme protein. Thus, during the course of evolution there is clustering on the chromosomes of genes that have related functions and in some cases there is gene fusion leading to the production of multienzyme polypeptides.

7.11 Multienzyme polypeptides: aspartate kinase–homoserine dehydrogenase, fatty-acid synthase, and the *arom* complex

In recent years it has become apparent that if enzymes are to be isolated with their complete polypeptide chains intact it is necessary to be particularly careful to avoid proteolysis occurring during isolation (see Chapter 2, Section 2.8.6). Many cells and tissues contain proteases that may be present in a latent form, e.g. within lysosomes, and these are often released during isolation. (For review of the problems associated with proteolysis during enzyme isolation see reference 95). On a simple statistical basis, proteolysis is more likely to cause a single break in a large polypeptide chain than in a small one. Thus a number of multienzyme proteins originally considered as multienzyme complexes have turned out to be multienzyme polypeptides. The examples of tryptophan synthase in *Neurospora* and *Saccharomyces* were mentioned in Section 7.9.1, but the most striking examples are the fatty-acid synthases and the complex of enzymes involved in aromatic amino-acid biosynthesis known as the *arom* complex; these are discussed in this section.

7.11.1 Aspartate kinase–homoserine dehydrogenase

The earliest reported example of a multienzyme polypeptide is aspartate kinase I—homoserine dehydrogenase I from *E. coli.*[96, 97] This multienzyme polypeptide is composed of four identical chains each having an M_r of 86 000. Each polypeptide chain has two active sites catalysing the reactions indicated below.

$$\xleftarrow{\qquad\qquad} M_r\ 86\,000 \xrightarrow{\qquad\qquad}$$

H_2N —— Aspartate kinase —— Homoserine dehydrogenase —— COOH

$$\text{ATP} + \text{L-Aspartate} \rightleftharpoons \text{ADP} + \text{4-Phospho-L-aspartate}$$

$$\text{L-Homoserine} + \text{NAD}^+ \rightleftharpoons \text{L-Aspartate } \beta\text{-semialdehyde} + \text{NADH}$$

These two reactions are the first and third steps in the biosynthesis of leucine, threonine, methionine, and lysine from aspartate.

The multienzyme polypeptide comprises 820 amino-acid residues. It has been shown that there is a region between residues 293 and 300 that is particularly sensitive to protease attack.[96] Two fragments that can be separated, and can be unfolded and refolded independently, contain the aspartate kinase and homoserine dehydrogenase activities, respectively. The C-terminal fragment, in addition to containing homoserine dehydrogenase, also has the ability to self-associate. A model[97] has been proposed (Fig. 7.15) in which there are three compact domains linked by more flexible regions.

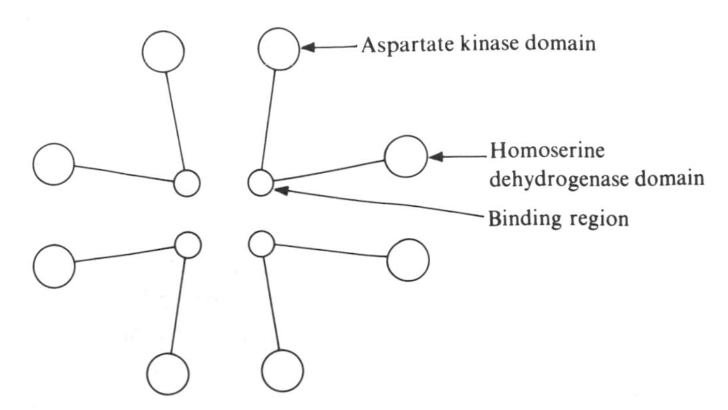

FIG. 7.15. Proposed tetrameric structure of aspartate kinase–homoserine dehydrogenase.[97]

7.11.2 Fatty-acid synthase

The ability to synthesize long-chain fatty acids is an attribute of all organisms and those in which fatty-acid synthesis has been examined in any detail show the same basic metabolic pathway, though with certain minor modifications, as illustrated in Fig. 7.16. Fatty-acids are synthesized from acetyl-CoA and malonyl-CoA and the overall process of extending an acyl chain by two methylene groups entails transacylations, two reductions, a dehydration, and a condensation and thus involves six catalytic activities as shown in Fig. 7.16. An additional enzyme acetyl-CoA carboxylase is necessary to catalyse the formation of malonyl-CoA but this is not part of the complex.

7.11.2.1 Organization of the fatty-acid synthase system

There are differences between organisms both in the predominant chain length of fatty acids they synthesize and in the mechanism of release of the long-chain acyl carboxylate, but the sequence of enzyme reactions is similar in all cases.[98] In spite of this similarity, the organization of the enzyme systems may differ. In *E. coli* six separate enzymes ([ACP] acetyltransferase, [ACP] malonyltransferase, 3-oxo-acyl-[ACP] synthase, 3-oxoacyl-[ACP] reductase, crotonyl [ACP] hydratase, and enoyl-[ACP] reductase), together with the acyl-carrier protein (ACP) have been isolated and purified. These enzymes together catalyse fatty-acid synthesis and there is no evidence that they are physically associated with one another. This non-aggregated type of fatty-acid synthase system (referred to as type II) has also been found in other bacteria, in the blue-green alga *Phormidium lunidum*, in *Euglena*, *Chlamydomonas*, Avocardo mesocarp, and in the chloroplasts from lettuce and spinach.

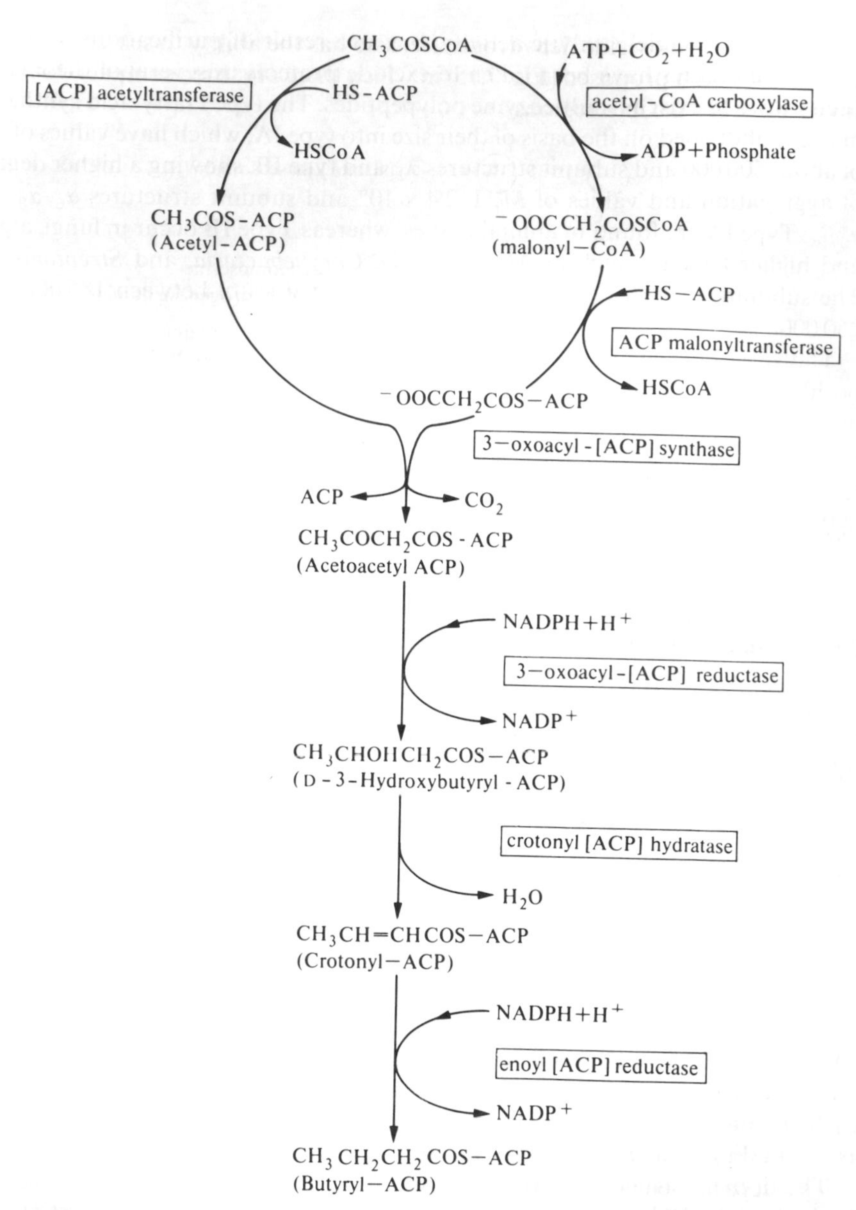

FIG. 7.16. The fatty-acid biosynthetic pathway. All reactions except the carboxylation of acetyl-CoA are catalysed by the fatty-acid synthase multienzyme system.

By contrast, in many animal tissues, avian tissues, yeast, *Neurospora*, and in some prokaryotes, the fatty-acid synthase system is found to exist as an aggregated multienzyme protein (type I). It was originally thought that these multienzyme proteins comprised several tightly associated polypeptide chains each having a single catalytic activity, but as a result of purifications in which rigorous measures have been taken to exclude proteolysis several of these have been shown to exist as multienzyme polypeptides. The type I fatty-acid synthases may be subdivided on the basis of their size into type IA, which have values of M_r of about 500 000 and subunit structures α_2, and type IB, showing a higher degree of aggregation and values of M_r 1–2.4 × 10^6 and subunit structures α_6–α_8, or $\alpha_6\beta_6$. Type IA are found in animal tissues, whereas Type IB occur in fungi, algae and higher bacteria such as *Mycobacteria*, *Corynebacteria*, and *Streptomyces*. The subunits of both types IA and IB have M_r values of between 185 000 and 250 000.

The yeast fatty-acid synthase system has been most thoroughly investigated both from a biochemical and genetic standpoint. By methods of genetic mapping[100] it has been shown that there are two unlinked genetic loci on the chromosome designated *fas 1* and *fas 2*, which code for the whole fatty acid synthase system. The two different subunits and the assignment of the catalytic activities to particular subunits is based on biochemical[101] and genetic[100] evidence. Fatty-acid synthase can be dissociated reversibly after modification of its free amino groups using maleic anhydride* (see Chapter 3, Section 3.4.2) and then the catalytic activities can be assigned to the separated polypeptide chains. The arrangement of catalytic centres on the α and β polypeptide chains is shown in Fig. 7.17, together with a possible model for the three-dimensional arrangement of the α and β subunits. The α chains are arranged in the form of a disc, and the β chains are looped across pairs of α chains above and below the plane of the disc.

In the smaller type IA fatty-acid synthases found in animal tissues, those that have been examined in detail have been found to be homodimers.[103–105] In the yeast fatty-acid synthase, the activities of [ACP]malonyltransferase, which is necessary to initiate fatty-acid chain growth (see Fig. 7.16), and of [ACP]palmityltransferase, which is necessary to terminate chain growth, are in identical regions of the polypeptide chain.[106, 107] On the other hand, in the mammalian multienzymes there are two different catalytic centres, making a total of seven different catalytic centres altogether. The fatty-acid synthases from chicken liver,[103] pigeon liver,[104] and rabbit mammary gland[105] each contain all seven catalytic activities within a single polypeptide chain and the multienzyme appears to be a homodimer arranged in head-to-tail fashion. Consistent with this is the finding of a single mRNA for all seven catalytic activities.[108]

The domain structure of the animal fatty-acid synthase has been studied by limited proteolysis using a variety of proteolytic enzymes. The presence of domains can readily be discerned by electron microscopy. By limited proteolysis the α chain is split into three domains of M_r 127 000, 107 000, and 33 000. The larger two can be further split. By locating the catalytic steps on particular domain fragments, it is possible to propose a model for the organization of the α_2 structure (Fig. 7.18). It can be seen that the growing chain is attached to the thiol

* Modification of free amino groups facilitates dissociation of the complex.[102]

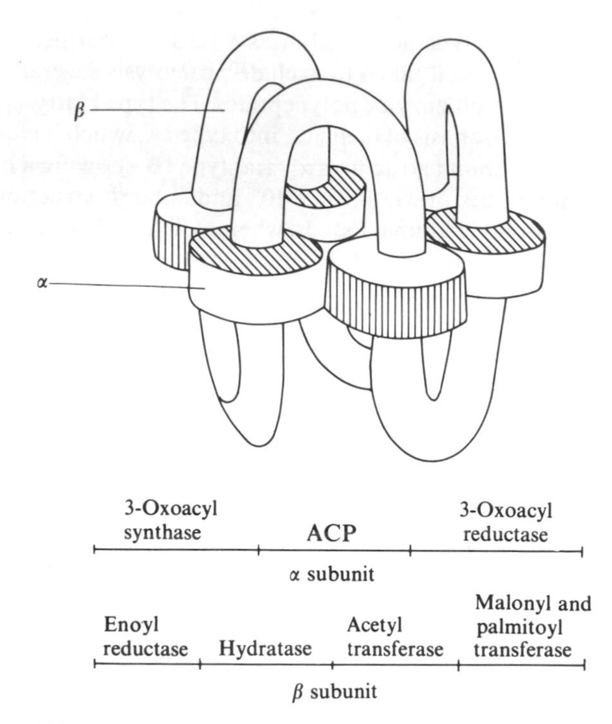

FIG. 7.17. A possible arrangement of the (α) and [β] subunits of yeast fatty-acid synthase ($\alpha_6\beta_6$), together with the location of enzyme activities on the polypeptide chains.

group on the end of the pantotheine moiety; it has to interact with the thiol on the synthase on the opposite α chain. The evidence for this stems from the observation that when the dimers are dissociated, $\alpha_2 \rightarrow 2\alpha$, the only activity lost is 3-oxoacyl-[ACP]synthase.[109]

A common feature of both type I and type II synthases is that the growing acyl chain is covalently attached to a protein through a phosphopantotheine

$$\text{Protein (Ser)OPO}_3^-\text{CH}_2.\text{C(CH}_3)_2.\text{CHOH.CONH(CH}_2)_2\text{CONH(CH}_2)_2\text{S}\underbrace{\text{COR}}_{\text{acyl group}}$$

prosthetic group. The acyl carrier protein has been isolated from several prokaryotes and sequenced. Its size varies between species, ranging from M_r values of 8600 to 16 000.[110] In eukaryotes it is clear that the equivalent of an acyl-carrier protein exists, since phosphopantotheine becomes covalently bound to the fatty-acid synthase and thus the 'acyl-carrier protein' is a section of the multienzyme polypeptide.

7.11.2.2 Control of the chain length of synthesized fatty-acids

The process of fatty-acid synthesis is often described as a spiral because a sequence of reactions occurs that leads to the addition of —CH_2CH_2— to a

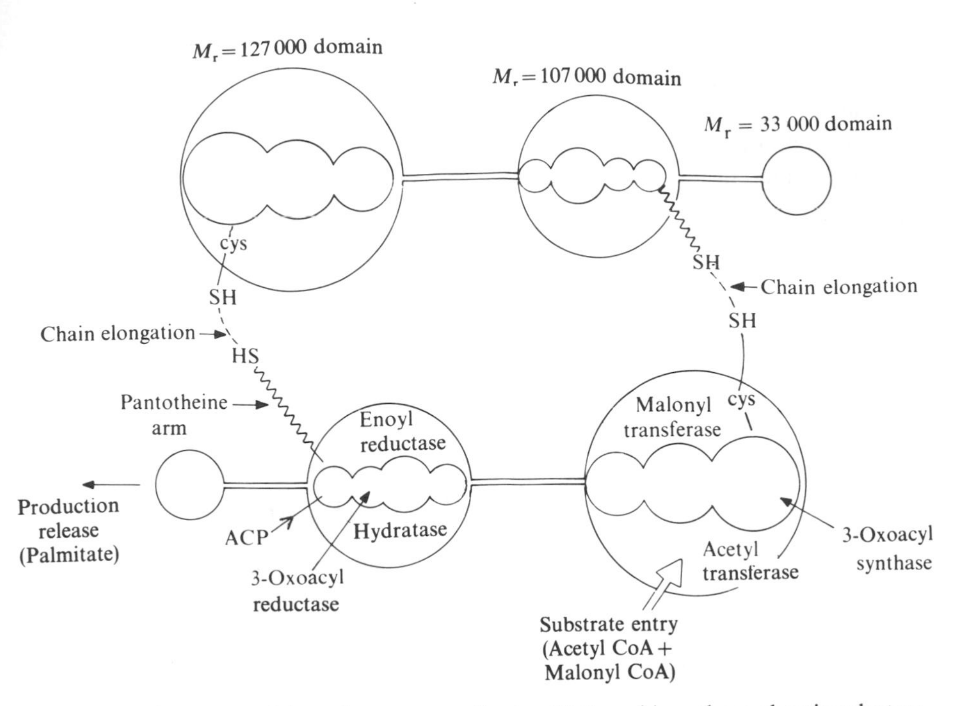

FIG. 7.18. The proposed domain structure of animal fatty-acid synthase, showing the two α subunits aligned head-to-tail.

growing acyl chain and then the sequence is repeated. During this growth the acyl group is attached to either the acyl-carrier protein or part of the multi-enzyme polypeptide. What determines the length of chain to be synthesized and how is it released from the enzyme? There is a fairly well-defined pattern for both the chain length and the degree of saturation of fatty acids synthesized by different species. In *E. coli* the predominant fatty acids synthesized are palmitic

Palmitic acid	$CH_3(CH_2)_{14}COOH$
Palmitoleic acid	$CH_3(CH_2)_5CH{=}CH(CH_2)_7COOH$
cis-Vaccenic acid	$CH_3(CH_2)_5CH{=}CH(CH_2)_9COOH$

acid, palmitoleic acid, and *cis*-vaccenic acid and the ACP derivatives of these acids transfer their acyl groups directly to glycerol-1-phosphate during phospholipid biosynthesis. The type of fatty acid synthesized in *E. coli* seems to be determined largely by the specificity of 3-oxoacyl[ACP] synthase, which catalyses the following two partial reactions:

$$\text{R CO-S-ACP} + \text{HS-enzyme} \rightleftharpoons \text{R-CO-S-enzyme} + \text{ACP—SH}$$

$$\text{R CO-S-enzyme} + \text{malonyl-S-ACP} \rightleftharpoons \text{R-CO-CH}_2\text{CO-S-ACP} + \text{HS-enzyme} + \text{CO}_2$$

where R = saturated or unsaturated alkyl group

The acyl group undergoing elongation is transferred to a cysteine side-chain on 3-oxoacyl[ACP] synthase.

When the maximum velocities of this synthase are compared using acyl-ACPs having different chain length of saturated hydrocarbon as substrates, there is very little difference comparing C_2 to C_{10} acyl ACPs, but the velocity decreases markedly for C_{14} acyl ACP and the enzyme is inactive towards C_{16} acyl ACP. When a similar comparison is made using substrates having unsaturated chains, the activity is low when hexadecenyl-ACP is used and inactive with *cis*-vaccenyl-ACP. Thus, the specificity of the synthase towards different acyl-ACPs sets an upper limit to the number of C_2 units that may be added to the growing chain.[111]

In yeast, the chain lengths of the fatty acids synthesized are predominantly C_{16} and C_{18} and control is probably similar to that in *E. coli*, namely, the specificity of 3-oxoacyl[ACP] synthase. The final step in the sequence is the transfer of palmityl-ACP or stearyl-ACP to coenzyme A catalysed by an acyltransferase.

$$\text{acyl-SACP} + \text{CoA} \rightleftharpoons \text{acyl-S-CoA} + \text{ACP}.$$

In yeast, palmityltransferase and malonyltransferase activities are associated with the same catalytic site on the multienzyme polypeptide, whereas in the mammalian fatty-acid synthases the two catalytic sites are distinct. A further difference is that in the mammalian system the sequence terminates with the release of free fatty acid, whereas, in yeast, acyl-CoA is released. This will greatly affect the rate of release of fatty-acid chains, since free fatty acids are readily released from the multienzyme complex, whereas acyl-CoAs, particularly of long-chain acids, are only slowly released. (For a review, see reference 112).

The sequence in yeast and in mammalian and avian systems[113] is as follows.

(i) *Yeast system*

1. E_1-Pantotheine-S-acyl + E_2-OH $\rightleftharpoons$ E_1-Pantotheine-SH + E_2-O-acyl.
2. E_2-O-acyl + CoASH $\rightleftharpoons$ Acyl-S-CoA + E_2-OH.

(ii) *Mammalian and avian systems*

1. E_1-Pantotheine-S-acyl + E_2-OH $\rightleftharpoons$ E_1-Pantotheine-SH + E_2-O-acyl
2. E_2-O-acyl + H_2O $\rightleftharpoons$ E_2-OH + acyl-OH

where E_1 and E_2 are different catalytic centres on the multienzyme polypeptides. Phosphopantotheine is covalently attached to E_1. The acyl group is transferred from the sulphydryl on pantotheine to a serine hydroxyl group on E_2.

7.11.2.3 Control of fatty-acid synthesis in Mycobacterium smegmatis

The control of fatty-acid synthesis in *Mycobacterium smegmatis* differs from that of *E. coli*, yeast, or mammalian systems and will be described here briefly. (For a detailed review see reference 114.) Fatty acids synthesized by *E. coli* are used primarily for incorporation into the phospholipids of the cell membrane. In mammals, they are, in addition, incorporated into triglycerides to be stored as an energy reserve. *Mycobacterium smegmatis*, which is a highly evolved prokaryote, synthesizes fatty acids both for incorporation into the cell membrane and into the cell wall. C_{16} and C_{18} fatty acids are used for membrane phospholipids, whereas C_{24} and C_{26} fatty acids are incorporated into the cell walls. Both C_{16}–C_{18} and C_{24}–C_{26} fatty acids are synthesized using a type I fatty acid synthase system. The fatty acids are synthesized whilst attached to the multienzyme protein via a phosphopantotheine residue and they are eventually transferred to CoA as in the yeast system (Fig. 7.19). Release of C_{16} and C_{18} acyl-CoAs from the enzyme is fairly slow (compare with yeast fatty-acid synthase, Section 7.11.2.2), but release of the C_{24} and C_{26} acyl-CoAs is very slow indeed, diffusion taking of the order of minutes.

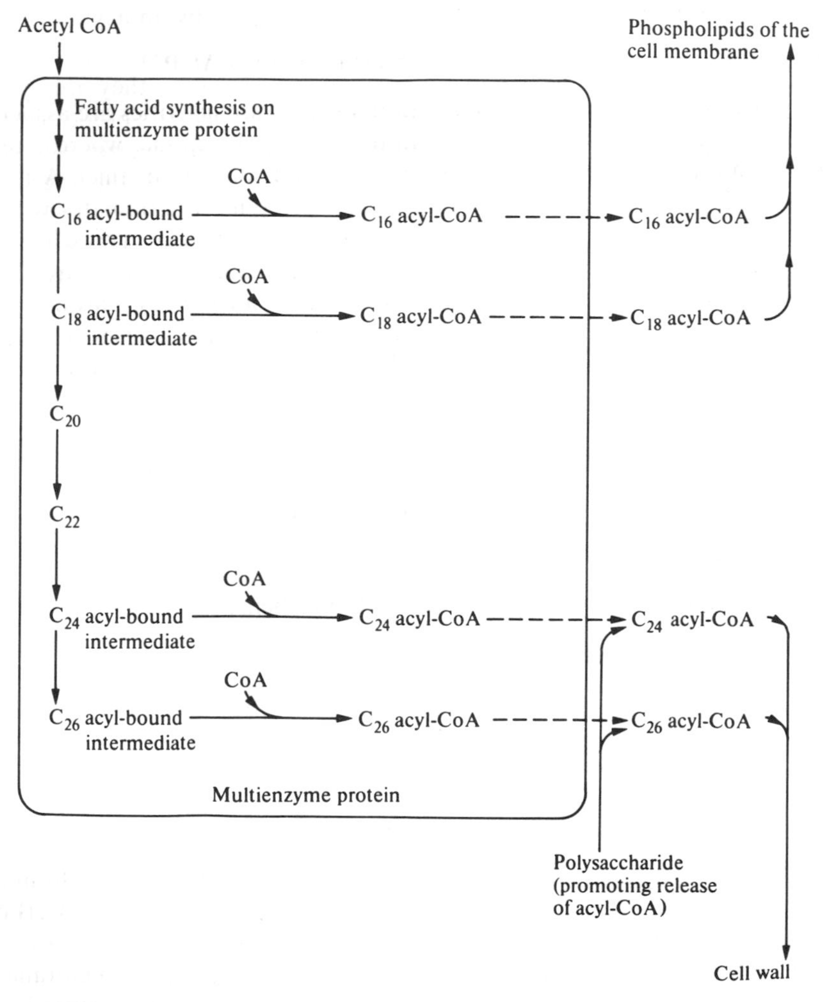

FIG. 7.19. Synthesis of fatty acids in *Mycobacterium smegmatis*.

Two factors regulate the balance between synthesis of C_{16} and C_{18} fatty acids and of C_{24} and C_{26} fatty acids: (i) high concentrations of acetyl-CoA stimulate the formation of C_{16}–C_{18} fatty acids; (ii) two classes of polysaccharides that contain decasaccharide segments of 6-*O*-methylglucose and 3-*O*-methylmannose residues promote the release of C_{24}–C_{26} acyl-CoAs from the fatty-acid synthase. Thus, in the presence of the polysaccharides, C_{24}–C_{26} fatty-acid synthesis is favoured, but when the polysaccharide concentration is low, slow release of C_{24}–C_{26} acyl-CoAs from the enzyme retards fatty-acid synthesis. This causes a build up of acetyl-CoA that in turn stimulates the synthesis of C_{16}–C_{18} acyl-CoAs.

7.11.3 Multienzyme proteins involved in the biosynthesis of aromatic amino acids

The aromatic amino acids tryptophan, tyrosine, and phenylalanine are synthesized *de novo* in most bacteria, fungi, and plants, whereas they are dietary requirements in many mammals. The biochemistry and genetics of the enzymes involved in this pathway have been studied in detail, particularly in certain prokaryotes and in fungi. The same pathway operates in all the organisms studied (as illustrated for *Neurospora crassa* in Fig. 7.20), but there is considerable variation between species in the organization of the genes, and hence for the enzymes for which they code.[115] When considering the conversion of chorismate to tryptophan, which is catalysed by enzymes E_8–E_{12} (Fig. 7.20), it is possible to identify seven different catalytic domains that can be completely separated in certain organisms. These are designated by the following letter code.

Domain E Catalyses the anthranilate synthase reaction with ammonia as amino group donor.
Domain G interacts with the E domain and provides the glutamine amidotransferase activity for the glutamine-dependent anthranilate synthase reaction.
Domain D is the catalytic site of E_9.
Domain F is the catalytic site of E_{10}.
Domain C is the catalytic site of E_{11}.
Domain A catalyses the conversion of indole glycerol phosphate to indole.
Domain B catalyses the conversion of indole to tryptophan.

The way in which these domains are linked shows considerable variation when different species are considered. Table 7.5 shows the linkage of polypeptide chains and subunit organisation for the best studied cases. Other fungi that have been studied, such as *Aspergillus nidulans*, *Coprinus lagopus*, and *Schizosaccharomyces pombe*, show the same pattern as *Neurospora crassa*.

In *Neurospora crassa* the biosynthetic pathway from erythrose 4-phosphate to tryptophan contains three multienzyme complexes (Fig. 7.20). There are also

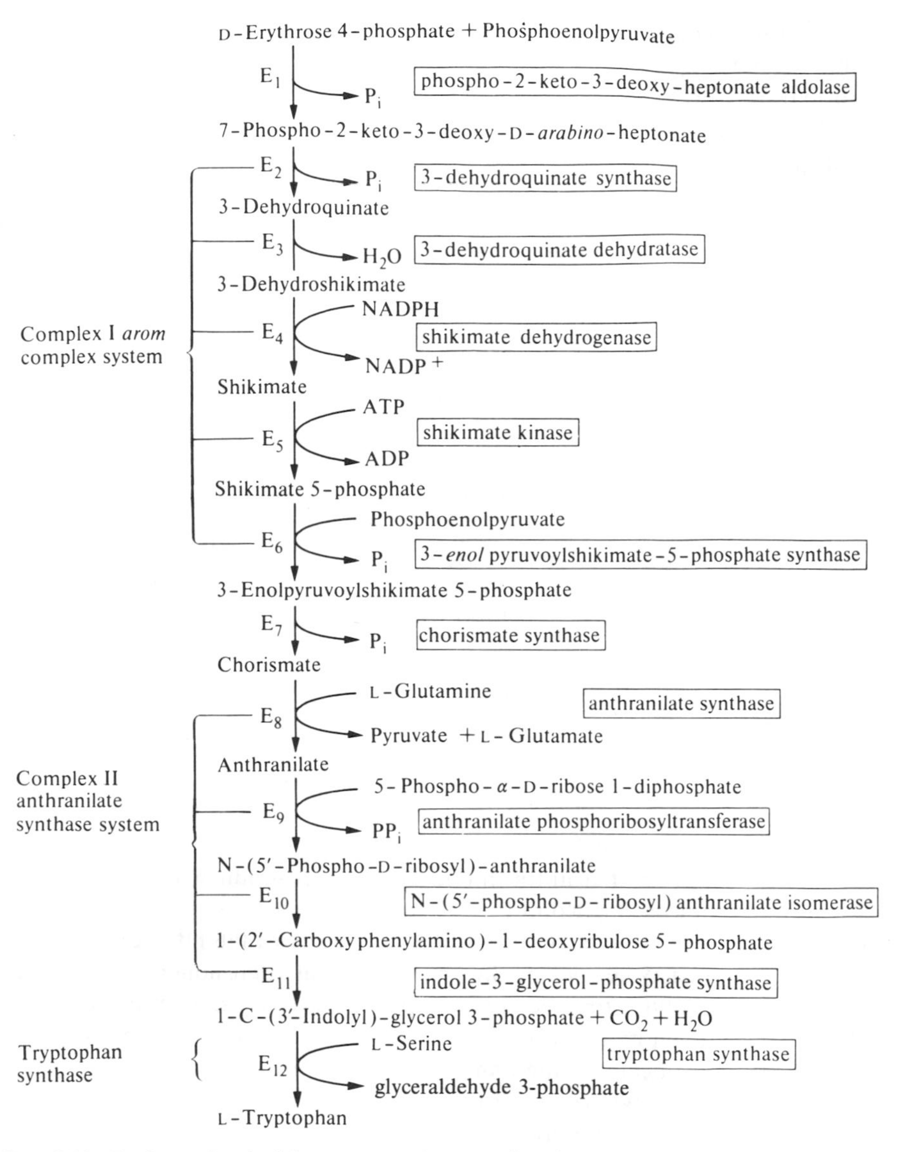

FIG. 7.20. Pathway for the biosynthesis of tryptophan in *Neurospora crassa*, showing the multienzyme proteins involved.

P_i = orthophosphate; PP_i = pyrophosphate

indications that weaker interactions may hold all three multienzyme proteins together with the remaining enzymes of the pathway in a 'super-multienzyme' protein.[116] In this section we discuss the properties of the *arom* complex, since certain features have been demonstrated with this multienzyme protein which have not been demonstrated in other multienzyme proteins.

TABLE 7.5
Organization of the domains for the conversion of chorismate to tryptophan

Organism			Domains			
Prokaryotes						
Escherichia coli	E–G–D		F–C		A, B	$(\alpha_2\beta_2)$
Pseudomonas putida	E, G $(\alpha\beta)$	D	F	C	A, B	$(\alpha_2\beta_2)$
Serratia marcescens	E, G $(\alpha\beta)$	D	F–C		A, B	$(\alpha_2\beta_2)$
Eukaryotes						
Neurospora crassa	E, G–F–C	$(\alpha_2\beta_2)$	D		A–B	(α_2)
Saccharomyces cerevisiae	E–G–C		D	F	A–B	(α_2)

'–'signifies a covalent link, ',' signifies a non-covalent link.

There are parallels between the history of the isolation and purification of the *arom** multienzyme protein and that of fatty-acid synthase type I. In 1972 the *arom* complex isolated from *Neurospora crassa*, having an M_r of 293 000, was shown to be separable into four distinct subunits having M_r of 54 000, 63 000, 84 000, and 95 000.[117] However, an isolation (Chapter 2, Section 2.8.6) in which rigorous attempts were made to inhibit proteolysis shows that the multienzyme protein comprises two polypeptide chains each having an M_r of 165 000.[118] From a study of the peptides obtained from hydrolysing the multienzyme protein, it is clear that the two chains of M_r 165 000 are identical or nearly identical[119] and each possesses the five catalytic activities (Fig. 7.20). The genes for all five catalytic activities are located in a cluster and may be transcribed as a polycistronic message, since the catalytic activities are transcribed and translated in an ordered sequence (compare with the tryptophan operon from *E. coli*, Fig. 7.8). It is probable that in the earlier isolation procedure some proteolysis occurred and that this resulted in rupturing the multienzyme polypeptide in the regions between the catalytic domains. It is noteworthy that in the early days of isolation of multienzymes, when the importance of exclusion of endogenous proteases was not fully realized, limited proteolysis occurred, preventing the isolation of the intact multienzyme polypeptide. Now that several multienzyme polypeptides have been isolated intact, limited proteolysis under carefully controlled conditions with exogenous proteases is being used as a method of exploring the domain structure.[120] In the case of the *arom* complex from *N. crassa*, limited proteolysis has been used to separate two fragments, one containing E_3/E_4 activity (Fig. 7.20) and the other E_6 activity. Different organizations occur in other species; in *E. coli* the catalytic activities appear to be on separate polypeptide chains, whereas, in plants, enzyme E_3 and E_4 copurify while E_2, E_5, and E_6 are separable (see reference 121).

A number of kinetic studies have been made to elucidate the working of the *arom* complex. It is thought that this complex may bring about a channelling of

* The *arom* complex was so designated because the catalytic domains are coded for by a contiguous region of the chromosome named by geneticists as the *arom* cluster.

the intermediates so as effectively to separate the biosynthetic from the degradative pathways for aromatic amino acids. There is one enzyme activity—

3-Dehydroquinate ⇌ (3-dehydroquinate dehydratase) 3-Dehydroshikimate + H_2O

3-dehydroquinate dehydratase, which is common to both biosynthetic and degradative pathways, and thus without a multienzyme polypeptide—which channels bound 3-dehydroshikimate on to the next catalytic domain there would be direct competition between the two pathways for these intermediates.

Calculations have been made to determine the expected transient time (see Section 7.6) for the reaction sequence from 7-phospho-2-keto-3-deoxy-D-*arabino*-heptonate to 3-enolpyruvoylshikimate 5-phosphate (Fig. 7.20) on the assumption that each step was catalysed by a separate enzyme and that the product of each step was released from each catalytic centre into the medium and was subsequently taken up by the next enzyme. When this calculated transient time is compared with the experimentally observed transient time for this multienzyme polypeptide, the latter is found to be approximately 1/10 to 1/15 of the former[122] (Fig. 7.21). Thus, when catalytic centres are in close proximity the metabolic rate can change much more rapidly to a new steady-state rate, and this

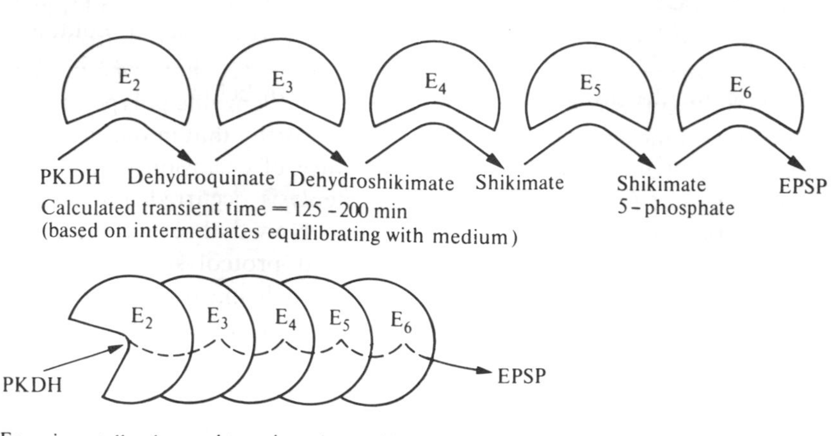

PKDH = 7-Phospho-2-keto-3-deoxy-D-*arabino*-heptonate
EPSP = 3-Enolpyruvoylshikimate 5-phosphate

FIG. 7.21. A comparison of the calculated and experimentally observed transient times for the enzyme-catalysed conversion of 7-phospho-2-keto-3-deoxy-D-*arabino*-heptonate to 3-enolpyruvoylshikimate 5-phosphate (see reference 122).

is probably due to the fact that the intermediates do not appreciably diffuse out into the medium but move directly from one catalytic centre to another.

Another related property of this multienzyme polypeptide is that of coordinate activation. It is found that the *arom* multienzyme polypeptide can convert 7-phospho-2-keto-3-deoxy-D-*arabino*-heptonate to 3-enolpyruvoylshikimate 5-phosphate (this requires E_2, E_3, E_4, E_5, and E_6; see Fig. 7.20) at approximately ten times the rate at which it can convert shikimate to 3-enolpyruvoylshikimate 5-phosphate (which requires E_5 and E_6 only). This is believed to be due to a process of coordinate activation of catalytic centres of the multienzyme polypeptide, with 7-phospho-2-keto-3-deoxy-D-*arabino*-heptonate possibly bringing about a conformational change in the whole multienzyme protein. There is evidence that 7-phospho-2-keto-3-deoxy-D-*arabino*-heptonate causes a reduction in K_m values of E_3, E_4, and E_6 for their respective substrates.[137] Also it is found that when 7-phospho-2-keto-3-deoxy-D-*arabino*-heptonate is bound to the *arom* complex it gives all five catalytic centres some protection against attack by proteases, suggesting possible conformational changes.

The organization of the anthranilate synthase complex (domains E, G–F–C) has been studied in *Neurospora crassa* using the technique of limited proteolysis to identify domains.[123] It comprises four polypeptide chains $\alpha_2\beta_2$. The α chain contains the catalytic site for anthranilate synthase capable of using ammonia as amino donor (domain E) and the β chain the catalytic site for phosphoribose anthranilate isomerase (domain F) and for indole glycerol phosphate synthase (domain C). When the multienzyme complex ($\alpha_2\beta_2$) is incubated with elastase, it cleaves the β chain between residues 237 and 238, splitting the multienzyme into two fragments. One of these fragments (M_r 98 000), comprising the complete α chain (M_r 70 000) and a fragment (M_r 29 000) of the β chain, contains the anthranilate synthase activity together with its glutamine-binding site (domain G). A model for the probable arrangement is shown below[123] (Fig. 7.22).

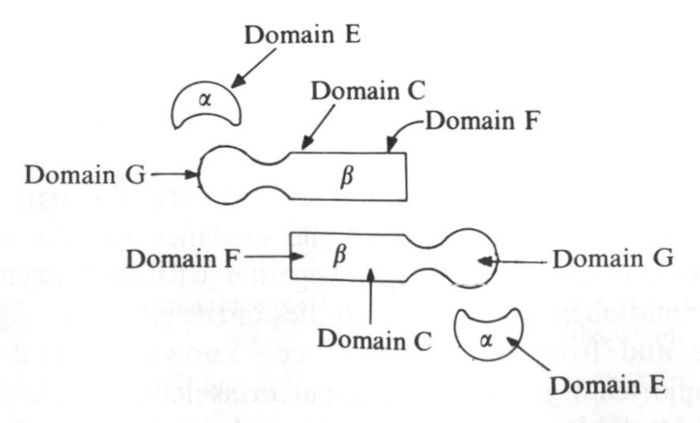

FIG. 7.22. A model for the arrangement of the anthranilite synthase complex ($\alpha_2\beta_2$) from *N. crassa*.

In *E. coli*, the phosphoribosyl anthranilate isomerase (domain F) and indole glycerol phosphate synthase (domain C) exist as a bifunctional monomeric enzyme. X-ray crystallographic studies have been carried out on this enzyme using an iodinated analogue of 1-(2′-carboxyphenylamino)-1-deoxyribulose 5-phosphate, the substrate of E_{11} (domain C) and the product of E_{10} (domain F).[124] This provided not only the three-dimensional structure of the enzyme but also a clear indication of the positions of the two catalytic sites. It is particularly interesting that the two catalytic sites are located back-to-back (Fig. 7.23) and this would appear to preclude the possibility of channelling of the reactants between the two catalytic sites. This is in contrast to another enzyme having two catalytic sites, namely, the DNA nucleotidyltransferase fragment known as the Klenow fragment[133] (see. Chapter 1, Section 1.3.2). This fragment contains both DNA nucleotidyltransferase activity and also 3′–5′ exonuclease activity. The latter serves an editing function (see Chapter 1, Section 1.3.2) in excising mismatched deoxyribonucleotides. In this case the two catalytic sites are juxtaposed so that the nascent chain from DNA nucleotidyltransferase activity passes directly through the second catalytic site for editing.[125]

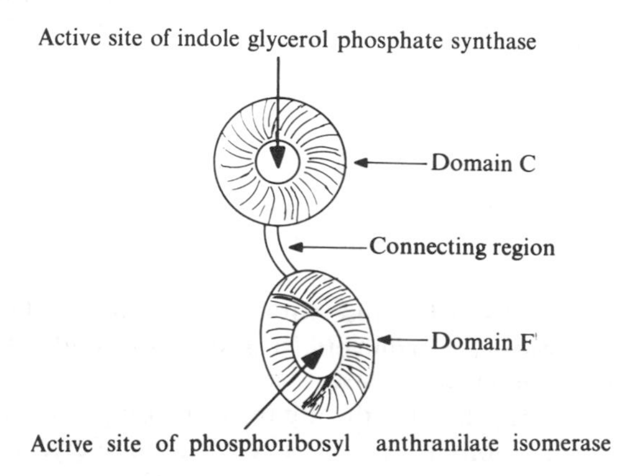

FIG. 7.23. Diagram showing the two catalytic sites in the bifunctional enzyme phosphoribosyl anthranilite isomerase and indole glycerol phosphate synthase from *E. coli*.

7.12 The glycogen particle

The multienzyme proteins or aggregates described in the last part of this chapter are those that are least well defined. The evidence for the existence of a subcellular particle containing glycogen together with the enzymes concerned with glycogen metabolism comes from studies of the isolation of glycogen from skeletal muscle and from liver. It has been known for many years from histological studies that glycogen is present in skeletal muscle in the form of particles (Fig. 7.24). If glycogen is isolated from skeletal muscle homogenates by procedures that do not cause protein denaturation, e.g. precipitation from the

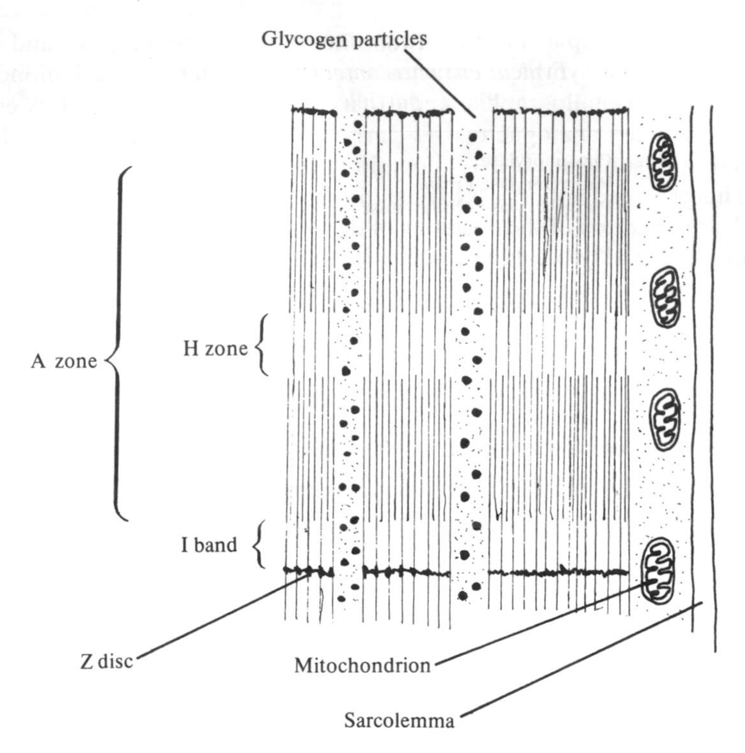

FIG. 7.24. The structure of a muscle fibre.

homogenate at pH 6.1 followed by high-speed centrifugation, then the particles containing glycogen also contain most of the glycolytic enzymes and they are of a similar size to those observed by histological studies. The evidence that these particles are not artefacts generated during the isolation procedure is supported by the following observations. (i) A high concentration of glycolytic enzymes is present after washing the particles. (ii) When three different methods are used to isolate the particles their composition is similar in each case.

The isolated glycogen particles have a reasonable well-defined composition, although not approaching the stoichiometric composition of the other multi-enzyme complexes. The enzymes present in glycogen particles are shown in Table 7.6. When the enzymes present in the particles are separated by poly-acrylamide gel electrophoresis it is found that over 95 per cent of the total enzyme protein is accounted for by phosphorylase, glycogen synthase, debranching enzyme (see Chapter 6, Fig. 6.25), and phosphorylase kinase. Although phosphorylase phosphatase is tightly associated with the complex, it represents only 0.2 per cent of the protein of the complex.

Isolated washed glycogen particles are capable of catalysing the conversion of glycogen to lactate at a reasonable rate, e.g. 1.3 μmol per min per 100 mg protein compared to 40 μmol lactate per min per 100 mg protein in skeletal muscle *in*

TABLE 7.6
Typical enzyme content of glycogen particles

Phosphorylase
Phosphorylase kinase
Phosphorylase phosphatase
Glycogen synthase
Protein kinase
Adenosinetriphosphatase
AMP deaminase
Adenylate kinase
Creatine kinase
Phosphoglucomutase

Reference 126.

vivo. The glycogen particles contain many of the enzymes that control glycogen synthase and phosphorylase. They also contain AMP deaminase, which may control AMP levels. Much of the interest in the glycogen particle has been in the

$$\text{AMP} + \text{H}_2\text{O} \xrightleftharpoons{\text{AMP deaminase}} \text{IMP} + \text{NH}_3$$

study of control of glycogen metabolism. In a number of situations the responses to ligands observed with the glycogen particle are similar to those observed in intact tissue. For example, the concentrations of Ca^{2+} and ATP required for phosphorylase kinase to phosphorylate phosphorylase, and the concentration of AMP required to inhibit purified phosphorylase phosphatase, in the glycogen particle are similar to those observed *in vivo*. Glycogen particles are polydisperse, varying in M_r from 6×10^6 to 1.6×10^9. Turnover of glucose moieties occurs more rapidly on the outside of the particles than on the inside. There is a different enzyme distribution in the particles depending on their size, e.g. phosphorylase activity is highest in small particles, whereas glycogen synthase activity is highest in the large particles.[127] Further study of the particle should help in understanding regulation of glycogen metabolism. The control of glycogen metabolism is discussed further in Chapter 6, Section 6.4.2 and also in references 126 and 128.

7.13 DNA synthesis

There is also evidence that the enzymes required for DNA synthesis are organized in a complex that has been referred to as a replication apparatus.[129] Six of the enzymes required for catalysing the conversion of ribonucleoside diphosphates into DNA (DNA nucleotidyltransférase, thymidine kinase, cytidylate kinase, nucleosidediphosphate kinase, thymidylate synthase, and tetrahydrofolate dehydrogenase) have been found to cosediment during sucrose density-gradient centrifugation. The complex or replication apparatus can be extracted from the nuclei of S-phase* fibroblasts but not from G-phase fibroblasts, and a similar complex has been detected in yeast.[130] There is also evidence that the several of the aminoacyl t-RNA synthetases from eukaryotes exist as a multienzyme complex.[131]

7.14 Conclusions

In this chapter we have described a number of different multienzyme proteins, less well-defined aggregates of enzymes, and an enzyme in which the different polypeptide chains operate in a concerted manner to catalyse a complex enzyme reaction (RNA nucleotidyltransferase). Multienzyme proteins are generally regarded as providing a more efficient method of catalysing a sequence of reactions than separate enzymes. The increase in efficiency is thought to stem from (i) reduction in transit time, (ii) reduction in transient time, (iii) channelling of intermediates so as to reduce competition from other pathways, (iv) coordinate activation of a group of catalytic centres, (v) protection of unstable intermediates from non-enzymatic degradation. (These factors were discussed in Section 7.6.) Examples in support of points (i) to (iv) have been given throughout this chapter. As yet there appears to be no well-documented example on an unstable intermediate that would be degraded spontaneously if it were not efficiently transferred from one catalytic centre to the next.

However, although it is not difficult to see that multienzyme proteins may improve efficiency, it is not clear how important or essential this is *in vivo*. For example, it has been shown that the transient time for the *arom* complex may be reduced by 10–15-fold as a result of being organized as a multienzyme protein (see Section 7.11.3), and thus the system can respond more rapidly to a change in substrate concentrations. Whether or not this is important in the growth of the organism and its ability to respond to changing conditions has not yet been established. The organization of the fatty-acid synthase as a multienzyme protein in yeast and mammalian systems means that a growing acyl group on the enzyme complex is free from competition from other growing chains. Only one fatty acid can be synthesized at a time per fatty-acid synthase. There is no evidence to suggest that fatty-acid synthesis is less efficient, either in terms of the maximum rates achievable or of the efficiency with which the initial precursor is converted to the final products, in the organisms in which the enzymes appear to be completely separate, e.g. *E. coli* and plants, than in the multienzyme polypeptides from yeast or mammalian tissues.

References

1. Srere, P. A., *Trends Biochem. Sci.* **5**, 120 (1980).
2. Karlson, P. and Dixon, H. B. F., *Trends Biochem. Sci.* **4**, N275 (1980).

* The nuclear division cycle can be divided into the phases G_1 (presynthetic phase), S (synthetic phase during which DNA synthesis occurs), G_2 (postsynthetic phase), and the M phase (mitotic phase). For details, see reference 3.

3. Adams, R. L. P., Knowler, J. T., and Leader, D. P., *The biochemistry of nucleic acids* (10th edn). Chapman Hall, London (1986).
4. Lewin, B., *Genes III*. Wiley, London (1987).
5. Kumar, S. A., *Prog. Biophys. molec. Biol.* **38**, 165 (1981).
6. Chamberlin, M. J., *Enzymes* (3rd edn). **15B**, 61 (1983).
7. Greenblatt, J. and Li, J., *Cell* **24**, 421 (1981).
8. Bujard, M., *Trends Biochem. Sci.* **5**, 274 (1980).
9. McClure, W. R., *A. Rev. Biochem.* **54**, 171 (1985).
10. Hansen, U. M. and McClure. R., *J. biol. Chem.* **255**, 9564 (1980).
11. Losick, R. and Pero, J., *Cell* **25**, 502 (1981).
12. Doi, R. H., *Archs. Biochem. Biophys.* **214**, 772 (1982).
13. Grossman, A. D., Erickson, J. W., and Gross, C. A., *Cell* **38**, 383 (1984).
14. von Hippel, P. H., Bear, D. G., Morgan, W. D., and McSwiggen, J. A., *A. Rev. Biochem.* **53**, 389 (1984).
15. Platt, T. A., *A. Rev. Biochem.* **55**, 339 (1986).
16. Yura, T. and Ishihama, A., *A. Rev. Genet.* **13**, 59 (1979).
17. Coggins, J. R., Lumsden, J., and Malcolm, A. D. B., *Biochemistry* **16**, 1111 (1977).
18. Hillel, Z. and Wu, C.-W., *Biochemistry* **16**, 3334 (1977).
19. Stöckel, P., May, R., Strell, I., Cejka, Z., Hoppe, W., Henman, H., Zillig, W., and Crespi, H. L., *Eur. J. Biochem.* **112**, 419 (1977).
20. Gilles, R. J., *Trends Biochem. Sci.* **8**, 301 (1985).
21. Van Schaftingen, E., *Adv. Enzymol.* **59**, 316 (1987).
22. Laporte, D. C. and Chung, T., *J. biol. Chem.* **260**, 15291 (1985).
23. Nimmo, H. G., in *Molecular aspects of cellular regulation* (Cohen, P., ed.), Vol. 3, p. 155. Elsevier, Amsterdam (1984).
24. Mori, M. and Tatibana, M., *Methods Enzymol.* **51**, 111 (1978).
25. Gaertner, F. H., *Trends Biochem. Sci.* **3**, 63 (1978).
26. Malley, M. I., Grayson, D. R., and Evans, D. R., *J. biol. Chem.* **255**, 11372 (1980).
27. Meijer, A. J., *FEBS Lett.* **191**, 249 (1985).
28. Bresters, T. W., De Abreu, R. A., Dekok, A., Visser, J., and Veeger, C., *Eur. J. Biochem.* **59**, 335 (1975).
29. Henderson, C. E. and Perham, R. N., *Biochem. J.* **189**, 161 (1980).
30. Jeyaseelan, K., Guest, J. R., and Visser, J., *J. gen. Microbiol.* **120**, 393 (1980).
31. Reed, L. J., *Acc. chem. Res.* **7**, 40 (1974).
32. Dixon, M., Webb, E. C., Thorne, C. J. R. and Tipton, K. F., *Enzymes* (3rd edn) Longman, London (1979). See p. 482.
33. Stephens, P. E., Darlison, M. G., Lewis, H. M., and Guest, J. R., *Eur. J.Biochem.* **133**, 155 (1983).
34. Stephens, P. E., Darlison, M. G., Lewis, H. M., and Guest, J. R., *Eur. J. Biochem.* **133**, 481 (1983).
35. Dennert, G. and Höglund, S., *Eur. J. Biochem.* **12**, 502 (1970).
36. Vogel, O., Hoehn, B., and Henning, U., *Eur. J. Biochem.* **30**, 354 (1972).
37. Eley, M. H., Namihara, G. Hamilton, L., Munk, P., and Reed, L. J., *Archs. Biochem. Biophys.* **152**, 655 (1972).
38. Danson, M. J., Hale, G., Johnson, P., Perham, R. N., Smith, J., and Spragg, P., *J. molec. Biol.* **129**, 603 (1979).
39. CaJacob, C. A., Frey, P. A., Hainfield, J. F., Wall, J. S., and Yang, H., *Biochemistry* **24**, 2425 (1985).
40. Schmidt, B. and Cohen, R., *Biochem. Biophys. Res. Commun.* **93**, 709 (1980).

41. Collins, J. H. and Reed, L. J., *Proc. natn. Acad. Sci. USA* **74**, 4223 (1977).
42. Angelides, K. J. and Hammes, G. G., *Proc. natn. Acad. Sci. USA* **75**, 4877 (1978).
43. Bates, P. L., Danson, M. J., Hale, G., Hooper, E. A. and Perham, R. N., *Nature, Lond.* **268**, 313 (1977).
44. Reed, L. J. and Mukherjee, B. B., *Methods Enzymol.* **13**, 55 (1969).
45. Koike, M., Reed, L. J., and Carroll, W. R., *J. biol. Chem.* **238**, 30 (1963).
46. Reed, L. J. and Yeaman, S. J., *Enzymes* (3rd edn.) **18B**, 77 (1987).
47. Bates, P. L., Danson, M. J., Hale, G., Hooper, E. A., and Perham, R. N., *Nature, Lond.* **268**, 313 (1977).
48. Hale, G. and Perham, R. N., *Biochem. J.* **177**, 129 (1979).
49. Guest, J. R., Lewis, H. M., Graham, L. D., Packman, L. C. and Perham, R. N., *J. molec. Biol.* **185**, 743 (1985).
50. Packman, L. C. and Perham, R. N., *Biochem. J.* **242**, 531 (1987).
51. Shepherd, G. B. and Hammes, G. G., *Biochemistry* **15**, 311 (1976).
52. Scouten, W. H., Visser, A. J. W. G., Grande, H. J., DeKok, A., and Graaf-Hess, A. C., *Eur. J. Biochem.* **112**, 9 (1980).
53. Stepp, L. R., Bleile, D. M., McRorie, D. K., Petit, F. H., and Reed, L. J., *Biochemistry* **20**, 4555 (1981).
54. Akiyama, S. K. and Hammes, G. G., *Biochemistry* **19**, 4208 (1980).
55. Danson, M. J., Fersht, A. R., and Perham, R. N., *Proc. natn. Acad. Sci. USA* **75**, 5386 (1978).
56. Henderson, C. E. and Perham, R. N., *Cell* **17**, 85 (1979).
57. Petit, F. H., Roche, T. E. and Reed, L. J., *Biochem. biophys. Res. Commun.* **49**, 563 (1972).
58. Randle, P. J., *Trends Biochem. Sci.* **3**, 217 (1978).
59. Sugden, P. H., Hutson, N. J., Kerbey, A. L., and Randle, P. J., *Biochem. J.* **169**, 433 (1978).
60. Cohen, P., *Trends Biochem. Sci.* **1**, 38 (1976).
61. Davis, P. F., Petit, F. H., and Reed, L. J., *Biochem. biophys. Res. Commun.* **75**, 541 (1977).
62. Wieland, O. H., Hartmann, U., and Siess, E. A., *FEBS Lett.* **27**, 240 (1972).
63. DeMarcucci, O. G. L., Gibb, G. M., Dick, J., and Lindsay, J. G., *Biochem. J.* **251**, 817 (1988).
64. Spencer, M. E., Darlison, M. G., Stephens, P. E., Duckenfield, I. K., and Guest, J. R., *Eur. J. Biochem.* **141**, 361 (1984).
65. Kresze, G-B., Ronft, H., Dietl, B. and Steber, L., *FEBS Lett.* **127**, 157 (1981).
66. Randle, P. J., Patston, P. A. and Espinal, J., *Enzymes* (3rd edn) **18B**, 97 (1987).
67. Bertrand, K., Korn, L., Lee, F., Platt, T., Squires, C. L., Squires, C., and Yanofsky, C., *Science NY* **189**, 22 (1975).
68. Yanofsky, C. and Crawford, I. P., *Enzymes* (3rd edn) **7**, 1 (1972).
69. Bartolomes, P., Böker H. and Jaenicke, R., *Eur. J. Biochem.* **102**, 167 (1979).
70. Dettwiler, M. and Kirschner, K., *Eur. J. Biochem.* **102**, 159 (1979).
71. Zalkin, H. and Yanofsky, C., *J. biol. Chem.* **257**, 1491 (1982).
72. Hankins, C. N. and Mills, S. E., *J. biol. Chem.* **252**, 235 (1977).
73. Miles, E. W., *Adv. Enzymol.* **49**, 127 (1979).
74. Kirschner, K. and Wiskocil, R., in *Protein–protein interactions* (Jaenicke, R. and Helmreich, E., eds.), pp. 245–268. Springer, New York (1972).
75. Miles, E. W., *Adv. Enzymol.* **49**, 127 (1979). See pp. 157–166.
76. Creighton, T. E. and Yanofsky, C., *J. biol. Chem.* **241**, 980 (1966).
77. Berger, F. G. and Hermann, K., *J. Bacteriol.* **124**, 800 (1975).

78. Miles, E. W., *J. biol. Chem.* **245**, 6016 (1970).
79. Lane, A. N. and Kirschner, K., *Eur. J. Biochem.* **129**, 571 (1983).
80. Matchett, W. H., *J. biol. Chem.* **249**, 4041 (1974).
81. Creighton, T. E., *Eur. J. Biochem.* **13**, 1 (1970).
82. Lane, A. N. and Kirschner, K., *Eur. J. Biochem.* **129**, 675 (1983).
83. Crawford, I. P., *Bacteriol. Rev.* **39**, 87 (1975).
84. Makoff, A. J. and Radford, A., *Microbiol. Rev.* **42**, 307 (1978).
85. Pierard, A., *Science NY* **154** 1572 (1966).
86. Lacroute, F., Pierard, A., Greenson, M. and Wiame, J. M., *J. gen. Microbiol.* **40**, 127 (1965).
87. Davis, R. H., in *Organizational biosynthesis* (Vogel, H. J., Lampen, J. O., and Bryson, V., eds.), pp. 303–332. Academic Press, New York. (1967).
88. Pierard, A. and Schröter, B., *J. Bacteriol.* **134**, 167 (1978).
89. Davis, R. H., Ristow, J. L. and Hanson, B. A., *J. Bacteriol.* **141**, 144 (1980).
90. Hoogenraad, N. J., Levine, R. L. and Kretchmer, N., *Biochem. biophys. Res. Commun.* **44**, 981 (1971).
91. Mori, M. and Tatibana, M. *J. Biochem. Tokyo* **78**, 239 (1975).
92. Coleman, P. F., Suttle, D. P. and Stark, G. R., *J. biol. Chem.* **252**, 6379 (1972).
93. Mori, M. and Tatibana, M., *Eur. J. Biochem.* **86**, 381 (1978).
94. Jones, M. E., *A. Rev. Biochem.* **49**, 253 (1980).
95. Pringle, J. R., *Methods Cell. Biol.* **12**, 149 (1975).
96. Cohen, G. N. and Dautry-Varsat, A., in *Multifunctional proteins* (Bisswanger, H. and Schmincke-Oft, E., eds.), pp. 47–121. Wiley, New York (1980).
97. Umbarger, H. E., *A. Rev. Biochem.* **47**, 533 (1978).
98. Sibilli, L., LeBras, G., LeBras, G., and Cohen, G. N., *J. Biol. Chem.* **256**, 10228 (1981).
99. Fazel, A., Muller, K., LeBras, G., Gard, J-L., Veron, M., and Cohen, G. N., *Biochemistry* **22**, 158 (1983).
100. Schweitzer, R., Kniep, B., Castorph, H., and Holzner, U., *Eur. J. Biochem.* **39**, 353 (1973).
101. Wieland, F., Renner, L., Verfürth, C., and Lynen, F., *Eur. J. Biochem.* **94**, 189 (1979).
102. Dixon, H. B. F. and Perham, R. N., *Biochem. J.* **109**, 312 (1968).
103. Stoops, J. K. and Wakil, S. J. *J. biol. Chem.* **256**, 5128 (1981).
104. Katiyar, S. S. and Porter, J. W., *Eur. J. Biochem.* **130**, 177 (1983).
105. McCarthy, A. D. and Hardie, D. G., *Eur. J. Biochem.* **130**, 185 (1983).
106. Engeser, H., Hübner, K., Straub, J., and Lynen, F., *Eur. J. Biochem.* **101**, 407 (1979).
107. Engeser, H., Hübner, K., Straub, J., and Lynen, F., *Eur. J. Biochem.* **101**, 413 (1979).
108. Mattick, J. S., Zehner, Z. E., Calabro, M. A. and Wakil, S. J., *Eur. J. Biochem.* **114**, 643 (1981).
109. Wakil, S. J. and Stoops, J. K., *Enzymes* (3rd edn), **16**, 3 (1983).
110. Volpe, J. J. and Vagelos, P. R., *A. Rev. Biochem.* **42**, 21 (1973).
111. Greenspan, M. D., Birge, C. H. Powell, G. L., Hancock, W. S., and Vagelos, P. R., *Science*, **170**, 1203 (1970).
112. Bloch, K. and Vance, D., *A. Rev. Biochem.* **46**, 263 (1977).
113. Kumar, S., *J. biol. Chem.* **250**, 5150 (1975).
114. Bloch, K., *Adv. Enzymol.* **45**, 1 (1977).
115. Hütter, R., Niederberger, P., and DeMoss, J. A., *A. Rev. Microbiol.* **40**, 55 (1986).
116. Welch, G. R. and DeMoss, J. A., in *Microenvironment and metabolic compartmentation* (Srere, P. and Easterbrook, R. W., eds.), p. 345. Academic Press, New York (1978).
117. Gaertner, F. H., *Archs. Biochem. Biophys.* **151**, 277 (1972).

118. Lumsden, J. and Coggins, J. R., *Biochem. J.* **161**, 599 (1977).
119. Lumsden, J. and Coggins, J. R., *Biochem. J.* **169**, 441 (1978).
120. Price, N. C. and Johnson, C. M. in *Proteolytic enzymes: a practical approach* (Beynon, R. J. and Bond, J. S., eds.) p. 163. IRL Press, Oxford (1989).
121. Coggins, J. R., Boocock, M. R., Campbell, M. S., Chaudhuri, S., Lambert, J. M., Lewendon, A., Mousdale, D. M. and Smith, D. D. S., *Biochem. Soc. Trans* **13**, 299 (1985).
122. Gaertner, F. H., in *Microenvironment and metabolic compartmentation* (Srere, P. and Easterbrook, R. W., eds), p. 345. Academic Press, New York (1978).
123. Walker, M. S. and DeMoss, J. A., *J. biol. Chem.* **261**, 16073 (1986).
124. Priestle, J. P., Grutter, M. G., White, J. L., Vincent, M. G., Kaina, M., Wilson, E., Jardetzky, T. S., Kirschner, K., and Jansonius, J. N., *Proc. natn. Acad. Sci. USA.* **84**, 5690 (1987).
125. Ollis, D. L., Brick, P., Hamlin, R., Xnong, N. G., and Steitz, T. A., *Nature, Lond.* **313**, 762 (1985).
126. Busby, S. J. W. and Radda, G. K., *Curr. Top, Cell. Reg.* **10**, 89 (1976).
127. Srere, P., *A. Rev. Biochem.* **56**, 89 (1987).
128. Ottaway, J. H. and Mowbray, J., *Curr. Top. Cell. Reg.* **12**, 108 (1977).
129. Reddy, G. P. and Pardee, A. R., *Proc. natn. Acad. Sci. USA.* **77**, 3312 (1980).
130. Jazwinski, S. M., and Edelman, G. M., *J. biol. Chem.* **259**, 6852 (1984).
131. Dang, C. V. and Dang, C. V., *Biochem. J.* **239**, 249 (1986).
132. Yeaman, S. J. *Trends. Biochem. Sci.* **11**, 293 (1986).
133. Joyce, C. M. and Steitz, T. A., *Trends. Biochem. Sci.* **12**, 288 (1987).
134. Helman, J. D. and Chamberlin, M. J.. *A. Rev. Biochem.* **57**, 839 (1988).
135. Carrey, E. A., *Biochem. J.* **236**, 327 (1986).
136. Perham, R. N., Duckworth, H. W., and Roberts, G. C. K., *Nature, Lond.* **292**, 474 (1981).
137. Welch, G. R. and Gaertner, F. H., *Archs. Biochem. Biophys.* **172**, 476 (1976).

8
Enzymes in the cell

8.1. Introduction

In order to make measurements of enzyme activity, whether of single enzymes or a sequence of enzymes, the organism or tissue concerned is disrupted and the enzyme activity measured in crude homogenates or partially purified enzyme preparations. Disruption is necessary in order that the enzymes have free access to the added substrates. In an intact cell the membrane often either prevents the substrate from entering the cell or limits the rate at which it enters. In whole tissues, only the cells on the surface would have unrestricted access to the substrates. Much or all of the subcellular organization is lost in this process of disruption and the assays for enzyme activity are usually performed under conditions which differ considerably from those *in vivo*. In this chapter we consider how the process of cell disruption, and the changed environment *in vitro* compared with that *in vivo*, may effect single enzymes, and also their interactions with other enzymes and cell structures, since this is important in assessing how far results obtained from enzyme measurements *in vitro* (cf. Chapters 4–6) may be used to understand enzyme action *in vivo*.

The principal differences between enzyme reactions measured *in vitro* and those occurring *in vivo* are as follows.

(i) Loss of organization and compartmentation
In an intact cell, enzymes, substrates, cofactors, or effectors are not distributed uniformly throughout the cell. There are often considerable differences between the concentrations of any of these in subcellular organelles; there is even evidence that concentration gradients exist within the soluble phase of the cytoplasm i.e. the cytosol, (see Section 8.2.6), but the differences may not exist once a cell or its constituent organelles have been disrupted.

(ii) The dilution factor
During the course of preparation of either a crude homogenate or a purified enzyme, the suspending medium used causes dilution of the solutes present within the cell. Protein concentrations within cell compartments are at the lower end of the range of protein concentrations that exist in protein crystals.[1] For example, muscle cells have 23% protein by weight; red blood cells, 35 per cent, and most dividing cells 17–26 per cent, whereas protein crystals range between 20 and 90 per cent protein. In contrast, the protein concentration in an enzyme assay using a liver homogenate might be 5 mg cm^{-3}, and with a highly purified enzyme the protein concentration might be of the order of 5 μg cm^{-3}. The dilution of the cell constituents may affect a number of interactions between

macromolecules and between macromolecules and small molecules and ions.

(iii) Relative concentrations of enzyme and substrate
In enzyme assays carried out under steady-state conditions (see Chapter 4, Section 4.1) there is a vast excess of substrate compared with that of enzyme. Typical concentrations might be in the region of 10^{-3} mol dm^{-3} for substrates and 10^{-6} mol dm^{-3} or below for enzymes. Measurements *in vivo* suggest that there are many instances in which enzymes are present in concentrations comparable with or greater than their substrates[2] and thus the kinetics observed under steady-state conditions may be somewhat different from those *in vivo*.

(iv) Closed and open systems
This aspect is related to the previous one. The enzyme assay carried out *in vitro* is essentially a closed system. The reactants are provided at the beginning of the assay and no replacement usually occurs during assay. The living cell, on the other hand, is an open system in which metabolites are continually entering and leaving. Thus, the concentrations of both the substrate and product of an enzyme reaction may be unchanged for a considerable period of time *in vivo* but nevertheless there may be considerable flux through the enzyme pathway (see Chapter 6, Section 6.3.1).

These four differences between enzymes in the cell and enzymes in the cuvette or test-tube will now be discussed in more detail. It will become apparent that certain aspects of this situation, e.g. compartmentation, have been explored extensively, whereas other areas, e.g. the high concentrations of protcins within organelles and their effects on reactions, have hardly been touched.

8.2 Intracellular compartmentation

Cells vary considerably in size, shape, and the amount of subcellular organization contained within them. A 'typical' cell illustrating subcellular organization is given in Fig. 8.1. Much of the subcellular organization can be discerned from light and electron microscopic observations, but in addition to this there are many examples of metabolic 'pools', where for a given metabolite there is kinetic evidence for compartmentation but for which no morphological counterpart can be discerned. A typical prokaryote cell has a volume in the range 0.5–500 μm^3 and well-defined subcellular structures have not been identified and isolated. A eukaryote cell is generally larger,[3] in the range 200 μm^3 to 15 000 μm^3 and many well-defined subcellular organelles have been characterized. The larger size probably necessitates a greater degree of subcellular compartmentation. In the following section we briefly describe the isolation and properties of the principal subcellular organelles of eukaryote cells. The main emphasis will be to give an indication of the size of each organelle, the selectivity of its surrounding membrane where appropriate, and the nature of the substances comprising the

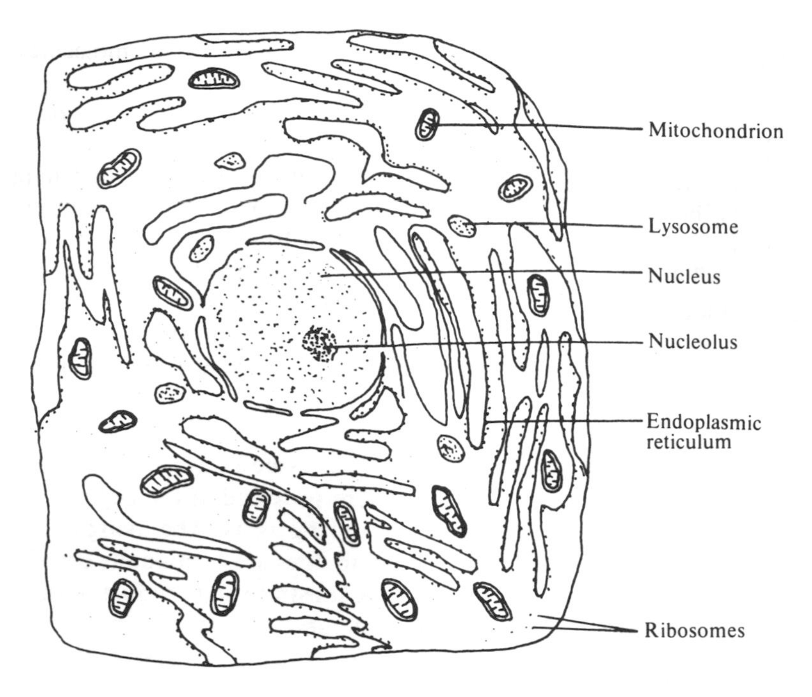

FIG. 8.1. A typical eukaryote cell showing the intracellular structure (subcellular organelles not drawn to scale).

matrix. Since the liver parenchyma cell is perhaps the most studied eukaryotic cell, the information given refers to the liver cell unless otherwise stated.

8.2.1 Isolation of subcellular organelles

The aim of subcellular fractionation procedures is to isolate subcellular organelles as free from contaminants as possible and with as little disruption as possible. The procedure involves two principal stages: (i) disruption of the tissue or cells to release the subcellular organelles and (ii) separation of the organelles from one another. It is necessary also to have methods for assessing the integrity of the organelles and the extent of contamination by other components. There is no single procedure suitable for all cells and tissues, and methods have to be adapted for the tissue concerned.

In the first stage, cells and tissues are disrupted by a variety of different methods according to the tissue involved (see Chapter 2, Section 2.5). The second step in the procedure is then to separate components from the homogenate usually by differential centrifugation. The rate of sedimentation in a centrifugal field increases with the size of a particle, the difference in density between the particle and the solvent, and the centrifugal field. There are small differences in the densities of the subcellular organelles, nuclei being denser than mitochondria, but the main factor influencing the separations are the differences in size of the organelles. The organelles are thus usually separated by centrifugation at increasing speeds, as indicated in Fig. 8.2. In order to obtain fairly pure preparations, the sedimented fractions are usually resuspended in buffer and then resedimented once or twice more. This procedure gives a fairly good separation of the main organelles: nuclei, mitochondria, and microsomes (disrupted endoplasmic reticulum). It is difficult to separate mitochondria from lysosomes because they have similar sizes and densities. An enriched lysosomal pellet may be prepared if a number of resuspensions and resedimentations are carried out, but this is unlikely even then to contain more than 10 per cent lysosomes. To obtain a complete separation of rat-liver lysosomes from mitochondria, the rats may be injected prior to sacrifice with Triton WR1339 (an anionic detergent, the sodium salt of an alkyl aryl polyether sulphate) or with colloidal gold. The Triton WR1339 or colloidal gold is phagocytosed by the

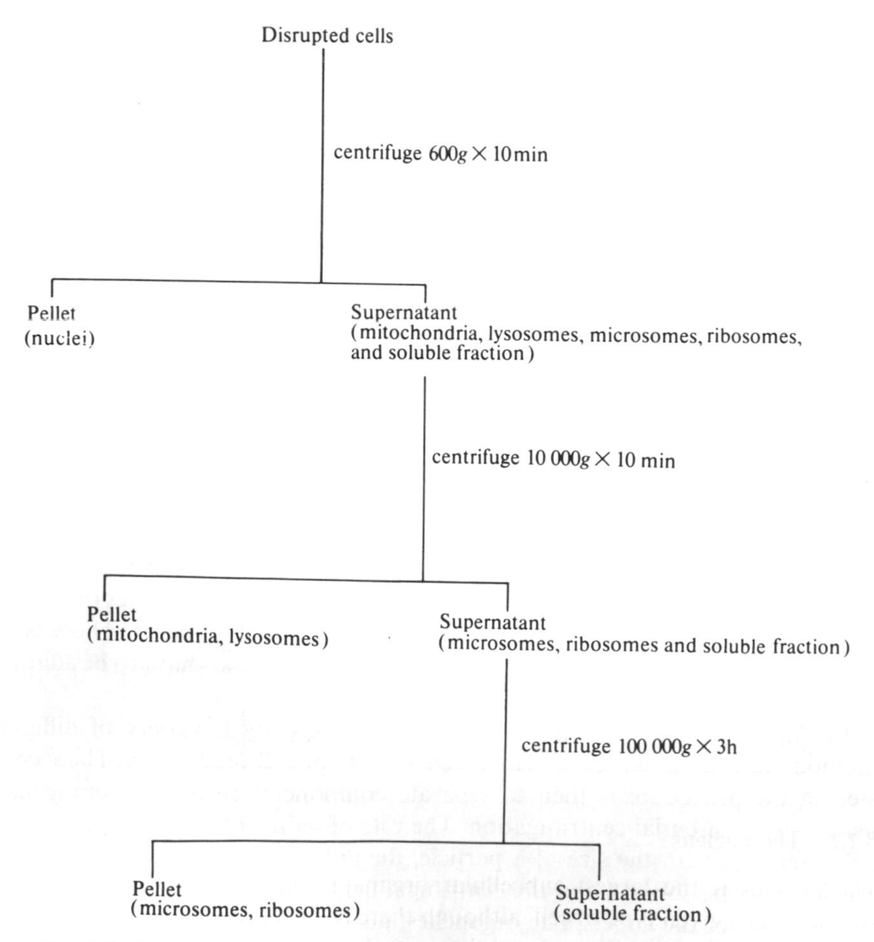

FIG. 8.2. Procedure for subcellular fractionation by differential centrifugation.

lysosomes, causing decreases or increases in their densities, respectively. They may then be separated from mitochondria.[4]

Further refinements to subcellular fractionations may be carried out; for example, smooth microsomes (disrupted endoplasmic reticulum lacking ribosomes) can be separated from rough microsomes (endoplasmic reticulum containing ribosomes) by centrifugation through layers of sucrose solution having different densities. The denser sucrose solutions retard the smooth microsomes, which are less dense than rough microsomes, because the former lack the more dense RNA.[5]

The purity of subcellular organelles can be assessed by biochemical and histological methods. Certain enzymes that have been found to be present either exclusively or predominantly in one subcellular organelle are used as marker enzymes to assess the extent of contamination of a particular subcellular fraction. For example, succinate dehydrogenase is a mitochondrial enzyme, and thus by measuring succinate dehydrogenase activity in a preparation of nuclei, the extent of contamination by mitochondria can be assessed. A list of marker enzymes for different subcellular fractions is given in Table 8.1. Besides the biochemical methods, it is useful to examine a preparation by light or electron microscopy, in order to check that the organelles are undamaged. Histochemical reagents can also be used. e.g. histochemical assay for acid phosphatase, or Feulgen's reagent for DNA (a reaction which depends upon the fact that partial hydrolysis products of DNA restore the colour of basic fuchsin that has been decolorized by sulphurous acid).

TABLE 8.1
Marker enzymes used in subcellular fractionation

Enzyme	Subcellular fraction
DNA nucleotidyltransferase	Nuclei
Nicotinamide-nucleotide adenylyltransferase	Nuclei
Succinate dehydrogenase	Mitochondria
Cytochrome *c* oxidase	Mitochondria
Glucose 6-phosphatase	Endoplasmic reticulum
Acid phosphatase	Lysosomes
Ribonuclease	Lysosomes
Catalase	Peroxisomes
Urate oxidase	Peroxisomes
5′-Nucleotidase	Plasma membrane
Glucose 6-phosphate dehydrogenase	Cytosol
Lactate dehydrogenase	Cytosol
6-Phosphofructokinase	Cytosol

8.2.2 The nucleus

The nucleus is the largest subcellular organelle. Its size does not vary substantially in a given type of cell, although there is a wide variation in the sizes of nuclei from different cell types and this variation is not related to the volume of the cytoplasm. For example, nuclei from certain fungi may be only 4 μm^3 in volume and occupy less than 1 per cent of the volume of the cytoplasm, whereas the nuclei of cells from the thymus gland are 300–400 μm^3 and occupy 70 per cent of the cell volume. The nuclei from liver cells have a volume of 300 μm^3 which is about 5 per cent of the total cell volume. The nucleus is bounded by two membranes which show some properties similar to those of the endoplasmic reticulum, e.g. all enzymes found on the nuclear membrane except cytochrome *c* oxidase are also found in the endoplasmic reticulum. However, they differ from the endoplasmic reticulum in buoyant density.[6, 7] There is a polymeric protein network lining the nucleoplasmic surface of the nuclear membrane.[8]

A characteristic feature of the nuclear membrane is the presence of pores. They vary in size and number according to the type of nucleus. In many mammalian nuclei the pores occupy 10–20 per cent of the surface area of the nucleus and the pore diameter is about 70 nm. Many small molecules and ions and also macromolecules are able to enter the nucleus by a diffusion process not requiring a source of metabolic energy, e.g. ATP. It seems likely that transport is through the nuclear pores. Larger proteins can enter through the pores by an active process that entails recognition by the pore of a signal sequence on the polypeptide chain.[9] There are differences in the concentrations of Na^+ and K^+ in the nucleus compared with outside the nucleus but these are probably due to different proportions of ions bound to macromolecules. It is thought that ribosomes (less than 50 nm wide from the outer edge of 40 S subunit to the outer edge of 60 S subunit) and RNA pass through the pores. The space within the nucleus into which sucrose and other small molecules cannot penetrate, is estimated at less than 10 per cent of the total volume.

The nucleus is the densest subcellular organelle (density approximately 1.35 g cm^{-3}) and this reflects the high concentrations of macromolecules within it (approximately 34 g protein, 9.5 g DNA, and 1.5 g RNA per 100 g nuclei).[10] The large amount of DNA in the nucleus necessitates a high degree of condensation from that of the double helix by as much as 50 000-fold in order to fit within the dimensions of the nucleus. The enzymes of the nucleus can be grouped according to their location (Table 8.2). The enzyme content of the soluble compartment of the nucleus is similar to that of the cytosol, especially in regard to that of the enzymes of the glycolytic pathway. The second and third groups (Table 8.2) of enzymes are bound to insoluble components of the nucleus which are not readily washed out of the preparations of isolated nuclei, but they may be extracted using concentrated salt solutions, e.g. 1 mol dm^{-3} NaCl. They comprise mainly the enzymes concerned with DNA and RNA synthesis and are attached to chromatin. The fourth group are only extracted by the use of detergents and are membrane-bound enzymes such as glucose 6-phosphatase and acid phosphatase. It is possible to isolate a nuclear matrix that comprises an insoluble skeletal framework connected to the nuclear membrane, and which has an appearance similar to that of the nucleus under the electron microscope. It comprises about 10% of the nuclear proteins, and processes such as replication and transcription are believed to occur when the enzymes concerned are associated with this

TABLE 8.2.
Enzymes present in the nucleus

1. Enzymes in the soluble space	**3. Enzymes concentrated in the nucleolus**
Glycolytic enzymes	RNA nucleotidyltransferase I
Pentose phosphate pathway enzymes	RNA methyltransferases
Lactate dehydrogenase	Ribonuclease
Malate dehydrogenase	
Isocitrate dehydrogenase	
Arginase	
2. Enzymes bound to chromatin	**4. Enzymes bound to membranes**
RNA nucleotidyltransferase II	Glucose 6-phosphatase
RNA nucleotidyltransferase III	Acid phosphatase
Nucleoside triphosphatase	
DNA nucleotidyltransferase	
Nicotinamide-nucleotide adenylyltransferase	

For further details see references 6, 7, 10–12.

framework.[13] The role of the nucleolus is principally the biogenesis of ribosomes and it develops and degenerates with each cell cycle.[14]

8.2.3 The mitochondrion

The mitochondrion is typically a sausage-shaped organelle having dimensions of approximately 0.5 μm diameter, 4 μm length, and 1 μm^3 volume. Although it is a relatively small organelle, there are usually a large number of mitochondria per cell and collectively they occupy a significant proportion of the cell volume, e.g. liver parenchymal cells have 1000–1600 mitochondria per cell and they occupy 10–20 per cent of the volume of the cytoplasm. The mitochondrion has two membranes that have markedly different properties. The outer membrane is smooth in appearance whereas the inner membrane has a number of infoldings called cristae (Fig. 8.3). The inner mitochondrial membrane encloses the matrix, which is 60–70 per cent of the total mitochondrial volume. On the inner surface of the inner mitochondrial membrane are small mushroom-shaped particles containing mitochondrial adenosinetriphosphatase, which plays an important role in oxidative phosphorylation (see Chapter 2, Section 2.8.3). The number of cristae per mitochondrion varies with the tissue, being highest in those tissues having high potential metabolic rates, e.g. flight muscle (see Chapter 6, Section 6.4.1.1).

The mitochondrial matrix is granular in appearance and has a very high concentration of protein, usually about 500 mg cm^{-3}.[15] About 70 per cent of the mitochondrial protein is in the matrix, 20 per cent is part of the inner mitochondrial membrane and only 4 per cent is part of the outer membrane. The inner mitochondrial membrane has a ratio of protein:lipid of 75 per cent to 25 per cent, which is high compared to other cell membranes. It is estimated, using

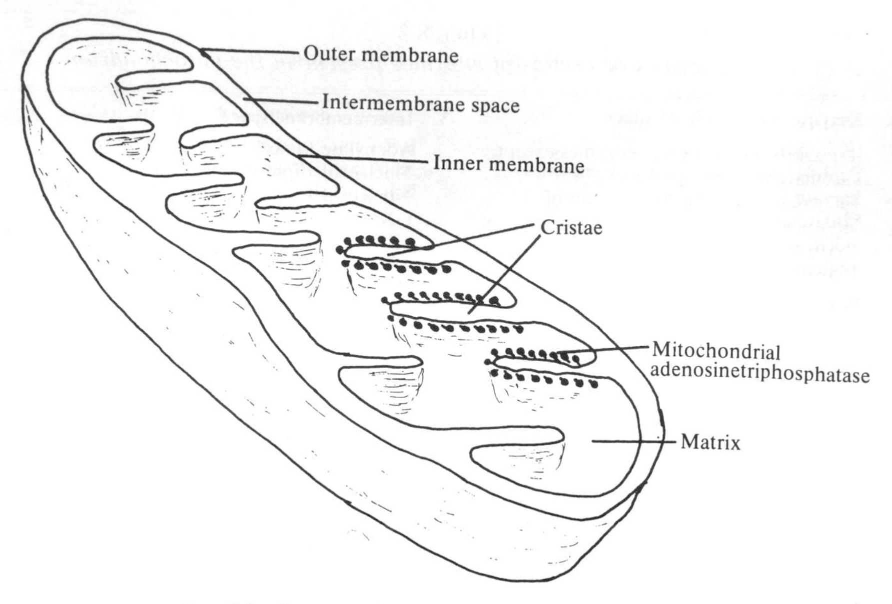

FIG. 8.3. Cross-section of a typical mitochondrion.

differential scanning calorimetry combined with freeze-fracture electron microscopy, that between one-third and one-half of the surface area is occupied by proteins,[16] most of which are involved in the electron-transport chain.

The chemical composition, permeability, and enzyme composition of the inner and outer membranes differ considerably. The outer membrane has a much higher proportion of lipid than the inner membrane, the latter having a very low content of cholesterol. Several of the enzymes present in the outer membrane are also present in the endoplasmic reticulum, e.g. cytochrome b_5 reductase and glycerophosphate acyltransferase, and it is often considered that the outer membrane may be derived from the endoplasmic reticulum. The outer membrane is generally permeable to small molecules having M_r up to 10 000, e.g. ATP, coenzyme A, and carnitine, whereas the inner membrane is generally permeable only to uncharged molecules having M_r up to 150. There are, however, a number of specific proteins in the inner membrane, called translocators, that are responsible for the transport of specific metabolites across the membrane. For example, translocators exist for inorganic phosphate, malate, succinate, citrate, isocitrate, 2-oxoglutarate, glutamate, aspartate, and adenine nucleotides (these are discussed further in Section 8.3.1.4). A list of the principal enzymes located in different regions of the mitochondrion is given in Table 8.3 and some are discussed later in the chapter. A number of the enzymes in the matrix have been shown to associate with one another and with proteins of the inner mitochondrial membranes. These are discussed in Section 8.5.3.2.

Although the concentration of protein is high in mitochondria, particularly in the matrix, the overall density of mitochondria is lower than that of the nucleus

TABLE 8.3
Principal enzymes or groups of enzymes present in the mitochondrion

Location	Enzymes
1. Matrix	Tricarboxylic acid cycle enzymes except succinate dehydrogenase
	Enzymes catalysing β-oxidation of fatty acids
	Pyruvate carboxylase
	Phosphoenolpyruvate carboxykinase (pigeon liver)
	Carbamoyl-phosphate synthase
	Ornithine carbamoyltransferase
	Glutamate dehydrogenase
2. Inner membrane	Succinate dehydrogenase, NADH dehydrogenase } +associated respiratory chain
	Adenosinetriphosphatase
	3-Hydroxybutyrate dehydrogenase
	Carnitine palmitoyltransferase
	Glycerol 3-phosphate dehydrogenase (FAD enzyme)
	5-Aminolaevulinate synthase
	Hexokinase
	Cytochrome *c* oxidase
3. Intermembrane space	Adenylate kinase
	Nucleosidediphosphate kinase
	Nucleosidemonophosphate kinase
	L-Xylulose-reductase
4. Outer membrane	NADH dehydrogenase (rotenone insensitive)
	Cytochrome b_5 reductase
	Amine oxidase (flavin-containing)
	Kynureninase
	Acyl-CoA synthetase
	Glycerophosphate acyltransferase
	Cholinephosphotransferase
	Adenylate kinase
	Hexokinase
	Phospholipase A_2

For further details see reference 17

(1.1 g cm^{-3} compared with 1.35 g cm^{-3}), largely owing to the absence of high concentrations of nucleic acids.

8.2.4 The lysosome

Lysosomes are roughly spherical organelles or vacuoles slightly smaller than mitochondria and having diameters between 0.2 μm and 0.8 μm. They were first discovered by De Duve[18] as particles containing hydrolytic enzymes that sedimented with the mitochondrial fraction in subcellular fractionation. Not only their sizes but their densities (about 1.2 g cm^{-3}) are similar to those of the mitochondria. The characteristic feature of lysosomes is that they contain about 60 hydrolytic enzymes, most of which have acid pH optima (Table 8.4).

Lysosomes are bounded by a single membrane[19] enclosing a dense granular matrix. The membrane shows limited permeability to solutes, e.g. very little sucrose can penetrate the lysosomal membrane. Transport of many solutes is by carrier-mediated processes. In liver cells the lysosomes only occupy 0.3 per cent of the total cell volume and on average there are about 20 lysosomes per cell. About 35 per cent of the total protein is in the membranes and the remainder is in the matrix. The protein concentration in the matrix is about 200 mg cm^{-3} (calculated from reference 20 assuming 35 per cent of lysosomal protein is membrane protein) and the intralysosomal pH is about 1.5 units below that of

TABLE 8.4
Enzymes present in lysosomes

1. Proteolytic enzymes	**3. Hydrolysis of nucleic acids**
Cathepsins B, D, G. L	Deoxyribonuclease II
Elastase	Ribonuclease II
Collagenase	
2. Hydrolysis of glycosides	**4. Hydrolysis of lipids**
β-D-Glucuronidase	Phospholipases A_1 and A_2
β-*N*-Acetyl-D-hexosaminidase	Cholesterol esterase
Hyaluronoglucosaminidase	
Lysozyme	
Neuraminidase	
5. Others	
Acid phosphatase	
Aryl sulphatase	

For a more comprehensive list, see references 4 and 25.

the cytosol. Several of the proteins both of the membrane and matrix are glycoproteins and many are acidic. The low intralysosomal pH is thought to be maintained mainly by Donnan equilibrium[21] (an equilibrium that results in the asymmetric distribution of diffusible ions across a membrane, brought about by the presence of a larger ion that is unable to diffuse across the membrane), but there is also some evidence for a membrane-bound adenosinetriphosphatase that may act as a proton pump[22] (a process by which protons are transported across a membrane against the concentration gradient at the expense of metabolic energy). The proteins on the inner surface of the lysosomal membrane contain about 16 μg sialic acid per mg protein and this, together with other glycoproteins in the matrix, is thought to be responsible for the maintenance of the low pH.

Lysosomes are morphologically heterogeneous and this may reflect the way in which they function. Their mode of action is still uncertain, but De Duve and others have proposed the following theory.[23, 24] Primary lysosomes that have a full complement of hydrolytic enzymes but that have not participated in intracellular digestion are formed by pinching off portions of the smooth endoplasmic reticulum at the Golgi apparatus. The primary lysosomes fuse with other membranous structures containing substrates to be degraded and thereby form secondary lysosomes, i.e. lysosomes in which active digestion is occurring. Some of the low M_r products may diffuse out of the secondary lysosomes.

8.2.5 The endoplasmic reticulum

Although the endoplasmic reticulum is not generally regarded as an organelle it forms an extensive network throughout the cytoplasm and effectively separates the cytosol into two components; that which is enclosed within the endoplasmic reticulum (the lumen) and that which is outside it (Fig. 8.4). It is not possible to

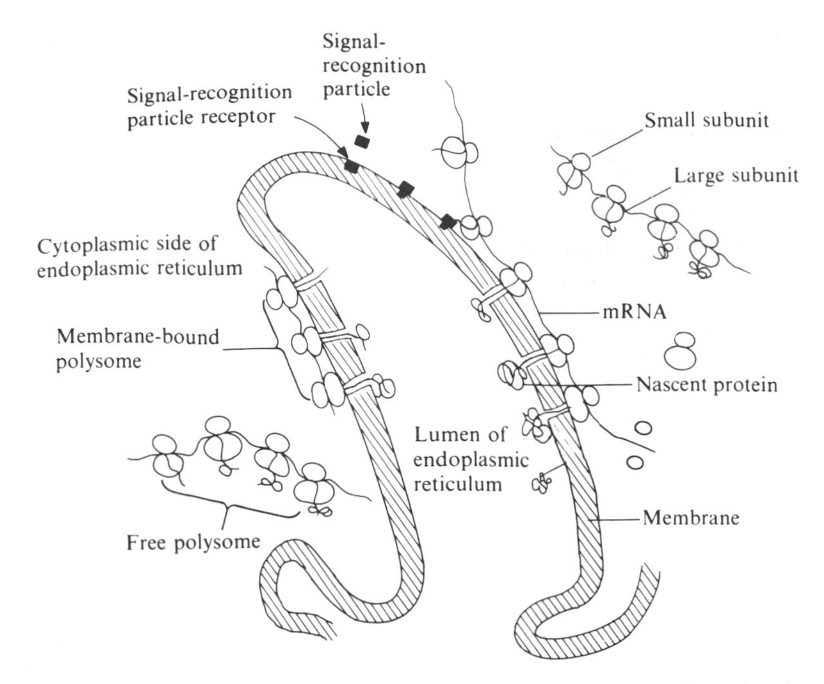

FIG. 8.4. The endoplasmic reticulum showing free and membrane-bound polysomes.

separate these two compartments by present biochemical techniques because when a cell is disrupted prior to carrying out subcellular fractionation, the endoplasmic reticulum is broken down and the membranous fragments form vesicles that constitute the 'microsome fraction' in subcellular fractionations. The amount of endoplasmic reticulum varies between different cell types but there is a good correlation between the density of reticulum and the extent to which the cell secretes proteins. The principal type of liver cell, the hepatocyte, and the cells of the exocrine pancreas (the cells concerned with its digestive functions rather than its hormonal functions) secrete a number of proteins and both have dense endoplasmic reticula (in the hepatocyte the endoplasmic reticulum occupies about 15 per cent of the total cell volume; 1 g of liver contains about 10 m^2 area of endoplasmic reticulum), whereas a non-secreting cell such as the erythrocyte is devoid of endoplasmic reticulum.

The endoplasmic reticulum, which is composed mainly of lipids and proteins, is organized into channels and it is assumed that proteins are collected and secreted into these channels. The proteins of the endoplasmic reticulum have both structural and catalytic roles. It is estimated that there are about 25–50 molecules of phospholipid per polypeptide chain in the endoplasmic reticulum. The membrane has limited permeability: uncharged molecules with M_r values up to about 600 can pass through freely, but charged species with M_r values greater than 90 and macromolecules, cannot pass freely. The amounts and distribution of many of the proteins have been estimated; cytochrome P450 is the most abundant, comprising 4 per cent of the total protein. The majority of the proteins appear to be on the cytoplasmic surface of the endoplasmic reticulum and those that are located on the luminal surface may be proteins of the cytosol that have been adsorbed on to the membrane (Table 8.5).

TABLE 8.5
Principal enzymes of the endoplasmic reticulum

Enzyme or group of enzymes	Location
Cholesterol biosynthetic enzymes	* Smooth endoplasmic reticulum
Steroid hydroxylation enzymes	Smooth endoplasmic reticulum
Carnitine acyltransferase	
Fatty acid elongating enzymes (C_{16}–C_{24})	
Glycerolphosphate acyltransferase	
Drug-metabolizing enzymes (aromatic ring hydroxylation, side-chain oxidation, deamination, dealkylation, dehalogenation)	Smooth endoplasmic reticulum
Protein synthesis	* Rough endoplasmic reticulum, cytoplasmic side
Glucose 6-phosphatase	Rough endoplasmic reticulum, luminal side
Adenosinetriphosphatase	Cytoplasmic side
Cytochrome b_5 reductase	Cytoplasmic side
NADPH-cytochrome reductase	Cytoplasmic side
GDPmannose α-D-mannosyltransferase	Cytoplasmic side
Nucleosidediphosphatase	Luminal side
β-D-Glucuronidase	Luminal side
UDP glucuronosyltransferase	
5′ Nucleotidase	Cytoplasmic side
Cholesterol acyltransferase	Cytoplasmic side

For further details see references 26, 27 and 28.
* Endoplasmic reticulum may be subdivided into rough endoplasmic reticulum and smooth endoplasmic reticulum. The areas of rough endoplasmic reticulum are those in which polysomes are present, whereas smooth endoplasmic reticulum lacks polysomes.

A large number of different processes occurs on the endoplasmic reticulum including protein synthesis and protein transport, biosynthesis of fatty acids, sterols, phospholipids, and the metabolism of foreign compounds, and this is reflected in the enzyme composition (Table 8.5). Protein synthesis occurs both on polysomes (Fig. 8.4), which are free in the cytosol, and on polysomes bound to the endoplasmic reticulum. The ribosomes that comprise the free and bound polysomes arise from the same ribosome pool. Whether a polysome remains free or bound is determined by the nature of mRNA that is being translated. Proteins that are destined either to become integral membrane proteins or to be transported across the endoplasmic reticulum on the lumenal side contain a signal sequence of between 16 and 30 amino acids, a preponderance of which are hydrophobic. The signal sequence is recognized by a signal-recognition particle which ensures that it becomes membrane bound. The growing polypeptide is extruded through the membrane eventually either becoming an integral membrane protein or being cleaved to remove the signal peptide and released into the lumen. These proteins may either become secreted out of the cell or be contained within vesicles or dispersed to the lysosomes. The endoplasmic reticulum is connected to the Golgi apparatus, which sorts these proteins in a manner dependent on their recognition regions for the lysosomes or for secretion. The enzymes destined for the lysosomes contain a recognition marker in the form of a mannose 6-phosphate residue that is recognized by a receptor (M_r 215 000) that transports the enzyme from the Golgi to the lysosomes (for details, see references 29 and 135).

8.2.6 The cytosol

The cytosol is the aqueous phase in which the cell organelles are suspended.* Operationally it is the soluble phase that remains as the supernatant after a tissue homogenate has been centrifuged for sufficient '$g \cdot$min' (Fig. 8.2) to sediment the microsome fraction and free ribosomes and is referred to as the 'high-speed' supernatant. It thus comprises most of the soluble phase from both the cytoplasmic and luminal sides of the endoplasmic reticulum. It will also contain soluble contaminants from any disrupted subcellular organelles. The 'high-speed' supernatant will also differ from the cytosol in that the former will be appreciably diluted with the buffer used for homogenization.

In recent years evidence has mounted that suggests that the cytosol is not simply a random mixture of proteins, nucleic acids, and other molecules. Certain groups of enzymes are known to form multienzyme complexes, e.g. fatty-acid synthase (see Chapter 7, Section 7.11.2), and in the case of glycolytic enzymes there is evidence for a type of association that is less well defined (see Section 8.5.3). It is possible that many other weak associations of macromolecules may occur that have not yet been detected either because they have not been sought or because the association may be so weak that the dilution of the cytosol during the isolation of a high-speed supernatant may be sufficient to cause dissociation.

In most tissues a substantial proportion of the protein of the cell is in the cytosol. The viscosity of the cytosol is approximately equivalent to that of a 15 per cent solution of sucrose. The cytosol may be regarded as a 'very crowded' protein solution. This causes the rate of diffusion of small molecules to be low (see Section 8.5.2). Measurements of the rates of diffusion of small molecules in the cytosol show that the rates are appreciably lower than in a protein solution of equivalent concentration. This has been interpreted as evidence that the characteristic macromolecules of the cytosol do form defined structures that retard the rates of diffusion of metabolites and other solutes.[30] The importance of diffusion rates in the cytosol is highlighted by evidence that ATP and O_2 gradients exist within the cytosol.[31] The presence of ATP gradients within the

* The term *cytoplasm* refers to the entire contents of the cell outside the nucleus, excluding the cell membrane. The cytoplasm therefore includes subcellular organelles such as mitochondria, lysosomes, etc.

cytosol has been demonstrated by showing that the catalytic activities of two ATP-requiring enzymes, one in the cytosol and one on the inside of the plasma membrane, differ in intact cells because their average distances from the sites of ATP generation (mitochondria) differ.

In addition to protein, the cytosol also contains most of the tRNA of the cell. The principal monovalent cations present are Na^+ and K^+ and these are thought to exist predominantly as hydrated ions in free solution, but the principal divalent cations, Mg^{2+} and Ca^{2+} are thought to be predominantly bound to nucleotides, nucleic acids, and acidic polysaccharides. A large number of enzymes are present in the cytosol, particularly those responsible for glycolysis, gluconeogenesis, fatty-acid synthesis, nucleotide biosynthesis, and aminoacyl-tRNA synthesis (Table 8.6).

TABLE 8.6
Enzymes or groups of enzymes present in the cytosol

1. Carbohydrate metabolism

Glycolytic enzymes including phosphorylase, phosphorylase kinase, and protein kinase
Glycogen synthase
Fructose-bisphosphatase
Phosphoenolpyruvate carboxykinase (rat liver)
Enzymes of the pentose phosphate pathway
Malate dehydrogenase
Isocitrate dehydrogenase
Lactate dehydrogenase
Citrate (*pro*-3*S*)-lyase
Malate dehydrogenase (oxaloacetate-decarboxylating) ($NADP^+$)
Glucose-1-phosphate uridylyltransferase

2. Lipid metabolism

Acetyl-CoA carboxylase
Fatty-acid synthase complex
Glycerol-3-phosphate dehydrogenase (NAD^+)

3. Amino-acid and protein metabolism

Aspartate aminotransferase
Alanine aminotransferase
Arginase
Argininosuccinate lyase
Argininosuccinate synthase
Aminoacyl-tRNA synthetases

4. Nucleic-acid synthesis

Nucleoside kinase
Nucleotide kinase

For further details see reference 36.

Other organelles that are not described here are discussed in the general reference 32, and in the following specific references: Chloroplasts, reference 33; Golgi apparatus, reference 34; Peroxisomes, reference 35; and Glycosomes, reference 137.

8.3 Compartmentation of metabolic pathways

In the intact cell the individual enzymes of a metabolic pathway function together and it is unusual for high concentrations of intermediates to build up. In addition, there are many connections between the main metabolic pathways, since some of the substrates, cofactors, and regulatory molecules, and sometimes the enzymes, are common to more than one pathway. These complex interactions can only be fully understood when, in addition to studying the isolated systems, some knowledge has been acquired concerning the intracellular locations and concentrations of the enzymes and substrates involved and any permeability barriers separating the components. The functions of some enzymes, e.g. malate dehydrogenase, isocitrate dehydrogenase, and ATP citrate (*pro*-3*S*)-lyase, cannot be understood fully without a knowledge of their location. We illustrate the importance of compartmentation using two examples: (i) the interrelationships between glycolysis, gluconeogenesis, fatty-acid oxidation and biosynthesis, and the tricarboxylic-acid cycle in the liver cell; and (ii) arginine and ornithine metabolism in the fungus *Neurospora*.

8.3.1 Compartmentation of carbohydrate and fatty-acid metabolism in liver

The processes of glycolysis, gluconeogenesis, β-oxidation of fatty acids, fatty-acid synthesis, and the tricarboxylic-acid cycle are of major importance in many cells. The enzymes concerned constitute a major portion of the protein of the cell. In this section we are primarily concerned with the intracellular location of the processes and their interconnections and we shall not consider the individual reactions in detail.

8.3.1.1 Location of the enzymes catalysing the pathways

The pathways discussed are illustrated in Fig. 8.5. The sequence of glycolytic enzymes from phosphorylase to pyruvate kinase and lactate dehydrogenase is located in the cytosol and in the soluble phase of the nucleus. It is not clear whether the soluble phase of the nucleus is continuous with that of the cytosol, but the concentrations of glycolytic enzymes are slightly higher in the nucleus.[10] In skeletal muscle there is evidence that the glycolytic enzymes are organized in a so-called 'glycogen particle' (see Chapter 7, Section 7.12) present in the cytosol. A similar structure may exist in liver and in other types of cell in which glycogen is stored, though they have not been investigated. Gluconeogenesis, the reversal of glycolysis, uses the same enzymes except for the three steps indicated below.

Step 1

$$\text{ATP} + \text{pyruvate} + CO_2 + H_2O \xrightleftharpoons{\text{pyruvate carboxylase}} \text{ADP} + \text{orthophosphate} + \text{oxaloacetate.}$$

$$\text{GTP} + \text{oxaloacetate} \xrightleftharpoons{\text{phosphoenolpyruvate carboxykinase}} \text{GDP} + \text{phosphoenolpyruvate} + CO_2.$$

Step 2

$$\text{D-Fructose 1,6-bisphosphate} + H_2O \xrightleftharpoons{\text{fructose bisphosphatase}} \text{D-fructose 6-phosphate} + \text{orthophosphate.}$$

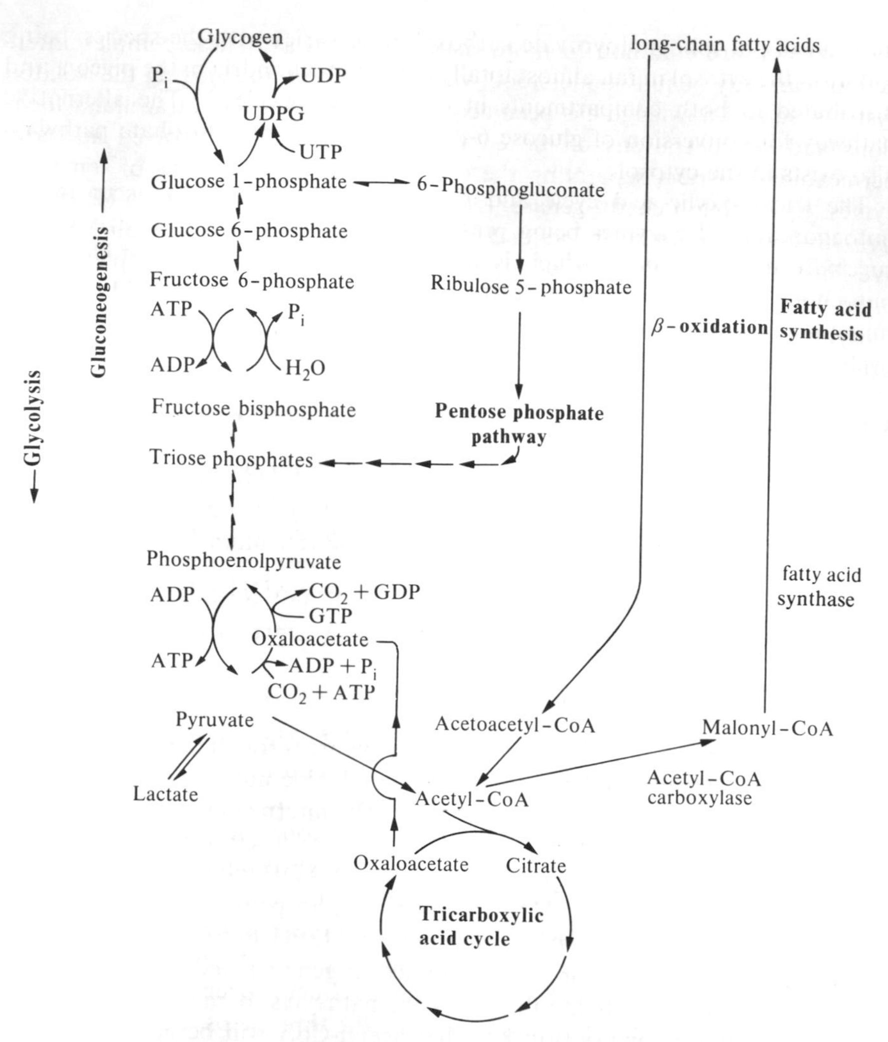

FIG. 8.5. Metabolic pathways for the biosynthesis and degradation of glycogen and of fatty acids. (P_i = orthophosphate.)

Step 3

$$\text{UTP} + \alpha\text{-D-glucose 1-phosphate} \xrightleftharpoons[]{\text{glucose 1-phosphate uridylyl transferase}} \text{UDPglucose} + \text{pyrophosphate}.$$

$$\text{UDPGlucose} + (1,4\text{-}\alpha\text{-D-glucosyl})_n \xrightleftharpoons[]{\text{glycogen synthase}} \text{UDP} + (1,4\text{-}\alpha\text{-D-glucosyl})_{n+1}.$$

Fructose-bisphosphatase and glycogen synthase are located in the cytosol, whereas pyruvate carboxylase is exclusively in the mitochondrial matrix. The location of phosphoenolpyruvate carboxykinase varies with the species, being totally in the cytosol in rat, almost totally in the mitochondria in the pigeon, and distributed in both compartments in many other species.[37] The alternative pathway for conversion of glucose 6-phosphate (pentose-phosphate pathway) also exists in the cytosol.

The tricarboxylic-acid cycle and β-oxidation of fatty acids occur in the mitochondria, all enzymes being present in the mitochondrial matrix except succinate dehydrogenase, which is bound to the inner surface of the inner mitochondrial membrane. The electron-transport chain is located on the inner mitochondrial membrane. It now seems probable that other tricarboxylic-acid cycle enzymes form a loose association with the membrane. The fatty-acids are synthesized from acetyl-CoA using acetyl-CoA carboxylase and fatty-acid synthase present in the cytosol. Palmitic acid is the main product of mammalian fatty-acid synthases.[38] Palmitic acid may then act as the substrate for the malonyl-CoA-dependent elongase present on the endoplasmic reticulum, where the chain may be extended up to C_{24}. The desaturating enzymes that catalyse the following reaction also occur on the endoplasmic reticulum.[39]

$$CH_3.(CH_2)_x.CH_2.CH_2.(CH_2)_y.CO.SCoA + NADH + H^+ + O_2 \rightleftharpoons$$
$$CH_3.(CH_2)_x.CH{=}CH.(CH_2)_y.CO.SCoA + 2H_2O + NAD^+.$$

8.3.1.2 *Interconnections between pathways*

The principal connections between the pathways are through the redox cofactors NAD^+, NADH and $NADP^+$, NADPH; the adenine nucleotides AMP, ADP, ATP; and the metabolite acetyl-CoA. Metabolic intermediates of one pathway may also be regulators of other pathways, e.g. citrate, fructose bisphosphate, and acetyl-CoA. The degradative pathways generate NADH (glycolysis, β-oxidation, tricarboxylic-acid cycle) and NADPH (pentose-phosphate pathway) and ATP, whereas the biosynthetic pathways require NADPH (fatty-acid biosynthesis, cholesterol biosynthesis) or NADH (gluconeogenesis) and ATP and GTP. Acetyl-CoA is the metabolite that links the pathways. It can be seen from the locations of the pathways (Fig. 8.6) that acetyl-CoA will be generated in the mitochondria by either pyruvate dehydrogenase (mitochondrial enzyme) or by β-oxidation of fatty acids, and it will be utilized either by the tricarboxylic-acid cycle in the mitochondria or by fatty-acid synthesis in the cytosol. It can also be appreciated that ATP and NADH will be generated principally by oxidative reactions in the mitochondria whereas ATP, NADH and NADPH will be required for biosynthesis in the cytosol.

8.3.1.3 *Location of metabolites involved in the pathways*

Using methods described in Section 8.2.1 it is a relatively straightforward matter to determine the subcellular location of the enzymes catalysing metabolic pathways, especially when compared with the problems of determining the intracellular location and concentrations of metabolites and cofactors. The

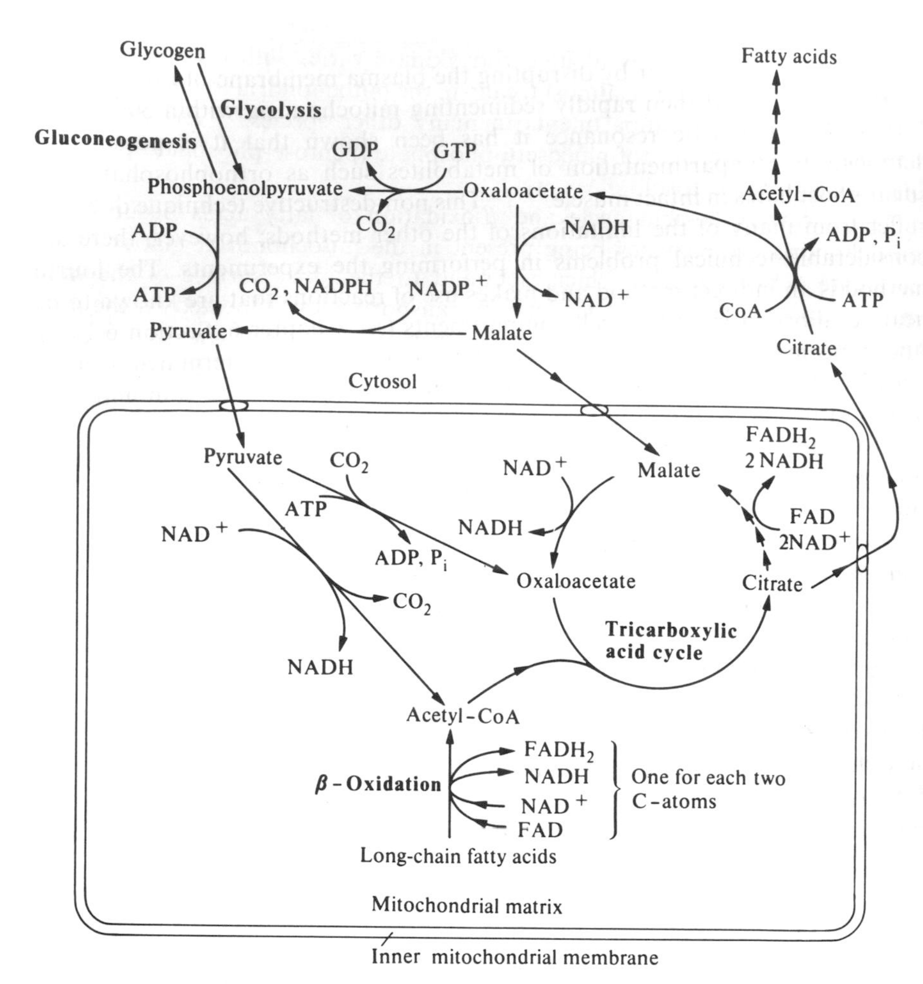

FIG. 8.6. The intracellular location of the principal metabolic pathways leading to the biosynthesis and degradation of carbohydrates and of fatty acids (P_i = orthophosphate).

direct determination of metabolite concentrations is difficult for two reasons: (i) the concentrations often change rapidly during the time required for isolation of subcellular organelles and (ii) substances having low values of M_r readily diffuse out of organelles and redistribute during isolation. Most attention has focused on the distribution of adenine nucleotides, the nicotinamide coenzymes, and the intermediates of the tricarboxylic-acid cycle in the mitochondrial compartments and in the cytosol. Little work has been carried out on other organelles. Four methods to determine these distributions have been generally used,[40] three direct and one indirect. The direct methods involve isolating mitochondria under conditions that minimize any change in the metabolite concentration or distribution, either by rapidly freeze-drying the tissue and then fractionating using organic solvents, or by disrupting the plasma membrane of isolated cells using digitonin and then rapidly sedimenting mitochondria within 30s. Using ^{31}P nuclear magnetic resonance it has been shown that it is possible to demonstrate compartmentation of metabolites such as orthophosphate and sugar phosphates in intact muscle.[41, 42] This non-destructive technique does not suffer from many of the limitations of the other methods; however, there are considerable technical problems in performing the experiments. The fourth method is an indirect method and makes use of reactions that are known to be near equilibrium in certain cell compartments (see Chapter 6, Section 6.3.3.1). The concentration of certain metabolites may then be determined from a knowledge of the concentrations of other metabolites that are common substrates of reactions at equilibrium (see reference 43).

All the methods have shortcomings. The method using non-polar solvents assumes that the metabolites are deposited during the freeze-drying process in the same location that they have *in vivo* and that they are not extracted by the non-polar solvent. The 'rapid' isolation method is still slow enough for certain changes to have taken place, e.g. it is estimated that a molecule of ATP can diffuse the length of a mitochondrion in <0.3s.[44] ^{31}P nuclear magnetic resonance is restricted to studying phosphate esters. It is less sensitive than other methods and the equipment necessary is not widely available. The indirect method assumes certain reactions are at equilibrium, whereas they may only be close to equilibrium and some may be closer than others. Nevertheless, in spite of these shortcomings there is general qualitative agreement about the distribution of many of the metabolites in the mitochondria and cytosol, but in some cases there is substantial disagreement about the actual magnitudes of the concentration gradients.

8.3.1.4 Location of nicotinamide cofactors, adenine nucleotides, and coenzyme A

One of the most important groups of substances about which information is necessary is the nicotinamide cofactors, since these cofactors participate in a large number of reactions. It has been known since the 1950s that NAD^+, NADH, $NADP^+$, and NADPH cannot cross the inner mitochondrial membrane. This became apparent when it was found that intact mitochondria oxidized exogenously supplied NADH very slowly compared with the rate of oxidation of NADH by damaged mitochondria.[45] The concentrations of these coenzymes in the mitochondria and the cytosol have been determined by the indirect method (see Section 8.3.1.3). In this method certain enzymes that are present in high concentrations and whose subcellular distribution is known are assumed to catalyse reactions that are near to equilibrium, whereas reactions catalysed by other enzymes or transport processes are assumed to be not at equilibrium. A list of some of the reactions that are considered to be near equilibrium is given in Table 8.7. The determination of the free

TABLE 8.7

Enzymes catalysing near equilibrium reactions in vivo *which have been used to estimate metabolite compartmentation in rat liver*

Enzyme	Location	Equilibrium definition	K_{eq}	Activity (katal kg^{-1})
Lactate dehydrogenase	Cytosol	$K = \frac{[\sum \text{pyruvate}][\text{NADH}][\text{H}^+]}{[\sum \text{lactate}][\text{NAD}^+]}$	1.1×10^{-11} mol dm^{-3}	4.1×10^{-3}
Malate dehydrogenase	Cytosol	$K = \frac{[\sum \text{OAA}][\text{NADH}][\text{H}^+]}{[\sum \text{malate}][\text{NAD}^+]}$	2.9×10^{-2} mol dm^{-3}	6.6×10^{-3}
Glycerol-3-phosphate dehydrogenase	Cytosol	$K = \frac{[\sum \text{DHAP}][\text{NADH}][\text{H}^+]}{[\sum \text{G3P}][\text{NAD}^+]}$	1.4×10^{-11} mol dm^{-3}	
Isocitrate dehydrogenase	Cytosol	$K = \frac{[\sum \text{oxogl}][\text{CO}_2][\text{NADPH}]}{[\sum \text{isocitrate}][\text{NADP}^+]}$	1.2 mol dm^{-3}	3.7×10^{-4}
Phosphogluconate dehydrogenase	Cytosol	$K = \frac{[\sum \text{ribulose 5P}][\text{CO}_2][\text{NADPH}]}{[\sum \text{phosphogluconate}][\text{NADP}^+]}$	0.17 mol dm^{-3}	4.7×10^{-5}
3-Hydroxybutyrate dehydrogenase	Inner mitochondrial Membrane	$K = \frac{[\sum \text{acetoacetate}][\text{NADH}][\text{H}^+]}{[\sum \text{3HOB}][\text{NAD}^+]}$	4.9×10^{-9} mol dm^{-3}	
Glutamate dehydrogenase	Mitochondrial Matrix	$K = \frac{[\text{oxogl}][\text{NH}_4^+][\text{NADH}][\text{H}^+]}{[\text{glutamate}][\text{NAD}^+]}$	3.9×10^{-13} mol^2 dm^{-6}	1.97×10^{-3}

Values of K are defined at 311 K (38°C), ionic strength, 0.25, and measured *in vitro* near pH 7.0.
References 44,46
OAA = oxaloacetate; DHAP = dihydroxyacetone phosphate; G3P = glycerol 3-phosphate; oxogl = 2-oxoglutarate; and 3HOB = 3-hydroxybutyrate.

$[NAD^+]/[NADH]$ ratio in the cytosol requires the following assumptions and measurements.

1. Lactate dehydrogenase is present exclusively in the cytosol and is present in sufficiently high concentration that the reaction is near equilibrium.

$$\text{Lactate} + NAD^+ \rightleftharpoons \text{pyruvate} + NADH + H^+$$

$$K = \frac{[\text{pyruvate}][NADH][H^+]}{[\text{lactate}][NAD^+]}.$$

2. The total concentrations of lactate and pyruvate in the tissue are determined after freeze clamping (see Chapter 6, Section 6.3.3.1).
3. It is assumed that the total concentration ratio of pyruvate/lactate is approximately the same as that in the cytosol. The mitochondria occupy about 20 per cent of the cell volume in liver and it is assumed that pyruvate and lactate freely diffuse into and out of the mitochondria.
4. It is also assumed that the ratio of the total concentrations of pyruvate and lactate is equal to the ratio of their free concentrations, i.e. no preferential binding occurs.

A further assumption is that the equilibrium constant measured *in vitro* with low enzyme concentrations is applicable *in vivo* (see discussion in Section 8.5.4).

By this method the $[NAD^+]/[NADH]$ ratio has been determined in the cytosol. A similar ratio is obtained using the other enzymes catalysing reactions that are near equilibrium, namely, glycerol 3-phosphate dehydrogenase and malate dehydrogenase, and these findings give support to the validity of this approach.

The same procedure is used to determine the $[NAD^+]/[NADH]$ ratio within the inner membrane of the mitochondria using reactions catalysed by the two enzymes 3-hydroxybutyrate dehydrogenase which is attached to the inner mitochondrial membrane, and glutamate dehydrogenase, which is in the mitochondrial matrix. It is necessary to measure the tissue contents of 3-hydroxybutyrate, acetoacetate, glutamate, 2-oxoglutarate, and ammonia. The procedure can be extended to determination of the ratio $[NADP^+]/[NADPH]$ in the cytosol using glucose 6-phosphate dehydrogenase, isocitrate dehydrogenase, or the 'malic' enzyme (L-malate: $NADP^+$ oxidoreductase (oxaloacetate-decarboxylating), EC 1.1.1.40). The mitochondrial $[NADP^+]/[NADPH]$ ratio is more difficult to determine because there is no $NADP^+$ -requiring enzyme that is exclusively mitochondrial. Calculations[46] have been made on the assumption that the mitochondrial glutamate dehydrogenase that reacts with either NAD^+ or $NADP^+$ *in vitro* (it is unusual for a dehydrogenase to use NAD^+ or $NADP^+$, interchangeably; see discussion in Chapter 5, Section 5.5.4.3) must react likewise *in vivo* and establish a near equilibrium as shown in the equations, but the results are still regarded as controversial.

(i) $NAD^+ + H_2O + $ L-glutamate $\rightleftharpoons$ 2-oxoglutarate $+ NH_3 + NADH$.

(ii) $NADPH + NH_3 +$ 2-oxoglutarate $\rightleftharpoons$ L-glutamate $+ H_2O + NADP^+$.

Sum $NAD^+ + NADPH \rightleftharpoons NADH + NADP^+$.

The results of determinations of the ratios $[NAD^+]/[NADH]$ and $[NADP^+]/[NADPH]$ are given in Table 8.8.

TABLE 8.8
The ratios of the concentrations of oxidized and reduced nicotinamide cofactors in rat liver determined by use of 'near equilibrium' enzymes

	Cytosol	Mitochondria
$[NAD^+]/[NADH]$	700–1000	7–8
$[NADP^+]/[NADPH]$	0.012–0.014	—

For further details see reference 43.

Although the exact values of these ratios vary with the method used and are sensitive to hormonal and nutritional changes, it is clear both that the ratio $[NAD^+]/[NADH]$ is much higher in the cytosol compared with the mitochondria and that in the cytosol the ratio of $[NADP^+]/[NADPH]$ is much lower than the ratio of $[NAD^+]/[NADH]$.

The principal significance of the different ratios $[NAD^+]/[NADH]$ and $[NADP^+]/[NADPH]$ is that it permits a further level of compartmentation within the cytosol. Spatial compartmentation occurs between pathways operating in the mitochondria and in the cytosol, but compartmentation based on enzyme specificity also occurs within the cytosol between NAD^+ and $NADP^+$-requiring dehydrogenases. The principal reactions generating NADPH are those of the pentose phosphate pathway,

$$\text{Glucose 6-phosphate} + NADP^+ \rightleftharpoons \text{glucono-}\delta\text{-lactone 6-phosphate} + NADPH,$$

$$\text{6-Phospho-D-gluconate} + NADP^+ \rightleftharpoons \text{D-ribulose 5-phosphate} + CO_2 + NADPH,$$

and the principal reactions utilizing NADPH are those of fatty-acid and cholesterol biosynthesis. There is a correlation between tissues synthesizing large amounts of fat, e.g. adipose and mammary tissue, and the activity of the pentose-phosphate pathway. Glycolysis and gluconeogenesis, on the other hand, require NAD^+ and NADH, respectively. Although the ratios of the NAD^+–NADH redox couple and of the $NADP^+$–NADPH redox couple are very different, they are nevertheless 'linked' to one another in the cytosol in such a way that changes in one couple will affect the other. The link between them is through enzymes that have common substrates, e.g. lactate dehydrogenase and the 'malic' enzyme, which are both present in the cytosol in large enough amounts to ensure equilibration.

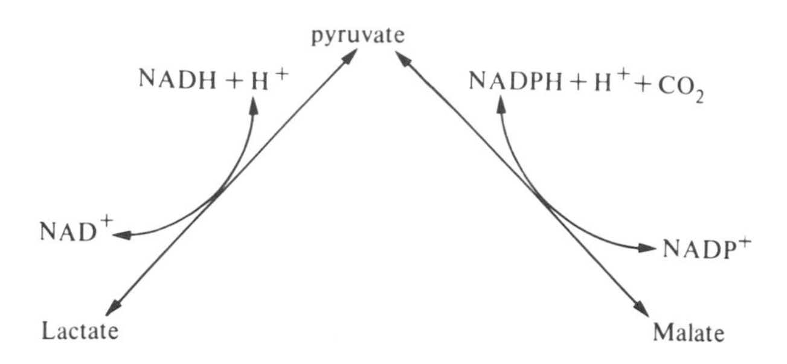

Since the principal oxidative reactions, e.g. oxidation of NADH, occur in the mitochondria and nicotinamide cofactors do not cross the inner mitochondrial membrane, it is necessary for the reducing power generated by glycolysis to be transferred to the mitochondria. This is brought about by two shuttles, the malate-aspartate shuttle and the glycerol 3-phosphate shuttle (Fig. 8.7). The malate-aspartate shuttle uses two malate dehydrogenases, two aspartate amino-transferases, and two translocators.* NADH in the cytosol is oxidized in the reaction catalysed by malate dehydrogenase and the malate formed enters the mitochondria in exchange for 2-oxoglutarate. There malate is reoxidized to oxaloacetate, which is then transaminated to aspartate that leaves the mitochondria in exchange for glutamate entering. The cycle is completed by transamination of aspartate in the cytosol. The malate-aspartate shuttle seems to be the principal shuttle in liver and heart for the transport of reducing equivalents into the mitochondria. A second shuttle that has been studied principally in insect flight muscle is the glycerol 3-phosphate shuttle. As can be seen in Fig. 8.7, it differs from the malate-aspartate shuttle in that *sn*-glycerol 3-phosphate does not enter the mitochondrial matrix, since the flavin-linked glycerol 3-phosphate dehydrogenase present in the inner mitochondrial membrane appears to have its substrate-binding site on the outer face of the membrane but the reduced flavin is generated on the inner face of the membrane (see Section 8.4.3).

Equally important in linking the biosynthetic and degradative pathways are the adenine nucleotides. The principal site of generation of ATP is the mitochondrion and of ATP utilization is the cytosol. Adenine nucleotides do not cross the inner mitochondrial membrane freely but are controlled by the adenine nucleotide translocator (Fig. 8.8), which allows the passage of one molecule of ATP in one direction in exchange for one molecule ADP in the other direction (for a review see reference 47). AMP is not transported by the translocator and cannot

* Specific proteins on the inner mitochondrial membrane allowing selective transport of particular ions across the membrane. (See Section 8.2.3.)

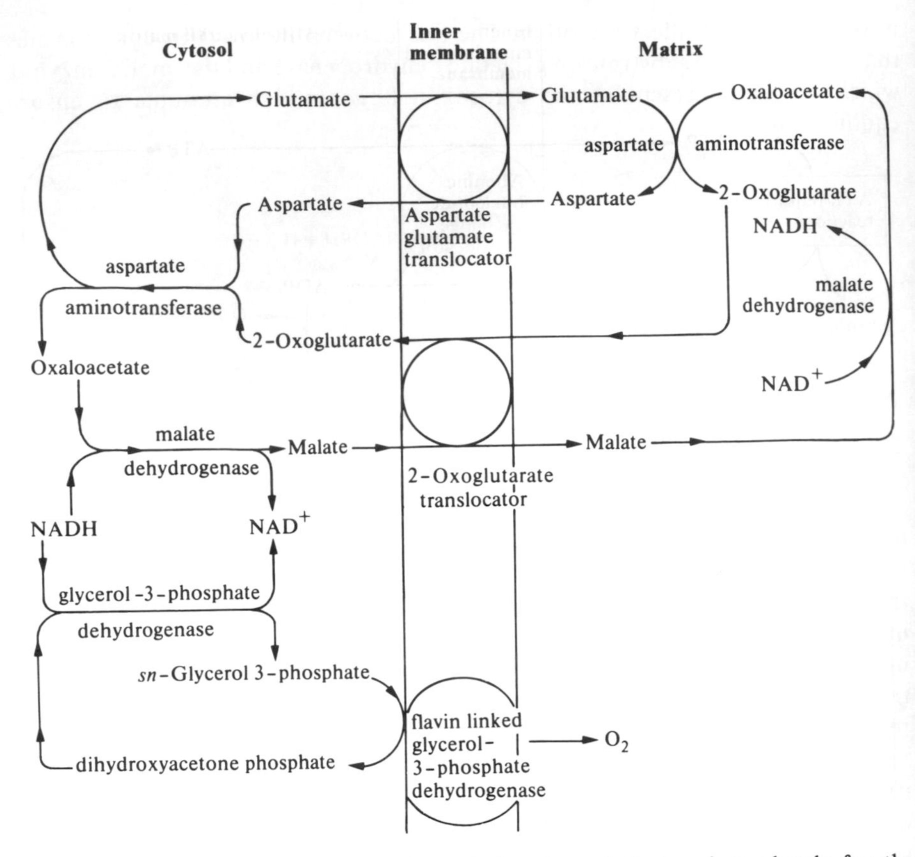

FIG. 8.7. The malate–aspartate shuttle and the glycerol 3-phosphate shuttle for the transport of reducing equivalents into the mitochondria.

cross the inner mitochondrial membrane without first being phosphorylated. Although the adenine nucleotide translocator constitutes as much as 6 per cent of the mitochondrial membrane protein, it appears to be the rate-limiting step in oxidative phosphorylation limiting the influx of ADP to the site of oxidative phosphorylation.[48] The ratio [ATP]/[ADP] in the cytosol is regulated by two enzymes present in large quantities and operating near equilibrium, namely glyceraldehyde-3-phosphate dehydrogenase and 3-phosphoglycerate kinase

(i) Glyceraldehyde 3-phosphate + NAD$^+$ + orthophosphate $\rightleftharpoons$

3-phosphoglyceroylphosphate + NADH + H$^+$.

(ii) 3-Phosphoglyceroylphosphate + ADP $\rightleftharpoons$ 3-phosphoglycerate + ATP.

Since these enzymes catalyse reactions near equilibrium, it can be seen that the ratio [NAD$^+$]/[NADH] in the cytosol is linked to the phosphorylation state of the adenine nucleotides in the cytosol.

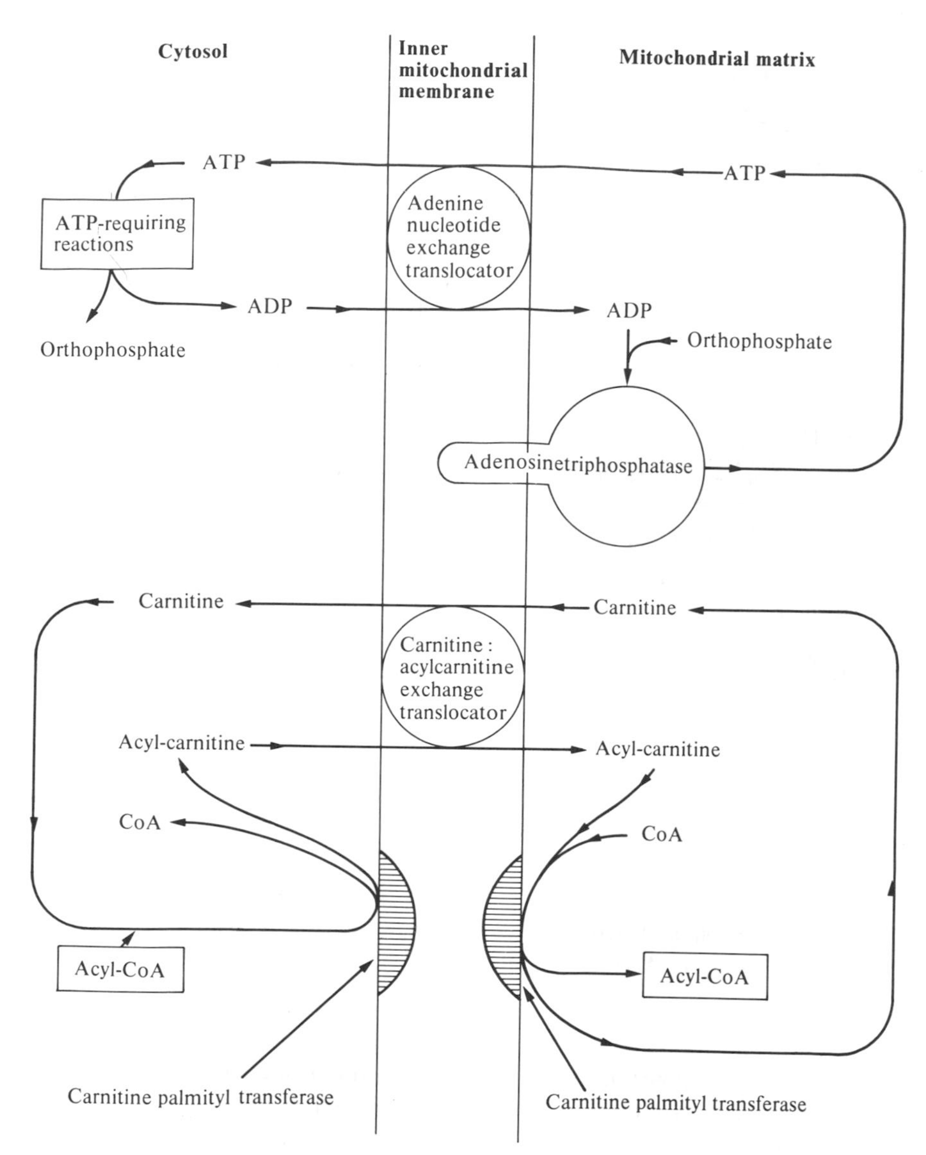

FIG. 8.8. Translocation of adenine nucleotides and of fatty acids across the inner mitochondrial membrane (see references 47, 49, and 138).

Besides nicotinamide cofactors and adenine nucleotides, the third important link between metabolism of fat and carbohydrate is acetyl-CoA, which can be either the end product of one pathway (β-oxidation or glycolysis) or the initial substrate of another (tricarboxylic-acid cycle or fatty-acid synthesis). The enzymes involved in acetyl-CoA metabolism, together with their intracellular locations, are shown in Fig. 8.9. Coenzyme A and its acylated derivatives are

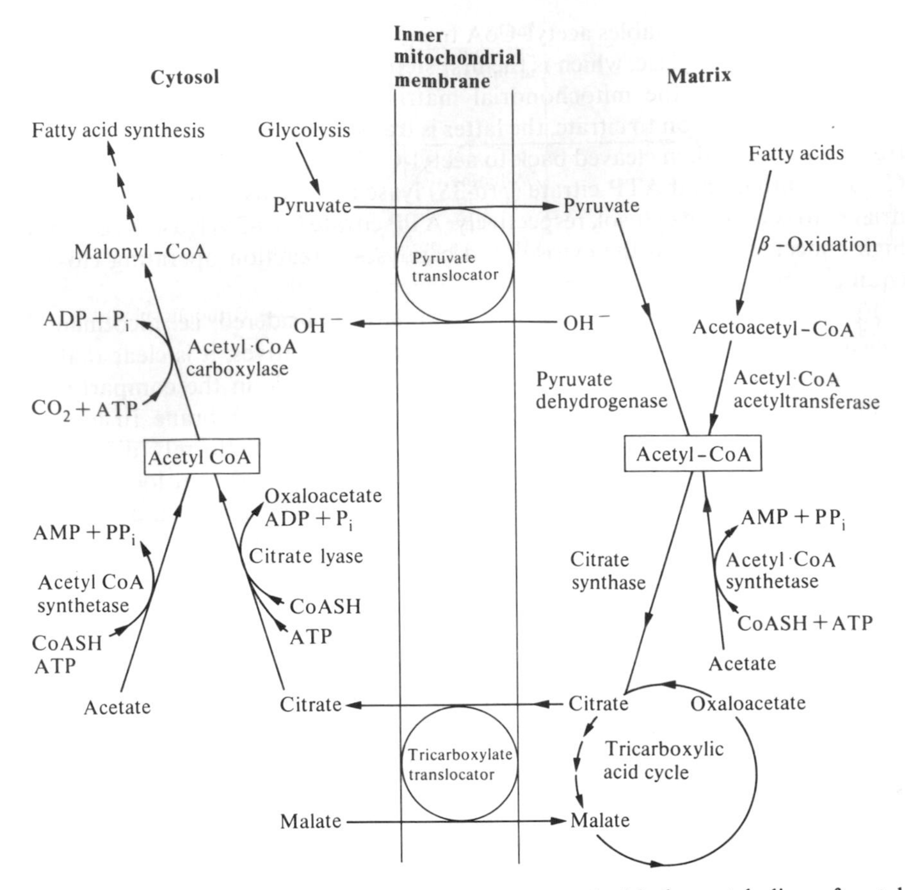

FIG. 8.9. The intracellular location of enzymes concerned with the metabolism of acetyl-coenzyme A. (P_i = orthophosphate; PP_i = pyrophosphate.)

unable to cross the inner mitochondrial membrane. Long-chain fatty acids are converted to their carnitine[50, 138] derivatives, which can then cross the membrane (Fig. 8.8)

$$(CH_3)_3N^+.CH_2.CHOH.CH_2COO^- + \text{Acyl-CoA} \rightleftharpoons$$
(Carnitine)
$$(CH_3)_3N^+.CH_2.\underset{\underset{\text{Acyl}}{|}}{\underset{O}{|}}{CH}.CH_2.COO^- + \text{CoASH}.$$

As can be seen from Fig. 8.9, most of the important steps generating acetyl-CoA, i.e. pyruvate dehydrogenase and acetyl-CoA acetyltransferase, are present in the mitochondrial matrix. The principal enzymes using acetyl-CoA are citrate (*si*) synthase, which enables acetyl-CoA to enter the tricarboxylic acid cycle, and acetyl-CoA carboxylase, which is the first step in fatty-acid biosynthesis. Acetyl-CoA generated in the mitochondrial matrix is effectively transferred to the cytosol by conversion to citrate, the latter is transported using the tricarboxylate translocator and then cleaved back to acetyl-CoA by ATP citrate (*pro*-3*S*) lyase. Citrate synthase and ATP citrate (*pro*-3*S*) lyase are exclusive to the mitochondrial matrix and the cytosol, respectively. ATP citrate (*pro*-3*S*) lyase is present in high concentrations in the cytosol and catalyses a reaction operating close to equilibrium.

From the three groups of substances we have considered, i.e. nicotinamide cofactors, adenine nucleotides, and coenzyme A derivatives, it is clear that the inner mitochondrial membrane plays an important role in the compartmentation of these compounds, since none can freely cross the membrane. In addition to these coenzymes, several other metabolites are not uniformly distributed across the mitochondrial membrane and there are several translocators that control the movement of intermediates of the tricarboxylic-acid cycle and certain amino acids. It is necessary to know the distribution of these compounds in order to understand the regulation of intermediary metabolism so that calculations of reaction velocities can be made from knowledge of the kinetic properties measured *in vitro*.

It is difficult to measure the distribution of these metabolites directly, for the reasons discussed with regard to nicotinamide cofactors. Indirect measurements have been made using the estimated values for the $[NAD^+]/[NADH]$ ratio. For example, the concentrations of malate and oxaloacetate in the cytosol are estimated as follows.

(1) Total cell malate and oxaloacetate is measured.
(2) The ratio of free $[NAD^+]/[NADH]$ is calculated as described earlier in this section. Now, for the reaction catalysed by malate dehydrogenase (MDH):

$$\text{Malate} + \text{NAD}^+ \rightleftharpoons \text{oxaloacetate} + \text{NADH} + \text{H}^+.$$

$$K_{\text{MDH}} = \frac{[\text{oxaloacetate}][\text{NADH}][\text{H}^+]}{[\text{malate}][\text{NAD}^+]}.$$

Therefore

$$\frac{[\text{Malate}]_{\text{cytosol}}}{[\text{Oxaloacetate}]_{\text{cytosol}}} = \frac{[\text{NADH}][\text{H}^+]}{[\text{NAD}^+]K_{\text{MDH}}}.$$

If $[H^+]$ and the ratio $[NADH]/[NAD^+]$ are known for the cytosol and also K_{MDH}, the ratio [malate]/[oxaloacetate] in the cytosol can be calculated. A similar equation may be written for malate and oxaloacetate in the mitochondria.

Total cell malate = $\text{malate}_{\text{cytosol}} + \text{malate}_{\text{mitochondria}}$.
Total cell oxaloacetate = $\text{oxaloacetate}_{\text{cytosol}} + \text{oxaloacetate}_{\text{mitochondria}}$.

Thus, the distribution of both malate and oxaloacetate can be calculated. There is, however, less agreement between workers concerning the concentrations of dicarboxylic and tricarboxylic acids in the two compartments than about the $[NAD^+]/[NADH]$ ratios. One reason for this is that the determinations are more indirect, relying on more than one reaction being at equilibrium, and thus discrepancies are likely to increase. There is also only qualitative agreement between results from the direct and indirect methods (see Section 8.3.1.3) of determining metabolite distribution. However, the data so far available do suggest that citrate, isocitrate, and possibly 2-oxoglutarate are more concentrated in the mitochondria, whereas glutamate, aspartate, and possibly oxaloacetate are more highly concentrated in the cytosol,[44, 51] but the actual concentration gradients obtained by different methods may differ by at least an order of magnitude and the results have to be treated with caution.

8.3.2 Compartmentation of arginine and ornithine metabolism in *Neurospora*

The metabolism of ornithine and arginine in *Neurospora* illustrates the importance of knowing not only the intracellular location of both enzymes and substrates but also the kinetic properties of the enzymes in order to understand their function *in vivo*. *Neurospora*, like many other fungi, is able to grow when supplied with glucose as the sole source of carbon. It is therefore readily able to synthesize both ornithine and arginine. *Neurospora* possesses all the urea-cycle enzymes (see Fig. 8.10) and also an active urease. When it is grown in a medium supplemented with arginine or ornithine it is readily able to assimilate these amino acids and use them in the biosynthesis of other amino acids. Table 8.9 shows some of the activities and K_m values of the principal enzymes present in *Neurospora* that utilize ornithine or arginine. Even when grown on minimal medium, *Neurospora* has substantial intracellular pools of ornithine and arginine, estimated at about 12 $mmol\,dm^{-3}$ and 8 $mmol\,dm^{-3}$, respectively.

At first sight it would seem that with a high intracellular pool of arginine and such an active arginase, arginine would be continually degraded at a high rate. It would also appear that provided the supply of carbamoyl phosphate was not limiting, with the presence of all the urea-cycle enzymes (Fig. 8.10), there would be a continuous high rate of synthesis of arginine and hence a substrate cycle (see Chapter 6, Section 6.3.4.1), but this does not occur.

However, a number of additional factors have to be taken into consideration before extrapolations from *in vitro* to *in vivo* conditions can be made. Arginase activity is normally assayed at pH 9.5 after a preincubation with Mn^{2+}. Although this treatment results in optimum catalytic activity, it is non-physiological and if the preincubation is omitted and assays are carried out at pH 7.5 the activity is appreciably lower[52] and appears sigmoidal with respect to arginine (for the significance of sigmoidal kinetics see Chapter 6, Section 6.2.2.1). However, the intracellular location of the enzymes and substrates is perhaps the most important factor, as seen in Fig. 8.11. Part of the urea cycle is located in the mitochondria, but the steps from citrulline to urea are located in the cytosol,

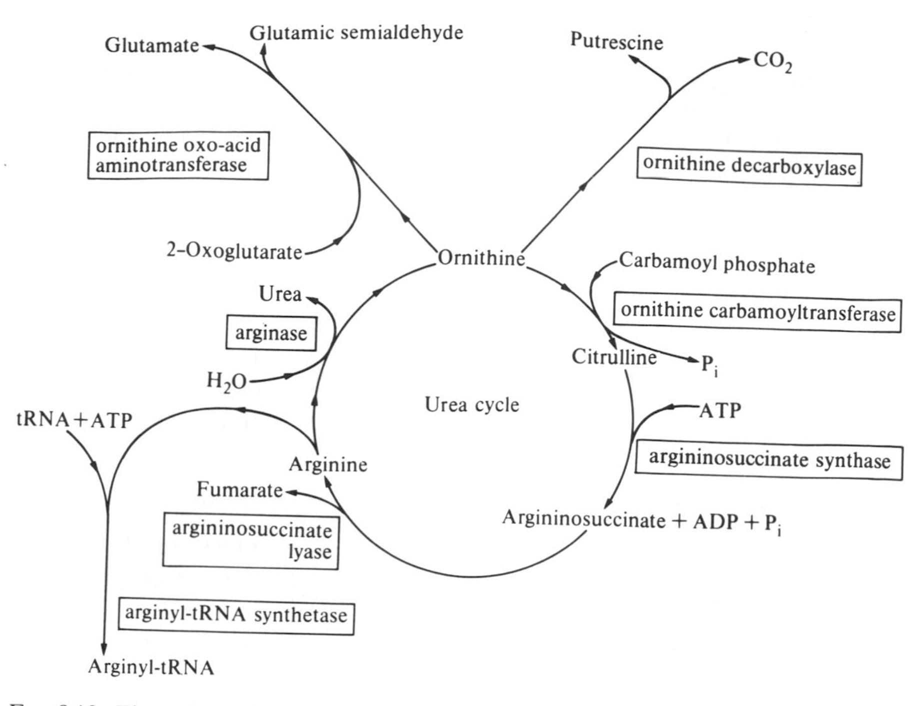

FIG. 8.10. The metabolic pathways for arginine and ornithine metabolism in the fungus *Neurospora crassa* (P_i = orthophosphate).

TABLE 8.9
The activities and Michaelis constants for ornithine- and arginine-metabolizing enzymes in Neurospora

Enzyme	K_m ornithine (mmol dm^{-3})	K_m arginine (mmol dm^{-3})	Activity (katal kg^{-1} protein)
Ornithine carbamoyltransferase	1.9		4.3
Ornithine oxyacid aminotransferase	2.0		0.23
Ornithine decarboxylase	0.2		8.3×10^{-4}
Arginase*		20	17.2
Arginyl-tRNA synthetase		0.02	1.3×10^{-5}
Argininosuccinate lyase		0.8	0.18
Argininosuccinate synthase			0.06

* Activity after Mn^{2+} activation.
For further details see references 55, 59–62.

where arginase and arginyl-tRNA synthetase are also located. It has been shown that approximately 98 per cent of the arginine and ornithine are located in a separate compartment known as a vacuole (Fig. 8.11) and thus the concentrations in the cytosol are much lower. This vacuole, which has been isolated and

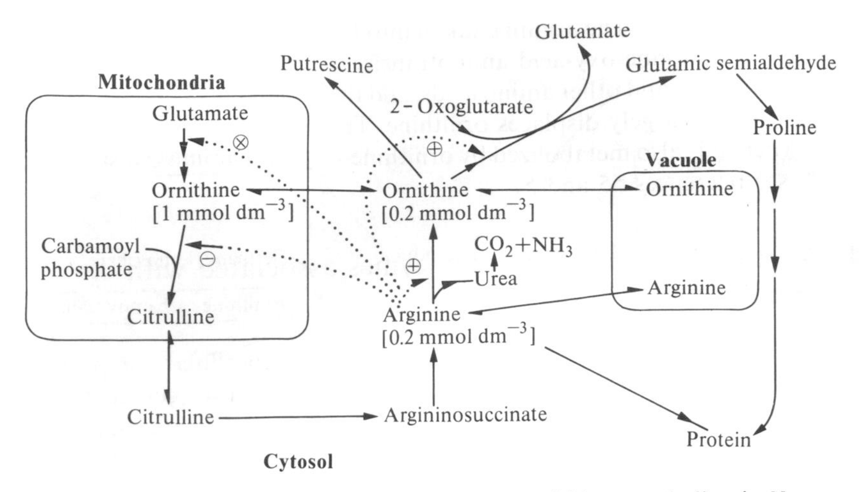

FIG. 8.11. The intracellular location of arginine and ornithine metabolism in *Neurospora crassa*: [ ] = concentrations in compartments indicated. Enzymes that are regulated by the intracellular concentrations of arginine are indicated ⊕ by induction, ⊖ by repression, and ⊗ by feedback inhibition (see references 53, 54 and 55).

partially characterized,[55] is capable of storing amino acids, particularly basic amino acids, and similar vacuoles have been found in other fungi and plants.[56]

The pathways are thus controlled to a large extent by the transport of substrates across both vacuole and mitochondrial membranes. In the intact organism the complete urea cycle does not operate to any appreciable extent, instead different halves of the cycle are operative under different physiological conditions. When a mutant deficient in urease is grown on a minimal medium and a small amount of [^{14}C]*guanidino*-arginine of high specific activity is added, most of the label is found in urea. The specific activities of the proteins labelled in short time periods show that the added [^{14}C]arginine does not equilibrate with the bulk of the arginine pool.[58] The low cytosolic concentration of arginine enables arginyl-tRNA synthetase to compete successfully with arginase (on the basis of their K_m values (Table 8.9) arginase will be operating at about 1 per cent maximum, whereas arginyl-tRNA synthetase will be at more than 90 per cent maximum provided $tRNA^{Arg}$ and ATP are not limiting). Thus, when grown on minimal medium the intracellular arginine, mainly in the vacuole, is synthesized using the urea-cycle enzymes to convert ornithine to arginine, whilst the arginase catalyses very little arginine degradation. On the other hand, when arginine is added to the growth medium it is the degradative part of the urea cycle that is active. The pool of arginine in the cytosol increases rapidly, e.g. from 0.2 mmol dm^{-3} to 15 mmol dm^{-3},[36] and, in addition, higher levels of ornithine-oxy-acid aminotransferase are induced and urea production is readily detectable. Also, the higher intracellular arginine inhibits further ornithine biosynthesis, probably by acting on acetylglutamate kinase, an enzyme on the ornithine biosynthetic pathway. The arginine taken into the cells is partly metabolized via arginase and ornithine-oxy-acid aminotransferase, leading to the formation of glutamate, proline, and other amino acids, and the remainder is taken up into the vesicles, where it largely displaces ornithine. The latter is transported into the cytosol where it is also metabolized by ornithine-oxy-acid aminotransferase. For reviews, see references 55 and 57.

8.4 Vectorial organization of enzymes associated with membranes

The intracellular membranes form the boundaries of subcellular compartments separating enzymes and metabolites in different regions of the cell, as has been discussed in the previous section. Associated with the intracellular membranes are a number of enzymes that are assymetrically arranged about the membranes. Several of these are concerned with the transport of metabolites and ions between compartments and their functioning can only be understood from studies of the intact membrane or organelles or in reconstituted systems where the enzyme, once isolated, has been reincorporated into membranes or membrane-like structures. In this section we discuss the methods used to study enzymes vectorially arranged in membranes and the progress that has been made in understanding how these enzymes function.

8.4.1 Enzymes in membranes

Much progress has been made in recent years in our understanding of the structures of membranes.[63] The 'fluid mosaic' model for membrane organization is generally accepted in which the phospholipids form a bilayer orientated with their polar head groups outside and the hydrocarbon chains towards the centre. The protein components of the membrane may be either on the surface of the lipid bilayer (peripheral proteins) or embedded in the lipid (integral proteins) as seen in Fig. 8.12. These two groups of proteins can be distinguished by the ease with which they are extracted from the membranes. Peripheral proteins form mainly electrostatic bonds with the charged phospholipid head groups and can be extracted by changes in ionic strength. The integral proteins bind to the hydrocarbon moiety of the phospholipids mainly by hydrophobic interactions and they are extracted from membranes using detergents or organic solvents. The extent to which an integral protein is embedded in the lipid bilayer varies from protein to protein, and this is reflected in the ease with which the protein is extracted. In addition, many of the integral proteins, once extracted, will aggregate once the detergent is removed because they contain a number of non-polar amino-acid side-chains that will take part in hydrophobic interactions. The range of interactions between proteins and the phospholipid bilayer is illustrated in Fig. 8.12.

Some proteins, such as fructose-bisphosphate aldolase and glyceraldehyde-3-phosphate dehydrogenase attached to the inner surface of the plasma membrane

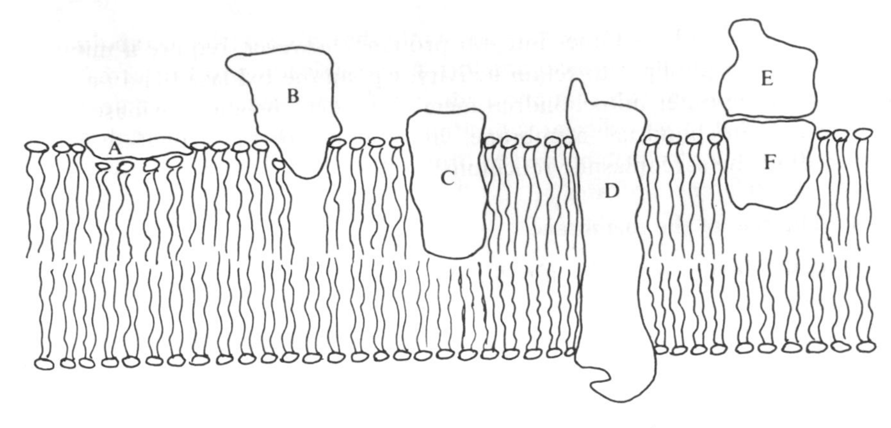

FIG. 8.12. Disposition of enzymes in a membrane.

A. A peripheral protein that can readily be extracted from the membrane by salt solutions, e.g. fructose-bisphosphate aldolase, glyceraldehyde-3-phosphate dehydrogenase in erythrocytes.
B. An integral protein with only a small portion of the polypeptide chain embedded in the bilayer, e.g. α-D-glucosidase and aminopeptidase in the brush border.
C. An integral protein with a large portion of the polypeptide chain embedded in the bilayer, e.g. acyl-CoA desaturase
D. An integral protein spanning the phospholipid bilayer, e.g. ion transporters such as adenosinetriphosphatase (Ca^{2+}-activated).
E. A protein attached to the membrane by a second protein (F) that is embedded in the bilayer, e.g. β-D-glucuronidase from the endoplasmic reticulum attached to a specific integral protein, egasin.

See references 39, 64–67.

of erythrocytes, can readily be extracted by 0.1 mol dm^{-3} EDTA or 1 mmol dm^{-3} ATP.[64] The integral proteins range from those in which only a small portion of the polypeptide chain is embedded in the bilayer, e.g. aminopeptidase from the intestinal brush border where a fragment of polypeptide chain of M_r weight 8000–10 000 is embedded within the membrane, whereas the remainder (M_r 280 000) is exposed in the lumen of the intestine.[65, 66, 68] to those in which the bulk of the polypeptide chain is presumed to lie in the bilayer, e.g. C_{55}-isoprenoid alcohol phosphokinase.[69] The latter enzyme from *Staphylococcus aureus* catalyses ATP-dependent phosphorylations of C_{55}-isoprenoid alcohols and functions in the biosynthesis of lipopolysaccharide. The enzyme has 58 per cent hydrophobic amino acids and when shaken with *n*-butanol–water mixtures it partitions in the butanol phase.

For some integral proteins the hydrophobic regions embedded in the bilayer merely act as an anchoring point on the enzyme. The enzymes may then be completely separated from phospholipid without any appreciable change in their kinetic properties, e.g. α-D-glucosidase and aminopeptidase[65, 66] from intestinal brush border, and nucleotide pyrophosphatase[70] from plasma membrane and endoplasmic reticulum. Other integral proteins, however, require a minimum amount of phospholipid to retain activity, e.g. amine oxidase (flavin-containing)[71] from the outer mitochondrial membrane, cytochrome *c* oxidase[72] from the inner mitochondrial membrane, and adenosinetriphosphatase (Ca^{2+}-activated) of the sarcoplasmic reticulum.

8.4.2 The role of the membrane

The role the membrane plays in relation to the bound enzymes falls into one of four categories.

1. The membrane can act as an anchoring point. For example, the functions of aminopeptidase and α-D-glucosidase in the brush border are to hydrolyse peptides and maltose prior to their uptake across the brush-border membrane. Anchoring the enzymes presumably makes more economical use of the enzymes, since they are retained at the required location (see Section 8.4.3). This is analogous to the use of immobilized enzymes in industrial processes; Chapter 11, Section 11.5.

2. The substrates and products of an enzyme reaction may have only limited solubility in water and may be more soluble in the lipid bilayer, and thus the enzyme will operate more efficiently when lipid bound, e.g. enzymes involved in lipid and glycolipid biosynthesis, in redox reactions involving long-chain ubiquinones, or in glycosyl-transfer reactions involving polyisoprenyl carrier lipids.[73]

3. The phospholipid bilayer may act as the medium in which certain multienzyme complexes are organized. An example of this is the acyl-CoA desaturase complex,[39] which comprises at least three proteins—cytochrome b_5 reductase, cytochrome b_5, and acyl-CoA desaturase—which act together to desaturate fatty acids. All three proteins are localized on the cytoplasmic face of the endoplasmic reticulum (Fig. 8.13).

4. Membranes may act to separate substrates, products, and effectors in an enzyme reaction, or may allow controlled transport processes to occur. These separations and transport processes cannot readily be studied once the enzyme has been separated from the associated phospholipid membrane components and are usually studied using intact membranes or vesicles obtained from them, or using reconstituted systems. Three examples, each differing in the way in which the enzyme is disposed in relation to the membrane are considered in the next sections.

8.4.3 Hydrolases of the microvillar membrane

Enzymes of this group are good examples with which to illustrate the role of the membrane as an anchoring point for enzymes. Microvillar membranes occur on the lumenal surfaces of the epithelium of the small intestine and the proximal tubule of the kidney. The microvillar membranes are highly convoluted to increase their surface area, so as to facilitate efficient absorption—in the case of the small intestine, absorption of nutrients, and in the case of the proximal tubule, reabsorption of solutes filtered by the kidney. The role of hydrolases in

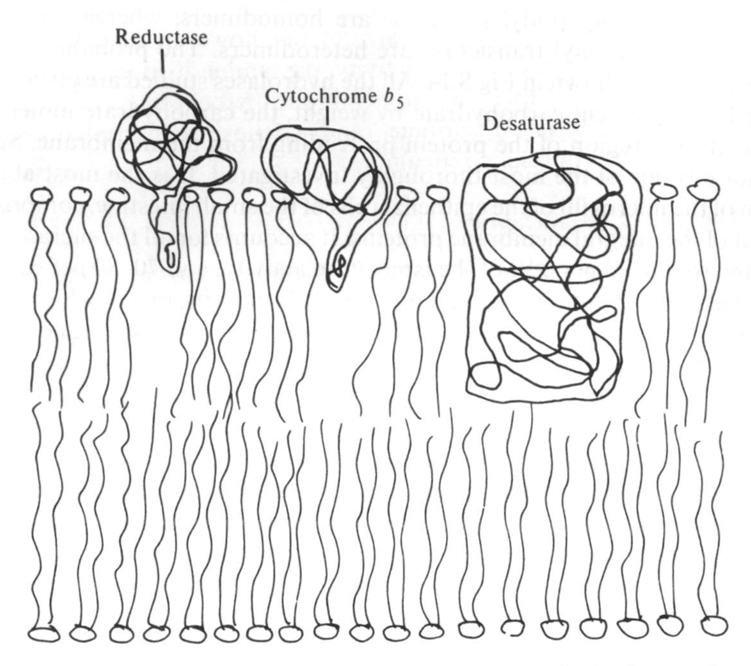

FIG. 8.13. The organization of the acyl-CoA desaturase complex in the membrane of the endoplasmic reticulum.

the small intestine is obviously to fulfil a digestive function. The proximal tubule contains a similar range of hydrolases but their function is not clear. The hydrolases form the major proportion of the integral membrane proteins of the microvillar membrane. A method that has been important in studying these enzymes is to compare the detergent extracted enzymes, usually using Triton X-100, with the enzymes extracted without detergent but after protease treatment to split off the hydrophobic anchoring region.

$$CH_3-C(CH_3)_2-CH_2-C(CH_3)_2-C_6H_4-(OCH_2CH_2)_nOH$$

Triton X-100 (isooctyl phenoxy polyethylene oxide)

With all these hydrolytic enzymes, the protease-extracted enzymes show full activity when compared with the detergent-extracted enzyme. The size of the hydrophobic anchor can be determined from the difference in M_r of the detergent-extracted hydrolase and the protease-extracted hydrolase. In determining the M_r of the detergent-extracted form by gel filtration, allowance has to be made for the micelle of detergent attached to the hydrophobic region necessary for solublization. Most anchors range in M_r from 3000 to 10 000. The structural arrangement of the hydrolases so far studied falls into one of two types, either homodimers (α_2) or heterodimers ($\alpha\beta$). Aminopeptidase, alkaline phosphatase and dipeptidyl peptidase are homodimers, whereas sucrase-isomaltase and γ-glutamyl transferase are heterodimers. The probable structural arrangements are shown in Fig 8.14. All the hydrolases studied are glycoproteins having 13–15 per cent carbohydrate by weight, the carbohydrate moiety being attached to the region of the protein protruding from the membrane. Sucrase-isomaltase is one of the most thoroughly investigated. It is the most abundant protein of the microvilli of the epithelial cells of the small intestine, comprising 10 per cent of the integral membrane proteins. It accounts for all the sucrase activity of the microvilli, 90 per cent of the isomaltase activity, and 70–80 per cent of the maltase activity. It comprises two non-identical subunits, one of which contains the isomaltase catalytic site and is anchored directly to the membrane, the other contains the sucrase activity and also accounts for the maltase activity and is bound to the isomaltase polypeptide, but is not anchored directly to the membrane (Fig 8.14). The sucrase-isomaltase is synthesized as a single polypeptide precursor that is cleaved during post-translational processing. It is generally assumed that the role of the membrane with all these hydrolases is to ensure efficient hydrolysis coupled with absorption through the associated membrane. For further details, see references 65, 66, 74, and 75.

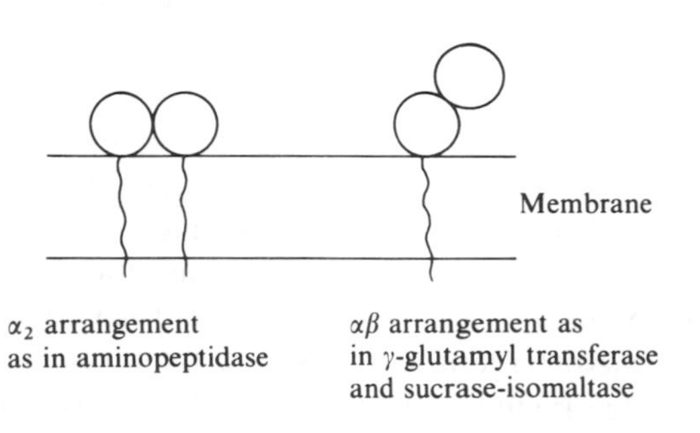

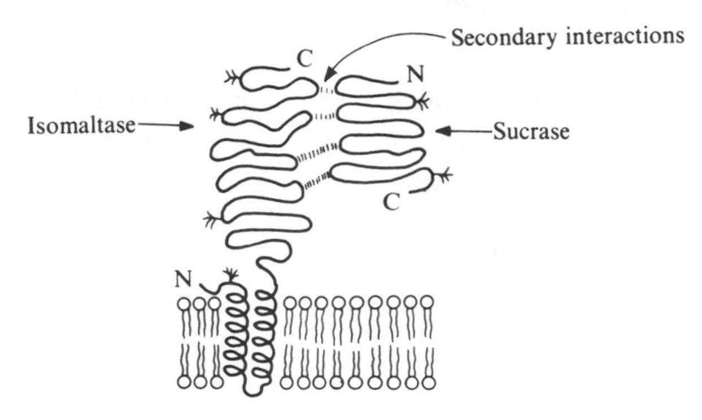

More detailed arrangement proposed for sucrase-isomaltase
(⚘ = carbohydrate residues)

FIG. 8.14. The attachment of microvillar enzymes to the cell membrane.[74,75]

8.4.4 Adenosinetriphosphatase (Ca^{2+}-activated) of the sacroplasmic reticulum

In contrast with the enzymes described in the previous section, for which the membrane acts primarily as an anchoring point, adenosinetriphosphatases (ATPases) are good examples of transmembrane proteins involved in the movement of ions through the membrane. A number of different adenosinetriphosphatase enzymes are known in which the hydrolysis of ATP is coupled to the transport of ions across a membrane, e.g. adenosinetriphosphatase (Na^+, K^+-activated) of the plasma membrane, mitochondrial adenosinetriphosphatase associated with oxidative phosphorylation (linked to H^+-movement), and adenosinetriphosphatase (Ca^{2+}-activated) of the endoplasmic reticulum and sarcoplasmic reticulum. Of these, the adenosinetriphosphatase (Ca^{2+}-activated) is perhaps the one of the best understood in molecular terms and therefore is the example chosen here. The sarcoplasmic reticulum is the intracellular membrane surrounding each myofibril in the muscle fibres. Calcium ions are released from the sarcoplasmic reticulum in response to a nervous impulse. This triggers the process of muscle contraction. Relaxation occurs when Ca^{2+} is removed and this is achieved by adenosinetriphosphatase (Ca^{2+}-activated), which pumps Ca^{2+} back into the sarcoplasmic reticulum at the expense of ATP hydrolysis. Vesicles of the sarcoplasmic reticulum can be prepared and from these adenosinetriphosphatase (Ca^{2+}-activated) can be isolated. There are only four main protein components of this membrane and adenosinetriphosphatase makes us about 90 per cent of this protein. The abundance and relative ease of isolation of this protein has made it a good model system for studying ion transport.

The adenosinetriphosphatase (Ca^{2+}-activated) requires an annulus of 30 molecules of phospholipid for activity. Removal of the phospholipids causes aggregation of the adenosinetriphosphatase; however, the phospholipids can readily be exchanged with added phospholipids if they are mixed in the presence of detergent and the detergent is later removed. Alternatively, the phospholipids may be replaced by certain non-ionic detergents, e.g. dodecyl octaethylene glycol monoether, which provide the amphiphilic environment necessary for maintenance of activity.[76] The adenosinetriphosphatase can be incorporated into vesicles or liposomes (structures composed of multiple phospholipid bilayers completely enclosing an aqueous phase)[77] containing internal phosphate ions. These reconstituted systems catalyse ATP-dependent Ca^{2+} transport. The stoichiometry of a Ca^{2+} transport associated with ATP hydrolysis is well established but the detailed mechanism by which the Ca^{2+} is transported across the membrane is not yet clear.

The sum of the steps is probably as follows:

$$\mathrm{ATP} + \underset{\text{outside}}{2\mathrm{Ca}^{2+}} + \underset{\text{inside}}{\mathrm{Mg}^{2+}} + 2\mathrm{K}^+ \rightleftharpoons \mathrm{ADP} + \text{orthophosphate} + \underset{\text{inside}}{2\mathrm{Ca}^{2+}} + \underset{\text{outside}}{\mathrm{Mg}^{2+}} + 2\mathrm{K}^+.$$

Under suitable conditions it is possible to demonstrate the reversibility of this process, i.e. a Ca^{2+} concentration gradient across the membrane can drive the synthesis of ATP. During the process of Ca^{2+} transport, the adenosinetriphosphatase becomes phosphorylated by ATP and then dephosphorylated.[78]

The ATPase comprises a single polypeptide chain (1001 amino-acid residues), the sequence of which has been deduced from the sequence of the cDNA.[79] There are two genes that encode the Ca^{2+} ATPases of fast-twitch and slow-twitch muscles and the sequences show them to be highly conserved.[80] The secondary structure and topography of the ATPase has been predicted on the basis of the likely folding pattern, together with the charged residues involved in the Ca^{2+}-binding and transmembrane regions. In addition, electron micrographs show a globule attached to the membrane by a stalk and helices are observed on freeze-fracture as intramembranous particles. A probable topology of the enzyme is given in Fig. 8.15. As can be seen, there are three distinct domains on the cytoplasmic side of the membrane, eight helices running through the membrane, and five loops on the lumenal side of the membrane. The steps in the mechanism are considered as follows:

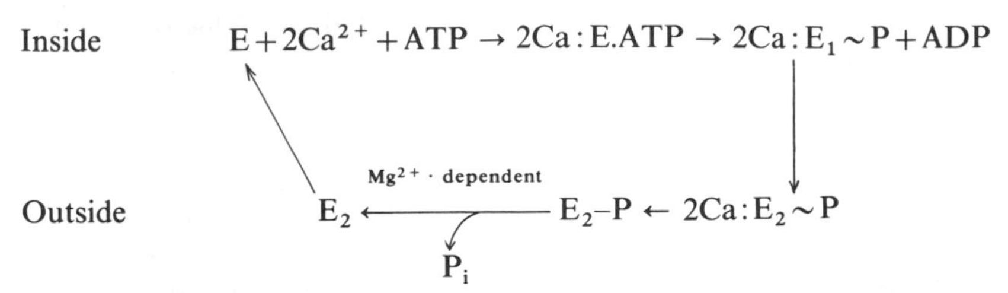

(The enzyme-bound calcium is shown without positive charge, since this is neutralized by the carboxyl groups on the enzyme.)

The initial step involves the binding of two Ca^{2+} and one ATP to the enzyme. The binding is cooperative and induces a conformational change (shown as

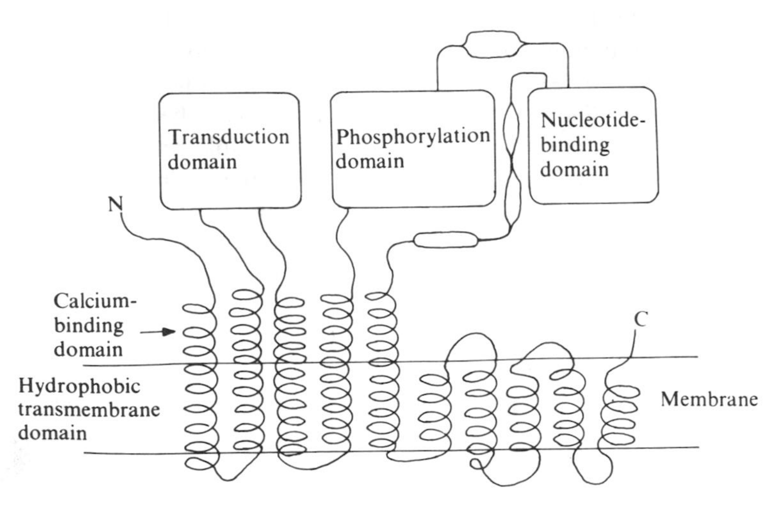

FIG. 8.15. The proposed structure of Ca^{2+}-ATPase.[80]

$E_1 \rightarrow E_2$) in the enzyme, which can be detected as a change in the fluorescence of tryptophan. The ATP is cleaved and phosphorylates Asp351 in the phosphorylation domain. The $2Ca:E_2{\sim}P$ has a lower affinity for Ca^{2+} than $2Ca:E_1ATP$ and is able to release Ca^{2+} on the lumenal side of the membrane, via the channel formed by the helices of the transmembrane region of the protein. The Ca^{2+} binding sites are believed to be the glutamic acid residues on the stalk of the helices close to the membrane.[80] It is not yet clear how many of the helices are involved in forming the channel, or how they are arranged in relation to one another.

8.4.5 Adenylate cyclase

3′:5′-Cyclic AMP is perhaps the best-understood of the second messengers that are released inside cells in response to an external signal, generally a hormone (see Chapter 6, Section 6.4.2.6). Other intracellular messengers include Ca^{2+} ions, phosphoinositides, and diacylglycerol. Cyclic AMP is generated by the reaction catalysed by adenylate cyclase:

$$\text{ATP} \rightarrow 3':5'\text{cyclicAMP} + \text{pyrophosphate}.$$

The intracellular concentrations of cAMP range from 10^{-7} to 10^{-6} mol dm^{-3}. The rises are generally short-lived and the cAMP is degraded by a phosphodiesterase. In this section we focus on the mechanism by which the signal from the hormone is relayed across the membrane to stimulate adenylate cyclase activity. In the early studies of this system it was not clear whether there existed a single protein spanning the membrane and having both adenylate cyclase activity and a hormone-binding site, or whether more than one protein was involved. The question was resolved by some ingenious cell-fusion experiments.[81] Turkey erythrocytes, which possess an adenylate cyclase activity responsive to catecholamines, when treated with *N*-ethyl maleimide lose their catalytic activity. A second type of cell, a mouse erythroleukaemia cell, lacks catecholamine receptors but possesses adenylate cyclase activity. These two cell types were fused and the resultant hybrid cell could be stimulated by catecholamines to increase cAMP levels. It was possible to show that the coupling between the activities exhibited a temperature dependence that related closely to the fluidity of the cell membrane. Clearly, the receptor and the adenylate cyclase were distinct entities able to associate by lateral movement through the membrane.

The involvement of a third protein became evident as a result of showing that the efficient signalling was dependent on GTP, the latter being degraded to GDP and orthophosphate. A prolonged stimulation could be obtained by using a non-hydrolysable analogue of GTP, guanyl-5′-ylimidodiphosphate. Having demonstrated that three proteins were involved, in order to fully understand the mechanism it was necessary to isolate the components and to reconstitute the system and show that it functioned normally.

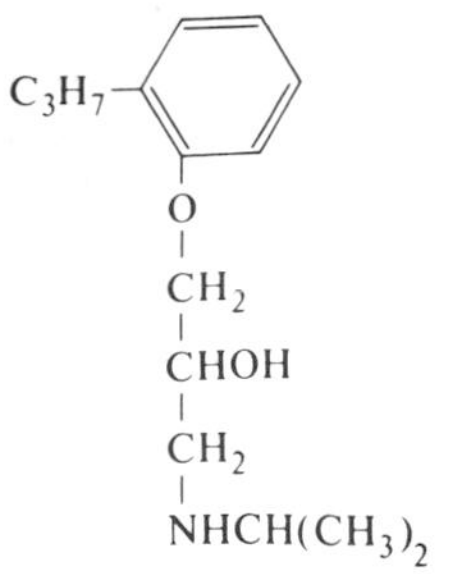

Dihydroalprenolol

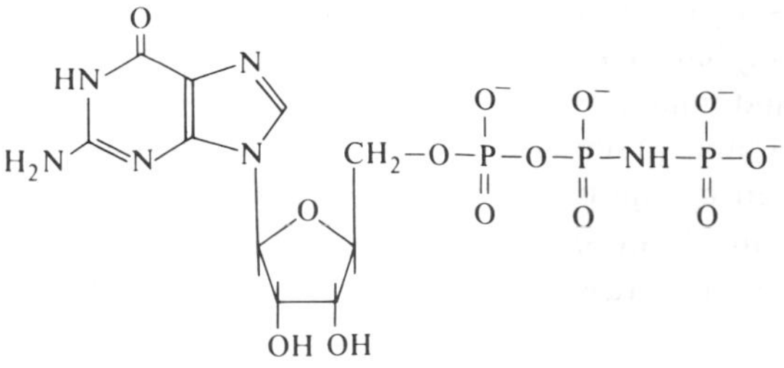

Guanyl-5′ylimidodiphosphate

As a result of more recent work, it has become apparent that there are a number of examples of intracellular signalling that are triggered by binding of a hormone to the outside of a cell. They each comprise three proteins—a receptor, a transducer (G protein), and an effector. In the case of cAMP signals, the effector is adenylate cyclase, but with phosphoinositides and diacylglycerol the effector is phospholipase C. The most thoroughly studied system is that involving the β-adrenergic receptor and adenylate cyclase. There are four types[82] of catecholamine receptor, α_1, α_2, β_1 and β_2, which differ in their relative affinities towards agonists* and antagonists* and also in the effector to which they are coupled. In the case of the system involving the β_1 type, it has now been possible to isolate and reconstitute the components. These will be described briefly. All three proteins are present in very small amounts in the cell membrane, and since they are membrane proteins it is necessary to extract them from the lipid of the cell membrane. It is also necessary to have sensitive methods for detection. In the case of the receptor this requires a suitable binding assay to measure $H + R = HR$. Hormones show very high affinities for their receptors, since most circulating hormones are present in nanomolar concentrations. Binding must be measured at physiological concentrations of hormones, or equivalent concentrations of their agonists or antagonists, otherwise non-specific binding to other proteins will confuse the picture. This generally means using radioactively labelled agonists or antagonists having very high specific radioactivity. In the case of the β-receptor,[81] ^{3}H-labelled dihydroalprenolol was used; this has $K_d = 6 \times 10^{-9}$ mol dm^{-3}. The β-receptor has been purified from turkey erythrocytes. The receptors are first solublized using the detergent digitonin, since the receptor is an integral membrane protein, and it is then purified using a combination of ion-exchange chromatography and affinity chromatography. The latter is performed on alprenolol–agarose and elution is with alprenolol,[83]; a 12 000-fold purification can be achieved. The high purification factor is an indication of the small proportion of the total protein in the extract it represents. Since the β-receptor has been isolated, a number of other receptors have been isolated and

* Agonists are structural analogues that mimic the effect of the hormone; antagonists are structural analogues that block the effect of the hormone.

the genes for a number of them have been cloned and sequenced. It is now possible to compare the sequences of the β-receptor, cholinergic receptors, various drug receptors, and light receptors such as rhodopsin and bacteriorhodopsin.[82,84,85] All of these receptors have a number of features in common in their sequences, in particular seven stretches of between 20 and 28 hydrophobic amino-acid residues, which are consistent with the arrangement in the membrane shown in Fig 8.16.

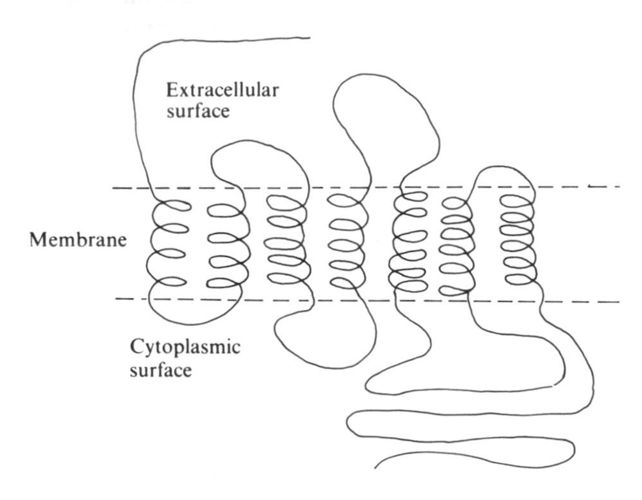

FIG. 8.16. Proposed model for receptor structure.

Although this arrangement has not been proven directly, it is analogous to that of the purple membrane in *Halobacterium*, which has been studied by high-resolution electron diffraction.[136] The first two cytoplasmic domains are the most conserved and may be involved in the interaction with the G-protein. The hormone binds into a pocket formed by the seven helices spanning the membrane.

The G-proteins have also been purified and there is a family of homologous transducing proteins.[86] They are peripheral membrane proteins and therefore more easily extracted. They are sometimes refered to as N-proteins (**N**ucleotide, as opposed to **G**uanine nucleotide binding proteins). Different types of G-proteins exist that may act positively (G_s, stimulating) on the effector protein or negatively (G_i, inhibitory). All the G-proteins that have been isolated have been found to be heterotrimers ($\alpha\beta\gamma$). Of the three polypeptide chains, the α chain differs most among members of the family, whereas the β and γ chains can be exchanged within a family. Two toxins have been useful in studying the mode of action of the G-proteins, namely cholera toxin and pertussis toxin. Both toxins are able to catalyse the ADP ribosylation of the α subunit of G-proteins. In the case of cholera toxin this leads to inhibition of GTP hydrolysis and thus to prolonged activation of adenylate cyclase. Pertussis toxin causes inhibition of hormone-mediated activation of adenylate cyclase.

The third component, adenylate cyclase, has been purified from rabbit myocardial membranes[87] and from bovine brain.[88] In both cases there is an initial solubilization and the main purification step is affinity chromatography on forskolin–agarose. Forskolin is a diterpene isolated from the roots of the aromatic herb *Coleus forskohlii* that causes stimulation of adenylate cyclase.

Having purified the three components, it is then possible to reconstitute the system. This has been done in stages.[89] The receptor and G_s protein can be combined in phospholipid vesicles. Addition of an agonist then stimulates the binding of guanine nucleotides and causes the hydrolysis of GTP to GDP and orthophosphate. A complete reconstitution has been effected in which the proteins R, G, and C are combined in the proportions 1:10:1 and the phospholipid vesicles comprising phosphatidylethanolamine and phosphatidylserine in the ratio 3:2. It was then shown that isoproterenol, an agonist, was able to stimulate an increase in adenylate cyclase activity.[89] A number of possible mechanisms have been proposed; the mechanism most generally favoured[90] is shown in Fig. 8.17.

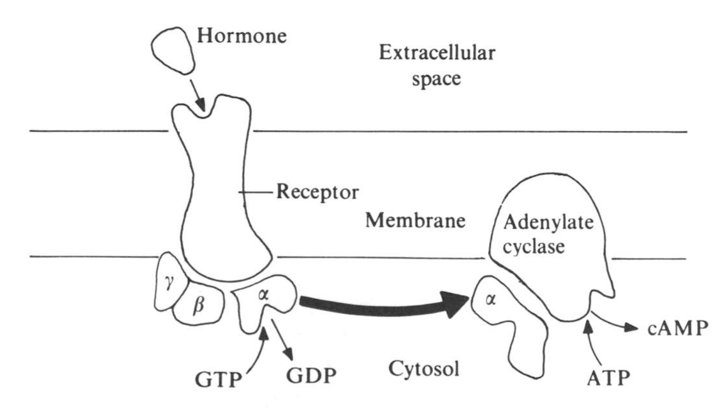

FIG. 8.17. Coupling between receptor, transducer, and adenylate cyclase.

8.5 The concentrations of enzymes and substrates *in vivo*

When cells or tissues are disrupted and extracts are prepared prior to assay, the suspension containing the enzymes is considerably diluted compared with that existing in the intact cell. This undoubtedly affects many aspects of an enzyme-catalysed reaction. Although many biochemists are aware of this problem when comparing measurements made *in vitro* and *in vivo*, there is little quantitative information that might enable a better extrapolation from *in vitro* to *in vivo* conditions. In this section we highlight some of these differences and discuss what progress has been made in understanding them.

8.5.1 The high concentrations of macromolecules within cells

The protein concentration in all of the intracellular compartments is high. In the mitochondrial matrix the concentration is estimated to be about 500 mg cm^{-3},[15] and in the cytosol concentrations may be in the region of 200 mg cm^{-3}. This has a number of effects on the properties of the medium in which enzymes act. The medium is much more viscous: a solution of albumin at a concentration of 200 mg cm^{-3} has 2–3 times the viscosity of water and one having a concentration of 500 mg cm^{-3} has 5–10 times the viscosity of water.[91] This will affect the rate of diffusion of solutes, particularly those of macromolecules (see Section 8.5.2) . A protein concentration of 500 mg cm^{-3} means that individual protein molecules will be in close proximity to one another and this will tend to promote interactions between protein molecules that would not occur in dilute solutions. A good example of this is given in Section 8.5.3.3. It will also have a profound effect on the state of water molecules in the cell. The water molecules in close proximity to protein molecules differ in some of their properties from those of pure solvent. Apart from a lower freezing point, it can be shown by various physical methods, e.g. n.m.r. spectroscopy, i.r. spectroscopy,[92] that these water molecules do not rotate as freely as those in pure water. These differences are due to various types of secondary bonds that bind water molecules to protein either directly or indirectly. The water molecules in solutions of protein have been classified into three types[93] based on the differences in their physical properties, as summarized in Table 8.10. The actual contents of type II and III water will depend on the particular protein, since a higher proportion of polar amino-acid residues will lead to a higher proportion of bound water. The number of molecules of internal water (type III) for various proteins that have been studied by X-ray crystallography[94] is as follows: papain, 15; actinidin (a related protease), 15; lysozyme, 4; carboxypeptidase, 24; cytochrome *c*, 3; and penicillopepsin, 13. In the case of lysozyme, where there are four

TABLE 8.10
Physical properties of water in protein solutions

Type		Rotational correlation time[a]	Average number of grams of water per gram of protein
I	as in pure water	3×10^{-12} s	
II	loosely bound to protein; freezing point lowered	approx. 10^{-9} s	0.3–0.5
III	irrotationally bound; motion same as protein molecule, integral part of protein structure	10^{-5}–10^{-7} s	0.003

[a] Rotational correlation time can be considered as the time taken for a molecule to rotate through an angle of 360°. For a more detailed explanation see reference 95.
For review on water in biological systems, see reference 93.

molecules of internal water tightly bound, it has been shown that when lysozyme is gradually hydrated, enzyme activity is just detectable when these are between 150 and 250 water molecules per lysozyme. This corresponds approximately to the sum of the type II and type III water molecules. If the mitochondrial proteins behave in the same way as detailed in Table 8.10 then in the mitochondrial matrix with a concentration of 500 mg cm^{-3} protein, 25 per cent of water would be type II and 1.5 per cent type III. Thus, a high proportion of the water in the mitochondrial matrix will have properties different from those of pure water. The same will apply to other subcellular compartments, though in some cases to a lesser degree. Although the nucleus has a slightly lower protein content (approximately 34 g protein per 100 g nuclei)[10] than the mitochondrial matrix, it has a much higher content of nucleic acids, particularly DNA (9 g DNA per 100 g nuclei).[10] Isolated DNA has been shown to bind approximately ten water molecules per nucleotide residue (approximately 5.6 g water per g nucleotide)[96] and thus DNA binds proportionally more water than does protein. It is unlikely that DNA in the nucleus will bind this high proportion of water, since much of its surface will be covered with nucleoprotein. Nevertheless, it would appear that much of the water in the nucleus will be bound.

8.5.2 Diffusion rates within cells

When considering the flux through a metabolic pathway *in vivo*, it is important not only to know the activities of all the enzymes involved in order to determine the rate-limiting steps but also to know the 'transit times' of each step. The 'transit time' is the time required for the product of one enzyme-catalysed reaction to diffuse to the active site of the next enzyme. There is at present insufficient information available to make accurate estimates of the 'transit time' although a number of approximate estimates have been made.[97–99] The two principal parameters required for these calculations are (i) the distance separating two sequential enzymes in a metabolic pathway and (ii) the diffusion coefficient of the product of the first enzyme in the medium of which the cell is composed. The average distance separating two enzymes may be estimated if the number of molecules of each enzyme per cell is known, together with the volume of the relevant subcellular compartment. The diffusion coefficient of the enzyme, substrates, and products can easily be measured in aqueous solution, but it is clear that the diffusion coefficients are significantly lower in the cytosol compared with those in pure water. Various methods have been used to estimate the diffusion coefficients in the cytosol.[30] The ratio D_{H_2O}/D_{cells} varies from 2 to 5 for molecules of $M_r = 200$–400, to 200-fold for protein of $M_r = 40\,000$–$150\,000$.

From the calculations[97,98] made so far it seems that the 'transit time' for the steps in certain metabolic pathways could be rate limiting, particularly in the larger cell compartments, unless the enzymes are not randomly distributed but are organized. For example, the calculated 'transit time' for fructose 6-phosphate to diffuse to phosphofructokinase in the cytosol is too long in relation to the known rate at which glycolysis can occur. This suggests that the enzymes may not be randomly arranged in the cytosol but that some kind of ordering exists. There is also kinetic evidence that suggests that an enzyme (aspartate aminotransferase) and its substrate (aspartate) may not be uniformly distributed throughout the mitochondrial matrix, but that the concentration of aspartate may be highest in the centre and the aspartate aminotransferase highest at the periphery of the matrix.[100]

8.5.3 Enzyme associations *in vivo*

It seems probable that in living cells many weak interactions occur that are not easily detected once a cell has been disrupted and the components suspended in buffer solution. In the equilibrium

$$\text{Protein}_A + \text{protein}_B \rightleftharpoons \text{protein}_A . \text{protein}_B$$

dilution will favour the dissociation of the $\text{protein}_A . \text{protein}_B$ complex, and in many cases when cell disruption is carried out prior to subcellular fractionation, then the concentrations of proteins may be lowered by at least one order of magnitude. Thus, the multienzyme complexes discussed in Chapter 7 probably represent only the stronger protein interactions. In recent years attempts have been made to try to demonstrate other weaker interactions. Most attention has focused on the two major metabolic pathways, namely the glycolytic pathway occurring in the cytosol, and the tricarboxylic-acid cycle occurring in the mitochondrial matrix. The enzymes catalysing the steps in both these pathways are usually present in cells in high concentrations, because of their central importance in metabolism. Skeletal muscle is a particularly rich source of glycolytic enzymes since most of the ATP generation in white muscle is through glycolysis. It has been shown that a number of glycolytic enzymes bind to fibrous-actin,[101] one of the major proteins concerned in muscle contraction. The capacity of actin to bind glycolytic enzymes varies with the enzyme; the capacities being greatest with 6-phosphofructokinase, pyruvate kinase, fructose-bisphosphate aldolase, glyceraldehyde-3-phosphate dehydrogenase, glucose-6-phosphate isomerase, and lactate dehydrogenase. The kinetic properties of some of the enzymes change on binding and this, together with the fact that binding to actin is reversible, could be an additional means of regulation, but further work is required before the relevance of these observations to the situation in intact muscles can be assessed.

8.5.3.1 Fructose-bisphosphate aldolase and glyceraldehyde-3-phosphate dehydrogenase

Several attempts have also been made to demonstrate associations between individual enzymes of the glycolytic pathway. We have discussed in Chapter 7, Section 7.12, the evidence for the existence of a 'glycogen particle' that contains a number of glycolytic enzymes. Fructose-bisphosphate aldolase and glyceraldehyde-3-phosphate dehydrogenase catalyse sequential steps in glycolysis:

$$\text{Fructose-bisphosphate} \rightleftharpoons \text{dihydroxyacetone phosphate} + \text{glyceraldehyde 3-phosphate.}$$

$$\text{Glyceraldehyde 3-phosphate} + NAD^+ + \text{orthophosphate} \rightleftharpoons \text{3-phospho-D-glyceroyl phosphate} + NADH.$$

(Triosephosphate isomerase converts dihydroxyacetone phosphate to glyceraldehyde 3-phosphate.)

So far direct methods to demonstrate association between the two enzymes, e.g. by ultracentrifugation or gel filtration, have yielded negative results; however, there is indirect evidence for some form of association.[102] When glyceraldehyde 3-phosphate is present in aqueous solution, there is normally an equilibrium between the aldehyde and the diol forms (see also Chapter 5, Section 5.5.2).

$$\begin{array}{l} H-C=O \\ \quad | \\ H-C-OH \\ \quad | \\ H_2C-O-PO_3^{2-} \end{array} + H_2O \rightleftharpoons \begin{array}{l} \quad OH \\ \quad | \\ H-C-OH \\ \quad | \\ H-C-OH \\ \quad | \\ H_2C-OPO_3^{2-} \end{array}$$

If the results of kinetic studies on glyceraldehyde-3-phosphate dehydrogenase alone are compared with those from fructose-bisphosphate aldolase coupled with glyceraldehyde-3-phosphate dehydrogenase, the rate of the coupled reaction is such as to suggest that the aldehyde form of glyceraldehyde 3-phosphate is used directly by the dehydrogenase and does not equilibrate with the diol. This in turn suggests a close association of the enzymes and is supported by measurements of polarization of fluorescence.

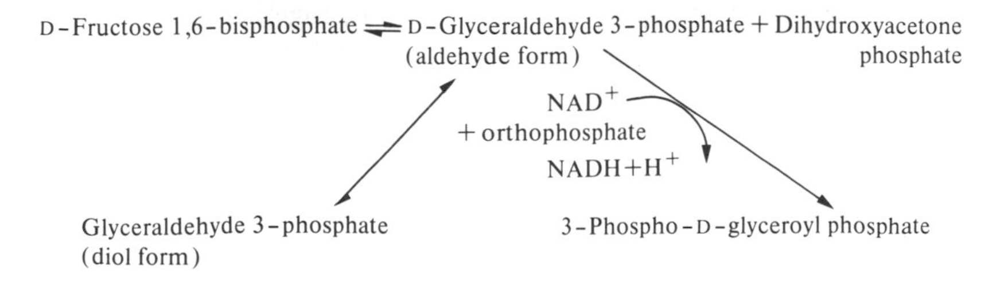

If fluorescein is coupled covalently to either fructose-bisphosphate aldolase or glyceraldehyde-3-phosphate dehydrogenase and the fluorescein-labelled enzyme mixed with varying amounts of the unlabelled form of the second enzyme, a change in fluorescence is observed, suggesting an interaction between the enzymes; the binding constant may be determined from the titration curve. This interaction between glyceraldehyde-3-phosphate dehydrogenase and fructose-bisphosphate aldolase has also been detected by active-enzyme centrifugation[104] (see Chapter 3, Section 3.2.4) and has a dissociation constant in the micromolar range[105]. There is also evidence for a weak association between fructose-bisphosphate aldolase and fructose-bisphosphatase isolated from rabbit muscle.[106]

8.5.3.2 The tricarboxylic-acid cycle enzymes and related metabolic pathways

The enzymes catalysing the tricarboxylic-acid cycle, together with aspartate aminotransferase and glutamate dehydrogenase, occur in the matrix of the mitochondria. Apart from succinate dehydrogenase that is covalently linked to inner mitochondrial membrane, all other enzymes are readily released when the inner mitochondrial membrane is disrupted. A number of investigations have been carried out to see whether there is any evidence for associations between the enzymes involved.

There is evidence for a weak interaction between mitochondrial malate dehydrogenase and citrate (*si*) synthase[107] and between malate dehydrogenase and aspartate aminotransferase.[108] During the purification of citrate (*si*) synthase it is difficult to remove the last traces of malate dehydrogenase activity, suggesting a type of interaction or similarity in properties.

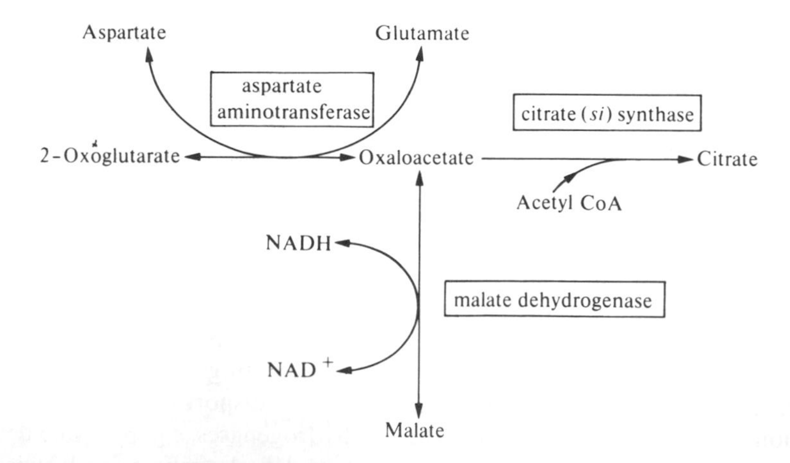

In the mitochondrial matrix, because of the very high protein concentration, there is a limitation on free water and this could promote interaction between the two enzymes. In order to bring about water limitation *in vitro*, experiments have been carried out in which some of the water has been replaced by poly(ethylene glycol). It has been demonstrated that under these conditions citrate (*si*) synthase will co-precipitate with mitochondrial malate dehydrogenase but not with cytosolic malate dehydrogenase. In a similar set of experiments, the partitioning of malate dehydrogenase and aspartate aminotransferase in a biphasic system of water and dextran–trimethylaminopropyl poly(ethyleneglycol) was studied. If either malate dehydrogenase and aspartate aminotransferase (both mitochondrial), or if malate dehydrogenase and aspartate aminotransferase (both cytosolic) were used, the enzymes in each pair appeared in the same fraction after counter current distribution. If, on the other hand the pair of enzymes were not from the same subcellular fraction, e.g. mitochondrial malate dehydrogenase and cytosolic aspartate aminotransferase, then they appeared in different fractions, suggesting no interaction. Thus, the interaction between the enzymes seems to be specific. Further evidence for an association between these enzymes and other enzymes of the tricarboxylic-acid cycle comes from studies on immobilized enzymes.[109] When fumarate hydratase and mitochondrial malate dehydrogenase were immobilized on a gel filtration column, it was possible to demonstrate that the immobilized enzymes were able to bind enzymes catalysing related steps in metabolism, e.g. immobilized fumarate hydratase binds malate dehydrogenase and citrate synthase whereas immobilized malate dehydrogenase binds citrate synthase and immobilized aspartate aminotransferase binds malate dehydrogenase.

Carbamoyl phosphate synthase I is the most abundant liver mitochondrial protein,[110] being present in concentrations between 0.4 and 1.0 mmol dm^{-3}. In addition to playing a key role in the urea cycle, it appears to act as a linking enzyme by binding glutamate dehydrogenase and also enhancing the stability of the complex between glutamate dehydrogenase and aspartate aminotransferase. The association between glutamate dehydrogenase and carbamoyl phosphate synthase I has been demonstrated by cross-linking the two in the mitochondrial matrix.[110]

Although associations have been demonstrated only between pairs of tricarboxylic-acid cycle enzymes in mitochondria, there is evidence[111] for a multienzyme aggregate in bacteria containing several of the tricarboxylic-acid cycle enzymes. By preparing spheroplasts and then lysing them in suitable medium, it was possible to demonstrate by gel filtration that a multienzyme aggregate existed that contained citrate synthase, aconitase, isocitrate dehydrogenase, succinate-CoA ligase, and malate dehydrogenase, and was thus able to catalyse the steps from fumarate to oxoglutarate. It is possible that it may also contain oxoglutarate dehydrogenase. The processes of β-oxidation of fatty acids and the tricarboxylic-acid cycle involve a number of dehydrogenases that have to transfer their reducing equivalents to the electron-transport chain on the inner mitochondrial membrane. Several of these dehydrogenases, e.g. pyruvate dehydrogenase, oxoglutarate dehydrogenase, malate dehydrogenase, and hydroxyacyl-CoA dehydrogenase are able to bind to complex I, which catalyses the electron-transport chain from NAD^+ to coenzyme Q,[112]. Thus there is evidence for a physical link between the tricarboxylic-acid cycle and the electron-transport chain. The concentrations of the tricarboxylic-acid cycle enzymes in the mitochondrial matrix[113] are greater than 1 mmol dm^{-3} so that high concentrations of enzymes might be expected to be necessary to demonstrate interactions *in vitro*. These experiments illustrate the point that many weak interactions may occur *in vivo* but that special methods may have to be used to detect them: neither of these complexes has been detected in the analytical ultracentrifuge.

8.5.3.3 Arginase and ornithine carbamoyltransferase

Another interaction between two enzymes that is not easily demonstrated in dilute solutions is that between arginase and ornithine carbamoyltransferase from yeast.[114,115] The function of ornithine carbamoyltransferase is in the biosynthesis of arginine from ornithine (see Fig. 8.10). When arginine is added to the growth medium it acts as a good source of both carbon and nitrogen for the synthesis of many compounds, e.g. other amino acids, proteins, and carbohydrates and also is a preferred source of arginine compared to that synthesized endogenously. Two mechanisms operate to suppress ornithine carbamoyltransferase activity in this situation. One is the suppression of synthesis of ornithine carbamoyltransferase and is relatively slow acting. The second mechanism, which is fast acting, involves interaction between ornithine carbamoyltransferase and arginase. This latter mechanism is demonstrated in concentrated suspensions by comparing the activity of ornithine carbamoyltransferase in cell extracts. (Fig. 8.18, curve A) with that in permeabilized cells (curve B). In the latter, yeast cells are made permeable to low-M_r substances by use of nystatin (a polyene fungicide produced by *Streptomyces noursei*), but the proteins remain within the cells and thus the protein concentration is unchanged and therefore high.

In both situations A and B in Fig. 8.18 the synthesis of ornithine carbamoyltransferase is suppressed after arginine is added to the medium, but additionally in B, ornithine carbamoyltransferase becomes bound to arginase in

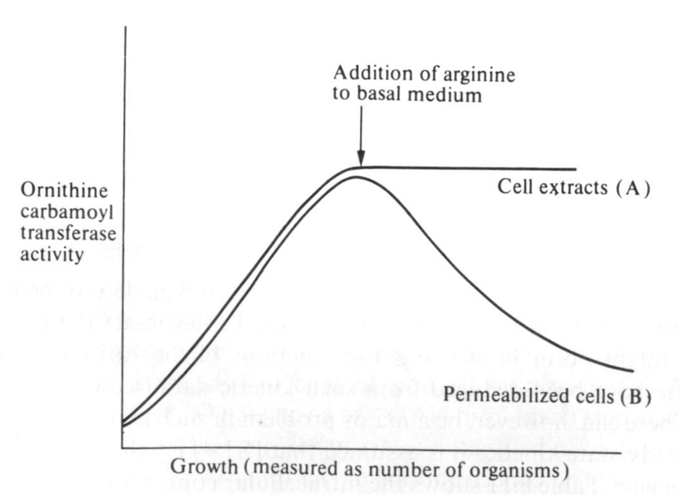

FIG. 8.18. The effects on ornithine carbamoyltransferase activity in yeast of adding arginine to the growth medium (see reference 114).

the presence of ornithine and arginine and the ornithine carbamoyltransferase activity is inhibited. Both arginase and ornithine carbamoyltransferase are trimeric, and the complex formed between them has been shown to have a stoichiometry $\alpha_3\beta_3$, and a dissociation constant of between 10^{-8} and 10^{-10} mol dm^{-3}:

$$\alpha_3\beta_3 = \alpha_3 + \beta_3.$$

The substrates ornithine and arginine promote the association of the complex $\alpha_3\beta_3$. It was originally proposed[114,115] that ornithine bound both to the catalytic site on ornithine carbamoyltransferase and also to a regulatory site. However, recent studies[116] using a bisubstrate analogue (transition-state analogue), δ-*N*-(phosphonacetyl)-L-ornithine,[117] have shown that a conformational change can be demonstrated on binding the catalytic site but give no evidence for a separate regulatory site. It seems probable that binding to the active sites of both enzymes induces the necessary conformational changes for their association.

8.5.3.4 *Membrane-bound hexokinase*

The binding of enzymes to other cell components may modify the properties of the enzyme concerned. In a number of mammalian tissues, hexokinase is distributed between the mitochondria and the cytosol. It has been suggested that an equilibrium exists between hexokinase that is free in the cytosol and that bound to the mitochondria and that this may act as a form of control mechanism. In one of the most studied tissues, the brain, up to 80 per cent of the hexokinase may be bound to the mitochondria and much of this may be released *in vitro* by increasing concentrations of glucose 6-phosphate. There is evidence that the kinetic properties of hexokinase are changed when it is released from the mitochondria. A higher proportion of hexokinase binds to mitochondria when ATP concentrations are diminished and this may enable hexokinase to compete more favourably with other ATP-requiring processes. There is also evidence that ATP released from the mitochondria is used by the hexokinase in preference to exogenous ATP. This in turn suggests that the initiation of the glycolytic pathway is directly coupled to ATP synthesis in mitochondria.[113]

8.5.4 The relative concentrations of enzymes and substrates

The constants, K_m and V_{max} (see Chapter 4, Section 4.3), determined by steady-state kinetic studies *in vitro*, are frequently used to estimate the rates at which reactions might occur *in vivo*; e.g. the functions of the various isoenzymes of hexokinase have been deduced from such kinetic data (see Chapter 4, Section 4.3.1.3). There can, however, be a major problem in such extrapolations, namely, that in steady-state kinetics it is assumed that $[S] \gg [E]$; this may not always be the case *in vivo*. Table 8.11 shows the intracellular concentrations of some of the glycolytic and tricarboxylic-acid cycle enzymes and their substrates. Although these are often quite wide variations in the calculated concentrations (compare

TABLE 8.11
Intracellular concentrations of enzymes and substrates in rat tissues

Enzyme	Enzyme concentration (μmol dm^{-3})	Substrate concentrations (μmol dm^{-3})	References
1. Phosphoglucomutase	5	G6P = 450	120, 124
2. Fructose-bisphosphate aldolase (muscle)	66	FBP = 65, DHAP = 28, G3P = 3	120, 125
3. Glyceraldehyde-3-phosphate dehydrogenase (muscle)	35	G3P = 3, NAD^+ = 600, P_i = 2800	120, 124, 125
4. Triosephosphate isomerase (muscle)	8	DHAP = 28, G3P = 3	120, 125
5. Pyruvate carboxylase (liver mitochondria)	10–40	Pyruvate = 50–1000 ATP = 1000–6000	120, 126, 128
6. Phosphoenolpyruvate carboxykinase (liver cytosol)	15	Oxaloacetate = 5, GTP = 100–600	37, 120
7. Citrate synthase (liver mitochondria)	35	Oxaloacetate = 0.01–0.1 Acetyl-CoA = 50–1000	120, 126, 128
8. Malate dehydrogenase (liver mitochondria)	70	Malate = 80–340 NAD^+ = 18 000	120, 129
9. Malate dehydrogenase (liver cytosol)	30	Malate = 300–450 NAD^+ = 500	120, 129
10. Aspartate aminotransferase (liver mitochondria)	50	Glutamate = 380–4800 Oxaloacetate = 0.01–0.1	120, 129
11. Aspartate aminotransferase (liver cytosol)	1.6	Glutamate = 2500–4400 Oxaloacetate = 5	120, 129

G6P = glucose 6-phosphate; FBP = fructose-bisphosphate; DHAP = dihydroxyacetone phosphate; G3P = glyceraldehyde 3-phosphate; and P_i = orthophosphate.

Table 8.11 with for example reference 113), it can be seen that for many of these the enzyme concentration is of the same order of magnitude as that of the substrate concentration. Glycolytic and tricarboxylic-acid cycle enzymes are of major importance in liver and muscle and are therefore present in higher concentrations than the majority of other enzymes in these tissues, but other enzymes are known to be present in high concentrations, e.g. 0.78 mmol dm^{-3} cathepsin D, and 1.54 mmol dm^{-3} cathepsin B in the lysosomal matrix.[23]

One of the assumptions in steady-state kinetics is that, since $[S] \gg [E]$, the total substrate concentration is assumed to approximate to the free substrate concentration. Clearly this will not apply for enzymes present in high concentrations as an appreciable amount of the substrate will be enzyme bound. Various equations have been derived for the rate of an enzyme reaction when the substrate is not in great excess.[120–123] Perhaps the most important factor is that knowledge of free substrate concentration *in vivo* is required and that this cannot easily be determined.

It is not only in the case of enzymes that are present in high concentrations that it is unsafe to assume that the total substrate and free substrate approximate to the same value. An enzyme present in small amounts in a tissue may be in competition with an enzyme present in high concentrations for a common substrate that may be largely bound to the latter enzyme. Thus, the free substrate concentration may, in several cases, be appreciably lower than the total substrate concentration. This situation is well illustrated in the case of oxaloacetate in mitochondria. A number of reactions compete for oxaloacetate and the enzymes concerned are present in high concentrations (compared with substrate); thus a very high proportion of the total oxaloacetate in mitochondria must be enzyme-bound (Fig. 8.19). The kinetic properties (K_m and V_{max}) of citrate (*si*) synthase are such that one would expect any oxaloacetate formed to be converted to citrate, and not to phosphoenolpyruvate for gluconeogenesis; clearly this does not always happen. It has been suggested that there must be some type of macromolecular organization (multienzyme complex) that prevents the competition between the enzymes for oxaloacetate.

However, this overlooks another aspect of the difference between experiments carried out *in vitro* and the situation *in vivo*, namely that the latter behaves more

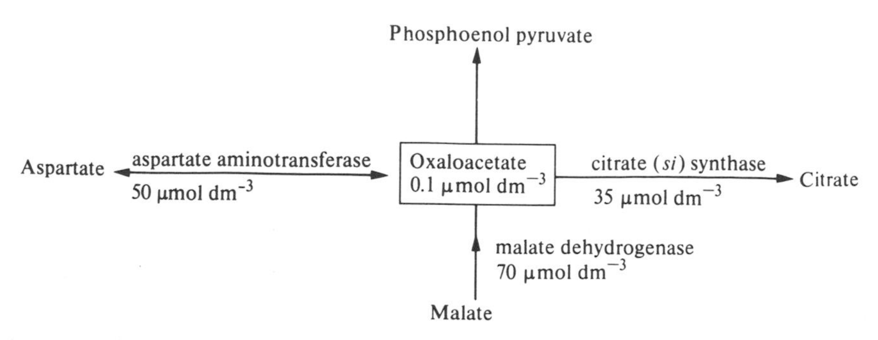

FIG. 8.19. The concentrations of enzymes metabolizing oxaloacetate in mitochondria.

like an open system rather than the closed system used *in vitro*. Although the intermediates of most metabolic pathways in cells are present in low concentrations, they are continuously being replaced by new source material and the end products are often removed from the cell. Thus, although the intracellular concentration of oxaloacetate is very low and undoubtedly much of it is enzyme bound, changes in the amounts of the enzymes utilizing oxaloacetate do not appreciably affect the intracellular concentration of free oxaloacetate in the 'near-open' system existing in the cell. Mathematical modelling with the aid of computers[130] has been made to determine the effect of a sharp increase in the amount of a protein that binds oxaloacetate (i.e. comparable with an oxaloacetate-metabolizing enzyme). This is found to cause only a transient fall in the intracellular concentration of free oxaloacetate. The reason is that oxaloacetate is rapidly replenished because it is linked metabolically to the other tricarboxylic-acid cycle intermediates that are present in much higher concentrations. In particular, the relatively large malate pool becomes slightly diminished to replenish oxaloacetate. Mathematical modelling techniques for studying the control of metabolic activity (see Chapter 6, Section 6.3.3.1) are still being developed and they require accurate kinetic data and knowledge of pool sizes to give reliable simulations, but they will play an increasing role in our understanding of metabolism in the whole cell, where the complexity of the situation makes computer simulations essential.

Another example in which the proportion of bound ligand is important *in vivo* is that of cAMP-dependent protein kinase.[131] This enzyme is present in muscle and when activated by a sufficiently high concentration of cAMP phosphorylates a number of proteins, amongst them phosphorylase kinase (as shown below).

$$R_2C_2(\text{inactive}) + 2\text{cAMP} \rightleftharpoons R_2(\text{cAMP})_2 + 2\text{C (active)}$$

R_2C_2 = inactive cAMP dependent protein kinase (R = regulatory, C = catalytic)

$$\text{ATP} + \text{protein} \xrightleftharpoons{\text{active protein kinase}} \text{ADP} + \text{phosphoprotein}$$

(e.g. inactive phosphorylase kinase) (e.g. active phosphorylase kinase)

The role of phosphorylase kinase in the control of glycogen metabolism is discussed in Chapter 6 (Section 6.4.2.1). The kinetics of activation of cAMP-dependent protein kinase by cAMP are complex,[132] but the concentration of cAMP that gives 50 per cent activation (apparent K_a) of the enzyme is reported to be in the range 10^{-8} to 10^{-7} mol dm^{-3}. The cAMP content of most tissues is in the range 0.1 to 1.0 μmol kg^{-1} wet weight (i.e. if equally distributed throughout the cell would give 10^{-7} to 10^{-6} mol dm^{-3}) and this would appear to be a sufficient concentration to activate substantially the phosphorylase kinase. However, it is known that in tissues such as resting muscle the rate of glycogenolysis is very low. This led initially to the suggestion that cAMP must be compartmentalized within the muscle so that the concentration to which the protein kinase is exposed is much lower. It is now clear that this explanation is unnecessary. The comparison between the *in vitro* experiment and the *in vivo* situation is false because the concentration of protein kinase in muscle is high (about 0.23 μmol dm^{-3})—approximately the same as that of cAMP. The apparent K_a will be dependent on the concentration of protein kinase, since high concentrations of the latter will favour association of the R and C subunits. Thus, at physiological concentrations of the protein kinase the apparent K_a is 1.5×10^{-6} mol dm^{-3}, i.e. higher concentrations of cAMP are required to effect activation. In addition, an inhibitor has been isolated from skeletal muscle that binds to the C subunit and inactivates it. Both these factors clearly have to be taken into account in explaining the activation *in vivo*.

Another factor to be considered, which arises from the high concentrations of certain enzymes in tissues, is the equilibrium position. When the enzymes in a reaction are present in only catalytic amounts, the equilibrium position for the reaction is unchanged, although in practice it may be difficult to study the equilibrium in the absence of the appropriate enzyme:

$$A + B \rightleftharpoons C + D$$

$$K_{eq} = \frac{[C][D]}{[A][B]}.$$

However, if the enzyme is present in concentrations similar to that of the substrates, so that appreciable amounts of the substrates and products are enzyme bound, then the equilibrium may be that between the enzyme–substrate complex and the enzyme–product complex. If the enzyme–catalysed reaction proceeds, for example by ternary complexes (Chapter 4, Section 4.3.5.1), the relevant equilibrium is

$$EAB \rightleftharpoons ECD$$

$$K'_{eq} = \frac{[ECD]}{[EAB]}.$$

A number of examples are given in Table 8.12. It can be seen that most of the values for K_{eq}, for the bound enzyme are near unity, and that this makes a very significant difference in a number of the examples cited. Thermodynamic arguments have been proposed (see reference 113) to account for the values of K'_{eq} near unity. These suggest that for the evolution of ideal catalysts the free-energy change should be near zero ($K_{eq} \approx 1$) (compare triose phosphate isomerase, Fig. 5.29), since this would give balanced rates of diffusion, chemical transformation, and desorption of steps in a metabolic sequence.

8.6 Conclusions

It can be seen from the preceding section that it has been possible to identify a number of factors that affect enzyme reactions and metabolic pathways *in vivo*,

TABLE 8.12
Changes in equilibrium by high concentrations of enzymes

	Enzyme	$K^{\dagger}_{eq}$	K'_{eq} enzyme bound
1.	Lactate dehydrogenase	0.37×10^{-4}	0.25
2.	Alcohol dehydrogenase	0.5×10^{-4}	5–10
3.	Pyruvate kinase	3×10^{-4}	1.0–2.0
4.	Arginine kinase	0.1	1.2
5.	Creatine kinase	0.1	≈1.0
6.	Adenylate kinase	0.4	1.6
7.	Phosphoglycerate kinase	8×10^{-4}	0.5–1.5
8.	Hexokinase	2000	≈1.0
9.	Adenosine triphosphatase	1.3×10^{5}	9
10.	Triose phosphate isomerase	2.2	0.6
11.	Phosphoglucomutase	1.7	0.4

References 113, 133, 134. † Equilibrium constant when catalytic amounts of enzymes are present, in contrast to when enzyme and substrate are present in comparable amounts.

but that are not always apparent in many of the better defined *in vitro* situations. Although many factors have been identified, in many cases it is not possible to be sure of the magnitude of these effects *in vivo*. In some cases this is because research interest in this area has only recently developed, in other cases it is a question of devising satisfactory techniques, for example there is still much controversy about the intercellular concentrations and distribution of certain metabolites because none of the methods used are ideal. Even for a metabolic pathway, such as glycolysis that has been actively studied since the turn of the century and in which the three-dimensional structures of many of the enzymes are known, there is still much to be learned about its organization in the cytosol and about its control *in vivo*.

References

1. Fulton, A. B., *Cell* **30**, 345 (1982).
2. Srivastava, D. K. and Bernhard, S. A., *Science* **234**, 1081 (1986).
3. De Robertis, E. D. P. and De Robertis, E. M. F., *Cell and molecular biology* (7th edn). Saunders, Philadelphia (1980). See p 16.
4. Dean, R. J., *Cellular degradative processes*. Chapman and Hall, London (1978). See Ch. 4.
5. Dallner, G., *Methods Enzymol.* **31A**, 191 (1974).
6. Zbarsky, I. B., *Int. Rev. Cytol.* **54**, 295 (1978).
7. Richardson, J. C. W. and Agutter, P. S., *Biochem. Soc. Trans.* **8**, 459 (1980).
8. Gerace, L., *Trends Biochem. Sci.* **11**, 443 (1986).
9. Dingwell, C., *Trends Biochem. Sci.* **10**, 64 (1985).
10. Siebert, G., in *Comprehensive biochemistry* (Florkin, M. and Stotz, E. H., eds), Vol 23, p. 1. Elsevier, Amsterdam (1968).

11. Fambrough, D. M., in *Handbook of cytology* (Lima de Faria, A., ed), Ch. 18, North Holland, Amsterdam (1969).
12. Siebert, G. and Humphrey, G. B., *Adv. Enzymol.* **27**, 239 (1965).
13. Nelson, W. G., Pienta, K. J., Barrack, E. R., and Coffey, D. S., *A. Rev. Biophys. Bioeng.* **15**, 457 (1986).
14. Sommerville, J., *Trends Biochem. Sci.* **11**, 438 (1986).
15. Srere, P. A., *Trends Biochem. Sci.* **5**, 120 (1980).
16. Hackenbrock, C. R., *Trends Biochem. Sci.* **6**, 151 (1981).
17. Tzagoloff, A., *Mitochondria*, Ch. 2. Plenum Press, New York (1982).
18. De Duve, C. and Wattiaux, R., *A. Rev. Physiol.* **28**, 435 (1966).
19. Lloyd, J. B. and Forster, S., *Trends Biochem. Sci.* **11**, 365 (1986).
20. Goldman, R. and Rottenberg, H., *FEBS Lett.* **33**, 233 (1973).
21. Price, N. C. and Dwek, R. A., *Principles and problems in physical chemistry for biochemists* (2nd edn). Oxford University Press (1979). See p 66.
22. Metzler, D. E., *Biochemistry: the chemical reactions of living cells.* Academic Press, New York (1977). See p. 604.
23. Dean, R. T. and Barrett, A. J., *Essays Biochem.* **12**, 1 (1976).
24. Schneider, Y. J., Octave, J. N., and Trouet, A., *Curr. Top. Membrane Transport* **24**, 413 (1985).
25. Barrett, A. J. and Heath, M. F., in *Lysosomes: a laboratory handbook* (Dingle, J. T., ed), pp. 19–45. North-Holland, Amsterdam (1977).
26. De Pierre, J. W. and Ernster, L., *A. Rev. Biochem.* **46**, 201 (1977).
27. De Pierre, J. W. and Dallner, G., *Biochim. biophys. Acta* **415**, 411 (1975).
28. Lichtenstein, A. H. and Breecher, P., *J. biol. Chem.* **255**, 9098 (1980).
29. von Figura, K. and Hasilik, A., *A. Rev. Biochem* **55**, 167 (1986).
30. Mastro, A. M., Babich, M. A., Taylor, W. D., and Keith, A. D., *Proc. natn. Acad. Sci., USA* **81**, 3414 (1984).
31. Aw, T. and Jones, D. P., *Am. J. Physiol.* **249**, C385 (1984).
32. Reid, R. A. and Leech, R. M., *Biochemistry and structure of cell organelles.* Blackie and Sons, Glasgow (1980).
33. Halliwell, B., *Chloroplast metabolism.* Clarendon Press, Oxford (1984).
34. Dunphy, W. G. and Rothman, J. E., *Cell* **42**, 13 (1985).
35. Tolbert, N. E., *A. Rev. Biochem.* **50**, 133 (1981).
36. Anderson, N. and Green, J. G., in *Enzyme cytology* (Roodyn, D. B., ed) Ch. 8. Academic Press, New York (1967).
37. Denton, R. H. and Halestrap, A. P., *Essays Biochem.* **15**, 37 (1979).
38. Bloch, K. and Vance, D., *A. Rev. Biochem.* **46**, 263 (1977).
39. Jeffcoat, R., *Essays Biochem* **15**, 1 (1979).
40. Zuurendonk, P. F., Akerboom, T. P. M., and Tager, J. M., in *Use of isolated liver cells and kidney tubules in metabolic studies* (Tager, J. M. and Williamson, J. R., eds), p. 17. North-Holland, Amsterdam (1976).
41. Seeley, P. J., Sehr, P. A., Gadian, D. G., Garlick, P. B., and Radda, G. K., in *NMR in biology* (Dwek, R. A., Campbell, I. D., Richards, R. E. and Williams, R. J. P., eds). Academic Press, London (1977). See p. 249.
42. Moore, G. R., Radcliffe, R. G., and Williams, R. J. P., *Essays Biochem.* **19**, 142 (1983).
43. Gumaa, K. A., McLean, P., and Greenbaum, A. L., *Essays Biochem* **7**, 39 (1971).
44. Veech, R. L., in *Microenvironments and metabolic compartmentation* (Srere, P. A. and Estabrook, R. W., eds), p. 17. Academic Press, New York (1978).
45. Lehninger, A. L., *J. biol. Chem.* **190**, 345 (1951).

46. Krebs, H. A. and Veech, R. L., in *Energy levels and metabolic control in mitochondria* (Papa, S., Tager, J. M., Quagliariello, E., and Slater, E. C., eds), p. 329. Adriatica Editrice, Bari, Italy (1969).
47. Klingenberg, *M*., *Trends Biochem. Sci.* **4**, 249 (1979).
48. Wilson, D. B., *A. Rev. Biochem.* **47**, 933 (1978). See p. 953.
49. Van Golde, L. M. G. and Van den Bergh, S. G., in *Lipid metabolism in mammals* (Snyder, F., ed), Vol. 1, pp. 37–42. Plenum Press, New York (1977).
50. Bieber, L. L. and Farrells, S., *The Enzymes* **16**, 627 (1983).
51. Williamson, J. R., in *Gluconeogenesis: its regulation in mammalian species* (Hanson, R. W. and Mehlman, M. A., eds). Ch. 5. Wiley, New York (1976).
52. Davis, R. H., Weiss, R. L., and Bowman, B. J., in *Microenvironment and metabolic compartmentation* (Srere, P. A. and Estabrook, R. W., eds). p. 197. Academic Press, New York (1978).
53. Bowman, B. J. and Davis, R. H., *J. Bact.* **130**, 274 (1977).
54. Goodman, I. and Weiss, R. L., *J. Bact.* **141**, 227 (1980).
55. Davis, R. H., *Microbiol. Rev.* **50**, 280 (1986).
56. Wiemken, A. and Durr, M., *Arch. Microbiol.* **101**, 45 (1974).
57. Davis, R. H., *Trends Biochem. Sci.* **13**, 101 (1988).
58. Weiss, R. L., *J. Bact.* **126**, 1173 (1976).
59. Weiss, R. L. and Davis, R. H., *J. biol. Chem.* **248**, 5403 (1973).
60. Vogel, R. H. and Kopac, M. J., *Biochim. biophys. Acta* **37**, 539 (1960).
61. Nazario, M., *Biochim. biophys. Acta* **145**, 146 (1967).
62. Cohen, B. B. and Bishop, J. O., *Genet. Res.* **8**, 243 (1966).
63. Houslay, M. D. and Stanley, K. K., *Dynamics of biological membranes.* Wiley, Chichester (1982).
64. Shin, B. C. and Carraway, K. L., *J. biol. Chem.* **248**, 1436 (1973).
65. Maroux, S. and Louvard, D., *Biochim. biophys. Acta* **419**, 189 (1976).
66. Kenny, A. J. and Booth, A. G., *Essays Biochem.* **14**, 1 (1978). See section VII, B.
67. Tomino, S. and Paigen, K., *J. biol. Chem.* **250**, 1146 (1975).
68. Benajlba, A. and Maroux, S., *Eur. J. Biochem.* **107**, 381 (1980).
69. Sandermann, H. and Strominger, J. L., *Proc. natn. Acad. Sci. USA* **68**, 2441 (1971).
70. Bischoff, E., Tran-Thi, T., and Decker, K. F. A., *Eur. J. Biochem.* **51**, 353 (1975).
71. Erwin, V. G. and Hellerman, L., *J. biol. Chem.* **242**, 4230 (1967).
72. Tzagoloff, A. and Maclennan, D. H., *Biochim. biophys. Acta* **99**, 476 (1965).
73. Sandermann, H., *Biochim. biophys. Acta* **515**, 209 (1978).
74. Kenny, A. J. and Maroux, S., *Physiol. Rev.* **62**, 91 (1982).
75. Brunner, J., Wacker, H., and Semenza, G., *Methods Enzymol.* **96**, 386 (1983).
76. Dean, W. L. and Tanford, C., *Biochemistry* **17**, 1683 (1978).
77. Bangham, A. D., Hill, M. V., and Miller, N. G. M., in *Methods in membrane biology* (Korn, E., ed), Vol. 1, Ch. 1. Plenum Press, London (1974).
78. De Meis, L. and Vianna, A. L., *A. Rev. Biochem.* **48**, 275 (1979).
79. Maclennan, D. H., Branbl, C. J., Korczak, B., and Green, N. M., *Nature, Lond.* **316**, 696 (1985).
80. Brandl, C. J., Green, N. M., Korczak, B., and Maclennan, D. H., *Cell* **44**, 597 (1986).
81. Strosborg, A. D., Vauguelin, G., Trautman, O., Klutchko, C., Bottari, S., and Andre, C., *Trends Biochem. Sci.* **5**. 11 (1980).
82. Leftkowitz, R. J. and Caron, M. G., *J. Biol. Chem.* **263**, 4993 (1988).
83. Brandt, D. R. and Ross, E. M., *J. biol. Chem.* **261**, 1656 (1986).

84. Gocayne, J., Robinson, D. A., FitzGerald, M. G., Chung, F.-Z., Kerlavage, A. R., Lentes, K.-U., Lai, J., Wang, C-D., Fraser, C. M., and Venter, J. C., *Proc. natn. Acad. Sci. USA* **84**, 8296 (1987).
85. Sibley, D. R., Benovic, J. L., Caron, M. G., and Leftkowitz, R. J., *Cell* **48**, 917 (1987).
86. Gilman, A. G., *A. Rev. Biochem.* **56**, 615 (1987).
87. Pfeuffer, E., Drehev, R-M., Metzger, H., and Pfeuffer, T., *Proc. natn. Acad. Sci. USA* **82**, 3086 (1985).
88. Smigel, M. D., *J. biol. Chem.* **261**, 1976 (1986).
89. May, D. C., Ross, E. M., Gilman, A. G., and Smigel, M. D., *J. biol. Chem.* **260**, 15829 (1985).
90. Rodbell, M., *Trends Biochem. Sci.* **10**, 461 (1985).
91. Ansari, M. H., MSc Thesis, University of Stirling (1980). See pp. 84–5.
92. Vandermeulen, D. L. and Ressler, N., *Archs Biochem. Biophys.* **199**, 197 (1980).
93. Cooke, R. and Kuntz, I. D., *A. Rev. Biophys. Bioeng.* **3**, 95 (1974).
94. Saenger, W., *A. Rev. Biophys. Bioeng.* **16**, 93 (1987).
95. Knowles, P., *Essays Biochem.* **8**, 79 (1972). See p 88.
96. Falk, M., Poole, A. G., and Goymour, C. G., *Can. J. Biochem.* **48**, 1536 (1970).
97. Weisz, P. B., *Nature, Lond.* **195**, 772 (1962).
98. Hubscher, G., Mayer, R. J., and Hansen, H. J. M., *Bioenergetics* **2**, 115 (1971).
99. Welch, G. R., *Prog. Biophys. molec. Biol.* **32**, 103 (1977).
100. Duszynski, J., Mueller, G., and La Noue, K., *J. biol. Chem.* **253**, 6149 (1978).
101. Masters, C. J., *Trends Biochem. Sci.* **3**, 206 (1978).
102. Ovadi, J. and Keleti, T., *Eur. J. Biochem.* **85**, 157 (1978).
103. Ovadi, J., Slaerno, C., Keleti, T., and Fasella, P., *Eur. J. Biochem.* **90**, 499 (1978).
104. Batke, J., Askoth, G., Lakatos, S., Schmitt, B. and Cohen, R., *Eur. J. Biochem.* **107**, 389 (1980).
105. Tompa, P., Bar, J., and Batke, J., *Eur. J. Biochem.* **159**, 117 (1986).
106. MacGregor, J. S., Singh, V. N., Davoust, S., Melloni, E., Pontremoli, S., and Horecker, B. L., *Proc. natn. Acad. Sci. USA* **77**, 3889 (1980).
107. Srere, P. A., Halper, L. A., and Finkelstein, M. B., in *Microenvironment and metabolic compartmentation* (*Srere, P. A. and Estabrook, R. W., eds*), p. 419. Academic Press, New York (1978).
108. Backman, L. and Johansson, G., *FEBS Lett.* **65**, 39 (1976).
109. Beeckmans, S. and Kanarek, L., *FEBS Lett.* **117**, 527 (1981).
110. Fahien, L. A., Kmiotek, E. H., Woldegiorgis, G., Evenson, M., Shrago, E., and Marshall, M., *J. biol. Chem.* **260**, 6069 (1985).
111. Barnes, S. J. and Weitzmann, P. D. J., *FEBS Lett.* **201**, 217 (1986).
112. Sumegi, B. and Srere, P. A., *J. biol. Chem.* **259** 15040 (1984).
113. Srivastava, D. K. and Bernhardt, S. A., *Curr. Top. Cell. Regul.* **28**, 1 (1986).
114. Wiame, J. M., *Curr. Top. Cell. Regul.* **4**, 1 (1971).
115. Messenguy, F., Penninckx, M., and Wiame, J. M., *Eur. J. Biochem.* **22**, 277 (1971).
116. Eisenstein, E. and Hensley, P., *J. biol. Chem.* **261**, 6192 (1986).
117. Mori, M., Aoyagi, K., Tatibana, M., Ishikawa, T., and Ishii, H., *Biochem. Biophys. Res. Commun.* **76**, 900 (1977).
118. Ottaway, J. H. and Mowbray, J., *Curr. Top. Cell. Regul.* **12**, 108 (1977).
119. Purich, D. L., Fromm, H. J., and Rudolph, F. B., *Adv. Enzymol.* **39**, 249 (1973).
120. Sols, A. and Marco, R., *Curr Top. cell. Regul.* **2**, 227 (1970).
121. Cha, S., *J. biol. Chem.* **245**, 4814 (1970).

122. Griffiths, J. R., *Biochem. Soc. Trans.* **7**, 15 (1980).
123. Chaplin, M. F., *Trends Biochem. Sci.* **6**, VI (1981).
124. Fersht, A. R., in *Enzyme structure and mechanism* (2nd edn). Freeman, New York (1985). See p. 328.
125. Veech, R. L., Rayman, L., Dalziel, K., and Krebs, H. A., *Biochem. J.* **115**, 837 (1969).
126. Barritt, G. J., Zander, G. L., and Utter, M. F., in *Gluconeogenesis: its regulation in mammalian species* (Hanson, R. W. and Mehlman, M. A., eds), p. 3. Wiley, New York (1976).
127. Greenbaum, A. L., Gumaa, K. A., and McLean, P., *Archs. Biochem. Biophys.* **143**, 617 (1971).
128. Srere, P. A., in *Gluconeogenesis: its regulation in mammalian species* (Hanson, R. W. and Mehlman, M. A., eds). p. 153. Wiley, New York (1976).
129. Sobboll, S., Scholz, R., Freisl, M., Elbers, R., and Heldt, H. W., in *Use of isolated liver cells and kidney tubules in metabolic studies* (Tager, J. M. and Williamson, J. R., eds), p. 29. North-Holland, Amsterdam (1976).
130. Ottaway, J. H., *Biochem. Soc. Trans.* **7**, 1161 (1979).
131. Nimmo, H. G. and Cohen, P., *Adv. cyclic Nucleotides Res.* **8**, 146 (1977).
132. Swillens, S., Van Cauter, E., and Dumont, J. E., *Biochim. biophys. Acta.* **364**, 250 (1974).
133. Nageswara, B. D., Cohn, M., and Scopes, R. K., *J. biol. Chem.* **253**, 8056 (1978).
134. Gutfreund, H., *Prog. Biophys. molec. Biol.* **29**, 161 (1975).
135. Pfeffer, S. R. and Rothman, J. E., *A. Rev. Biochem.* **56**, 829 (1987).
136. Henderson, R. and Unwin, P. N. T., *Nature, Lond.* **257**, 28 (1975).
137. Opperdoes, F. R., *Trends Biochem. Sci.* **13**, 255 (1988).
138. Bieber, L. L., *A. Rev. Biochem.* **57**, 261 (1988).

9
Enzyme turnover

9.1 Introduction

Since the early 1940s when Schoenheimer *et al.*[1] first used isotopically labelled precursors to study protein synthesis, it has been realized that most proteins in living cells are continually being replaced whether the organism is fully grown and in nitrogen balance or is growing and increasing its total protein (and hence nitrogen) content. However, it was not until the late 1960s that much information was gained about the rate at which individual enzymes are replaced or turned over in particular tissues and cells and the general importance of enzyme turnover as a control mechanism was appreciated (for reviews see references 2–10).

In any given tissue or cell type there is a wide range of turnover rates for different enzymes, e.g. in rat liver the half-lives of enzymes range from about 15 minutes to over 100 hours.[4,10] The underlying mechanisms that account for this wide range of turnover rates are not known. Mechanisms have been proposed to account for the degradation of certain enzymes or groups of enzymes, and some general correlations have been made between the properties of enzymes and their rates of turnover (see Sections 9.5 and 9.6), but more information is required before a general theory to account for the turnover of enzymes can be formulated.

At first sight the process of enzyme turnover might seem to be inefficient and wasteful of energy, since every peptide bond formed requires the hydrolysis of ATP and GTP, whereas proteolysis is not coupled to the generation of ATP. However, turnover is important if a cell is to be able to adapt to changes in its environment or if it becomes necessary to remove abnormal enzyme or protein molecules that may arise by mutation or by errors in gene expression. In general, the longer the life of an individual cell the more important is the process of intracellular enzyme turnover. For example, in a bacterium such as *Escherichia coli* growing under optimal conditions, cell division may occur every 20 min. Adaptation occurs largely by induction or repression of enzyme synthesis. If lactose is added to the growth medium, induction of β-D-galactosidase occurs. If lactose is then removed from the medium, existing enzyme molecules will rapidly be diluted out within the cells provided they continue to divide rapidly and β-D-galactosidase is no longer synthesized. Early experiments aimed at detecting turnover of proteins in exponentially growing bacteria suggested that the proteins were completely stable.[11] More recent experiments have shown that turnover does occur but that it is very limited. Increases in the rate of turnover occur when bacteria approach the stationary phase and the supply of nutrients becomes limiting. In contrast to that of rapidly dividing bacteria, the average life of an adult liver cell is between 160 and 400 days and many enzymes are completely replaced every few days. This turnover is necessary, for example, if a liver cell is to be able to adapt to changes in nutrients that it may encounter throughout its life.

When discussing enzyme turnover in this chapter we are referring to the intracellular processes by which individual enzyme molecules are degraded and replaced by synthesis of new enzyme molecules. The steady-state level of an enzyme in a cell depends on the rates of two opposing processes, namely the rate of enzyme synthesis and the rate of enzyme degradation. The general mechanism by which enzyme synthesis occurs is better understood than that of enzyme degradation, although, on the whole, enzyme degradation has a more important regulatory influence on enzyme turnover, particularly in eukaryotes. Therefore, we shall now consider in outline the mechanism of enzyme synthesis before considering the kinetics of enzyme turnover and possible mechanisms of enzyme degradation.

9.2 Mechanism of enzyme synthesis

The mechanism by which enzymes are synthesized is no different from that of protein synthesis in general. There are good accounts of this in many textbooks[12–13] and monographs[10, 14, 15] and only the salient features are given here. The information that determines the primary sequence of an enzyme is contained in the order of the deoxyribonucleotide residues of DNA. This information is transferred to a sequence of ribonucleotide residues in mRNA in a process known as transcription; specificity is achieved by complementary base pairing between DNA and mRNA. In eukaryotes, the primary transcript generally undergoes processing to excise sections of the RNA before forming mRNA.[16] The sequence of ribonucleotide residues in mRNA is then expressed in the sequence of amino-acid residues in the enzyme in a process known as translation. The steps in these processes are shown in Fig. 9.1. The amino acids are activated by conversion to aminoacyl adenylates; these then react with species of tRNA to form aminoacyl-tRNAs (see Chapter 5, Section 5.5.5). For each different L-amino acid there exists at least one distinct tRNA. The tRNAs differ from one another in their nucleotide sequence and in particular in the trinucleotide sequence known as the anticodon, which recognizes a complementary sequence (by base pairing) on mRNA known as a codon. The conversion of amino acids to aminoacyl adenylates and thence to aminoacyl-tRNA is catalysed by a class of enzymes known as aminoacyl-tRNA synthetases (see Chapter 5, Section 5.5.5.) Each of these enzymes recognizes a specific amino acid and a specific tRNA.

L-Amino acid + ATP = L-aminoacyl-AMP + pyrophosphate.

L-Aminoacyl-AMP + tRNA = L-aminoacyl-tRNA + AMP.

Enzyme specificity is thus crucial in accurately relaying information contained in a ribonucleotide sequence to that in an amino-acid sequence during the process of translation. The ribosomes, together with a number of other specific proteins, constitute the machinery that in the presence of GTP enables aminoacyl-tRNAs, aligned on the mRNA, to react to form polypeptides and release the tRNAs.

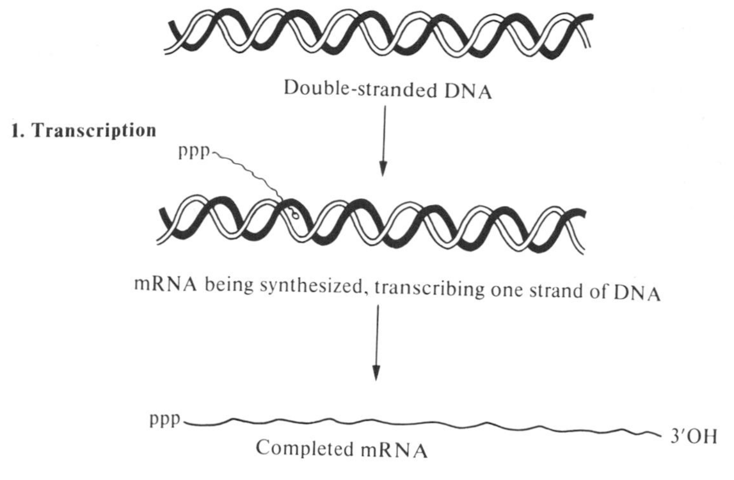

FIG. 9.1. Diagram to illustrate the principal features of the processes of mRNA and protein synthesis, as it occurs in prokaryotes. In eukaryotes the primary transcript is spliced to form mRNA.[16] (For details see references 14 and 15.)

The enzymatic steps that lead to the synthesis of an enzyme are thus well understood. However, the factors that regulate the rate of synthesis of the majority of enzymes are not clear. A limited number of inducible* and

* Inducible and repressible enzymes are those whose synthesis in cells is regulated by inducer and/or repressor molecules of low M_r.

repressible enzymes from *E. coli* have been studied in great depth (see reference 17), but the factors regulating the expression of the genes for the majority of enzymes, particularly those from eukaryotes, are not yet fully understood.

9.3 Kinetics of enzyme turnover

In order to understand the importance of enzyme turnover as a regulatory mechanism, it is necessary to make a kinetic analysis of the process. Equations that govern the steady-state levels of enzymes, and the rates of changes to new steady-state levels, have been derived by Schimke *et al.*[2, 10, 18] From the study of a number of enzymes in which the amounts of enzyme protein present in cells or tissues have been measured, it has been shown that the rate of degradation of an enzyme normally obeys first-order kinetics, i.e. the rate of enzyme degradation $=k[\mathrm{E}]$, where [E] is the enzyme concentration.* This means that the degradation of an enzyme is a random process; a newly synthesized enzyme molecule is as likely to be degraded as an old one. The process of enzyme synthesis has been shown generally to obey zero-order kinetics, i.e. the rate of synthesis of an enzyme is independent of the enzyme concentration within the cell. Thus, the rate of change of level of an enzyme in a cell ($\mathrm{d}[\mathrm{E}]/\mathrm{d}t$) is given by

$$\frac{\mathrm{d}[\mathrm{E}]}{\mathrm{d}t}=k_{\mathrm{s}}-k_{\mathrm{d}}[\mathrm{E}],$$

where k_{s} is the rate constant for enzyme synthesis and k_{d} the rate constant for enzyme degradation. In the steady-state, $\mathrm{d}[\mathrm{E}]/\mathrm{d}t=0$ and therefore $k_{\mathrm{s}}=k_{\mathrm{d}}[\mathrm{E}]$.

We now consider the expression for the rate of approach to a new steady-state level if either the rate of synthesis or the rate of degradation changes. If the rate constants for the new rates of synthesis and degradation are k'_{s} and k'_{d}, respectively, $[\mathrm{E}_0]$ is the initial enzyme concentration, and $[\mathrm{E}_t]$ is the enzyme concentration at time t after the change in rates of synthesis or degradation, then the equation describing the time course of approach to a new steady-state level is given by

$$\frac{[\mathrm{E}_t]}{[\mathrm{E}_0]}=\frac{k'_{\mathrm{s}}}{k'_{\mathrm{d}}[\mathrm{E}_0]}-\left(\frac{k'_{\mathrm{s}}}{k'_{\mathrm{d}}[\mathrm{E}_0]}-1\right)\mathrm{e}^{-k'_{\mathrm{d}}t}.$$

For a derivation of this equation, see Appendix 9.1 and reference 18. This equation is useful for analysing changes in enzyme concentrations resulting from hormonal, nutritional, or other physiological changes. When the rate of degradation of an enzyme is measured, it is more common to express the result in terms of the half-life of the enzyme rather than in terms of the rate constant, k_{d}. For a first-

* There are, however, some exceptions; e.g. some proteins present in nerve and muscle show biphasic kinetics, that is, one fraction of the protein is degraded at a different rate from the remainder of the same protein in that tissue.[19]

order reaction the relationship is:

$$t_{\frac{1}{2}} = \frac{\ln 2}{k_d} = \frac{0.69}{k_d}.$$

Figure 9.2 illustrates the effect of differences in the rate of degradation of three enzymes on the rate of attainment of new steady states, when a tenfold increase in the rate of synthesis of the enzyme is induced by a stimulus. The effect of withdrawal of the stimulus 5 hours later is also shown. It can be seen that the enzyme having the shortest half-life, ornithine decarboxylase, responds very rapidly to the changed rate of synthesis and has almost reached the new steady state within 2–3 hours, similarly when the stimulus is removed the old steady-state level is reached within a further 2–3 hours. In contrast, the amount of the enzyme having the longest half-life, pyruvate kinase, is still increasing almost linearly after five hours and would require over 70 hours to achieve 90 per cent of the new steady-state level.

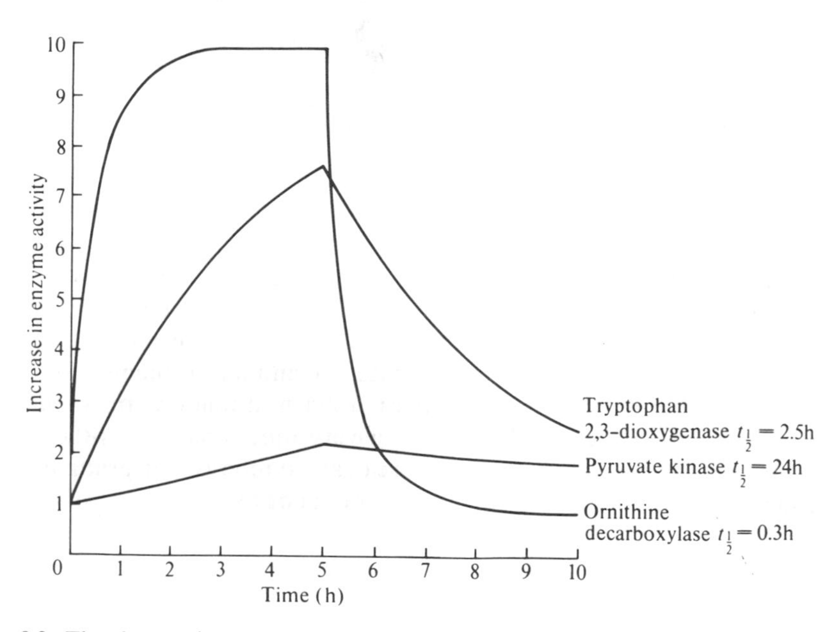

FIG. 9.2. The changes in enzyme activities resulting from a 10-fold increase in the rate of synthesis of each enzyme at zero time, and a 10-fold reduction in the rate of synthesis at 5 hours. Note how the differences in half-lives affect the rates of attainment of new steady states.

Thus, it is the rate of degradation of an enzyme that primarily determines how rapidly the amount of enzyme changes in response to a given stimulus. The shorter the half-life, the quicker the response, and thus the better the control of the level of enzyme. In deriving these equations it is assumed that the rate of growth of the cells is slow in comparison with the rate of synthesis of the enzymes so that changes in the total volume of the cells can be neglected.

9.4. Methods for measurement of rates of enzyme turnover

There are several methods that can be used to measure the rates of protein synthesis and protein degradation; the most appropriate will depend on the type of information required. In this section we outline the principal methods; there are several reviews and papers in which more detailed treatment is given.[3, 20–22] In order to understand how the concentration of a particular enzyme is maintained within a cell or tissue and how the concentration is changed in various physiological conditions, it is most useful to measure k_s and k_d. Knowledge of these rate constants will not give any insight as to the mechanism of synthesis or degradation, e.g. as to which are the rate-limiting steps in these processes, but they will indicate whether the change in concentration is due to a change in the rate of synthesis or of degradation, or of both.

A change in the enzyme content of a tissue or organism will usually first be observed from the measurement of enzyme activity. It is then necessary to establish whether this change represents a change in the amount of enzyme protein as opposed to some conformational change or covalent modification of the enzyme that alters its activity (see Chapter 6, Section 6.2). To distinguish these it is necessary to purify the enzyme and then measure the amount of enzyme protein. If previously purified enzyme is available it may be used to raise a specific antibody by injecting the enzyme into another species. The serum fractions containing the antibody may then be collected and used to titrate enzyme protein. Alternatively, monoclonal antibodies may be raised against the enzyme (for details see reference 13). The complex between the enzyme and the antibody precipitates and can be collected by centrifugation and then estimated. Isolation of the enzyme from the tissues at a number of different times, to prove that the amount of enzyme protein is changing, is a very time-consuming process and may require large amounts of tissue. The use of an antibody greatly simplifies the procedure and can be used with smaller amounts of tissue.

Once it has been established that a change in the amount of enzyme protein occurs, the rates of synthesis and degradation are measured. Isotopically labelled precursors are most frequently used for this, as outlined in the following sections.

9.4.1 Measurement of k_s

The rate of enzyme synthesis is usually measured either after giving a single pulse of a suitable radioactively labelled precursor, e.g. an amino acid, or by giving a constant infusion of the labelled precursor. In the first method (involving a single pulse), incorporation of labelled precursor into the enzyme is measured at short time intervals. The specific activity of the amino-acid pool in the tissue or cells concerned is also measured. If short time intervals are used, the initial rate of incorporation is determined (i.e. when the specific activity of label in the enzyme is low and the loss of radioactivity by degradation is negligible). From these data k_s may be determined. One problem arises particularly in experiments with whole animals. The radioactive amino acids enter the plasma and thence pass into the cells, where they equilibrate with endogenous amino acids. The amino acids are then converted to aminoacyl-tRNAs, which are the ultimate precursors of the enzyme. Therefore, to calculate accurately the rates of synthesis, the specific radioactivity of the isolated aminoacyl-tRNAs should be used.[21] However, in rat liver, the tissue in which many measurements of k_s have been made, there is rapid equilibration between the free amino-acid pool within the cells and the aminoacyl-tRNAs and therefore in this particular tissue the specific radioactivity of the free amino acids may be used.

The second method is that of constant infusion of radioactively labelled amino acids. The amino acids are either infused intravenously or given in the diet until the specific radioactivity of the amino-acid pool within the cells has reached a plateau level. Measurements are then made of the incorporation of the label into the enzyme at the end of the infusion period and of the changes in specific radioactivity of the free amino acid within the tissue. From these measurements k_s may be determined; the mathematical relationships are complex and are described in references 20 and 21.

9.4.2 Measurement of k_d

The rate of degradation can also be measured by use of radioactivity labelled precursors. If a pulse of labelled precursor is given and the decay of the specific radioactivity of the labelled enzyme is measured at fixed time intervals, k_d can be calculated. Measurements are made for a longer period than those required for measurement of k_s. Ideally k_d is measured when the specific radioactivity of the enzyme is high, and the specific radioactivity of the amino-acid pool has declined (Fig. 9.3).

An assumption made in the calculation of k_d is that no significant re-utilization of the precursors released by protein degradation takes place. If re-utilization takes place, k_d may be considerably underestimated. This has been a major problem in the accurate determination of k_d, but re-utilization can often be minimized by a suitable choice of precursor; for example [*guanidino*-^{14}C]arginine is found to be a better label for this purpose than uniformly labelled arginine or other labelled amino acids when liver protein degradation is measured.[23] This is because the guanidino group is rapidly removed by arginase present in high concentrations in the liver. [^{14}C]carbonate has also been found to be a good precursor for measurement of k_d; much of the [^{14}C]carbonate is incorporated into the carboxyl groups of glutamate and aspartate.[24] The radioactivity of the latter amino acids declines rapidly because of decarboxylations that occur when glutamate and aspartate become transaminated and subsequently oxidized in the tricarboxylic-acid cycle. The $H^{14}CO_3^-$ produced in this way is diluted by the large intracellular pool of HCO_3^- and lost as $^{14}CO_2$.

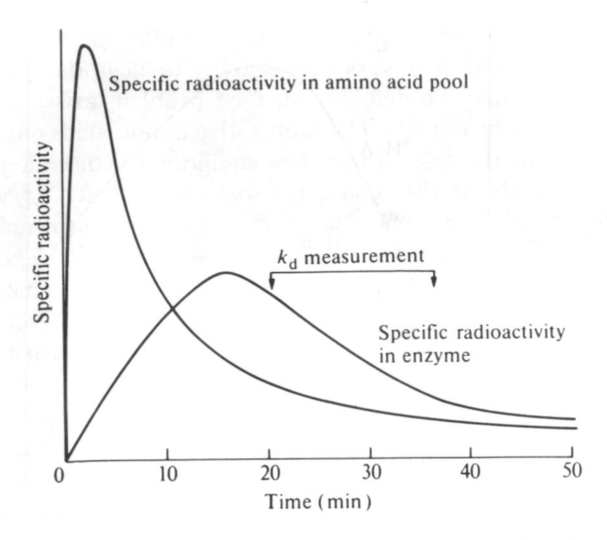

FIG. 9.3. Precursor-product relationship after a single pulse of radioactively labelled amino acid is given to a tissue or cells. Diagram shows the specific radioactivities in the amino-acid pool (precursor) and in the enzyme (product).

Another example in which reincorporation does not occur is that of 5-aminolaevulinic acid,[25] which is incorporated specifically into haem-containing proteins, e.g. cytochromes, catalase. Haem is degraded by a pathway that differs from that of its synthesis and thus reincorporation does not take place. The validity of this method depends on the assumption that the haem prosthetic group is degraded at the same rate as the protein moiety, which may or may not be the case.

The double isotope technique[20] is a convenient method of comparing the rates of turnover of proteins having the same cellular origin. It entails giving two successive pulses of a radioactively labelled amino acid, the first being ^{3}H-labelled and the second ^{14}C-labelled; these are shown as ^{3}H-A and ^{3}H-B, and ^{14}C-A and ^{14}C-B, respectively, in Fig. 9.4. where A and B are two different proteins. In the example in Fig. 9.4, the turnover of A is twice that of B. The proteins that are being studied are isolated at time Y. Typically the ^{3}H pulse would have been given 24 h before isolation of the proteins, whereas the ^{14}C pulse would have been given 30 min. before isolation (time X), although the precise times would depend on the proteins and tissue concerned. The times are chosen so that the specific activities of the first label (^{3}H) in the proteins being studied are declining, whereas the specific activities of the second label (^{14}C) are increasing (Fig. 9.4). The proteins are isolated at time Y and the ratio ^{14}C/^{3}H (b/a and d/c) in each is measured; these ratios are directly related to their turnover rates. The method has two advantages, namely, (i) that the proteins only have to be isolated at one time point, and (ii) that the differences in turnover rates between proteins are more readily apparent than when a single isotope is used.

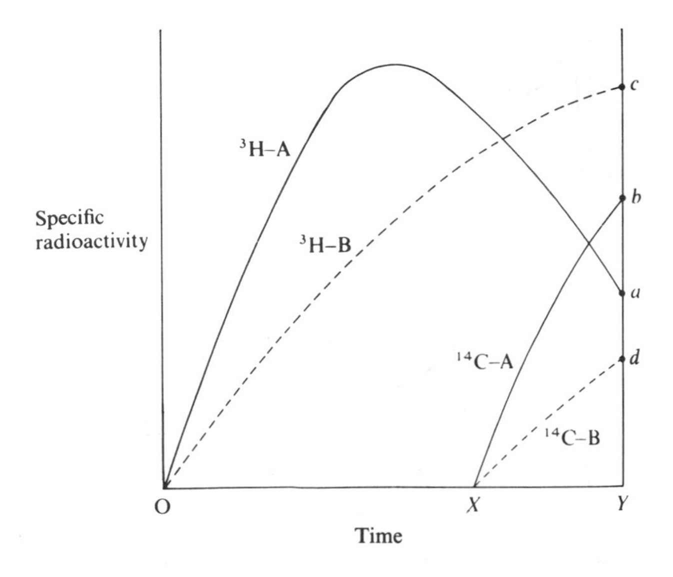

FIG. 9.4. Double isotope technique for measuring protein turnover.

A method that has been widely used to determine rates of degradation is to inhibit protein synthesis in the tissue concerned by administration of cyclohex-imide or puromycin and then to measure the subsequent decay of enzyme activity. This method has the advantage of simplicity in that it does not require the isolation of the enzyme, but it has the disadvantages (a) that it cannot distinguish enzyme inactivation from enzyme degradation and (b) that cyclohex-imide and puromycin, in addition to inhibiting protein synthesis, may also affect protease activity, thereby affecting the rate of enzyme degradation.[3] Evidence suggests that the latter complication is less serious when studying enzymes that have short half-lives.[27]

9.5 Results from measurements of rates of enzyme turnover

The rates of turnover of a number of enzymes have now been measured in a variety of tissues and organisms and it is possible to correlate these results with the structures and properties of the enzymes concerned. The most widely measured parameter is the enzyme's half-life. Some of the results obtained are given in Table 9.1. It is apparent that a wide range of turnover rates can occur within one tissue, e.g. the half-lives of liver proteins range from about 15 min (ornithine decarboxylase) to 7 days (6-phosphofructokinase). The half-life of a given enzyme may differ from one tissue to another (Table 9.2) or even between different compartments within the same cell, e.g. 5-aminolaevulinate synthase in the cytosol has a half-life of 20 min, whereas in the mitochondria the half-life is 60 min.[32] Although some results suggest that in certain organelles or membranes

TABLE 9.1
The half-lives of enzymes in rat liver

Enzyme	Half-life (hours)
Ornithine decarboxylase	0.3
5-Aminolaevulinate synthase (mitochondria)	1.2
RNA nucleotidyltransferase I	1.3
Tyrosine aminotransferase	1.5
Tryptophan 2,3-dioxygenase	2.0
Thymidine kinase	2.6
Hydroxymethylglutaryl-CoA reductase	4.0
Serine dehydratase	5.2
Phosphoenolpyruvate carboxykinase	8.0
Dehydro-orotase	12.0
RNA nucleotidyltransferase II	13.0
Glucose-6-phosphate dehydrogenase	24
Glucokinase	30
Catalase	33
Acetyl-CoA carboxylase	50
Glyceraldehyde-3-phosphate dehydrogenase	74
Pyruvate kinase	84
Arginase	108
Fructose-bisphosphate aldolase	117
Lactate dehydrogenase (LDH-5)	144
6-Phosphofructokinase	168

See references 2, 4, 28–31.

TABLE 9.2
The half-lives (in days) of enzymes in different tissues of the rat

Enzyme	Liver	Heart	Skeletal muscle	Kidney	Brain
Pyruvate kinase	3.5	4.2	21.6	—	—
Lactate dehydrogenase[a]	1.8–3.8	3.7–9.9	3.0–8.5	3.0–5.0	3.0–7.2
Ornithine oxo-acid aminotransferase	0.95	—	—	4.0	—
Fatty-acid synthase	6.4	—	—	—	2.8

See references 2, 28, 33, and 34.
[a] The half-life of lactate dehydrogenase varies with the particular isoenzyme.[35]

within an organelle a number of enzymes turnover at similar rates,[36] e.g. cytochrome *c* oxidase and adenosinetriphosphatase from the inner mitochondrial membrane of yeast turnover at similar rates during sporulation, and also ribosomal proteins turnover at similar rates, other data indicate that there is a considerable range of turnover rates of enzymes within a single organelle or membrane[38, 39] (Table 9.3) and it seems likely that in general organelles do not turnover as units. From the limited data available so far, it seems that the polypeptide chains within a multienzyme complex, e.g. pyruvate dehydrogenase,

TABLE 9.3
Variation in half-lives of enzymes within the same organelle of rat liver

	Half-life (hours)
1. Microsomal membrane	
Cytochrome b_5 (outside)	100
Cytochrome b_5 reductase (outside)	140
NADPH-cytochrome reductase (outside)	70
Nucleosidediphosphatase (inside)	30
Carboxylesterase (inside)	95
Hydroxymethylglutaryl-CoA reductase	4
NAD^+ nucleosidase	430
2. Mitochondria	
Carbamoyl phosphate synthase	185
Malate dehydrogenase	62
Glutamate dehydrogenase	24
Carnitine acetylase transferase	43
Ornithine oxo-acid aminotransferase	46
Cytochrome *c* oxidase	>190
Amine oxidase (flavin-containing)	67
Pyruvate dehydrogenase system	
$E_1\alpha$	108
$E_1\beta$	120
E_2	134
E_3	120

See references 39–43.

and within an oligomeric enzyme, e.g cytochrome *c* oxidase and amine oxidase (flavin-containing), turnover at similar rates (Table 9.3).

9.6 Possible correlations between the rates of turnover and the structure and function of enzymes

A number of enzymes having short half-lives catalyse either the first or the rate-limiting steps in a metabolic pathway, e.g. ornithine decarboxylase, 5-aminolaevulinate synthase, tyrosine aminotransferase, tryptophan 2,3-dioxygenase, serine dehydratase, hydroxymethylglutaryl-CoA reductase, and phosphoenolpyruvate carboxykinase. This finding may be important, since the amounts of enzymes that turnover rapidly can be more precisely controlled than the amounts of those that turnover more slowly (see Section 9.3).

Although the list of characterized intercellular proteases* is growing, it seems highly improbable that there exists within each cell such a wide range of

* All proteolytic enzymes fall within the class of 'proteases'. These may be subdivided into 'endopeptidases' and 'exopeptidases'. The term 'proteinases' can be used interchangeably with endopeptidases.[44]

proteases that each is able to control the rate of degradation of either individual enzymes or small groups of enzymes; this would require an enormous number of proteases, the half-lives of which would in turn have to be regulated. What, therefore, are the characteristics of those enzymes that are degraded most rapidly? Do they have any common features? The outline of a pattern is beginning to emerge. The pattern may be sought from either a structural or a functional standpoint. From the structural standpoint, results obtained from a large number of proteins from the cytosol of liver suggest that there is a correlation between the subunit size of the protein and its rate of degradation, the larger subunits being degraded faster.[45, 46] The explanation for this could simply be that larger polypeptides have more protease-sensitive sites and are thus more susceptible to proteolysis. It is not yet clear how general is this correlation, since exceptions have been noted in some mitochondrial[41] and lysosomal[47] enzymes.

As well as the size correlation, some data support a correlation between the turnover rate and hydrophobicity, the most hydrophobic enzymes being degraded fastest.[48] Enzymes with low p*I* are more labile.[49] However, there are exceptions to these general correlations.[50]

Interesting correlations can be made between the rate of degradation of enzymes and their ligand-binding properties. There are a number of examples in which the binding of a substrate, cofactor, or competitive inhibitor has been known to stabilize an enzyme against degradation (Table 9.4). The group of enzymes that has been most studied in this respect is the pyridoxal phosphate-requiring enzymes. A group of proteases were isolated that preferentially degraded the apoenzymes of pyridoxal phosphate-requiring enzymes. These proteases were believed to be responsible for the intracellular turnover of apoenzymes in liver and skeletal muscle.[57] However, later evidence throws doubt on this hypothesis, since the group-specific proteases in question were found to originate in the mast cells associated with skeletal muscle and liver.[58] Also, if rats were fed on a pyridoxal-(vitamin B_6) deficient diet so that they became pyridoxal deficient, no increase was observed in the rate of degradation of pyridoxal phosphate-requiring apoenzymes.[59] Thus, although there is good evidence for the destabilization of enzymes *in vitro* when depleted of ligands, it is doubtful whether this is an important mechanism *in vivo*.

The explanation for the wide variation in the rates at which many enzymes are turned over may be found by referring to the structures of the enzymes, i.e. those enzymes that have most protease-sensitive sites, that bind ligands least strongly, and that are denatured most readily are those most likely to be turned over most rapidly. One of the difficulties in making comparisons of structure with turnover is that the enzymes that are most interesting for studying turnover are the enzymes present in the cell in small amounts and about which least structural information is available. By contrast, the enzymes about which there is detailed structural information, including three-dimensional structure, are often those that are most stable *in vivo*. The situation has partly been remedied by Rechsteiner *et al.*,[60] who have developed a method for microinjection of purified

TABLE 9.4
Examples of ligands that stabilize enzymes against degradation

Enzyme	Ligand	Effect
Hexokinase	Glucose	Reduces rate of degradation by trypsin
Thymidine kinase	Thymidine	Stabilizes against degradation in cell extracts
Tryptophan 2,3-dioxygenase	Tryptophan or tryptophan analogues	Induces tryptophan 2,3-dioxygenase by stabilizing it against degradation
Serine dehydratase Tyrosine aminotransferase	Pyridoxal phosphate	Protects against group specific proteases that degrade pyridoxal phosphate-requiring apoenzymes
Aspartate carbamoyltransferase	Aspartate CTP or ATP	Increased susceptibility to trypsin Decreased susceptibility to trypsin
Tetrahydrofolate dehydrogenase	Methotrexate	Decreases rate of intracellular degradation
Ornithine decarboxylase	α-Methylornithine or diaminobutanone	Decreases rate of intracellular degradation, protects against chymotrypsin
Acetyl-CoA carboxylase	Palmityl-CoA citrate	Increased susceptibility to lysosomal protease Decreased susceptibility to lysosomal protease
Hydroxymethylglutaryl-CoA reductase	Mevinolin	Decreases intracellular degradation

See references 3, 51–56.

enzymes into cells. Using this method in a study of 35 proteins of known sequence and X-ray structure they found that the half-lives did not correlate with size, charge, or proportion of hydrophobic residues. More recently they have tried to relate the abundance of particular amino acids in a protein to turnover rate.[61] A correlation was found between the half-life and the abundance of four particular amino acids, namely proline, glutamate, serine, and threonine. All the proteins studied with a half-life less than 2 hr had regions rich in some of, or a combination of, these four amino acids.

In addition to making comparisons on purely structural grounds, it is useful to relate turnover to function. A number of workers have divided enzymes into long-lived and short-lived. This is a convenient functional division, since it appears to relate to the mode of degradation. Many long-lived enzymes are the so-called 'house-keeping' enzymes. These include enzymes catalysing non-rate-limiting steps in the major metabolic pathways. They are generally degraded by lysosomal proteases. The short-lived enzymes are present in much smaller amounts and are degraded by extralysosomal routes. Four categories of short-lived proteins can be distinguished at present. The first are enzymes catalysing regulatory steps in metabolic pathways; examples include all those cited at the beginning of this section. The second are enzymes required only at particular stages in the life of the cell, e.g. enzymes required for thymidine triphosphate synthesis, which is required only when the cell enters the presynthetic (G_1) phase of the cell cycle, when the total amount of DNA is doubled. The third type are the enzyme precursors that contain the signal peptide required for entry into an organelle. These are converted into the functional enzymes after entry to the organelle. The fourth are a rather different category, namely the abnormal or non-functional enzymes.

Abnormal proteins, e.g. mutant proteins or proteins containing chemically modified amino acids are degraded much faster than their normal counterparts. Mutations in β-D-galactosidase[62] from *E. coli* cause the enzyme to be degraded rapidly, and the incorporation of azetidine-2-carboxylate in place of proline into haemoglobin reduces its half-life from 120 days to 20 min.[63] The mechanism by which abnormal proteins are selectively degraded at a faster rate is not known.

9.7 The mechanisms of protein degradation

Enzymes degraded by lysosomal and non-lysosomal routes can often be distinguished by two methods. Chloroquine and ammonia are able to penetrate the lysosomal membranes and raise the internal pH,[64] thereby inhibiting lysosomal proteases and so affecting the breakdown of enzymes degraded by the lysosomal pathway. Secondly, the Q_{10} (see Chapter 4, Section 4.3.4.1) for degradation of short-lived proteins degraded by the non-lysosomal route is about 2,[65] whereas degradation via the lysosomes has a Q_{10} of about 4.[66] It is estimated that in the liver about 60 per cent of proteins are degraded in the cytosol and about 40 per cent in the lysosomes.[67]

The mechanisms by which enzymes are degraded are still only poorly understood. A few enzymes have been studied in detail and certain degrading systems have been examined, but how far one can generalize from these limited examples is not clear. It is generally assumed that enzymes are eventually degraded to small peptides or their constituent amino acids. This may involve, particularly in the case of short-lived enzymes, a highly selective initial step followed by further degradation by less specific proteases. The initial step is assumed to recognize specific features in the enzyme to be degraded, but the nature of these might be quite varied. The recognition might involve a highly specific protease that then partially degrades the enzyme; it might entail some kind of modification or tagging of the enzyme by addition of some covalent group; or it might involve binding to a membrane.

9.7.1 Intracellular proteases

In the past decade the number of intracellular proteases that have been characterized has increased considerably. Much of this has resulted from the use of highly specific substrates, e.g. peptides linked through an amide bond to a chromophoric group. The general method of detection of a protease would be that to incubate a cell extract with a protein such as casein and measure the degradation to peptides that are soluble in trichloracetic acid. Alternatively, a protein that has been chemically modified to contain a chromophore (e.g. azocoll, a chemically modified collagen or azocasein) is incubated with a cell extract, and the soluble, coloured peptides released are detected. These methods are satisfactory for detecting endoproteases of low or unknown specificity. The use of peptides modified with chromophores detects more highly specific proteases. The introduction of this method dramatically increased the number of proteases detected in yeast from 8 in 1979 to 29 in 1984.[68]

The lysosomes, which are the richest source of intracellular proteolytic activity in many tissues, contain a variety of exopeptidases and endopeptidases called cathepsins. The concentrations of cathepsins D and B within the lysosomes are estimated to be as high as 0.78 mmol dm^{-3} and 1.54 mmol dm^{-3}, respectively.[69] Lysosomal proteases are thought to be involved in the degradation of longer-lived proteins. It is now apparent that there are proteases present in all cell compartments.

Peptide-bond cleavage is an exergonic process and thus might not be expected to require energy coupling. However, it is now clear that in both eukaryotes and prokaryotes there exist ATP-dependent proteases.[44] Those found in the cytosol of eukaryotes are ATP stimulated, i.e. ATP is required for activity although it is not degraded. Those proteases found in bacteria and in mitochondria hydrolyse ATP during protein degradation. This ATP requirement may be the 'price' for specificity, in the same way that there is net degradation of ATP during substrate cycles (see Chapter 6, Section 6.2.1.2). The intracellular proteases known range in

M_r from 20 000 to 800 000. The sizes of these very large proteases may reflect the possession of a high specificity and complex regulatory mechanism.

9.7.2 How enzymes or proteins are selected for intracellular degradation

The key question in protein degradation is understanding the selection mechanism. Various methods of 'tagging' proteins for degradation have now been identified. These include the formation of a peptide bond between the protein and another protein called ubiquitin, the modification of the protein by phosphorylation or by oxidation, or the binding of the protein to a membrane. There may be other mechanisms yet to be discovered. The different methods will be considered in turn. The degradation of proteins in reticulocytes has been extensively studied and it has been shown by Herscho and co-workers[70, 71] that degradation involves tagging with ubiquitin (M_r 9000). Ubiquitin is so called because of its widespread distribution throughout eukaryotes, and it is possibly also present in prokaryotes. Its role in protein turnover has only been established in a few tissues. Although it is present in kidney and liver, it has not been shown to be involved in degradation in those tissues. Proteins become covalently attached to ubiquitin via their $-NH_2$ groups either the N-terminal α-NH_2 or ε-NH_2 of lysine. A protein may become multiply tagged. ATP is required for the conjugation, together with three other enzymes designated E_1, E_2, and E_3.

$$E_1\text{–SH} + \text{ATP} + \text{Ub} \rightarrow E_1\text{–S–Ub} + \text{AMP} + \text{PP}_i .$$

$$E_1\text{–S–Ub} + E_2\text{–SH} \rightarrow E_2\text{–S–Ub} + E_1\text{–SH}.$$

$$E_2\text{–S–Ub} + \text{protein} \rightarrow E_2\text{–SH} + \text{protein–Ub} \quad (E_3 \text{ catalysed}).$$

The ubiquitinated protein is thus tagged for degradation by proteolytic enzymes. Bachmair *et al.*[72] examined the specificity of the degrading system. They constructed a chimeric gene consisting of the genes for ubiquitin linked to β-galactosidase. When this gene is expressed it produced a conjugate protein with ubiquitin linked to the N-terminus of β-galactosidase via a peptide bond. The protein became de-ubiquitinated and degraded by the cell's proteolytic system. Site-directed mutagenesis (see Chapter 5, Section 5.4.5) was then used to replace the normal N-terminal amino acid of β-galactosidase (Met) in turn by other amino acids. It was then possible to study the degradation of the fusion products. The half-lives of the fusion products ranged from 2 min to over 20 h depending on the N-terminal amino acid as shown below:

$$\underset{\text{ubiquitin}}{H_2N\text{------------}}\text{Gly-Met}\underset{\beta\text{-galactosidase}}{\text{------------COOH}}$$

Replacement of Met by Ser, Ala, Gly, Thr, or Val has a stabilizing effect. Replacement of Met by Ile, Gln, Glu or Tyr has a destabilizing effect. Replacement of Met by Leu, Phe, Asp, Lys, or Arg has a strongly destabilizing effect. The N-terminal residue is therefore an important determinant in protein degradation through the ubiquitin pathway.

Another method of tagging an enzyme prior to its degradation is by phosphorylation. Two examples of this are fructose-bisphosphatase[73] from *Saccharomyces cerevisiae* and the NAD^+-requiring glutamate dehydrogenase from *S. cerevisae* and *Candida utilis.*[74, 95] When *S. cerevisiae* is grown with acetate as the source of carbon, it carries out gluconeogenesis to provide it with adequate hexoses and their derivatives. On the other hand, when it is supplied with glucose as the source of carbon, gluconeogenesis is suppressed. These differences are reflected in the stability of fructose-bisphosphatase, which catalyses a rate-limiting step in gluconeogenesis (see Chapter 6, Section 6.4.1). When grown on acetate, the k_d for fructose-bisphosphatase breakdown is 0.008 h^{-1}, whereas when grown on glucose it is 0.42 h^{-1}.[75] The role of phosphorylation is best seen when the acetate-grown cells are switched to a glucose medium. Within 2 min the activity of fructose-bisphosphatase drops very dramatically. This initial fall can be shown to be due to phosphorylation of a serine residue on fructose-bisphosphatase, although no proteolysis occurs. However by 30 min the fructose-bisphosphatase protein is degraded. Thus, it appears that the initial inactivation is through phosphorylation but that this then labels the protein for degradation. A similar phenomenon is observed with NAD^+-requiring glutamate dehydrogenase in *S. cerevisiae* and *C. utilis.* The NAD^+-requiring glutamate dehydrogenase in these organisms is principally concerned with oxidative deamination of glutamate:

$$\text{Glutamate} + NAD^+ \rightarrow \text{2-oxoglutarate} + NADH + NH_4^+ .$$

When *S. cerevisiae* is grown with glutamate as the source of carbon and nitrogen, the NAD^+-requiring glutamate dehydrogenase activity is high. If the organism is switched to glucose and ammonia as the sources of carbon and nitrogen, then oxidative deamination of glutamate is not required and the NAD^+-requiring glutamate dehydrogenase activity rapidly declines. The mechanism is similar to that in the previous example, namely the glutamate dehydrogenase initially becomes phosphorylated and is then degraded by proteolysis. Two other enzymes that become more susceptible to proteolysis following phosphorylation are hydroxymethylglutaryl-CoA reductase[76] and pyruvate kinase.[77]

The method of tagging in the case of glutamine synthetase from bacteria is by oxidation.[78] The enzyme is oxidized by mixed-function oxidases and it has been shown that there is loss of a single histidine residue per subunit. The oxidized enzyme is then more susceptible to attack by proteases. Several other enzymes may also be tagged by oxidation, including glutamine phosphoribosyl pyrophosphate amidotransferase, phosphoenolpyruvate carboxykinase, tyrosine aminotransferase, fructose bisphosphate aldolase, and ornithine decarboxylase.[78]

There is some evidence that certain enzymes become associated with membranes prior to degradation.[5] The enzyme hydroxymethylglutaryl-CoA reductase (HMG-CoA reductase) contains two domains, a catalytic domain (548 residues) and an anchoring domain (339 residues). The latter anchors the enzyme to the endoplasmic reticulum. HMG-CoA reductase has a short half-life (Table 9.1). It has been shown[79] that if the catalytic domain is cleaved from the complete enzyme, then the rate of degradation of the truncated enzyme (catalytic domain only) is only one-fifth the rate of degradation of the complete enzyme. The truncated enzyme is no longer able to associate with the membrane, which is believed to be involved in initiating the breakdown.

Thus it can be seen that there are a number of possible methods of tagging enzymes to mark them for degradation, but it is not yet clear how generally each route is used.

9.8 The significance of enzyme turnover

The turnover of specific enzymes has been studied most extensively in mammalian tissues in response to nutritional and hormonal changes, and to a lesser extent during development. In microorganisms, enzyme turnover has been studied in response to changes in nutrients, particularly carbon and nitrogen sources. These aspects will be considered in turn.

9.8.1 Changes in enzyme-turnover rates in animals in response to changes in diet

Arginase and serine dehydratase present in the liver are concerned indirectly with protein breakdown. If an animal is fed on a high-protein diet, then the surplus protein is not stored as such; it becomes deaminated and the carbon skeletons of the amino acids are converted to precursors of lipid or carbohydrate before being stored in those forms. Non-essential dietary amino acids, of which serine is one, are degraded before the essential amino acids. Serine dehydratase catalyses the first step in the breakdown of serine:

$$CH_2OH.CH(NH_2)COOH \rightarrow CH_2{=}C(NH_2)COOH \rightarrow$$
$$CH_3C({=}NH)COOH \rightarrow CH_3CO.COOH + NH_3.$$

The ammonia resulting from deamination of the amino acids eventually enters the urea cycle to be converted to urea. Arginase catalyses the final step in urea formation. The levels of these two enzymes change in response to changing dietary intake. An increase in protein intake causes increases in the activities of both arginase and serine dehydratase in the liver. Both of these increases have been shown to be due to increases in total enzyme protein and not due to activation of pre-existing enzyme. A decrease in dietary protein causes a decrease in the amount of arginase. As can be seen from Table 9.5, changes in the levels of enzymes can be affected by changes in the rate of synthesis, in the rate of degradation, or of both. In the case of arginase a further adaptation has been observed by starving animals that had previously been fed on a low-protein diet. Under these conditions the arginase content of the liver increases in spite of there being no protein intake. Endogenous proteins are now degraded as a source of energy. In contrast to the increase in arginase activity when placed on a high-protein diet, under fasting conditions the increase in arginase activity is brought about in a more economical fashion by a decrease in k_d but no change in k_s. An enzyme that resides in the plasma membrane and is believed to be important in

TABLE 9.5
The effect of diet on the turnover rates of enzymes from rat liver

Enzyme	Change of diet	Effect on enzyme
Arginase	Low-protein → high-protein	Increased activity mainly due to increased k_s
	High-protein → low-protein	Decreased activity, k_s decreased, k_d increased
	Low-protein → fasting	Increased activity, k_d decreased
Serine dehydratase	Protein-free → high-amino-acid	Large increase in activity due to increased k_s
Ornithine oxoglutarate aminotransferase	Basal → high-protein	Increased activity, increase in k_s k_d unchanged
Acetyl-CoA carboxylase	Basal → fat-free diet	Increased activity, increased k_s
	Basal → fasting	Decreased activity, k_s decreased, k_d increased
	Basal → high-fat diet	Decreased activity, k_s decreased, k_d increased
Fatty acid synthase	Basal → fat-free diet	Increased activity, k_s increased k_d unchanged
	Basal → fasting	Decreased activity, k_s decreased, k_d increased
Pyruvate kinase	Basal → fasting	Decreased activity, k_s decreased, k_d slight increase
	Fasting → refeeding	Increased activity, k_s increased, k_d unchanged
Phosphoenol pyruvate carboxykinase	Fed → starved	Increased activity, k_s increased, k_d slightly decreased
	Starved → refed	Decreased activity, k_s decreased, k_d slightly increased

See references 2, 33, 42, 80–84.

amino-acid uptake into cells is γ-glutamyltransferase. The activity of this enzyme is much higher in hepatoma cells than in normal liver cells, and this probably reflects the need for a greater rate of uptake to sustain the more rapid growth of the tumour cells. The increase is brought about by a decrease in the rate of degradation of the enzyme. Its half-life in normal cells is 3 h compared to 24 h in hepatoma cells.[85]

Another important enzyme to be studied in response to dietary change is acetyl-CoA carboxylase. This enzyme, which catalyses the first (rate-limiting) step in fatty-acid synthesis, is sensitive to changes in dietary fat.

$$\text{Acetyl-CoA} + CO_2 + \text{ATP} = \text{malonyl-CoA} + \text{ADP} + \text{orthophosphate.}$$

High dietary fat lessens the need for endogenous fatty-acid synthesis and *vice versa*. Again, the changes in enzyme level can be brought about by changing either the rate of synthesis or the rate of degradation (Table 9.5). When animals are fasted, not only acetyl-CoA carboxylase but also fatty-acid synthase and L-malate-$NADP^+$ oxidoreductase (oxaloacetate-decarboxylating) are degraded more rapidly than in fed animals. L-Malate-$NADP^+$ oxidoreductase (oxaloacetate-decarboxylating) is one of the enzymes involved in lipogenesis, since it catalyses the formation of NADPH.

9.8.2 Changes in enzyme stability in microorganisms in response to changes in carbon- and nitrogen-containing nutrients

Changes in the carbon and nitrogen sources in the growth medium cause adaptation in microorganisms; this adaptation may involve synthesis of new enzymes and degradation or inactivation of existing enzymes. An enzyme that is no longer necessary for the growth of the microorganisms on a new medium may be lost either passively, i.e. diluted out as the organism continues to grow, or in an active fashion by inactivation or degradation. The active loss of an enzyme that is no longer required may involve reversible or irreversible steps. Reversible inactivation may occur by covalent modification (see Chapter 6, Section 6.2.1). Examples from microorganisms include the following.

(1) Phosphorylation of the pyruvate dehydrogenase complex in *Neurospora crassa*[86] which occurs in the presence of ATP and inactivates the complex.
(2) Phosphorylation of yeast NAD^+-dependent glutamate dehydrogenase to form a much less active enzyme.[74] This enzyme is responsible in yeast for glutamate breakdown and is phosphorylated when glutamate in the medium is exchanged for ammonia and glucose as sources of nitrogen and carbon.
(3) Inactivation of enzymes required for gluconeogenesis, i.e. phosphoenolpyruvate carboxykinase, malate dehydrogenase, and fructose-bisphosphatase in *S. cerevisiae* when the source of carbon is changed from acetate to glucose (see Section 9.7.2).

(4) Inactivation of ornithine carbamoyltransferase in yeast by binding arginase. This occurs when arginine is added to the growth medium and is described more fully in Chapter 8, Section 8.5.3.3.

9.8.3 Changes in enzyme turnover in response to hormonal stimuli

The systems most studied under this heading are various liver enzymes concerned with the catabolism of amino acids and their response to hormones, particularly the glucocorticoids. Some studies have been made with intact animals but others have used a strain of hepatoma cells grown in culture which responds very similarly to that of intact liver and which has the advantage that the hormone concentration can be more easily controlled. Glucocorticoids include cortisol, corticosterone, and cortisone and their principal effects are to raise blood glucose levels, mobilize amino acids, and stimulate gluconeogenesis in the liver. Trytophan 2,3-dioxygenase, the first enzyme in the tryptophan catabolic pathway, can be induced in liver by administration of either cortisol or of tryptophan or analogues of tryptophan. In each case the total amount of enzyme protein is increased, but in the case of cortisol administration it is due to an increase in the rate of enzyme synthesis, whereas in response to tryptophan or its analogues the rate of enzyme degradation is decreased, possibly as a result of stabilization of the enzyme in the presence of the substrate. The activity of tyrosine aminotransferase, the first step in tyrosine breakdown, is also induced by cortisol which increases the rate of its biosynthesis. Unlike tryptophan 2,3-dioxygenase, tyrosine aminotransferase is also induced by the pancreatic hormones insulin and glucagon.

Other liver enzymes that have been studied are alanine aminotransferase and ornithine oxo-acid aminotransferase. The former increases in response to glucocorticoids, which affect only the rate of synthesis, while the latter has been shown to increase in response to oestrogen, again by an increase in the rate of synthesis; there is no change in the rate of degradation. The effects are summarized in Table 9.6.

The mechanisms by which these hormones bring about increases in enzyme activity are still unclear. The steroid hormones have been shown to penetrate the cells of the target tissue, where they bind to receptors and ultimately bring about a change in gene expression.[87] In the case of certain enzymes, e.g. tryptophan 2,3-dioxygenase, there is evidence for increased production of mRNA, which is assumed to be due to changes in gene expression. Other hormones, such as glucagon and adrenalin, bind to receptors on the cell membrane and their effects are thought to be mediated through changes in the intracellular concentrations of cyclic nucleotides (see Chapter 6, Section 6.4.2.1).

9.8.4 Changes in enzyme turnover during development

Although changes in the levels of many enzymes have been studied in a number of different developmental systems and in some cases groups of enzymes have

TABLE 9.6

Changes in enzyme turnover in response to hormonal treatment

Tissue	Enzyme	Hormonal change	Effect on enzyme
Rat liver	Tryptophan 2,3-dioxygenase	Glucocorticoid administration	Increased activity, k_s increased, k_d unchanged
Rat liver	Tyrosine aminotransferase	Glucocorticoid, glucagon or insulin administration	Increased activity, increased k_s, k_d unchanged
Rat kidney	Ornithine oxo-acid aminotransferase	Oestrogen	Increased activity, k_s increased
Rat liver	Carbamoyl phosphate synthase	Thyroidectomy	Decreased k_d
Rat liver	Malate dehydrogenase	Thyroidectomy	Decreased k_d
Rat liver	Arginase	Glucocorticoid	Increased activity, k_s increased
Rat liver	Serine dehydratase	Glucagon	Increased activity, k_s increased

References 2, 34, 41.

been shown to appear rapidly at particular stages in development,[88] none of these changes has been analysed in as great detail as the dietary effects described in Section 9.8.1. The most widely used method of study is to measure enzyme activity during different stages of a particular developmental process, in some cases perturbing the system by the use of inhibitors. In more detailed studies, enzymes have been purified and the enzyme protein has been measured. These studies taken together allow us to build up an overall picture of gene expression during development. As yet very few studies[89] have been made to determine changes in k_s and k_d for particular enzyme systems during development. A number of simple systems have been used as models to study development, e.g. sporulation of yeast and bacteria and the life cycle of the cellular slime mould.

The sporulation of yeast illustrates the type of studies made. Certain strains of *Saccharomyces cerevisiae* can be induced to sporulate by starvation on a nitrogen-free medium. During the process of sporulation, modified cells or ascospores are formed within an ascus. Since this occurs during nitrogen starvation, all new proteins are made from existing nitrogen soruces. A large amount of protein turnover occurs during sporulation and the total protein of the culture remains fairly constant. Extensive degradation occurs to provide a source of amino acids for the synthesis of new proteins. By differential labelling of vegetative and sporulating yeast cells it has been shown that proteins in both the vegetative cells and the spores are degraded.[90] Two endoproteases, (A and B), one carboxypeptidase (Y) and two or more aminopeptidases are present in yeast cells. There are also specific inhibitors for some of the proteases but these are located in the cytosol, whereas the proteases are contained within vacuoles. The activities of proteases A and B increase several-fold during sporulation; the activity of proteases A appears essential for sporulation, since mutants lacking this enzyme are unable to sporulate. A number of enzymes have been shown to decrease their specific activities to less than 50 per cent of their initial value,[90] e.g. glutamate dehydrogenase ($NADP^+$-requiring), isocitrate lyase, malate dehydrogenase, cytochrome *c* oxidase, succinate dehydrogenase, fructose-bisphosphatase, aspartate aminotransferase, alcohol dehydrogenase, and hexokinase. It is not yet clear whether these changes are due to proteolysis, although this seems most probable. As can be seen, many enzymes that decrease in activity are those one might expect to accompany a decrease in metabolic activity.

Proteolysis occurs in bacteria during sporulation. This has been studied in *Bacillus subtilis*, where it has been shown that selective inactivation of certain enzymes occurs, e.g. aspartate carbamoyltransferase and amidophosphoribosyltransferase. This is due to inactivation followed by proteolysis.[91] (For review see reference 92.)

Large increases in proteolytic activity have been found associated with particular stages of development in many organisms, e.g. during germination and sporulation in the water mould,[93,94] *Blastocladiella emersonii*, in the germination of macroconidia of *Microsporum gypseum*, and in the regression of the salivary glands of the insect larval instar of *Chironomus tentans*, but in the developmental systems studied up until now the rates of synthesis and degradation of individual enzymes have not yet been determined. Thus, more research is needed before the control of protein turnover during development is understood.

9.9 Conclusions

Enzyme turnover is an important method of regulation of enzyme activity and is controlled by regulation of either enzyme synthesis or enzyme inactivation/degradation, or in some cases by regulation of both steps. The general mechanism of enzyme synthesis is well understood, although its control has only been studied in depth in a limited number of prokaryote systems. The mechanism of enzyme inactivation is less well understood. A number of correlations with structure and function and with particular sequences of individual enzymes have been made. Many enzymes appear to become tagged before becoming susceptible to protease attack. The methods of tagging include conjugation with ubiquitin, phosphorylation, oxidation, or binding to membranes, but it is still not generally clear precisely which features of the enzymes are recognized by the tagging system, although some clues are beginning to emerge.

References

1. Schoenheimer, R., *The dynamic state of body constituents.* Harvard University Press, Cambridge, Massachusetts (1942).
2. Schimke, R. T., *Adv. Enzymol.* **37**, 135 (1973).
3. Goldberg, A. L. and Dice, J. F., *A Rev. Biochem.* **43**, 835 (1974).
4. Goldberg, A. L. and St. John, A. C., *A. Rev. Biochem.* **45**, 747 (1976).
5. Ballard, F. J., *Essays Biochem.* **13**, 1 (1977).
6. Beynon, R. J. and Bond, J. S., *Am. J. Physiol.* **251**, C 141 (1986).
7. Mayer, R. J. and Doherty, F. *FEBS Lett.* **198**, 181 (1986).
8. Rivett, A. J., *Curr. Top. Cell. Regul.* **28**, 291 (1986).
9. Rechsteiner, M., Roger, S., and Rote, K., *Trends Biochem. Sci.* **12**, 390 (1987).
10. Walker, R., *The molecular biology of enzyme synthesis*, Ch. 1. Wiley, New York (1983).
11. Goldberg, A. L. and St. John, A. C., *A. Rev. Biochem.* **45**, 760 (1976).
12. Stryer, L., *Biochemistry* (3rd edn), Ch. 5. Freeman, New York (1988).
13. Darnell, J., Lodish, H., and Baltimore, D., *Molecular cell biology*, Ch. 4. Scientific American Books Inc., New York (1986).
14. Sezekely, M., *From DNA to protein.* MacMillan, London (1980).
15. Adams, R. L. P., Knowler, J. T., and Leader, D. P., *The biochemistry of nucleic acids* (10th edn), Ch. 11. Chapman and Hall, London (1986).
16. Adams, R. L. P., Knowler, J. T., and Leader, D. P., *The biochemistry of nucleic acids* (10th edn), pp. 289–292. Chapman and Hall, London (1986).
17. Lewin, B., *Gene expression—III*, Ch. 10. Wiley, London (1987).
18. Berlin, C. M. and Schimke, R. T., *Molec. Pharmacol.* **1**, 149 (1965).
19. Chiu, F. C. and Goldman, J. E., *J. Neurochem.* **42**, 166 (1984).

20. Zak, R., Martin, A. F., and Blough, R., *Physiol. Rev.* **59**, 407 (1979).
21. Garlick, P. J., in *Comprehensive biochemistry* (Florkin, M. and Stotz, E. H., eds.), Vol. 19B, Ch. 2. Elsevier, Amsterdam (1980).
22. Wilde, C. J., Paskin, N., Saxton, J., and Mayer, R. J., *Biochem. J.* **192**, 311 (1980).
23. Schimke, R. T., *J. biol. Chem.* **239**, 3808 (1964).
24. Millward, D. J., *Clin. Sci.* **39**, 577 (1970).
25. Druyan, R., De Bernard, B., and Rabinowitz, M., *J. biol. Chem.* **244**, 5874 (1969).
26. Doyle, D. and Tweto, J. *Methods Cell Biol.* **10**, 235 (1975).
27. Gunn, J. M., Ballard, J. F., and Hanson, R. W., *J. biol. Chem.* **251**, 3586 (1976).
28. Schimke, R. T. and Katanuma, N., *Intracellular protein turnover*, p. 143. Academic Press, New York (1975).
29. Schimke, R. T. and Katanuma, N., *Intracellular protein turnover*, p. 181. Academic Press, New York (1975).
30. Schimke, R. T. and Katanuma, N., *Intracellular protein turnover*, p. 283. Academic Press, New York (1975).
31. Hopgood, M. F. and Ballard, F. J., *Biochem. J.* **144**, 371 (1974).
32. Goldberg, A. L. and St. John, A. C., *A. Rev. Biochem.* **45**, 751 (1976).
33. Volpe, J. J., Lyles, T. O., Runcari, D. A. K., and Vagelos, P. R., *J. biol. Chem.* **248**, 2502 (1978).
34. Kobayashi, K., Kito, K., and Katunuma, N., *J. Biochem.* **79**, 787 (1976).
35. Masters, C., *Curr. Top. Cell. Regul.* **21**, 205 (1982).
36. Gear, A. R. L., *Biochem. J.* **120**, 577 (1970).
37. Jacobs, F. A., Bird, R. C., and Sells, B. H., *Eur. J. Biochem.* **150**, 255 (1985).
38. Walker, J. H., Burgess, R. J., and Mayer, R. J., *Biochem. J.* **176**, 927 (1978).
39. Russell, S. M., Burgess, R. J., and Mayer, R. J., *Biochem. J.* **192**, 321 (1980).
40. Schimke, R. T. and Katanuma, N., *Intracellular protein turnover*, p. 325. Academic Press, New York (1975).
41. Nicoletti, M., Guerri, C. and Grisolia, S., *Eur. J. Biochem.* **75**, 583 (1977).
42. Ip, M. M., Chee, P. Y., and Swick, R. W., *Biochem. biophys. Acta* **354**, 29 (1974).
43. Miyazawa, S., Ozasa, H., Furuta, S., Osumi, T., Hashimoto, T., Miura, S., Mori, M., and Tatibana, M., *J. Biochem.* **93**, 453 (1983).
44. Bond, J. S. and Butler, P. E., *A. Rev. Biochem.* **56**, 333 (1987).
45. Dice, J. F., Dehlinger, P. J., and Schimke, R. T., *J. biol. Chem.* **248**, 4220 (1973).
46. Dice, J. F. and Goldberg, A. L., *Archs. Biochem. Biophys.* **170**, 213 (1975).
47. Dean, R. T., *Trans. Biochem. Soc.* **3**, 250 (1975).
48. Bohley, P., Wollert, H-G., Riemann, D., and Riemann, S., *Acta Biol. Med. Ger.* **40**, 1655 (1981).
49. Dice, J. F. and Goldberg, A. L., *Proc. natn. Acad. Sci. USA* **72**, 3893 (1975).
50. Russell, S. M., Wilde, C. J., White, D. A., Hasan, H. R., and Mayer, R. J., *Acta Biol. Med. Ger.* **40**, 1397 (1981).
51. McLintock, D. K. and Markus, G., *J. biol. Chem.* **243**, 2855 (1968).
52. Hillcoat, B. L., Swett, V., and Bertino, J. R., *Proc. natn. Acad. Sci. USA* **58**, 1632 (1967).
53. McCann, P. P., Tardiff, C., Duchesne, M., and Mamont, P. S., *Biochem. Biophys. Res. Commun.* **76**, 893 (1977).
54. Stevens, L. and McKinnon, I. M., *Biochem. J.* **166**, 635 (1977).
55. Taneke, T., Wade, K., Ogiwara, H., and Numa, S., *FEBS Lett.* **82**, 95 (1977).
56. Sinensky, M. and Logel, J., *J. biol. Chem.* **258**, 8547 (1983).
57. Katunuma, N., Kominami, E., Banno, Y., Kito, K., Aoki, Y., and Urata, G., *Adv. Enz. Regul.* **14**, 325 (1976).

58. Woodbury, R. G., Everett, M., Sanada, Y., Katunuma, N., Lagunoff, D., and Neurath, H., *Proc. natn. Acad. Sci. USA* **75**, 5311 (1978).
59. Perry, S. T., Lee, K., and Kenney, F. T., *Archs. Biochem. Biophys.* **195**, 362 (1979).
60. Rechsteiner, M., Chin, D., Hough, R., McGarry, T., Rogers, S., Rote, K. V., and Wu, L., *Cell Fusion* (Ciba Symposium 103) (Evered, D. and Whelan, J., eds), pp. 181–201 (1984).
61. Rogers, S., Wells, R., and Rechsteiner, M., *Science* **234**, 364 (1986).
62. Lin, S., and Zabin, I., *J. biol. Chem.* **247**, 2205 (1972).
63. Ballard, F. J., *Essays Biochem.* **13**, 22 (1977).
64. Poole, B., and Wibo, M., *J. biol. Chem.* **248**, 6221 (1973).
65. Neff, N. T., De Martino, G. N., and Goldberg, A. L., *J. Cell Physiol.* **101**, 439 (1979).
66. Hough, R., and Rechsteiner, M. *Proc. natn. Acad. Sci. USA* **81**, 90 (1984).
67. McElligott, M. A., Miao, P., and Dice J. F., *J. biol. Chem.* **260**, 11986 (1985).
68. Wolf, D. H., *Biochem. Soc. Trans.* **13**, 279 (1985).
69. Dean, R. T. and Barrett, A. J., *Essays Biochem.* **12**, 1 (1976).
70. Hershko, A. and Ciechanover, A., *A. Rev. Biochem.* **51**, 335 (1982).
71. Hershko, A. and Ciechanover, A., *Prog. Nucl. Acid Res. Molec. Biol.* **33**, 19 (1986).
72. Bachmair, A., Finley, D., and Varshavsky, A., *Science* **234**, 179 (1986).
73. Purwin, C., Leidig, F. and Holzer, H., *Biochem. Biophys. Res. Commun.* **107**, 1482 (1982).
74. Hemmings, B. A., *Biochem. Soc. Trans.* **10**, 328 (1982).
75. Funayama, S., Gancedo, J. M., and Gancedo, C., *Eur. J. Biochem.* **109**, 61 (1980).
76. Parker, R. A., Miller, S. J., and Gibson, D. M., *Biochem. Biophys. Res. Commun.* **125**, 629 (1984).
77. Bergström, G., Ekman, P., Humble, E., and Engström, L., *Biochem. Biophys. Acta* **532**, 259 (1978).
78. Rivett, A. J. and Levine, R. L., *Biochem. Soc. Trans.* **15**, 816 (1987).
79. Gill, G., Faust, J. R., Chiu, D. J., Goldstein, J. L., and Brown, M. S., *Cell*, **41**, 249 (1985).
80. Schimke, R. T. and Katunuma, N., *Intracellular protein turnover*, p. 127. Academic Press, New York (1975).
81. Cladaras, C. and Cottam, G. L., *Archs Biochem. Biophys.* **200**, 426 (1980).
82. Gunn, J. M., Hanson, R. W., Meynhas, O., Reshef, L., and Ballard, F. J., *Biochem. J.* **150**, 195 (1975).
83. Hopgood, M. F., Ballard, F. J., Reshef, L., and Hanson, R. W., *Biochem. J.* **134**, 445 (1973).
84. Numa, S. and Yamashita, S., *Curr. Top. cell. Regul.* **8**, 197 (1974). see pp. 206–20.
85. Ding, J. L., Smith, G. D., and Peters, T. J., *FEBS Lett.* **142**, 207 (1982).
86. Wieland, O. H., Hartmann, U., and Siess, E. A., *FEBS Lett.* **27**, 240 (1972).
87. King, R. J. B., *Essays Biochem.* **12**, 41 (1976).
88. Greengard, O., *Essays Biochem.* **7**, 159 (1971).
89. Ballard, F. J., in *Principles of metabolic control in mammalian systems* (Herman, R. H., Cohn, R. M., and McNamara, P. D., eds.), p. 245. Plenum Press, New York (1980).
90. Betz, H., in *Cell differentiation in microorganisms, plants and animals* (Nover, L. and Mothes, K., eds.), p. 243. North Holland, Amsterdam (1977).
91. Switzer, R. L., Maurizi, M. R., Wong, J. Y., Brabson, J. S., and Meyer, E., in *Limited proteolysis in microorganisms* (Cohen, G. N. and Holzer, H., eds.), p. 103. NIH Bethesda, Maryland (1978).
92. Maurizi, M. R. and Switzer, R. L., *Curr. Top. cell. Regul.* **16**, 163 (1980).

93. Leaver, C. J. and Lovett, J. S., *Cell Differentiation* **3**, 165 (1974).
94. Correa, J. U., Rodrigues, V., and Lodi, W. R., in *Limited proteolysis in microorganisms* (Cohen, G. N. and Holzer, H., eds.), p. 127. NIH Bethesda, Maryland (1978).
95. Hemmings, B. A., in *Molecular aspects of cellular regulation* (Cohen, P., ed.), Vol. 3, p. 155. Elsevier, Amsterdam (1984).

Appendix 9.1.

Derivation of equation describing the time course of approach to a new steady-state level of enzyme after a change in its rate of synthesis and/or degradation has occurred.

The steady-state level of an enzyme in a cell or tissue is given by the equation:

$$\frac{d[E]}{dt} = k_s - k_d[E] \qquad \text{(see Section 9.3)}$$

where k_s is the rate constant for its synthesis (zero order with respect to [E]) and k_d is the rate constant for its degradation (first order with respect to [E]).

If there is a change in the rate of synthesis and/or degradation of the enzyme as a result of some hormonal, dietary, etc. influence then the enzyme concentration will change to a new steady-state level.

Let

k_s' be the new rate constant for enzyme synthesis,
k_d' be the new rate constant for enzyme degradation,
$[E_0]$ be the initial steady-state concentration,
$[E_t]$ be the enzyme concentration at time t after change in rate constants from k_s to k_s' and k_d to k_d'.

$$\frac{d[E]}{dt} = k_s' - k_d'[E] \tag{A.9.1}$$

$$\frac{d[E]}{(k_s' - k_d'[E])} = dt. \tag{A.9.2}$$

By integration of eqn (A.9.2),

$$\ln(k_s' - k_d'[E]) = -k_d't + c. \tag{A.9.3}$$

At time $t=0$, $[E]=[E_0]$, therefore $c = \ln(k_s' - k_d'[E_0])$. Thus at time t,

$$\ln(k_s' - k_d'[E_t]) - \ln(k_s' - k_d'[E_0]) = -k_d't. \tag{A.9.4}$$

Taking antilogarithms of eqn (A.9.4):

$$\frac{k_s' - k_d'[E_t]}{k_s' - k_d'[E_0]} = e^{-k_d't}. \tag{A.9.5}$$

This can be rearranged to give

$$\frac{[E_t]}{[E_0]} = \frac{k_s'}{k_d'[E_0]} - \left(\frac{k_s'}{k_d'[E_0]} - 1\right) e^{-k_d't}.$$

See also reference 18.

10
Clinical aspects of enzymology

10.1 Introduction

Over the last thirty years there has been a considerable increase in both the measurement of enzyme activities and in the use of purified enzymes, in clinical practice. Before 1940 the only enzymes whose activities were measured for clinical diagnoses were the hydrolytic enzymes: lipase, amylase, phosphatases, trypsin, and pepsin and their measurement constituted less than 5 per cent of the analyses carried out in the average clinical chemistry laboratories. At the present time up to 25 per cent of the work of an average clinical chemistry laboratory may consist of enzyme assays for diagnoses and up to about 20 different enzymes are assayed routinely. The reason for the change in emphasis is largely due to increased understanding and awareness of the molecular details of metabolism and to the development of rapid and reliable methods of enzyme assay. The discovery in the 1950s of the relationship between the activities of serum aminotransferases and dehydrogenases and heart and liver diseases gave the impetus to search for other enzymes present in serum that could be of diagnostic value. In addition, during the last twenty-five years many enzymes have been isolated and purified (see Chapter 2) and this has made it possible to use enzymes to determine the concentrations of substrates of clinical importance. The availability of highly specific antibodies, often from single clones (monoclonal antibodies), in conjunction with purified enzymes to amplify detection has enabled a wide range of biological compounds to be measured by enzyme immunoassay. A further development arising from the increased availability of purified enzymes is the potential for enzyme therapy (see Section 10.8).

At the present time the major importance of enzymes in clinical practice is in the measurement of enzyme levels in serum as an aid to diagnosis and we shall therefore discuss this first (Sections 10.2–10.6). In Section 10.7 we shall describe the use of enzymes as a means of estimating concentrations of substrates, and finally in Section 10.8 we consider enzyme-replacement therapy. There are a number of books and reviews on clinical applications of enzymology in which more detailed information can be obtained.[1–7]

10.2 Determination of enzyme activities for clinical diagnosis

For the measurement of an enzyme activity to be useful as a routine diagnostic clinical method the following conditions should be fulfilled.

1. The enzyme should be present in blood, urine, or some readily available tissue fluid. Tissue biopsies should not be used routinely, although they may be used if the diagnostic value is sufficiently important.
2. The enzyme should be easy to assay and it is advantageous if the method can readily be automated.
3. The differences in the ranges of enzyme activities obtained from normal and diseased subjects should be diagnostically significant and there should be a good correlation between the level of enzyme activities and the pathological state.
4. It is also useful if the enzyme is sufficiently stable so that the sample may be stored at least for a limited time.

Serum is the fluid in which most measurements of enzyme activities are made. Urine can be used for a few enzymes that are cleared by the kidney. Red blood cells and white blood cells, although they are relatively accessible, have not so far been used extensively in diagnosis. The enzymes present in serum can be considered in one of two categories: (i) plasma-specific enzymes and (ii) non-plasma-specific enzymes. The former are enzymes whose normal function is in the plasma, e.g. enzymes concerned with blood coagulation,[8,42] complement activation,[9,43] and lipoprotein metabolism. The latter are enzymes that have no physiological function in the plasma, enzymes for which the cofactors or even substrates may be lacking. This category includes enzymes that are secreted by tissues, e.g. amylase, lipase, and phosphatases, and also enzymes associated with cellular metabolism, whose presence in normal serum at low levels may be due to turnover of cells within the tissue causing release of the enzyme content. The rationale of enzyme measurement for diagnosis is that if a tissue is broken down or if it is producing an abnormally high amount of intracellular enzyme that is then released, there is an elevation of enzyme activity in the serum. For several enzymes there is a very high concentration gradient (a factor of the order of 10^3 or greater) across the cell membrane to the extracellular fluid and thus a small amount of tissue damage may affect considerably the serum concentration of certain enzymes.[6]

In the diseased state a tissue may become inflamed or it may become necrotic.* If the latter occurs there is likely to be fairly complete release of the enzyme content from the dead cells. However, the pattern of enzymes detectable in the serum may not completely resemble that of the tissue from which it arises, since enzymes may be inactivated at different rates. Inflammation of a tissue may result in a change in the permeability of the cells, so that release of enzymes may occur from the cytosol but not from organelles,[10] e.g. glutamate dehydrogenase, which is present in high concentrations in the mitochondrial matrix, is released only when cell destruction is fairly complete.[11] The enzymes released into the serum seem to be removed at a fairly rapid rate. The mechanism of their removal has not been studied appreciably: it may be due to inactivation and degradation occurring in the serum or it may be due to clearance by the kidney. Some

* Necrosis is the death of a portion of tissue or organ.

experiments, involving intravenous injections of ^{125}I-labelled lactate dehydrogenase (LDH-5) into rabbits, have been carried out to study the fate of serum enzymes. The results suggest that LDH-5 undergoes denaturation in the plasma and that the products are excreted into the small intestine where they are further degraded and then reabsorbed, as amino acids and small peptides, back into the circulation.[12] The quite rapid removal of most released enzymes means that monitoring particular enzymes present in the serum of a diseased subject gives an up-to-date picture of the release of enzymes from diseased tissue. Monitoring enzymes in the serum is thus useful not only in the initial diagnosis but also in studying the course of the disease and the response to treatment.

Ideally it would be desirable to study tissue-specific enzymes for diagnostic purposes, since these would identify the tissue from which they arose. However, there are relatively few enzymes that are entirely tissue specific, although there are several that are much more abundant in one tissue than another, e.g. acid phosphatase in the prostate and acetylcholinesterase (EC 3.1.1.7) in erythrocytes. Although a particular enzyme activity may not be specific to one tissue, there may exist isoenzymes (see Chapter 1, Section 1.5.2) that show a different pattern in different tissues. The best-studied example is that of lactate dehydrogenase (further information on lactate dehydrogenase is given in Chapter 5, Section 5.5.4). Lactate dehydrogenase is composed of four subunits. There are two types of subunit, which combine in lactate dehydrogenase to give five possible combinations: α_4, $\alpha_3\beta$, $\alpha_2\beta_2$, $\alpha\beta_3$, and β_4. These five types may be separated by electrophoresis and it has been found that their distribution varies with the tissue (Fig. 10.1). Thus, although by measurements of lactate dehydrogenase activity in serum the tissue of origin could not be identified, identification might be possible

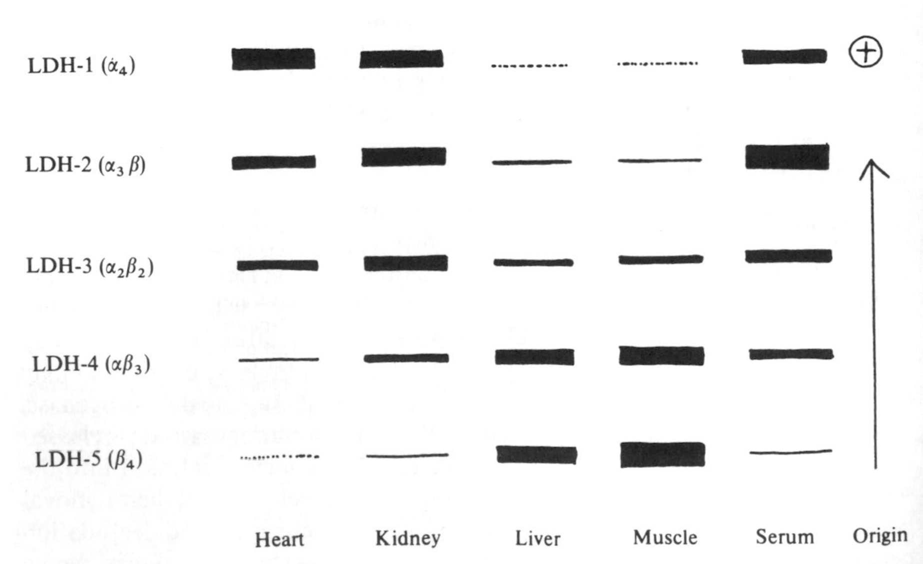

FIG. 10.1. Electrophoretic separation of isoenzymes of lactate dehydrogenase from different tissues.

if the isoenzyme distribution were determined by electrophoresis. Several enzymes that may be present in serum are known to exist in multiple forms, e.g. alkaline phosphatase, amylase, creatine kinase, ceruloplasmin, glucose 6-phosphate dehydrogenase, and aspartate aminotransferase, but these isoenzymes have not yet been as well characterized as those of lactate dehydrogenase. Some isoenzymes may be differentiated by criteria other than electrophoretic mobility, e.g. by substrate specificity, heat stability, or sensitivity to inhibitors (Table 10.1). It is now possible in some cases to distinguish isoenzymes by use of monoclonal antibodies. This method has been used to distinguish the different isoenzymes of human phosphofructokinase, and to identify which form is absent in inherited deficiencies of phosphofructokinase.[42] Monoclonal antibodies against specific isoenzymes may in future be used for the detection and quantitation of isoenzymes used as tumour markers, such as prostatic acid phosphatase, placental alkaline phosphatase, and terminal deoxynucleotide transferase.

In many cases where measurements of enzyme activity are used as an aid to diagnosis, more than one activity is measured in order to permit a greater degree of discrimination. Perhaps the two examples in which enzyme monitoring has been most useful and widely applied are those of liver and heart diseases. It must be emphasized that the measurement of enzyme activity itself is not sufficient evidence on which to make a diagnosis but must be taken in conjunction with the

TABLE 10.1
Methods used to determine the tissue of origin of serum enzymes

Enzyme	Differentiation by electrophoresis	Differentiation by other methods
Lactate dehydrogenase	LDH-1 predominant in heart, erythrocytes, brain and kidney	Differential heat stability, LDH-4 and LDH-5 show greater heat lability.
	LDH-3 predominant in leucocytes, adrenal and thyroid	LDH-1 is more sensitive to inhibition by excess pyruvate than LDH-5
	LDH-5 predominant in skeletal muscle and liver	
Acid phosphatase		Prostate enzyme inhibited by tartrate. Erythrocyte enzyme inhibited by formaldehyde. Differential activities towards substrates, phenyl phosphate, glycerol 2-phosphate and thymolphthalein phosphate
Alkaline phosphatase	Complex electrophoretic patterns. Liver enzyme most cationic	Placental enzyme shows greater heat stability. Enzymes from both placenta and intestine are sensitive to phenylalanine
Creatine kinase	Brain, lung and thyroid mostly CK-1 (or BB). Skeletal muscle almost exclusively CK-3. Cardiac tissue CK-3+ about 15% CK-2	

See references 1, 2.

clinical symptoms and other types of evidence, e.g. electrocardiographic evidence in the case of heart disease.

Increased activities of certain enzymes in serum are commonly found in malignant diseases, but no enzyme is specific for cancer and so far none has been very useful in the early detection of cancer.[13]

Besides the use of isoenzyme patterns to identify the tissue of origin, isoenzyme patterns are also used in forensic science. Since a number of enzymes in serum and red blood cells occur as different isoenzymes, the particular pattern in a blood sample may provide supporting evidence as to its origin. Isoenzyme patterns that are currently used by the Metropolitan Police Science Laboratory are adenosine deaminase, adenylate kinase, carbonate dehydratase, acid phosphatase, esterase, phosphoglucomutase, aminopeptidase and lactoylglutathione lyase.[14]

10.3 Clinical enzymology of liver disease

The enzymes most used in diagnosis of liver disease are serum aspartate aminotransferase (often referred to as serum glutamate oxaloacetate transaminase, SGOT), alanine aminotransferase (often referred to as serum glutamate pyruvate transaminase, SGPT), alkaline phosphatase, and γ-glutamyltransferase. Others that are also sometimes measured are lactate dehydrogenase and isocitrate dehydrogenase. Ornithine carbamoyltransferase is an enzyme that is exclusive to the liver, but it has not been widely used as a diagnostic aid because the assay is complex compared to those of other liver enzymes. The aminotransferases that catalyse the following reactions are usually assayed by coupling to the reactions catalysed by malate and lactate dehydrogenases so that the change in A_{340} due to oxidation of NADH can be measured continuously (see Chapter 4, Section 4.2).

1. Aspartate + 2-oxoglutarate $\rightleftharpoons$ oxaloacetate + glutamate

   Oxaloacetate + NADH + H^+ $\rightleftharpoons$ malate + NAD^+.

2. Alanine + 2-oxoglutarate $\rightleftharpoons$ pyruvate + glutamate.

   Pyruvate + NADH + H^+ $\rightleftharpoons$ lactate + NAD^+.

Alkaline phosphatase is assayed by measuring the hydrolysis of *p*-nitrophenyl phosphate. γ-Glutamyltransferase is assayed by measuring the release of *p*-nitroaniline when γ-glutamyl-*p*-nitroanilide is used as one of the substrates. *p*-Nitroaniline absorbs at 400 nm.

$$HOOC.CH(NH_2)CH_2CH_2CO(NH\text{-}C_6H_4\text{-}NO_2) + \text{glycylglycine} \longrightarrow \gamma\text{-glutamylglycylglycine} + H_2N\text{-}C_6H_4\text{-}NO_2$$

The most commonly occurring liver diseases are viral or toxic hepatitis, cirrhosis, primary and secondary liver tumours, and obstruction of the bile-duct

due to stones, tumours, or stricture. The conditions may be acute, int
or chronic. The two aminotransferases are in the category of non-plasm
enzymes, i.e. they are enzymes concerned with intracellular metabolism and have no known function in the plasma. They are released from the liver when parenchymal cells (the principal cells of the liver) become necrotic as in, for example, viral or toxic hepatitis or cirrhosis; the increase in their activity in the plasma is related to the extent of cell breakdown. Aminotransferases in serum may also arise from cardiac tissue after myocardial infarction, but the proportions of asparate aminotransferase and alanine aminotransferase differ. This ratio was used to differentiate between various conditions, although it is only rarely used now. In cases of chronic hepatitis or cirrhosis, aminotransferases are often monitored over a period of serveral months to follow the progress of the disease and to provide an indication of the prognosis.

Measurements of serum alkaline phosphatase are also important in diagnosis of liver disease. Serum alkaline phosphatase arises from two main sources, namely, the liver and bone; the enzymes from these sources differ in their isoenzyme patterns. Liver alkaline phosphatase is a non-plasma-specific enzyme that is secreted from the sinusoidal surface (i.e. that adjacent to the blood space) of the liver cell and thus is present in the serum at low levels in the absence of liver cell damage. It is possible to differentiate obstructive jaundice from hepatitis by measurement of aminotransferases and alkaline phosphatase, since there is a proportionately greater increase in alkaline phosphatase activity in obstructive jaundice than in hepatitis, so that the ratios of activities of serum alkaline phosphatase/serum alanine aminotransferase one week after the onset of viral hepatitis might be 0.2–0.6 compared with 10–20 one week after the onset of obstructive jaundice.[4] Whether the bile-duct obstruction is intermittent or complete can be assessed by monitoring the alkaline phosphatase activity over a period of a few days (Fig. 10.2). Glutamate dehydrogenase can also be used in place of alkaline phosphatase to distinguish hepatic and biliary disease. The ratio of glutamate dehydrogenase/alanine aminotransferase is higher in cases of obstructive jaundice than, for example, in viral hepatitis.[11]

γ-Glutamyltransferase is now also being used extensively in the diagnosis of liver disease, since most of this enzyme activity detectable in the serum is of hepatic origin. The activity is greatly elevated in the serum in cases of chronic alcoholism, when it causes cirrhosis of the liver, and in cases of metastatic invasion of the liver. The increased serum level in cases of chronic alcoholism is due to both increased synthesis in the liver and increased release into the serum. γ-Glutamyltransferase activity follows a similar pattern to that of alkaline phosphatase in obstructive jaundice.

10.4 Clinical enzymology of heart disease

Enzymological methods have been particularly useful in providing supporting evidence for myocardial infarction (the process of formation of an area of dead heart muscle) and also monitoring the course of the infarct. In myocardial

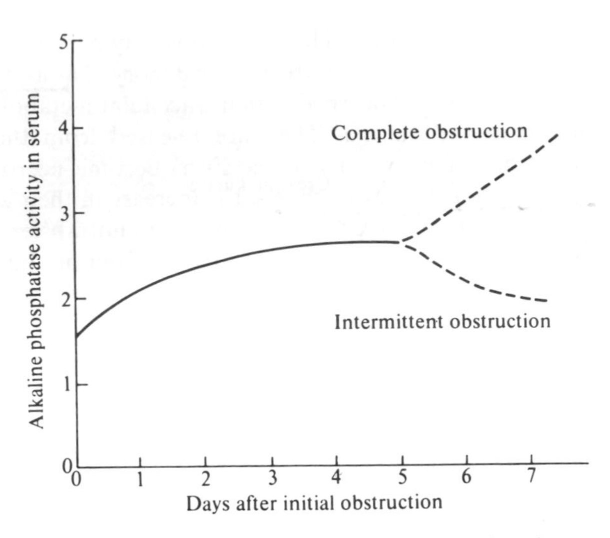

FIG. 10.2. Typical changes in the activity of serum alkaline phosphatase after complete and intermittent obstructive jaundice (see reference 4.)

infarction a coronary artery becomes obstructed and this leads to irreversible damage and necrosis of the heart tissue. Enzymes are released from the necrotic tissue into the plasma; the three enzymes most commonly assayed are creatine kinase, aspartate aminotransferase, and lactate dehydrogenase. Each enzyme shows a different time course for release into the plasma and subsequent disappearance. These differences depend on the concentration of each present in cardiac tissue, their rate of release, and their subsequent rate of clearance or degradation.[1,4] Their half-lives in the plasma are creatine kinase $\approx$ 15 h, asparate aminotransferase $\approx$ 17 h, and lactate dehydrogenase $\approx$ 110 h. Creatine kinase is the earliest to be detectable, rising 4–6 h after the onset of pain, reaching a peak at 24–36 h and then rapidly declining. Aspartate aminotransferase reaches a peak between 48 and 60 h and lactate dehydrogenase between 48 and 72 h; the latter declines more slowly than the former (Fig. 10.3). The prognosis and progress can be assessed by the increase in the activities of these enzymes. Creatine kinase, which does not occur in the liver unlike aspartate aminotransferase and lactate dehydrogenase, can therefore be used to differentiate heart disease from liver disease. However, since the level of creatine kinase rises and falls rapidly it is sometimes safer to measure aspartate aminotransferase or lactate dehydrogenase, but in other situations creatine kinase activity can be used in the early detection and prognosis of myocardial infarction.

Creatine kinase can be particularly useful if its estimation is accompanied by determination of the isoenzyme pattern. Creatine kinase derived from skeletal muscle, a common source of creatine kinase in serum, consists almost exclusively of the CK-3(MM) type, whereas creatine kinase derived from heart tissue,

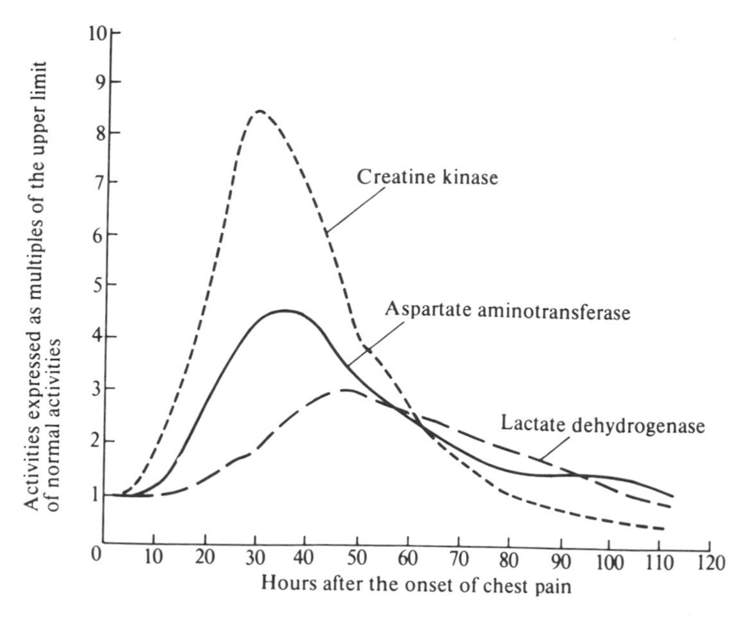

FIG. 10.3. Typical changes in the activities of serum creatine kinase, aspartate aminotransferase (SGOT), and lactate dehydrogenase following myocardial infarction. Activities are expressed as multiples of the upper limit of the normal level (see reference 1).

although predominantly CK-3(MM), also contains about 15 per cent CK-2(MB). Negligible CK-2 activity occurs outside heart tissue and thus it can be used as a specific indicator of damage to heart tissue in normal subjects.[15,16] An antibody against the M subunit is sometimes used to inhibit activity of the M subunit, and any remaining creatine kinase activity in serum will then be due to the B subunit of CK-2 and the CK-1 (BB). Lactate dehydrogenases from liver and heart may also be distinguished by their isoenzyme patterns (Fig. 10.1).

Measurement of both serum aspartate aminotransferase and alanine aminotransferase can be used to differentiate between liver and heart disease. The ratio of activities of aspartate aminotransferase/alanine aminotransferase in heart tissue (20–25) is higher than the ratio in liver (3–5) and this is reflected in the ratios of enzymes released into the serum. However, this ratio of activities in the serum is not now widely used, because the measurement of, for example, creatine kinase or γ-glutamyltransferase is generally regarded as a more reliable method of distinguishing heart and liver diseases.

10.5 Other enzyme activities that become elevated in serum in disease

A number of other enzymes are assayed in serum and urine for diagnostic purposes; the more frequently measured ones are discussed below.

10.5.1 α-Amylase

α-Amylase is an endoamylase that hydrolyses the 1,4-α link in amylose and amylopectin. In humans the enzyme occurs in a variety of tissues, but the highest concentrations are in the pancreas and in the salivary glands. Low amylase activities can be detected in the serum and urine of normal healthy subjects. The most pronounced increases in serum and urinary amylase occur in acute pancreatitis, when the activities may rise to 20–30 times the normal levels. In other pancreatic diseases, e.g. chronic pancreatitis or carcinoma of the pancreas, serum and urinary amylase activities may be raised, but only slightly. Although the highest concentration of amylase is in the pancreas, the increased excretion of amylase may occur in a number of intra-abdominal disorders in which the tissue affected is in close proximity to the pancreas, e.g. gastric or duodenal ulcers, peritonitis, and mesenteric thrombosis. Amylase activity in urine may also be increased in mumps when the salivary glands become inflamed.

An unusual feature of amylase secretion compared with other enzymes is that it is readily detected in urine. Only enzymes having M_r values less than about 55 000 are cleared by the kidney and appear in urine. Most enzymes of importance in clinical diagnosis have M_r values greater than 70 000; two exceptions are uropepsinogen and α-amylase (49 000).[17]

During an attack of acute pancreatitis the level of α-amylase in the serum is usually raised after one day and falls on the second or third day; the level of the the enzyme in urine, on the other hand, may be raised earlier and persist longer.

10.5.2 Creatine kinase and fructose-bisphosphate aldolase

These two enzymes are useful as indicators of skeletal muscle disorders. Creatine kinase occurs predominantly in skeletal muscle, cardiac muscle, and brain, but in very low activities in all other tissues. The serum activities are raised in all forms of muscular dystrophy, the level depending on the mass of muscle that is becoming dystrophic. Although creatine kinase activity is also raised in myocardial infarction, the rise is short-lived and there would be no likelihood of confusing it with skeletal muscle defects; in addition, the isoenzyme patterns differ (see Section 10.4 and references 15, 16).

Fructose-bisphosphate aldolase is also present in high concentrations in skeletal muscle and thus could be used to support the conclusions reached from the assay of creatine kinase. However, fructose-bisphosphate aldolase is also present in many other tissues, though not at such high concentrations. Particularly important in this respect is the high fructose-bisphosphate aldolase content of red blood cells; it is therefore essential that a sample of serum to be assayed does not contain lysed red blood cells.

10.5.3 Alkaline phosphatase

Alkaline phosphatase is a phosphatase having low specificity, hydrolysing a variety of organic phosphate esters, and having a pH optimum between 9 and 10.

Its activity is detectable in a number of different tissues notably intestinal epithelium, kidney tubules, bone (osteoblasts), leucocytes, liver, and placenta. The true substrates of the enzyme are not known and it is usually assayed with *p*-nitrophenyl phosphate. Apart from its application in the diagnosis of obstructive jaundice, it is also very useful in the detection of bone disease. Its activity in serum is raised in those bone diseases in which there is increased activity of the bone-forming cells or osteoblasts as in, e.g. osteomalacia, rickets, Paget's disease, and hyperparathyroidism. The alkaline phosphatase present in normal serum is believed to be derived mainly from the liver cells. It is useful to be able to differentiate between increased activities in the serum due to secretion from liver, from bone, or from other tissues. The isoenzyme patterns of alkaline phosphatase from various tissues are quite complex, multiple bands being obtained from most tissues. For this reason differentiation by electrophoresis is not an easy method to use in a routine investigation. Other properties of the alkaline phosphatases from different tissues that have been useful are the effects of heat and various inhibitors (see Table 10.1). The results suggest that alkaline phosphatases are of three distinctive types, namely, placental, intestinal, and those from other tissues. There is an increase in alkaline phosphatase activity during the third trimester of pregnancy and this is due to release of placental enzyme. The latter can readily be differentiated from the bone and liver enzymes because it is quite stable to heating for 30 min at 329 K (56 °C) while the others are labile.[41]

10.5.4 Acid phosphatase

This enzyme, like alkaline phosphatase, is a phosphatase of low specificity and the natural substrates are unknown. The pH optimum is between 4.0 and 5.5. The enzyme occurs in a variety of tissues, e.g. liver, spleen, erythrocyte, and prostate. The highest concentration is present in the prostate and detection of prostatic carcinoma is the main purpose of clinical assays of this enzyme. Approximately one-third of the normal circulating acid phosphatase in adult males is derived from the prostate. The enzyme is very labile and must be assayed as soon as possible after collecting the serum. It is clearly important to have an early and unequivocal diagnosis of prostatic carcinoma; however, small deviations in the level of acid phosphatase in the serum could be due to release from other tissues or lysis of erythrocytes during serum collection. The erythrocyte and prostatic enzymes can be differentiated by their differing sensitivities towards formaldehyde and tartrate: 0.5 per cent formaldehyde almost completely inhibits the erythrocyte enzyme but hardly affects that of the prostatic enzyme, whereas 0.01 mol dm^{-3} tartrate has almost no effect on the erythrocyte enzyme whilst inhibiting the prostatic enzyme. Most recently, however, it has been found that the best substrate for the prostatic enzyme is thymolphthalein phosphate, which is only very poorly attacked by the erythrocyte enzyme and thus can be used to distinguish these enzymes.

10.6 The detection and significance of enzyme deficiencies

The previous section has been concerned with the detection of elevated enzyme activities in serum and urine and the diagnosis of the associated diseases. These represent the large majority of enzyme assays performed for clinical diagnosis. Much less frequently encountered are those subjects who suffer from enzyme deficiencies due to inborn errors in metabolism. More than 140 different inborn errors in metabolism are now known although none of them occur very frequently, e.g. phenylketonuria, 1 in 15 000 births (about 40 births per year in the UK);[18] cystinuria, 1 in 14 000; galactosaemia, 1 in 33 000; maple syrup disease, 1 in 330 000; alcaptonuria, 1 in 200 000.[19] In many cases the enzyme deficiency has been established (Table 10.2), although it is not generally known whether it is a mutation in the genome that has led to the production of a defective enzyme or failure to produce enzyme at all. The initial identification of the disease is rarely the result of an enzyme assay and in many cases the presence of a metabolite is used in detection of the disease; thus, phenylketonuria is usually detected by measurement of phenylpyruvate in urine or phenylalanine in blood. The ultimate determination of the enzyme defect may require a tissue biopsy. Some inborn errors in metabolism are relatively harmless, e.g. albinism, alcaptonuria, but others must be detected early if the defect is to be circumvented. Two important ones that fall in the latter category are phenylketonuria and galactosaemia. In phenylketonuria there is a lack of phenylalanine 4-monooxygenase, an enzyme required for the breakdown of phenylalanine:

$$C_6H_5-CH_2CH(NH_2)COOH \rightleftharpoons HO-C_6H_4-CH_2CH(NH_2)COOH$$

$$+ \text{Tetrahydropteridine} + O_2 \qquad\qquad + \text{Dihydropteridine} + H_2O$$

Phenylalanine is being produced continuously in a normal subject as a result of protein turnover and dietary intake and is largely broken down via tyrosine to homogentisic acid. In the phenylketonuric subject, phenylalanine has to be degraded by a minor pathway via phenylpyruvic acid (Fig. 10.4).

Phenylketonuria is associated with mental retardation: it is not clear whether this is due to the build up of high concentrations of phenylalanine or its metabolites. The mental disturbance can be prevented if the subjects from birth to 7 years are fed a diet with a considerably reduced phenylalanine content, although there is evidence that the diet must be maintained continuously throughout adulthood to prevent the accumulation of deleterious effects.[21] Special diets with low-protein breads, cereals and vegetable fats have been compiled.[22] Early detection is thus very important and infants are now screened at birth in many countries for phenylketonuria.

TABLE 10.2
Some examples of inborn errors in metabolism due to enzyme deficiencies

Inborn error	Enzyme deficiency
Alcaptonuria	Homogentisate 1,2-dioxygenase
Phenylketonuria	Phenylalanine 4-monooxygenase
Cystinuria	(Renal and intestinal transport defect of cysteine, ornithine and lysine)
Maple syrup disease	Branched-chain oxo-acid oxidative decarboxylases
Galactosaemia	UDPglucose-hexose-1-phosphate uridylyltransferase
Glycogen storage diseases	Glucose-6-phosphatase, α-D-glucosidase, debranching enzyme, liver or muscle phosphorylase
Pentosuria	L-Xylulose reductase
Tay–Sachs disease	β-*N*-acetyl-D-hexosaminidase
Nieman–Pick disease	Phospholipase *C*
Gangliosidosis (generalized)	β-D-Galactosidase
Acatalasia	Catalase

For a more complete list see references 18–20.

In galactosaemia the deficient enzyme is galactose 1-phosphate uridylyltransferase. As can be seen from Fig. 10.5, this results in a block in the conversion of galactose 1-phosphate to glucose 1-phosphate. The deficiency results in a build up of high intracellular concentrations of galactose 1-phosphate, a high level of galactose in body fluids, and galactose excretion in urine. The pathological consequences, mental retardation, slow weight gain, enlarged liver, and cataracts are generally attributed to the high intracellular concentrations of galactose 1-phosphate. The harmful effects can be avoided by reducing the galactose content of the diet from birth. (Lactose present in milk is the principal source of galactose in infants.) Galactose 1-phosphate uridylyltransferase is normally present in many human tissues including erythrocytes. Thus, galactosaemia may be detected by assaying lysed erythrocytes for the enzyme.

The most frequently occurring hereditary disease known is glucose-6-phosphate dehydrogenase deficiency, in which the erythrocytes are affected. It is more common in Negros and in Sephardic Jews than in other races. The disease is manifested as haemolytic anaemia and occurs after the subjects have been treated with antimalarial drugs such as pamaquine or primaquine, or have received other drugs such as sulphonamides or aspirin, or have eaten fava beans. Erythrocytes metabolize glucose using the glycolytic and pentose phosphate pathways. The agents that cause haemolysis appear to inhibit glucose-6-phosphate dehydrogenase, which is present only in low levels in these subjects. This in turn causes reduction in the intracellular NADPH concentration and the latter is required for glutathione reductase. It is thought that haemolysis is

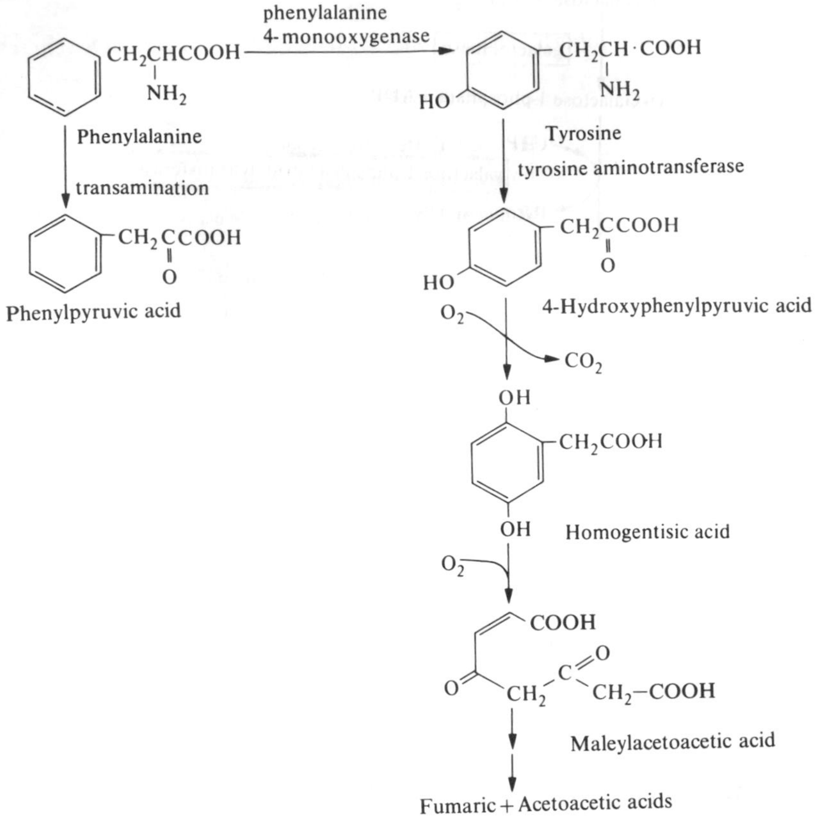

phenylalanine 4-monooxygenase

$CH_2CHCOOH$ (NH_2) → HO– $CH_2CH\cdot COOH$ (NH_2)

Phenylalanine | Tyrosine

transamination | tyrosine aminotransferase

CH_2CCOOH (=O) | HO– CH_2CCOOH (=O)

Phenylpyruvic acid | 4-Hydroxyphenylpyruvic acid

O_2 → CO_2

OH, CH_2COOH, OH Homogentisic acid

O_2

COOH, O=, CH_2, C=O, CH_2–COOH

Maleylacetoacetic acid

Fumaric + Acetoacetic acids

FIG. 10.4. Metabolic pathway for the degradation of phenylalanine in humans.

induced by lowering of the levels of glutathione:

$$\text{D-Glucose 6-phosphate} + \text{NADP}^+ \rightleftharpoons \text{D-glucono-}\delta\text{-lactone 6-phosphate} + \text{NADPH} + \text{H}^+.$$

$$\text{H}^+ + \text{NADPH} + \text{oxidized glutathione} \rightleftharpoons \text{NADP}^+ + 2 \text{ glutathione}.$$

Another example of drug sensitivity being associated with a genetically inherited deficiency occurs in the case of serum cholinesterase (EC 3.1.1.8) deficiency. Cholinesterase is normally present in serum at quite high levels. The enzyme is considered to be synthesized by the liver and lower levels of activity are detected in serum in certain liver diseases, e.g. acute hepatitis and advanced carcinoma of the liver, when the rate of synthesis of the enzyme is reduced. The enzyme is sometimes referred to as pseudocholinesterase since it hydrolyses a variety of esters of choline, in contrast to the acetylcholinesterase (EC 3.1.1.7) of nervous tissue and erythrocytes, which has a more restricted specificity. The

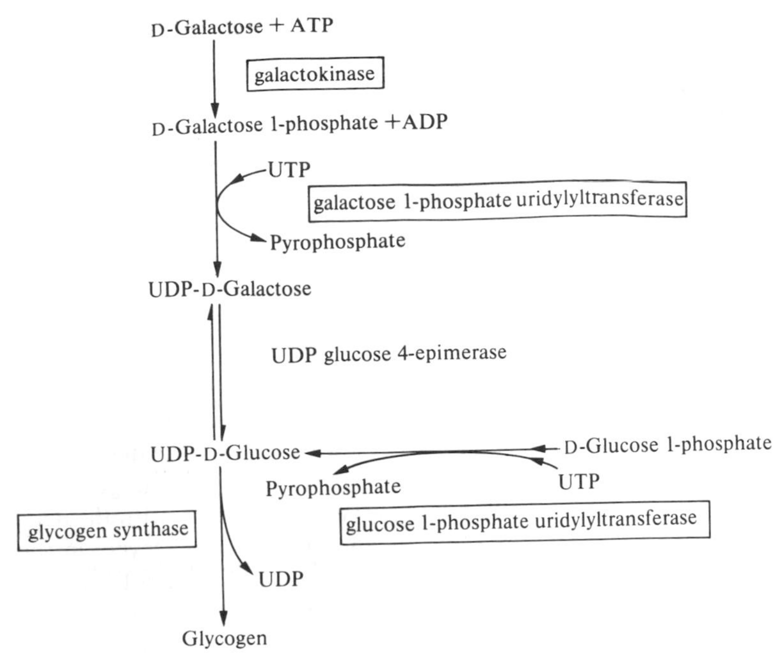

FIG. 10.5. The pathways of galactose metabolism.

$$\text{Acylcholine} + H_2O \xrightleftharpoons{\text{cholinesterase}} \text{choline} + \text{carboxylate anion}$$

relative deficiency of cholinesterase in some individuals became apparent as a result of the use of suxamethonium (succinyl choline) as a muscle relaxant in surgery.

$$(CH_3)_3N^+(CH_2)_2O_2C(CH_2)_2CO_2(CH_2)_2N^+(CH_3)_3$$

Succinyl choline

In normal subjects the effects of suxamethonium are short-lived and this is due to its breakdown by serum cholinesterase. However, in certain individuals the effects of the drug were much more long-lasting; this was associated with low cholinesterase levels. When the cholinesterase from suxamethonium-sensitive individuals was further studied it became apparent that the enzyme differed from that of normal individuals in having a high K_m with all substrates tested. This enabled a simple screening test to be carried out on serum to determine whether a subject is likely to be sensitive to suxamethonium. The test involves determining the percentage inhibition of cholinesterase by a particular concentration of dibucaine measured under standard conditions. The cholinesterase from suxamethonium-sensitive individuals is inhibited by only about 20 per cent under conditions that reduce the normal enzyme activity by about 80 per cent.[18]

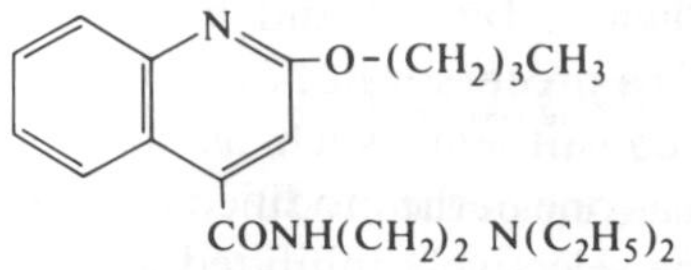

Dibucaine

The diseases described so far in this section are all genetically inherited diseases in which there is either a deficiency or a change in properties of a particular enzyme or protein. An important genetically inherited disease that involves a protein defect, though not an enzyme defect, is familial hypercholesterolaemia. It is included in this section because one of the successful treatments for the disease involves the use of a competitive inhibitor. Familial hypercholesterolaemia is characterized by high circulating cholesterol and cholesterol ester levels in blood. Subjects homozygous for the disease may have 4–6 times the normal circulating levels of cholesterol, and generally have symptoms of coronary disease in childhood or adolescence, with the deposition of arterial plaques. The manifestation of the disease is less severe in the more common heterozygous state, with cholesterol and cholesterol ester levels of about twice the normal range and increased frequency of coronary disease at middle age. The condition has been shown to arise from a deficiency in the low-density lipoprotein receptors that are responsible for the uptake of cholesterol and cholesterol esters by the tissues: hence the higher circulating levels of cholesterol and cholesterol esters.

It has been found that the circulating levels of cholesterol and cholesterol esters can be reduced, and therefore so can the tendency to form arterial plaques, if the subjects are given a competitive inhibitor of hydroxymethylglutaryl-CoA reductase (HMG CoA reductase). HMG CoA reductase catalyses the rate-limiting step in the formation of cholesterol. It is an enzyme having a short half-life (see Chapter 9, Section 9.6).

$$\text{HMG CoA} + 2\text{NADPH} + 2\text{H}^+ \rightarrow \text{mevalonate} + \text{CoASH} + 2\text{NADP}^+.$$

Three competitive inhibitors[23] of HMG CoA reductase have been found, namely, mevinolin, compactin, and monacol K, which can be used to inhibit the

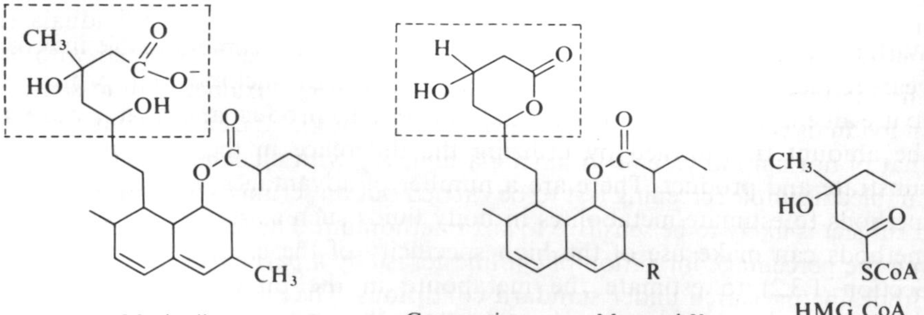

Mevinolin Compactin Monacol K HMG CoA

R = H R = CH_3

Competitive inhibitors of HMG CoA reductase; the regions resembling HMG CoA are outlined.

enzyme. All three are fungal products and have a structural resemblance to HMG CoA. Mevalonate is a precursor not only of steroids but also of a number of other important cell constituents such as ubiquinone, dolichol, and isopentenyl adenine (present as one of the modified bases in tRNA). However, sterol biosynthesis appears to be selectively inhibited at low concentrations.

The last example in this section is rather different since it is a case where an enzyme assay is used to detect a vitamin deficiency. The enzyme transketolase, which catalyses the following reaction, requires thiamin diphosphate as a cofactor for activity:

$$\text{D-Ribose 5-phosphate} + \text{D-xylulose 5-phosphate} \rightleftharpoons \text{sedoheptulose 7-phosphate} + \text{D-glyceraldehyde 3-phosphate.}$$

Normally thiamin deficiency rarely occurs in developed countries, where there is very little malnutrition. However, the deficiency is common in areas where there is severe malnutrition and it also occurs in the case of Wernicke's encephalopathy. This syndrome occurs in alcoholics, who become thiamin-deficient as a result of consuming little other than alcohol. The early stages of this syndrome may be detected as thiamin deficiency. Thiamin is normally converted into thiamin diphosphate, the cofactor for transketolase and also pyruvate and oxoglutarate dehydrogenases:

$$\text{Thiamin} + \text{ATP} \rightleftharpoons \text{Thiamin monophosphate} + \text{ADP.}$$

$$\text{Thiamin monophosphate} + \text{ATP} \rightleftharpoons \text{thiamin diphosphate} + \text{ADP.}$$

The erythrocytes are a good and readily available source of enzymes of the pentose-phosphate pathway such as transketolase. The transketolase activity is thus measured in the presence and absence of added thiamin diphosphate. In normal subjects the activity is not increased by addition of the cofactor, in contrast to the situation in thiamin-deficient subjects.

For reviews on genetically inherited enzyme deficiencies see references 18, 19, 24, 25.

10.7 The use of enzymes to determine the concentrations of metabolites of clinical importance

With the wider availability of purified enzymes it has become possible in recent years to measure a variety of metabolites by enzymatic methods. The principle is to use an enzyme to transform a metabolite into its product and then to estimate the amount transformed by utilizing the difference in the properties of the substrate and product. There are a number of advantages in using enzymatic methods to estimate metabolites in body fluids such as serum or urine. These methods can make use of the high specificity of the enzyme (see Chapter 1, Section 1.3.2) to estimate the metabolite in the presence of many other substances, thus avoiding purification steps that may be necessary prior to a chemical analysis. Enzymatic methods can sometimes be used to measure concentrations of labile substances that would be degraded by harsher chemical methods. There may also be certain disadvantages. The cost of purified enzymes may be too high for routine use. This will depend on the amount required to carry out each assay in a reasonably short time. In some cases it is now becoming possible to use immobilized enzymes that can then be reused (see Chapter 11, Section 11.5). Enzymes are susceptible to inactivation by a variety of agents. An inhibitor present in serum or urine might, for example, completely inhibit enzyme activity or it may only cause partial inhibition. In the latter case we might obtain a false result unless the time course of the reaction were studied to check that complete reaction had occurred.

A number of points have to be checked before an enzyme can be reliably used in the measurement of the concentration of a metabolite. (For a detailed discussion see reference 26.) The equilibrium position of the reaction is important and it is best, if possible to use a reaction in which the equilibrium is far to the right in order to obtain a quantitative conversion of the metabolite to the product of the reaction. If the equilibrium is not so favourable, this can sometimes be overcome either by removing the products by chemical trapping or by coupling to a second enzyme reaction. For example, the equilibrium constant for glucose-6-phosphate isomerase is close to unity,

$$K = \frac{[\text{fructose 6-phosphate}]}{[\text{glucose 6-phosphate}]} = 0.3,$$

but by coupling this to glucose-6-phosphate dehydrogenase the product of the isomerase is effectively removed.

1. D-Fructose 6-phosphate $\rightleftharpoons$ D-glucose 6-phosphate.
2. D-Glucose 6-phosphate + $NADP^+$ $\rightleftharpoons$ D-glucono-δ-lactone 6-phosphate + NADPH + H^+.
3. D-Glucono-δ-lactone 6-phosphate + H_2O $\rightleftharpoons$ 6-phospho-D-gluconate.

The third step, which is the spontaneous non-enzymic hydrolysis of D-glucono-δ-lactone 6-phosphate, ensures that reactions 1 and 2 proceed to virtual completion.

The equilibrium for the glutamate dehydrogenase reaction lies to the left, but glutamate can be estimated in terms of the NADH produced,

$$\text{L-Glutamate} + NAD^+ + H_2O \rightleftharpoons \text{2-oxoglutarate} + NADH + NH_4^+,$$

provided the 2-oxoglutarate is trapped by reaction with hydrazine:

$$\begin{array}{l} COO^- \\ | \\ C=O \\ | \\ CH_2 \\ | \\ CH_2 \\ | \\ COO^- \end{array} + H_2NNH_2 \rightarrow \begin{array}{l} COO^- \\ | \\ C=NNH_2 \\ | \\ CH_2 \\ | \\ CH_2 \\ | \\ COO^- \end{array} + H_2O$$

A sufficiently large amount of enzyme must be used so that the substrate is almost completely (≥99 per cent) converted to the product in a short time (usually a few minutes). This will depend on the V_{max} for the enzyme and also on the K_m in relation to the amount of metabolite to be estimated. If the latter is, for example, approximately equal to the K_m, then when the substrate has become 99 per cent converted to the products the rate will only be 1 per cent of V_{max}.

The specificity of the enzyme should be known, since some enzymes that do not show absolute specificity will convert related substrates or homologues to their products, often at a reduced rate. If the enzyme preparation is not pure, then side-reactions will occur. Some examples of the use of enzymes to assay metabolites that are important in clinical chemistry are given below.

10.7.1 Blood glucose

The oral glucose tolerance test is still the most widely used diagnostic procedure in the investigation of glucose metabolism in man, e.g. in the investigation of diabetes. A subject is usually given an oral test dose of glucose and the blood glucose concentration is measured for the subsequent three to four hours. In a normal subject the blood glucose concentration rises during the initial 30–60 min and then falls rapidly back to the basal level due to the increased utilization by the tissues. In a diabetic subject the increase in the concentration of blood glucose is more pronounced and the decline to normal levels is much slower. The extent of the rise and the rate of the subsequent fall depends on the severity of diabetes. The test requires several measurements of blood glucose. Chemical methods that are used depend on the reducing properties of glucose and therefore are not specific for glucose. There are three enzymatic methods that are generally available: hexokinase coupled with glucose-6-phosphate dehydrogenase, glucose oxidase coupled with peroxidase, and glucose dehydrogenase.

1. $$\begin{cases} \text{D-Glucose} + \text{ATP} \rightleftharpoons \text{D-glucose 6-phosphate} + \text{ADP}. \\ \text{D-Glucose 6-phosphate} + \text{NADP}^+ \rightleftharpoons \text{D-glucono-}\delta\text{-lactone 6-phosphate} + \text{NADPH} + \text{H}^+. \end{cases}$$

2. $$\begin{cases} \text{D-Glucose} + H_2O + O_2 \rightleftharpoons \text{D-gluconic acid} + H_2O_2. \\ H_2O_2 + \text{dye}_{\text{reduced}} \rightleftharpoons H_2O + \text{dye}_{\text{oxidized}}. \end{cases}$$

3 $$\text{D-Glucose} + \text{NAD}^+ \rightleftharpoons \text{D-glucono-}\delta\text{-lactone} + \text{NADH} + \text{H}^+.$$

In the first method the NADPH produced would be measured by the change in A_{340}. Although hexokinase is not specific for glucose, glucose-6-phosphate dehydrogenase is highly specific for glucose 6-phosphate and thus the specificity is built into the second step. The second method using glucose oxidase is more frequently used because the enzymes and substrates are less expensive and the final absorption is in the visible range. Glucose oxidase specifically oxidizes β-D-glucose, and α-D-glucose is oxidized 150 times slower. Blood glucose contains an equilibrium mixture of both isomers. Most glucose oxidase preparations contain some aldose 1-epimerase as an impurity that enables rapid epimerization to occur and the total glucose to be estimated. The hydrogen peroxide produced in the first step oxidizes an acceptor, which then has a changed absorption spectrum. The most commonly used acceptor was *o*-dianisidine; however, on account of its mild carcinogenic properties it has been replaced by 2,2′-azino-di-(3-ethylbenzthiazoline)-6-sulphonate, which absorbs at 400–450 nm. This replacement also increases the sensitivity of the assay. The concentrations of glucose determined by glucose oxidase assays are usually lower than the true concentrations; this is due to the presence of substances in serum that interfere with the peroxidase reaction by acting as alternative acceptors.

The third method entails glucose dehydrogenase (EC 1.1.1.47) from *Bacillus cereus* and has the advantages that it does not require any additional coupling enzymes and also that the glucose dehydrogenase can be immobilized on nylon tubing and therefore can be reused several times (see reference 27; for further discussion on immobilized enzymes, see Chapter 11, Section 11.5.4).

10.7.2. Uric acid

Uric acid is the end product of purine metabolism in humans. Measurement of uric acid in blood and urine is important in the diagnosis of gout, when the levels of uric acid are raised. They are also raised in other conditions, e.g. in leukaemia and after ingestion of food rich in nucleoproteins. Uric acid may be estimated by either chemical or enzymatic methods. The chemical method involves the oxidation of uric acid to allantoin using phosphotungstic acid, whereas in the enzymatic method urate oxidase is used to catalyse a similar oxidation

Chemical

$$\text{Uric acid} + H_3PW_{12}O_{40} \rightarrow \text{allantoin} + CO_2 + \text{tungsten blue.}$$

Enzymatic

$$\text{Uric acid} + 2H_2O + O_2 \rightleftharpoons \text{allantoin} + CO_2 + H_2O_2.$$

The chemical method has the disadvantage of lack of specificity, whereas the enzymatic method has the advantage that urate oxidase shows absolute specificity but the possible disadvantage of requiring measurement of absorption in the ultraviolet region. The enzyme measurement uses the difference in absorbance at 293 nm between uric acid and allantoin.

The hydrogen peroxide formed could be coupled to peroxidase as in the glucose oxidase assay to enable measurements to be made in the visible part of the spectrum. The enzymatic method can be adapted for automated uric acid determination, whereas this is not easily possible with the chemical method.[28]

10.7.3 Urea

Determinations of the levels of urea in serum are frequently carried out as a test of renal function, especially in conjunction with urinary urea estimation, which enables the glomerular filtration rate to be assessed. Urea is hydrolysed by urease to ammonium carbonate. The ammonia released is reacted with sodium hypochlorite and phenol under alkaline conditions to form indophenol that can be estimated spectrophotometrically.

$$\text{urea} + 2H_2O \xrightleftharpoons{\text{urease}} (NH_4)_2CO_3$$

$$NH_3 + NaOCl + 2C_6H_5OH \longrightarrow O{=}C_6H_4{=}N{-}C_6H_4{-}O^-$$

Indophenol

10.7.4 Cholesterol, cholesterol esters, and triglycerides

The measurement of the concentrations of cholesterol and cholesterol esters in the serum is important in assessing the risk factors in arteriosclerosis and in myocardial infarction. High circulating levels of cholesterol and cholesterol esters are indicative of high risk. Cholesterol can be measured by a colorimetric method that is not highly specific and thus not very satisfactory. It also involves the use of concentrated sulphuric acid, which can be a nuisance in routine analysis. The introduction of straightforward enzymatic methods of high specificity has greatly simplified procedures. Cholesterol can be measured using a single enzyme, cholesterol oxidase, and cholesterol esters can be measured by using this enzyme together with cholesterol esterase. The product of cholesterol oxidase action, 4-cholesten-3-one, can be estimated by its absorption at 240 nm.

$$\text{Cholesterol ester} + H_2O \rightleftharpoons \text{cholesterol} + RCOO^-.$$

$$\text{Cholesterol} + O_2 \rightleftharpoons \text{4-cholesten-3-one} + H_2O_2.$$

The measurement of serum triglycerides in the plasma or serum is useful in following the course of diabetes mellitus, nephrosis, and biliary obstruction. It can also be measured enzymatically, usually after alkaline hydrolysis of the triglyceride. The glycerol formed can then be assayed using the following coupled assays:

$$\text{ATP} + \text{glycerol} \xrightleftharpoons{\text{glycerol kinase}} \textit{sn}\text{-glycerol 3-phosphate} + \text{ADP}.$$

$$\text{ADP} + \text{phosphoenolpyruvate} \xrightleftharpoons{\text{pyruvate kinase}} \text{ATP} + \text{pyruvate}.$$

$$\text{Pyruvate} + \text{NADH} + H^+ \xrightleftharpoons{\text{lactate dehydrogenase}} \text{L-lactate} + \text{NAD}^+.$$

10.7.5 Other metabolites

A number of other metabolites that are measured less frequently can also be measured by enzymatic methods.

Creatine may be measured by the decrease in A_{340} using creatine kinase coupled to pyruvate kinase and lactate dehydrogenase:

$$\text{Creatine} + \text{ATP} \xrightleftharpoons{Mg^{2+}} \text{phosphocreatine} + \text{ADP}.$$

$$\text{ADP} + \text{phosphoenolpyruvate} \xrightleftharpoons{Mg^{2+}} \text{ATP} + \text{pyruvate}.$$

$$\text{Pyruvate} + \text{NADH} + \text{H}^+ \rightleftharpoons \text{L-lactate} + \text{NAD}^+.$$

The measurement of creatinine is important in the creatinine-clearance test, which is used to test glomerular filtration occurring in the kidney. The colorimetric method that has been used for many years is relatively non-specific and so the recent development of an enzymatic method using creatininase is a considerable improvement. It is used in conjunction with the coupling enzymes just mentioned for the assay of creatine:

$$\text{Creatinine} + \text{H}_2\text{O} \rightleftharpoons \text{creatine}.$$

Ammonia may be estimated using glutamate dehydrogenase:

$$\text{NH}_4^+ + \text{NADH} + \text{2-oxogluturate} \rightleftharpoons \text{L-glutamate} + \text{NAD}^+.$$

Ethanol may be estimated using alcohol dehydrogenase,

$$\text{CH}_3\text{CH}_2\text{OH} + \text{NAD}^+ \rightleftharpoons \text{CH}_3\text{CHO} + \text{NADH} + \text{H}^+,$$

and lactate in a similar way by using lactate dehydrogenase.

10.7.6 Enzyme immunoassay

In addition to the direct use of enzymes to measure the concentrations of their substrates in body fluids, there is also an indirect method in which enzymes are used as a type of amplification system in what is known as enzyme immunoassay. Since the first description of the method of radioimmunoassay by Yalow and Berson in 1960, a large number of methods have been developed for measurement of hormones, drugs, and a variety of small biological molecules that are based on antigen–antibody interactions. These methods rely on the specificity of antigen–antibody interactions for binding the species to be assayed, together with some method of increasing the sensitivity of detection by use of either a radioactive label or an enzyme label.

Small molecules such as hormones and drugs do not induce antibody formation when injected into a foreign species. However, if they are chemically combined to a protein such as albumin, and then are injected into an animal, they will induce antibody formation. The antibodies so formed will combine specifically with the antigen, forming an insoluble antigen–antibody complex. The antibodies will also combine with the original small molecule (known as a hapten) although they will not give rise to a precipitate. Immunoassays can thus be devised for any molecule (acting as a hapten) that, when coupled to a protein, is capable of inducing antibody formation. The method thus has potentially very wide application.

The second stage of the method is to detect the antigen–antibody or antibody–hapten complexes, preferably when they are present in low concentrations. Two of the methods used are either to incorporate a radioactive label on to the antigen or antibody (radioimmunoassay), or to incorporate an enzyme molecule on to the antigen or antibody (enzyme immunoassay). Radioimmunoassays do not in general entail the use of enzymes* and thus are outside the scope of this book (a good review of the method is given in reference 29). Radioimmunoassays are, however, being used in the detection of prostatic acid phosphatase and also the isoenzyme CK-2 of creatine kinase because they provide a more sensitive means of detection.[30]

The procedure of enzyme immunoassay is outlined in Fig. 10.6. The drug, hormone, etc. (designated compound L) to be assayed is chemically coupled to a protein and this conjugate is injected into a rabbit to raise the necessary antibodies. The rabbit serum is fractionated to obtain the fraction that is to be used as the reagent. A second conjugate is now made, this time between compound L and a purified enzyme. This conjugate must retain enzyme activity. A number of enzymes have been used in this type of coupling, e.g. alkaline phosphatase, β-galactosidase, horseradish peroxidase, lysozyme, malate dehydrogenase, and glucose-6-phosphate dehydrogenase. The solution containing an unknown concentration of compound L can now be assayed. The solution is mixed with a fixed concentration of the antibodies, containing an excess of antibody combining sites. A known concentration of enzyme-labelled compound L is then added and a portion of this combines with the remaining antibody-combining sites. It is then necessary to be able to measure the amount of free enzyme-labelled compound L. How this is done depends on whether the enzyme-labelled compound L–antibody complex shows enzyme activity (Fig. 10.6). In some cases the combination with the antibody blocks and the active site inhibits enzyme activity. The free and bound enzyme-labelled compound L can then be quantitated without prior separation. This is sometimes referred to as homogeneous enzyme immunoassay. In other cases, when the active site is not blocked, it is necessary to separate the free and bound enzyme-labelled compound L prior to measurement of enzyme activity (heterogeneous enzyme immunoassay).

There are a number of variants of this type of immunoassay. In one the enzyme is covalently linked to an antibody that is able to bind L (immuno-enzymetric assay). Another involves a sandwich technique in which an antibody–antigen complex involving the ligand is detected by a second antibody against the first. The second antibody has an enzyme covalently linked to it. The variants on the basic techniques are described in reference 31.

The procedure is thus very simple once a suitable antibody and enzyme-labelled compound L has been obtained. This method has been developed to

* Lactoperoxidase (iodide peroxidase EC 1.11.1.8) is sometimes used to catalyse the incorporation of ^{131}I or ^{125}I into the antigen.

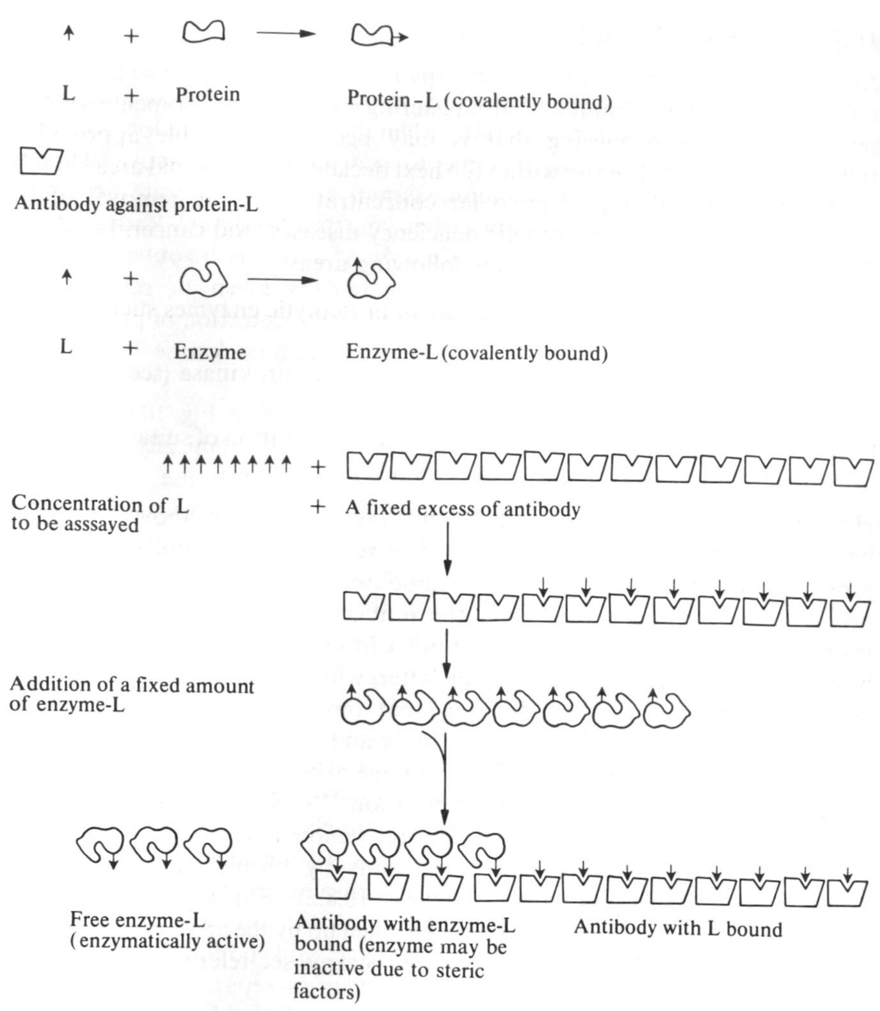

FIG. 10.6. The method of enzyme immunoassay. The diagram illustrates how this method is used to assay compound L after formation of a covalent derivative of compound L with a purified enzyme.

measure a variety of drugs, including opiates, barbiturates, amphetamine, cocaine, anticancer drugs, and hormones such as thyroxine, and to screen antigens from pathogenic organisms, and is being used increasingly in clinical laboratories. It has two advantages over radioimmunoassays in that the reagents are stable and it avoids the need to use radioactive materials. (For a review, see reference 32.) The method is likely to be improved in specificity by use of high-affinity monoclonal antibodies.

10.8 Enzyme therapy

Although the attempts at enzyme therapy in clinical trials have so far met with limited success, the technique of administering enzymes for therapeutic purposes seems sufficiently promising that is may become a feasible approach for treatment of certain diseases within the next decade. The principal areas in which work on enzyme therapy has so far concentrated are the removal of toxic substances from the blood, genetic deficiency diseases, and cancer. In addition, enzyme therapy may be used in the following areas:

(1) degradation of necrotic tissue by use of proteolytic enzymes such as trypsin, chymotrypsin, or bromelain;
(2) removal of blood clots using streptokinase or urokinase (see Chapter 11, Section 11.4.3.8); and
(3) treatment of pancreatic insufficiency by administration of suitably entrapped enzymes, e.g. proteolytic enzymes, as in cystic fibrosis.

How the enzyme is administered depends on which of the above objectives is being considered. If the enzyme is used to remove or metabolize a substance present in the blood, e.g. a toxic metabolite or a blood clot, then it is not necessary for the enzyme to enter the intracellular compartments; it is only necessary for the enzyme to be present in the blood. This type of application may be either intra- or extracorporeal. The latter, which involves the use of a bypass as in kidney dialysis, has the advantage that immobilized enzymes may be used and these can be retained outside the body and are not cleared by the kidney. Immobilized enzymes also have the advantage of being less immunogenic. There are several examples of this type of application,[33] such as removal of circulating urea by urease in cases of kidney failure, removal of bilirubin by bilirubin oxidase, removal of fibrin from blood clots by fibrolysin, and removal of asparagine by asparaginase (see Section 10.8.2). For intracellular enzyme therapy, it is necessary for the enzyme to be taken up by the appropriate target cells.

For more detailed accounts of enzyme therapy, see references 34–36.

10.8.1 Treatment of genetic deficiency diseases

There are now a large number of genetic deficiency diseases in which the enzyme deficiency has been identified (Table 10.2), although these only afflict small proportions of the population. The only treatment that is at present available for a number of these diseases is careful regulation of the diet to reduce the substrate loading and thereby minimize the harmful effects of the defect, e.g. for phenylketonuria, a diet low in phenylalanine; for galactosaemia, a diet low in galactose; for maple syrup disease, a diet low in leucine, isoleucine, and valine. The ultimate cure for these diseases would be to insert the gene that codes for the deficient enzyme into chromosomes. Although there are still many problems to be solved before this may be feasible, significant progress has been made. For example, phenylalanine hydroxylase, the enzyme deficient in phenylketonuria has been cloned and has been introduced into cultured human cells using a retrovirus.[21]

The introduction of the gene in somatic cells (somatic replacement), as opposed to germ cells, may thus become a feasible proposition. A more immediate prospect is the replacement of the deficient enzyme. Although purified enzymes are available for many of these deficiency diseases, there are formidable problems in delivering the enzyme to the required site in such conditions that it will remain stable and functional for a reasonable time.

When considering enzyme therapy to alleviate genetic deficiency disease, the prospects of a successful treatment are likely to be best in cases in which the enzyme can be most easily targeted to the tissue or organelle required, and also in cases in which an enzyme treatment for a limited period is likely to be of value. The latter is a consideration because any enzyme applied is likely to be turned over and eventually cleared from the system. These restrictions have made lysosomal storage diseases the prime candidates for enzyme therapy.[36] These are a group of diseases in which one of the normal lysosomal enzymes is deficient. Various substances are normally taken up into the lysosomes by the process of endocytosis, in which a portion of the membrane surrounds the substance and is then involuted and pinched off within the lysosome, and are subsequently degraded. However, in lysosomal storage diseases the substances are endocytosed but then accumulate. Because the lysosomes have the capability to endocytose foreign material, it would seem reasonable that they should take up added enzymes. There is also quite a good understanding of the signals that cause lysosomal proteins made in the Golgi apparatus to be taken up by the lysosomes. The presence of a mannose-6-phosphate residue on the protein targets it to the lysosomes. Work has been in progress since 1964 on attempts at replacement therapy in lysosomal storage diseases and a number of technical problems have been identified. Quite large amounts of the appropriate enzyme are required in a high state of purity and in a non-immunogenic form. Some of the early problems arose from immunological reactions, especially when the enzyme was obtained from a different organism. Many enzymes, when administered, are either inactivated or degraded fairly rapidly, although some experiments carried out using fibroblasts grown in culture have shown that enzymes can be taken up from the culture medium and will function within the cells. The lysosomal storage disease most studied is Gauchers disease. In this disease a glucocerebroside accumulates largely in macrophages. In this case a single treatment with a glucocerebrosidase might be expected to clear the lysosomes of glucocerebroside, but results so far show only marginal effects. The enzyme, although reaching the required site, has a very short half-life.

Perhaps the most promising developments are in the area of enzyme stabilization and in the vehicle used to supply the enzyme. In a number of cases it has been shown that enzymes may be stabilized against degradation either by their encapsulation within a matrix or by cross-linking to other substances; thus, cross-linking human α-galactosidase A to an anti-α-galactosidase antibody using hexamethyl isocyanate increased the half-life of the enzyme under physiological conditions from 9 min to 46 min.[37] A method that provides protection of the enzyme and may also improve its uptake is entrapment in erythrocyte ghosts or in liposomes.[38] The former may be prepared by lysing erythrocytes (these could be the subject's own erythrocytes) in dilute saline containing the enzyme and then resealing them in isotonic saline, thereby entrapping some of the enzyme. Liposomes containing enzyme are prepared from a mixture of cholesterol, phosphatidyl choline, and phosphatidic acid that is sonicated in aqueous suspension together with the enzyme. They have been prepared containing, for example, amyloglucosidase, and have been used in the treatment of a glycogen storage disease with encouraging results.[39]

10.8.2 Cancer therapy

The enzyme therapy that has been most tested as potential anticancer treatment has involved administration of asparaginase:

$$\text{L-Asparagine} + H_2O \rightleftharpoons \text{L-aspartate} + NH_3$$

it was observed that injection of normal guinea-pig serum caused complete regression of a lymphosarcoma in mice and delay in the appearance of other tumours in experimental animals. These effects occurred only when guinea-pig serum was used. Extensive analysis showed that the active factor present in guinea pig was asparaginase. This enzyme is widely distributed in plants, animals, and bacteria but it is absent from the serum of most common mammals. Other sources of asparaginase were tested and also found to cause regression of certain tumours. Bacteria represent one of the best sources of asparaginase and much subsequent experimentation has been carried out using asparaginase from *E. coli*. A number of different types of tumour, but particularly lymphosarcomas, have been found to be sensitive to asparaginase. The precise mechanism of action of asparaginase in causing tumour regression is not completely clear but it appears that the sensitive tumours, in contrast to normal host tissues, lack the ability to synthesize asparagine. Administration of asparaginase thus reduces the circulating level of asparagine. Asparaginase treatment has been given in a number of clinical trials and has been found most effective in relief of acute lymphocytic leukaemia. Unfortunately, there are a number of adverse side-effects such as liver damage and immune reactions, and in addition, repeated treatment can lead to the development of resistance. Attempts are being made to try to improve the treatment, i.e. to reduce the side-effects by administering the enzyme in microcapsular or other immobilized forms[33] or by giving asparaginase in combination with other chemotherapy. Other enzymes, e.g. glutaminase, serine dehydratase, arginase, phenylalanine ammonia-lyase, and leucine dehydrogenase, are also being tested as possible anticancer agents,[40] the rationale being that certain types of cancer cells may have lost the ability to synthesize amino acids and are thus dependent on uptake from the serum for supply. Normal cells can synthesize all but the eight essential amino acids. Thus, enzymes that degrade any of the non-essential amino acids from the serum might be effective therapeutic agents against a cancer cell having a particular amino-acid requirement.

10.9 Conclusions

Measurements of enzyme activity are being used increasingly in clinical diagnoses and enzyme preparations are being used increasingly to estimate metabolites and in trials for enzyme therapy. These trends are likely to continue to increase for several reasons. Sophisticated equipment is becoming more widely available in clinical laboratories and many assays can now be performed on automated equipment. A wider variety of pure enzymes is becoming commercially available and many of these have been successfully immobilized to support materials. This should enable the enzymes to be used more economically both in the assay of metabolites and in enzyme-replacement therapy.

In the field of clinical diagnosis, the assay of more than one enzyme or the combination of enzyme assay and isoenzyme distribution can give a reasonable degree of discrimination between different diseases. Many enzymes which are at present difficult to assay routinely may, in the future, be measured by radioimmunoassays or enzyme immunoassays, and this will greatly increase the range of possibilities. This, in turn, could widen the scope of enzyme diagnoses.

References

1. Horder, M. and Wilkinson, J. H., in *Chemical diagnosis of disease* (Brown, S. S., Mitchell, F. L., and Young, D. S., eds.), Ch. 7. Elsevier/North-Holland, Amsterdam (1979).
2. Wilkinson, J. H., *The principles and practice of diagnostic enzymology*. Edward Arnold, London (1979).
3. Tietz, N. W., *Fundamentals of clinical chemistry* (3rd edn). Ballière Tindall (1987).
4. Schmidt, E. and Schmidt, F. W., *Brief guide to practical enzyme diagnosis* (2nd edn). Boehringer Mannheim GmbH, Mannheim (1976).
5. Curtius, H. C. and Roth, M., *Clinical biochemistry: principles and methods,* Vol. 2, Ch. 11. Walter de Gruyter, Berlin (1974).
6. Schmidt, E. and Schmidt, F. W., *FEBS Lett.* **62**, Suppl. E62 (1976).
7. Bergmeyer, H., *Methods in enzymatic analysis* (3rd English edn). Verland Chemie Weinheim, Academic Press, New York (1983). See Vol. 1.
8. Furie, B. and Furie, B. C., Cell **53**, 505 (1988).
9. Reid, K. B. M., *Essays Biochem.* **22**, 27 (1986).
10. Horder, M. and Wilkinson, J. H., in *Chemical diagnosis of disease* (Brown, S. S., Mitchell, F. L., Young, D. S., eds.), Ch. 7, pp. 378–9. Elsevier/North-Holland, Amsterdam (1979).
11. Schmidt, E. S. and Schmidt, F. W., *Clin. Chim. Acta* **43**, 43 (1988).
12. Wilkinson, J. H., *The principles and practice of diagnostic enzymology*, pp. 225–9, Edward Arnold, London (1976).
13. Schwartz, M. K. and Young, D. S., in *Chemical diagnosis of disease* (Brown, D. S., Mitchell, F. L., and Young, D. S., eds.), Ch. 25. Elsevier/North Holland, Amsterdam (1979).
14. Divall, G. B., *Electrophoresis* **6**, 249 (1985).
15. Horder, M. and Wilkinson, J. H., in *Chemical diagnosis of disease* (Brown, S. S.,

Mitchell, F. L., and Young, D. S., eds.), Ch. 7, p. 383. Elsevier/North-Holland, Amsterdam (1979).

16. Konttinginen, A., in *Enzymes in health and disease, Soc. clin. Enzymol.* p. 149. Karger, Basel (1978).
17. Hall, C. L. and Hardwicke, J., *A. Rev. Med.* **30**, 199 (1979).
18. Harris, H., *The principles of human biochemical genetics* (3rd edn), Chs. 5 and 6. Elsevier/North-Holland, Amsterdam (1980).
19. Raine, D. H., in *Chemical diagnosis of disease* (Brown, S. S., Mitchell, F. L., and Young, D. S., eds.), Ch. 18. Elsevier/North-Holland, Amsterdam (1979).
20. Dean, R. J., *Cellular degradative processes.* Chapman and Hall, London (1978). See p. 38.
21. Ledley, F. D., Dilella, A. G., and Woo, S. L. C., *Trends in Genetics* **1**, 309 (1985).
22. Fitzpatrick, C., *N. Staffs. med. Inst. J.* **8**, 35 (1976).
23. Endo, A., *Trends Biochem. Sci.* **6**, 10 (1981).
24. Cotton, R. G. H., *Int. J. Biochem.* **8**, 333 (1977).
25. Huijing, F., *Physiol. Rev.* **55**, 609 (1975).
26. Bergmeyer, H. U., *Methods in enzymatic analysis* (2nd English edn). Verland Chemie Weinheim, Academic Press, New York (1974). See Vol. 1, Section A. II. 3.
27. Hornby, W. E., Noy, G. A., and Salleh, A. B. B., in *Biotechnological applications of proteins and enzymes* (Bohak, Z. and Sharon, N. eds.), Ch. 16. Academic Press, New York (1977).
28. Ziegenhorn, J., Brandhuber, M., and Bartl, K., in *Enzymes in health and disease, Soc. clin. Enzymol.*, p. 131. Karger, Basel (1978).
29. Orth, D. N., *Methods Enzymol.* **37**, 22 (1975).
30. Rider, C. C. and Taylor, C. B., *Isoenzymes*, pp. 68–70. Chapman and Hall, London (1980).
31. Oellerich, M., *J. Clin. Chem. Clin. Biochem.* **22**, 895 (1984).
32. O'Sullivan, M. J. and Bridges, J. W., *Ann. clin. Biochem.* **16**, 221 (1979).
33. Klein, M. D. and Langer, R., *Trends Biotech.* **4**, 179 (1986).
34. Beutler, E., *Trends Biochem. Sci.* **6**, 95 (1981).
35. Holcenberg, J. S., *A. Rev. Biochem.* **51**, 795 (1982).
36. Brady, R. O., *Enzymes* **16**, 679 (1983).
37. Desnick, R. J., Thorpe, S. R., and Fiddler, M. B., *Physiol. Rev.* **56**, 57 (1976).
38. Colley, C. M. and Ryman, B. E., *Trends Biochem. Sci.* **1**, 203 (1976).
39. Ryman, B. E., in *Enzymes in health and disease, Soc. clin. Enzymol.*, p. 222. Karger, Basel (1978).
40. Chong, E. D. S. and Chang, T. M. S., in *Biomedical applications of immobilized enzymes and proteins* (Chang, T. M. S., ed.), Vol. 1, Ch. 9. Plenum Press, New York (1977).
41. Wilkinson, J. H., *The principles and practice of diagnostic enzymology*, p. 136. Edward Arnold, London (1976).
42. Vora, S., *Analyt. Biochem.* **144**, 307 (1985).
43. Mann, K. G., Jenny, R. J., and Krishnaswamy, S., A. *Rev. Biochem.* **57**, 915 (1988).
44. Müller-Eberhard, H. J., *A. Rev. Biochem.* **57**, 321 (1988).

11
Enzyme technology

11.1 Introduction

Enzymes have remarkable catalytic properties (see Chapter 1, Section 1.3.1), especially when compared with other types of catalysts. In recent years they have been exploited on an increasing scale in food, pharmaceutical, and chemical industries. The advantages of high catalytic activity, lack of undesirable side-reactions, and operation under mild conditions are often highly desirable. Also, with the advent of genetic engineering, a wider range of enzymes will become available on a larger scale and thus increase the scope of enzyme technology. Genetic engineering itself creates a demand for certain highly purified enzymes (see Section 11.5.5) and the whole area of enzyme technology is expanding rapidly.

The enzymes present in a number of bacteria and fungi have been used for many years by man, but it is only in recent years that isolated enzymes, as opposed to enzymes present in whole organisms, have been used in industrial processes. We first consider the main applications of reactions that are catalysed by enzymes in intact organisms, then consider the use of isolated enzymes, and follow by discussion of the recent and future applications of immobilized enzymes. This short chapter serves to illustrate some of the new and traditional technological applications of enzymology. There are many articles and books on enzyme technology (see references 1–5).

11.2 Use of microorganisms in brewing and cheesemaking

The oldest industries in which enzymes in living organisms have been used are those of brewing and cheesemaking. These processes were carried out centuries before there was any understanding of their biochemical basis. In fact, much of the emphasis in biochemistry at the end of the last century and in the early part of this century was aimed at understanding these fermentation processes.

As can be seen from Table 11.1, enzymes may be involved in either hydrolysis to monosaccharides or in glycolysis in the process of ethanol production. In the case of beer making, the starting material is the polysaccharide starch, whereas in wine making the starting material is mainly mono- or disaccharides. The largest carbohydrates that most yeasts are able to ferment are trisaccharides. Therefore, when polysaccharides are used as the principal source of carbohydrates they must be degraded before fermentation can occur. The principal raw material used in brewing is the starch obtained from barley. Barley seeds are allowed to germinate and this leads to release of amylases that cause the breakdown of starch. The starch grains are contained within the cells of the endosperm, the

TABLE 11.1
The use of whole organisms as a source of enzymes in the food industry

Process	Organism used	Enzyme step or steps involved
Brewing and winemaking	Barley (*Hordeum*)	α-Amylase and β-amylase, endo-1,3-β-D-glucanase
	Saccharomyces spp	Limited proteolysis; oligosaccharidases (oligosaccharides → monosaccharides) Glycolytic enzymes (monosaccharides → ethanol)
Cheesemaking	*Streptococcus lactis* *Streptococcus cremoris*	Lactose → lactate + limited lipolysis and proteolysis
	Propionobacteria spp	Conversion of lactate → propionate + acetate $+ CO_2 + H_2O$
	Penicillium camemberti and *Penicillium roqueforti*	Limited lipolysis and proteolysis plus other undefined activities
	Calf stomach	Chymosin (rennin); caseinogen → casein
Vinegar production	*Acetobacter*	Ethanol to acetate

walls of which are surrounded by a number of polymers including 1,3-β-D-glucans. It is necessary for these glucans to be degraded so that the amylases can reach the starch granules (Fig. 11.1). Anaerobic glycolysis is the second stage in the process in which ethanol is the end-product. (For further details of the brewing process see references 6–9). New strains of yeasts that possess amylolytic

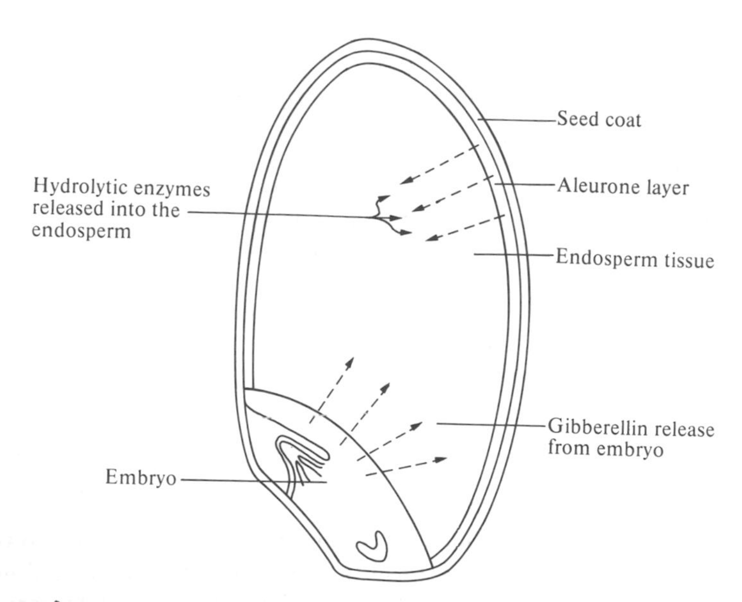

FIG. 11.1 The release of hydrolytic enzymes from a germinating barley seed.

ng constructed by recombinant DNA technology. These are able
de starch to glucose and to ferment glucose and may replace
sts in the future.[10]
emembered that ethanol is not only produced as a beverage but is
... important solvent and fuel. There are considerable differences between the production of alcoholic beverages and of industrial ethanol. For the former it is important that the composition of the product is maintained constant and thus all components contributing to the flavour and appearance must be maintained. On the other hand for industrial ethanol, the main consideration is low-cost production. Much industrial alcohol is produced by catalytic hydration of ethene derived from crude oil. However, with the increased cost of petrochemicals there is a greater interest in the fermentation process, particularly if a cheap source of raw material is available. Thus, much effort is being expended on devising efficient methods of degrading the most abundant polysaccharide, cellulose. It is also important to use a strain of yeast that will allow maximum fermentation to occur. For example if sugar beet is used as the raw material, since it contains the sugar raffinose, it is important to use a raffinose-metabolizing strain of yeast. The possibility of production of ethanol using fermentative bacterium such as *Zymomonas* is being explored. It has the advantage of catalysing a faster fermentation[11].

For cheesemaking the main process is that of fermentation of lactose to lactic acid and this is most frequently carried out by *Streptococcus lactis* or *Streptococcus cremoris*. These organisms are used primarily because they have limited metabolic diversity and do not therefore produce large quantities of other products. However, it is the small quantities of other metabolities that are partially responsible for the characteristic flavours of cheeses. Other species of bacteria are used additionally in the case of certain cheeses that have gaseous cavities. These are the result of decarboxylations occurring late in the fermentation process, usually brought about by *Streptococcus diacetylactis* and *Leuconostoc* spp.

In Swiss cheese the flavour is partly due to propionic acid formation by *Propionobacteria* spp. The latter species only begins to thrive once appreciable lactic acid formation has occurred. This second fermentation, in which lactic acid is converted to propionic acid, does not result in a high energy yield compared to the energy yield in lactate formation and thus only develops in the later stages of fermentation. The ATP is generated in the decarboxylation of pyruvate.

$$\underset{\text{Lactate}}{3CH_3.CHOH.COO^-} \xrightarrow{-6[H]} \underset{\text{Pyruvate}}{3CH_3.CO.COO^-} \xrightarrow{+6[H]}$$

$$\rightarrow \underset{\text{Propionate}}{2CH_3.CH_2.COO^-} + CH_3COO^- + CO_2 + H_2O$$

The second stage in the production of Camembert, Brie, or blue cheeses is the ripening brought about by fungi *Penicillium camemberti* and *Penicillium roqueforti*. The *P. camemberti* grows on the surface of the ripening Camembert or Brie, whereas *P. roqueforti* has the ability to grow under fairly anaerobic conditions within the cheese. Although chemical analyses of the blue cheeses have shown a variety of carboxylic acids, alcohols, and esters present, very little is known about the enzymes involved. A more detailed account of the organisms used in cheese making is given in references 9, 12 and 13.

11.3 Use of microorganisms in the production of organic chemicals

Microorganisms are also used to produce a variety of complex organic compounds. In many cases the enzymes involved have not been characterized. The most widely used organisms for these processes are fungi, and several organic acids are synthesized by species of *Aspergillus*. The conditions of growth are arranged so as to produce the maximum amount of acids concerned, e.g. growth of *Aspergillus niger* on high concentrations of sugars and low concentrations of Mn^{2+} and Fe^{3+} gives a high yield of citric acid. In general it is not economical to produce simple organic compounds using microorganisms, since they can be produced more cheaply by chemical syntheses. However, this is not the case with more complex compounds such as antibiotics, for which microorganisms are used. Table 11.2 summarizes the main processes in which the enzyme complements of certain microorganisms are used to synthesize organic compounds; further details are given in references 14 and 15.

11.4 Use of isolated enzymes in industrial processes

Although several industrial processes are carried out using whole organisms as a source of enzymes, the efficiency of these processes can often be improved by using isolated enzymes. In addition, enzymes are now being used in a wider variety of processes. The advantages of using isolated enzymes compared to intact organisms are as follows: (i) it may be possible to obtain higher catalytic activity; (ii) undesirable side-reactions may be avoided; (iii) the processes may be able to be carried out more reproducibly. The principal disadvantage is likely to be the higher cost of using isolated enzymes and also that they may be more sensitive to inactivation than the enzymes within an intact organism. It is advantageous to use enzymes from thermophilic bacteria or fungi (when a suitable enzyme is available) so as to reduce the problem of heat inactivation.

11.4.1 Sources of isolated enzymes

The most important sources of isolated enzymes are bacteria and fungi, although enzymes from plant and animal tissues are also used. Fungi and bacteria have the advantage that they can be easily grown and it is usually not difficult to scale up a production process (for reviews, see references 16–18). In addition, the sources are not subject to seasonal or other factors, e.g. chymosin (rennin) production is

TABLE 11.2

The production of some organic compounds by use of microorganisms

Compound	Organism used for synthesis	Enzyme steps if known	Uses of compound
Citric acid	*Aspergillus niger*	Monosaccharides → citrate Glycolysis and tricarboxylic acid cycle involved	Soft drinks, jams, jellies, flavouring, blood transfusion, Sequestering agent in electroplating and leather tanning
Itaconic acid $H_2C{=}C(COOH)-CH_2-COOH$	*Aspergillus* spp	As with citrate formation plus citrate ⇌ aconitate → CO_2 + itaconate	Copolymer in acrylic resins
Gluconic acid	*Aspergillus niger*	Glucose → gluconolactone → gluconate (glucose oxidase)	Calcium gluconate is used in Ca therapy, sequestering agent, food additive, pharmaceutical industry
Amphotericin B Chloramphenicol Cycloheximide Erythromycin Novobiocin Streptomycin	*Streptomyces* spp		Antibiotics
Griseofulvin Penicillin	*Penicillium* spp		Antibiotics
Bacitracin Gramicidin	*Bacillus subtilis* *Bacillus brevis*		Antibiotics
Carotenoid pigments	*Blakeslea trispora*	Synthesis from acetate	Colouring agent
Riboflavin	*Ashbya gossypi*	Synthesis from glycine, formate and CO_2	Nutrient

seasonal because it is dependent on the supply of calf stomachs. The majority of enzymes that have so far been used are hydrolytic enzymes and many of these are produced extracellularly by fungi. The possibility of producing larger quantities of enzymes from animal and plant sources by use of tissue culture methods is now being explored. Some proteins such as vaccines are already being produced in tissue culture.

With microbial enzymes it is often possible to increase the yields by changes in the growth conditions, addition of inducers, or strain selection, including increasing the number of gene copies by genetic engineering.[19-21]

With enzymes from animal and plant sources, the yields may be increased by the introduction of the appropriate genes and their promoter regions into the more rapidly growing microorganisms. However, there are often problems to be overcome before the gene is satisfactorily expressed and extracted from the microorganism as, for example in the case of chymosin in *E. coli*.[22,41]

11.4.2 Isolation of enzymes

The methods available for isolating enzymes on laboratory scale (Chapter 2) or an industrial scale are the same. However, the criteria for selection of a particular method vary according to use. Enzymes required for food-processing and related industries and for detergents are generally required in large quantities at relatively low purity. Those enzymes required for clinical diagnosis and related areas are generally required in smaller quantities at a higher level of purity. Although a high state of purity is not generally required in food processing, it may be necessary to exclude certain contaminating enzymes; e.g. in biscuit manufacture, where proteases are used it is necessary that the extract containing protease is low in amylase activity. Many of the basic methods of enzyme purification require some adaptation for scaling up.[40]

As mentioned previously, many of the fungal enzymes used are extracellular enzymes and they can readily be separated from the mycelium by filtration. It is then usually necessary to concentrate the extract and this is done by spray drying, which is less costly than precipitation by ammonium sulphate or organic solvents. If the enzyme is intracellular, the mycelium has to be disrupted. Enzymatic methods of breaking the cell wall would be too costly and difficult to scale up, so high-pressure homogenization methods are usually used.

11.4.3 Enzymes isolated on an industrial scale and their applications

The principal isolated enzyme preparations that are used industrially are those concerned with carbohydrate metabolism (Table 11.3) and protein hydrolysis (Table 11.4) and these account for over three-quarters of current sales. Other enzymes that have been used less extensively include AMP deaminase, steroid 11β-isomerase, aminoacylase, streptokinase, penicillinase, hyaluronoglucosaminidase, hyaluronoglucuronidase, triacylglycerol lipase, catalase, keratinase, and phosphodiesterase I.

TABLE 11.3

Carbohydrate-metabolizing enzymes used in industry

Enzyme	Reaction or substrate	Sources of enzyme	Utilization
Amylase	Endohydrolysis of 1,4-α-D-glucosidic linkages in polysaccharides	*Bacillus subtilis* *Aspergillus niger* *Aspergillus oryzae*	Production of sugars and oligosaccharides from starch Enables starch to be fermented
Exo-1,4-α-D glucosidase (amyloglucosidase)	Hydrolysis of terminal 1,4-linked α-D-glucose residues successively from non-reducing end of polysaccharides	*Aspergillus niger* *Aspergillus oryzae* *Rhizopus* sp	Production of glucose from starch
Cellulase	Endohydrolysis of 1,4-β-D-glycosidic linkages in cellulose and β-D-glucans	*Trichoderma viride* *Aspergillus niger*	Cellulose → cellobiose for fermentation (cellulase preparations having very high activity have yet not been obtained)
Polygalacturonase	Random hydrolysis of 1,4-α-D-galactosiduronic linkages in pectin	*Mucor*, *Botrytis*, *Pencillium*, and *Aspergillus*	Extraction of fruit juice from pulp, clarification of wines and fruit juices
β-D-Galactosidase	Lactose → glucose + galactose	*Aspergillus niger* *Aspergillus oryzae*	Hydrolysis gives sweeter, more-soluble sugars
β-D-Fructofuranosidase	Sucrose → glucose + fructose	*Aspergillus oryzae* *Saccharomyces*	Hydrolysis gives sweeter, more-soluble sugars
Glucose oxidase	Glucose + O_2 → gluconolactone + H_2O_2	*Aspergillus niger* *Penicillum* spp	Analytical reagent for glucose; desugaring egg products, removing O_2 from mayonnaise and fruit juices susceptible to oxidation
Xylose isomerase (usually referred to as glucose isomerase)	Glucose ⇌ fructose	*Streptomyces* spp *Lactobacillus brevis*	Production of high-fructose syrup

TABLE 11.4
Protein-metabolizing enzymes used in industry

Enzyme	Reaction or substrate	Source of enzyme	Utilization
Papain	Peptide-bond cleavage	*Papaya* latex	Removal of turbidity (due to protein) in beer, meat tenderizing
Chymosin (rennin)	Clotting of milk	Calf stomach	production of cheese
Trypsin	Peptide-bond cleavage	Animal pancreas	Meat tenderizers, medical uses
Chymotrypsin	Peptide-bond cleavage	Animal pancreas	
Fungal proteases	Peptide-bond hydrolysis	*Aspergillus oryzae* *Aspergillus niger*	Meat tenderizers, breadmaking to improve viscoelastic properties of doughs, chill-proofing of beer (to prevent haze due to protein–tannin interaction)
		Mucor miehei *Edothia parasitica*	Used as a substitute for chymosin, since causes only limited proteolysis
Bacterial proteases		*Bacillus subtilis*	Detergents, removal of gelatin from films

In some instances isolated enzymes are replacing whole organisms in the processing of foods, but in many cases enzymes are being used in completely new processes. We consider briefly the principal uses of enzymes in industry.

11.4.3.1 Alcoholic beverages

In the traditional brewing process, where the starch from barley is hydrolysed prior to fermentation, the barley is allowed to germinate to an extent that ensures sufficient α- and β-amylases are produced for the hydrolytic steps not to be rate-limiting. The addition of hydrolytic enzymes prior to fermentation in brewing has not been extensively practised. Enzyme supplementation has been used more extensively in the production of spirits, where sorghum, potatoes, wheat, or rye may be used as a source of starch. The enzymes most used are amylases from *Bacillus subtilis, Aspergillus oryzae* or *A. niger*. The α-amylase from *B. subtilis* is more heat stable and is used when harsh conditions are required to liquefy particular cereals. However, the extract lacks α-1,6 glucosidase activity and therefore only degrades starch (amylopectin*) to oligosaccharides. The enzymes from *Aspergillus* are also able to cleave α-1,6 glucoside links and therefore degrade the starch more completely into fermentable sugars. When used in brewing, these enzymes result in the production of a low-calorie beer since there is very little oligosaccharide remaining (important for slimming drinkers!).

* Amylopectin is a polysaccharide containing D-glucose units linked α-1,4 and α-1,6.

11.4.3.2 Breadmaking

In breadmaking the carbohydrate for fermentation is also derived from the hydrolysis of starch. The aeration (CO_2 production) of bread during its manufacture will depend on adequate α- and β-amylase activity. In some countries, where harvesting conditions do not promote germination, amylase activity becomes a limiting factor. In this case, addition of the fungal amylases (see Section 11.4.3.1) is preferable on account of their greater heat lability. It is important that only limited starch hydrolysis occurs. The fungal enzymes are rapidly inactivated during baking. Fungal proteases may also be used in breadmaking to improve the viscoelastic properties of the dough.

11.4.3.3 Cheesemaking

The increased demand for cheese coupled with the reduced availability of chymosin (rennin), which is obtained from calf stomach, has led to a search for a substitute for chymosin. Chymosin causes the clotting of milk, a process which involves cleavage of a single peptide bond in casein κ between Phe 105 and Met 106, releasing the acidic C-terminal peptide. This release causes the micelles containing casein (which normally have a negative charge and so are mutually repelled) to aggregate and clot. A detailed account is given in reference 23. Proteolytic enzymes such as trypsin also cause clotting, but then further degrade casein. If this occurs in cheesemaking it leads to undesirable flavours. In the search for a substitute for chymosin, enzymes that cause clotting but only limited proteolysis are desirable. The fungal proteases from *Mucor pusillus*, *Mucor miehei*, and *Edothia parasitica* can substitute for chymosin, although they do not have quite such a good clotting/proteolysis ratio. They are now used on a commercial scale and account for more than one-third of the cheese production world wide.[24]

11.4.3.4 Meat tenderizing

Toughness in meat is due primarily to collagen and elastin and also actomyosin. Lower-quality cuts of meat are just as nutritious as prime cuts. Thus, to make maximum use of the carcass, efforts have been made to tenderize the lower-quality cuts. This is done by injecting proteolytic enzymes into the vascular system either before or after slaughter. The enzymes most used are papain, trypsin, chymotrypsin, and also microbial proteases from *Aspergillus*. Over 5 per cent of United States beef is subject to tenderizing procedures by packers.

11.4.3.5 Sweeteners

A quantitatively very important use of enzymes is in the production of glucose syrups to be used as sweetening agents in a variety of processed foods. A glucose–fructose mixture is sweeter than glucose alone (glucose is about 70 per cent as sweet as sucrose, whereas fructose is about 60 per cent sweeter than sucrose), and cheaper to produce than sucrose that is extracted from sugar cane or sugar beet. The glucose–fructose mixture is produced from starches, particularly corn starch. Two enzymes are used in the process: exo-1,4-α-D-glucosidase obtained from *Aspergillus niger* and xylose isomerase from *Streptomyces* spp. (The enzyme used for isomerization is xylose isomerase, EC 5.3.1.5, although it is usually referred to as glucose isomerase, since it will also isomerize glucose;[25] $K_{m(\text{D-glucose})} = 0.17$ mmol dm^{-3} and $K_{m(\text{D-xylose})} = 0.01$ mmol dm^{-3}.)

$$\text{Starch} \rightarrow \text{glucose} \rightleftharpoons \text{fructose.}$$

This process is discussed further in Section 11.5.1 on immobilized enzymes.

Another example of the use of an enzyme to cause an increase in sweetness is the use of β-D-galactosidase in ice-cream manufacture. Since the mixture of glucose and galactose is sweeter than lactose, this enzyme can be used to catalyse an increase in sweetness of a product.

11.4.3.6 Clarification of beers, wines, and fruit juices

Turbidity in beers, wines, and fruit juices is undesirable because turbid drinks are less acceptable to the consumer. Enzymes have been used to clarify a variety of beverages. In the case of fruit juices they are more easily prepared from extracted pulp if the viscosity is not too high. In the case of beer a haze or turbidity may develop on cooling and this appears to be due to a complex between tannin and protein. It can be successfully removed by addition of small amounts of protease. In the case of wines and fruit juices the turbidity may be due to starch or pectin and is removed by addition of amylases or polygalacturonases, which also cause a decrease in viscosity. A detailed discussion of the enzymes used is given in reference 26.

11.4.3.7 Detergents

Detergents containing enzymes were developed extensively during the 1960s, so that by 1969 about half of the detergents marketed contained enzymes. Following the indication of possible health hazards, the use of such detergents declined sharply in the early 1970s; since then methods have been developed for encapsulation of enzyme particles using non-ionic surfactants and this has reduced dust inhalation problems with powdered enzymes. The principal enzymes used in 'biological' detergents are mixtures of amylase and neutral and alkaline proteases that are active in the pH range 6.5 to 10 and at temperatures from 303 K (30°C) to 333 K (60° C). Detergents often contain oxidizing and chelating agents and so the enzymes must withstand these also. *Bacillus* neutral protease and *Bacillus* alkaline protease are the most suitable to date, but others from thermophilic organisms having high thermal stability are also being developed.[27] A mixture of carbohydrate-hydrolysing and protein-hydrolysing enzymes of low specificity is most useful, since polysaccharides and proteins require partial degradation before they are removed from garments. By contrast, lipids can be removed without any prior degradation by ionic detergents at alkaline pHs.

11.4.3.8 Medical applications

Isolated enzymes are used in medicine both as reagents in analysis and in therapy. We discussed the use of enzymes, e.g. glucose oxidase, in clinical analysis in Chapter 10, Section 10.7. As a therapeutic agent, trypsin has been used as an anti-inflamatory agent and as a wound cleanser. Hydrolytic enzymes have also been used in clinical trials to liquefy blood clots. Blood clots are liquefied under physiological conditions by the action of plasmin. Plasmin is produced slowly in the plasma by the action of fibrinokinase on plasminogen (plasmin precursor). Streptokinase (an enzyme from *Streptococcus haemolyticus*) or urokinase (an enzyme isolated from human urine) are effective activators of plasminogen and

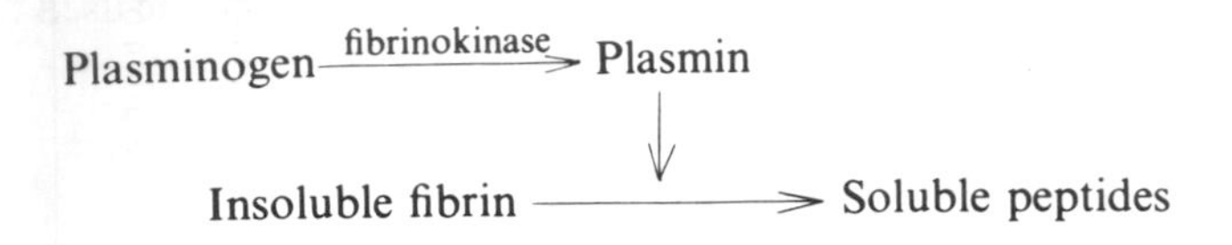

appear to be useful in relieving peripheral thrombosis. Plasminogen activators are also produced in other tissues. Attempts are being made to develop cell lines that produce larger quantities of these activators. The tissue plasminogen activators are often tissue specific.

Hyaluronidase* obtained from beef testes is a useful agent to aid diffusion when coinjected with other drugs such as antibiotics, adrenalin, heparin, and local anaesthetics. Hyaluronidase hydrolyses polyhyaluronic acid, the main polysaccharide component of connective tissue, and thus aids the diffusion of other substances by reducing the viscosity in the locality of the injection.

We have discussed the possible use of asparaginase in cancer chemotherapy in Chapter 10, Section 10.8.2. For further details of the medical applications of isolated enzymes, see references 28 and 29.

11.5 Immobilized enzymes

Attaching an enzyme to an insoluble support will permit its reuse and continuous use without a difficult recovery process. The attachment to the support may also stabilize the enzyme. If two or more enzymes catalysing a sequence of reactions are immobilized in close proximity to each other, an efficient immobilized multienzyme protein analogue can be produced. There are therefore potential advantages in immobilizing enzymes that are required on a large scale. Much research effort has been devoted during the last decade to producing satisfactory immobilized enzymes for large-scale application. For details of the methods used for immobilization, see references 30–33. A few immobilized enzymes are being used currently in industry (Table 11.5) but a large number are still at the research and development stage.

* The recommended names for the two enzyme activities commonly referred to as hyaluronidase are hyaluronoglucosaminidase and hyaluronoglucuronidase.

TABLE 11.5

Industrial applications of immobilized enzymes and whole cells

Enzyme or whole cells	Carrier	Methods of immobilization	Application
Aminoacylase	DEAE-Sephadex	Adsorption	Preparation of L-amino acids from a racemic DL-amino acid mixture
Xylose isomerase (glucose isomerase)	Duolite A7 (ceramic beads)	Adsorption	Preparation of fructose–glucose mixture
	Amberlite IRA904 exchange resin	Adsorption	
E. coli containing asparate ammonia-lyase	Polyacrylamide	Entrapment	Preparation of L-aspartate
Pseudomonas putida containing arginine deaminase	Polyacrylamide	Entrapment	Preparation of citrulline
Achromobacter liquidum containing histidine ammonia-lyase	Polyacrylamide	Entrapment	Preparation of uraconic acid
E. coli containing penicillin amidase	Polyacrylamide, cellulose	Entrapment	Preparation of 6-aminopenicillanic acid
Brevibacterium ammoniagenes containing fumarate hydratase	Polyacrylamide	Entrapment	Preparation of malate and fumarate
Lactase (β-galactosidase)	Silica particles	Adsorption	Preparation of lactose-free milk

There are a number of ways in which an enzyme may be immobilized, by adsorption, covalent linkage, matrix entrapment, or encapsulation, as summarized in Fig. 11.2. The uses of immobilized enzymes can be broadly subdivided into preparative and analytical. Examples of the former are large-scale isomerization of glucose and production of L-amino acids from racemic mixtures. Use of immobilized glucose oxidase is an example of the latter. We discuss in the following sections the current applications of immobilized enzymes.

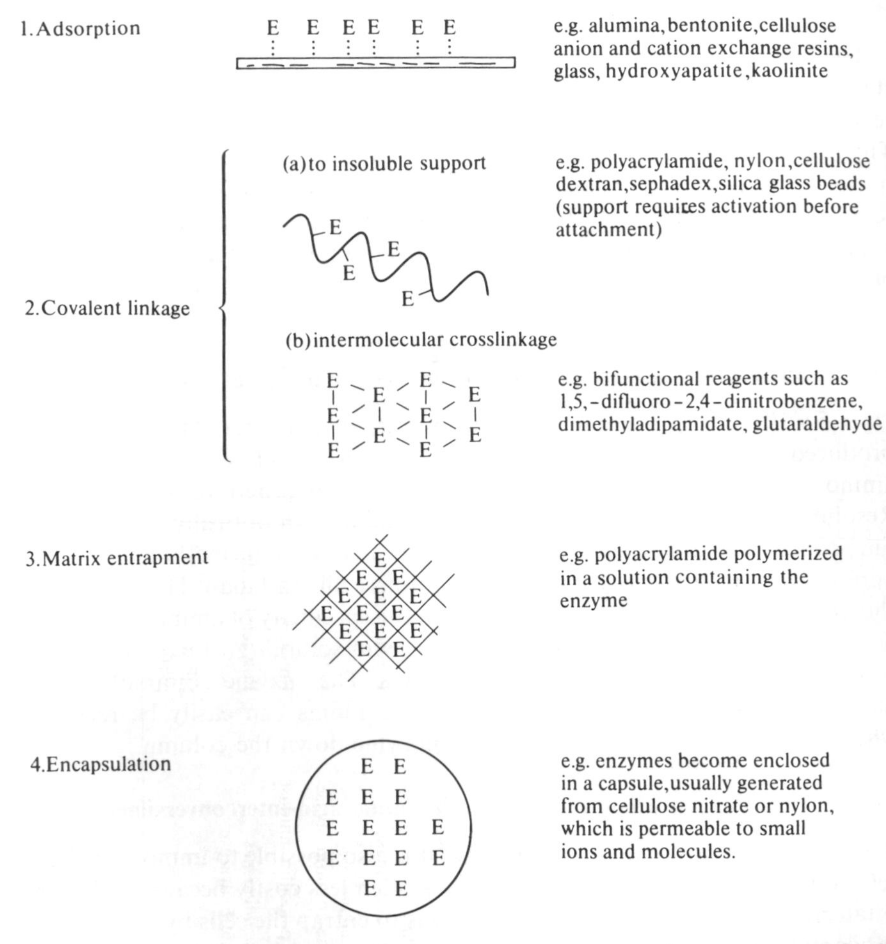

FIG. 11.2. Methods used for enzyme immobilization. (For details see references 31–33).

11.5.1 Production of syrups from corn starch

The production of syrups from starches was mentioned in Section 11.4.3.5. Syrups (sugar solutions) are used extensively as sweetening agents in a variety of manufactured foods. Several million tonnes of high-fructose corn syrup are now produced annually and this is displacing sucrose in traditional applications. It represents one of the biggest developments in food science since 1974. Fructose is more satisfactory than glucose because it is sweeter and does not crystallize so readily from concentrated solutions. The overall process requires two main enzyme-catalysed steps:

$$\text{Starch} \xrightarrow{\text{exo-1,4-}\alpha\text{-D-glucosidase}} \text{glucose} \xrightleftharpoons{\text{xylose isomerase}} \text{fructose.}$$

The first of these steps is catalysed by exo-1,4-α-D-glucosidase obtained from *Aspergillus niger*. This enzyme has been immobilized on porous silica. The second step is catalysed by xylose isomerase obtained from *Streptomyces* spp. This enzyme has been immobilized by attachment of DEAE-cellulose or to a porous ceramic carrier. The latter method has proved more successful in large-scale columns because of the non-compressible nature of the support. The equilibrium constant for the enzyme at 323 K (50°C), the operating temperature of the process, is approximately unity, so that approximately equal amounts of fructose and glucose are present in the product.

11.5.2 Production of L-amino acids from racemic mixtures

Over 5×10^7 kg of DL-amino acids are produced each year. Most of these are produced by synthetic methods and since most biological systems utilize only L-amino acids, a method for resolving the isomers is required for many purposes. Resolution of optically active acids by combining with optically active bases is time-consuming and expensive. Resolution by an enzymatic method that uses an immobilized enzyme is now carried out commercially in Japan. The procedure is shown in Fig. 11.3. The method is based on the specificity of aminoacylase for L-*N*-acetylated amino acids and the differences in solubility of free amino acids compared with their *N*-acetylated derivatives. The enzyme is immobilized by binding to DEAE-Sephadex columns. The columns can easily be recharged when necessary simply by passing more enzyme down the column.

11.5.3 Immobilized whole cells for use in amino-acid interconversions

Besides immobilization of single enzymes it is also possible to immobilize whole cells (for a review, see reference 31). This is often less costly because it does not entail enzyme isolation. The usual method is to entrap the cells by polymerizing acrylamide and *N,N'*-methylenebisacrylamide around them. This has been successfully used in Japan for the production of aspartate, citrulline, phenylalanine, tryptophan, serine, urocanic acid, and 6-aminopenicillanic acid by entrapment of strains of *E. coli* and *Pseudomonas putida* containing the appropriate enzymes.[34] Immobilized *E. coli* cells have been found to have an operational half-life of up to two years.[42] Aspartate and citrulline are used in the pharmaceutical industry, urocanic acid is used as an ultraviolet filter in suntan preparations, and 6-aminopenicillanic acid is used as a precursor of the

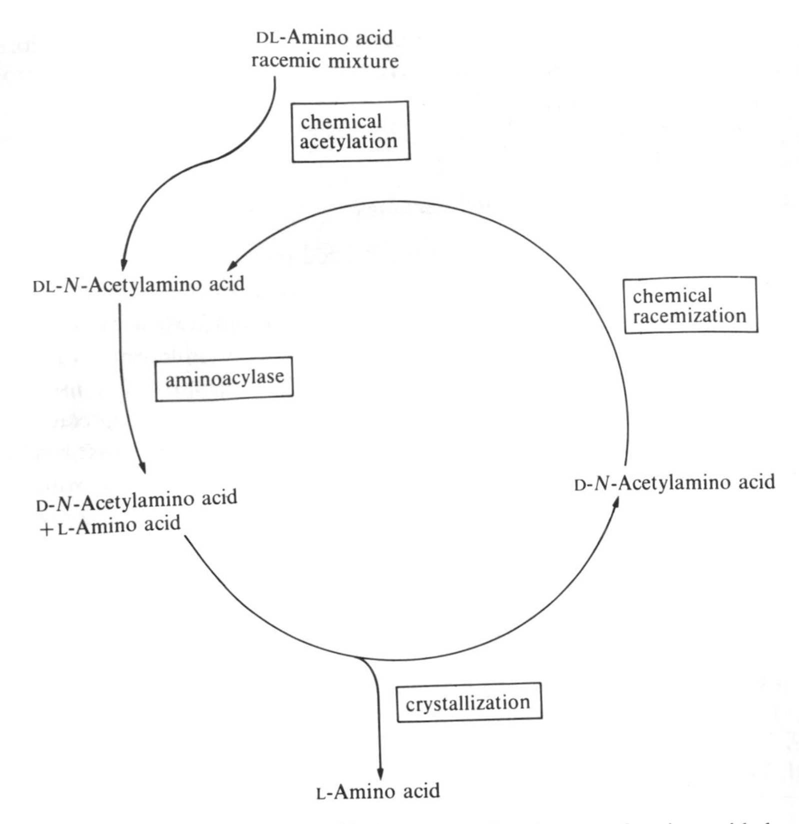

FIG. 11.3. Production of L-amino acids from racemic mixtures of amino acids by use of aminoacylase.

semisynthetic penicillins, e.g. methicillin, oxacillin, carbenicillin, and ampicillin. Aspartate, together with phenylalanine, is used in the sweetening agent Aspartame (*N*-L-α-aspartyl-L-phenylalanine methyl ester). The enzyme catalysed reactions for some of the steps are given below.

$$\underset{\text{Fumarate}}{{}^{-}OOC\,CH{=}CH{-}COO^{-}} + NH_3 \underset{}{\overset{\text{aspartate ammonia-lyase}}{\rightleftharpoons}} \underset{\text{Aspartate}}{{}^{-}OOCCH_2CH(NH_3^{+})COO^{-}}$$

$$\underset{\text{Arginine}}{H_2N{-}C({=}NH_2^{+}){-}NH(CH_2)_3CH(NH_3^{+})COO^{-}} \overset{\text{L-arginine deiminase}}{\rightleftharpoons} \underset{\text{Citrulline}}{H_2N{-}C({=}O){-}NH(CH_2)_3CH(NH_3^{+})COO^{-}} + NH_4^{+}$$

$$\underset{\text{Histidine}}{HC{=}C(\text{–}NH{-}CH{=}N\text{–}){-}CH_2CH(NH_3^{+})COO^{-}} \overset{\text{histidine ammonia-lyase}}{\rightleftharpoons} \underset{\text{Urocanic acid}}{HC{=}C(\text{–}NH{-}CH{=}N\text{–}){-}CH{=}CHCOO^{-}} + NH_4^{+}$$

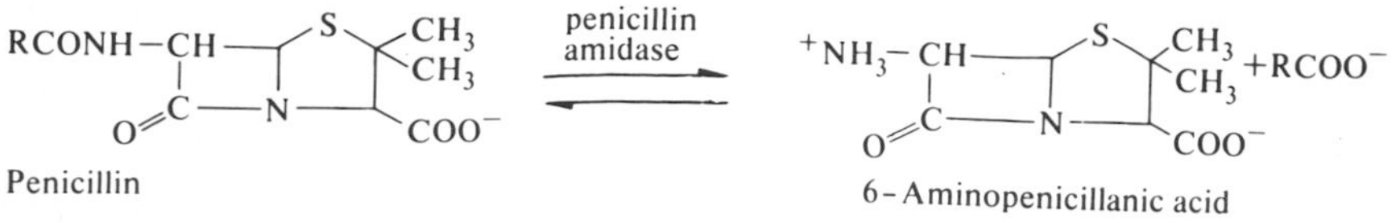

Penicillin $\xrightarrow{\text{penicillin amidase}}$ 6-Aminopenicillanic acid + $RCOO^-$

11.5.4 Other uses of immobilized enzymes

A number of other applications of immobilized enzymes are being developed. Lactose is a major byproduct of cheese manufacture, since much of the lactose occurs in the whey. A small-scale process in operation in Italy uses immobilized β-galactosidase to hydrolyse lactose to yield a mixture of glucose and galactose, the mixture being more useful as a sweetening agent. Cortisol is a useful drug for arthritis treatment and it can be made from the cheap precursor 11-deoxycortisol. Trials with immobilized steroid 11β-monooxygenase and a Δ'-dehydrogenase are being carried out to produce prednisolone, which is a superior therapeutic agent to cortisone.

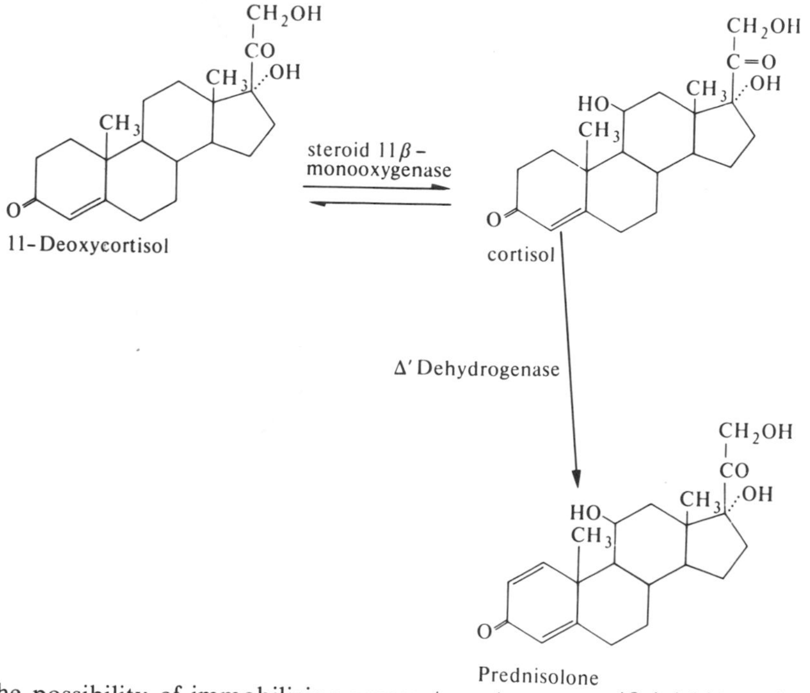

11-Deoxycortisol $\xrightarrow{\text{steroid } 11\beta\text{-monooxygenase}}$ cortisol

cortisol $\xrightarrow{\Delta' \text{ Dehydrogenase}}$ Prednisolone

The possibility of immobilizing urease to act as an artificial kidney in the removal of urea is being tested:

$$\text{Urea} + H_2O \overset{\text{urease}}{\rightleftharpoons} CO_2 + 2NH_3$$

This would be coupled to an absorbing device to remove the ammonia produced.

Immobilized enzymes can also be used in analytical procedures. The most developed application to date is in the analysis of blood glucose using immobilized glucose dehydrogenase. The glucose dehydrogenase from *Bacillus cereus* oxidizes only D-glucose and can be used specifically in the determination of this sugar:

$$\text{D-Glucose} + \text{NAD}^+ \rightleftharpoons \text{D-glucono-}\delta\text{-lactone} + \text{NADH} + \text{H}^+.$$

The NADH formed is measured spectrophotometrically or spectrofluorimetrically. The enzyme has been immobilized on to nylon tubing and thus can be reused, so that as little as 0.18 mg can be used for 1500 glucose determinations.

Another analytical development is that of an enzyme electrode that can be used for the continuous monitoring of specific substrates. It has the advantage over spectrophotometric methods that it can be used with fluids that are optically opaque, such as blood or culture media. An example of an enzyme electrode is that used to measure urea. Urease is immobilized within a gel layer in close proximity to cation-sensitive electrode that measures NH_4^+ concentration. As the urea is hydrolysed in close proximity to the electrode, the electrode detects the high local concentration of NH_4^+. A similar type of electrode can be used to monitor glutamate (Fig. 11.4). The arrangement is that NAD^+ is immobilized to

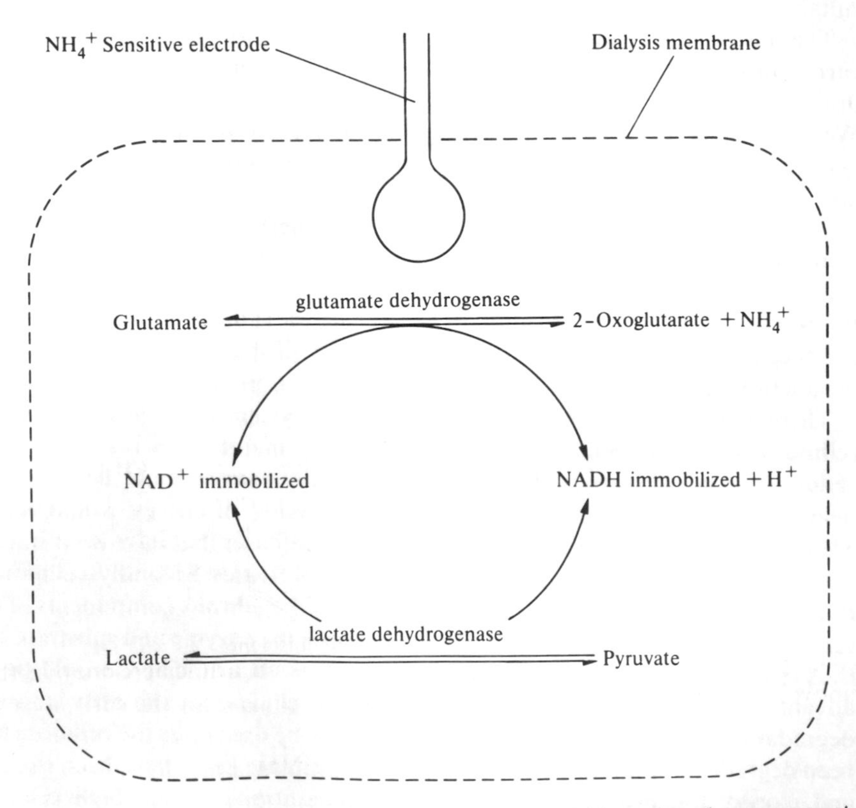

FIG. 11.4. Diagrammatic arrangement of an electrode used to monitor the concentration of glutamate (see reference 35).

an insoluble dextran, this and glutamate dehydrogenase and lactate dehydrogenase are all entrapped in a dialysis membrane surrounding the ammonium-ion-sensitive electrode. Again the NH_4^+ generated in close proximity to the electrode results in an electrode potential proportional to the logarithm of the glutamate concentration. Enzyme electrodes have been devised for measurement of a wide range of amino acids, lactic acid, acetic acid, formic acid, penicillin, and alcohols.[36]

11.5.5 Future developments

Several processes using immobilized enzymes have been tested at the pilot plant level, and can be expected to be developed on an industrial scale in the next few years. In addition, there are a number of areas where new developments are hoped for. These include immobilized enzymes for the oxidation of hydrocarbons, for the synthesis of steroids, vitamins, and antibiotics, for nitrogen fixation, for removal of pesticides from water supplies, and for the photolysis of water by a suitable photosynthetic cell.

The oxidation of hydrocarbons to specific alcohols and acids is expensive to carry out chemically. There are microbial cells capable of oxidizing carbohydrate, but the enzymes or the cells have not yet been successfully immobilized. We mentioned in Section 11.5.4 the production of prednisolone from 11-deoxycortisol. Similar procedures should permit the production of more steroids and antibiotics by the same type of method. An encapsulated cell capable of fixing nitrogen would enable some of the atmospheric nitrogen to be used as a source of nitrogen to be incorporated into amino acids and proteins. Similarly, stabilization of a cell or components of a cell capable of carrying out photosynthesis would permit the utilization of solar energy for the photolysis of water to produce hydrogen as a fuel. There are still many problems associated with the production of a stable immobilized system that can photolyse water.[37]

There is still a great need for an enzymatic system capable of degrading cellulose: 40–60 per cent of waste disposed of by major cities is composed of cellulose or its derivatives. If cellulose and its derivatives could be efficiently degraded to monosaccharides a considerable saving of energy would result. There are two main difficulties here. Firstly, the cellulases that have been studied and isolated so far do not have very high specific activities. Secondly, cellulose is insoluble and occurs interwoven with other insoluble fibrous components of the cell. This greatly limits the area of contact between the enzyme and substrate and thus enzymatic degradation is likely to be very inefficient. There would be no advantages in this case in using an immobilized cellulase for the early stages of degradation, though immobilized cellulase might be used after the cellulose had been degraded to a soluble form. A number of cellulase genes have been isolated and cloned; the prospect of commercial preparations having high cellulase activity is now good.[38]

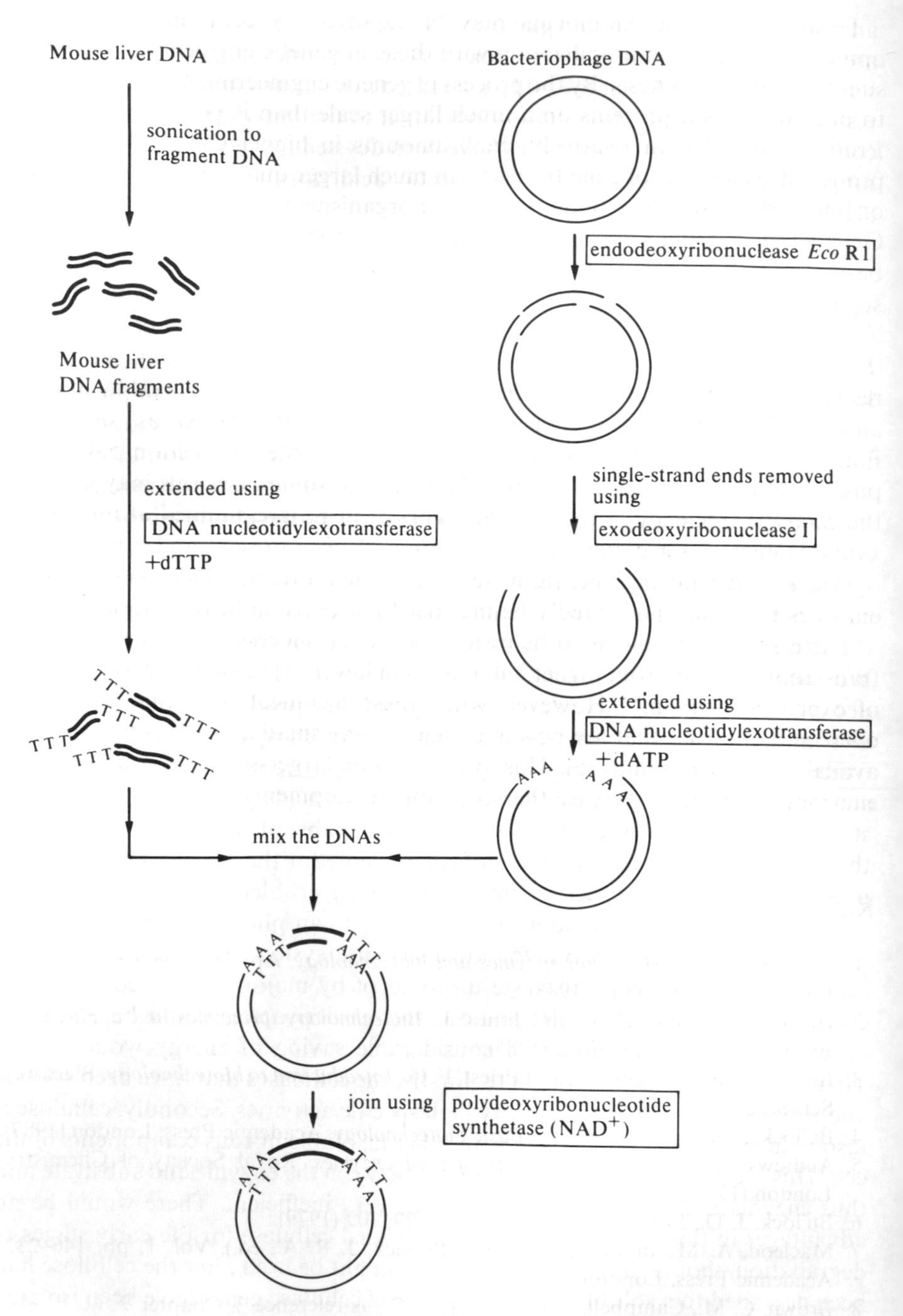

FIG. 11.5. Diagram to illustrate the use of enzymes for the production of recombinant DNA using mouse liver DNA and bacteriophage DNA.

Finally, further developments may be expected by combining the developments in enzyme immobilization with those in genetic engineering, including site-directed mutagenesis. By the process of genetic engineering it will be possible to produce certain proteins on a much larger scale than is possible at present. Proteins produced in relatively small amounts in higher eukaryotes may be produced, as a result of gene transfers, in much larger quantities in prokaryotes or lower eukaryotes, e.g. fungi, since these organisms can multiply much faster. Genetic engineering entails fragmenting DNA from one organism at specific base sequences using restriction endonucleases. The DNA fragments so obtained have to be separated and then joined to a suitable vector, such as plasmid or virus, in order to introduce the DNA into a second species. (For details, see references 19–21) The gene transfer thus entails the use of restriction endonucleases, polydeoxyribonucleotide synthetases (DNA ligases), and DNA nucleotidylexotransferases (terminal transferases) as shown in Fig. 11.5. Using site-directed mutagenesis (see Chapter 5, Section 5.4.5) it is possible to replace specific amino-acid residues; in some cases this may improve the catalytic efficiency of the enzyme. Work is in progress aimed at improving xylose isomerase for commercial applications.[39] The present and future developments in genetic engineering have thus created a demand for these purified enzymes that will undoubtedly be increased by future industrial applications.

There are still problems to be overcome for the successful transcription and translation of genes from higher eukaryotes in lower organisms and for recovery of expressed protein.[41] However, when these technical problems have been overcome, enzymes that are now available in only small quantities will become available in larger quantities. This availability on larger-scale a wider variety of enzymes will further increase the scope for development.

References

1. Fogarty, W. M., *Microbial enzymes and biotechnology.* Applied Science Publishers, London and New York (1983).
2. Higgins, I. J., Best, D. J., and Jones, J., *Biotechnology: principles and applications.* Blackwell Scientific, Oxford (1985).
3. Brown, C. M., Campbell, I., and Priest, F. G., *Introduction to biotechnology.* Blackwell Scientific, Oxford (1987).
4. Bu'lock, J. and Kristiansen, B., *Basic biotechnology.* Academic Press, London (1987).
5. Andrews, A. T., *Chemical aspects of food enzymes.* Royal Society of Chemistry, London (1987).
6. Bu'lock, J. D., *Symp. Soc. gen. Microbiol.* **29**, 309 (1979).
7. Macleod, A. M., in *Brewing science* (Pollock, J. R. A., ed.), Vol. 1, pp. 146–232. Academic Press, London (1979).
8. Brown, C. M., Campbell, I. and Priest, F. G., as reference 3, Chapter 9.
9. Beech, G. A., Melvin, M. A., and Taggart, J., as reference 2, Chapter 3.
10. Tubb, R. S., *Trends Biotech.* **4**, 98 (1986).
11. Buchholz, S. E., Dooley, M. M., and Eveleigh, D. E., *Trends Biotech.* **5**, 199 (1987).

12. Lawrence, R. C. and Thomas, T. D., *Symp. Soc. Gen. Microbiol.* **29**, 187 (1979).
13. Kirby, C. J., and Law, B., as reference 5, p 106 ff.
14. Meers, J. L., and Milson, P. E., as reference 4, Chapter 13.
15. Corbett, K., as reference 4, Chapter 16.
16. Lilly, M. D., in *Applied Biochemistry and Bioengineering* (Wingard, L. B., Katchalski-Katzir, E., and Goldstein, L. eds.), Vol. 2, p. 1. Academic Press, New York (1979).
17. Aunstrup, K., as reference 16, Vol. 2, p. 28.
18. Bucke, C., in *Principles of Biotechnology* (Wiseman, A., ed.), Ch. 6, Surrey University Press, New York, 1983.
19. Morrow, J. F., *Methods Enzymol.* **68**, 3 (1979).
20. Old, R. W. and Primrose, S. B., *Principles of gene manipulation: an introduction to genetic engineering* (3rd edn). Blackwell, Oxford (1985).
21. Brown, T. A., *Gene cloning: an introduction.* Van Nostrand and Reinhold (U.K.) 1986.
22. Cullen, D. and Leong, S., *Trends Biotech.* **4**, 285 (1986).
23. Andrews, A. T., as reference 5, p. 203 ff.
24. Adler-Nissen, J., *Trends Biotech.* **5**, 170 (1987).
25. Antrim, R. L., Colilla, W., and Schnyder, R. J., as reference 16, Vol. 2, p. 97.
26. Fogarty, W. M. and Kelly, C. T., as reference 1, Ch. 3.
27. Cowan, D., Daniel, R., and Morgan, H., *Trends Biotech.* **3**, 68 (1985).
28. Chang, T. M. S., *Biomedical applications of immobilised enzymes and proteins*, Vol. 1, Chs. 9, 10, and 18. Plenum Press, London (1977).
29. Bohak, Z. and Sharon, N., *Biotechnological applications of proteins and enzymes*, Chs. 16 and 18. Academic Press, New York (1977).
30. Mosbach, K., *Methods Enzymol.* **44** (1976).
31. Woodward, J., *Immobilised cells and enzymes.* IRL Press, Oxford (1985).
32. Mosbach, K., *Methods Enzymol.* **135** (1987).
33. Mosbach, K., *Methods Enzymol.* **136** (1987).
34. Hamilton, B. K., Hsiao, H-Y., Swann, W. E., Anderson, D. M., and Delente, J. J., *Trends Biotech.* **3**, 64 (1985).
35. Bohak, Z. and Sharon, N., *Biotechnological applications of proteins and enzymes*, p. 141. Academic Press, New York (1977).
36. Guilbault, G. G. and de Olivera Neto, G., as reference 31, Ch. 5.
37. Hall, D. O., Coombs, J., and Higgins, J., as reference 2, Chapter 2.
38. Schreirer, W., *Trends Biotech.* **5**, 55 (1987).
39. Hartley, B. S., Anderton, T. and Shaw, P. C., as reference 5, p. 120 ff.
40. Bruton, C. J., *Phil. Trans. R. Soc. Lond.* **B300**, 249 (1983).
41. Marston, F. A. O., *Biochem. J.* **240**, 1 (1986).
42. Wood, L. L., *Biotechnology* **2**, 1081 (1984).

Appendix

Enzymes referred to in Chapters 1–11

The enzymes referred to in the 11 chapters of this book are listed below in alphabetical order generally under their recommended names as in *Enzyme nomenclature* (Recommendations (1984) of the Nomenclature Committee of the International Union of Biochemistry), except in a few cases where trivial names are more frequently used. A few of the enzymes listed have not yet been classified, usually because insufficient information is at present available for this to be done. In these cases we have indicated what is known about the probable nature of the reaction concerned. The specificities of a number of proteases are given in an abbreviated form, e.g. arg- means the amide bond cleaved is that on the carboxyl side of arginine, whereas -leu means that cleavage occurs on the amino side of leucine.

	EC number
Acetylcholinesterase Acetylcholine + H_2O ⇌ choline + acetate	3.1.1.7
Acetyl-CoA acetyltransferase Acetyl-CoA + Acetyl-CoA ⇌ CoA + acetoacetyl-CoA	2.3.1.9
Acetyl-CoA carboxylase ATP + Acetyl-CoA + CO_2 + H_2O ⇌ ADP + orthophosphate + malonyl-CoA	6.4.1.2
Acetylglutamate kinase ATP + *N*-acetyl-L-glutamate ⇌ ADP + *N*-acetyl-L-glutamate 5-phosphate	2.7.2.8
β-N-acetyl-D-hexosaminidase R-2-Acetamido-2-deoxy-β-D-hexoside + H_2O ⇌ ROH + 2-acetamido-2-deoxy-hexose	3.2.1.52
Acid phosphatase An orthophosphoric monoester + H_2O ⇌ an alcohol + orthophosphate	3.1.3.2
Aconitase Citrate ⇌ *cis*-aconitate + H_2O The preferred name is aconitate hydratase	4.2.1.3
[Acyl-carrier-protein] acetyltransferase Acetyl-CoA + [acyl-carrier protein] ⇌ CoA + acetyl-[acyl-carrier protein]	2.3.1.38
[Acyl-carrier-protein] malonyltransferase Malonyl-CoA + [acyl-carrier protein] ⇌ CoA + malonyl-[acyl-carrier protein]	2.3.1.39

[Acyl-carrier-protein] palmityltransferase
A component of certain eukaryote fatty-acid synthase systems
Palmityl-CoA + [acyl-carrier protein] ⇌ CoA + palmityl-[acyl-carrier protein]

Acyl-CoA desaturase
Stearyl-CoA + AH_2 + O_2 ⇌ oleyl-CoA + A + $2H_2O$ 1.14.99.5

Acyl-CoA synthetase
ATP + an acid + CoA ⇌ AMP + pyrophosphate + An acyl-CoA 6.2.1.3

Adenosinetriphosphatase
ATP + H_2O ⇌ ADP + orthophosphate 3.6.1.3

Adenylate cyclase
ATP ⇌ 3′:5′ cyclic AMP + pyrophosphate 4.6.1.1

Adenylate kinase
ATP + AMP ⇌ ADP + ADP 2.7.4.3

Alanine aminotransferase
L-Alanine + 2-oxoglutarate ⇌ pyruvate + L-glutamate 2.6.1.2

Alcohol dehydrogenase
Alcohol + NAD^+ ⇌ aldehyde or ketone + NADH 1.1.1.1

Aldose 1-epimerase
α-D-Glucose ⇌ β-D-glucose 5.1.3.3

Alkaline phosphatase
An orthophosphate monoester + H_2O ⇌ an alcohol + orthophosphate 3.1.3.1

Amidophosphoribosyltransferase
5-Phospho-β-D-ribosylamine + pyrophosphate + L-glutamate ⇌ L-glutamine + 5-phospho-α-D-ribose 1-diphosphate + H_2O 2.4.2.14

Amine oxidase (flavin-containing)
$RCH_2NH_2 + H_2O + O_2$ ⇌ $RCHO + NH_3 + H_2O_2$ 1.4.3.4

Amino-acid acetyltransferase
Acetyl-CoA + L-glutamate ⇌ CoA + *N*-acetyl-L-glutamate 2.3.1.1

Aminoacylase
An *N*-acyl-aminoacid + H_2O ⇌ a fatty acid anion + an aminoacid 3.5.1.14

Aminoacyl-tRNA synthetases (*ligases*)
One enzyme for each of the amino acids
ATP + L-amino acid + $tRNA^{AA}$ ⇌ AMP + pyrophosphate + L-aminoacyl-$tRNA^{AA}$ 6.1.1.(1–7, 9–12, 14–22)

5-*Aminolaevulinate synthase*
Succinyl-CoA + glycine ⇌ aminolaevulinate + CoA + CO_2 2.3.1.37

Aminopeptidase (cytosol)
Aminoacyl-peptide + H_2O ⇌ amino acid + peptide 3.4.11.1

AMP deaminase
AMP + H_2O ⇌ IMP + NH_3 3.5.4.6

α-Amylase
Endohydrolysis of 1,4-α-D-glucosidic linkages in polysaccharides containing three or more 1,4-α-linked D-glucose units — 3.2.1.1

β-Amylase
Hydrolysis of 1,4-α-D-glucosidic linkages in polysaccharides so as to remove successive maltose units from the non-reducing ends of the chains — 3.2.1.2

Amylo-1,6-glucosidase
Endohydrolysis of 1,6-α-D-glucosidic linkages at points of branching in chains of 1,4-linked α-D-glucose residues. Together with EC 2.4.1.25, these two activities constitute the glycogen *debranching* system — 3.2.1.33

Anthranilate phosphoribosyltransferase
N-(5′-Phospho-D-ribosyl)-anthranilate + pyrophosphate ⇌ anthranilate + 5-phospho-α-D-ribose 1-diphosphate

In many organisms this enzyme exists as a complex with anthranilate synthase (EC 4.1.3.27) — 2.4.2.18

Anthranilate synthase
Chorismate + L-glutamine ⇌ anthranilate + pyruvate + L-glutamate

In many organisms this enzyme exists as a complex with anthranilate phosphoribosyltransferase (EC 2.4.2.18) — 4.1.3.27

Arginase
L-Arginine + H_2O ⇌ L-ornithine + urea — 3.5.3.1

Arginine deiminase
L-Arginine + H_2O ⇌ L-citrulline + NH_3 — 3.5.3.6

Argininosuccinate lyase
L-Argininosuccinate ⇌ fumarate + L-arginine — 4.3.2.1

Argininosuccinate synthase
ATP + L-citrulline + L-aspartate ⇌ AMP + pyrophosphate + L-argininosuccinate — 6.3.4.5

Arginyl-tRNA synthetase (ligase)
ATP + L-arginine + $tRNA^{Arg}$ ⇌ AMP + pyrophosphate + L-arginyl-$tRNA^{Arg}$ — 6.1.1.19

Aromatic-L-amino acid decarboxylase
L-Tryptophan ⇌ tryptamine + CO_2 — 4.1.1.28

Arylsulphatase
A phenol sulphate + H_2O ⇌ a phenol + sulphate — 3.1.6.1

Asparaginase
L-Asparagine + H_2O ⇌ L-aspartate + NH_3 — 3.5.1.1

Aspartate aminotransferase
L-Aspartate + 2-oxoglutarate ⇌ oxaloacetate + L-glutamate — 2.6.1.1

Aspartate ammonia-lyase
L-Aspartate ⇌ fumarate + NH_3 — 4.3.1.1

Aspartate carbamoyltransferase
Carbamoylphosphate + L-aspartate ⇌ Orthophosphate + *N*-carbamoyl-L-aspartate — 2.1.3.2

Aspartate kinase
ATP + L-aspartate ⇌ ADP + 4-phospho-L-aspartate — 2.7.2.4

ATP citrate (pro-3S)-*lyase*
ATP + citrate + CoA ⇌ ADP + orthophosphate + acetyl-CoA + oxaloacetate.
Often referred to as citrate cleavage enzyme — 4.1.3.8

Branched-chain 2-oxoacid dehydrogenase
3-Methyl-2-oxobutanoate + lipoamide ⇌ *S*-(2-methylpropanoyl) dihydrolipoamide + CO_2
Part of multienzyme complex — 1.2.4.4

Branching enzyme
See 1,4-α-glucan branching enzyme

Bromelain
A thiol protease, cleaving preferentially at lys-, ala-, tyr-, and gly- — 3.4.22.4

Carbamoyl-phosphate synthase (ammonia)
2ATP + NH_3 + CO_2 + H_2O ⇌ 2ADP + orthophosphate + carbamoyl phosphate
This enzyme is present in the livers of ureotelic vertebrates. — 6.3.4.16

Carbamoyl-phosphate synthase (glutamine-hydrolysing)
2ATP + glutamine + CO_2 + H_2O ⇌ 2ADP + orthophosphate + glutamate + carbamoyl phosphate
This enzyme is present in bacteria, fungi, and mammals. Both carbamoyl phosphate synthases from fungi are EC 6.3.5.5, as is the mammalian enzyme that participates in pyrimidine biosynthesis. — 6.3.5.5

Carbonate dehydratase
H_2CO_3(or H^+ + HCO_3^-) ⇌ CO_2 + H_2O
Trivial name often used is carbonic anhydrase — 4.2.1.1

Carboxylesterase
A carboxylic ester + H_2O ⇌ an alcohol + a carboxylic acid anion — 3.1.1.1

Carboxypeptidase A
Peptidyl-L-amino acid + H_2O ⇌ peptide + L-amino acid — 3.4.17.1

Carboxypeptidase B
Peptidyl-L-lysine (-L-arginine) + H_2O ⇌ peptide + L-lysine (or L-arginine) — 3.4.17.2

Carnitine palmitoyltransferase
Palmitoyl-CoA + L-carnitine ⇌ CoA + L-palmitoylcarnitine — 2.3.1.21

Catalase
H_2O_2 + H_2O_2 ⇌ O_2 + $2H_2O$ — 1.11.1.6

Cathepsin B
Hydrolyses proteins with a specificity resembling that of papain — 3.4.22.1

Cathepsin D
Protease with specificity similar to, but narrower than that of pepsin A — 3.4.23.5

Cathepsin G
Protease with specificity similar to that of chymotrypsin — 3.4.21.20

Cathepsin L
Hydrolysis of proteins: no action on acylamino acid esters — 3.4.22.15

Cellulase
Endohydrolysis of 1,4-β-D-glycosidic linkages in cellulose, lichenin and cereal β-D-glucans — 3.2.1.4

Chitinase
Random hydrolysis of 1,4-β-acetamido-2-deoxy-D-glucoside linkages in chitin and chitodextrin — 3.2.1.14

Chitin synthase
UDP-2-Acetamido-2-deoxy-D-glucose + $[1,4\text{-}(2\text{-acetamido-2-deoxy-}\beta\text{-D-glucosyl})]_n \rightleftharpoons$ UDP + $[1,4\text{-}(2\text{-acetamido-2-deoxy-}\beta\text{-D-glucosyl})]_{n+1}$ — 2.4.1.16

Cholesterol acyltransferase
Acyl-CoA + cholesterol $\rightleftharpoons$ CoA + cholesterol ester — 2.3.1.26

Cholesterol esterase
A cholesterol ester + $H_2O \rightleftharpoons$ cholesterol + a fatty-acid anion — 3.1.1.13

Cholesterol oxidase
Cholesterol + $O_2 \rightleftharpoons$ 4-cholesten-3-one + H_2O_2 — 1.1.3.6

Cholinephosphotransferase
CDPcholine + 1,2-diacylglycerol $\rightleftharpoons$ CMP + a phosphatidyl-choline — 2.7.8.2

Cholinesterase
An acylcholine + $H_2O \rightleftharpoons$ choline + a carboxylic-acid anion — 3.1.1.8

Chorismate synthase
3-Phospho-5-enolpyruvoylshikimate $\rightleftharpoons$ chorismate + orthophosphate — 4.6.1.4

Chymosin
Protease, cleaves a single bond in casein κ

Trivial name often used is rennin — 3.4.23.4

Chymotrypsin
Protease, preferential cleavage: tyr-, trp-, phe-, leu- — 3.4.21.1

Citrate (si) *synthase*
Citrate + CoA $\rightleftharpoons$ acetyl-CoA + H_2O + oxaloacetate — 4.1.3.7

Clostripain
Protease, with preferential cleavage of peptides: arg–, especially arg–pro bond — 3.4.22.8

Collagenase (vertebrate)
Cleaves preferentially one bond in native collagen leaving a *N*-terminal (75%) and C-terminal (25%) fragment — 3.4.24.7

Creatine kinase
ATP + Creatine $\rightleftharpoons$ ADP + phosphocreatine — 2.7.3.2

Creatininase
Creatinine + $H_2O \rightleftharpoons$ creatine — 3.5.2.10

Crotonyl-[acyl-carrier-protein] hydratase
D-3-Hydroxybutyryl-[acyl-carrier protein] $\rightleftharpoons$ crotonyl-[acyl-carrier protein] + H_2O — 4.2.1.58

3′:5′-Cyclic-nucleotide phosphodiesterase
Nucleoside 3′:5′-cyclic phosphate + $H_2O \rightleftharpoons$ nucleoside 5′-phosphate — 3.1.4.17

Cystathionine-γ-lyase
L-Cystathionine + $H_2O \rightleftharpoons$ L-cysteine + NH_3 + 2-oxobutyrate — 4.4.1.1

Cytidylate kinase
ATP + (d)CMP $\rightleftharpoons$ ADP + (d)CDP — 2.7.4.14

Cytochrome b$_5$ *reductase*
NADH + 2 ferricytochrome $b_5 \rightleftharpoons$ NAD^+ + 2-ferrocytochrome b_5 — 1.6.2.2

Cytochrome c *oxidase*
4 Ferrocytochrome $c + O_2 \rightleftharpoons$ 4 ferricytochrome $c + 2H_2O$ — 1.9.3.1

Debranching enzyme
Consists of two activities: amylo-1,6-glucosidase and 4-α-D-glucanotransferase

3-Dehydroquinate dehydratase
3-Dehydroquinate $\rightleftharpoons$ 3-dehydroshikimate + H_2O — 4.2.1.10

3-Dehydroquinate synthase
7-Phospho-3-deoxy-D-*arabino*-heptulosonate $\rightleftharpoons$ 3-dehydroquinate + orthophosphate — 4.6.1.3

Deoxyribonuclease I
Endonucleolytic cleavage of DNA to 5′-phospho-dinucleotide and 5′-phospho-oligonucleotide end-products — 3.1.21.1

Deoxyribonuclease II
Endonucleolytic cleavage of DNA to 3′-phospho-mononucleotides and 3′-phospho-oligonucleotides — 3.1.22.1

Dihydrolipoamide acetyltransferase
Acetyl-CoA + dihydrolipoamide $\rightleftharpoons$ CoA + S^6-acetylhydrolipoamide

A component of the pyruvate dehydrogenase multienzyme system — 2.3.1.12

Dihydrolipoamide dehydrogenase
Dihydrolipoamide + $NAD^+ \rightleftharpoons$ lipoamide + NADH

A component of the pyruvate, oxoglutarate and branched chain 2-oxoacid dehydrogenase multienzyme complexes — 1.8.1.4

Dihydrolipoamide succinyltransferase
Succinyl-CoA + dihydrolipoamide $\rightleftharpoons$ CoA + S^6-succinyl-dihydrolipoamide

A component of the oxoglutarate dehydrogenase multienzyme system — 2.3.1.61

Dihydro-orotase
L-5,6-Dihydro-orotate + $H_2O \rightleftharpoons$ *N*-carbamoyl-L-aspartate — 3.5.2.3

Dipeptidyl peptidase
A number of enzymes of differing specificities catalysing the reaction
Dipeptidyl-polypeptide + $H_2O \rightleftharpoons$ dipeptide + polypeptide — 3.4.14.1–5

DNA nucleotidylexotransferase
n Deoxynucleoside triphosphate + $(\text{deoxynucleotide})_m \rightleftharpoons$ n pyrophosphate + $(\text{deoxynucleotide})_{m+n}$ — 2.7.7.31

DNA nucleotidyltransferase
n Deoxynucleoside triphosphate ⇌ *n* pyrophosphate + DNA_n

Often referred to as DNA polymerase 2.7.7.7

Elastase (pancreatic)
Protease preferentially cleaving bonds involving the carbonyl groups of amino acids bearing uncharged non-aromatic side chains. Hydrolysis of elastin 3.4.21.36

Endodeoxyribonuclease

A group of endodeoxyribonucleases often referred to as restriction endonucleases. The Type II enzymes (EC 3.1.21.4) catalyse sequence-specific cleavage. There are over 400 enzymes of differing specificity known.

Endo-1,3-β-D-glucanase
Hydrolysis of 1,3-β-D-glucosidic linkages in 1,3-β-D-glucans 3.2.1.39

Endoproteases A and B (yeast)
These enzymes are classified as EC 3.4.23.6 (*Saccharomyces* aspartic protease) and EC 3.4.21.48, respectively

Endoprotease Glu-C
A serine protease from *Staphylococcus aureus* with preferential cleavage glu-, asp- 3.4.21.19

Enolase
2-Phospho-D-glycerate ⇌ phosphoenolpyruvate + H_2O 4.2.1.11

3-enol*pyruvoylshikimate-5-phosphate synthase*
Phosphoenolpyruvate + shikimate 5-phosphate ⇌ orthophosphate + 3-enolpyruvoylshikimate 5-phosphate 2.5.1.19

Enoyl-[acyl-carrier-protein] reductase (NADPH)
Acyl-[acyl-carrier protein] + $NADP^+$ ⇌ 2,3-dehydroacyl-[acyl-carrier protein] + NADPH 1.3.1.10

Enteropeptidase
Selective cleavage of lys^6-ile^7 bond in trypsinogen; often referred to as enterokinase 3.4.21.9

Exodeoxyribonuclease I
Exonucleolytic cleavage of DNA in the 3′ to 5′ direction to yield 5′-phosphomononucleotides 3.1.11.1

Exo-1,4-α-D-glucosidase
Hydrolysis of terminal 1,4-linked α-D-glucose residues successively from the non-reducing ends of the chains with release of β-D-glucose 3.2.1.3

Fatty-acid synthase system
A multienzyme protein; for reactions catalysed, see individual catalytic activities by reference to Fig. 7.16. This is now known as Fatty-acyl–CoA synthase; the enzymes from animal sources and from yeast are given EC numbers 2.3.1.85 and 2.3.1.86, respectively

Ficin
A thiol protease preferentially cleaving lys-, ala-, tyr-, gly-, asn-, leu-, val- 3.4.22.3

β-D-Fructofuranosidase
Hydrolysis of terminal non-reducing β- D-fructofuranoside residues in β-D-fructofuranosides 3.2.1.26

Fructokinase
ATP + D-fructose ⇌ ADP + D-fructose 6-phosphate — 2.7.1.4

Fructose-bisphosphatase
D-Fructose 1,6-bisphosphate + H_2O ⇌ D-fructose 6-phosphate + orthophosphate — 3.1.3.11

Fructose-2,6-bisphosphatase
D-Fructose 2,6-bisphosphate + H_2O ⇌ D-fructose 6-phosphate + orthophosphate — 3.1.3.46

Fructose-bisphosphate aldolase
D-Fructose 1,6-bisphosphate ⇌ dihydroxyacetone phosphate + D-glyceraldehyde 3-phosphate — 4.1.2.13

Fumarate hydratase
L-Malate ⇌ fumarate + H_2O
Often referred to as fumarase — 4.2.1.2

Galactokinase
ATP + D-galactose ⇌ ADP + α-D-galactose 1-phosphate — 2.7.1.6

Galactose 1-phosphate uridylyltransferase
UTP + α-D-galactose1-phosphate ⇌ pyrophosphate + UDPgalactose — 2.7.7.10

α-D-Galactosidase
Hydrolysis of terminal, non-reducing α- D-galactose residues in α-D-galactosides, including galactose oligosaccharides, galactomannans and galactolipids — 3.2.1.22

β-D-Galactosidase
Hydrolysis of terminal non-reducing β-D-galactose residues in β-D-galactosides — 3.2.1.23

GDPmannose α-D-mannosyltransferase
GDPmannose + heteroglycan ⇌ GDP + 1,2(1,3)-α-D-mannosylheteroglycan — 2.4.1.48

1,4-α-Glucan branching enzyme
Transfers a segment of 1,4-α-D-glucan chain to a primary hydroxyl group in a similar glucan chain — 2.4.1.18

4-α-D-Glucanotransferase
Transfers a segment of 1,4-α-D-glucan to a new 4-position in an acceptor which may be glucose or a 1,4-α-D-glucan
This activity forms part of the glycogen debranching system — 2.4.1.25

3-Glucanase (now known as Glucan endo-1,3-glucosidase)
See endo-1,3-β-D-glucanase — 3.2.1.39

Glucokinase
ATP + D-glucose ⇌ ADP + D-glucose 6-phosphate — 2.7.1.2

Glucose dehydrogenase
β-D-Glucose + $NAD(P)^+$ ⇌ D-glucono-δ-lactone + NAD(P)H — 1.1.1.47

Glucose oxidase
β-D-Glucose + O_2 ⇌ D-glucono-δ-lactone + H_2O_2 — 1.1.3.4

Glucose-6-phosphatase
D-Glucose 6-phosphate + H_2O ⇌ D-glucose + orthophosphate — 3.1.3.9

Glucose-6-phosphate dehydrogenase
D-Glucose 6-phosphate + $NADP^+$ ⇌ D-glucono-δ-lactone 6-phosphate + NADPH — 1.1.1.49

Glucose-6-phosphate isomerase
D-Glucose-6-phosphate ⇌ D-fructose 6-phosphate — 5.3.1.9

Glucose-1-phosphate uridylyltransferase
UTP + α-D-glucose 1-phosphate ⇌ pyrophosphate + UDPglucose — 2.7.7.9

α-D-Glucosidase
Hydrolysis of terminal, non-reducing 1,4-linked α-D-glucose residues with release of α-D-glucose — 3.2.1.20

Often referred to as maltase

β-D-glucuronidase
A β-D-glucuronide + H_2O ⇌ an alcohol + D-glucuronate — 3.2.1.31

Glutamate dehydrogenase
L-Glutamate + H_2O + NAD^+ ⇌ 2-oxoglutarate + NH_3 + NADH — 1.4.1.2

Glutamate dehydrogenase ($NAD(P)^+$)
L-Glutamate + H_2O + $NAD(P)^+$ ⇌ 2-oxoglutarate + NH_3 + NAD(P)H — 1.4.1.3

Glutaminase
L-Glutamine + H_2O ⇌ L-glutamate + NH_3 — 3.5.1.2

Glutamine phosphoribosyl pyrophosphate amidotransferase
5-Phospho-β-D-ribosylamine + pyrophosphate + L-glutamate ⇌ L-glutamine + 5-phospho-α-D-ribose 1-diphosphate + H_2O — 2.4.2.14

Glutamine synthetase
ATP + L-Glutamate + NH_3 ⇌ ADP + Orthophosphate + L-Glutamine — 6.3.1.2

γ-Glutamyltransferase
(5-L-Glutamyl)-peptide + An amino acid ⇌ Peptide + 5-L-Glutamylamino acid — 2.3.2.2

Glutathione-insulin transhydrogenase [recommended name: Protein-disulphide reductase (glutathione)]
2 Glutathione + protein-disulphide ⇌ oxidized glutathione + protein-dithiol — 1.8.4.2

Glutathione reductase (NAD(P)H)
NAD(P)H + oxidized glutathione ⇌ $NAD(P)^+$ + 2 glutathione — 1.6.4.2

Glyceraldehyde-3-phosphate dehydrogenase
D-Glyceraldehyde 3-phosphate + orthophosphate + NAD^+ ⇌ 3-phospho-D-glyceroyl phosphate + NADH — 1.2 1.12

Glycerol kinase
ATP + glycerol ⇌ ADP + *sn*-glycerol 3-phosphate — 2.7.1.30

Glycerol-3-phosphate dehydrogenase
sn-Glycerol 3-phosphate + acceptor ⇌ dehydroxyacetone phosphate + reduced acceptor — 1.1.99.5

Glycerol-3-phosphate dehydrogenase (NAD^+)
sn-Glycerol 3-phosphate + NAD^+ ⇌ dihydroxyacetone phosphate + NADH — 1.1.1.8

Glycerol-3-phosphate acyltransferase
Acyl-CoA + *sn*-glycerol 3-phosphate ⇌ CoA + 1-acylglycerol 3-phosphate — 2.3.1.15

Glycogen synthase
UDPglucose + (1,4-α-D-glycosyl)$_n$ ⇌ UDP + (1,4-α-D-glucosyl)$_{n+1}$ — 2.4.1.11

Hexokinase
ATP + D-hexose ⇌ ADP + D-hexose 6-phosphate — 2.7.1.1

Histidine ammonia-lyase
L-Histidine ⇌ urocanate + NH_3 — 4.3.1.3

Homogentisate 1,2-dioxygenase
Homogentisate + O_2 ⇌ 4-maleylacetoacetate — 1.13.11.5

Homoserine dehydrogenase
L-Homoserine + NAD(P)$^+$ ⇌ L-aspartate β-semialdehyde + NAD(P)H — 1.1.1.3

Hyaluronoglucosaminidase
Random hydrolysis of 1,4-linkages between 2-acetamido-2-deoxy-β-D-glucose and D-glucuronate residues in hyaluronate — 3.2.1.35

Hyaluronoglucuronidase
Random hydrolysis of 1,3-linkages between β-glucuronate and 2-acetamido-2-deoxy-D-glucose residues in hyaluronate — 3.2.1.36

Hydrogenase
2 Reduced ferredoxin + $2H^+$ ⇌ 2 oxidized ferredoxin + H_2 — 1.18.99.1

3-Hydroxyacyl-CoA dehydrogenase
(*S*)-3-hydroxyacyl-CoA + NAD^+ ⇌ 3-oxoacyl-CoA + NADH — 1.1.1.35

3-Hydroxybutyrate dehydrogenase
D-3-Hydroxybutyrate + NAD^+ ⇌ acetoacetate + NADH — 1.1.1.30

Hydroxymethylglutaryl-CoA reductase
Mevalonate + CoA + $2NAD^+$ ⇌ 3-hydroxy-3-methylglutaryl-CoA + 2NADH — 1.1.1.88

Indole-3-glycerol-phosphate synthase
1-(2′-Carboxyphenylamino)-1-deoxyribulose 5-phosphate ⇌ 1-*C*-(3′-indolyl)-glycerol 3-phosphate + CO_2 + H_2O — 4.1.1.48

Iodide peroxidase
Iodide + H_2O_2 ⇌ iodine + $2H_2O$
Used to iodinate tyrosine residues — 1.11.1.8

Isocitrate dehydrogenase (NAD$^+$)
threo-D_S-Isocitrate + NAD^+ ⇌ 2-oxoglutarate + CO_2 + NADH — 1.1.1.41

Isocitrate lyase
threo-D_S-Isocitrate ⇌ succinate + glyoxylate — 4.1.3.1

Isoleucyl-tRNA synthetase (ligase)
ATP + L-isoleucine + $tRNA^{Ile}$ ⇌ AMP + pyrophosphate + L-isoleucyl-$tRNA^{Ile}$ — 6.1.1.5

Isomaltase
Hydrolysis of 1,6-α-D-glucosidic linkages in isomaltose and dextrins produced from starch and glycogen by α-amylase.

Also known as oligo-1,6-glucosidase.

The enzyme from intestinal mucosa also catalyses the hydrolysis of sucrose and maltose (see Chapter 8, Section 8.4.3) 3.2.1.10

C_{55} isoprenoid-alcohol kinase
ATP + C_{55}-isoprenoid-alcohol ⇌ ADP + C_{55}-isoprenoid-alcohol phosphate 2.7.1.66

Kynureninase
L-Kynurenine + H_2O ⇌ anthranilate + L-alanine 3.7.1.3

Lactate dehydrogenase
L-Lactate + NAD^+ ⇌ pyruvate + NADH 1.1.1.27

Lactose synthase
UDPgalactose + D-glucose ⇌ UDP + lactose 2.4.1.22

Lactoylglutathione lyase
(*R*)-*S*-Lactoylglutathione ⇌ glutathione + methylglyoxal 4.4.1.5

Leucine dehydrogenase
L-Leucine + H_2O + NAD^+ ⇌ 4-methyl-2-oxopentanoate + NH_3 + NADH 1.4.1.9

Lysozyme
Hydrolysis of 1,4-β-linkages between *N*-acetylmuramic acid and 2-acetamido-2-deoxy-D-glucose residues in a mucopolysaccharide or muropeptide 3.2.1.17

Malate dehydrogenase
L-Malate + NAD^+ ⇌ oxaloacetate + NADH 1.1.1.37

Malate dehydrogenase (oxaloacetate-decarboxylating) ($NADP^+$)
L-Malate + $NADP^+$ ⇌ pyruvate + CO_2 + NADPH

Often referred to as the 'malic' enzyme 1.1.1.40

Malonyl-CoA dependent elongase system
A microsomal membrane enzyme system that catalyses the elongation of acyl-CoAs (C_{16}–C_{22}) using malonyl-CoA and NADPH to acyl-CoAs (C_{18}–C_{24}).

Methionine adenosyltransferase
ATP + L-methionine + H_2O ⇌ orthophosphate + pyrophosphate + *S*-adenosyl-L-methionine 2.5.1.6

NADH dehydrogenase
NADH + acceptor ⇌ NAD^+ + reduced acceptor 1.6.99.3

NAD^+ nucleosidase
NAD^+ + H_2O ⇌ nicotinamide + ADPribose 3.2.2.5

$NADP^+$-cytochrome reductase
$NADP^+$ + 2 ferrocytochrome ⇌ NADPH + 2 ferricytochrome 1.6.2.4

Neuraminidase
Hydrolysis of 2,3-, 2,6-, and 2,8-glycosidic linkages joining terminal non-reducing *N*- or *O*-acylneuraminyl residues to galactose, *N*-acetylhexosamine, or *N*- or *O*-acylated neuraminyl residues in oligosaccharides, glycoproteins, glycolipids, or colominic acid 3.2.1.18

Nicotinamide-nucleotide adenylyltransferase
ATP + nicotinamide ribonucleotide ⇌ pyrophosphate + NAD^+ 2.7.7.1

Nitrogenase (flavodoxin)
6 Reduced flavodoxin + 6H^+ + N_2 + nATP ⇌ 6 oxidized flavodoxin + 2NH_3 + nADP + n orthophosphate 1.19.6.1

Nucleosidediphosphatase
A nucleoside diphosphate + H_2O ⇌ a nucleotide + orthophosphate 3.6.1.6

Nucleosidediphosphate kinase
ATP + nucleoside diphosphate ⇌ ADP + nucleoside triphosphate 2.7.4.6

Nucleosidemonophosphate kinase
ATP + nucleoside monophosphate ⇌ ADP + nucleoside diphosphate 2.7.4.4

Nucleoside triphosphatase
NTP + H_2O ⇌ NDP + orthophosphate 3.6.1.15

5′-Nucleotidase
A 5′-ribonucleotide + H_2O ⇌ A ribonucleoside + orthophosphate 3.1.3.5

Nucleotide pyrophosphatase
A dinucleotide + H_2O ⇌ 2 mononucleotides
Substrates include NAD^+, $NADP^+$, FAD, CoA, ATP, and ADP 3.6.1.9

Ornithine carbamoyltransferase
Carbamoylphosphate + L-ornithine ⇌ orthophosphate + L-citrulline 2.1.3.3

Ornithine decarboxylase
L-Ornithine ⇌ putrescine + CO_2 4.1.1.17

Ornithine-oxo-acid aminotransferase
L-Ornithine + a 2-oxoacid ⇌ L-glutamate γ-semialdehyde + an L-amino acid 2.6.1.13

Orotate reductase (NADH)
L-5,6-Dihydroorotate + NAD^+ ⇌ orotate + NADH 1.3.1.14

Orotidine-5′-phosphate decarboxylase
Orotidine 5′-phosphate ⇌ UMP + CO_2 4.1.1.23

3-Oxoacyl-[acyl-carrier-protein] reductase
D-3-Hydroxyacyl-[acyl-carrier protein] + $NADP^+$ ⇌ 3-oxoacyl-[acyl-carrier protein] + NADPH 1.1.1.100

3-Oxoacyl-[acyl-carrier-protein] synthase
Acyl-[acyl-carrier protein] + malonyl-[acyl-carrier protein] ⇌ 3-oxoacyl-[acyl-carrier protein] + CO_2 + acyl-carrier protein 2.3.1.41

Oxoglutarate dehydrogenase (also known as oxoglutarate decarboxylase)
2-Oxoglutarate + lipoamide ⇌ S^6-succinyldihydrolipoamide + CO_2
A component of the oxoglutarate dehydrogenase multienzyme system 1.2.4.2

Papain
Protease preferentially cleaving arg-, lys-, phe-X- (the peptide bond next but one to the carboxyl group of phenylalanine); limited hydrolysis of native immunoglobulins 3.4.22.2

Penicillin amidase
Penicillin + H_2O ⇌ 6-aminopenicillinate + a carboxylic acid anion 3.5.1.11

Penicillinase (recommended name β-lactamase)
Penicillin + H_2O ⇌ penicilloate — 3.5.2.6
Other β-lactam substrates are also acted on by the enzyme.

Penicillopepsin
An aspartic protease from *Penicillum janthinellum* which resembles pig pepsin structurally. — 3.4.23.6

Pepsin A
Protease causing preferential cleavage at phe-, leu- usually referred to as pepsin. There are also pepsins B and C, which show more restricted specificity — 3.4.23.1

Peroxidase
Donor + H_2O_2 ⇌ oxidized donor + $2H_2O$ — 1.11.1.7

Phenylalanine ammonia-lyase
L-Phenylalanine ⇌ *trans*-cinnamate + NH_3 — 4.3.1.5

Phenylalanine 4-monooxygenase
L-Phenylalanine + tetrahydropteridine + O_2 ⇌ L-tyrosine + dihydropteridine + H_2O — 1.14.16.1

Phosphodiesterase 1
Hydrolytically removes 5′-nucleotides successively from the 3′-hydroxy termini of 3′-hydroxy-terminated oligonucleotides — 3.1.4.1

Phosphoenolpyruvate carboxykinase (GTP)
GTP + oxaloacetate ⇌ GDP + phosphoenolpyruvate + CO_2 — 4.1.1.32

6-Phosphofructokinase
ATP + D-fructose 6-phosphate ⇌ ADP + D-fructose 1,6-bisphosphate — 2.7.1.11

6-Phosphofructo-2-kinase
ATP + D-fructose 6-phosphate ⇌ ADP + D-fructose 2,6-bisphosphate — 2.7.1.105

Phosphoglucomutase
α-D-Glucose 1,6-bisphosphate + α-D-glucose 1-phosphate ⇌ α-D-glucose 6-phosphate + α-D-glucose 1,6-bisphosphate
This enzyme was previously classified as EC 2.7.5.1 — 5.4.2.2

Phosphogluconate dehydrogenase (decarboxylating)
6-Phospho-D-gluconate + $NADP^+$ ⇌ D-ribulose 5-phosphate + CO_2 + NADPH — 1.1.1.44

Phosphoglycerate kinase
ATP + 3-phospho-D-glycerate ⇌ ADP + 3-phospho-D-glyceroyl phosphate — 2.7.2.3

Phosphoglycerate mutase
2,3-Bisphospho-D-glycerate + 2-phospho-D-glycerate ⇌ 3-phospho-D-glycerate + 2,3-bisphospho-D-glycerate — 5.4.2.1
This enzyme was previously classified as EC 2.7.5.3

Phospho-2-keto-3-deoxy-gluconate aldolase
6-Phospho-2-keto-3-deoxy-D-gluconate ⇌ pyruvate + D-glyceraldehyde 3-phosphate — 4.1.2.14

Phospho-2-keto-3-deoxy-heptonate aldolase
7-Phospho-2-keto-3-deoxy-D-*arabino*-heptonate + orthophosphate ⇌ phosphoenolpyruvate + D-erythrose 4-phosphate + H_2O — 4.1.2.15

Phospholipase A_1
A phosphatidylcholine + H_2O ⇌ 2-acylglycerophosphocholine + a fatty-acid anion — 3.1.1.32

Phospholipase A_2
A phosphatidylcholine + H_2O ⇌ 1-acylglycerophosphocholine + an unsaturated fatty-acid anion — 3.1.1.4

Phospholipase C
A phosphatidylcholine + H_2O ⇌ 1,2-diacylglycerol + choline phosphate — 3.1.4.3

Phosphorylase
$(1,4\text{-}\alpha\text{-D-Glycosyl})_n$ + orthophosphate ⇌ $(1,4\text{-}\alpha\text{-D-glucosyl})_{n-1}$ + α-D-glucose 1-phosphate — 2.4.1.1

Phosphorylase kinase
4ATP + 2 phosphorylase *b* ⇌ 4ADP + phosphorylase *a* — 2.7.1.38

Phosphorylase phosphatase
Phosphorylase *a* + $4H_2O$ ⇌ 2 phosphorylase *b* + 4 orthophosphate — 3.1.3.17

Plasmin
Protease cleaving preferentially lys- >arg-; higher selectivity than trypsin — 3.4.21.7

Polydeoxyribonucleotide synthetase (NAD^+)
NAD^+ + (deoxyribonucleotide)$_n$ + (deoxyribonucleotide)$_m$ ⇌ AMP + NMN + (deoxyribonucleotide)$_{n+m}$

Often referred to as DNA ligase — 6.5.1.2

Polynucleotide 5′-hydroxyl-kinase
ATP + 5′-dephospho-DNA ⇌ ADP + 5′-phospho-DNA — 2.7.1.78

Polygalacturonase
Random hydrolysis of 1,4-α-D-galactosiduronic linkages in pectate and other galacturonans — 3.2.1.15

Proline racemase
L-Proline ⇌ D-proline — 5.1.1.4

Protein disulphide-isomerase
Catalyses the rearrangement of —S—S— bonds in proteins — 5.3.4.1

Protein kinase
ATP + a protein ⇌ ADP + a phosphoprotein

The enzyme from rat tissues is stimulated by cyclic AMP and will activate phosphorylase kinase. Other enzymes will phosphorylate other proteins. Some enzymes are activated by cyclic AMP, some by cyclic GMP and some by neither (see Chapter 6, Section 6.4.2). — 2.7.1.37

Protein phosphatase
A phosphoprotein + nH_2O ⇌ a protein + *n* orthophosphate

See Chapter 6, Section 6.4.2. for an account of some of the different protein phosphatases — 3.1.3.17

Protein-tyrosine kinase
ATP + protein-tyrosine ⇌ ADP + protein-tyrosine phosphate — 2.7.1.112

Pyruvate carboxylase
ATP + pyruvate + CO_2 + H_2O ⇌ ADP + orthophosphate + oxaloacetate — 6.4.1.1

Pyruvate dehydrogenase (lipoamide) (also known as pyruvate decarboxylase)
Pyruvate + lipoamide ⇌ S^6-acetyldihydrolipoamide + CO_2

A component of the pyruvate dehydrogenase multienzyme system — 1.2.4.1

[*Pyruvate dehydrogenase (lipoamide)*] *kinase*
ATP + [pyruvate dehydrogenase (lipoamide)] ⇌ ADP + [pyruvate dehydrogenase (lipoamide)] phosphate — 2.7.1.99

[*Pyruvate dehydrogenase (lipoamide)*]*-phosphatase*
[Pyruvate dehydrogenase (lipoamide)] phosphate + H_2O ⇌ [pyruvate dehydrogenase (lipoamide)] + orthophosphate — 3.1.3.43

Pyruvate kinase
ATP-pyruvate ⇌ ADP + phosphoenolpyruvate — 2.7.1.40

Restriction endonuclease
See Endodeoxyribonuclease

Reverse transcriptase
n Deoxynucleoside triphosphate ⇌ *n* pyrophosphate + DNA_n

Also known as DNA nucleotidyltransferase (RNA-directed)) — 2.7.7.49

Ribonuclease II
Exonucleotic cleavage of RNA in the 3′ to 5′ direction to yield 5′ phosphomononucleotides — 3.1.13.1

Ribonuclease (pancreatic)
Endonucleolytic cleavage of RNA to 3′-phosphomonucleotides and 3′ phosphooligonucleotides ending in Cp or Up with 2′,3′-cyclic phosphate intermediates — 3.1.27.5

Ribulosebisphosphate carboxylase
D-Ribulose 1,5-bisphosphate + CO_2 ⇌ 2 3-phospho-D-glycerate — 4.1.1.39

RNA-methyltransferases
S-Adenosyl-L-methionine + tRNA ⇌ *S*-adenosyl-L-homocysteine + tRNA containing methyl groups

A group of enzymes which methylate tRNA on different positions on the purine or pyrimidine bases — 2.1.1.29–2.1.1.36

RNA nucleotidyltransferase
n Nucleoside triphosphate ⇌ *n* pyrophosphate + RNA_n

Usually referred to as RNA polymerase — 2.7.7.6

L-Serine dehydratase
L-Serine + H_2O ⇌ pyruvate + NH_3 + H_2O — 4.2.1.13

Shikimate dehydrogenase
Shikimate + $NADP^+$ ⇌ 3-dehydroshikimate + NADPH — 1.1.1.25

Shikimate kinase
ATP + shikimate ⇌ ADP + shikimate 5-phosphate — 2.7.1.71

Steroid 11β-monooxygenase
A steroid + reduced adrenal ferredoxin + O_2 ⇌ an 11β-hydroxysteroid + oxidized adrenal ferredoxin + H_2O — 1.14.15.4

Streptokinase
Although this protein has been purified, it is not clear whether it possesses a catalytic site. Streptokinase interacts with plasminogen and the complex formed shows plasmin activity (see under plasmin). The nature of activation process has to be clarified.

Subtilisin
Hydrolysis of proteins and peptide amides: isolated from *Bacillus subtilis* — 3.4.21.14

Succinate-CoA ligase (GDP-forming)
GTP + succinate + CoA ⇌ GDP + orthophosphate + succinyl-CoA
Sometimes misleadingly referred to as succinate thiokinase — 6.2.1.4

Succinate dehydrogenase
Succinate + acceptor ⇌ fumarate + reduced acceptor — 1.3.99.1

Sucrose α-D-glucohydrolase
Hydrolysis of sucrose and maltose by an α-D-glucosidase-type action often referred to as sucrase or invertase — 3.2.1.48

Terminal deoxyribonucleotidyltransferase
n Deoxynucleoside triphosphate + $(\text{deoxynucleotide})_m$ ⇌ n pyrophosphate + $(\text{deoxynucleotide})_{m+n}$
Also known as terminal transferase — 2.7.7.31

Tetrahydrofolate dehydrogenase
5,6,7,8-Tetrahydrofolate + $NADP^+$ ⇌ 7,8-dihydrofolate + NADPH — 1.5.1.3

Thermolysin
A metalloprotease cleaving preferentially -leu > -phe
The recommended name for this enzyme is *Bacillus thermoproteolyticus* neutral protease but it is usually referred to as thermolysin — 3.4.24.4

Thiosulphate sulphurtransferase
Thiosulphate + cyanide ⇌ sulphite + thiocyanate — 2.8.1.1

Threonine dehydratase
L-Threonine + H_2O ⇌ 2-oxobutyrate + NH_3 + H_2O — 4.2.1.16

Thrombin
Protease cleaving preferentially arg-: activates fibrinogen to fibrin — 3.4.21.5

Thymidine kinase
ATP + thymidine ⇌ ADP + thymidine 5′-phosphate — 2.7.1.21

Thymidylate synthase
5,10-Methylenetetrahydrofolate + dUMP ⇌ dihydrofolate + dTMP — 2.1.1.45

Transketolase
Sedoheptulose 7-phosphate + D-glyceraldehyde 3-phosphate ⇌ D-ribose 5-phosphate + D-xylulose 5-phosphate — 2.2.1.1

Triacylglycerol lipase
Triacylglycerol + H_2O ⇌ diacylglycerol + a fatty-acid anion — 3.1.1.3

Triosephosphate isomerase
D-Glyceraldehyde 3-phosphate ⇌ dihydroxyacetone phosphate 5.3.1.1

Trypsin
A protease cleaving preferentially arg-, lys- 3.4.21.4

Tryptophan 2,3-dioxygenase
L-Tryptophan + O_2 ⇌ L-formylkynurcnine 1.13.11.11

Tryptophan synthase
L-Serine + indoleglycerol 3-phosphate ⇌ L-tryptophan + glyceraldehyde 3-phosphate 4.2.1.20

Tyrosine aminotransferase
L-Tyrosine + 2-oxoglutarate ⇌ 4-hydroxyphenylpyruvate + L-glutamate 2.6.1.5

Tyrosine 3-hydroxylase
L-Tyrosine + tetrahydropteridine + O_2 ⇌ 3,4-dihydroxy-L-phenylalanine + dihydropteridine + H_2O

Also known as tyrosine 3-monooxygenase 1.14.16.2

Tyrosyl-tRNA synthetase (ligase)
ATP + L-tyrosine + $tRNA^{tyr}$ ⇌ AMP + pyrophosphate + L-tyrosyl-$tRNA^{tyr}$ 6.1.1.1

UDPglucose 4-epimerase
UDPglucose ⇌ UDPgalactose 5.1.3.2

UDPglucose-hexose-1-phosphate uridylyltransferase
UDPglucose + α-D-galactose 1-phosphate ⇌ α-D-glucose 1-phosphate + UDPgalactose 2.7.7.12

UDPglucuronosyltransferase
UDPglucuronate + acceptor ⇌ UDP + acceptor β-D-glucuronide 2.4.1.17

Urate oxidase
Urate + O_2 + $2H_2O$ ⇌ allantoin + CO_2 + H_2O_2

The identity of the products of the reaction is uncertain 1.7.3.3

Urease
Urea + H_2O ⇌ CO_2 + $2NH_3$ 3.5.1.5

Urokinase (recommended name: Plasminogen activator)
Protease cleaving preferentially arg-val in plasminogen 3.4.21.31

Uropepsinogen
The enzyme, pepsin, which is secreted in the stomach in mammals, is subsequently absorbed by the intestine into the blood, whence it is excreted by the kidney as an inactive protease, uropepsinogen. Uropepsinogen is activated to uropepsin at pH 1–2

Xanthine oxidase
Xanthine + H_2O + O_2 ⇌ urate + supcroxide 1.1.3.22

Xylose isomerase
D-Xylose ⇌ D-xylulose 5.3.1.5

Sometimes referred to as glucose isomerase, since it will isomerise D-glucose to D-fructose, but D-xylose is the preferred substrate

L-Xylulose reductase
Xylitol + $NADP^+$ ⇌ L-xylulose + NADPH 1.1.1.10

Index